Internal Combustion Engine Principles
with Vehicle Applications

Spencer C. Sorenson

Automatic Press / VIP

Automatic Press / VIP
Information on this title: www.vince-inc.com

© Automatic Press / VIP 2017
This publication is in copyright. Subject to statuary exception
and to the provisions of relevant collective licensing agreements,
no reproduction of any part may take place without
the written permission of the publisher.

First published 2017

Printed in the United States of America
and the United Kingdom

ISBN-13 978-87-92130-57-0 paperback

The publisher has no responsibilities for
the persistence or accuracy of URLs for external or
third party Internet Web sites referred to in this publication
and does not guarantee that any content on such
Web sites is, or will remain, accurate or appropriate.

Internal Combustion Engine Principles

with Vehicle Applications

Professor Emeritus Spencer C. Sorenson
Department of Mechanical Engineering
Technical University of Denmark
DK 2800, Kgs. Lyngby, Denmark
August 28, 2017
Cover picture courtesy of Ford Motor Co. Used by permission.

Contents

Preface

The internal combustion engine has been one of the major factors in changing the world in the 20th century and beyond. Nearly all forms of transport rely, or have relied on them for propulsion. Engines have changed in many ways from the low speed devices of the late 19th century, to the high speed, computer controlled engines we see today. Yet most of the basic principles remain unchanged. As seen in a problem in this text, Nicolas Otto's first engine achieved an indicated thermal efficiency of about 80% of that allowed by thermodynamic principles, a value that is typical of modern spark ignition engines today. Improved fuel quality, control systems, lubricants, materials etc, make it possible to run engines at much higher speeds and compression ratios than Otto could, thus increasing the thermodynamic limits to which engine designers aspire.

One of the main goals of this book is to present the basic principles and concepts involved in determining engine performance and efficiency. In the modern age, with complex simulations, and a wealth of information it is easy to lose sight of some these basics. The main focus, then, is on principles, and not so much on details. The book is not intended to present all the latest details, (a nearly hopeless task) but to provide a foundation on which to understand them and relate them to how much (or little) they affect engine operation. In the years that have elapsed since this book was started, many changes have occurred in engine technology and the requirements placed on them. Turbocharging has become prevalent in diesel engines, and almost all diesel engines produced today are turbocharged, there has been a substantial increase in the number of turbocharged spark ignition engines, and methods to prevent them from knocking have been developed.

The environment poses the largest challenge for engines today. Though emissions reductions of well over 90% have been achieved since control started in the late 60's, recent proposals have been made to ban IC engines in some areas, with a major focus on particulate emissions. The ultimate challenge, though, is the greenhouse effect and the production of CO_2 from fossil fuels. Transport is responsible for 23% of global CO_2 emissions (17% of greenhouse gas emission), and 72% of the 23% is due to road traffic, essentially all of which is due to IC engines. So engines are a logical target for control initiatives. Since CO_2 is the desired product of combustion of fossil fuels, and removal of CO_2 from exhaust gas of engines appears extremely difficult, if at all possible, alternate fuels or fuel production methods look to be important areas. Efficiency improvements have been achieved, and one of the goals of this book is to focus on the factors that determine efficiency of engines. It is not easy to see how efficiency improvements alone will reduce CO_2 emissions to levels in climate change proposals.

In response to the CO_2 challenge, the number of electric vehicles has increased greatly in the past few years, and will increasingly prove to be an alternative to fossil fuel power plants, since they can be fueled by renewable energy - currently wind and solar power. The challenges to internal combustion engines in the future will be immense.

The book is written in American English, and uses "conventional" US notation for the most part, though a few changes from the author's experience in Europe have crept in. The SI unit system is used, and the American system of periods for decimal points, and commas for 1000's separators is used.

The chapters on vehicle loading and gearing are based to a large extent on material provided by my former colleague, Prof Arne Boyhus of the Technical University of Denmark (DTU). Through the years, the author has had inspiration, guidance and help from many individuals. A few of those are: Professors Phil Myers and Otto Uyehara of the University of Wisconsin, who opened the doors to engine study and research, and showed how the search for truth can be challenging and intense without compromising personal values. My former colleague at the University of Illinois at Urbana-Champaign Prof William L.

Hull showed me many new ways to look at engines and how to teach them to students. Two of my other colleagues at UIUC, Professors Lester D. Savage Jr and Roger A Strehlow, were inspiring and taught me many valuable lessons in the search to better understand combustion. Dr Patrick F. Flynn late of Cummins Engine Company, was a good friend for many years, and never failed to start an interesting discussion of new developments in engine technology. Dr Gustav Winkler showed the value of the Wilans line, also gave many insights into turbochargers. Professor Elbert Hendricks of DTU, with whom I have worked closely for many years, inspired me to look at engines new ways. Though we never worked together, I also owe a debt of appreciation to Prof Gordon P. Blair, of Queens University, Belfast, for showing me that more complicated is not always better. The continuing encouragement of Prof. Duane Abata is greatly appreciated.

I also am indebted to the many, many students I have had the privilege to teach through the years, many of whom have become good friends as well. Their provocative and interested questions have forced me to relearn many things. As Professor Myers always said, "Those at the university who need education most are the professors, and it is the students who teach them". I would also like to thank Professors Bjørn Qvale and Jesper Schramm of DTU for providing an inspiring working environment. Thank you to all the rest who helped me through the years.

This book is dedicated to Hans and George

Spencer C. Sorenson

Kgs. Lyngby, Denmark,

October 2017

Nomenclature

Symbol	Meaning	Units
a	Acceleration	m/s^2
A	Area	m^2
A_f	Frontal area, flame area	m^2
A_D	Drag area	m^2
$[A_i]$	Concentration of specie i	mol/m^3
A_n	Normalized throttle opening area	m^2
AF	Air Fuel ratio	kg/kg
B	Cylinder bore	m
BP, P_b	Brake Power	kW
$bmep$	Brake Mean Effective Pressure	bar, kPa
$BSFC, bsfc$	Brake Specific Fuel Consumption	g/kW-h
C_r	Radial clearance	m
C_D, C_W	Drag coefficient	
C_f	Flow coefficient	
c_p	Specific heat at constant pressure	kJ/kg-K
c_v	Specific heat at constant volume	kJ/kg-K
D	Diffusion coefficient	m^2/s
D	Diameter	m
E	Activation energy	kJ/mol
F, f	Force	N
f	Residual fraction	kg/kg
f_r	Rolling resistance coefficient	
FA	Fuel Air ratio	kg/kg
FBP	Final Boiling Point	°C, K
$FMEP, fmep$	Friction Mean Effective Pressure	bar, kPa
FP, P_f	Friction Power	kW
$fumep$	Fuel Mean Effective Pressure	kPa, bar
H	Enthalpy	kJ
H_u	Lower heating value	kJ/kg
h	Specific enthalpy	kJ/kg
h	Absolute humidity	kg/$kg_{dry\ air}$
h	Heat transfer coefficient	$W/m^2, kJ/m^2, kJ/m^2-°$
h	Bearing clearance	m
I	Moment of inertia	kg-m^2
IP	Indicated Power	kW
$imep$	Indicated Mean Effective Pressure	bar, kPa
$isfc$	Indicated Specific Fuel Consumption	g/kW-h
K	Bulk modulus	Pa/bar
K_p	Equilibrium constant	for partial pressures
l	Valve lift	m
L	Connecting rod length	m
Le	Lewis number	
m	Mass	kg

Symbol	Meaning	Units
m_a	Air mass	kg
$\dot{m}_f$	Air flow, Fuel flow at the intake valve	kg/s
m_f	Fuel mass	kg
$\dot{m}_f$	Fuel flow	kg/s
$\dot{m}_{fi}$	Injected fuel rate	kg/s
$\dot{m}_{ff}$	Fuel film flow rate	kg/s
$\dot{m}_{fv}$	Evaporated fuel flow rate	kg/s
$\dot{m}_{at}$	Air flow through the throttle	kg/s
$\dot{m}_{ap}$	Air flow through the intake value	kg/s
M	Moment	Nm
MW	Molecular weight	kg/kmol
n_{cyl}	number of cylinders	-
N	Rotation speed	Rev/min
p	Pressure	bar, kPa
P	Power	kW
q	Specific heat transfer	kJ/kg
Q	Heat transfer, fuel energy	kJ
$\dot{Q}$	Volume flow	m^3/s
r	radius	m
R	Gas Constant	kJ/kg-K
R	Radius	m
R	Crankshaft arm radius	m
R_o	Universal gas constant	kJ/kmol-K
R_d	Delivery ratio	-
R_s	Scavenging ratio	-
RVP	Reid Vapor pressure	kPa
S	Stroke	
S	Entropy	kJ/K
s	Specific entropy	kJ/kg-K
S_l	Laminar flame speed	m/s
S_p	Mean piston speed	m/s
t	time	s
T	Torque	N-m
T	Temperature	, K,°C
U	internal energy	kJ
U	velocity	m/s, km/h
u	Specific internal energy	kJ/kg
v	Specific volume	m^3/kg
v	Velocity	m/s, km/h
V	Volume	m^3
V_d, V_s	Displacement (swept) Volume	m^3
W	Work	Nm
W	Road force	N
w	Specific work	Nm/kg
X	pressure amplitude ratio	-
x	Distance of piston from TDC	m
x	Mole fraction	-
x	Number of revolutions per cycle	-
X	Fraction of fuel on manifold wall	-
y	distance	m

Symbol	Meaning	Units
y	Mass burned fraction	-
α	Angle of inclination, throttle angle	-
α	Angular acceleration	$1/m^2$
ε	Compression ratio	-
ε	Relative eccentricity	-
η	Efficiency	-
γ	Specific heat ratio	-
κ	Thermal conductivity	W/m-K
λ	Excess air ratio	-
ω	Angular frequency	s^{-1}
ϕ	Fuel Air equivalence ratio	-
ϕ	Angle at small end of con. rod	-
Φ	Compressor flow coefficient	-
ρ	Density	kg/m^3
Psi	Head coefficient	
τ	Ignition delay	sec, deg
τ	Shear stress	MPa
θ	Gear ratio	-
θ	Crank angle degree	Degree
μ	Viscosity	kg/m-s
ν	Kinematic viscosity	m/s^2

Subscripts

a	air, areodynamic
ac	acceleration
amb	ambient
b	brake
c	clearance, combustion, compressor, compression
CV, cv	control volume
d	displacement, delivery
e	engine
ex	exhaust, expansion
f	fuel, filling/charging, filter, formation
fg	evaporation
g	gradient
HR	Heat Release
i	indicated, substance i
id	ideal
in	intake, into cycle
inj	injector, injected
man	manifold
o	initial
o	atmosphere
p	particle
$prod$	products
$react$	reactants
r	rolling
R	reaction
s	stoichiometric, isentropic, sensible, sonic, superposition
sc	scavenging, supercharged
t	turbine
tr	trapping, transmission
u	unburned
v	volumetric, valve
w	water

Terms and Abbreviations

ASI	After Start of Injection
ATDC, ABDC	After Top Dead Center, After Bottom Dead Center
BTDC , BBDC	Before Top Dead Center, Before Bottom Dead Center
BDC	Bottom Dead Center
CI	Compression Igntion
CN	Cetane Number
CVT	Continuously Variable Transmission
DI	Direct Injection
DPF	Diesel Particle Filter
EAD	Equilibrium Air Distillation
EGR	Exhaust Gas Recirculation
GDI	Gasoline Indirect Injection
HCCI	Homogeneous Charge Compression Ignition
IDI	InDirect Injecion
MBT	Minimum Spark Advance for Maximum Brake Torque
MON	Motor Octane Number
NA	Naturally Aspirated (no pressure charging)
RON	Research Octane Number
SC	Super Charged
SI	Spark Ignition
SOI	Start of Injection
TC	Turbocharged
TDC	Top Dead Center
WOT	Wide Open Throttle

Chapter 1

Engine Basics

1.1 Introduction

1.1.1 General

A piston engine produces work from high pressure gasses that act against a piston, which moves up and down in a cylinder. The high pressure gasses are formed by burning a mixture of fuel and air. In the very first engines, a mixture of fuel and air was drawn into a cylinder, then burned, without compressing it prior to combustion. The combustion was followed by an expansion, where the work was produced, and then an air exchange process at atmospheric pressure, where the cylinder was recharged.

Nikolaus Otto is credited as being the first to use the compression cycle, that is he developed a cycle where the fuel air charge was drawn into the cylinder, then compressed before ignition. After the combustion, the gasses expanded to produce work, the net work being equal to the difference between the work required to compress the gasses, and the work delivered during expansion of the gasses. A simple thermodynamic analysis shows that a greater amount of compression allows a greater amount of expansion, improving the efficiency of the process. This is presented in Section 2.1 on ideal cycles in more scientific terms.

The standard air cycle with constant volume heat addition (simulated combustion) has been named the Otto cycle in honor of this achievement. Otto's original engines were close to the constant volume process, and had efficiency as close to the thermodynamic limits as modern engines, about 80% of the ideal efficiency at the low compression ratios possible with the fuel he had available. Otto's main contribution, though, was not the constant volume process, but rather the concept of compression of gasses before combustion. This gave an immediate step increase in efficiency over the previous engines. Otto's engines used a homogeneous mixture of fuel and air, with some kind of positive ignition, eventually a spark plug. Thus, modern spark ignition (SI or Otto) engines are direct descendants of Otto's work.

Another major milestone in engine development was the work performed by Dr. Rudolph Diesel. In his efforts to achieve Carnot like conditions (heat addition at constant temperature), Diesel found it necessary to wait to inject the fuel into the air until the compression of the air raised the temperature and pressure sufficiently to ignite the fuel. Diesel was actually closer to achieving constant pressure combustion than constant temperature, and the constant pressure ideal air cycle is named the Diesel cycle in his honor. The concept of compression ignition is Diesel's main contribution, even though it was not his initial objective. But in order to accomplish ignition, a high degree of compression was necessary, and as the standard analysis of ideal cycles will show, a high compression ratio gives a high efficiency. Diesel engines are also called compression ignition (CI) engines.

All modern piston engine cycles have the same basic elements as Otto's and Diesel's engines, and an engine cycle consists of :

- *A Charging Process* - where fresh air (Diesel) or fresh fuel-air mixture (Spark Ignition) is introduced into the cylinder.

- *Compression* - where the volume of the mixture resulting from the charging process is reduced.

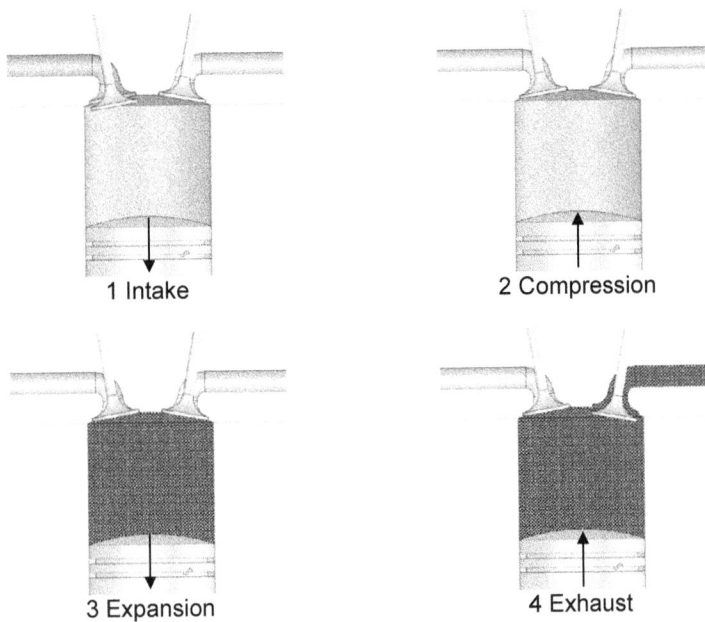

Figure 1.1: The 4 strokes of the piston in the 4-stroke cycle. Combustion occurs between compression and expansion.

- *Combustion* - where the chemical energy in the fuel is liberated to raise the temperature and pressure inside the combustion chamber.

- *Expansion* - where work is produced by the force of the high pressure gasses moving a piston to set a connected mechanism in motion.

- *An Air Exchange Process* - where the expended combustion products are exhausted to the atmosphere and fresh charge brought in.

The differences in engine types derive mainly from the manner in which the combustion is performed and how the air exchange is accomplished. These choices have many implications as to efficiency, power density, fuel requirements, exhaust emissions, *etc.*, that will be discussed throughout the book. The Otto (spark ignition/SI) process involves positive ignition of a "ready to burn" fuel air charge. The Diesel (compression igntion/CI) process involves the "auto-ignition" of fuel as it is injected into high temperature compressed air. In principle, a spark ignition engine could be converted to a diesel engine by exchanging the spark plug with a fuel injector. From the inside of the cylinder, that is all that would be needed. For a myriad of practical reasons this is not a good idea, as many changes in piston design, compression ratio, fuel type and many other parameters would be necessary to provide reliable, long-term efficient service, but in principle the changes are simple.

The other major distinction between engine types can be made with respect to the number of engine revolutions required to perform one work producing cycle. The first compression cycle engines operated using a process where the air exchange process occurred over an entire engine revolution. That is at the end of expansion, the cylinder was opened to the environment and the combustion products sent to the surroundings as the piston moved from bottom to top of the cylinder. When this was done, the exhaust opening was closed, an intake valve opened and a new charge drawn in as the piston moved again to the bottom of the cylinder. This cycle is known as the 4-stroke cycle because of the 4 strokes of the piston, intake, compression, expansion and exhaust. The cycle is shown schematically in Figure 1.1.

This process involves positive displacement of the gasses, and is very effective in removing old cylinder gasses and introducing new. The problem with it is that it requires one entire revolution of the engine for it to be completed. There is essentially no positive work output produced during the air exchange process. In fact, a small amount of work is required to perform it in non-turbocharged engines, such work being called *pumping* work, since the intake pressure is normally lower than the exhaust pressure.

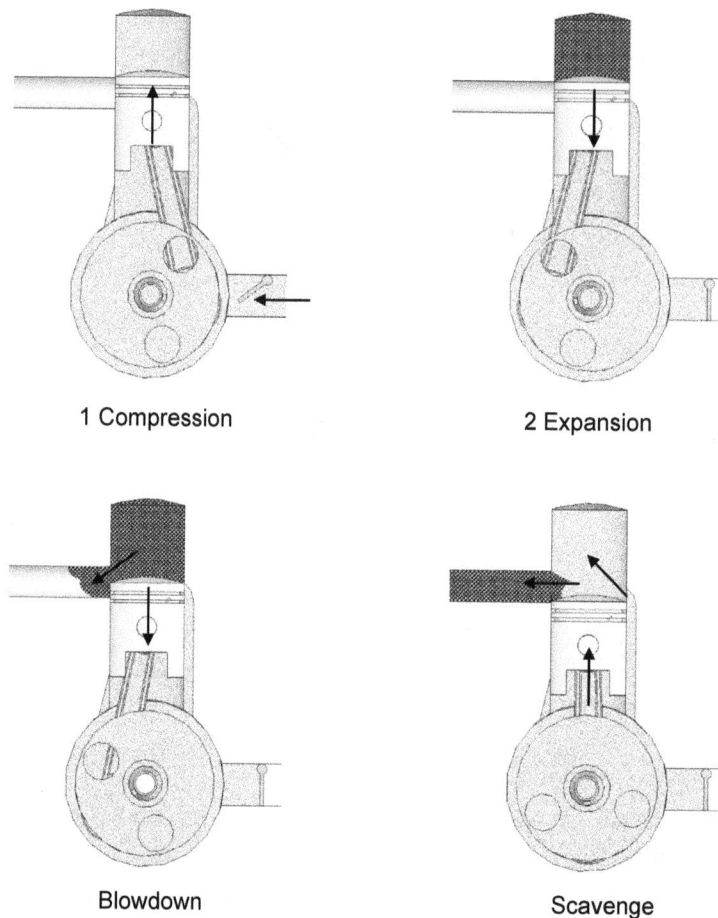

1 Compression	2 Expansion
Blowdown	Scavenge

Figure 1.2: The 2-stroke cycle. Combustion occurs between compression and expansion.

In order to give more work producing processes per engine revolution, the 2-stroke cycle was developed. The basic idea here is to limit the air exchange process to a short period of time between the end of expansion and the start of compression, such that positive work is produced every time the piston moves from its position at the very top of the cylinder, *Top Dead Center* (TDC) to its postion at the very bottom of the cylinder *Bottom Dead Center* (BDC)[1]. If the air exchange can be made as effective as in a 4-stroke cycle, the 2-stroke engine can produce twice as much power as the 4-stroke engine for an engine of the same size and speed. The same amount of work produced twice as often gives twice the power. The cycle is shown schematically in Figure 1.2.

A major obstacle in the 2-stroke engine is overcoming the situation that the air exchange process (Blowdown and Scavenge) must occur rapidly while the piston is near its BDC position, and without the benefits of positive displacement. It is more difficult to get a good air exchange with the 2-stroke engine, but it is possible by a careful design and optimization of the flows in and out of the engine. With some 2-stroke diesel engines, auxiliary blowers are used to help this situation.

In the situations where an effective air exchange process is obtained, engines with quite high power density can be produced. For a variety of reasons, the 2-stroke SI engine has significant problems meeting modern emissions standards, which has prevented it from being used in automobile applications. However, the most efficient engines produced today are large 2-stroke marine diesel engines, with efficiencies of over 50%.

The topic of this book is what is commonly called internal combustion engines, although as the following will show, only a portion of all possible types of internal combustion engines will be discussed. Engines

---

[1]In engine terminology, the term "top" refers to the position where the cylinder volume is at a minimum, and the term "bottom" to the position where the volume is a maximum, regardless of orientation of the cylinder axis.

or motors are devices that convert other forms of energy to mechanical work. The mechanical work can be in used for many things, to move a vehicle, to drive an electric generator or pump or a variety of types of tools. Combustion engines are driven using the release of chemical energy from the combustion of a fuel as the primary source. Vehicle applications will be of major interest in this book, although the engine principles are valid for engines built for a variety of other purposes.

What can be defined as an *internal* combustion engine is a device that combusts a fuel with an oxidizer, and uses the combustion products directly to produce work. In the ideal world, an internal combustion engine can work without the requirement of heat transfer to the working fluid, since the combustion products *are* the working fluid. Though in practice there is some heat transfer taking place, it is not required for the operation of the device. On the other hand, one can refer to an *external* combustion engine. In this device, the combustion products are not the working fluid, and heat transfer to the work producing substance is *required* for the device to work. An example is the Rankine cycle, where a boiler is used to transfer heat from the combustion products to the actual working fluid, in this case usually water/steam. The Stirling engine is also an example of an external combustion engine when powered by combustion products, though it can actually be operated on solar energy, or other non-combustion thermal sources.

Using the above definition of an internal combustion engine, the classic gas turbine is also an internal combustion engine, since the combustion gasses are used directly to produce power by expansion over a turbine, and no heat transfer to the working fluid is *required* in order for the engine to operate. The turbine is a steady flow device, whereas in the conventional piston engine, the expansion is an intermittent process. The thermodynamics of the two expansion and compression processes are, therefore, somewhat different, and so gas turbines and piston engines are normally discussed separately[2]. The intermittency of operation in a piston engine has major implications for efficiency. A piston engine operates with high temperatures, sometimes approaching 3000K. But this high temperature is only present for a very short time, and therefore does not result in such severe cooling problems as in a gas turbine, where the turbine is exposed to the highest cycle temperatures continuously.

The distinction between internal and external combustion engines does not depend on the expansion device as such. The classic steam railway locomotive is an external combustion engine, even though it uses pistons to produce work. The important point is that the working fluid, steam, requires heat transfer from the combustion gasses in order for the engine to operate, therefore, it is an external combustion engine.

The engines to be discussed in this book will be internal combustion engines that involve expansion of the combustion gasses in a closed system with a variable volume. For the most part, this means piston engines, but the thermodynamic principles of non-piston engines are the same as those of piston engines. Of the non-piston engine type, the Wankel engine is the most well-known, though there are other types of rotary engines that have been developed through the years. In practice it is the classic piston-crank mechanism that is all-dominating in the "closed-volume" or positive displacement internal combustion engine category today. It is effective, relatively simple to produce and has proven to be extremely reliable, and has an advantage in that the piston moves very little at the end of the compression stroke while combustion occurs. This mechanism is used with all types of engines, 2- and 4-stroke, spark ignition and compression ignition. The differences between these engine types arise in the configuration of the valves and ports for letting gasses in and out of the cylinder, and in the equipment used to get the fuel into the cylinder and ignited (carburetors, injectors, spark plugs, *etc.*).

1.1.2   Engine Configuration

Given the above, then consider an internal combustion engine with a piston-in-cylinder arrangement. Figure 1.3 shows a sketch of such a system. The piston is enclosed in a cylinder, where sealing between the walls and the moving piston is accomplished through a series of metal rings placed in grooves around the periphery of the piston. The rings expand and press against the cylinder wall, providing a good seal. Especially on the top rings, cylinder pressure acts on the inside of the ring and forces it outward toward the cylinder wall. This helps to maintain a good seal even with very high pressures. Maximum cylinder pressures in excess of 100 atmospheres are not unusual in heavy duty engines. For mechanical reasons,

---

[2]The turbocharger discussed in Chapter 11 is basically a gas turbine, driven by the high temperature exhaust gas.

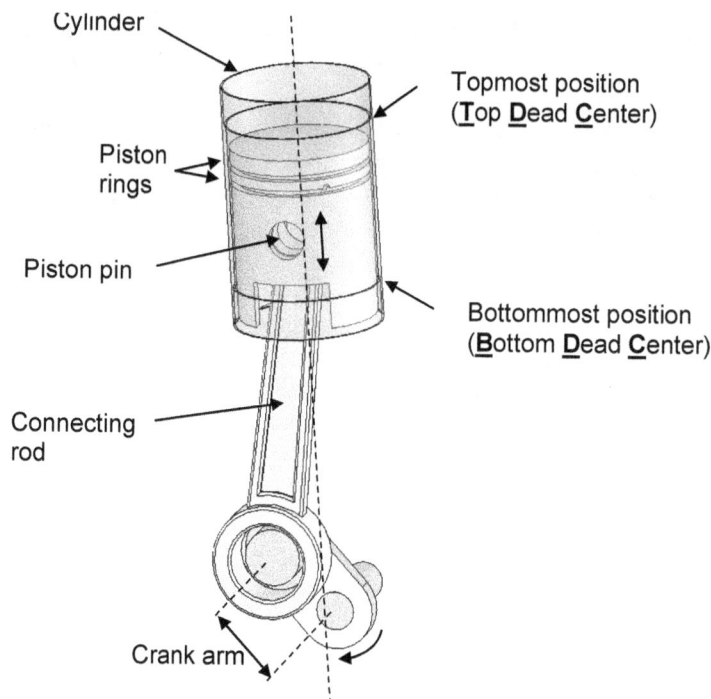

Figure 1.3: The basic geometry of a piston engine.

there is a small gap between the ends of the rings where there is a very small amount of gas leakage. An additional ring, furthest away from the combustion chamber, is used to control the flow/distribution of lubricating oil on the cylinder wall, to limit wear of the the components to an acceptable level while minimizing the amount of oil consumed. The number of rings can vary, but the use of two compression rings and one oil control ring is a very common configuration.

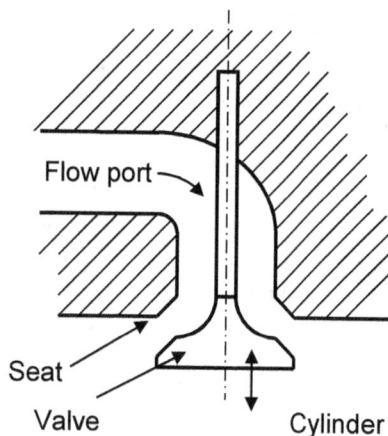

Figure 1.4: A poppet valve.

The top of the cylinder is often closed by a separate piece called the cylinder head. This is bolted to the construction containing the cylinders, pistons and crankshaft apparatus, called the cylinder block. Valves for control of the flow into and out of the engine are normally installed in the cylinder head (overhead valves), though in some smaller or older 4-stroke engines they can be found in the block (flathead, or side valve engine). Camshafts to drive the valves are located in either the block or the cylinder head (overhead camshaft). 2-stroke engines can operate without the valves in the cylinder head that 4-stroke engines need. In the 2-stroke engine, air and exhaust flows often occur through ports in the cylinder walls. This makes it possible for the head and block can be cast in one piece, which is a cheaper construction. In this case, there is no need for machining at the top of the engine to make facilities for the valves. All that is needed is a hole for the spark plug in a spark ignition engine or an injector in a diesel engine. In a 4-stroke engine, the cylinder head is made removable to facilitate installation of the valves, among other reasons.

In order for a combustion process to take place, the cylinder must be filled with air and fuel, so provisions must be made to open and close the cylinder at appropriate times to fill it with fresh charge prior to combustion, and to expel the combustion products after they have been expanded to perform work. There are two common physical methods for this. The first is that of the poppet valve shown in

Figure 1.4, which is opened and closed by the thrust from a cam on a camshaft. An area around the head of the valve is ground to a size matching that of a corresponding area in the cylinder head, called the valve seat. When the valve rests on the seat, there is a good seal, and the opening is effectively closed. The camshaft is physically connected to the crankshaft such that valve openings and closings occurs at fixed crankshaft (piston) positions. Overhead valves have become dominant for 4-stroke engine, since they result in lower flow restriction, and the area exposed to the cylinder pressure is smaller than the outdated flathead design. This also reduces forces acting on the cylinder head. Lately, advances have been made in the area of variable valve timing, such that valve openings and closings can be varied somewhat with respect to piston position at different operating conditions to improve performance and fuel economy.

The other method for air exchange is that of cylinder ports. This is common practice in 2-stroke engines. The ports are normally at the bottom of the cylinder, and whether they are open or blocked is determined by the piston position. If the piston rings are below the port opening, the cylinder is open to flow, and the port then is closed as the piston and rings on their way to the top of the cylinder pass the port and block its flow path. Ports can be used for both intake and exhaust gases, and some engines have a combination of ports and poppet valves. This is found in large marine 2-stroke engines, where there are intake ports in the cylinder walls near BDC and an exhaust poppet valve is located in the cylinder head. Intake flows in 2-stroke engines are basically always controlled by ports.

Figure 1.5: A cut-away view of a 3 cylinder, 4-stroke Spark Ignition (SI) engine. Picture courtesy of Ford Motor Company, used by permission.

The piston moves linearly in the cylinder, and this motion is normally conveniently converted to a rotary motion for driving vehicles or other rotating equipment such as electric motors, pumps, compressors. This is accomplished by the classic slider-crank mechanism shown in Figure 1.3. Many other mechanisms have been proposed, but the slider crank is still far and away the most dominant mechanism for piston powered engines today. The piston is the sliding part, as it reciprocates inside the cylinder. There is a pin in the piston, that connects to a connecting rod, which can rotate around the piston pin. At the other end of the connecting rod, there is a facility for connection to the arm of a crankshaft. The crankshaft rotates around its center, and as the piston moves down in the cylinder, the force on the piston is transmitted through the connecting rod, and applies a torque to the crankshaft, causing it to rotate. Similarly, a torque applied to the crankshaft can be converted to a force on the piston, which can be used to compress the cylinder gasses. Piston motion as a function of crank position is described in Chapter 13.

The torque applied to the crankshaft varies throughout the cycle. It is positive on the expansion, negative on the compression. This could result in a variation of the engine speed throughout the cycle, and give rise to vibrations or other undesirable effects. Therefore, there is a flywheel attached to the crankshaft of an engine. The flywheel is a disc that has an inertia that is normally greater than that of the rotating and reciprocating parts of the engine, and is used to maintain a nearly constant rotational speed of the crankshaft. The greater the inertia of the flywheel, the smaller the variation of rotational speed through the engine cycle. There is a compromise when selecting a flywheel, as a heavy flywheel, which gives very stable operation, gives a slower transient response. The intended application must be considered when designing a flywheel. These elements can be seen in a cut-away view of a spark ignition engine, as shown in Figure 1.5, where the gears on the flywheel indicate another use of the flywheel, as it is connected to an electric motor when the engine is to be started.

An important part of the intake system of a spark ignition engine is the throttle, located near the inlets of the cylinder. Since the SI engine operates with an intake mixture of fuel and air, regulation of the engine requires that the air and fuel flows be regulated to provide a nearly constant ratio of fuel and air. At part load, the throttle opening area is reduced, giving a flow restriction and resulting in a low pressure at the intake. The low pressure means less mixture is filled in the cylinder, and consequently less fuel is burned and less power produced.

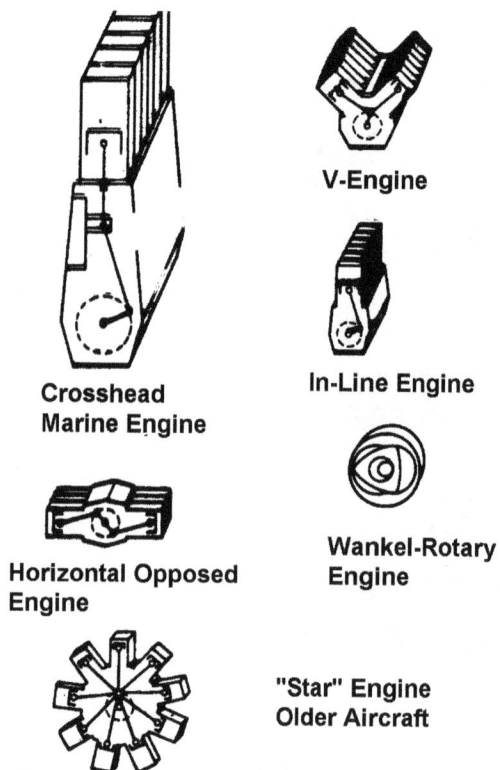

V-Engine

In-Line Engine

Wankel-Rotary Engine

Crosshead Marine Engine

Horizontal Opposed Engine

"Star" Engine Older Aircraft

Figure 1.6: Common cylinder arrangements for internal combustion engines.

Most vehicle engines have more than one cylinder connected to a common crankshaft. There are many possible cylinder arrangements, the most common at the time of writing being either the in-line configuration, where all cylinders lie in a single plane, or the V-arrangement, where two banks of cylinders form a V. There are many aspects to cylinder arrangement choice, some of the more important being balancing and physical space requirements. Some of the implications of the configuration on engine balancing are presented in Chapter 13. Figure 1.6 shows a few of the geometric configurations of modern engines. The large marine diesel engines have a crosshead construction, where the piston is connected to the crosshead, which is better suited for carrying the heavy side thrust from the crank mechanism. This may be recognized from old steam railroad locomotives.

The discussion above has been concerned with the geometrical aspects of piston engines. The geometries can be used with different types of combustion systems. At the moment, there are two major types of combustion systems used for internal combustion engines. The first is the Spark Ignition (SI) engine, also called the gasoline/petrol engine or the Otto engine. As the first name indicates, this is an engine where the combustion process is started through the application of a spark. In order for this to occur, there must be a combustible mixture located near the spark plug at the time the spark is discharged.

A mixture of typical fuel and air can only be ignited by a spark at engine conditions if the mixture ratio is somewhere near the chemically correct proportions. For a typical gasoline this means a ratio of air to gasoline vapor from approximately 8 kg/kg to 25 kg/kg. Outside of these conditions, ignition is uncertain at best.

To achieve this kind of combustion, SI engines traditionally have mixed the fuel and air outside of the engine, and drawn a homogeneous charge of fuel and air in the proper proportions through the intake valves. In older engines, carburetors were used, but today the use of electronic fuel injection has become predominant in automobile engines as well as other vehicle engines. Carburetors are still found in some applications, typically those were strict exhaust emissions standards have not been imposed, and engine price has a high relative importance. The use of a homogeneous mixture of fuel and air has important implications on the operation and efficiency of SI engines, as power regulation requires both the flow of fuel and air be regulated at the same time. As will be seen later, this leads to pumping losses and reduced engine efficiency at part loads. The compression of a mixture of fuel and air can also cause problems if abnormal ignition occurs at an inopportune time in the cycle (knocking).

The other type of combustion system is what is called Compression Ignition (CI) or more commonly the diesel engine. The CI engine compresses air alone to a high temperature and pressure as the piston nears the TDC position. Then, fuel is injected into the air shortly before combustion is desired. After a short time to prepare a suitable amount of combustible air-fuel vapor mixture, this portion of the fuel ignites due to the high temperature and pressure. The flame then spreads throughout the cylinder, and consumes the fuel and air in a process where the mixing between the two components is the dominant factor in determining how combustion occurs. Figure 1.7 shows a cutaway view of a compression ignition engine. The engine is very similar to the spark ignition engine of Figure 1.5, the differences consisting of fuel injectors instead of spark plugs, and a different geometry of the combustion chamber/piston bowl.

The CI engine shown is called a Direct-Injection (DI) engine, as the fuel is injected into a single chamber where the combustion is completed and the gasses expand. This type of engine is replacing the InDirect-Injection (IDI) engine, in which fuel is injected into a preliminary chamber, where it ignites and then the burning fuel and air flow into the main combustion chamber, where combustion is completed. The DI engine is more efficient, and due to modern fuel injection and computer technologies, is now able to meet the speed demands of small vehicles.

There are two advantages to the CI system regarding efficiency. The first is that a high degree of compression is needed to make the air suitable for ignition. This in turn allows for more expansion after combustion and greater efficiency. This will be discussed in more detail in the combustion and ideal cycles sections. The second advantage is that the fuel "finds" its own air during combustion, and it is not necessary to mix the fuel and air outside the engine. So there is no need to control the air flow by restricting the intake air flow. This leads to reduced flow (pumping) losses and greater efficiency, especially at part load. As a result of the above two factors, CI engines have higher efficiencies than their SI counterparts.

Another type of engine has been produced recently, called Gasoline Direct Injection (GDI), which is a kind of hybrid between SI and CI combustion. In the GDI engine, an injector sprays fuel into the combustion chamber as in a diesel engine, but a spark plug is used to start the combustion of the spray, giving direct control of the start of combustion, independent of the type of fuel. The idea is to be able to avoid pumping losses by mixing the fuel with only a portion of the air in the cylinder, eliminating the need to throttle the intake air. With modern computer and injection technology, this type of engine switches between combustion with a spray, and combustion with a completely mixed fuel air charge when the spray combustion is ineffective. This engine is discussed in Chapter 16.

Developments are also under way at the time of writing to develop what is called a Homogeneous Charge Compression Ignition or HCCI engine. This is also a combination of the SI and CI concepts, in that a homogeneous charge of fuel vapor and air is compressed as in a spark ignition engine, but ignition occurs due to the high temperature encountered under compression as in a diesel engine. Since the fuel and air are pre-mixed, they tend to ignite in a very short time as opposed to CI spray combustion. The problem to date has been controlling the ignition process, as it is extremely sensitive to changes in cylinder temperature and pressure. The system is of great interest since it produces very low levels of the emissions of oxides of nitrogen. At the current time, HCCI engines can operate in this mode over a part of the operating region, and must switch between CI and HCCI operation to obtain a suitable range of output.

Figure 1.7: A cut-away view of a 6 cylinder, 4-stroke Direct Injection (DI) compression ignition engine. Picture Courtesy of Ford Motor Co., used by permission.

The different kinds of ignition system pose different demands on the type of fuel used. Therefore, there are significant differences between the physical and chemical properties of gasoline and diesel fuel. These will be discussed later in Chapter 4.

1.1.3 Choice of Engine

Internal combustion engines are used in a wide variety of applications, ranging from small model aircraft engines with power under 1 kW to large multi-cylinder diesel engines for ships, with power on the order of 50 MW. In terms of number, there are hundreds of millions of automobiles throughout the world, and with automobile engine power on the order of 50 to 150 kW, they represent an enormous amount of power generating capacity. Other applications are hand powered tools, heavy-duty road vehicles, portable and emergency electric generators, railway locomotives, light private aircraft, pumping stations, all sizes of ships and a variety of other applications.

Each of these applications has its set of requirements that determines the choice of engine. A few of the important factors that determine selection of engine design are:

- Economy - including purchase, operation and repairs. One of the most important factors in the choice of an engine is the economical aspect. It is important to weigh all the various economic aspects for an application. Prime considerations are the purchase price of the engine, the amount of money used for fuel and other consumable items (mainly lubricating oil), inspection and maintenance, including repairs, possibly even disposal costs. There is a wide range in the relative importance of these parameters as the applications for engines range from small tools used a few hours per year to large engines in operation almost around the clock for many years. Engines that use a lot of fuel can be designed to operate on very cheap fuel, large marine engines that run on bunker fuel being a classic example.

- Size and Weight - These two parameters are normally closely related. Again, there is a wide range of requirements where size and weight are very important such as hand tools and the like, to cases where there is an abundance of room such as in larger ships. In modern passenger cars, the desire for aerodynamically efficient chassis design often sets tight limits on the allowable size and shape of the engine. Weight is a very important factor in piston engines for light aircraft.

- Lifetime and durability - Since durability is often accomplished by more expensive materials, larger parts and special design, costs can be limited by making the engine no more durable than needed. Small power tools normally have lower requirements for durability than commercial transport engines, which must operate reliably for 10's of thousands of hours. Certain applications require high reliability, such as aircraft engines or emergency generators. This also has an impact on the design and ultimate choice of the engine for an application.

- Environment - One of the most important environmental aspects related to internal combustion engines has become that of air pollution. Especially in the field of transport, engine emissions have played a large role in air pollution problems. This has primarily been the case concerning the local air quality in cities, but vehicle emissions have also contributed to acidification of freshwater lakes in certain areas. Great strides have been taken in this area since about 1970, and engines at the time of writing exhibit exhaust emission reductions of more than an order of magnitude from the engines before emission limits were implemented. A side benefit of environmental requirements for engines has been a more rapid use of electronics and computer technology. In the 21st century, greenhouse gas emissions of CO_2 will be a serious consideration in engine and vehicle design.

  Noise is also an environmental factor, and certain engine types are well known for generating noise. Engine choice and installation have a large impact on the noise generation. Other environmental aspects that may be considered are the disposal of waste products such as used lubricating oil and filters or coolants.

1.2   Engine Parameters

In the following, a review is given of some general parameters of engine operation. There are parameters used to describe performance as well as parameters determining performance. The development of the basic power equation gives an important basis for understanding the interrelation between engine parameters and performance.

1.2.1   Measuring Power

The purpose of an engine is to produce Work/Power. From basic physics, mechanical work is produced when a force is exerted through a distance:

$$Work = F \cdot Distance \quad [N-m]. \tag{1.1}$$

Power is the rate at which work is performed in N-m/s or J/s (Watt). This can be calculated by the product of a force and a velocity in linear systems.

Since the primary purpose of the engine is to function as a power source, the power output of the engine is the most important experimental parameter to be measured. A machine called a *dynamometer* is used to measure the power output of an engine. One typically uses a *"brake"* or dynamometer to apply a load to an engine. Figure 1.8 shows the general principle of a dynamometer. There are two basic parts, the stator and the rotor. The *stator*

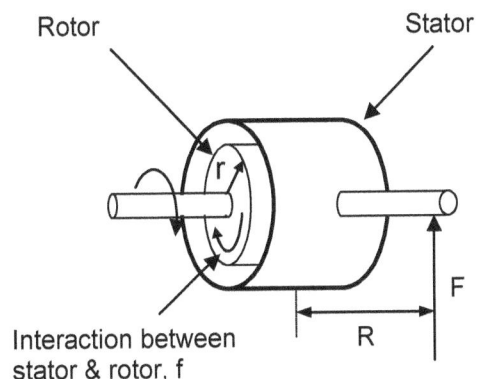

Figure 1.8: Schematic diagram of an engine dynamometer.

is mounted in a set of outer (trunnion) bearings that allows it to rotate without friction. The *rotor* is directly connected to the engine, and rotates inside of the stator. By electrical, mechanical or hydraulic means, there is an interactive force, f, between the rotor and the stator which acts at the inner radius of the stator, (outer radius of the rotor) r. The methods of doing this in practice will be discussed shortly. But the stator is kept from rotating by an arm with a radius from the center of the axle R, which is attached to a scale or force transducer that measures the force applied to the stator, F. By a moment balance on the stator, $rf = RF$.

The work done through one rotation of the rotor is the force, f, times the distance through which it reacts, which is the circumference of the rotor $2\pi r$,

$$Work/rev = 2\pi rf = 2\pi RF \tag{1.2}$$

The product RF is called the torque of the engine, denoted here by T and has units of N-m.

The power is then equal to the rate of work per unit time:

$$P = \frac{2\pi RFN}{60000} = \frac{2\pi TN}{60000} \quad \frac{N-m}{rev}\frac{rev}{min}\frac{min}{s}\frac{kW}{1000W} \Rightarrow \quad kW \tag{1.3}$$

The power measured using a dynamometer is called brake power because the device used to measure it is often called a brake. The term brake comes from old devices, where a mechanical friction brake was used to squeeze the output shaft of the motor. In modern usage, it is more common to use the term dynamometer for the power absorption device. The terms are interchangeable. Equation (1.3) is valid for any type of engine or power producing device, since it is based on an analysis of the dynamometer and not of the engine.

There are several dynamometer types used to measure engine performance. Each has its advantages and disadvantages.

1. **Water Brake** - This type of brake uses water to transmit a force from the rotor to the stator. In both elements, there are specially shaped buckets that receive water from the other element, change its direction, and then send it back to that element. From basic fluid mechanics we know that changing the momentum of a fluid (changing at least the direction of its momentum vector) results in a force. In addition to this basic force, there are hydraulic losses that occur due to friction that also result in forces to the elements. The load applied by a water brake can be adjusted by regulating the water level in the housing.

Since fluid mechanics are involved, the load attained for a given water level is proportional to the square of the velocity; that is to say, the load (torque) applied by the dynamometer is proportional to the square of the rotor speed, as show in the characteristics of Figure 1.9. Applying Equation (1.3), the power absorbed by the dynamometer is then proportional to the cube of the engine speed. This makes the water brake a natural device for testing to simulate practical types of conditions in applications where the load follows this pattern (load proportional to the square of the speed): Ships, propeller driven aircraft and land vehicles where aerodynamic loading is important. Typically with a water brake, the loading is set for a given condition, and the load "follows" along as the speed

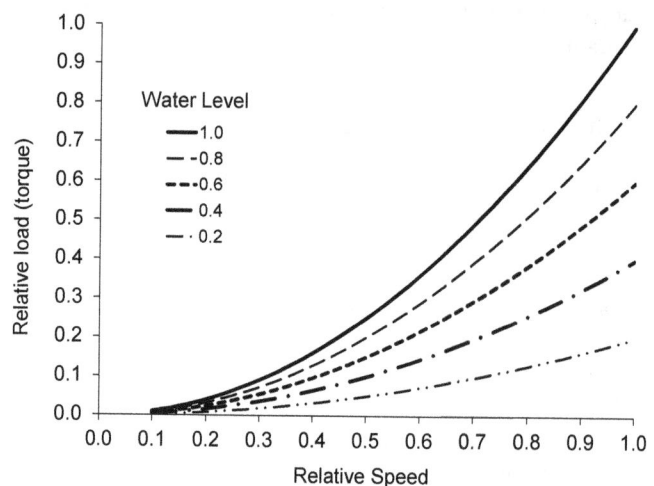

Figure 1.9: The schematic characteristic of a water brake, showing the increase in the load with the square of the increasing speed.

changes. That is, the torque is proportional to the square of the engine speed and the power is then proportional to the cube of the engine speed, the speed in essence determining the load. A water brake is normally smaller than the other common brakes/dynamometers for a given power absorption, which is an advantage, especially when measure marine engines with power in the megawatt range. While they are convenient for the loading types mentioned above, water brakes are less suited for engine operation at low loads than the eddy current dynamometer, which will be discussed in the following.

2. **Eddy Current Dynamometer** - A second type of brake/dynamometer is that which uses the eddy current principle. It is known that currents are induced in a conductor moving through a magnetic field. In this type of dynamometer, these currents are established in the rotor and are called eddy currents. They establish a magnetic field that opposes the motion of the metal rotor by interacting with the field that established them. In an eddy current dynamometer, the rotor is made of a suitable metal, and there are coils in the stator. The coils are energized by a dc current, establishing a magnetic field through which the rotor moves. This establishes the eddy currents in the rotor, and the strength of the eddy currents and their reaction with the stator, is regulated by regulating the current through the coils.

A typical characteristic for an eddy current dynamometer is shown in Figure 1.10 As opposed to the water brake, large loads are easily established at low speeds. Therefore, this type of dynamometer is well suited for engine testing over the entire speed and load range, in that its characteristic loading fits the basic nature (torque curve) of a motor with regards to speed. It is also easy to regulate an eddy current dynamometer with a fairly small DC current to the coils. Eventually, the power absorption of an eddy current dynamometer is limited by the amount of power that can be removed by the cooling system. This gives the hyperbolic limiting curve in the high speed and load regions of Figure 1.10, since the product of speed and torque are constant in this region.

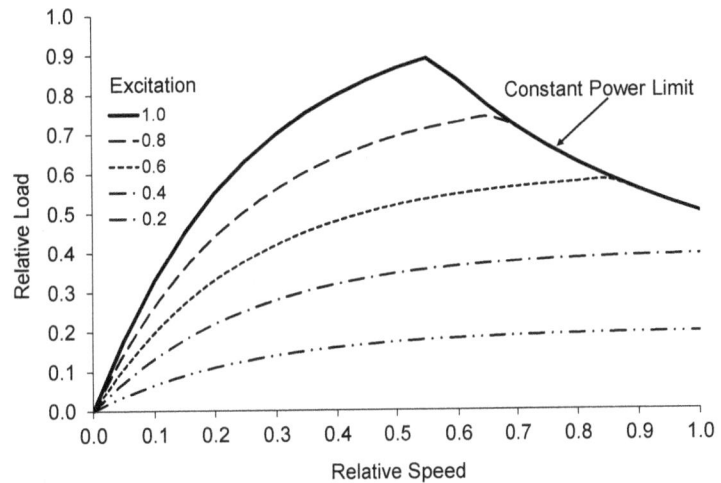

Figure 1.10: The schematic characteristic of an eddy current dynamometer, showing a nearly constant load with increasing speed.

Electric Dynamometer - A third type of dynamometer is the electric dynamometer. An electric generator provides the load on the engine in this case. The generator is mounted in trunnion bearings, and the load measured by a load cell on the dynamometer arm. In principle, any kind of generator can be used here. One application that is very useful is that of an electric motor-generator set. This has the advantage that the operating mode can be switched, so that when the unit runs as an electric motor, the combustion engine can be turned over for starting purposes. But more important is the possibility of turning the engine over at higher speeds in order to obtain a measure of the friction of the engine, through the so-called *motoring* method. This method has some disadvantages, but has been used for many years to estimate engine friction. The motoring method is an alternative to the Willans line method, and the method of using cylinder pressure measurements, which will be described later. From a practical point of view, the electric dynamometer is very advantageous in situations where a large amount of engine testing is done, in that it is possible to return the engine power to the electrical grid and receive payment for it. With the water brake and eddy current dynamometers, the power generated is merely dissipated in cooling water. On the other hand, these devices are normally cheaper to purchase and install than the electric motor-generator unit.

In modern applications, large motor-generator sets are used to provide transient engine test facilities. Transient test cycles are used for emissions testing of heavy-duty engines, and the dynamometer and its control system must be able to rapidly switch between motoring and absorbing states, to simulate acceleration and deceleration of vehicles, a challenging application.

1.2.2 Indicated Power

In an engine, all the mechanical work originates from the pressure of the gasses in the combustion chamber acting on the top of the piston, as is shown in a pressure - volume $(p - V)$ diagram. The volume change is prescribed by the motion of the piston. For most of the time, it is assumed that the engine is operating in a steady state, that is, that any given engine cycle is identical to the preceding cycle. This is not true in the case of transient operation, but the details of cylinder conditions have been of less interest for transient operation. In determining the indicated power with this method, it is necessary to measure the instantaneous cylinder pressure throughout the cycle.

Historically, several kinds of sensors have been used to measure cylinder pressure in engines. Cylinder pressure measurement technology has evolved to the stage that nearly all engine cylinder pressure measurements today are made using a piezoelectric quartz transducer. The principle of this transducer is that a force applied to a certain kind of quartz crystal causes an electric charge on the crystal. The charge is proportional to the force, and the crystals are mounted in adapters that can be inserted into the engine combustion chamber. A schematic picture of a transducer is shown in Figure 1.11. The cylinder pressure is transferred to the crystal through a thin metal diaphragm, in such a way that the charge produced has a very linear response with respect to the pressure. The charge is converted to a voltage using a charge amplifier, based on operational amplifiers. The electric signal is an ideal form for modern data acquisition methods. The transducers have adequate frequency response to provide an accurate measurement of the pressure.

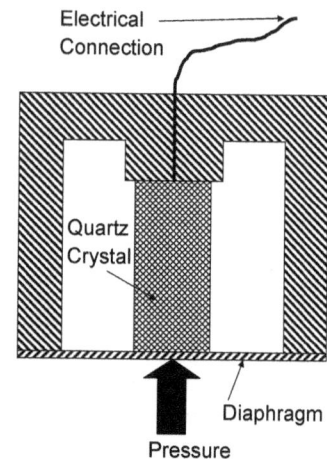

Figure 1.11: Schematic diagram of a piezoelectric pressure transducer.

Piston position is necessary to determine cylinder volume, and the most appropriate method for determining it is through the use of a rotary shaft encoder. The latter is connected to the crankshaft at a known position, and produces a known number of pulses per revolution. Since the pulses are produced at fixed degree intervals, they are used to send a signal to the data acquisition system "telling" it to measure the pressure. The pulses act like a clock that does not necessarily operate with a constant frequency. This is important, since the engine may experience slight changes in speed during the measurement period. The key point is that a value of the pressure is sampled at a known crank angle position for each pressure measurement. The corresponding cylinder volume can then be easily calculated for each pressure measurement, using Equation (2.106).

A rotary shaft encoder which produces 360 pulses per revolution, for example, is suitable for collecting the cylinder pressure data every crank angle degree. If a higher degree of accuracy is desired, encoders can readily be obtained which give more than 360 pulses per revolution, though extreme care must be taken to accurately determine the position of the encoder relative to top dead center.

The crank angle shaft encoders used normally have more than one channel, and it is common to have an additional channel with one pulse per revolution. This pulse is normally associated with top dead center, or some well known, accurately determined crank shaft position. This serves as a reference value from which all the other crank angle positions can be determined by simple counting, and is typically used to initiate the overall data collection process. For a 4-stroke engine, a sensor mounted on the crankshaft would give two pulses per cycle, hence, extra handling of the data is necessary, or else the crankshaft pulse should be logically compared with a signal on the camshaft, which rotates at half the crank angle speed. It would not be a good idea to put the crank angle sensor on the camshaft as there is a mechanical connection between it and the crankshaft, and position errors could occur due to clearances and the like.

Typical data of this kind is shown in Figure 1.12, which is a plot of cylinder pressure and volume as a function of crank angle degree. The engine is the CFR[3] fuel testing engine, which has been used for decades for measuring fuel knock (the original purpose for its design) and fuel combustion characteristics. Here, the engine is operating at a compression ratio of 7:1, with a spark timing of 20° Before Top Dead Center (BTDC), with a fuel air ratio of 1.05 times the chemically correct mixture of gasoline and air at

---

[3]CFR denotes Cooperative Fuel Research.

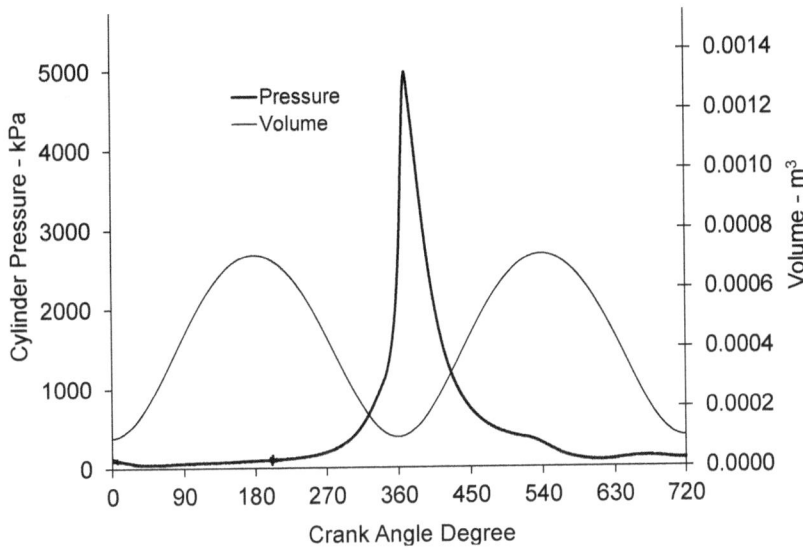

Figure 1.12: Cylinder pressure and volume as functions of crank angle degree for a CFR, a 4-stroke, single cylinder spark ignition engine.

a speed of 1200 revolutions per minute.

These data have been plotted again in Figure 1.13, where measured cylinder pressure is shown as a function of cylinder volume. The resulting pressure-volume diagram is called an *indicator diagram*. The

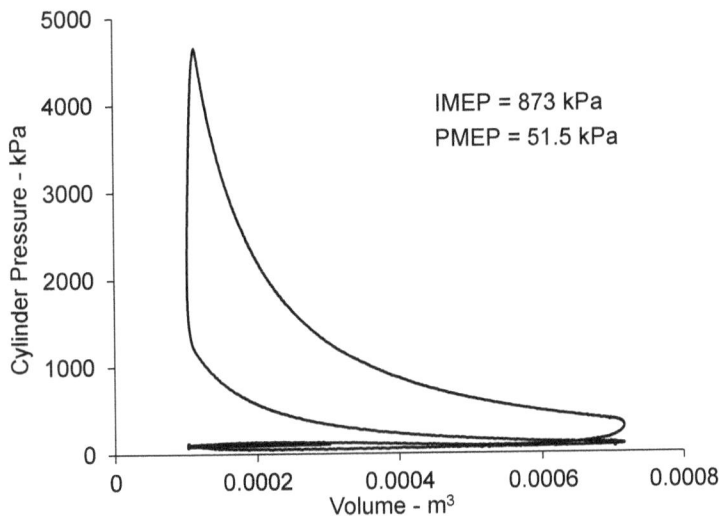

IMEP = 873 kPa
PMEP = 51.5 kPa

Figure 1.13: A pressure volume diagram for a CFR, single cylinder, 4-stroke spark ignition engine.

work produced is called the indicated work. The work produced for the cycle, W_i, is then equal to:

$$W_i = \int_{BDC_{comp}}^{BDC_{exp}} p \cdot dV \tag{1.4}$$

and is called the indicated work, since it comes from the indicator diagram.[4] The work from the start of compression to the end of expansion is the net positive work-producing portion of the cycle in

---

[4]The term derives from steam engines where the pressure was measured with a device called the pressure indicator.

4-stroke engines. The remainder of the cycle, from the end of expansion to the start of compression, is the portion associated with pumping the air through the engine. This work is normally associated with the friction of the engine, since it is work that must performed in order to operate the engine. There is no universally accepted standard on this subject, so one needs to be careful, as some authors include pumping in the indicated work and others do not. Some authors distinguish the gross indicated work (that from the compression and expansion strokes) from the the net indicated work (all four strokes of a 4-stroke engine). In this text, pumping work is understood to be part of the friction, and is not included in the indicated work, the latter being the work on only the compression and expansion strokes.

In 2-stroke engines, the integral in Equation (1.4) covers the complete cycle of the engine. However, in many 2-stroke spark ignition engines, the bottom of the piston is used to pump the gasses into the cylinder, and a compression process occurs below the piston. This is not visible from a standard p-V diagram, since it is on the other side of the piston from the combustion chamber, and must be measured separately. This crankcase pumping work is considered part of the engine friction. It can be measured in the same way as the cylinder pressure, using an appropriate pressure transducer in the crankcase. Typical data for a crankcase compression of a small 2-stroke spark ignition engine is shown in Figure 1.14.

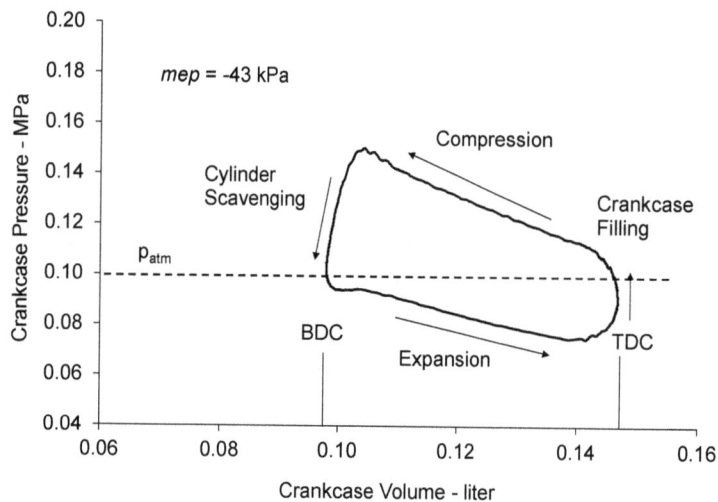

Figure 1.14: Crank case pressure and volume for a small 2 stroke single cylinder, spark ignition engine.

In connection with the indicated work just discussed, the power produced on the top of the piston is called the indicated power, IP. When the friction is overcome by using some of the indicated work, the remainder is available for brake, i.e. useful, work, that is the friction power, FP, is subtracted from the indicated power to give the brake power:

$$BP = IP - FP \qquad (1.5)$$

The mechanical efficiency is defined as a way to characterize the relationship between the piston work and the brake work, that is the relative loss due to friction:

$$\eta_m = \frac{BP}{IP} = \frac{IP - FP}{IP} \qquad (1.6)$$

1.2.3 Mean Effective Pressure

Engines are produced in a wide range of sizes, from small engines for model aircraft to large marine diesel engines, spanning approximately 6 orders of magnitude in power. Yet when viewed in the proper perspective, the engines are remarkably similar. Equation (1.4) shows that the work produced by an engine is proportional to its volume. Therefore, it is natural to normalize this work with respect to the engine's displacement volume. The displacement volume, is the volume displaced by the piston in one stroke between TDC and BDC. It is calculated:

$$V_d = \frac{\pi}{4} B^2 \cdot S \cdot n_{cyl} \tag{1.7}$$

where: B is the cylinder bore diameter, S is the length of the stroke of the piston, and n_{cyl} is the number of cylinders in the engine.

It can be seen that the units of the normalized work will be pressure, and in fact this term is defined as the *indicated mean effective pressure*:

$$imep \equiv \frac{W_i}{V_d} = \frac{\int p \cdot dV}{V_d} \tag{1.8}$$

While the units of *imep* are those of pressure, in reality it refers to a different quantity, which is the work performed per engine cycle per unit of displacement. Given this, it is straightforward to calculate the indicated engine power from the *imep*. For *imep* in kPa, volume in m^3 and power in kW:

$$IP = imep \cdot V_d \cdot \frac{N}{60 \cdot x} \quad \left[\frac{kJ}{m^3 \cdot cycle} \cdot m^3 \cdot \frac{rev}{min} \cdot \frac{min}{s} \cdot \frac{cycle}{rev} = \frac{kJ}{s} = kW \right]$$

$$= \frac{imep \cdot V_d \cdot N}{60 \cdot x} \tag{1.9}$$

Here, x is the number of revolutions of the engine output shaft per working cycle of the engine, hence $x = 1$ for a 2-stroke engine, and $x = 2$ for a 4-stroke engine.

The indicated mean effective pressure for the engine in Figure 1.13 was determined by numerical integration of the p-V diagram and found to be 873 kPa. This is the portion of the diagram from the start of compression to the end of expansion. For the rest of the cycle, the pumping mean effective pressure is found to be 51.5 kPa. The pumping work must be subtracted from the indicated work to determine what is left over to overcome the mechanical friction and be available for useful work.

Note that absolute level of the cylinder pressure does not directly influence the *imep*. That is, it is the integral of the product of the pressure and the volume change over the compression and expansion portions of the cycle that is important. For example, if all of the pressures in Figure 1.13 were arbitrarily increased by a constant value, the work and *imep* would remain unchanged, since the area of the diagram would be unaffected. Thus, while high *imep* producing engines often are connected with high cylinder pressures, high pressure itself is no guarantee of a high *imep*.

Note also that given the *imep*, the indicated work per cycle can easily be calculated by multiplying the *imep* by the displacement volume. In some texts, this idea is used to "define" the mean effective pressure as that pressure which when exerted on the piston over the expansion stroke (piston movement corresponding to V_d), would give the engine work produced. This is, in reality, a result of the true definition, Equation (1.8), and often serves to confuse rather than clarify.

Using the idea of a mean effective pressure as a work per cycle per displacement volume, any power can be converted to a corresponding mean effective pressure, though not with the same physical relation to the cylinder pressure as Equation (1.8). Thus, the brake and friction mean effective pressures are very commonly used in engines. The indicated, brake and friction mean effective pressures can be calculated as follows:

$$imep = \frac{60 \cdot x \cdot IP}{V_d \cdot N} \tag{1.10}$$

$$bmep = \frac{60 \cdot x \cdot BP}{V_d \cdot N} \tag{1.11}$$

$$fmep = \frac{60 \cdot x \cdot FP}{V_d \cdot N} \tag{1.12}$$

It is also obvious that:

$$imep = bmep + fmep \tag{1.13}$$

Throughout this text, the following units will typically be used: Displacement volumes in m^3 and pressures in kPa. These units substituted directly in Equations (1.10) to (1.12) will give power in kW. Volume in m^3 is equal to volume in liters divided by 1000.

An interesting result can be determined by equating the brake power from Equation (1.11) with that from the torque measured with the dynamometer, Equation (1.3):

$$BP = \frac{bmep \cdot V_d \cdot N}{60 \cdot x} = \frac{2\pi \cdot T \cdot N}{60000}$$
$$\Rightarrow bmep = \frac{2\pi \cdot T \cdot x}{1000 V_d} \tag{1.14}$$

The term $2\pi T$ was earlier seen to be the work produced per engine revolution (Equation (1.2)), the factor x is the number of revolutions per cycle, and so the $bmep$, naturally is the work per cycle per unit displacement. Equation (1.14) shows the torque and $bmep$ to be proportional, so discussion of trends in one automatically applies to the other.

There are different ways of calculating the indicated work. The correct method by definition is that from Equation (1.4). This requires a cylinder pressure as a function of volume. This is simple in computer simulations, as the pressure is the primary variable calculated. For an actual engine, the pressure must be measured in the cylinder, and pressure measurement is not trivial. Other approaches include the estimation of engine friction by the motoring method, as described with the electric motor generator dynamometer, or the Willans line procedure, described in Section 1.4. If the friction and brake powers are known, then the indicated power can be found by Equation (1.5)

1.2.4 Mean Piston Speed

Although the absolute values of power, torque, fuel flow rates, *etc.* are important, the use of parameters such as efficiency/specific fuel consumption and mean effective pressure allow us to compare engines of different sizes and speeds. A parameter which allows us to compare engine speed is the mean piston speed. Especially in the areas of friction, wear and lubrication, it is not necessarily the speed *per se* (RPM) that is important, but rather the velocity with which the various engine parts move relative to one another. To make this comparison, we use the concept of the mean piston speed.

The idea behind the mean piston speed is this: In a piston engine the piston moves up the length of the stroke and then down the same distance in the course of one engine revolution. The distance traveled in one revolution is then twice the stroke of the engine. The mean piston speed is calculated as the distance moved in this period of time:

$$S_p \equiv \frac{2 \cdot S \cdot N}{60} \tag{1.15}$$

The maximum instantaneous piston speed is higher than the mean piston speed, but basically proportional to it, since all piston engines use a similar geometry. The mean piston speed is, therefore, a readily available parameter that is useful for describing engine speed, especially when considering wear and durability.

The utility of mean effective pressure and mean piston speed can be seen by comparing specifications for a variety of arbitrarily chosen engines. Tables 1.1, 1.2 and 1.3 show specifications for some typical passenger car engines, typical diesel engines for road goods transport, and some miscellaneous engines with a wide range of powers, typical of engines of the late 1990's. For these engines, the power ranges from 2.5kW to 82MW, a factor of over 30000, the displacement varies by a factor of more than 40000 and the speed range is from 80 rpm to 6000 rpm, a factor of 75. Yet when looking at the bmep, the range is from 405 to 2767 kPa, a factor of 6.8.

The piston speed range is from about 8 to 16 m/s at rated power, with most diesel engines below 12, giving a range of about a factor of 2 as compared to the absolute speed range of 75. Piston speed is a very important parameter for engine wear and is a limiting factor in engine durability. Values of piston speed as above give acceptable wear and durability regardless of engine size.

This all suggests that there is a great similarity in engines, regardless of differences in speed and power. In this text, there is probably more focus on similarities between engines than their differences. The *bmep* ranges are predicted by simple arguments and the use of reasonable parameters. This is the topic of the next section.

Table 1.1: Specifications for a variety of 2017 4-stroke passenger car engines. TC denotes turbocharger, SC denotes supercharger, NA denotes Naturally Aspirated

Manufacturer:	Peugeot	Citroën	Kia	Ford	Audi	Hyundai	Peugeot	VW
Type	SI	SI	SI	SI	SI	CI	CI	CI
Air intake	NA	TC	NA	TC	SC	TC	TC	TC
Bore - mm	71.0	75.0	77.0	87.5	84.5	75.0	75.0	82.5
Stroke - mm	84.0	90.5	85.4	83.1	89.0	84.5	88.3	92.8
Cylinders	3	3	4	4	6	3	4	4
Displ. - liter	1.00	1.20	1.59	2.00	3.00	1.12	1.56	1.99
Rated power - kW	52	81	77	184	245	55	88	162
Rated rpm	6000	5500	5700	5500	5500	4000	3500	4500
Rated S_p - m/s	16.8	16.6	16.2	15.2	16.3	11.2	10.3	13.9
Max Torque - Nm	95	205	147	360	440	180	285	350
T_{max} rpm	4300	1500	4000	2000	2900	1750	1750	1500
Max $bmep$	1197	2148	1161	2263	1846	2020	2295	2217
S_p @ T_{max}-m/s	12.0	4.2	11.2	5.6	8.1	4.9	4.9	4.2

Table 1.2: Specifications for a variety of 2017 4-stroke engines for highway goods vehicles. TC denotes turbocharger, NA denotes Naturally Aspirated, T/B denotes truck, bus, all engines with charge after coolers.

Manufacturer:	Peugeot	Peugeot	GM	Volvo	Scania	MAN	Daimler	Volvo
Type	SI	CI	CI	CI	CI	CI	CI	CI
Air intake	TC	TC	TC	TC	TC	TC	TC	TC
Useage	Van	Van	Pickup	T/B	T/B	T/B	T/B	T/B
Bore-mm	77.0	75.0	94.0	110	134	126	139	144
Stroke- mm	88.5	88.3	100	135	140	166	171	165
Cylinders	4	4	4	4	5	6	6	6
Swept vol-liter	1.65	1.56	2.78	5.13	9.87	12.4	15.6	16.1
Rated Power-kW	89.5	85.8	147	177	236	373	460	552
Rated rpm	6000	3600	3600	2200	1800	1800	1700	1800
S_p @ BP_{max}-m/s	11.7	10.6	12.0	9.9	8.4	10	9.7	9.9
Max Torque	160	240	500	900	1370	2300	3000	3550
T_{max} speed - rpm	4250	1500	2000	1200	1100	1200	1100	950
Max $bmep$	1220	1933	2263	2204	1744	2327	2421	2767
S_p @ T_{max}-m/s	12.5	4.4	6.7	5.4	5.1	5.5	6.3	5.2

Table 1.3: Specifications for a variety of "other" engines from 2017. TC denotes turbocharger, all with after cooler, NA: Naturally Aspirated, DI: Direct Injection, IDI: InDirect Injection

Manufacturer:	TVS	Yanmar	Yanmar	Volvo	MAN	Deutz	MAN
Type	SI	PC-CI	DI-CI	DI-CI	SI-Gas	DI-CI	DI-CI
2/4 Stroke	2	4	4	4	4	4	2
Air intake	NA	NA	NA	TC	TC	TC	TC
Application	MoPed	Utility	Utility	Marine	Utility	Off-road	Ship
Bore-mm	46.0	72.0	88.0	110	108	132	950
Stroke-mm	42.0	74.0	90.0	135	125	145	3460
Cylinders	1	3	3	4	6	8	12
Swept vol-liter	0.07	0.90	1.64	5.13	6.82	15.9	29430
Rated Power-kW	2.6	14.0	26.2	160	184	440	82440
Rated rpm	6000	3600	3000	2200	2200	1900	80
S_p @ Pmax-m/s	8.4	8.9	9.0	9.9	9.2	9.2	9.2
Max Torque-Nm	5	51	105	910	850	2650	$9.84 \cdot 10^6$
T_{max} speed - rpm	3500	2400	1900	1450	900	1400	70
Max $bmep$ - kPa	450	709	803	2218	1555	2098	2100
S_p @ T_{max} - m/s	4.9	5.9	5.7	6.5	3.8	6.8	8.1

1.2.5 Basic Engine Power Equation

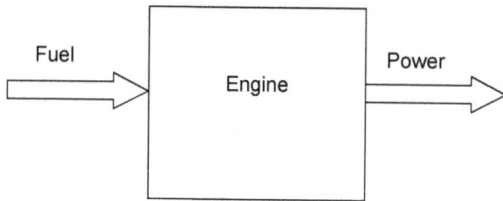

In this section, a general equation will be developed for the power output of an engine as a function of several parameters and variables. In the course of the development, several important engine parameters will be defined. The development starts with a very simple description. Consider an engine as a "black box" where we put in fuel and obtain an output power:

When we send fuel into the engine, we are really putting in chemical energy, which depends on the lower heating value of the fuel, H_u - kJ/kg, which is the energy released when the fuel is completely burned and the combustion products cooled to the original temperature. The units are typically kJ/kg fuel or MJ/kg. The input energy is then:

$$Q_{in} = \dot{m}_f H_u \tag{1.16}$$

We define the *brake engine (thermal) efficiency*, η_e such that:

$$BP = \dot{m}_f H_u \eta_e \tag{1.17}$$

Similarly, the indicated power can be related to the *indicated efficiency*:

$$IP = \dot{m}_f H_u \eta_i \tag{1.18}$$

The indicated thermal efficiency is determined by the compression ratio, the timing and speed of the combustion, and the amount of heat loss. The *compression ratio* is defined as the maximum cylinder volume divided by the minimum cylinder volume.

$$\varepsilon \equiv \frac{V_{max}}{V_{min}} = \frac{V_c + V_d}{V_c} \tag{1.19}$$

Where V_c is the clearance (minimum) volume, and V_d is the displacement volume.

One means of describing engine efficiency is the *specific fuel consumption*.

While the thermal efficiency is the most significant parameter from a thermodynamic point of view, in practical terms the efficiency is usually expressed through the brake specific fuel consumption, defined as:

$$bsfc \equiv \frac{\dot{m}_f}{BP} \tag{1.20}$$

The normal units for this are in grams per kilowatt hour. Obviously, the $bsfc$ and brake thermal efficiency are related.

Substituting Equation (1.17) into (1.20):

$$bsfc = \frac{\dot{m}_f}{\dot{m}_f \cdot H_u \cdot \eta_e} = \frac{1}{H_u \eta_e} \tag{1.21}$$

The units used in this text are normally heating value in kJ/kg and $bsfc$ in g/kWh, so using these units, with 1000g/kg and 3600 s/hour:

$$bsfc = \frac{3.6 \cdot 10^6}{H_u \eta_e} = \frac{3.6 \cdot 10^6}{H_u \eta_i \eta_m} \tag{1.22}$$

It can easily be shown that:

$$isfc = \frac{3.6 \cdot 10^6}{H_u \eta_i} \tag{1.23}$$

The *fuel air ratio*, denoted FA, is defined as the mass of fuel divided by the mass of air, and is an important engine variable. If we look at most engine fuels today, they are made up of many compounds consisting of mainly hydrogen and carbon, and the correct proportion for complete combustion (stoichiometric conditions) is about:

$$FA \equiv \frac{\dot{m}_f}{\dot{m}_a} = m_f/m_a \simeq 0.068 \tag{1.24}$$

If the fuels are liquid, then using typical densities in kg/m^3:

$$\frac{\rho_f}{\rho_a} \simeq \frac{800}{1.2} \simeq 670 \tag{1.25}$$

which means that the volume of the air required to burn a mass of fuel is about 10000 times larger than the volume of fuel. Even for most gaseous fuels (LPG, natural gas) the air volume is about 15 times larger than the fuel volume. Therefore, the air volume is a most important limiting factor for engine performance, and it is more useful to write the fuel flow in terms of the airflow and the fuel air ratio, since this focuses on the most important engine parameters.

The air flow is determined by the displacement (swept) volume, which we theoretically can fill with air at the intake density, ρ_{in}. Since the process is not perfect, we introduce another efficiency called the volumetric efficiency, η_v :

$$\eta_v \equiv \frac{\dot{m}_a}{\dot{m}_{air,ideal}} = \frac{\dot{m}_a}{\dfrac{\rho_{in} V_d N}{60x}} \tag{1.26}$$

In 2-stroke engines, the terms corresponding to the volumetric efficiency and the air density used in Equation (1.26) are defined in a slightly different way, but the principle is the same. That is, ideally we fill the cylinder volume with air that exists at a reference density. In the 4-stroke engine, the volume to be filled is the displacement volume, V_d.

Then we can use the fuel air ratio, FA, and the air flow to express the fuel flow, and for the four-stroke engine, the equation for the power becomes:

$$BP = \dot{m}_f \cdot H_u \cdot \eta_e \tag{1.27}$$
$$= \dot{m}_a \cdot FA \cdot H_u \cdot \eta_e \tag{1.28}$$
$$= \eta_v \cdot \rho_{in} \cdot \frac{V_d \cdot N}{120} \cdot FA \cdot H_u \cdot \eta_e \tag{1.29}$$

We are also interested in explicitly showing the friction losses in our equation, as well as showing the efficiency of the work produced on the piston, that is the indicated thermal efficiency, η_i. The latter is obtained from thermodynamic analysis of engine cycles, a topic of its own, which will be discussed at length later. The brake engine efficiency is the product of the indicated and mechanical efficiencies.

We can then write the power equation in either of the following equivalent forms:

$$BP = \eta_v \cdot \rho_{in} \cdot \frac{V_d N}{120} \cdot FA \cdot H_u \cdot \eta_i \cdot \eta_m \tag{1.30}$$

$$= \eta_v \cdot \rho_{in} \cdot \frac{V_d N}{120} \cdot FA \cdot H_u \cdot \eta_i - FP \tag{1.31}$$

The above equations are very important in showing us the connection between different engine operating and design variables. We can see many things, for example that we can produce a given power output by operating a large engine at low speed or a small engine at high speed. The primary objective being to send more air through the engine in order to use more fuel and, therefore, produce more power.

We can see different ways of regulating the engine. For example, traditional spark ignition engines operate with a well-mixed (homogeneous) fuel and air blend at a constant fuel air ratio. The combustion is started by discharging a spark through a gap in the spark plug in the cylinder at the proper time. A flame spreads uniformly from the spark. To change the amount of work produced by the engine, we must regulate the amount of input energy, that is, regulate the fuel input quantity.

Since the SI engine, in principle, operates with a fixed fuel air ratio, changing the fuel amount requires changing the air amount at the same time. This is done by adjusting the intake air density in Equation (1.30), that is, regulating the intake manifold pressure. The homogeneous combustion has other effects, since compression of the homogeneous fuel air mixtures can cause knock, which we will see later, can severely damage an engine. To avoid knock, we must use high quality fuel and are limited to as to how high a compression ratio we can use. The following chapter will show how this affects power and efficiency.

There is another consequence of this type of operation. Lowering the intake pressure results in the engine "seeing" a low pressure intake gas, and pushing it out at the (higher) environmental pressure. This is simply a pump, which requires energy to perform the air exchange process. The lower the load (intake pressure) the greater the difference between the intake and exhaust pressures, and thus the more pumping work required. This has consequences for the efficiency of spark ignition engines operating at low load. Figure 1.13 shows the "pumping loop" for an SI engine at full load. It is small, since the engine is operating at wide open throttle, WOT, where there is minimum restriction between the atmosphere and the cylinder on the intake stroke. At part load the magnitude of this pumping loss is larger.

A diesel engine uses a combustion process where the fuel is injected into the cylinder under high pressure just before combustion is to occur. If the cylinder temperature and pressure are high enough, the fuel will ignite with no further help. The amount of fuel consumed is basically regulated independently of the amount of air, such that, in principle, it is only the fuel amount that is changed when the power output changes. That is the fuel air ratio in Equation (1.30) is variable, and if the air flow does not change, the power is directly proportional to the fuel air ratio.

The air amount can change, and does especially in turbocharged engines, but this does not, in general, depend on the amount of fuel. Regulation of the fuel amount independently of the air amount (variable fuel air ratio) is possible because the flame in a diesel engine differs from that in a spark ignition engine, and the fuel burns as it mixes with the air. There are, of course limits to how much we can change fuel air ratios, if too much fuel is used in a diesel engine, smoke and efficiency will suffer.

The kind of combustion occurring in a diesel engine has two important consequences with regard to efficiency. The ignition of the fuel in this manner requires high cylinder temperature and pressure, which are achieved by using a high compression ratio. This leads to higher indicated efficiencies than typical spark ignition engines. The fact that we do not restrict the airflow almost eliminates the losses due to pumping the air through the engine. This is especially important at part load, where the intake pressure in the spark ignition engine is quite low. For a CI engine, the pumping loop usually remains on the magnitude of that in the $p - V$ diagram in Figure 1.13. Due to lower pumping losses, diesel engines have an even higher brake efficiency than spark ignition engines at part load.

Equation (1.29) can be converted to bmep by using Equation (1.11):

$$bmep = \eta_v \cdot \rho_{in} \cdot FA \cdot H_u \cdot \eta_e \tag{1.32}$$

Similarly for the indicated mean effective pressure, a very important relationship is;

$$imep = \eta_v \cdot \rho_{in} \cdot FA \cdot H_u \cdot \eta_i \tag{1.33}$$

These equations give an indication of relative values and limits for engines in general. Similar versions exist for the two-stroke engine, as seen in Chapter 8.

For diesel engines, it is not possible to operate at stoichiometric (chemically correct) mixture ratios without making excess smoke and suffering large efficiency losses. For standard diesel fuel, a maximum value of the fuel air ratio of about 0.05 is typical for engines without a turbocharger, depending on the specific details of any given engine, and an indicated thermal efficiency of about 0.45 is realistic for a modern engine. Using a typical diesel fuel with a heating value of 42500 kJ/kg, a volumetric efficiency of 0.8 and a mechanical efficiency of 0.8 as representative values, one obtains a maximum value of about 735 kPa for a direct injection diesel engine for a non-turbocharged engine. This is typical of engines found in the literature (See Table 1.2).

For spark ignition engines, operation with a stoichiometric mixture ratio is standard today. In practice at full load, a slightly rich ($FA > 0.068$) mixture is often used because it gives a little more power, while operation at part load for modern vehicle engines is at the chemically correct value since it permits operation with a 3-way catalyst and large reductions in exhaust emissions. The compression ratio for spark ignition engines is lower than diesel engines, and an indicated thermal efficiency of 0.36 is reasonable.

Using Equation (1.32) for a spark ignition engine with the same volumetric and mechanical efficiency, a stoichiometric mixture, a heating value of 44000 kJ/kg, and an indicated efficiency of 0.36, one obtains a maximum bmep of about 830 kPa. Modern automobile engines have special manifold arrangements that use pressure pulsations to increase the intake pressure at intake valve closing, and can often have a maximum *bmep* of over 1000 kPa (See Table 1.1).

For non-turbocharged engines, in spite of the lower efficiency, the spark ignition engine of a given size then produces more power than a diesel engine, in reality because it uses more of the air than the diesel engine, due to the combustion process (homogeneous in spark ignition, heterogeneous in diesel). The homogeneous combustion also enables the spark ignition engine to operate at higher speeds than the diesel. This further increases the amount of power produced per unit swept volume by spark ignition engines as compared to diesel engines.

Modern diesel engines use turbocharging extensively. This increases intake pressure, and Equation (1.32) shows that this will increase the brake mean effective pressure of the engine (See Table 1.2). Turbocharging also is applicable to spark ignition engines, but there is a limit to how high the pressure can be raised, since that promotes knock. On the other hand, the higher pressure and temperature are just what the diesel combustion process needs, so diesel engines can readily be turbocharged to a high degree.

1.3   Sample Calculations

This section takes the motor specifications from a given motor, and goes through a series of calculations to illustrate the use of the various definitions of the different parameters describing engine operation and performance. The engine described is the turbocharged Daimler-Benz OM611 engine and the data mentioned have been obtained from a technical publication describing the engine [1], [2]. Among the information shown in the paper are torque and power curves, full load power curves, and an engine map showing specific fuel consumption as a function of bmep and speed. The engine is a 4-stroke diesel engine with 4 cylinders.

1. Displacement volume: The first variable is the engine displacement volume, calculated from the bore of 88 mm and the stroke of 88.4 mm:

$$V_d = \frac{\pi}{4}B^2 \cdot S \cdot n_{cyl}$$

$$= \frac{\pi}{4}\, 0.088^2 \cdot 0.0884 \cdot 4$$

$$= 2.15 \cdot 10^{-3}m^3 = 2.15\ liter$$

2. Power-Torque: The maximum torque of the engine can be read to be 300 Nm at 2500 rpm. This corresponds to a power of:

$$BP = \frac{2\pi TN}{60000} = \frac{2\pi \cdot 300 \cdot 2500}{60000} = 78.5 kW$$

at the rated maximum torque condition.

3. Brake Mean Effective Pressure: The bmep can be calculated from the above condition:

$$bmep = \frac{120 \cdot BP}{V_d \cdot N}$$

$$= \frac{120 \cdot 78.5}{0.00215 \cdot 2500} = 1750 kPa$$

This should also be the same as calculated from the torque, using Equation (1.14):

$$bmep = \frac{2\pi \cdot T \cdot x}{1000V_d}$$

$$= \frac{2\pi \cdot 300 \cdot 2}{2.15} = 1750 kPa$$

Note that this value is over twice the estimate for a non-turbocharged diesel engine in the previous section. This indicates the effect of turbocharging. The bmep is higher than that for a non-turbocharged SI engine, and shows that turbocharging is a good way to increase the specific output of a CI engine, especially since the turbocharging is beneficial to the combustion process in CI engines.

4. Fuel consumption: From the power and torque from the data point in item 2., the $bsfc$ can be read as 221 g/kWh. This can be converted to the actual fuel flow:

$$\dot{m}_f = bsfc \cdot BP = 221 \cdot 78.5 \ = 1.74 \cdot 10^4 g/h = 4.83 \cdot 10^{-3} kg/s$$

5. Injected fuel mass: In the calculations concerned with fuel injection, it is often of interest to calculate the amount of fuel injected in one injection process, m_f. This can be determined from the fuel flow, number of cylinders and engine speed:

$$\dot{m}_f = \frac{m_f \cdot n_{cyl} \cdot N}{60x}$$
$$\Rightarrow \quad m_f = \frac{60 \cdot x \cdot \dot{m}_f}{n_{cyl} \cdot N}$$
$$= \frac{60 \cdot 2 \cdot 4.83 \cdot 10^{-3}}{4 \cdot 2500} = 5.79 \cdot 10^{-5} \ kg/injection$$

6. Fuel volume injected: The fuel volume (neglecting compressibility) is obtain from the fuel mass. A typical diesel fuel density is 0.84 g/cm^3 = 840 kg/m^3 :

$$V_f = \frac{m_f}{\rho_f} = \frac{5.79 \cdot 10^{-5}}{840} \cdot \frac{10^9 \ mm^3}{m^3} = 68.9 \ mm^3/injection$$

Note that the ratio of the fuel volume to the displacement volume per cylinder is

$$\frac{68.9 \cdot 10^{-9}}{\frac{2.15 \cdot 10^{-3}}{4}} = 1.28 \cdot 10^{-4}$$

Which emphasizes that the air volume is the limiting factor for power.

7. Friction Power:

In principle for a diesel engine, the data on an engine map can be used to obtain the friction through the Willans line method. This will be illustrated in a following section. For the time being, assume that it is found that the $fmep$ for the engine at 2500 rpm is 195 kPa. Then the friction power is:

$$FP = \frac{fmep \cdot V_d \cdot N}{60 \cdot x} = \frac{195 \cdot 0.00215 \cdot 2500}{120} = 8.73 \ kW$$

8. Indicated Power:

When the brake and friction powers are known, the indicated power can be obtained:

$$IP = BP + FP = 78.5 + 8.73 = 87.2 \ kW$$

9. Indicated Mean Effective Pressure:

The imep can be calculated from the indicated power or the bmep and the fmep:

$$imep = bmep + fmep = 1750 + 195 = 1945 \ kPa$$

10. Mechanical Efficiency:

The mechanical efficiency is calculated from the indicated and brake powers or mean effective pressures:

$$\eta_m = \frac{BP}{IP} = \frac{bmep}{imep} = \frac{78.5}{87.5} = \frac{1750}{1945} = 0.900$$

12. Brake Thermal Efficiency:

The brake thermal efficiency can be calculated from the brake power and the fuel flow and heating value, here assumed to be 42500 kJ/kg.

$$\eta_e = \frac{BP}{Q_{in}} = \frac{BP}{\dot{m}_f \cdot H_u} = \frac{78.5}{4.83 \cdot 10^{-3} \cdot 42500} = 0.382$$

13. Indicated Efficiency:

The indicated thermal efficiency can be calculated from the indicated power, the fuel flow and the heating value.

$$\eta_i = \frac{IP}{Q_{in}} = \frac{IP}{\dot{m}_f \cdot H_u} = \frac{87.2}{4.83 \cdot 10^{-3} \cdot 42500} = 0.425$$

14. Indicated Specific Fuel Consumption:

The most direct way to calculate the *isfc* is directly from the definition:

$$isfc = \frac{\dot{m}_f}{IP} = \frac{4.83 \cdot 10^{-3}}{87.5} \cdot 3.6 \cdot 10^6 = 199 \; g/kWh$$

Equation (1.23) could also be used:

$$isfc = \frac{3.6 \cdot 10^6}{\eta_i \cdot H_u} = \frac{3.6 \cdot 10^6}{0.425 \cdot 42500} = 199 \; g/kWh$$

The above data can be obtained directly from an engine map with no assumptions other than the heating value of the fuel, and that the Willans line analysis is valid.

It would be of interest in some cases, to have an estimate of some of the air flow parameters. In the published data set, there is a bit more information, which makes it possible to estimate air flow. In the following, some estimates will be made concerning air flow, the main purpose of which is to illustrate the calculation of basic engine parameters.

In the referenced article, intake pressure data are given, but there are no data for either air flow, intake temperature or volumetric efficiency. One good place to start would be to estimate the volumetric efficiency. This is a modern engine, with special attention given to the intake manifold design, and operating at an intermediate speed, where experience shows that volumetric efficiencies are near their maximum. Therefore, it will be assumed that the volumetric efficiency is on the order of 0.9. To calculate the air flow with the intake pressure and volumetric efficiency known, Equation (1.26) indicates that the intake manifold temperature should be known. This is not given either, but the manifold pressure ratio gives the opportunity to make a reasonable estimate, using experience with compressors and intercoolers. That is the approach that will be taken. In Chapter 11, Equation (11.7) shows that the temperature out of the compressor can be determined if the pressure ratio and the compressor efficiency are known. Here, only the pressure ratio is known, and one must rely on experience that indicates that a compressor of an engine of this size will have an optimum efficiency of 0.72 to 0.74, so a value of 0.73 will be assumed.

For the engine condition shown, the compressor pressure ratio is about 2.2. Then using a result to be obtained from the the chapter on turbocharging, Equation (11.7), the specific heat ratio for air, $\gamma = 1.4$, and an inlet temperature of 298K, the temperature after compression is estimated to be:

$$Y_c = \frac{p_2}{p_2}^{\frac{\gamma-1}{\gamma}} - 1 = 2.2^{\frac{1.4-1}{1.4}} - 1 = 0.253$$

$$T_2 = T_1 \left(1 + \frac{Y_c}{\eta_c}\right) = 298 \left(1 + \frac{0.253}{0.73}\right) = 401K$$

The engine has an intercooler, which uses atmospheric air to reduce the temperature of the compressed air. Experience indicates that such a device can be expected to cool with an effectiveness, α, of about 0.65. That is, the actual reduction in temperature is about 65% of the ideal reduction from the compressor temperature to the inlet temperature. (See Chapter 11). Then, the engine inlet air temperature is estimated to be:

$$T_{in} = T_{comp,out} - \alpha \left(T_{comp,out} - T_{atm}\right) = 401 - 0.65 \left(401 - 298\right) = 334K$$

It is now possible to estimate the engine air flow using Equation (1.26) and the ideal gas law:

$$\rho_{in} = \frac{p_{in}}{RT_{in}} = \frac{220}{0.287 \cdot 334} = 2.30 \frac{kg}{m^3}$$

$$\dot{m}_a = \eta_v \frac{\rho_{in} \cdot V_d \cdot N}{60x} = 0.9 \frac{2.30 \cdot 0.00215 \cdot 2500}{120} = 9.27 \cdot 10^{-2} kg/s$$

Is this reasonable? One way to check is to calculate the fuel air ratio. It is well known that diesel engines always run lean, and that experience says that a fuel air ratio greater than about 0.045 to 0.05 is not very likely. In the case at hand:

$$FA = \frac{\dot{m}_f}{\dot{m}_a} = \frac{4.83 \cdot 10^{-3}}{9.27 \cdot 10^{-2}} = 0.0521$$

The value is not totally unreasonable, but seems a bit high based on experience. The fuel flow is basically given in the data, and should be accurate, and any discrepancy would then be due to the air flow estimates. Since FA seems to be too large, the estimated air flow is expected to be too small. Factors that would increase the air flow would be 1. A higher volumetric efficiency, which is not impossible, since the manifold is carefully designed, 2. A lower intake temperature. This could result from a higher compressor efficiency, which does not seem likely, since the value used was taken from an engine of similar age, size and technology. It could also result from a better effectiveness for the intercooler, which is probably more realistic than a higher compressor efficiency.

Without additional information, it would be difficult to make a better estimate. But literature data is often incomplete, and lacks information needed to make more detailed calculations. Therefore, a good understanding of the principles of engine operation, and the use of fundamental parameters can be useful in "extracting" information from published data.

1.4   Willans Line

In a subsequent chapter, different methods of determining engine friction will be discussed. Among these methods is the Willans line method [3]. The method is discussed here because the information obtained from this method includes more than just engine friction. In particular, engine efficiencies can be obtained and simply demonstrated with the help of the Willans line[5].

The Willans line method is based on the assumption that for a constant indicated thermal efficiency, the indicated power of an engine is proportional to the fuel mass flow to the engine, and that the friction mean effective pressure is independent of engine load, and depends only on engine speed:

$$IP = \dot{m}_f \cdot H_u \cdot \eta_i \tag{1.34}$$

This equation is always valid, since it is in essence the definition of the indicated efficiency. To use the Willans line analysis to obtain friction, it is necessary to assume that the indicated efficiency is constant.

When the friction power is subtracted from the indicated power, the following equation is obtained for the brake power:

$$BP = \dot{m}_f \cdot H_u \cdot \eta_i - FP \tag{1.35}$$

These equations can also be converted to mean effective pressures.

$$imep = \frac{m_f \cdot H_u \cdot \eta_i}{V_d} \tag{1.36}$$

$$bmep = \frac{m_f \cdot H_u \cdot \eta_i}{V_d} - fmep \tag{1.37}$$

One can also define a *fuel mean effective pressure*, that is, fuel input energy per engine cycle per displacement volume:

$$fumep = \frac{m_f \cdot H_u}{V_d} \tag{1.38}$$

---

[5]The Willans line actually goes back to steam engines

where: m_f is the fuel injected per engine cycle to the volume V_d.

Using this definition, then:

$$imep = fumep \cdot \eta_i \tag{1.39}$$

$$bmep = fumep \cdot \eta_i - fmep = fumep \cdot \eta_e = fumep \cdot \eta_i \cdot \eta_m \tag{1.40}$$

On the basis of these equations, one can draw the Willans line, that is, a plot of brake mean effective pressure as a function of fuel mean effective pressure. In drawing the Willans line, it is important that a line be made only for constant speeds, since the extrapolated result gives only one value, and it is well known that friction in engines depends on the speed. Note that Equation (1.35) demonstrates that the Willans line can be drawn using fuel mass or mass flow rate instead of $fumep$. Willans lines can also be drawn using power and input energy in kW. However, the use of mean effective pressures allows the Willans lines for the entire range of engine operating speeds to be drawn on one set of axes, a compact depiction of the entire engine operating range.

From Equation (1.37), it can be seen that if the indicated efficiency is constant at low loads, and the data are extrapolated to the value of zero fuel mep, the value on the y axis will be equal to $(-fmep)$. Then in principle it is possible to obtain the engine friction with the Willans line by extrapolating to the point of $fumep = 0$. In addition, equations (1.39) and (1.40) indicate that the slope of the Willans line at any engine condition is equal to the indicated efficiency. Also, the slope of the line from the point $bmep = 0$ to an engine operating condition, will be equal to the brake thermal efficiency. For the point at $bmep = 0$, the brake efficiency is 0 because the mechanical efficiency is 0.

An example of a Willans line from experimental data is shown in Figure 1.15 for the OM611 engine for which the sample calculations were performed, operating at 2500 rpm, where it appears that the indicated efficiency is nearly constant for the whole load range. A straight line has been drawn through the data at the lower loads to find the fmep of 195 kPa. Even though the data looks nearly linear over the entire range, some curvature can be seen at heavy loads, and some uncertainty in the value of the friction must be admitted. The indicated thermal efficiency shown in the curve is the value for the linear portion of the curve at the lighter loads. Over a bmep of 1000 kPa, the indicated efficiency can be seen to fall. The sample calculation shown previously gave an indicated efficiency of 0.425, consistent with the Willans line at high loads.

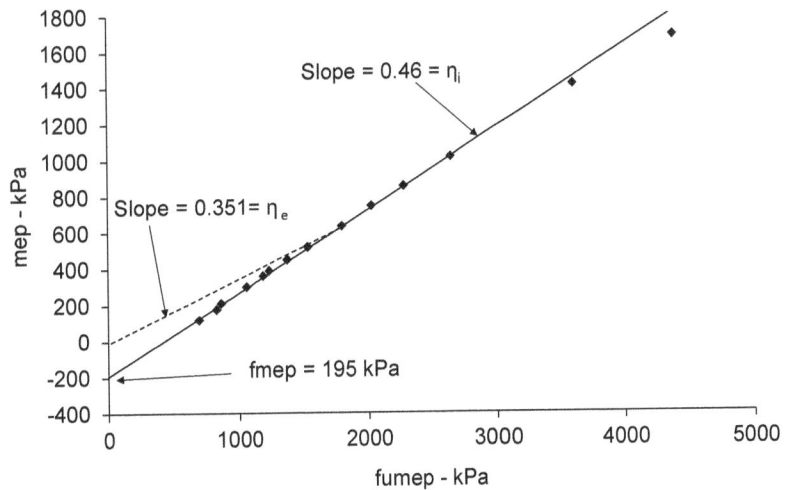

Figure 1.15: A Willans line generated from the engine map data for the OM611 engine at 2500 rpm. Note that the line is not generated from the trend line facility in the spread sheet for all the data points, since it has some curvature at the end.

A point for the brake efficiency has been shown, where the $bmep = 630$ kPa, and the $fumep = 1794$ kPa, giving a brake thermal efficiency of 0.351. As the output increases, the slope of the line for the brake thermal efficiency approaches that for the indicated thermal efficiency. This is another way of showing that at light loads, much of the engine indicated work goes to friction, giving a low brake efficiency. The brake efficiency can fall at high loads if the indicated efficiency falls sufficiently. This is common with diesel engines at high load (fuel air ratios), especially naturally aspirated engines, and is due to a long combustion period.

NB: It should be noted that this method is only suited for diesel engines or a gasoline direct injection engine, as long as the latter is unthrottled, since the method assumes that the friction is constant at lighter loads where the extrapolation is made. Throttled engines (some modern diesel engines use throttling as a way to control exhaust gas recirculation) have a pumping loss that is part of the friction. As was shown earlier, the variation of the pumping loss is a direct consequence of throttled operation, and in this case it cannot be assumed that the friction is constant as load varies.

Above all, it should always be remembered, that *extrapolation* of data is involved, which increases uncertainties significantly. The utility of the Willans line method for diesel engines depends on the quality of the data and on the operating tendencies of the engine. That is, if the engine operation at light loads is such that the thermal efficiency of the engine is not constant, then the extrapolation of the data to $fumep = 0$ becomes unusable. Timing adjustments and EGR for emission control for light loads can have an effect on the indicated efficiency and cause uncertainty in the friction values found.

In principle, one can obtain a Willans line from an engine map. The accuracy will depend strongly on the quality of the engine map bsfc contours and the resolution, especially at low loads, where the Willans line normally is most linear. At each point, the bsfc can be read off an engine map at a given speed and load (Torque or bmep). By reading off values at a constant speed, the fuel flows in g/hr are readily obtained by:

$$\dot{m}_f = bsfc \cdot BP = bsfc \cdot \frac{2\pi \cdot T \cdot N}{60000} = bsfc \cdot \frac{bmep \cdot V_d \cdot N}{60x} \tag{1.41}$$

This is the information required to perform the Willans line analysis.

1.5   A Few Conversion Factors

English units are still found in some literature, here are a few conversion factors for units commonly found:

1 pound (lb) = 0.4536 kg

1 hp (UK/US) = 0.7457 kW, 1hp (metric or PS) = 0.7355 kW

1 ft lb$_f$ = 1.356 Nm

1 cubic inch = 0.01639 liter

1 pound/sq. inch (psi) = 6.893 kPa

1 pound force (lb$_f$) = 4.448 N

1 Btu = 1.055 kJ

1 lb/hp-hr = 608.3 g/kWh

1° C = 1.8 ° F

1.6 Problems

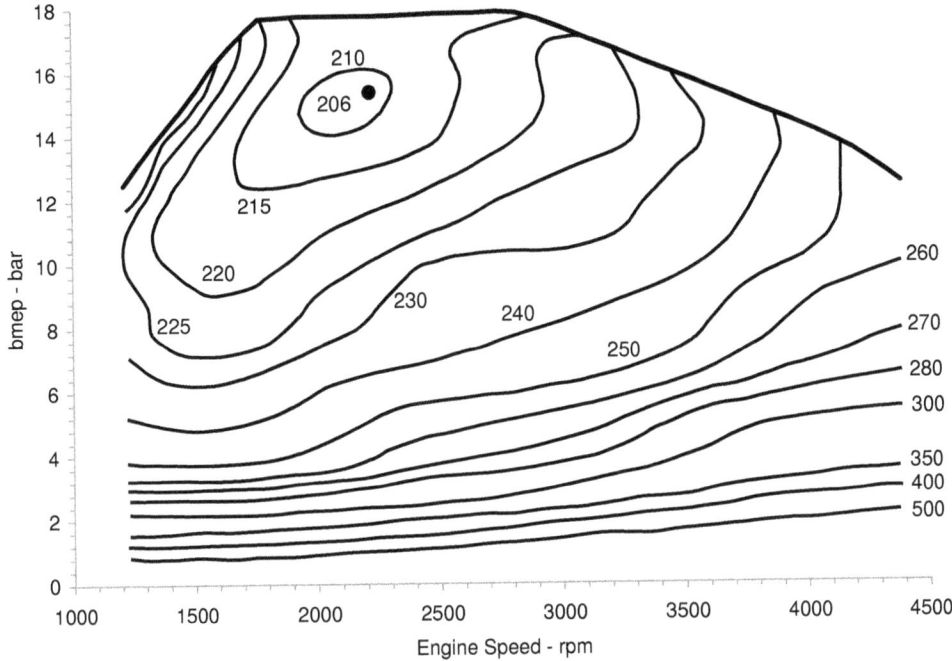

Figure 1.16: Engine map for a 4-stroke 3-cylinder turbocharged diesel.

Problem 1.1

Figure 1.16 shows the engine map for an engine with the following specifications: Cylinder diameter: 83 mm, Stroke: 92 mm, Compression ratio: 18:1, Fuel consumption in the curves is in g/kWh. Assume the fuel heating value is 42500 kJ/kg.

1. Find the displacement volume.

2. Find the clearance volume.

3. Draw a curve for the engine's maximum power and torque as a function of engine speed.

4. Calculate the mean piston speed at 4000 rpm.

5. At the minimum brake specific fuel consumption, calculate the fuel flow in g/s.

6. At 2500 rpm, the friction mean effective pressure is 220kPa. Calculate the indicated and brake thermal efficiencies and the mechanical efficiency at full load and also at a brake mean effective pressure of 6 bar.

7. From the literature, it is found that, at the compressor's maximum pressure ratio is 2.3:1, with an atmospheric pressure of 1.0 bar. The engine uses a charge air cooler after the compressor, which reduces the intake temperature after the compression. At full load, 2500 rpm the volumetric efficiency is estimated to be 0.88 and the air intake temperature 340K. Find the fuel air ratio.

(1. 1.493 liter, 2. 0.0878 liter, 3. $T_{max} \approx 215$ Nm, $BP_{max} \approx 70$ kW, 4. 12.3 m/s, 5. 2.43 g/s, 6. Full load: $\eta_e = 0.394$, $\eta_i = 0.443$, $\eta_m = 0.890$; 6 bar: $\eta_e = 0.342$, $\eta_i = 0.467$, $\eta_m = 0.732$, 7. $FA = 0.0516$)

Problem 1.2

An 8-cylinder 4-stroke spark ignition engine uses 1300 kg/hr of a mixture with mixture ratio of 1 kg gasoline to 12.5 kg air. The heating value = 43000 kJ/kg, the engine speed = 4000 rpm, the brake thermal efficiency = 30% and the stroke to bore ratio = 1.1:1

1. Find the brake power in kW.

2. Find cylinder diameter D and stroke S.

($BP = 345$ kW; $B/S = 129/142$ mm)

Problem 1.3

An engine produces a torque of 700 Nm at a speed of 3000 rpm. The air consumption is 970 kg/hour, the fuel/air ratio $= 0.069$, the heating value $= 43000$ kJ/kg and the indicated efficiency can be assumed to be $= 33\%$. Find the mechanical efficiency. ($\eta_m = 0.833$)

Problem 1.4

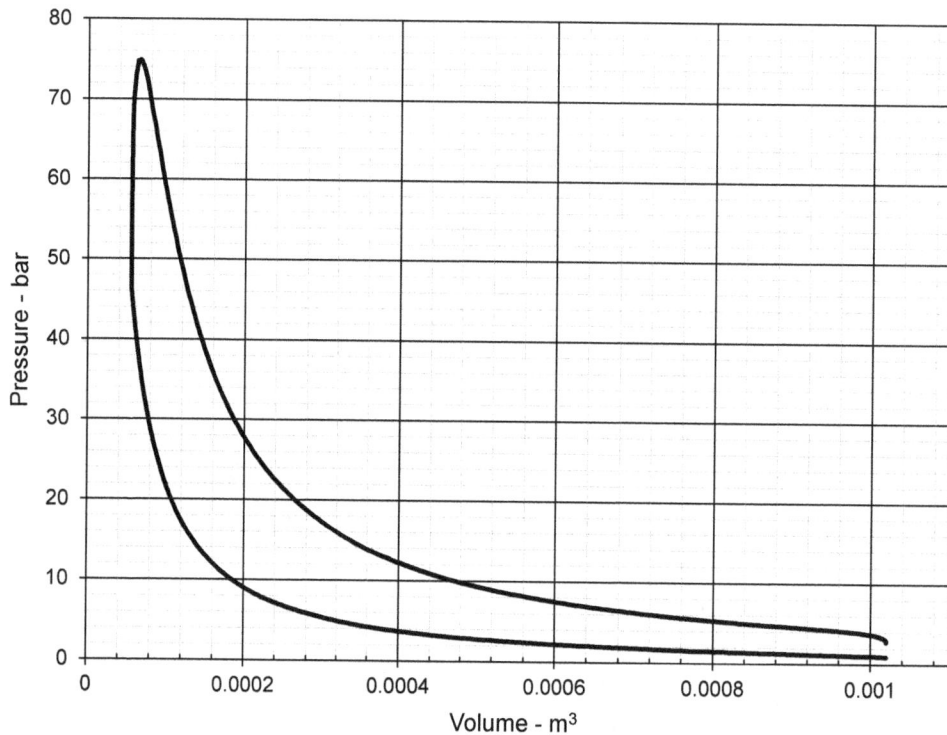

Figure 1.17: p-V diagram for Problem 1.4.

Figure 1.17 shows a pressure-volume diagram for a 2-cylinder, 4-stroke direct injection diesel engine, that has the following operating parameters: Bore= 85 mm, Stroke= 85 mm, Speed = 1992 rpm, Brake power= 12.6 kW.

Calculate:

1. Indicated mean effective pressure

2. Indicated power

3. Friction mean effective pressure

4. The mechanical efficiency

5. The friction mean effective pressure if the pressure transducer's calibration has been in error, such that the pressure measurements were 5 % too low

6. The polytropic coefficients, n, ($pV^n = const$) on the expansion and compression processes

($imep \approx 10.1$ bar; $IP \approx 16.2$kW; $fmep \approx 2.2$ bar; $\eta_m \approx 0.78$; $n_{comp} \approx 1.29$; $n_{exp} \approx 1.21$)

Chapter 2

Cycles

2.1 Ideal Air Cycles

2.1.1 Ideal Processes

While the actual processes that occur in real piston engines can be quite complex, one can obtain a good idea of the parameters that are important from a thermodynamic point of view by examining some idealized processes. These simplified processes are a first step in the range of simulation methods for engines, but they are useful, in that they give the basis for a fundamental understanding of engine cycles. The fundamental importance of certain variables will be established here. As more and more complicated engine simulations or calculation models are developed, a larger number of parameters can be included, but the relationships and understanding developed with the simple air cycles remain the fundamental basis for understanding the performance of engines. Predicted values using these simplified models may not be precise, but the qualitative trends established are valid.

The ideal air cycles are typically presented in thermodynamics courses, but are presented again here for the sake of completeness and as a basis for discussion. In the ideal air cycles, a series of simplifying assumptions are made:

- The cylinder is closed.

- The gasses in the cylinder are ideal (pV=mRT).

- The specific heat of the cylinder gasses is constant.

- Combustion is simulated with an appropriate heat addition.

- The air exchange process is simulated with a heat removal.

Typically, the ideal air cycles are calculated on the basis of a mass of 1 kg.

2.1.2 The Otto Cycle

The simplest process is the Otto cycle, named after the inventor of the 4-stroke compression cycle, Nicolaus Otto, and traditionally is used as the model for the spark ignition engine. It consists of 4 processes:

1. Isentropic compression

2. Heat addition at constant volume ("Combustion")

3. Isentropic expansion

4. Heat rejection at constant volume

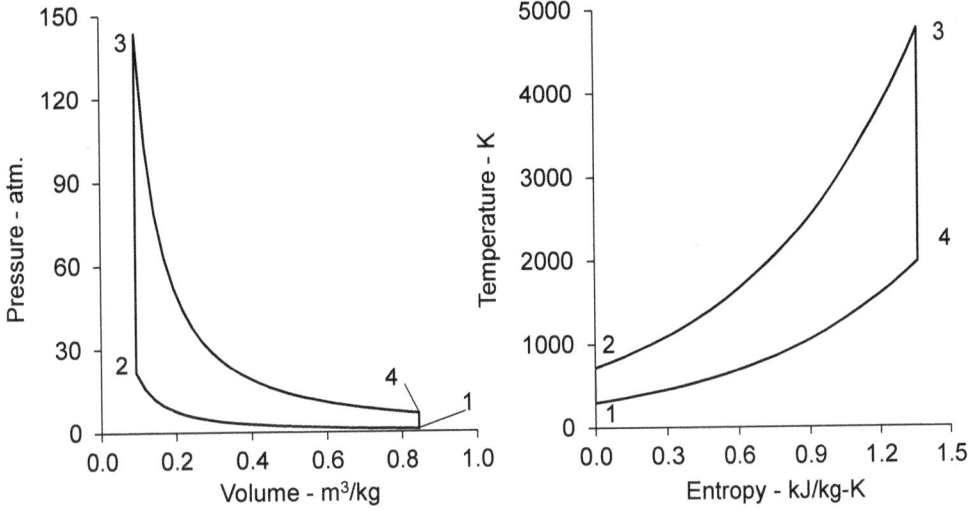

Figure 2.1: $p - v$ and $T - s$ diagrams for the ideal air Otto cycle, compression ratio $= 9{:}1$, $q_{in} = 2900$ kJ/kg.

The processes are shown on $p - v$ and $T - s$ diagrams in Figure 2.1. The conditions for the points of the cycle can be calculated using the following relationships. It is assumed that there is a mass of 1 kg and that p_1 and T_1 are known. Then the states throughout the cycle can be calculated as:

State 2:

$$v_2 = v_1 \varepsilon^{-1} \tag{2.1}$$
$$T_2 = T_1 \varepsilon^{\gamma - 1} \tag{2.2}$$
$$p_2 = p_1 \varepsilon^{\gamma} \tag{2.3}$$
$$s_2 = s_1 \tag{2.4}$$

where: ε is the compression ratio, defined as: v_1/v_2, and $\gamma = c_p/c_v$, and lower case variables denote mass specific quantities.

State 3:

$$v_3 = v_2 \tag{2.5}$$
$$T_3 = T_2 + \frac{q_{in}}{c_v} \tag{2.6}$$
$$p_3 = p_2 \frac{T_3}{T_2} \tag{2.7}$$
$$s_3 = s_2 + c_v \ln \frac{T_3}{T_2} + R \ln \frac{v_3}{v_2} \tag{2.8}$$

State 4:

$$v_4 = v_1 \tag{2.9}$$
$$T_4 = T_3 \frac{1}{\varepsilon^{(\gamma - 1)}} \tag{2.10}$$
$$p_4 = p_3 \left(\frac{1}{\varepsilon}\right)^{\gamma} \tag{2.11}$$
$$s_4 = s_3 \tag{2.12}$$

In the calculation of T_3, a value of q_{in} is needed. This can be estimated by the following:

$$Q_{in} = m_f \cdot H_u \tag{2.13}$$
$$= m_a \cdot FA \cdot H_u \Rightarrow q_{in} = FA \cdot H_u \tag{2.14}$$

Table 2.1: Stoichiometric fuel air ratios, heating values and their product for selected engine fuels

Fuel	FA_s - kg_f/kg_a	H_u - kJ/kg_f	$FA_s \cdot H_u$ - kJ/kg_a
Methane, CH_4	0.0581	50020	2910
Propane, C_3H_8	0.0637	46360	2950
Octane, C_8H_{18}	0.0661	44786	2960
Hexadecane, $C_{16}H_{34}$	0.0667	44307	2955
Propylene, C_3H_6	0.0676	45334	3065
Cyclohexane, C_6H_{12}	0.0676	43435	2936
Benzene, C_6H_6	0.0752	40580	3050
Toluene, C_7H_8	0.0741	40528	3003
Xylene, C_8H_{10}	0.0730	40800	2978
Methanol, CH_3OH	0.156	19195	2994
Ethanol, C_2H_5OH	0.111	26803	2975
Dimethyl Ether, CH_3OCH_3	0.111	28800	3197
Soybean Oil Methyl Ester, $C_{15}H_{27.1}O_{1.48}$	0.0795	37260	2960

Where m_a is the air mass, assumed 1 kg, and m_f the associated fuel mass using the fuel air ratio, FA. This procedure is acceptable as long as the fuel air ratio is less than or equal to the stoichiometric value, FA_s. The methods for calculating the stoichiometric fuel air ratio are described in Chapter 3. For $FA > FA_s$ not all of the chemical energy in the fuel can be released, as the air present is not sufficient to burn all the fuel to CO_2 and H_2O, and the full heating value, H_u is not released to the gas. (Only fuels containing C, H and O, are considered here).

Already here one can ask the question: how much difference do we expect the choice of fuel to make with respect to the power and efficiency of our engine? A first estimate can be made using Equation (2.14). To do this, let us consider a variety of potential fuels for engines. Since our ideal cycles will be based on a mass of 1 kg, we can look at the product $FA_s \cdot H_u$ for some guidance. This determines the basic energy that can be released from combustion of the fuel, which is available to produce work. The results are shown in Table 2.1

The results indicate that on the basis of 1 kg of air consumed, there is not all that great a difference in the energy available between the different fuels. Even fuels such as methanol or ethanol, whose stoichiometric fuel air ratios differ significantly from hydrocarbons, have about the same amount of energy available per kg of air in a stoichiometric mixture. Remember that air is more significant because we use more of it ($AF \simeq 15$ for most hydrocarbons) and liquid fuels have a much higher density.

From the properties of the states, it is possible to calculate the work and efficiency for the cycle. These quantities are denoted with the subscript 'i' for indicated, since they originate from the $p - V$ diagram, which has been called an indicator diagram as described in Chapter 1. The following calculations will be based on 1 kg of air.

The indicated work:

From the $p - v$ diagram it can be seen that $v_2 = v_3$, and that $v_1 = v_4$. Therefore, there is no work involved for the processes 2-3 and 4-1. From the first law for a closed adiabatic system:

$$w_{1 \to 2} = c_v(T_1 - T_2) \tag{2.15}$$
$$w_{3 \to 4} = c_v(T_3 - T_4) \tag{2.16}$$

The indicated work, w_i is therefore:

$$w_i = c_v(T_1 - T_2 + T_3 - T_4) \tag{2.17}$$

Since the heat addition $q_{in} = c_v(T_3 - T_2)$, the indicated efficiency η_i can be determined:

$$\eta_i \equiv \frac{w_i}{q_{in}} \tag{2.18}$$

$$\eta_i = \frac{(T_1 - T_2 + T_3 - T_4)}{T_3 - T_2} \tag{2.19}$$

Using the equations for the calculation of the states it can be shown that Equation (2.19) reduces to:

$$\eta_i = 1 - \left(\frac{1}{\varepsilon}\right)^{\gamma-1} \tag{2.20}$$

Note that the indicated efficiency is independent of the choice of the fuel.

Otto Cycle Example

Otto cycle, constant specific heat ratio of 1.4, heat addition of 2900 kJ/kg, and a compression ratio of 9.0:1. Initial conditions: $T_1 = 298K$ and $p_1 = 100kPa$. One can quickly determine thermal efficiency by using equation (2.20), but in the following, all the properties will be calculated and Equation(2.19) used instead. This will be useful when the air exchange process is considered later.

Using equations (2.1) to (2.12), values of all the properties can be calculated:

$$v_1 = \frac{RT_1}{p_1} = \frac{0.287 \cdot 298}{100} = 0.855 \ m^3/kg$$

$$T_2 = 298 \cdot 9^{(1.4-1)} = 717.5 \ K$$

$$v_2 = 0.8558/9.0 = 0.0951 \ m^3/kg$$

$$s_2 = s_1 \equiv 0 \ kJ/kg - K$$

$$p_2 = 100 \cdot 9^{1.4} = 2167 \ kPa$$

$$T_3 = 717.5 + \frac{2900}{0.717} = 4757 \ K$$

$$p_3 = 2167\frac{4757}{717.5} = 14370 \ kPa$$

$$s_3 = s_2 + c_v \ln\frac{4757}{717.5} + 0.287\ln\frac{0.0951}{0.0951} = 1.358 \ kJ/kg - K$$

$$v_3 = v_2 = 0.0951 \ m^3/kg$$

$$v_4 = 0.8558 \ m^3/kg$$

$$T_4 = 4757\frac{1}{9^{(1.4-1)}} = 1975 \ K$$

$$p_4 = 14370\left(\frac{1}{9}\right)^{1.4} = 663 \ kPa$$

$$s_4 = s_3 = 1.357 \ kJ/kg - K$$

The states of the cycles are shown as $p - v$ and $T - s$ diagrams in Figure 2.1.

Then the work terms can be calculated to give the indicated work:

$$w_{1\to2} = c_v(T_1 - T_2) = 0.717(298 - 717.5) = -301.3 \ kJ/kg$$

$$w_{3\to4} = c_v(T_3 - T_4) = 0.717(4757 - 1975) = 1997 \ kJ/kg$$

$$w_i = w_{comp} + w_{exp} = -301.3 + 1997 = 1696 \ kJ/kg$$

The indicated thermal efficiency can be calculated in different equivalent ways, either from its basic definition: (2.18), Equation 2.19 or with Equation (2.20).

$$\eta_i \equiv \frac{1696}{2900} = 0.585$$

$$\eta_i = \frac{(298 + 717.5 + 4757 - 1975)}{4757 - 717.5} = 0.585$$

The thermal efficiency naturally agrees with (2.20):

$$\eta_i = 1 - \left(\frac{1}{9}\right)^{1.4-1} = 0.585$$

The indicated mean effective pressure can also be calculated. Recall that it is the work performed per cycle per unit displacement volume. For our mass of 1 kg of air, the displacement is then $v_1 - v_2$, so:

$$imep = \frac{w_i}{v_d} = \frac{1691}{0.8558 - 0.0951} = 2229 kPa$$

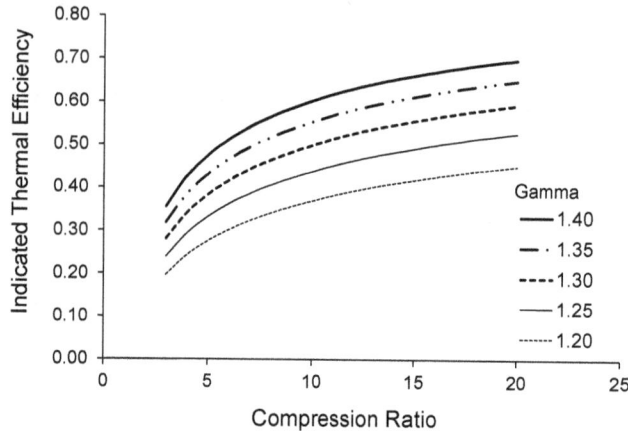

Figure 2.2: The effect of compression ratio and specific heat ratio, γ, on the thermal efficiency of an ideal air Otto cycle.

The indicated efficiency and mean effective pressure are important, since they are independent of engine size. So, already at this primitive stage of calculation, one could estimate the power and efficiency of an engine of a given displacement volume and speed. That is, of course, the reason that these calculations are performed.

The above indicates that for an ideal Otto process, the efficiency is only affected by the compression ratio and the specific heat ratio γ. Figure 2.2 shows the indicated thermal efficiency as a function of compression ratio for the Otto cycle, using different values of γ. The processes have higher efficiency for high values of ε. Since γ decreases with increasing temperature, we can see that a calculation with a constant value of $\gamma = 1.4$, (air at 25°C) will over estimate the thermal efficiency for a real process. Low values of γ come from a high specific heat, since:

$$\gamma = \frac{c_p}{c_v} = \frac{c_v + R}{c_v} = 1 + \frac{R}{c_v} \tag{2.21}$$

Thus, with lower values of γ, the temperatures, and especially pressures, are lower throughout the cycle, and then so are the integrals of $p \cdot dV$

Why is the compression ratio so important? We know from physics that we obtain work by moving a force through a distance: no movement - no work. In engines, the force is obtained by the cylinder pressure acting on the piston top, $F = p \cdot A$ and if the piston moves a distance δx, the volume change is $A \cdot \delta x = \delta V$. So the work done in the piston movement is:

$$\delta W = F \cdot \delta x = p \cdot \delta V \tag{2.22}$$

Then the greater the expansion, the greater amount of work we can extract from the gasses. When heat is added, the temperature and pressure increase. The more the gasses are expanded after that, the more work is obtained, and the higher the thermal efficiency. For understanding the cycle, it might be preferable to call ε the expansion ratio instead of the compression ratio, which has been the traditional definition. Since we have a thermodynamic cycle, that is the process returns to the initial state, one simplified way of looking at things is that compression is the "price" we have to pay in order to be able to expand.

2.1.3 Diesel Cycle

The next process is the Diesel cycle, named after the inventor of the compression ignition engine, Rudolf Diesel. The cycle also consists of 4 processes:

1. Isentropic compression

2. Heat addition at constant pressure ("Combustion")

3. Isentropic expansion

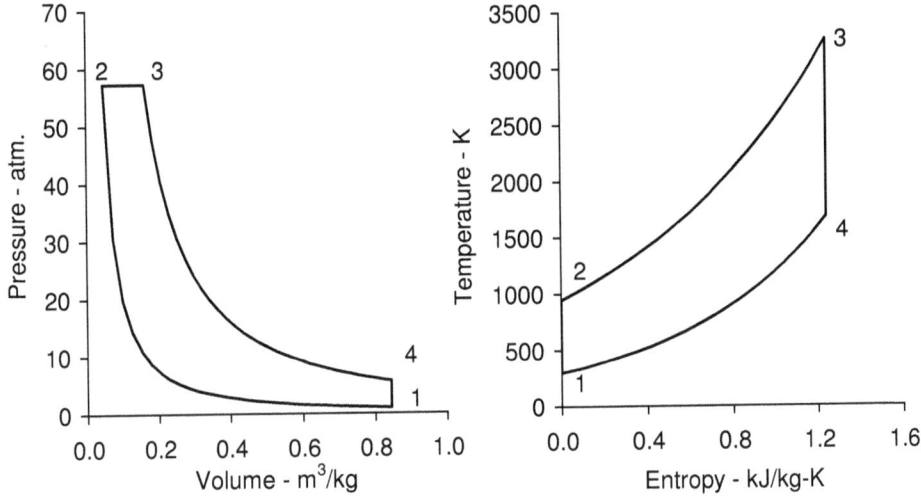

Figure 2.3: $p - v$ and $T - s$ diagrams for the ideal air Diesel cycle, compression ratio = 18:1, $q_{in} = 2320$ kJ/kg.

4. Heat rejection at constant volume

The processes are shown on $p - v$ and $T - s$ diagrams in Figure 2.3. The states throughout the cycle are calculated in a similar manner to the Otto cycle with small differences. Once again it is assumed that the mass = 1 kg, and that p_1 and T_1 are known. Then:

State 2 (Same as Otto cycle), use Equations (2.1) to (2.4)

State 3:

The heat addition process occurs at constant pressure now. For that process, the first law of thermodynamics gives:

$$\delta q - \delta w = \Delta u \tag{2.23}$$

$$\delta q - \int p \cdot dv = \Delta u \tag{2.24}$$

$$\delta q - p_3(v_3 - v_2) = \Delta u \tag{2.25}$$

Since $h \equiv u + pv$, and $p_2 = p_3$, then

$$\delta q = \Delta h = m \cdot c_p \Delta T = m \cdot c_p \left(T_3 - T_2 \right) = q_{in} \tag{2.26}$$

$$p_3 = p_2 \tag{2.27}$$

$$T_3 = T_2 + \frac{q_{in}}{c_p} \tag{2.28}$$

$$v_3 = v_2 \frac{T_3}{T_2} \tag{2.29}$$

$$s_3 = s_2 + c_p \ln \frac{T_3}{T_2} - R \ln \frac{p_3}{p_2} \tag{2.30}$$

State 4: v_4 is obtained from (2.9), and s_4 is obtained from (2.12)

$$T_4 = T_3 \left(\frac{v_3}{v_4} \right)^{(\gamma - 1)} \tag{2.31}$$

$$p_4 = p_3 \left(\frac{v_3}{v_4} \right)^{\gamma} \tag{2.32}$$

The indicated work:

The compression and isentropic expansion works are obtained from (2.15) and (2.16). However, there now an additional process to consider, as there is also some work performed on the piston top during the heat addition process, so an additional term is needed:

$$w_{2\to 3} = \int_2^3 p \cdot dv \tag{2.33}$$
$$= p_3(v_3 - v_2) = p_2(v_3 - v_2)$$
$$= R(T_3 - T_2)$$

The indicated work, w_i is then:

$$w_i = c_v(T_1 - T_2 + T_3 - T_4) + R(T_3 - T_2) \tag{2.34}$$

and since the heat addition, $q_{in} = c_p(T_3 - T_2)$, the indicated efficiency η_i can be found:

$$\eta_i = \frac{w_i}{q_{in}} \tag{2.35}$$

$$\eta_i = \frac{c_v(T_1 - T_2 + T_3 - T_4) + R(T_3 - T_2)}{c_p(T_3 - T_2)} \tag{2.36}$$

Through the use of the equations for the state conditions, it can be shown that:

$$\eta_i = 1 - \left(\frac{1}{\varepsilon}\right)^{\gamma-1} \left[\frac{\left[\left(\frac{T_3}{T_2}\right)^{\gamma} - 1\right]}{\gamma\left(\frac{T_3}{T_2} - 1\right)}\right] = 1 - \left(\frac{1}{\varepsilon}\right)^{\gamma-1} \left[\frac{\left[\left(\frac{v_3}{v_2}\right)^{\gamma} - 1\right]}{\gamma\left(\frac{v_3}{v_2} - 1\right)}\right] \tag{2.37}$$

The term in the square parentheses can be shown to be > 1 for $T_3/T_2 > 1$. Therefore the Diesel Cycle has a lower indicated thermal efficiency that the Otto cycle *IF* the compression ratio is the same.

Diesel Cycle Example

Determine the efficiency and indicated mean effective pressure for a Diesel cycle using the conditions of the previous example.

Diesel cycle, constant specific heat ratio of 1.4, heat addition of 2900 kJ/kg, and a compression ratio of 9.0:1. Initial conditions: $T_1 = 298K$ and $p_1 = 100kPa$.

Using the equations above, all the properties can be calculated:

$$v_1 = \frac{RT_1}{p_1} = \frac{0.287 \cdot 298}{100} = 0.8558 \ m^3/kg$$
$$T_2 = 298 \cdot 9^{(1.4-1)} = 717.5 \ K$$
$$v_2 = 0.8558/9.0 = 0.0951 \ m^3/kg$$
$$s_2 = s_1 \equiv 0 \ kJ/kg - K$$
$$p_2 = 100 \cdot 9^{1.4} = 2167 \ kPa$$
$$T_3 = 717.5 + \frac{2900}{1.005} = 3603 \ K$$
$$p_3 = p_2$$
$$s_3 = s_2 + 1.005 \ln\frac{3603}{717.7} - 0.287 \ln\frac{2167}{2167} = 1.621 \ kJ/kg - K$$
$$v_3 = 0.0951\frac{3603}{717.5} = 0.4774 \ m^3/kg$$
$$v_4 = 0.8558 \ m^3/kg$$
$$T_4 = 3603\frac{1}{\frac{0.4774}{0.8558}^{(1.4-1)}} = 2852 \ K$$
$$p_4 = 2167(\frac{0.4774}{0.8558})^{1.4} = 957.2 \ kPa$$
$$s_4 = s_3 = 1.358 \ kJ/kg - K$$

Then the work terms can be calculated:

$$w_{1\to2} = c_v(T_1 - T_2) = 0.717(298 - 717.5) = -301.3kJ/kg$$
$$w_{2\to3} = R(T_3 - T_2) = 0.287(3603 - 717.5) = 828.6kJ/kg$$
$$w_{3\to4} = c_v(T_3 - T_4) = 0.717(3603 - 1975) = 538.6kJ/kg$$
$$w_i = -301.3 + 828.6 + 538.6 = 1066kJ/kg$$

The indicated thermal efficiency can be calculated:

$$\eta_i \equiv \frac{(1066)}{2900} = 0.368$$

The indicated mean effective pressure can also be calculated.

$$imep = \frac{w_i}{v_d} = \frac{1066}{0.8558 - 0.0951} = 1402kPa$$

The conclusion that the diesel cycle is less efficient that the Otto cycle comes from the mathematical analysis of the cycles, but what is the physical reason? As mentioned before, work is obtained during the expansion process. In the Otto cycle, all the heat is added, then the expansion occurs. On the other hand, there is expansion in the Diesel cycle while the heat addition process is taking place. Then the last energy added during the heat addition process cannot be expanded as much as the first portion, and the efficiency of the Diesel cycle must be lower than that of the Otto cycle for a given compression ratio and heat addition. The longer the heat addition period, that is the greater q_{in}, the lower the efficiency. As q_{in} approaches 0, the efficiency approaches that of the Otto cycle, but the work done approaches 0 as well, so there is not much to gain here.

The idea of the effects of combustion extending into the expansion stroke can also be used in an actual engine. If the combustion process (modelled in the ideal cycles by the heat addition) occurs while the piston is on the way down in the cylinder, then it will give rise to a lower efficiency than if the combustion occurs close to the top dead center position. A similar effect arises if the combustion starts around top dead center, but is slow and continues into the expansion stroke. We expect (and find in practice) that late and slow combustion gives low efficiency. This is a very important concept when considering the timing and combustion duration in developing engines.

There is another significant difference between the two cycles, the cylinder pressure is much lower in the case of the Diesel cycle. Although the Otto cycle efficiency is much higher, the cylinder pressure is also 6.63 times higher with the Otto cycle. This obviously has implications with respect to engine construction.

Figure 2.4 shows the $p-v$ and $T-s$ diagrams for the Otto and Diesel cycles with the same compression ratio. These cycles have the same heat addition, and it can be seen from the $T-s$ diagram that the Otto cycle has a smaller heat rejection (the constant volume process at v_1). This means that there has been a greater conversion of the energy added through heat transfer to mechanical work, and therefore a higher efficiency.

But it is also clear that the Otto cycle results in a much higher maximum pressure than the Diesel cycle. In an actual engine the maximum cylinder pressure is very important. The higher the maximum cylinder pressure, the stronger, larger, heavier and more expensive an engine must be built. Therefore, it is interesting to compare the two cycles under a more practical restraint, i.e. that of a given maximum cylinder pressure. This comparison is shown in Figure 2.5 for $\gamma = 1.4$ and a maximum cylinder pressure of 8000 kPa. For the Otto cycle, the compression ratio which results in a maximum cylinder pressure of 8000 kPa is 5.175 with the given amount of heat addition. In order to achieve this pressure, a Diesel cycle needs a compression ratio of 22.87:1. With such a high compression ratio the Diesel cycle has a higher efficiency that the Otto cycle at the lower compression ratio. This can be seen from the diagram in Figure 2.4, where it can be seen that the Diesel cycle has the lowest heat rejection, and therefore the highest efficiency.

At a given compression ratio, the Otto pressure is very high, the Diesel efficiency low, so a third cycle is an obvious possibility.

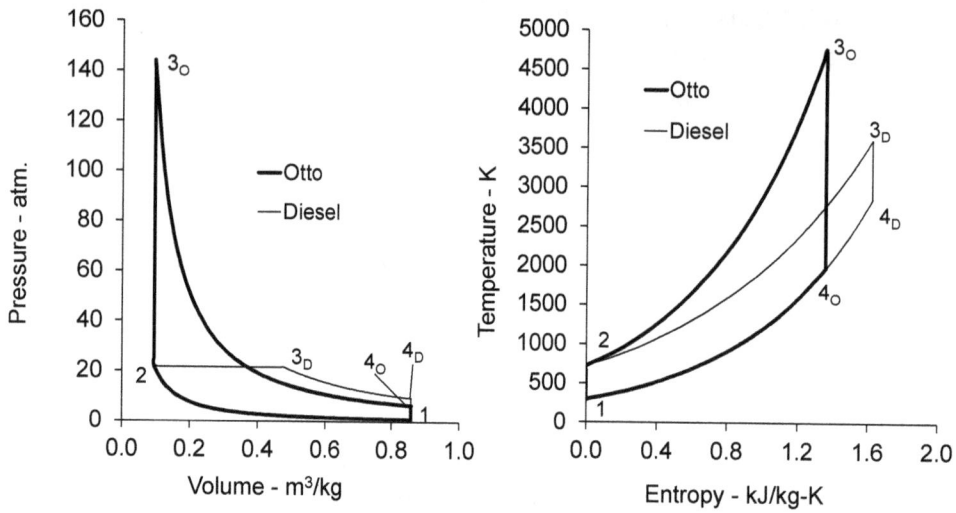

Figure 2.4: $p - v$ and $T - s$ diagrams for the Otto and Diesel ideal air cycles, with the same compression ratio. 9:1 and heat addition, 2900 kJ/kg. For the Otto cycle, $\eta_i = 0.585$, and $imep = 2229$ kPa. For the Diesel cycle, $\eta_i = 0.368$, and $imep = 1401$ kPa.

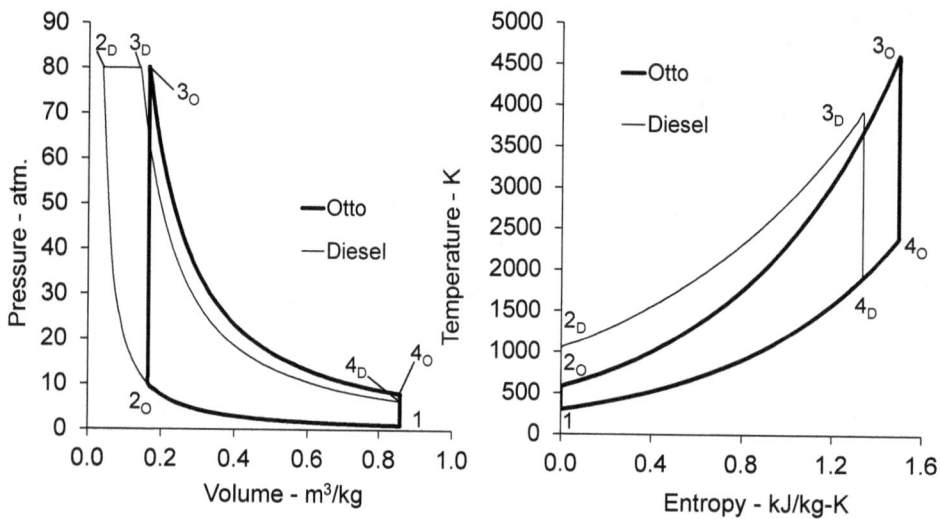

Figure 2.5: $p - v$ and $T - s$ diagrams for the Otto and Diesel ideal air cycles, with the same maximum cylinder pressure, 8000 kPa and heat addition, 2900 kJ/kg. For the Otto cycle, $\eta_i = 0.482$, and $imep = 2023$ kPa. For the Diesel cycle, $\eta_i = 0.601$, and $imep = 2130$ kPa.

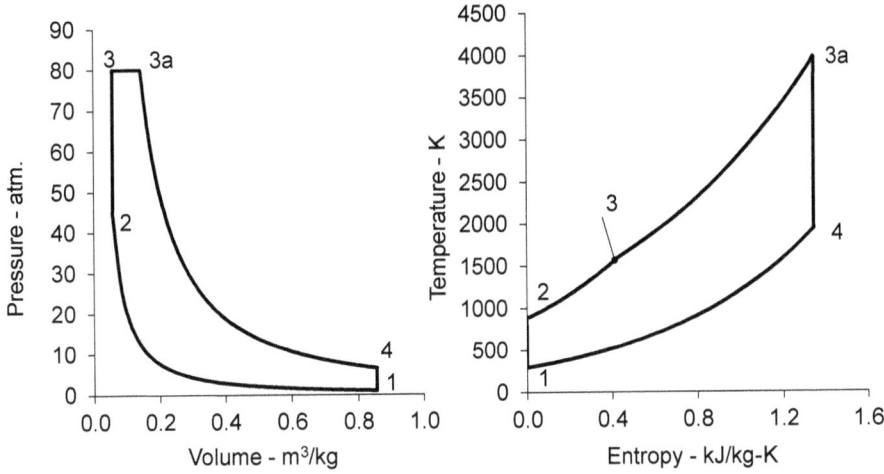

Figure 2.6: $p - v$ and $T - s$ diagrams for the ideal air Limited Pressure (Seiliger) cycle for a maximum pressure of 8000 kPa, heat addition of 2900 kJ/kg and compression ratio of 15:1.

2.1.4 Limited Pressure Cycle

The third cycle, called the Dual Cycle, Limited Pressure Cycle, or Seiliger Cycle, is a combination of the Otto and Diesel cycles. It is based on the concept that there is a maximum allowable cylinder pressure and that the heat addition process can be adjusted to accommodate it. This is accomplished by starting the heat addition at constant volume at the end of compression until the maximum pressure is attained. Thereafter, the remainder of the heat addition is continued at constant pressure. The cycle then consists of 5 processes:

1. Isentropic compression

2. Constant volume heat addition ("Combustion phase 1")

3. Constant pressure heat addition ("Combustion phase 2")

4. Isentropic expansion

5. Heat rejection at constant volume

The cycle is shown on $p - v$ and $T - s$ diagrams in Figure 2.6. The states are calculated in a similar manner as the Otto and Diesel cycles. Once again it is assumed that the mass is 1 kg, and that p_1 and T_1 are known. It follows that:

State 2:

$$v_2 = v_1 \varepsilon^{-1} \tag{2.38}$$
$$T_2 = T_1 \varepsilon^{(\gamma-1)} \tag{2.39}$$
$$p_2 = p_1 \varepsilon^{\gamma} \tag{2.40}$$
$$s_2 = s_1 \tag{2.41}$$

Here, the compression ratio must be less than that which gives the maximum allowable pressure at the end of compression:

$$\varepsilon < \varepsilon_{max} = \left(\frac{p_{max}}{p_1} \right)^{\frac{1}{\gamma}} \tag{2.42}$$

State 3:

In order to calculate state 3, it is necessary to find out how much heat can be transferred to the gasses to achieve the temperature that gives the maximum allowable pressure:

$$v_3 = v_2 \tag{2.43}$$

$$T_3 = T_2 \left(\frac{p_{max}}{p_2} \right) \tag{2.44}$$

The amount of heat transfer occurring at constant pressure is the amount that is left after the heat addition at constant volume:

$$q_{3-3a} = q_{in} - q_{2-3} = q_{in} - c_v \left(T_3 - T_2 \right) \tag{2.45}$$

State 3a is calculated in a similar manner to the Diesel cycle:

$$p_{3a} = p_3 \tag{2.46}$$

$$T_{3a} = T_3 + \frac{q_{3-3a}}{c_p} \tag{2.47}$$

$$v_{3a} = v_3 \frac{T_{3a}}{T_3} \tag{2.48}$$

$$s_{3a} = s_3 + c_p \ln \frac{T_{3a}}{T_3} - R \ln \frac{p_{3a}}{p_3} \tag{2.49}$$

State 4: v_4 and s_4 are obtained from

$$v_4 = v_1 \tag{2.50}$$

$$s_4 = s_{3a} \tag{2.51}$$

And the remaining states:

$$T_4 = T_{3a} \left(\frac{v_{3a}}{v_4} \right)^{(\gamma-1)} \tag{2.52}$$

$$p_4 = p_{3a} \left(\frac{v_{3a}}{v_4} \right)^{\gamma} \tag{2.53}$$

Indicated work:

As in the case of the Diesel cycle, the work performed during the heat addition must be included in the overall cycle calculation:

$$w_{3 \to 3a} = \int_3^{3a} p \cdot dv \tag{2.54}$$
$$= p_{3a}(v_{3a} - v_3) = p_{max}(v_{3a} - v_3)$$
$$= R(T_{3a} - T_3)$$

The indicated work, w_i is then:

$$w_i = c_v(T_1 - T_2 + T_{3a} - T_4) + R(T_{3a} - T_3) \tag{2.55}$$

and since the heat addition, $q_{in} = c_v(T_3 - T_2) + c_p(T_{3a} - T_3)$, the indicated efficiency η_i can be found:

$$\eta_i = \frac{w_i}{q_{in}} \tag{2.56}$$

$$\eta_i = \frac{c_v(T_1 - T_2) + c_v(T_{3a} - T_4) + R(T_{3a} - T_3)}{c_v(T_3 - T_2) + c_p(T_{3a} - T_3)} \tag{2.57}$$

Through the use of the equations for the state conditions, it can be shown that:

$$\eta_i = 1 - \left(\frac{1}{\varepsilon} \right)^{\gamma-1} \left[\frac{\left(\frac{p_3}{p_2} \right) \left(\frac{v_{3a}}{v_3} \right)^{\gamma} - 1}{\left(\frac{p_3}{p_2} - 1 \right) + \gamma \left(\frac{p_3}{p_2} \right) \left(\frac{v_{3a}}{v_2} - 1 \right)} \right] \tag{2.58}$$

From Equation (2.58), the results of (2.20) and (2.37) respectively can be obtained by assuming either $v_3 = v_{3a}$ or $p_2 = p_3$.

As expected, the limited pressure cycle has an indicated thermal efficiency that lies between those of the Otto and Diesel cycles for the same compression ratio, ε.

From these simple air cycles, we can conclude that one of the most important parameters for determining the efficiency of an engine is the compression ratio. For all the processes, the efficiency increases with increasing compression ratio. We have seen that the Otto cycle has the highest efficiency (and cylinder pressure) *for a given compression ratio and heat addition.*

Example for Limited Pressure Cycle:

In previous examples, a compression ratio of 9 was used, giving a very high pressure for the Otto cycle, and a low pressure for the Diesel cycle. In this example, assume that the maximum pressure that the engine can tolerate is 7500 kPa and that the heat addition is again 2900 kJ/kg. The compression ratio is assumed to be 16:1

$$v_1 = \frac{RT_1}{p_1} = \frac{0.287 \cdot 298}{100} = 0.8558 \ m^3/kg$$

$$T_2 = 298 \cdot 16^{(1.4-1)} = 903.4 \ K$$

$$v_2 = 0.8558/16.0 = 0.05349 \ m^3/kg$$

$$s_2 = s_1 \equiv 0 \ kJ/kg - K$$

$$p_2 = 100 \cdot 16^{1.4} = 4850 \ kPa$$

$$T_3 = = 903.4\frac{7500}{4850} = 1397 \ K$$

$$q_{2-3} = 0.717(1397 - 903.4) = 353.9 \ kJ/kg$$

$$q_{3-3a} = 2900 - 353.9 = 2546.1 \ kJ/kg$$

$$s_3 = s_2 + 1.005\ln\frac{1397}{903.4} - 0.287\ln\frac{7500}{4850} = 0.3130 \ kJ/kg - K$$

$$T_{3a} = 1397 + \frac{2546.1}{1.005} = 3930 \ K$$

$$v_{3a} = .05349\left(\frac{3930}{1397}\right) = 0.1505 \ m^3/kg$$

$$p_{3a} = p_3 = 7500 \ kPa$$

$$s_{3a} = 0.3130 + 1.005\ln\frac{3930}{1397} - 0.287\ln\frac{7500}{7500} = 1.353 \ kJ/kg - K$$

$$v_{3a} = 0.8558 \ m^3/kg$$

$$T_4 = 3930\frac{1}{\left(\frac{0.1505}{0.8558}\right)^{(1.4-1)}} = 1961 \ K$$

$$p_4 = 7500\left(\frac{0.1505}{0.8558}\right)^{1.4} = 658.0 \ kPa$$

$$s_4 = s_{3a} = 1.358 \ kJ/kg - K$$

Then the work terms can be calculated:

$$w_{1\to2} = c_v(T_1 - T_2) = 0.717(298 - 903.4) = -434.7 \ kJ/kg \tag{2.59}$$

$$w_{3\to3a} = R(T_{3a} - T_3) = 0.287(3930 - 1397) = 726.7 \ kJ/kg$$

$$w_{3a\to4} = c_v(T_{3a} - T_4) = 0.717(3930 - 1961) = 1412 \ kJ/kg$$

$$w_i = -434.7 + 726.7 + 1412 = 1704 \ kJ/kg$$

The indicated thermal efficiency can be calculated from the definition or Equation (2.58):

$$\eta_i \equiv \frac{(1704)}{2900} = 0.588$$

The indicated mean effective pressure can also be calculated.

$$imep = \frac{w_i}{v_d} = \frac{1704}{0.8558 - 0.05349} = 2124 \; kPa$$

Some of the conclusions gathered from the mathematical analysis of the cycles can be shown graphically with the use of $T - s$ and $p - v$ diagrams. These diagrams are really only useful in terms of visualizing processes and comparing ideal cycles under different assumptions. But, they do illustrate some basic concepts that are important in relation to engine operation and efficiency, so they will be presented and discussed here. The $p - v$ diagram is important because the work involved in the different processes is proportional to the area under the curve on the diagram (assuming that it is a reversible process):

$$\delta w = \int p \cdot dv \qquad \qquad (2.60)$$

In a real engine, the $p - v$ diagram is the only way to determine the actual indicated work performed, and these curves are integrated with numerical methods using actual engine data.

In a similar manner, heat transfer is proportional to the area under a curve on a $T - s$ diagram. Using the definition of entropy, (again assuming a reversible process) then:

$$\Delta q = \int T \cdot ds \qquad \qquad (2.61)$$

If we consider the $p - v$ diagram of an Otto cycle (Figure 2.1), we can see that there is work performed on the system for the compression process from point 1 to point 2. Note that in going from point 1 to 2, the volume is decreasing, that is, dv is negative. This means that the work is negative, consistent with the usual mechanical engineering sign convention that says work done on a system is negative. The greater the compression ratio, the more work required. No mechanical work is done during the heat addition process, since $dv = 0$. On the expansion stroke, positive work is performed from point 3 to point 4, since dv is positive. The area (amount of work) is greater than the compression work, since the pressure is higher than on the compression stroke.

Theoretically, it would be possible to determine the amount of net work done by measuring the area enclosed within the 4 process lines. This is, of course, very inconvenient, and the mathematical methods are much simpler. Note that there is no way to get information about the efficiency of the cycle from a $p - v$ diagram, since it contains no direct information about the energy added to the cycle in the process between points 2 and 3.

So we have to use the $T - s$ diagram (Figure 2.1). If we look at the processes one-by-one, we see that for the compression, there is no heat transfer, since by definition it is adiabatic. For the process from 2-3, there is heat addition and therefore an increase in entropy. The area under the curve is equal to the heat added (see Equation (2.14)). The expansion takes place between points 3 and 4, and again there is no heat transfer in the ideal process, so the line is vertical on the $T - s$ diagram.

In our ideal air cycles, the air exchange process is replaced by a heat rejection process, as the air is cooled back to the original state, 1. From either the $p - v$ diagram or the $T - s$ diagram, it can be seen that when the expansion is completed, the temperature and pressure of the gasses are above the the values at the start of the cycle. In order to get back to the original condition, we must "throw away" some energy. In the ideal cycle, this is an amount of cooling. In a actual engine, this energy is sent out of the engine when the exhaust valve is opened and the gases sent out into the atmosphere. The reason for the high temperature of the exhaust gasses is that given the fixed geometry of a piston engine, the cylinder gasses cannot be expanded adequately to cool them down to atmospheric temperature after the heat addition. On the $T - s$ diagram, the area under the line $4 \rightarrow 1$ can be used to indicate the amount of energy in the exhaust gas (exhaust temperature).

The $T - s$ diagram can then give us information about the efficiency of the process. First of all, we have an indication of the heat addition, the area under the line $2 \rightarrow 3$. In addition, we know from basic thermodynamics the for cycle:

$$\oint \delta w = \oint \delta q \tag{2.62}$$

This means that the net amount of work done is equal to the difference between the heat added and rejected, and that quantity is equal to the area enclosed on the $T - s$ diagram as well as enclosed area on the $p - v$ diagram. So for the cycle:

$$w_{net} = w_i \tag{2.63}$$
$$= q_{in} - q_{rejected} = q_{net} \tag{2.64}$$

and the indicated thermal efficiency is :

$$\eta_i = \frac{w_i}{q_{in}} \tag{2.65}$$
$$= \frac{q_{in} - q_{rejected}}{q_{in}} \tag{2.66}$$
$$= 1 - \frac{q_{rejected}}{q_{in}} \tag{2.67}$$

The last expression is the usually the most useful, as it allows comparisons of cycles under different conditions. Usually, the heat addition is kept constant, and then different processes are drawn on the $T - s$ and $p - v$ diagrams. The process with the smallest $q_{rejected}$ is then the most efficient.

A simple example is that of the effect of increasing the compression ratio on the efficiency of an Otto cycle. We know from the mathematical result, Equation (2.20), that the efficiency will go up. But we can ask another question: What happens to the exhaust temperature as the compression ratio increases?

To make answer the question, we make some conditions. First, the same amount of heat must be added to each cycle. Second, each cycle must start from the same temperature and pressure. The $T - s$ and $p - v$ diagrams can be drawn for a given Otto cycle. Here we are only interested in comparisons, and the diagrams can simply be sketched on a piece of paper, for comparison. The processes are shown in $p - v$ and $T - s$ diagrams in Figure 2.7.

To draw the second cycle, the line of the process $1 \to 2$ on the $p - v$ diagram must be extended along the same line to some arbitrary volume less than that of point 2, ending at a higher pressure. The heat addition process then takes place along a vertical line from the new end of compression, point 3', It is not yet known where to stop, but the new pressure will be higher than that of point 3. To get some guidance, attention is now turned to the $T - s$ diagram. The compression processes are vertical lines on the $T - s$ diagram, and the point 2' is higher up on the line than point 2. Heat is now added, and the area under the line (remember $\delta q = \int T ds$) $2 \to 3$ must be equal to the area under the line $2' \to 3'$. Lines of constant volume on a $T - s$ diagram are nearly parallel. This means that the heat addition must stop a at point on the line, where $s'_3 < s_3$, that is to the left of the original expansion line. The temperature of point 3' must be higher than that of point 3, since the same heat addition will give a higher final temperature when the initial temperature is higher. The $T - s$ diagram can now be completed by drawing a vertical line down from point 3' to intersect the line $4 \to 1$, defining a new point 4'. To complete the $p - v$ diagram, point 4' can be drawn between points 4 and 1, and then the line $3' \to 4'$ drawn, with such that $p_{3'} > p_3$.

Looking at the $T - s$ diagram, it can now be seen that when the compression ratio is increased, the amount of heat rejection decreases. This is clear, even though the diagrams may just be sketched. Therefore, according to Equation (2.65), the thermal efficiency is greater when the compression ratio is increased, as expected.

What about the second question? Since the amount of heat rejection is smaller, the exhaust temperature must be *lower*. This is often an unexpected result for the beginner, as high compression ratios are equated with high temperatures and pressures. The latter do occur, as $T_{3'} > T_3$ and $p_{3'} > p_3$. But because the process is more efficient, a larger portion of the heat added is converted to work, and thus less energy is rejected unused to the environment. In practice, changes in exhaust temperature can give an indication of changes in efficiency, though care must be taken, since there are other parameters that affect exhaust temperature in actual operating engines.

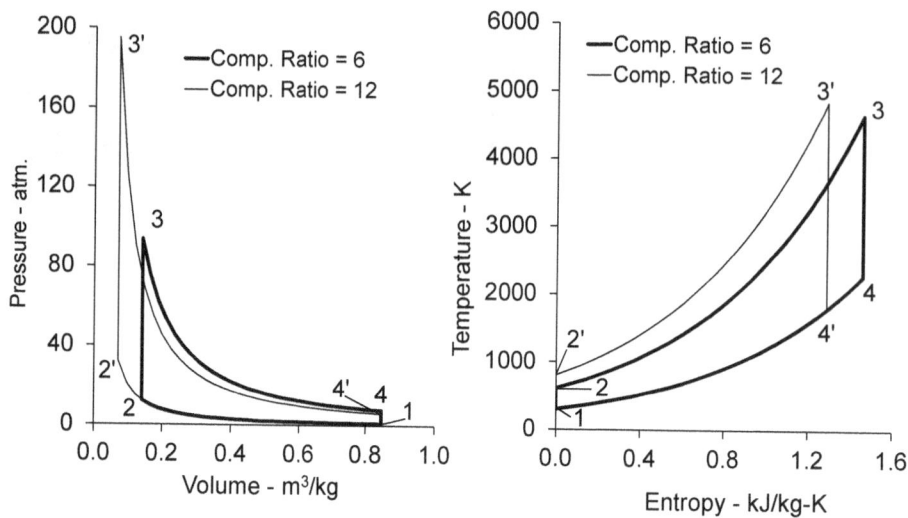

Figure 2.7: $p - v$ and $T - s$ diagrams for two Otto ideal air cycles with different compression ratios and the same heat addition, 2900 kJ/kg.

The above example shows that the $p - v$ and $T - s$ diagrams can be used as visualization tools. They can help to explain and understand fundamental concepts of thermodynamics and trends that determine engine operation and performance. For a more thorough discussion of ideal cycles, the reader is referred to other textbooks on internal combustion engines, for example references [4], [5], [6] and [7].

2.2 Simple Air Exchange in 4-Stroke Air "Engines"

In the discussion of the ideal air cycles, it was assumed that the system was closed, and that heat transfer to and from the system replaced the combustion and air exchange processes respectively. In this section, one of these assumptions will be removed. That is, the air exchange process will be looked at in a very simple fashion. The ideal air exchange process discussed here will only be that for the 4-stroke engine. The 2-stroke process is described in Chapter 8. While the methods are again simple, some important concepts will be discussed, including the residual fraction, *i. e.* the old gasses left over from the previous cycle and its influence on the subsequent cycle. The idea of pumping work will also be introduced, and the importance of intake pressure on the control and performance of an SI engine emphasized.

2.2.1 Exhaust Process

There are two separate process in the air exchange of a 4-stroke engine. The first to be considered is the exhaust process. When the work has been performed on the expansion process, and the piston is near the bottom of the cylinder, it is necessary to open the cylinder and remove the combustion products to make room for a fresh charge of either air or fuel-air mixture, depending on the engine type. There are two parts to the exhaust process, the first is the process where the exhaust valve is opened while the piston is near bottom dead center. The cylinder pressure here is above atmospheric, often high enough above the exhaust pressure that there is sonic flow through the exhaust valve for a period of time. A substantial portion of the exhaust gasses flow out of the cylinder through the exhaust port due to this pressure difference.

This process is commonly called the *blowdown* process, since the exhaust gasses are blown out of the cylinder and the pressure approaches the exhaust pressure. In spite of generating noise, there is an advantage to this process, in that a substantial fraction of the exhaust gasses leave the cylinder without any work being done on them. This means that there will be a smaller amount of exhaust gas to be forced out of the cylinder on the second part of the exhaust process, the actual *exhaust stroke*. Here, the exhaust valve is held open, and the piston moves from bottom dead center to top dead center, exhausting most of the remaining gasses. Notice that not all the gasses can be expected to be removed, because of the clearance volume needed to obtain the desired compression ratio. When the piston is at TDC, this volume can be expected to be filled with exhaust gasses at roughly the exhaust pressure. Pressure pulsations play a role here in real engines, and will be discussed later in Chapter 7. The remaining gasses are called the *residual gasses* or sometimes just the residual, since they are left over and will be present in the following cycle.

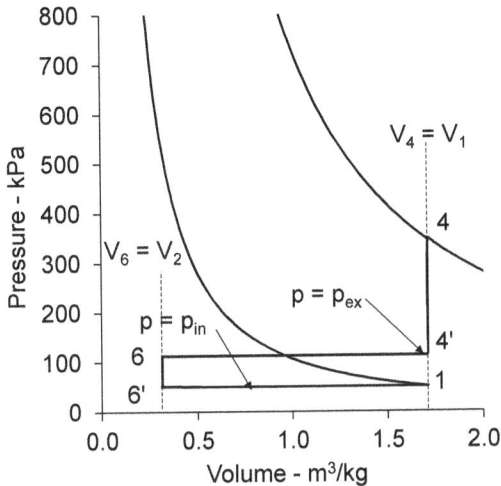

Figure 2.8: The intake and exhaust processes in an ideal engine with an early opening of the intake valve.

Two questions need to be answered: 1: What is the quantity of residual gas? and 2: What is the temperature of the residual gas? Consider first is simply the quantity. The more residual gas, the less room in the cylinder for a new charge, and so the less power the engine is capable of producing. Then with regard to the temperature, the residual gasses are also warm, since they are combustion products. This will influence the temperature of the following cycle, since the residual gets mixed with the incoming air. In the HCCI (Homogeneous Charge Compression Ignition) engine this is very important, and the residual amount is regulated in order to control the temperature of the cylinder gasses on the compression stroke and so, indirectly, the ignition timing.

Before going through the mathematics, it is clear that there can be a slight problem with calculations, since any given cycle is affected by the previous cycle. It will be shown that the open cylinder processes result in an iterative calculation process when the engine is assumed to be operating at steady state. The problem is that residual gasses are present at the start of a cycle, but their state is determined by the previous cycle, which is identical to that being analyzed. This iterative process is encountered with basically all simulations of engine operation where air exchange is encountered.

For the idealized cycles, the process will be considered to occur as shown in Figure 2.8. It is assumed that there is a constant cylinder pressure during the intake stroke, p_{in}. At bottom dead center, the intake valve is closed, the gasses trapped and the compression and expansion portions of the cycle occur as described in the previous section. The air exchange process is the same regardless of type of combustion.

It is further assumed that after heat addition (combustion), the gasses expand to BDC, at which time the exhaust valve is opened and a portion of the gasses flow out of the cylinder. The expansion occurs down to a pressure in the exhaust system, p_{ex}. This is assumed to occur instantaneously. After the pressure in the cylinder has fallen to p_{ex}, the piston is moved towards TDC, and the cylinder pressure held constant at p_{ex}. The figure shows the transition from exhaust to intake as occurring at constant volume. That is, it is assumed that the exhaust valve closes at TDC. Immediately after that, the intake valve is opened, and the pressure drops to p_{in}, at which time the intake process begins. There will be a flow of residual gasses back into the intake manifold

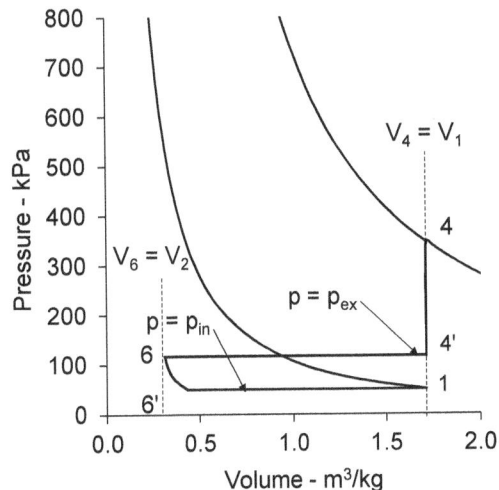

Figure 2.9: The intake and exhaust processes in an ideal engine with a late opening of the intake valve.

with this process, but it is assumed that they all are returned to the cylinder during the remainder of the intake process.

An alternative to the transition is the process where the opening of the intake valve is delayed to the time when the cylinder pressure is equal to the intake pressure as the piston moves away from TDC. In this way there is no back flow of residual gasses into the intake manifold. In this process, the residual gasses expand isentropically all the way down from the pressure after heat addition to the intake pressure before the intake valve opens. The process is shown in Figure 2.9

During the blowdown process, it is assumed that the expansion of the gasses inside the cylinder occurs isentropically. As the gasses pass through the exhaust valve, there is a throttling process, in which the entropy of the gasses leaving the cylinder increases, but these gasses are no longer relevant to the combustion process, since they will not return to the engine. Therefore, one considers a fictional process for the entire mass of cylinder gasses, where it is assumed that all the gasses, both in and out of the cylinder, assume a state that is the result of isentropic expansion to p_{ex}. This will give the correct thermodynamic state of the gasses in the cylinder, which is the only portion still of interest to us.

This fictive process results in a state of the entire cylinder gasses, which is called state 5. The volume and temperature of this state are[1]:

$$V_5 = \frac{m_5}{\rho_5} = \frac{1}{\rho_5} = V_4 \left(\frac{p_4}{p_{ex}} \right)^{\frac{1}{\gamma}} \tag{2.68}$$

$$T_5 = T_6 = T_4 \left(\frac{p_{ex}}{p_4} \right)^{\left(\frac{\gamma - 1}{\gamma} \right)} \tag{2.69}$$

It is further assumed that after the expansion, there is no heat transfer from the cylinder gasses, and that there is no change in pressure as the gasses are moved out of the cylinder by the upward motion of the piston. Then the mass of residual gas that cannot be removed from the cylinder is determined by the compression volume, and is equal to:

$$m_6 = \rho_6 V_6 = \rho_5 V_6 = \rho_5 V_2 \tag{2.70}$$

It is convenient to express the residual mass in terms of the total mass that is trapped in the cylinder after the air exchange is completed, and so the residual fraction is defined:

$$f \equiv \frac{residual\ mass}{total\ trapped\ mass} \tag{2.71}$$

This mass is constant from state 1 to state 4, and is the mass of the fictional volume V_5. Then:

$$f = \frac{residual\ mass}{total\ trapped\ mass} = \frac{\rho_5 V_2}{\rho_5 V_5} \tag{2.72}$$

$$f = \frac{V_2}{V_5} \tag{2.73}$$

Equation (2.73) is the important result that will be used in further calculations. It is also important to remember that it is determined by isentropic expansion from the state at the end of expansion, regardless of the cycle or working gasses.

2.2.2 Intake Process

In the ideal engine, the intake process begins when the exhaust process is completed. The process is simpler than the exhaust process, since the pressure differences are small. The flows are driven by the motion of the piston rather than by pressure differences. Figure 2.10 shows the states at the start and the end of the intake process. The initial state is assumed known, and the condition at the end of the process is needed, since it is the initial condition for the cycle calculations. The equations needed to determine the states are the equations of mass and energy conservation. In the following, some simplifying assumptions make it possible to obtain an analytic solution to the problem. For simulations in which the dynamic processes must be considered, such as the case of a detailed simulation where the instantaneous

---

[1]Capital letters are used to denote volumes in m^3, since the cylinder is open in this process and the mass may not be the same at all states

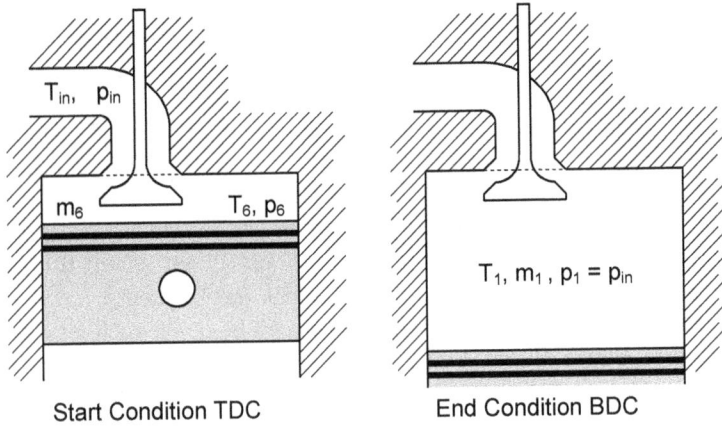

Figure 2.10: Schematic representation of the states at the start and end of in ideal intake process

flows through the valves are calculated, the equations must be solved in differential form, usually by numerical integration. The simplified analytic solutions shown here demonstrate the general concepts and tendencies.

The conservation of mass must be observed:

$$\frac{dm}{dt} = \dot{m}_{in} - \dot{m}_{out} \tag{2.74}$$

This can be simplified using the assumptions at hand and integrated over the intake stroke:

$$\int_{TDC}^{BDC} \frac{dm}{dt} dt = \int_{TDC}^{BDC} dm = m_1 - m_6 = \int_{TDC}^{BDC} \dot{m}_{in} dt \tag{2.75}$$

The first law for an open system neglecting kinetic energy is:

$$\dot{Q}_{cv} - \dot{W} = \frac{d(mu)}{dt} + \sum_{out} \dot{m}_{out} h_{out} - \sum_{in} \dot{m}_{in} h_{in} \tag{2.76}$$

In the case of the ideal intake stoke, there is no heat transfer and no flow out of the system. The rate of control volume work is:

$$\dot{W} = p\frac{dV}{dt} \tag{2.77}$$

Integrating Equation (2.76) from TDC to BDC, one obtains:

$$-\int_{TDC}^{BDC} p\frac{dV}{dt} dt = \int_{TDC}^{BDC} \frac{d(mu)}{dt} dt - \int_{TDC}^{BDC} \dot{m}_{in} h_{in} dt \tag{2.78}$$

The case is considered where the intake valve opens at TDC. During the intake stroke, the pressure is constant at p_{in} and the enthalpy of the intake mixture is regarded as constant, and can be taken out of the last integral in Equation (2.78). The resulting integral can be replaced by Equation (2.75). As a result, Equation (2.79) is obtained:

$$-p(V_{BDC} - V_{TDC}) = mu_{TDC} - mu_{BDC} - h_{in}(m_1 - m_6) \tag{2.79}$$

where point 1 corresponds to BDC and point 6 corresponds to TDC. Further, using $V = v \cdot m$, the ideal gas law, and the definitions of enthalpy and residual fraction:

$$h_1 = (1 - f) h_{in} + f\left(u_6 + \frac{p_{in}}{p_6} \cdot RT_6\right) \tag{2.80}$$

This is the sought result, as the temperature (enthalpy) at the start of the cycle is given as a function of the condition at the end of the exhaust stroke, described in Equations: (2.68) and (2.73) and the intake condition. There are two unknowns, f and T_1 and two Equations, (2.73) and (2.80). In principle, then, the cycle is determined. The normal procedure is to guess values of f and T_1 at the start of a calculation, conduct the calculations for the compression, heat addition and expansion, determine state 5 (note that $T_5 = T_6$), calculate f with (2.73), and then calculate T_1 with Equation (2.80). If both f and T_1 are the same (very close to) the original estimates, convergence is achieved. If not, the whole process must be conducted again until convergence on f and T_1 is achieved. This is efficiently done by simply using the estimates for f and T_1 for a cycle as the new estimates for the next cycle. Normally the process will converge within 3 or 4 iterations, for agrement better than about 0.1°C for T_1 and about 0.1 % for f.

Equation (2.80) is general, and can be used for temperature dependent specific heats and such. However, so far the rest of the cycle is idealized, and so some simplifications can be made without compromising too much accuracy. Since there is no chemical reaction occurring on the intake stroke, the enthalpies in Equation (2.80) can be replaced by the integral of $c_p dT$ for each substance, and Equation (2.80) then only involves some temperatures, the gas constant and some c_p values. For the ideal air process (the original reason for doing this analysis here), there is only one value of c_p, and it is normally assumed to be constant. Then:

$$T_1 = (1 - f) T_{in} + fT_6 \left(\frac{1}{\gamma} + \frac{p_{in}}{p_6} \frac{\gamma - 1}{\gamma} \right) \tag{2.81}$$

where T_6 is the temperature of the cylinder gasses expanded isentropically from p_4 to the exhaust pressure. For the case where the intake and exhaust pressures are equal, Equation (2.81) reduces to:

$$T_1 = (1 - f) T_{in} + fT_6 \tag{2.82}$$

The analysis will not be shown here, but for the case where the intake opening is delayed until the cylinder pressure is equal to the intake manifold pressure, Equation (2.78) becomes:

$$h_1 = (1 - f) h_{in} + fh_{6''} \tag{2.83}$$

In this case, the state, 6" refers to the condition where the cylinder gasses have been expanded isentropically from state 4 to the intake pressure. As with Equation (2.81), Equation (2.83) can be simplified for the ideal air process to:

$$T_1 = (1 - f) T_{in} + fT_{6''} \tag{2.84}$$

With Equation (2.80), the pumping portion of the $p - V$ diagram is simply a rectangle, so the pumping work is easy to calculate, and the equation for T_1 a little more complicated. For the late opening intake valve the situation is reversed, the equation for T_1 is simple, but the pumping work less convenient to determine. Relative to the overall accuracy, and all other simplifications, the error in assuming the that the pumping loop is a rectangle is of minor significance.

An important detail of operation of an SI engine can be seen by looking at the residual gas process. The residual fraction is determined by the fictive volume, V_5, through Equation (2.73), and $V_2 = V_1/\varepsilon$. Since the mass is constant between points 1 and 5, if can be shown that :

$$f = \frac{1}{\varepsilon} \frac{T_{in}}{T_5} \frac{p_{ex}}{p_{in}} \tag{2.85}$$

The value of T_5 does not change too much, since the largest drop from the combustion temperature is on the expansion from TDC to BDC, and the temperature ratio is determined only by the expansion (compression) ratio here. In a spark ignition engine, the power output is regulated by adjusting the intake pressure. When the engine is idling, the power output is 0, and the power produced on the piston is used only to overcome friction. Therefore, the intake pressure is very low, (typically on the order of about 0.25 bar) and the resulting pressures throughout the entire cycle much lower than at full load. Equation (2.85) then indicates a larger value of f when the intake pressure (load) decreases in an SI engine. The largest residual dilution is at idle, this will be shown later to have implications for stability and emissions when an SI engine is at idle.

In an actual engine, the physical volume is fixed and the amount of air changes. The residual mass is determined by the temperature and pressure at the end of the exhaust stroke. The dominant factor for determining the residual mass is the pressure at the end of the exhaust stroke, which in an engine without turbocharging remains near atmospheric, especially at the low flow condition with idle. On the intake side with an SI engine, the amount of air taken in is drastically reduced with respect to full load by lowering the intake pressure. Thus an approximately constant amount of residual gas becomes more and more significant when it is mixed with a smaller and smaller amount of fresh charge. If the residual fraction is too high, the flame will not propagate, a problem with operation at idle in SI engines.

2.2.3 Summary

This section has shown that when the air exchange process, even a highly idealized process, is considered, the calculation of an engine cycle becomes an iterative process, and the starting cycle temperature and residual fraction must be determined by iteration. The various equations for the temperature at the start of compression, while differing slightly in details are of the same general form, and show the influence of the amount of residual and its temperature on the starting fresh mass in the cylinder. This in turn determines how much work can be produced by the cycle. High residual amounts displace large amounts of fresh air, and high residual temperature results in a low density at the start of compression, which also reduces power.

2.3 Air Cycle with Air Exchange

With the basics of the air exchange process in hand, attention can now be paid to the calculation of engine processes with air exchange included. The thermodynamic processes are as used in the case of the ideal air cycles with the exception of the heat rejection process, which will now be replaced by the air exchange process. The result is an ideal air cycle with air exchange, or what might be called an ideal air engine.

Some adjustments to the calculation procedure must be made. The amount of heat addition to the cycle will now depend on the air exchange effectiveness. To take an extreme example, if there is 100% residual gas, there can be no combustion and the heat addition is zero. Under the assumptions that the combustion is complete, the total cylinder mass is 1 kg and that there is only air in the cylinder, the following equation can be used for the calculation of the heat addition:

$$Q_{in} = 1 \cdot H_u \cdot FA \cdot (1 - f) \tag{2.86}$$

It is implicitly assumed here that the residual mass is "old" mass and therefore cannot contribute to the combustion. It must also be remembered that this model is not valid for combustion of mixtures where there is more than the chemically correct amount of fuel.

The calculation process is then:

- Guess f and T_1. Without any other experience, a value of 0.05 is a reasonable starting value for f, and it is also reasonable to use $T_1 = T_{in} + 30°$.

- Calculate the value of Q_{in}

- Find the states 2,3 and 4 as was the case with the ideal air cycle, Otto, Diesel or Limited Pressure

- Find V_5 from state 4 and the exhaust pressure, using Equation (2.68)

- Use Equation (2.73) to determine f

- Use Equation (2.81), (2.82) or (2.84) to find T_1 using the value of f just calculated, and the value of T_6 determined from Equation (2.69). In the case of the late opening intake valve, use the intake pressure instead of the exhaust pressure in Equation (2.69) to find $T_{6''}$

- Compare the values of f and T_1 just found. If they agree to within the definition of acceptable agreement, convergence has been achieved. If not, use these values as new starting values and repeat the entire calculation until convergence is attained. Normally, if it takes more than 5 iterations, the problem is poorly posed or there is an error somewhere. Remember that when using the equations to determine T_1, the intake temperature is a constant parameter in Equations (2.81), (2.82) and (2.84). It is a common mistake to use T_1, instead of the intake temperature, but if that is done the problem will not converge.

When convergence is achieved, the final values can be calculated using the equations previously developed for the ideal air cycles for the processes between the start of compression and the end of expansion. The following uses the Otto cycle as an example. The use of a diesel or limited pressure cycle should be straightforward.

The indicated work:

$$W_i = 1 \cdot [c_v (T_1 - T_2) + c_v (T_3 - T_4)] \tag{2.87}$$

The pumping work:

$$W_p = (p_{in} - p_{ex}) (V_1 - V_2) \tag{2.88}$$

The indicated mean effective pressure:

$$imep = \frac{W_i}{(V_1 - V_2)} \tag{2.89}$$

The indicated thermal efficiency

$$\eta_i = \frac{W_i}{Q_{in}} \tag{2.90}$$

The indicated specific fuel consumption:

$$isfc = \frac{3.6 \cdot 10^6}{H_u \cdot \eta_i} \tag{2.91}$$

where $isfc$ is in g/kWh and the heating value in kJ/kg

The charging efficiency, here denoted η_f is referred to the situation where the engine is filled with air at atmospheric density:

$$\eta_f = \frac{1 - f}{\rho_{atm} (V_1 - V_2)} \tag{2.92}$$

The volumetric efficiency is referred to the situation where the engine is filled with air at the intake density:

$$\eta_v = \frac{1 - f}{\rho_{in} (V_1 - V_2)} \tag{2.93}$$

Air Engine Example

Consider an ideal engine with air as the working fluid, operating at a fuel air ratio of 0.065 on a fuel with a lower heating value of 44000 kJ/kg. The compression ratio is 9.5:1, the intake temperature 298K, the intake pressure 0.9 bar and the exhaust pressure 1.2 bar. Find the indicated mean effective pressure, the indicated thermal efficiency, the indicated specific fuel consumption and the charging efficiency.

To start, assume that T_1=329.9 K and $f = 0.03715$. These are the correct values and the example will verify this. The usual method is to assume a mass of 1 kg air. In determining the convergence, only the essential calculations will be performed. When convergence is obtained, the correct intermediate properties and final output variables can be calculated.

The initial volume is:

$$V_1 = \frac{mRT_1}{p_1} = \frac{1 \cdot 0.287 \cdot 329.9}{90} = 1.053 \ m^3$$

Using the compression ratio:

$$V_2 = \frac{V_1}{\varepsilon} = \frac{1.053}{9.50} = 0.1108 \ m^3$$

With isentropic compression:

$$T_2 = T_1 \cdot \varepsilon^{(\gamma-1)} = 329.9 \cdot (9.5)^{0.4} = 811.9 \ K$$

The temperature at the end of heat addition (combustion) is calculated from the first law. It is assumed that the mixture is leaner than stoichiometric and the entire heating value is released upon combustion, that is the combustion efficiency is 100%.

$$T_3 = T_2 + \frac{1 \cdot FA \cdot H_u \left(1 - f\right)}{c_v} = 811.9 + \frac{0.065 \cdot 44000 \cdot (1 - .03715)}{0.717} = 4652 \ K$$

The temperature at the end of expansion is:

$$T_4 = T_3 \left(\frac{1}{\varepsilon}\right)^{\gamma-1} = 4652 \left(\frac{1}{9.5}\right)^{0.4} = 1891 \ K$$

The pressure at the end of expansion is obtained from the ideal gas law:

$$p_4 = \frac{mRT_4}{V_4} = \frac{1 \cdot 0.287 \cdot 1891}{1.053} = 515.8 \ kPa$$

The isentropic expansion volume can now be calculated:

$$V_5 = V_4 \left(\frac{p_4}{p_5}\right)^{\frac{1}{\gamma}} = 1.053 \frac{515.8}{120} = 2.981 \ m^3$$

Now the value of f can be checked:

$$f' = \frac{V_2}{V_5} = \frac{0.1108}{2.981} = 0.03717$$

This is in good agreement with the assumed value. The temperature of the exhaust gas is found by expanding from state 4 to the exhaust pressure, either using isentropic relations or, as here, the ideal gas law with V_5 and the exhaust pressure:

$$T_6 = T_5 = \frac{p_{ex}V_5}{mR} = \frac{120 \cdot 2.981}{1 \cdot 0.287} = 1246 \ K$$

The temperature at the start of compression is found using the process where the exhaust valve closes at TDC, Equation (2.81):

$$T_1 = (1 - f') \, T_{in} + f' \left(\frac{1}{\gamma} + \frac{\gamma - 1}{\gamma} \frac{p_{in}}{p_{ex}}\right) T_6$$

$$= (1 - 0.03717) \, 298 + 0.03737 \left(\frac{1}{1.4} + \frac{(1.4 - 1)}{1.4} \frac{90}{120}\right) 1246 = 329.9 \ K$$

Since f and T_1 agree with the initial values, convergence has been obtained. If they do not agree, then the latest calculated values of f and T_1 should be used as new starting values and the procedure repeated.

Now the remaining pressures in the cycle can be calculated:

$$p_2 = \frac{mRT_2}{V_2} = \frac{1 \cdot 0.287 \cdot 811.9}{0.1108} = 2103 \ kPa$$

$$p_3 = \frac{mRT_3}{V_3} = \frac{1 \cdot 0.287 \cdot 4627}{0.1108} = 11981 \ kPa$$

The problem is completed by calculating the engine output parameters.

Indicated Work:

$$W_i = m \cdot c_v \left(T_1 - T_2 + T_3 - T_4 \right) = 1 \cdot 0.717 \left(329.9 - 811.9 + 4653 - 1891 \right) = 1635 \ kJ$$

Indicated Mean Effective Pressure:

$$imep = \frac{W_i}{V_1 - V_2} = \frac{1636}{1.052 - 0.1108} = 1737 \ kPa$$

Indicated Thermal Efficiency:

$$\eta_i = \frac{W_i}{Q_{in}} = \frac{W_i}{m_a FA \cdot H_u \left(1 - f \right)} = \frac{1635}{1 \cdot 0.065 \cdot 44000 \cdot \left(1 - 0.03717 \right)} = 0.5938$$

It is of interest to note that for the Otto ideal air cycle:

$$\eta_i = 1 - \left(\frac{1}{\varepsilon} \right)^{\gamma - 1} = 1 - \left(\frac{1}{9.5} \right)^{0.4} = 0.5936$$

The process is just as efficient as the Otto cycle, since it is a constant volume process, and the same thermodynamic properties have been assumed. Because of a higher T_1 and the presence of residual gasses, the *imep* will be less.

Indicated Specific Fuel Consumption:

$$isfc = \frac{3.6 \cdot 10^6}{H_u \cdot \eta_i} = \frac{3.6 \cdot 10^6}{44000 \cdot 0.5936} = 137.8 \ g/kWh$$

Charging Efficiency:

$$\eta_f = \frac{1 - f}{\rho_{atm} \left(V_1 - V_2 \right)} = \frac{1 - 0.03717}{\frac{100}{0.287 \cdot 298} \left(1.052 - 0.1108 \right)} = 0.8488$$

Volumetric Efficiency:

$$\eta_v = \frac{1 - f}{\rho_{in} \left(V_1 - V_2 \right)} = \frac{1 - 0.03717}{\frac{90}{0.287 \cdot 298} \left(1.052 - 0.1108 \right)} = 0.9721$$

The volumetric efficiency is quite high. It has been assumed that pressure in the cylinder at the start of compression is that same as the intake pressure. In an actual engine it will be lower due to pressure drops across the intake port and valve, and reverse flow at the end of the intake stroke. But the volumetric efficiency will normally be close to one, and the changing air flow with changing load is due to the change in the intake pressure as a result of throttling in an SI engine, or pressure charging in turbocharged or supercharged engines.

2.4 Fuel Air Cycles

This chapter started with the most simple model of engine cycles, that is, the ideal air cycle. They showed the importance of the compression ratio, the thermodynamic properties, and the placement and speed of the combustion process in a very simple way. A second improvement was the introduction of a simple air exchange process for the 4-stroke engine. This introduced the idea of the residual gas fraction, and the "carry over" from one cycle to the next. This lead to the need for iterative solutions for steady state engine cycle simulation. These are some fundamental concepts, that are valid no matter what the type of engine.

The assumption of constant specific heats in these models is unrealistic, and can be eliminated by using gas tables for variable specific heats, or modern simulation programs, such as EES, which has the thermodynamic properties included. This is left as an exercise for the reader to perform, and only some results shown here in Table 2.2. It can be seen that the temperatures and pressures are lower, when more accurate specific heats are used, as well as the output and efficiency.

Table 2.2: Comparison of Otto cycle variables with specific heat ratio of 1.4 with that for variable specific heat of air. Compression ratio = 9.5, heat addition=2548 kJ/kg, $p_1 = 100kPa$, $T_1 = 330K$

Variable	$\gamma = 1.4$	Variable γ
T_2 - K	812.1	781.2
T_4 - K	1773	1794
p_2 - kPa	2338	2249
p_3 - kPa	12560	9948
p_4 - kPa	537.2	543.5
η_i -	0.5936	0.5114
$imep$ - kPa	1785	1537

Attention will be given to a much greater step. That is the calculation of ideal cycles will be made in which all the specific heats are temperature and composition dependent, and the effects of temperature, pressure and fuel air ratio on the composition of the products of combustion are included. This is normally called the fuel air cycle, or equilibrium fuel air cycle.

One way of handling the composition for combustion products of fuel air mixtures is to assume that the composition is that resulting from ideal combustion. For lean mixtures of fuels containing C, H, and O, the ideal products are CO_2, H_2, N_2 and O_2. For rich mixtures, CO and H_2 will be present, and some methods to calculate their concentrations are given in Section 3.1. However, at temperatures and pressures encountered during engine combustion, these ideal combustion products are not stable, but react to form additional substances such as O, H, OH, and NO, In addition, even for lean mixtures, CO and H_2 can be formed during and after combustion. Many of these reactions involve significant amounts of energy, and, therefore, influence the temperatures and pressures of cycle.

Though there may be some exceptions in actual engines, reactions of the fuel and air mixture before ignition are ignored, and the composition of pre-combustion gasses is assumed to be that of a mixture of fuel, air and residual gas, the latter normally assumed to be of constant composition.

When dealing with the equilibrium systems, it is common to use the fuel air equivalence ratio instead of the fuel air ratio. The equivalence ratio is defined in Equation (3.12) as the ratio of the actual fuel air ratio to the fuel air ratio for a chemically correct mixture of fuel and air. For mixtures with an excess of fuel, the equivalence ratio, ϕ, is greater than one, while for an excess of oxygen, it is less than one. It is a convenient parameter for specifying the relative mixture strength, regardless of the absolute value of the chemically correct fuel air ratio.

2.4.1 Chemical Equilibrium

The composition of the combustion products when all these species are involved is determined by the assumption that the gasses are always in a state of chemical equilibrium. This state is the composition that gives the minimum Gibbs free energy of the system. Of interest with a reaction or system of reactions is the composition which results in a state of equilibrium. This is the condition where there is no thermodynamic tendency for the composition to change. This is the chemical equivalent of a mechanical system in thermal equilibrium, where there is no tendency for the temperature to change when the system is isolated.

The properties of the combustion products are determined through a calculation to determine the equilibrium composition. Studies show that to a large degree, the combustion products are close to the equilibrium composition once a flame has occurred, and remain so until a certain temperature is reached during expansion, where the composition "freezes". This temperature is on the order of 1500-1800 K. The transition from equilibrium to frozen composition normally occurs over a fairly small temperature range, so the assumption of fixed temperature at which the composition remains fixed is usually acceptable. In a system undergoing chemical reactions at a given temperature and pressure, the thermodynamic properties are dependent on the progress of the reactions, in addition to temperature and pressure.

In order to understand what is involved in the calculation, some background material is given here. For a more detailed discussion of chemical equilibrium and the solution of the equilibrium composition in reacting systems, the reader is referred to more advanced books on engine thermodynamics or combustion.

See for example references [8], [9], [7], [10], [11].

In order to find the equilibrium condition, an expression for the Gibbs free energy for the reacting system is written as a function of the degree of reaction, and the state of minimum Gibbs free energy obtained. The procedure can be complicated and will not be described here, but leads to the concept of the equilibrium constant. For ideal gasses, which is the norm for engines, and a given reaction, the equilibrium constant used is usually that written in terms of the partial pressures of the reactants and products. For a general reaction, where species A and B react to form species C and D with the relative number of moles ν_A, ν_B, ν_C, and ν_D:

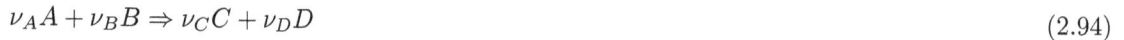

$$\nu_A A + \nu_B B \Rightarrow \nu_C C + \nu_D D \tag{2.94}$$

The equilibrium constant in terms of the partial pressures can be written:

$$K_p(T) = \frac{p_C^{\nu_C} \cdot p_D^{\nu_D}}{p_A^{\nu_A} \cdot p_B^{\nu_B}} \tag{2.95}$$

Where p_i is the partial pressure of gas i, normally in atmospheres in equilibrium calculations. For ideal gasses, the partial pressure is calculated as

$$p_i = x_i \cdot p \tag{2.96}$$

Where x_i is the mole fraction of specie i and p is the total pressure of the system.

The equilibrium constant can be shown to be a function of temperature alone, and values of $K_p(T)$ can be found in combustion and thermodynamics texts. Since the value of K_p depends only on the temperature, and the partial pressures depend on the composition (mole fraction) and the total pressure, Equation (2.95) represents a relation between the temperature, pressure and degree of reaction (composition). Equation (2.95) is often nonlinear, and when several reactions are involved, a number of expressions for equilibrium constants for these reactions must be solved simultaneously, along with the atomic balances. The solution of the reaction system for combustion products requires special techniques. It is possible to set up the equations for combustion products and solve the equations using existing equation solvers. However, several programs exist to solve the relevant equations for combustion systems, and it is not necessary to develop new programs to accomplish this. See for example references [8], [9], [10] and [11] for FORTRAN equilibrium programs.

For most engine work, the species considered in the equilibrium calculations are:

CO_2, H_2O, CO, H_2; N_2, O_2, H, O, OH, NO, N with Ar as an inert substance.

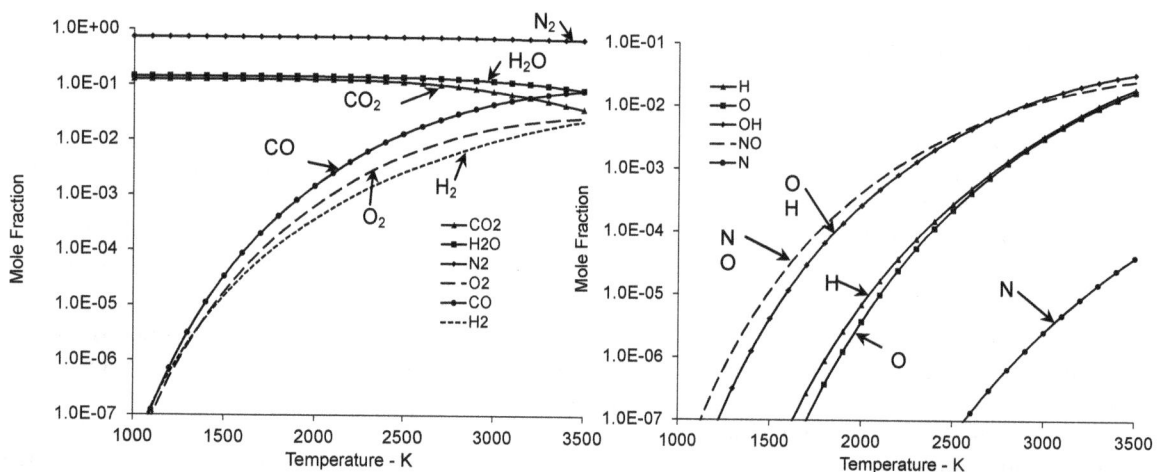

Figure 2.11: Equilibrium combustion products for a stoichiometric octane-air mixture at 20 atm pressure

To solve the reaction system at a given temperature, pressure and equivalence ratio, atomic balances for all the elements must be used in combination with a series of equilibrium constants. In equilibrium systems, a set of linearly independent reactions is utilized. For the calculations presented in the following,

Table 2.3: The set of equilibrium reactions and corresponding equilibrium constants used to determine the equilibrium composition of combustion products for hydrocarbon fuels

Reaction	Equilibrium constant
$CO_2 \rightleftharpoons CO + \frac{1}{2}O_2$	$K_p = \dfrac{p_{CO} \cdot p_{O_2}{}^{0.5}}{p_{CO_2}}$
$H_2O \rightleftharpoons H_2 + \frac{1}{2}O_2$	$K_p = \dfrac{p_{H_2} \cdot p_{O_2}{}^{0.5}}{p_{CO_2}}$
$H_2O \rightleftharpoons \frac{1}{2}H_2 + OH$	$K_p = \dfrac{p_{H_2}^{0.5} \cdot p_{OH}}{p_{H_2O}}$
$\frac{1}{2}H_2 \rightleftharpoons H$	$K_p = \dfrac{p_H}{p_{H_2}^{0.5}}$
$\frac{1}{2}O_2 \rightleftharpoons O$	$K_p = \dfrac{p_O}{p_{O_2}^{0.5}}$
$\frac{1}{2}N_2 \rightleftharpoons N$	$K_p = \dfrac{p_N}{p_{N_2}^{0.5}}$
$\frac{1}{2}N_2 + \frac{1}{2}O_2 \rightleftharpoons NO$	$K_p = \dfrac{p_{NO}}{p_{N_2}^{0.5} \cdot p_{O_2}^{0.5}}$

a simplified version of the reaction set used by Strehlow [9] was used. The possibility of formation of methane is not included, and for engine purposes, the formation of NO_2 is not considered. The reactions and corresponding expressions for the equilibrium constants are shown in Table 2.3

The set of chemical balance and equilibrium equations can be solved for a range of temperatures and pressures typical for engine combustion products. Figure 2.11 shows the variation of the exhaust gas composition from octane combustion as a function of temperature for a pressure of 20 atmospheres and a stoichiometric fuel air ratio. At temperatures below around 1500 K, the minor products are of little significance, equilibrium can be neglected, and the composition is essentially that predicted for ideal combustion products from the overall chemistry. This helps to reduce the importance of the assumption of a freezing temperature for exhaust gas composition.

At higher temperatures, secondary products begin to appear according to the reactions in Table 2.3. Several of these reactions have significance with respect to the temperature. For example, the second reaction in Table 2.3 involves the dissociation of water to form hydrogen and oxygen. The reverse reaction is, of course, the combustion of hydrogen and involves a significant release of chemical energy, that is, it is exothermic. The formation of hydrogen and oxygen from water thus requires an input of energy (is endothermic). In the high temperature combustion products, hydrogen is formed in increasing amounts as the temperature increases, as shown in Figure 2.11. As a result of the dissociation of the water and carbon dioxide, there is chemical energy stored in the combustion products, especially in the form of carbon monoxide and hydrogen, such that the equilibrium temperature is lower than that obtained with ideal products (no dissociation).

2.4.2   Equilibrium Cycle

In order to accurately predict the state of the combustion products with chemical equilibrium, the reaction system in Table 2.3 must be solved along with atomic conservation and the first law of thermodynamics. The fuel air cycle is a cycle consisting of the same processes as the ideal air engine discussed previously, but in which the correct thermodynamic properties are included. The same simple air exchange process as used for the ideal air engine is normally used. Earlier engine text books have included charts for temperature and equivalence ratio dependent properties of fuel air mixtures, and equilibrium charts for the properties of the combustion products. With these charts, it is possible to perform cycle calculations. Since the solution for chemical equilibrium in the combustion products is now conveniently performed using computers, the entire fuel air cycle can be quickly performed using a computer program. The details of such a calculation are discussed in Ferguson [11].

A series of calculations have been made for the fuel air cycle for the Otto process, using the WEIN equilibrium procedure developed by Strehlow [9]. Calculations were made for iso-octane, but the results are fairly similar for most hydrocarbon fuels or simple alcohols that are suited for SI engine use. The processes included are:

- Isentropic compression (the specific heat is a function of temperature and equivalence ratio), no chemical reaction on compression.

- Constant volume combustion, combustion products satisfy chemical equilibrium immediately after combustion.

- Isentropic expansion to BDC, combustion products satisfy chemical equilibrium at each state.

- Isentropic expansion to exhaust pressure, combustion products for stoichiometric or lean mixtures are CO_2, H_2O, O_2, N_2 and Ar, and for rich mixtures, CO_2, H_2O, O_2, N_2, CO, H_2 and Ar. The latter composition is calculated from the water gas reaction (see Section 3.1.2) at a temperature of 1500K.

- Air exchange process with constant exhaust pressure, intake opening at when intake pressure is equal to the cylinder pressure.

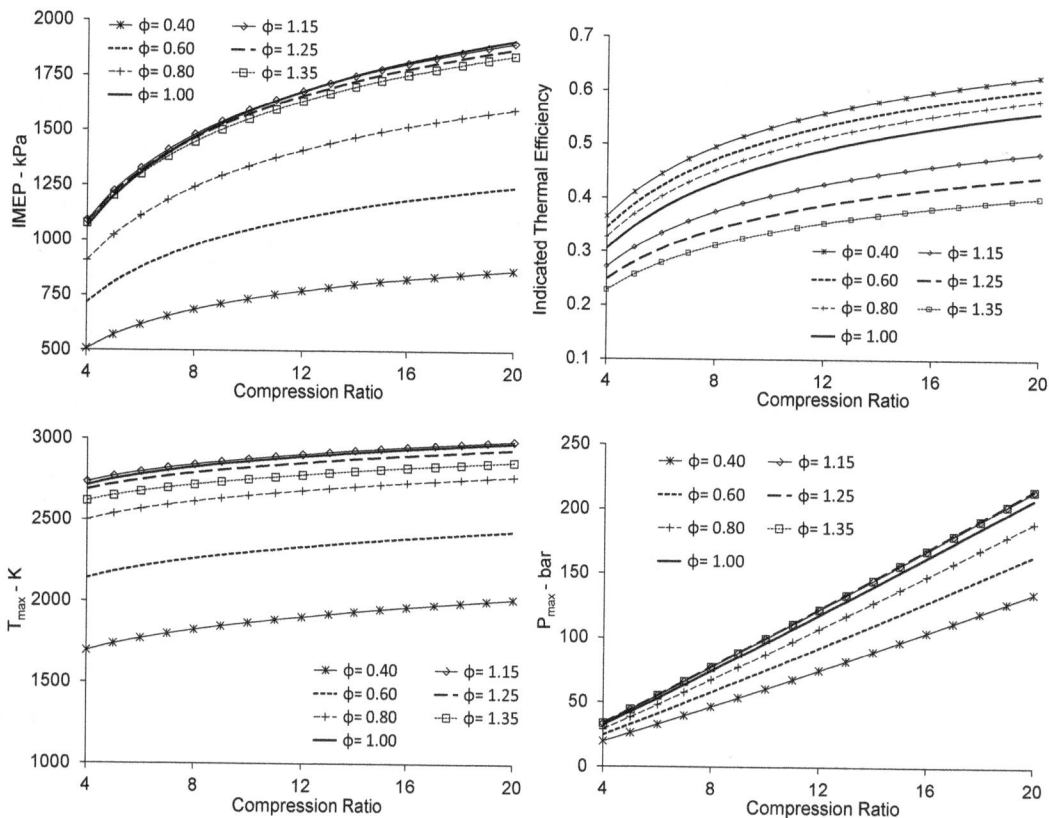

Figure 2.12: Indicated thermal efficiency, indicated mean effective pressure, maximum cycle temperature and pressure for the ideal fuel air cycle with constant volume combustion as a function of compression ratio and fuel air equivalence ratio with iso-octane fuel. Intake temperature and pressure 298K, 100 kPa, exhaust pressure 100 kPa.

Note that the fuel air equilibrium Otto cycle gives the maximum possible efficiency and mean effective pressure for a constant volume process for an spark ignition engine under a given set of conditions.

The estimates using even temperature dependent properties of air alone are actually too high, since the properties would not include fuel specific heats or chemical equilibrium. It is found that at the end of combustion, and throughout most of the expansion stroke, the chemical composition in a real engine is quite close to that predicted by the equilibrium calculation for a given temperature and pressure. The errors due to the equilibrium assumption concerning the composition and properties in the ideal fuel-air cycle are very small. Differences in the fuel air equilibrium cycle in comparison to real engines are due to other processes such as the time it take for combustion to occur (not at constant volume) and the heat loss to the cylinder walls. In other words, for an engine operating with a given compression ratio, fuel air ratio and intake pressure, it is not possible for the engine to produce more power or have a better efficiency than that predicted by the fuel air cycle with constant volume. A real engine will produce less power.

The indicated thermal efficiency, indicated mean effective pressure, maximum temperature and maximum pressure for the ideal fuel air cycle are shown in Figure 2.12 as a function of compression ratio for intake and exhaust pressures of one atmosphere, and intake air temperature of 298K at various fuel air equivalence ratios. The same results have been re-plotted with equivalence ratio as the abscissa in Figure 2.13. The following trends can be seen. First of all, the effect of the compression ratio on efficiency and mean effective pressure is similar to that of the ideal cycles, although the values are lower than the air cycles. That is, the efficiency and mean effective pressure increase with increasing compression ratio. This is due to the same effect, as more of work is extracted from the high temperature gasses as the compression (expansion) ratio increases.

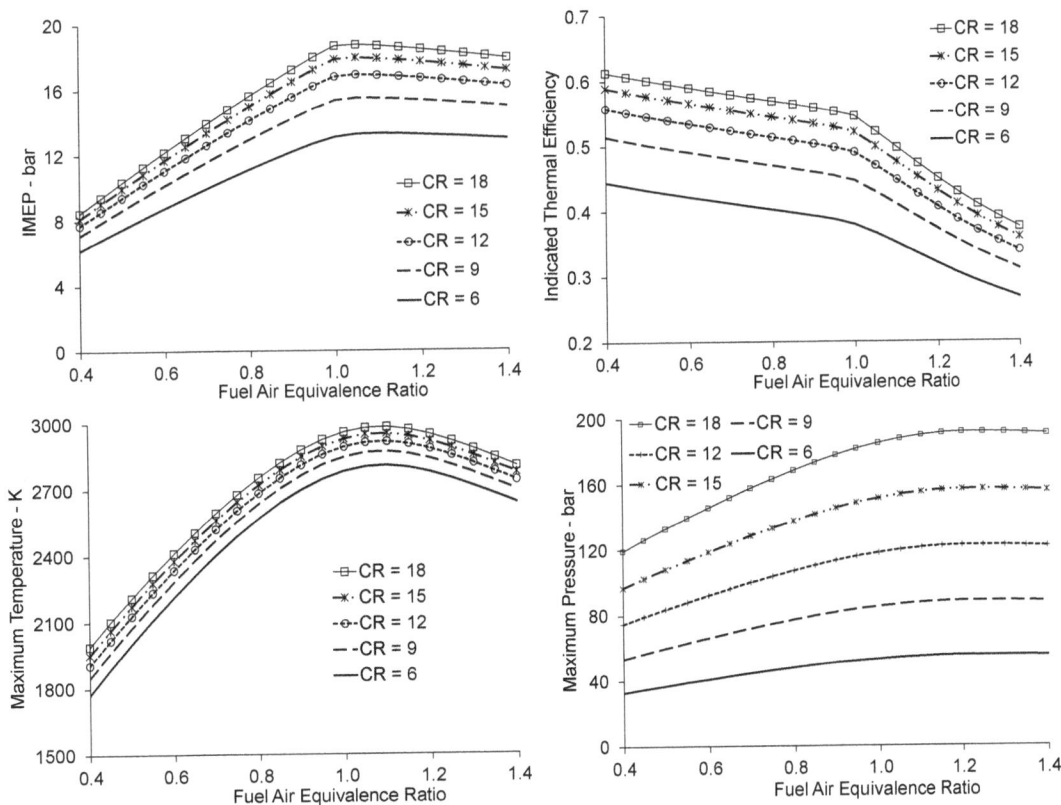

Figure 2.13: Indicated mean effective pressure, indicated thermal efficiency, maximum cycle temperature and pressure for the ideal Fuel Air cycle with constant volume combustion as a function of fuel air equivalence ratio and compression ratio with iso-octane fuel. Intake temperature and pressure 298K, 100 kPa, exhaust pressure 100 kPa.

The values are lower than the ideal air cycles because the presence of fuel increases the specific heat of the mixture, causing a lower value of the specific heat ratio, γ, which in turn lowers the efficiency. The tendency is the same as that seen for the ideal air cycle in Figure 2.2, the values, though are lower.

Additionally, the chemical equilibrium reactions result in lower pressures after combustion, which gives less work for the same fuel input. Figure 2.13 shows the highest thermal efficiency for the condition where there is the least fuel, that is for the lowest equivalence ratios. This is because the fuel has a higher specific heat than the air, and for an equivalence ratio of 0, the efficiency would be the same as that of an ideal air cycle with temperature dependent specific heats, though with no output! In addition, the dissociation of ideal reaction products due to chemical equilibrium becomes less important as temperatures after combustion decrease as the mixture becomes leaner.

Another effect regarding the the curve of indicated thermal efficiency as a function of equivalence ratio is that a sharp change occurs at stoichiometric mixtures. This is due to the lack of oxygen when the mixture becomes richer than the chemically correct ratio. (see Figure 2.15). As the equivalence ratio increases above 1.0, more and more CO and H_2 are formed, and exhausted to the atmosphere. The amount of energy liberated by combustion remains quite close to being constant near the stoichiometric value for these richer mixtures. Thus, there is a decrease in the thermal efficiency due to a poorer combustion efficiency, and the increased amount of chemical energy not used. This energy is found in higher exhaust emissions of CO and H_2. The indicated mean effective pressure remains approximately constant for mixtures richer than stoichiometric.

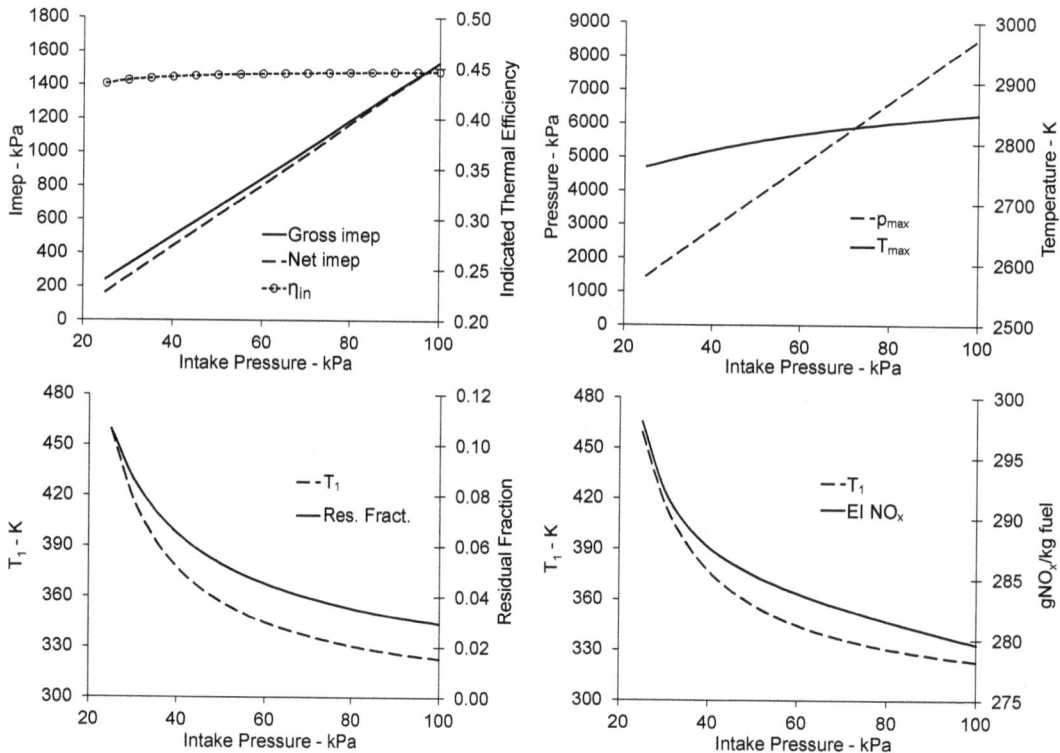

Figure 2.14: Indicated thermal efficiency, indicated mean effective pressure, maximum cycle temperature and pressure for the ideal Fuel Air cycle with constant volume combustion as a function of intake pressure for a stoichiometric mixture and compression ratio of 9:1 for iso-octane. Intake temperature 298K, exhaust pressure 100 kPa. Also shown are the temperature at the start of compression, T_1, residual fraction, and grams of NO_x produced per kg fuel.

Though the NO_x emissions will be discussed more in detail in Section 6.2, they are presented here, as they are a result of the equilibrium process, through the reaction between oxygen and nitrogen, as shown in Table 2.3. NO_x is an undesirable air pollutant. Figure 2.15 shows the emissions of NO_x as a function of fuel air equivalence ratio for various compression ratios for the constant volume fuel air cycle for isooctane. It is seen that the largest amounts of NO_x are formed with lean mixtures. This is because oxygen is required for its formation according to the reactions in Table 2.3. The oxygen concentration at the peak temperature and pressure is also shown in Figure 2.15. However, high temperature is also needed for the NO to be formed, and as the mixture becomes leaner, the temperature falls, and the NO_x emission decreases. Figure 2.13 shows the peak temperature decreasing as equivalence ratio decreases because there is less fuel to heat up the same mass of air.

Figure 2.14 shows effect of intake pressure on mean effective pressure, efficiency, maximum temperature and maximum pressure for the ideal fuel-air cycle. In addition, the residual fraction, temperature at the start of compression, and the NO_x produced per unit fuel at state 3 per fuel mass are also shown.

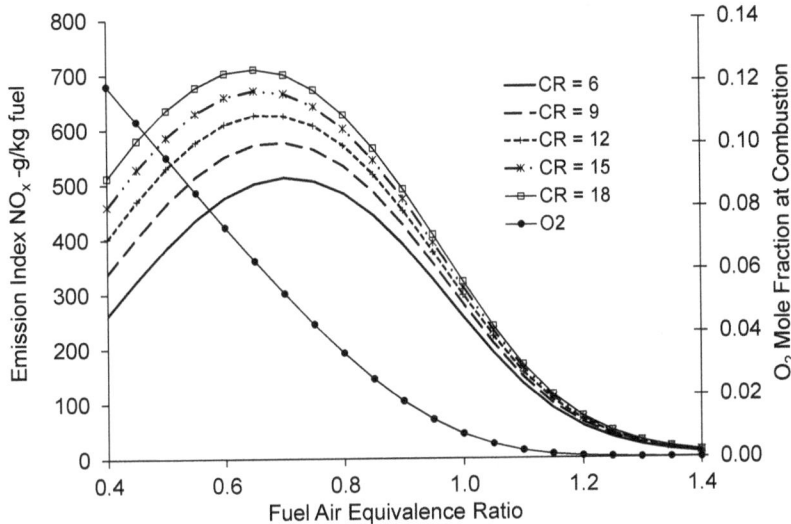

The pressure has little effect on the efficiency, but a major effect on the indicated mean effective pressure, similar to the trends shown with the ideal air cycles. Since the fuel air ratio is constant, lower pressure means less fuel and therefore less power. In Figure 2.14 both the gross and net imep are shown, the difference between these two being the pumping work caused by low intake pressure and higher exhaust pressure. This pumping loss is normally associated with friction, and is the cause of lower efficiency at light loads when using throttled engine operation with constant fuel air ratio.

Figure 2.15: NO_x, as gram NO_2 per kg of fuel, and mole fraction of oxygen at peak temperature and pressure in a constant volume, ideal Fuel Air cycle as a function mixture ratio and compression ratio for iso-octane. Intake temperature 298K, exhaust pressure = intake pressure = 100 kPa.

The residual fraction increases as the intake pressure (load) is decreased (see Equation (2.85)), since the mass of residual at the end of the ex-

haust process is relatively constant, and the inducted mass becomes smaller as the pressure decreases. High residual fraction at very light loads can give problems with stable engine operation, especially at idling, another consequence of the constant fuel air ratio operation.

The pressure has little effect on the thermal efficiency. So the variation of *imep* with intake pressure is almost wholly due to the change in the inducted mass of air and fuel in the cylinder. The effects of chemical equilibrium and property variation are very small when the pressure is changed. Load variation can easily be determined from these results by changing the intake pressure.

In Figure 2.13, the *imep* is seen to increase steadily with increasing equivalence ratio until the stoichiometric condition is reached. After that, it remains nearly constant until an equivalence ratio of 1.2 is reached whereafter it begins to decline. On closer examination of the curve, it can be seen that the *imep* is a little higher for some slightly rich mixtures, especially for lower compression ratios and lower intake pressures. This is one reason why most spark ignition engines operate with slightly rich mixtures when full power is needed. A second factor is that the flame speed is slightly higher for these mixtures, which also has a small effect on power, as can be shown with the simulation presented in the following section.

How does the fuel air cycle model compare with practical results? Figure 2.16 shows the ratio of the indicated thermal efficiency of an experimental CFR engine to that predicted by the fuel air cycle. The engine is operating at full throttle, with a spark timing of 20°BTDC, which is close to the maximum torque value, a compression ratio of 6.23:1 and a speed of 1200 rpm. The figure shows that there is an approximately constant ratio of about 0.7 for the thermal efficiencies of the CFR engine and the ideal cycle.

In practice, it is found that a modern operating spark ignition engine will have an indicated thermal efficiency and mean effective pressure in the vicinity of 80% that predicted by the constant volume fuel air cycle. This is due to the fact that combustion and air exchange processes require some time to occur, and that there are heat losses to the cylinder due to high temperature gasses, particularly on the expansion stroke. The next section will demonstrate this using a simple crank angle based simulation.

This result can be used in a predictive form. For a good first estimate of the indicated properties of a spark ignition engine, the indicated thermal efficiency and mean effective pressures can be obtained from equilibrium fuel air cycle charts or computer calculations and multiplied by about 0.8 to give a realistic estimate of cycle performance values. The indicated efficiency is best estimated in this way, since it is not very dependent on the intake pressure. The indicated mean effective pressure is directly dependent on the intake pressure, so the accuracy here will depend on the assumed value of the intake pressure at the desired load. For more precise estimates, more complicated programs are needed that include effects of valve timing, engine friction, pressure pulsations in intake systems, etc..

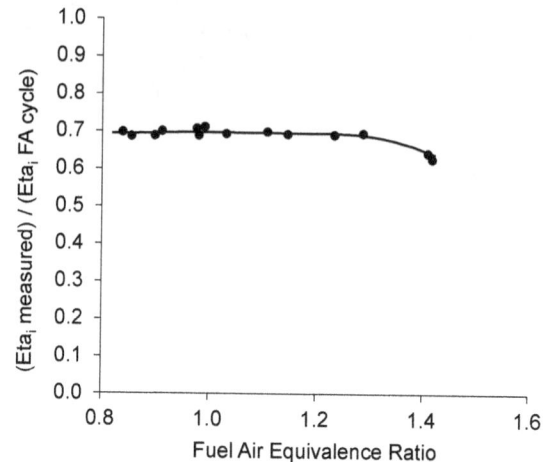

Figure 2.16: The ratio of indicated thermal efficiency of a CFR engine to that predicted by the constant volume ideal fuel air cycle.

2.5 Engine Simulations

Since the advent of digital computers in the late 1960's, computer simulations have increasingly become some of the most useful tools available to engineers. After steady development of a large range of simulation procedures, simulations are used to design all kinds and sizes of projects and products. The following discussion applies to a much broader scope than just engines. While the ideal air processes and the fuel air equilibrium process are in reality also simulations of engine performance, the term engine simulation usually refers to a time dependent calculation of engine processes. A variety of simulations have been developed and and are in use.

2.5.1 Advantages of Simulations

- Reduction in development time and cost - This is the prime driving force for the enormous increase in the number of simulations and other engineering software. By shortening development time, large savings are achievable through reductions in manpower and through less construction and testing.

- Determine variables not possible to measure - In many cases, it is possible to calculate properties of systems that are difficult or impossible to measure. One of the best examples of this is finite element software for determining internal stresses in materials. Through the years, the reliability and accuracy of these programs has improved such that the calculated values of many of the stresses are nearly as trustworthy as experimental measurements. Similarly, in engines better simulations are capable of calculating engine temperatures and instantaneous flow, quantities that are difficult and costly to measure.

- "Test" cases that cannot be built - An example of this is the calculation of the effects of heat transfer in engines. With a simulation, it is possible to determine the performance of a perfectly adiabatic engine, something that cannot be achieved in practice. Such a calculation could give the engineer an idea of the benefits of pursuing a method to perfectly insulate the combustion chamber, and is helpful in choosing fruitful development areas. But one must be careful that these conditions can be described accurately, the heat transfer case being a good example, as shown in the following list of disadvantages.

- Isolate effects of variables, that cannot be isolated in experiments - In experiments, it is often not possible to change one engine variable by itself, for example changing the residual temperature without changing the start temperature of compression. The can be readily done with a simple simulation, provided the user has access to the basic code, or that there are provisions in the simulation software to change the desired variables.

2.5.2 Disadvantages of Simulations

- Limited by the assumptions of the model - A simulation is no better than the sub-models used to describe the many complicated processes occurring in the device that is the object of the simulation. The limitations and basic assumptions of any simulation or software must be known before application: This is crucial! For example, a design exercise for the improvement of the combustion chamber of a spark ignition engine may give incorrect conclusions, if it is based on a simulation that does not consider engine knock. Not only is the inclusion of the phenomena at hand important, it is also necessary to understand the theoretical background for the calculation procedures. In some cases, inclusion alone is not sufficient. In the 1980's, a lot of work was performed on the development of low heat rejection engines, primarily through the application of ceramic components or coatings inside engines. All computer simulations of the time predicted improvements in performance with coatings, yet when engines were modified to give the low heat rejection, the improvements in performance were not found. Eventually it was found that the description of the boundary layer and heat transfer processes used for standard engines was insufficient for low heat rejection engines. This lead to improvements in engine heat transfer modelling procedures, but at the expense of a lot of time and effort. There is a positive side to this, though, in that it lead to a better understanding of engine heat transfer.

- "forget to think" - One of the greatest hazards of using computers and simulations is the implicit trust the user often has in the results from the calculations. Things tend to be accepted when they are the products of large, complicated computer programs. The huge amount of computing power available today, combined with a large amount of software enables the engineer to accumulate and process an immense amount of information in a very short time. In this process, the temptation is often great to look quickly at elegant plots from a plotting program, but without taking the time to look at the curves in detail, and making sure that the results are reasonable, and that they have been understood. Valuable information may be lost or overseen and incorrect results go unnoticed without taking the time to study the results in a more than superficial manner.

- Simulation development can easily becomes the goal instead of the problem to which it is to be applied. Sometimes the user or developer of computer software and simulations becomes involved in the business of preparing the simulations, and they become the goals in of themselves. While this may be enjoyable for the programmer, it can easily be considered to be a waste of time for an employee to develop simulations in excess of needs.

- Time to develop or learn the simulations - Simulation programs and computer software can often be very comprehensive and complex. It normally takes a user a certain amount of time to learn to use these tools, as one must learn input and output procedures, often special structures of input data, as well as the theoretical background for the program. If a simulation is only planned to be used for a limited number of special jobs, it is well worth considering if the user should spent the time involved in learning all aspects of a program. Simpler solutions may give adequate answers in less time at lower cost, or consultants may be used instead.

- Confirmation of tends reliability required.- The validity of the procedures used in any simulation MUST be verified by comparison with known and accurate results. No one should use a simulation in an important application without being absolutely sure that the model is successful at predicting trends in effects of important variables on output results. Basic trends and results should be verified. The confirmation of results can be performed either by the user himself, or it could be performed by the supplier or a technical consulting agency. Often the latter have experience with different kinds of programs, and are capable of verifying that the simulation procedures do behave appropriately in practice.

2.5.3 Choosing a Simulation

1. "Buy or build?" Before computers became so wide-spread and so much software became available, there was usually not a question. If a simulation was needed, it had to be developed by the user. Through the years, however, many simulations have been commercialized, or are publicly available from non-commercial developers such as governmental laboratories. Today, a huge amount of software is available to solve a vast array of engineering problems. Finite element and computational fluid dynamics programs are two examples of simulations/computers codes which have been highly developed.

It is still a frequent occurrence that no commercial simulation routines are available that can solve a given problem. In that case, simulation development is necessary. Since the early 1990's, many computer packages have become available for solving complex systems of algebraic and differential equation. This makes the programming of simulations much simpler that programming in the more "Traditional" programming languages like FORTRAN and PASCAL. There are also packages available, which contain accurate procedures for calculating the properties of materials in the calculation[2]. However, the fundamental questions that need to be asked when developing a new simulation are the same as those for a purchase.

2. What program should be purchased or developed? Some important questions to ask are:

1. "What is the goal of the simulation?" The first question to be asked is why is the simulation being used in the first place. That is, there are many reasons for choosing to use a simulation, each of which will have an impact of the choices made.

2. "What is the type of work to be performed?" Is the simulation to be used for preliminary investigations, or to make detailed calculations of products. For engines, the choice could be between relatively simple simulations consisting of ordinary differential equations and that of something like computational fluid mechanics (CFD) programs, that solve non-linear partial differential equations in three dimensional. One would hardly use the latter to find the general effects of spark timing on performance!

3. "What are the accuracy requirements?" In some cases, accuracy of calculations is not so important, provided that the general predicted trends are correct and can be used for comparisons. In other situations, particularly those where customers and sale of products are involved, the absolute accuracy of the simulation is very important. Often, the simulations with the highest accuracy have the least generality. That is, where accuracy is required, simulation models are often fit to known results, often by adjusting coefficients in equations, *etc.* This process limits the generality of the simulation models, but is acceptable as long as the known limitations are not exceeded.

Simulations that are not based on solid physical principles can be treacherous. For example, if a sub-model is fit to a polynomial of higher order, a slight excursion outside the area of validity can cause drastic errors, or even incorrect signs.

2.6 Types of Engine Simulations

The simplest models available for estimating the performance of an engine cycle are the air cycles discussed in Section 2.1.1, typically using a constant specific heat. These models, in spite of their simplicity, can be used to determine some important trends in engine performance, such as the effect of compression ratio and importance of maximum allowable cylinder pressure. They are the first step for understanding the fundamental thermodynamics of engine operation. There are many factors that the ideal air cycles cannot include, however, such as the air exchange process, effects of properties and chemistry, and time dependent processes, such as combustion rate, combustion timing and time dependent heat transfer.

---

[2]For example the program EES, developed by Prof Klein at the University of Wisconsin

The fuel air equilibrium cycle, based on temperature and fuel air ratio dependent properties, and using chemical equilibrium can be used to evaluate the effects of chemical properties without heat transfer. It typically estimates indicated performance parameters that are about 20% higher than found with actual engines. The differences are commonly attributed to losses associated with time dependent processes. That is, the combustion rate in actual engines is not instantaneous. There is also heat transfer throughout the entire cycle, and it depends on the instantaneous cylinder temperature and pressure.

"Complete" engine cycle simulation programs include time dependent flow and heat transfer, equilibrium combustion products, accurate thermodynamic properties and can be used to give good estimates of engine performance, and in the case of spark ignition engine, good estimates of CO and NO_x emissions. The development of these models is a time consuming task. In order to evaluate the relative effects of the time dependent processes, a simple simulation of a spark ignition engine will be presented in the following. The main emphasis will be on the time dependent aspects, that is combustion rate and timing as well as the heat loss to the cylinder wall. In order to simplify the problem, thermodynamic properties will be described through the use of a specific heat ratio. The model can also be used for a diesel engine by changing the heat release rate function.

The model presented in the following section is the simplest form of cycle simulation model, and is called a 0-dimensional model. This is because the combustion chamber is treated as consisting of a uniform volume of gasses, and no consideration is taken as to variations of any variables in a spatial direction. A 0-dimensional model is probably a better approximation to a diesel engine, since there is more mixing.

The next step in improving the description of the physical processes in the engine is the 2-zone, 0-dimensional model, where both a burned and an unburned zone are included, each zone being uniform in itself, and only the volumes of the zones are known, again with no consideration being taken of spatial variations [12].

An improvement over the zero dimensional, 2-zone model for spark ignition engines is the so-called "quasi-dimensional" model. This is really a 2-zone model, in which one assumes that the flame spreads in a spherical manner from the spark plug . In this way, it is possible to calculate the flame area from the burned volume, and some geometrical aspects of the combustion process can be considered [13]. This model is not valid for conditions where there is a large rotary air motion in the cylinder, since it distorts the flame. All of the above mentioned models involve the solution of ordinary differential equations. Various forms of diesel models have been used in a quasi-dimensional model, where knowledge or injector spray characteristics is "forced" upon the model, to determine spatial variations in the combustion chamber gasses, when the model still consists of time dependent, ordinary differential equations.

The next type of simulation is the multi-dimensional model. Currently, the only type of model used very much is the full 3-dimensional model. This type of model solves the complete set of partial differential equations for mass, momentum, energy and species conservation in 3 physical dimensions throughout the engine cycle, and includes a model for the calculation of turbulent flow. The CFD[3] programs used normally here require a large computer capacity. This type of model gives a tremendous amount of information, and needs special post-processing software in order to present the results in a form that is understandable for the user of the program. Such models are becoming more common in the total design of the flow and combustion systems of engines.

2.7 Simple Engine Simulation

In this section, a very simple simulation of an engine is presented. The purpose of this simulation is to illustrate some of the relative effects of timing and duration of combustion, and heat transfer throughout the cycle. In addition, it is intended to serve as introduction into the general area of time dependent cycle simulations. In line with this goal, only very simple empirical models of the processes will be employed.

The air exchange process will be simulated by the simple relationships used in the ideal air engine models as described in Section 2.3. That is, constant pressure on the intake and exhaust strokes with no heat transfer there. The model will not, therefore, be able to predict effects of engine speed on airflow and power, since the speed dependence of the engine's volumetric efficiency is related to the time dependent

---
[3]Computational Fluid Dynamics

flows through the values of the engine. Since a simple properties model is used, neither will it be possible to accurately calculate the details of the combustion process. Therefore, the same concept used with the air cycles is used, that is, the combustion process is simulated by an equivalent heat transfer, but in this case, it is written as a function of time using an empirical equation.

2.7.1 Basic Analysis

The system to be analyzed is considered to be the gasses in the cylinder, as enclosed by the top of the piston, the cylinder sidewalls, and the compression volume in the cylinder head. The basic equations for the simplified combustion analysis are the first law of thermodynamics for the cylinder and the equation of state for ideal gasses. The system is shown in Figure 2.17.

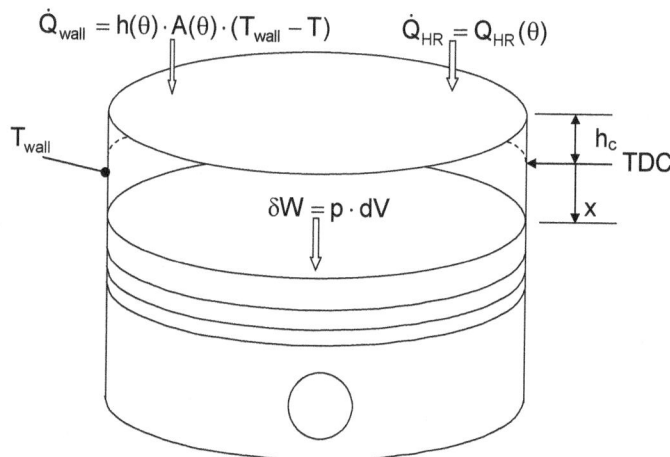

Figure 2.17: The system for the simulation consists of the gasses above the piston, enclosed by the cylinder head and piston at a constant, uniform wall temperature T_{wall}.

The first law for a time dependent open control volume is:

$$\dot{Q} - \dot{W} = \frac{d(m \cdot u)}{dt} + \sum \dot{m}_{out} h_{out} - \sum \dot{m}_{in} h_{in} \tag{2.97}$$

When calculating the portions of the engine cycle where the valves are open, this equation needs to be solved. The flows are given by isentropic relations, using time dependent valve areas as determined by the camshaft or ports. The enthaplies related to the flows are determined by the direction of the flow, and can be either cylinder or manifold conditions. The book by Ferguson [11] gives examples of how to determine intake manifold conditions in the case where cylinder gasses flow back into the intake manifold and are completely mixed with the fresh charge. In addition, one must keep track of the mass in the cylinder through the mass conservation equation:

$$\frac{dm}{dt} = \sum \dot{m}_{in} - \sum \dot{m}_{out} \tag{2.98}$$

It is also necessary to perform a mass balance on the residual gasses in the cylinder.

The use of cycle simulations for the complete engine cycle is a very useful tool, and allows the determination of valve parameters such as area, lift and timing on the air exchange process and, therefore, the performance and fuel consumption of an engine. Unfortunately, this calculation is too complicated to be covered within the scope of this book. There are many articles on cycle simulations in the literature, dating back to the 1960's. Two references that include cycle simulations and intake and exhaust tuning processes are the books by Blair [14] and [15].

In order to give an idea of how simulations are performed and to investigate some time dependent processes that ideal cycles cannot determine, a simple simulation will be developed.

2.7.2 Assumptions for the Simple Model

- 1. The properties in the cylinder are uniform. This means that one time dependent temperature and one time dependent pressure are used to describe the state of all the gasses in the combustion chamber. In addition, the chemical composition of the chamber is the same throughout. That is to say, this is a 0-dimensional, 1-zone model.

- 2. The in-cylinder mass is constant. This means that the piston rings seals are assumed to be perfect. In addition, the analysis presented here will only consider the portion of the engine cycle where the valves are closed. In the ideal case this means the time from the start of compression (180° before TDC) to the end of expansion (180° after TDC).

- 3. Combustion can be modelled using the idea of an equivalent rate of heat transfer to the cylinder gasses. This is the same idea as the air cycles, but here the heat addition will be made a function of time, so it will be possible to evaluate the effects of combustion rate and timing.

- 4. Heat losses to the cylinder gasses are a function of time, and are estimated by an effective convection to the surfaces of the engine. All surfaces are assumed to have the same, constant temperature. Only heat transfer by convection is considered.

- 5. The gasses are ideal gasses, that is they obey the ideal gas law $pV = mRT$, which also implies that c_v is a function of temperature alone.

- 6. The engine is operating at a constant rotational speed of N revolutions per minute.

In the development of the model, no distinction is made between spark ignition and diesel engines. The single zone model is best suited to diesel engines, but is applied to an SI engine here because the heat release functions are easier to describe. A diesel engine needs a different heat release model, and a model for the ignition delay, for example the form of Equation (4.1) for the latter.

Since this simulation is restricted to the portion of the engine cycle when the valves or ports are closed, the flow terms in Equation (2.97) and (2.98) are equal to 0. The first law reduces to the following form for a time dependent closed system:

$$\dot{Q} - \dot{W} = \frac{d(m \cdot u)}{dt} \tag{2.99}$$

Where: $\dot{Q}$ is rate of heat transfer to the system, $\dot{W}$ is the rate of work done by the system on the surroundings, m is the system mass, u is the mass specific internal energy of the gasses, and t is the time. The mass conservation equation becomes irrelevant. Since the engine speed is constant, there is a direct relation between the time and the angle of crankshaft rotation:

$$\delta t = \frac{\delta \theta}{6 \cdot N} \tag{2.100}$$

for time in seconds. Therefore, the equations can also be written in terms of the crank angle degree, θ, instead of the time t. Since the mass is constant in the closed system, and for an ideal gas,

$$\dot{Q} - \dot{W} = \frac{d(m \cdot u)}{d\theta} = m\frac{du}{d\theta} = mc_v\frac{dT}{d\theta} \tag{2.101}$$

where the rates of heat and work transfers are understood to be in units of J/crank angle degree.

In terms of measurable quantities, it is the pressure that is of most interest in our calculations, since it determines the work performed. Therefore, the temperature will be eliminated from Equation (2.101). That is done by differentiating the ideal gas law:

$$V\frac{dp}{d\theta} + p\frac{dV}{d\theta} = mR\frac{dT}{d\theta} \tag{2.102}$$

Using $\dot{W} = p\frac{dV}{d\theta}$ and the relationships: $R = c_p - c_v$, and $\gamma = \frac{c_p}{c_v}$:

$$\dot{Q}_{net} = \frac{\gamma}{\gamma - 1}p\frac{dV}{d\theta} + \frac{1}{\gamma - 1}V\frac{dp}{d\theta} \tag{2.103}$$

Where $\dot{Q}_{net}$ is the combined effect of the "combustion" and heat transfer. For modelling the combustion, the same idea of using an equivalent heat transfer is employed. This has been used in the ideal air cycles and the ideal air engine, but in the time dependent case, the heat addition is written as a function of time or crank angle degree. The integral of this function through the duration of the heat addition process will give the same total amount of heat addition as an ideal process with constant volume heat addition, for example. The term *apparent heat release* is often used in connection with this idea, and the energy released by the combustion of fuel can be regarded as an apparent heat release, or in other words, an amount or rate of heat addition that has the same effect as combustion of fuel.

Using the concept of apparent heat release, we can divide the net heat transfer into two portions, the first, $\dot{Q}_{HR}$, being the apparent heat release (combustion rate), and the second, $\dot{Q}_{wall}$, being the equivalent heat transfer to the chamber walls. While it is the combined effect of these two terms that determines the pressure development in the cylinder, it is useful to regard them separately, since they describe two different physical phenomena. Then:

$$\dot{Q}_{HR} + \dot{Q}_{wall} = \frac{\gamma}{\gamma - 1} p \frac{dV}{d\theta} + \frac{1}{\gamma - 1} V \frac{dp}{d\theta} \tag{2.104}$$

This is the basic equation that will be used in the simple simulation model. Note that V is a known function of the time/crank angle in the slider crank mechanism (see Equation (2.106)), and so its derivative is also known (see Equation (2.108)). For now, we will assume that the wall heat transfer can be determined as a function of engine operating conditions and the cylinder variables of temperature and pressure. The heat transfer area will be assumed to be a known function of the time/crank angle, and is a known function of time. In the single zone model, the heat transfer area is equal to the entire surface area enclosed by the piston top, cylinder walls, and combustion chamber in cylinder head. In multi-zone models, it is necessary to divide the total surface area into separate zones in contact with the respective gas zones.

If one then knows the heat release function, Equation (2.104) can be solved for the pressure derivative, which then can be integrated as a function of time or crank angle degree to give the cylinder pressure as a function of crank angle (or equivalently time). In any practical system, this requires the use of a numerical method to integrate the equations. There are many types of integration schemes available with commercial computer software today.

Another application of Equation (2.104) is as an analysis tool. If one measures the pressure-crank angle history of an engine, the pressure derivative can be numerically estimated. Knowing the volume-time characteristics and using the ideal gas law to determine the temperature, the apparent heat release rate from the engine can be directly calculated from Equation (2.104). This can be done as either the net neat release (where the combined contribution of the heat transfer and combustion are determined) or the gross heat transfer, where an estimate is made of the heat loss to the walls, and the contribution due to combustion alone can be estimated. In fact, Equation (2.104) is used in this form to investigate the combustion before making the simulation, so that simulation models for the combustion process can be fit and applied.

2.7.3 Cylinder Pressure

As indicated above, Equation (2.104) can be written in a different form to solve for the pressure:

$$\frac{dp}{d\theta} = \frac{\gamma - 1}{V} \left(\dot{Q}_{HR} + \dot{Q}_{wall} \right) - \gamma \frac{p}{V} \frac{dV}{d\theta} \tag{2.105}$$

Note that it has not been necessary to assume that γ is constant. In the results presented in the following simulation, this assumption has been made, but a temperature dependent γ would improve the predictive capabilities. It is also possible to include the effects of fuel and air mixtures, or simplified combustion products on the specific heat ratio by using relevant values of γ with chemical species calculations as shown in Section 3.1. Chemical equilibrium cannot be included here, though.

The volume is a known geometrical function of the crank angle or output shaft position in any positive displacement engine. Though the most common type of geometry found in engines is the conventional slider-crank mechanism, any other geometry where the volume is known as a function of the output shaft position can be used. For the slider crank mechanism, with a connecting rod length L, and the crank shaft arm length $R = S/2$, where S is the stroke of the engine and B is the bore:

$$V(\theta) = V_c + \frac{\pi B^2}{4}\left(R\left[1 - \cos\theta\right] + \frac{R^2}{2L}\sin^2\theta\right) \tag{2.106}$$

$$= V_c + \frac{\pi B^2}{4}x(\theta) \tag{2.107}$$

where x is the distance to the top of the piston from its top dead center position, and the piston is at TDC for $\theta = 0°$. V_c is the compression volume, that is the minimum cylinder volume occurring at Top Dead Center. By differentiating Equation (2.106), the volume derivative needed for Equation (2.105) is obtained:

$$\frac{dV}{d\theta}(\theta) = \frac{\pi B^2}{4}R\left(\sin\theta + \frac{R}{2L}\sin 2\theta\right) \tag{2.108}$$

It should be noted that in Equation (2.108) the units are $m^3/radian$. If the time units are chosen to be crank angle degrees, then it should be recalled that there are $180/\pi$ degrees per radian, and Equation (2.108) modified accordingly.

A simple combustion model can be used to calculate the heat release. It is based on the combustion results discussed in the Section 3.8.4. In this case, the trigonometric approximation, Equation (2.109), is used:

$$\dot{Q}_{HR} = \frac{\pi H_u m(FA)(1-f)\eta_c}{2\Delta\theta_c(1+FA)}\sin\left(\pi\frac{\theta - \theta_o}{\Delta\theta_c}\right) \tag{2.109}$$

where: H_u is the lower heating value in kJ/kg, m is the total cylinder mass, f is the residual gas fraction, FA is the fuel air ratio in kg/kg, θ_o is crank angle for the start of the heat release, $\Delta\theta_c$ is the duration of the heat release, and η_c is the combustion efficiency. For lean and stoichiometric mixtures, $\eta_c \approx 1$, while with richer mixtures, η_c decreases as the mixture becomes richer. Note that θ_o is not the time at which the spark is fired, but a later time, that compensates for the time it takes for the flame to build up to a suitable size to give significant energy release. This is discussed further in Chapter 3. It is easy to include other functions instead of the trigonometric function. The Wiebe function, also described in Section 3.8.4, is often used, as it can describe non-symmetrical combustion rates.

The heat transfer to the walls of the cylinder can be estimated using empirical correlations for the convective heat transfer coefficient. They are discussed in Section 10.4.1. Two correlations will be used in the simulation results, the simplest, which is the Eichelberg correlation [16], shown in Equation (10.21) and the most commonly used heat transfer correlation for spark ignition engines developed by Woschni [17], Equation (10.23). For diesel engines, a significant amount of heat transfer occurs by radiation, and the forms of the Eichelberg and Woschni equations are not suited to radiation. A correlation by Annand has been developed for diesel engines [18], and includes a radiation term, that is, a term which depends on temperature to the 4th power. This is not necessarily correct if an average gas temperature is calculated in the simulation, since the radiation in a diesel flame arises from the flame, where the local temperatures are close to the stoichiometric equilibrium flame temperature, even though the remainder of the gasses may be colder. As the size and shape of the radiating zone has an influence on the rate of heat transfer, it is difficult to find a simple diesel heat transfer model with proper dependencies on relevant variables.

The area for convective heat transfer is calculated from the surface area of the combustion chamber in the cylinder head, the top surface area of the piston, and time dependent cylinder area exposed by the piston between the cylinder head and the top of the piston. For the simple simulations, it is sufficient to consider that the combustion chamber in the cylinder head is cylindrical in form, as the correlation in Equation (10.21) is not sufficiently accurate to give greater detail. The area can then be calculated as follows:

$$A_{HT} = 2\left(\frac{\pi B^2}{4}\right) + \pi B\left[h_c + x(\theta)\right] \tag{2.110}$$

Where: h_c is the height of the cylindrical compression volume, and $x(\theta)$ is as used in Equation (2.107).

When including the heat transfer in Equation (2.105), it is important to remember the sign convention. Heat transfer to the system is defined as positive. Therefore:

$$\dot{Q}_{wall} = h \cdot A(\theta) \cdot (T_{wall} - T) \tag{2.111}$$

That is, if the cylinder wall is warmer than the cylinder gasses, heat is added to the system. Typically, the sign of the heat transfer changes throughout the cycle. At the start of compression, the cylinder gasses are normally colder than the walls. Later on in the compression, the gasses become warmer than the wall, and there is a heat loss to the walls. This normally continues throughout the rest of the cycle as calculated here.

For the simulation here, where the specific heat is constant, when implementing the Woschni correlation, it is suggested that the motoring pressure be calculated using Equation (10.25). The reference condition can be chosen at the start of compression at BDC. Since a simplified air exchange process will be used in the simulation, the Woschni parameters for that part of the cycle are not used. As will be seen below, there results with the Eichelberg or the Woschni correlation are not dramatically different in the cases simulated, though this may not be the case for different engine sizes, speeds and types.

2.7.4   Numerical Method

Since the equations for the volume and volume derivatives are normally complicated trigonometric functions and due to the nature of the heat transfer correlations, Equation (2.105) cannot normally be solved by conventional, analytical methods. Therefore, a numerical method must be used for an approximate solution. There are many methods that can be used for the solution of Equation (2.105). Many of these methods are available in commercial mathematical software packages, and can be used to integrate the above equation. A few examples are Matlab, EES and IMSL routines. For programming in a computer language such as FORTRAN, PASCAL, *etc.* one needs to know the numerical method used to solve the equations, and write it into the program.

One simple method for numerical integration that is usually sufficiently accurate is a second order Runge-Kutta method. Starting from a known condition at a known crank angle, a first estimate of the pressure after the integration time step, $\Delta\theta$, is made using the simple Euler method:

$$p_{\theta+\Delta\theta}^{+} = p_\theta + \left[\frac{dp}{d\theta}\right]_\theta \Delta\theta \tag{2.112}$$

The value of the derivative in Equation (2.112) is obtained by substituting the known conditions into Equation (2.105). The estimate of the new temperature is obtained from the use of the ideal gas law with the pressure from Equation (2.112) and the volume at the crank angle position $\theta + \Delta\theta$.

The value of the pressure derivative based on the estimated variables at $\theta + \Delta\theta$ is obtained by substitution of the estimates from Equation (2.112) into Equation (2.105). An improved value for the average slope over the time interval $\Delta\theta$ is taken to be the average of the initial and slope and that resulting from using the estimated pressure and temperature:

$$\left(\overline{\frac{dp}{d\theta}}\right)_\theta = \frac{1}{2}\left[\left(\frac{dp}{d\theta}\right)_\theta + \left(\frac{dp}{d\theta}\right)_{\theta+\Delta\theta}^{+}\right] \tag{2.113}$$

The final value of the pressure is then obtained by reapplication of the simple Euler technique with the corrected slope of Equation (2.113).

$$p_{\theta+\Delta\theta} = p_\theta + \left[\overline{\frac{dp}{d\theta}}\right]_\theta \Delta\theta \tag{2.114}$$

This procedure is continued throughout the period where the cylinder is closed. For the period when the valves are open, other procedures are needed to calculate the cylinder conditions, which include the flow in and out of the valves. This is outside the scope of these notes. Unless the combustion is very rapid, an integration time step of one crank angle degree gives adequate accuracy. An estimate of the adequacy of the step size can be obtained by reducing it and seeing the effects on the final results. If the results are not affected by a reduction is step size, it is normally assumed that the accuracy is sufficient. With sophisticated integration packages, actual error control is available. With the simple model used here, the errors from physical approximation are considerably greater than those from the numerical method.

2.7.5 Simple Intake and Exhaust Processes

For estimating effects of intake and exhaust pressure, the same simple intake and exhaust processes as the ideal air engine are used. They assume ideal flow with no losses through either the intake or exhaust valves. It is assumed that at BDC at the end of the expansion stroke, the pressure drops isentropically to that of the exhaust manifold, p_{ex}. In the following, p_4 and T_4, are the pressure and temperature of the gases at the end of the expansion stroke (just before the exhaust valve opens). If it is assumed that there is no pressure drop through the exhaust valve and no heat transfer on the exhaust stroke, then the residual fraction can be calculated via Equation (2.73):

$$f = \frac{V_{comp}}{V_5} \tag{2.115}$$

V_5, is given by:

$$V_5 = V_d \left(\frac{\varepsilon}{\varepsilon - 1} \right) \left(\frac{p_4}{p_{ex}} \right)^{\frac{1}{\gamma}} \tag{2.116}$$

Where ε is the compression ratio V_{max}/V_{min}, V_d is the displacement volume, and $V_{max} = \frac{V_d \cdot \varepsilon}{\varepsilon - 1}$ is the total cylinder volume at BDC.

The temperature at the start of the compression stroke is dependent on the residual fraction, and the temperature at the end of the exhaust stroke and can be approximately calculated by Equation (2.81):

$$T_1 = (1 - f) T_{in} + f \cdot T_{4''}$$

T_{in} is the intake manifold temperature, and $T_{4''}$ is the temperature of the gasses at the end of the exhaust stroke expanded isentropically to the intake pressure:

$$T_{4''} = T_4 \left(\frac{p_{in}}{p_4} \right)^{\frac{\gamma - 1}{\gamma}} \tag{2.117}$$

Since the conditions at the start of the compression stroke, that is the initial temperature and the residual fraction, depend on the values from the previous cycle, an iterative procedure must be utilized. Values for f and T_1, again must be assumed, the calculations carried out for a complete cycle, and the initially assumed values and finally calculated values compared. This procedure must be continued until convergence is obtained, as in the case of the ideal air engine. The procedure which has proved to yield the fastest convergence is again to substitute the final values from the previous iteration directly as the initial values for the next iteration. A reasonable convergence criteria is a change in T_1 of $< 0.1\ K$ and a change in the residual fraction $< 0.1\ \%$.

When convergence has been attained, the results include the pressure and (mass average) temperature of the uniform cylinder gasses as a function of crank angle throughout the integration period. From this, the engine output can be obtained. This is done by utilizing the fact that the work is equal to the integral of $p \cdot dV$ over the operating cycle. The integrated work, the indicated work, is obtained by a numerical integration of the pressure and volume values, using the trapezoidal rule:

$$W_i = \oint p \cdot dV \approx \sum_{i=1}^{i=M-1} \left(\frac{p_{i+1} + p_i}{2} \right) \cdot (V_{i+1} - V_i) \tag{2.118}$$

or the direct use of the pressure and the volume derivative:

$$W_i = \oint p \cdot dV \approx \sum_{i=1}^{i=M} p_i \cdot \frac{dV}{d\theta}_i \tag{2.119}$$

Where the index, i, corresponds to a crank angle position, M is the total number of crank angles in the cycle. The indicated thermal efficiency is the ratio of the indicated work performed to the total chemical energy released in the combustion process:

$$\eta_i = \frac{W_i}{m_f \cdot H_u} = \frac{W_i}{m_{tot}\,(1-f)\,\dfrac{FA}{1+FA} \cdot H_u} \tag{2.120}$$

Where: m_{tot} is the total mass in the cylinder, FA is the fuel air ratio, f, is the residual fraction, and H_u the lower heating value of the fuel.

The indicated mean effective pressure is the amount of indicated work performed per engine cycle per unit displacement volume, and is calculated as:

$$imep = \frac{W_i}{V_d} \tag{2.121}$$

Using these parameters, other engine output variable such as torque and power can be calculated using formulas found in the section on engine basics.

2.7.6 Results and Trends

The simulation described has been implemented and used to conduct some simple estimates of trends in engines. One of the primary purposes of the simulation is to investigate the effect of combustion time and heat transfer throughout the cycle. The first trend investigated is that of combustion duration and timing. The simple simulation was used to show the connection between the combustion duration, the timing of the combustion and the indicated thermal efficiency. Only the indicated thermal efficiency will be shown, since the imep is proportional to the indicated thermal efficiency and the intake density for constant fuel air ratio. Therefore, trends in imep will be related to trends in η_i for a given intake pressure. All the simulations shown are for a speed of 1800 RPM and the wall surface temperature of the combustion chamber is assumed to be 150°C.

Figure 2.18 shows the effect of combustion duration and timing on the thermal efficiency of an SI engine with the simple simulation model, with $\gamma = 1.35$, for stoichiometric

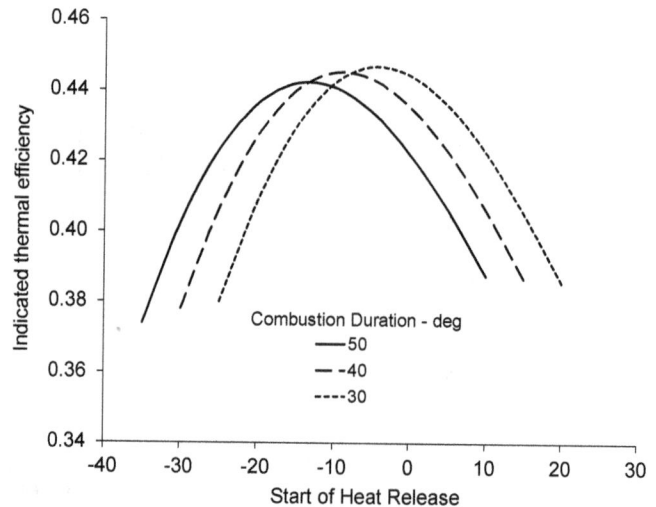

Figure 2.18: Indicated thermal efficiency as a function of the start of combustion for the CFR engine for three different combustion durations, as predicted by a simple simulation.

octane-air, compression ratio 9.0:1, intake pressure = 95 kPa and exhaust pressure = 105 kPa, 1800 RPM, Eichelberg heat transfer coefficient. First it should be noted that for any given duration of combustion, there is an optimum timing. If the timing is too early or too late, efficiency will be reduced. Late timing causes too much combustion on the expansion stroke, with lower efficiency as was seen in the Otto-Diesel cycle comparison at the same compression ratio in Section 2.1. Combustion which occurs too early results in an increase in the work required to compress the mixture, as it causes the pressure on the compression stroke to rise above the normal compression pressure before TDC, increasing the compression work. The optimum combustion timing is a compromise between these two limits.

Note also that the curves are somewhat flat near the optimum. In practice, a change in timing of a few degrees near the optimum timing is barely measurable on an operating engine. Therefore, the engineer often selects the latest timing that gives the maximum torque. The minimum is advantageous, as the maximum cylinder pressures are slightly lower, giving less mechanical stress on the engine and less tendency for knock. The condition is called , MBT timing which refers to Minimum Spark Advance for **M**aximum **B**rake **T**orque.

Another point of interest is that the optimum start of combustion is earlier in the cycle as the combustion duration increases. The MBT timing typically results in the condition where the maximum cylinder pressure occurs at about 15° ATDC. This is where the compromise mentioned in the previous paragraph is optimum. Remember that the start of combustion used in the simulation is not the actual spark timing in an engine, but a later time, determined by the very first buildup of the flame. This is discussed in more detail later in Section 3.5.

The results show that when the combustion timing is optimized, there is not a large effect of combustion duration on the maximum efficiency. This is because the piston moves very little in the time period around TDC. The effects can be illustrated in Figure 2.19, which shows the simulated maximum indicated thermal efficiency for the CFR engine as a function of combustion duration for the same conditions as Figure 2.18. Two cases are shown, the first being the adiabatic situation. Here, it is possible to calculate the efficiency with zero duration, since it is simply the efficiency of the ideal air engine described previously. For a combustion duration of 60 degrees, the indicated efficiency is still 96.6% of the value for instantaneous combustion. This is an advantage of the slider crank mechanism, and the small amount of movement around TDC. Thus, as long as the combustion duration is reasonable, there is not a large improvement in efficiency to be found by increasing combustion speed.

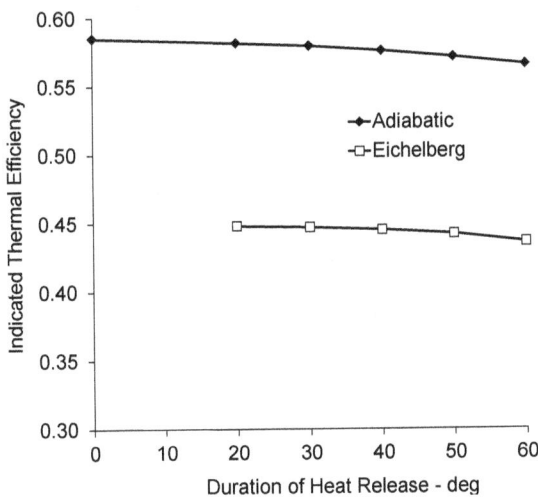

Figure 2.19: Maximum indicated thermal efficiency as a function of combustion duration for the CFR engine, as predicted by a simple simulation.

Another thing to be seen from Figure 2.19 is the relative effects of the heat transfer on the efficiency. Taking the case at a combustion duration of 50 degrees, the ratio of the cycle efficiency with Eichelberg heat transfer to the adiabatic cycle efficiency with constant volume is 0.755. For the Woschni heat transfer correlation the number is a bit higher at about 0.789. Thus, on this simple basis, it appears that there is about a 20% loss of efficiency/power due to the combined effects of heat losses and a finite duration of combustion. This same general relationship (the "80% rule of thumb") is also found with real engine operation compared to the fuel air cycle.

The effects of heat transfer can be seen on pressure and temperature diagrams. Figure 2.20 shows the pressure time and pressure volume diagrams for the adiabatic simulation and the two heat transfer correlations presented in the previous section. The heat transfer is most important on the expansion stroke where the temperatures are high. Very small differences could be seen on an expanded scale for the compression process, but are not visible here. Seen from a pressure-volume point of view, the loss of power and efficiency is more apparent, and the difference in the work on the expansion stroke is shown to be the major factor for the loss of power, as shown in Figure 2.20.

The simulation involves solving a differential equation for the cylinder pressure as a function of time, but the temperature is also available as a function of time by using the ideal gas law. Some care must be exercised here. A basic assumption of the simulation is that there is a uniform temperature. This is not the case during combustion in a spark ignition engine, where there are two separate zones, the burned and unburned gasses. Thus during the simulated combustion period, the temperature is a mass based average temperature for the two zones. Before and after the combustion, the assumption of a single temperature is a reasonable assumption, and results there can be considered useful. This emphasizes the need to understand the limitations of the models built into simulation programs. For example, a simulation with these assumptions is reasonable to use for the effects of timing, and other basic processes. However, in the case of something like NO_x emissions, where the process is extremely temperature dependent, such a model is not at all useful and could easily be misleading. A better combustion model including the temperature of a least two separate combustion zones in needed for this purpose.

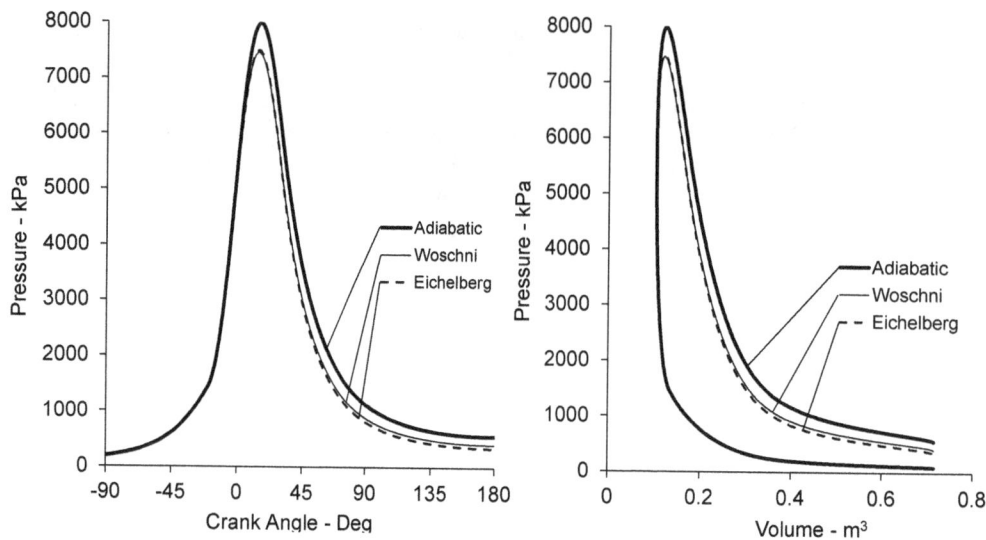

Figure 2.20: Cylinder pressure as a function of crank angle and volume for the CFR engine for three different heat transfer conditions, as predicted by a simple simulation with $\gamma = 1.35$, for stoichiometric octane-air, compression ratio 9.0:1, intake pressure = 95 kPa and exhaust pressure = 105 kPa. The start of heat release is 20° BTDC, and the duration 50°.

With that background, Figure 2.21 shows the simulated mass average cylinder temperature as a function of crank angle degree and cylinder volume. Note that the temperatures are quite high. This is a consequence of the chosen value of 1.35 for the specific heat ratio. When temperature dependent values for specific heats are used the temperature will fall. In an actual engine, the temperatures are on the order of 2500K, in reasonably good agreement with the Fuel Air cycle. The combustion temperature in the Fuel Air cycle is strongly affected by high specific heats and the chemical dissociation described in Section 2.4.1. As expected from the pressure figures, there is a big difference when the heat transfer is included, but not much difference between the two heat transfer correlations.

2.8 Two Zone Combustion Model

In the previous section, a simple model was presented, in which all the properties in the combustion chamber are assumed to be the same at any instant in time. This approximation gives a very simple model for both combustion and heat release calculations, but cannot be used for calculations which require a more accurate determination of the local temperatures or compositions in the combustion chamber - nitric oxide calculations, for example. In this section, a simplified two zone model will be presented, which is able to predict trends in temperature developments in both a burned and unburned zone. This is a good approximation to combustion in spark ignition engines. The model was developed by Blumberg and Kummer in connection with early studies concerned with the effects of engine variables on NOx emissions in spark ignition engines [12].

The model is valid for the duration of the combustion, and it is assumed that the combustion chamber is divided into two zones: a burned zone and an unburned zone. It is assumed that the pressure in the combustion chamber is the same throughout the entire combustion chamber at any instant, and that each zone is characterized by a time-dependent temperature which is the same at all locations in that zone. It is further assumed that the total mass, m, in the cylinder is constant throughout the combustion and is equal to the sum of the burned and unburned masses.

$$m = m_u + m_b \tag{2.122}$$

The burning rate is defined as the rate at which the unburned mass decreases:

$$\dot{m} = -\frac{dm_u}{d\theta} = \frac{dm_b}{d\theta} = m_{u,0}\frac{d\alpha}{d\theta} \tag{2.123}$$

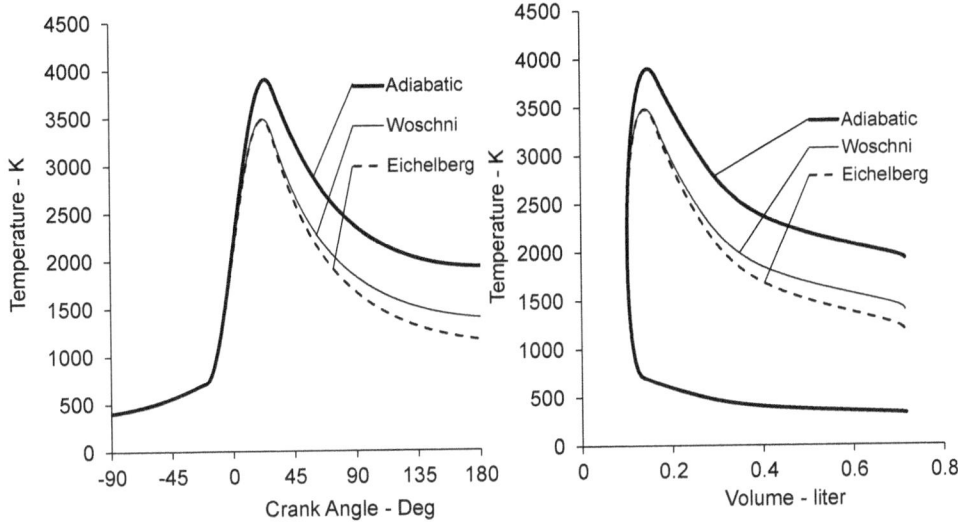

Figure 2.21: Mass average cylinder temperature as a function of crank angle degree and volume for the CFR engine for three different heat transfer conditions, as predicted by a simple simulation with $\gamma = 1.35$, for stoichiometric octane-air, compression ratio 9.0:1, intake pressure = 95 kPa and exhaust pressure = 105 kPa. The start of heat release is $20°$ BTDC, and the duration $50°$.

where α is the mass burned fraction and $m_{u,0}$ is the initial unburned mass, that is the total mass in the cylinder at the start of combustion. The ideal gas law is assumed to apply for each zone:

$$pV_u = m_u R_u T_u \tag{2.124}$$
$$pV_b = m_b R_b T_b \tag{2.125}$$

and the total cylinder volume is equal to the sum of the volumes of the two zones:

$$V = V_u + V_b = \frac{m(1-\alpha)R_u T_u}{p} + \frac{m\alpha R_b T_b}{p} \tag{2.126}$$

The first law of thermodynamics is written for the entire combustion chamber gas:

$$\frac{dQ}{d\theta} - p\frac{dV}{d\theta} = \frac{d(m \cdot u)}{d\theta} \tag{2.127}$$

In Equation (2.127), Q is the total heat transfer to the gasses from the cylinder walls and u is the total internal energy of the combustion chamber gasses, both burned and unburned, and is the weighted sum of the burned and unburned fractions:

$$u = (1-\alpha)u_u + \alpha u_b \tag{2.128}$$

The idea of an apparent heat release rate is not used here, as the energy associated with the change from unburned to burned mass gives the pressure rise due to combustion. In the combustion process with detailed models, the combustion products are normally assumed to be those determined by chemical equilibrium, that is they are a function of temperature and pressure, as well as fuel air ratio. In order to simplify the calculation process, Blumberg and Kummer assumed that the internal energy of unburned gasses and the combustion products can be approximated by a linear function of the temperature:

$$u_u = c_{v,u}T_u + a_u \tag{2.129}$$
$$u_b = c_{v,b}T_b + a_b \tag{2.130}$$

Where $c_{v,u}$ and $c_{v,b}$ are the average specific heats of the reactants and products per unit mass, and

$$a = \frac{N_{moles}}{MW}\sum x_i \left(h^o_{f,i} - (\bar{c}_{v,i} + R_o)\cdot T_o\right) \tag{2.131}$$

In Equation (2.131), x_i is the mole fraction of each specie, and the specific heat, gas constant and heat of formation are all on a mole basis. T_o is the reference temperature, 298.15 K, N_{moles} the number of moles in the mixture, and MW the molecular weight of the mixture. The units of a are then kJ per kg of mixture, either burned or unburned.

For the unburned gasses, it is assumed that there is no chemical reaction and the internal energy can be calculated for a given composition and fuel using the principles of Section 3.3, or an engineering calculation package such as the program EES. The internal energy can be calculated for a range of temperatures typical of those experienced under combustion and fit to equation (2.129).

In their model, Blumberg and Kummer assumed that the temperature development in the unburned zone is determined by an isentropic process throughout the combustion.

$$T_u = T_{u,0} \left(\frac{p}{p_0} \right)^{\frac{\gamma_u - 1}{\gamma_u}} \tag{2.132}$$

This basically assumes that the unburned zone is adiabatic and that all heat transfer is associated with the burned zone. This assumption eliminates the need for a separate differential equation for the unburned temperature and greatly simplifies the calculation.

The above equations can be combined to yield the following ordinary differential equation for the pressure development in the cylinder [4]:

$$\frac{dp}{d\theta} = \frac{m \frac{d\alpha}{d\theta} \left[c_{v,u} T_{u,0} \left(\frac{p}{p_0} \right)^{\frac{\gamma_u - 1}{\gamma_u}} (\gamma_b - \gamma_u) - (a_b - a_u)(\gamma_b - 1) \right] - \gamma_b p \frac{dV}{d\theta} + (\gamma_b - 1) \frac{dQ}{d\theta}}{V + V_0 \left(\frac{\gamma_b - \gamma_u}{\gamma_u} \right) \left(\frac{p}{p_0} \right)^{\frac{-1}{\gamma_u}} (1 - \alpha)} \tag{2.133}$$

In Equation (2.133), α, the mass burned fraction, V, the total cylinder volume, and p, the cylinder pressure are functions of the time or as here, crank angle position, θ. The subscripted quantities, $T_{u,0}, p_0,$ and V_0 refer to the conditions in the combustion chamber at the time of the start of the combustion process.

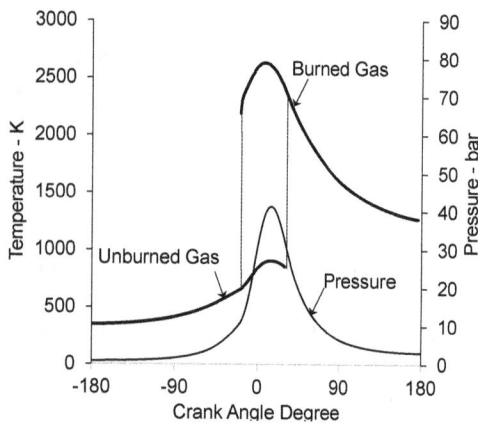

Figure 2.22: A two zone simulation of a CFR engine using the simulation model of Equation (2.133).

The instantaneous heat transfer rate, $\frac{dQ}{d\theta}$, can be determined from the heat transfer correlations described in Chapter 10. Since there are two zones during combustion, this is the sum of the heat transfer from both the burned and unburned zones. To determine these contributions, assumptions as to the flame geometry and exposed surface areas of these two zones during combustion are needed. As an alternative, one could use the mass averaged temperature of the entire combustion chamber to calculate an overall average heat transfer rate, as was done with the single zone model previously presented.

For the periods before and after combustion, Equation (2.133) reduces to Equation (2.105), since there is only one zone, and therefore one value of γ and $\frac{d\alpha}{d\theta} = 0$. The equation can be integrated to the start of combustion to find $T_{u,0}$ and p_0. During the combustion process, the temperature of the unburned mixture will change according to the pressure and it is assumed to be an isentropic compression as in Equation (2.132).

---

[4]It is useful to recall that $R = c_v \cdot (\gamma - 1)$

As an initial condition for the temperature of the burned gas, one can use the adiabatic flame temperature at constant pressure (the pressure change is infinitely small at the start of combustion). This means that the process is isenthalpic:

$$h_u = h_b \tag{2.134}$$

Using Equations (2.129) and (2.130), and $h = u + pv = u + RT$ for an ideal gas, one obtains:

$$T_{b,o} = \frac{a_u - a_b}{c_{v,b} + R_b} + T_{u,o} \frac{c_{v,u} + R_u}{c_{v,b} + R_b} \tag{2.135}$$

A sample calculation is shown in Figure 2.22, for a CFR engine operating at 1500 rpm, intake pressure of 85 kPa, iso-octane and air at an equivalence ratio of 0.8. Note the two different temperatures during combustion, and the compression of the unburned gasses in addition to that from a motoring cycle. The extra compression due to combustion increases the temperature of the unburned gas, and is important with respect to the undesirable process of knock in SI engines, to be discussed in Chapter 4. A two zone model is also needed for the calculation of nitrogen oxide emissions, which will be discussed in Chapter 6. The temperature used in single zone simulations is a mass average temperature during combustion, and is not actually found in the combustion chamber at this time.

Equation (2.133) can used for the heat release rate calculations by solving for $\frac{d\alpha}{d\theta}$:

$$m\frac{d\alpha}{d\theta} = \frac{\frac{dp}{d\theta}\left[V + V_0\left(\frac{\gamma_b - \gamma_u}{\gamma_u}\right)\left(\frac{p}{p_0}\right)^{\frac{-1}{\gamma_u}}(1-\alpha)\right] + \gamma_b p\frac{dV}{d\theta} - (\gamma_b - 1)\frac{dQ}{d\theta}}{c_{v,u}T_{u,0}\left(\frac{p}{p_0}\right)^{\frac{\gamma_u - 1}{\gamma_u}}(\gamma_b - \gamma_u) - (a_b - a_u)(\gamma_b - 1)} \tag{2.136}$$

The trapped mass can be calculated from the measured fuel and air flow rates and an estimate of the residual gas fraction, and $T_{u,0}$ from the ideal gas law.

The properties in Equations (2.133) and (2.136) have been determined for unburned iso-octane/air mixtures and the equilibrium properties for combustion of these mixtures. The temperature range is from 300 to 1500 K for the unburned mixtures, and from 1800 to 2700 K for the combustion products. For determining the equilibrium products, a pressure of 2500 kPa was assumed.

For unburned mixtures $0.7 \le \phi < 1.35$:

$$c_{v,u} = 0.834094 + 0.146378\phi \tag{2.137}$$
$$a_u = -359.109 - 166.173\phi \tag{2.138}$$

For equilibrium combustion products $0.7 \le \phi < 1$:

$$c_{v,b} = 2.9309 - 4.2104\phi + 2.7514\phi^2 \tag{2.139}$$
$$a_b = -4068.2 + 4685.8\phi - 4828.6\phi^2 \tag{2.140}$$

For equilibrium combustion products $1.0 \le \phi < 1.2$:

$$c_{v,b} = 10.522 - 15.694\phi + 6.6429\phi^2 \tag{2.141}$$
$$a_b = -21948 + 29916\phi - 12177\phi^2 \tag{2.142}$$

Equation (2.136) was used to analyze some pressure diagrams from a CFR engine, and the results are shown in Figure 2.23. It can be seen that by choosing a suitable value for γ in the single zone model, essentially the same results are obtained, thus the single zone model is quite acceptable for finding trends. It is not known beforehand, however, what value should be chosen, and the single zone model can, of course, give no indication of the difference between the burned and unburned temperatures.

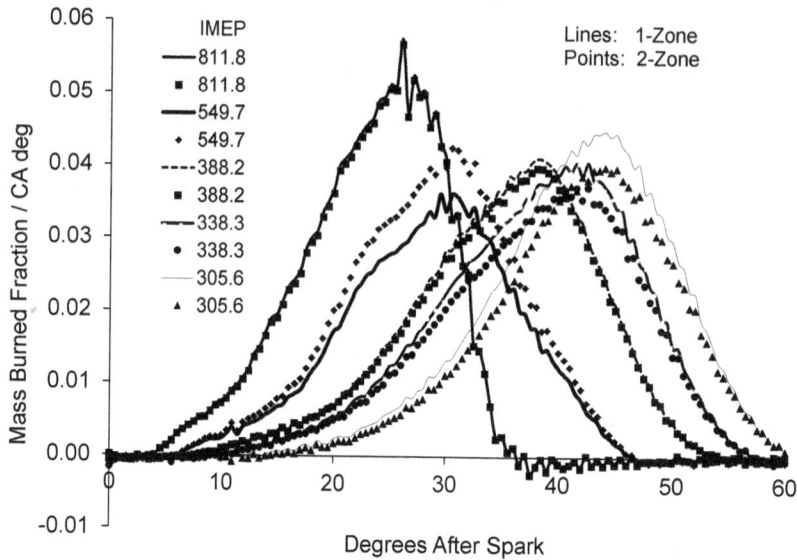

Figure 2.23: Heat release rate for varying *imep* for the CFR engine using two different heat release models: a single zone model with $\gamma = 1.26$ (lines), and a two zone model using Equation and the above correlations for $a_u, a_b, c_{v,u}$ and $c_{v,b}$ (symbols). 1200 rpm, stoichiometric fuel-air ratio, compression ratio = 7.0:1, the numbers give *imep* in kPa.

2.9 Full Cycle Simulation

So far, this chapter has presented time dependent cycle simulations based on the ideal intake and exhaust process described in 2.2. This was done in order to simplify the results, and concentrate on the time dependent effects of combustion timing and duration and instantaneous heat transfer. The ideal processes cannot calculate the effects of the instantaneous flow through the valves and thus cannot be used to investigate valve timing and speed effects on engine performance.

To do this, a complete cycle simulation is needed that has models for flow through the valves and the behavior of residual gases when they enter the intake manifold. Two of the equations needed for this have already been presented: The first law for an open system: Equation (2.97) and mass conservation for an open system: Equation (2.98). During the time the valves are closed, the simulation is the same as those discussed above. But when valves are open, these equations must be numerically integrated as well, since the assumptions of the ideal engine that allowed integration over the intake and exhaust strokes are no longer valid. In addition, an equation is needed for the flow of combustion products into and out of the cylinder in order to determine the residual fraction.

$$\frac{dm_r}{dt} = \sum f_{in} \cdot \dot{m}_{in} - \sum f_{out} \cdot \dot{m}_{out} \tag{2.143}$$

where m_r is the mass of residual (burned) gas in the cylinder

The reader is referred to other books on the subject, in particular, the books by Blair are good starting points [14], [15]. Heat release models with chemical equilibrium are presented in [19]. Some general considerations will be discussed briefly to give a better understanding of the considerations involved in developing these models.

The first consideration is that of flow through the valves. The valve flow is based on isentropic flow, corrected by empirical flow coefficients, where it is necessary to check if the pressure ratio is greater than that required for critical flow. The area can be take to be the product of the instantaneous valve lift and the perimeter of the valve head, as discussed in Section Section 7.5.1. In open cylinder simulations, one always needs to check the pressure differences between the manifolds and cylinder to determine flow directions (which define the flow properties in the first law) and whether the flows are choked or not. The most common assumption is that the valve area is the minimum of the product of the instantaneous valve lift and the perimeter of the valve head or the cross section of the port near the valve. Numerical problems often arise in low speed situations where the pressure difference between cylinder and manifold is small, and the valve area is relatively large. Small integration step sizes and/or stable numerical integration methods are needed to prevent the calculations from being unstable and giving meaningless results. Intake flows at low speeds typically give the most trouble.

When calculating instantaneous flows in engines, the situation often arises where the flow is reversed, that is exhaust gases flow back into the cylinder, and cylinder gases flow back into the intake manifold. The exhaust gas reverse flow is normally not a problem if one finds the average temperature of the gases that have flowed out of the cylinder and uses that for determining the properties. Compositions of the cylinder gases and the exhaust are usually the same when reverse exhaust flow occurs. Reverse intake flow is more troublesome, as it usually occurs just when the intake valve starts to open. The flow out of the cylinder here normally consists of combustion products and they flow into the intake manifold which contains either air, or an air fuel mixture. One must decide how to model what happens to the residual gases. There are two extreme assumptions. The first is that there is no mixing of the two, and that the residual that flows back into the intake remains intact as a kind of plug, and that it must be drawn back into the cylinder before fresh intake charge (containing no residual) can flow into the cylinder. This requires the calculation of the total amount and average temperature of the residual gases that have entered the intake manifold, and when the flow direction changes to that of entering the cylinder the properties are those of the residual until that amount of mass has been drawn into the cylinder. From then on, the intake flow consists of fresh charge. In addition, one must decide whether it is the fresh intake charge or the residual gases temperature that must be used in the isentropic flow equation during the time when the residual plug is being purged.

The other option is to assume that residual flow that enters the intake manifold from back flow is completely mixed with the intake gases, and that as soon as the flow reverses such that gases flow into the cylinder, the composition of the gases entering the cylinder consists of a mixture of fresh charge and residual, and that this fraction remains the same when the flow through the intake valve is into the cylinder. This model requires that one maintain a mass and energy balance of the intake manifold to determine temperature and residual fraction there. These are additional parameters that must be solved in an iterative manner until their change for successive iterations is acceptably small. A method of calculating these properties is presented in Ferguson [11].

2.10 Problems

Problem 2.1

Calculate the indicated efficiency and the indicated mean effective pressure for an ideal air Otto cycle for under the following conditions: $p_1 = 1.013$ bar, $T_1 = 333$K, $\gamma = 1,4$ $c_p = 1.001$ kJ/kg-K $q_{2-3} = 2900$ kJ/kg $\varepsilon = 9$. These conditions are typical for a spark ignition engine.($\eta_i = 0.585$; $imep = 20.24$ bar.)

Problem 2.2

Repeat Problem 2.1, for a Diesel cycle with $q_{2-3} = 2300$ kJ/kg and $\varepsilon = 20$. These conditions are typical for a pre-chamber diesel engine. ($\eta_i = 0.605$; $imep = 15.5$ bar)

Problem 2.3

Repeat problem 2.1, for an ideal air cycle with limited pressure, where $\varepsilon = 16$, $p_{max} = 68$ bar, and $q_{in} = 2300$ kJ/kg. These conditions are typical for a direct injection diesel engine. ($\eta_i = 0.603$ $imep = 15.69$ bar)

Problem 2.4

Calculate the indicated efficiency, the indicated mean effective pressure, the indicated specific fuel consumption, the volumetric efficiency (ideal air quantity based on intake conditions) and the charging efficiency (ideal air quantity based on atmospheric condition) for an ideal air engine under the following conditions: $\gamma = 1.4$ $c_p = 1.003$ kJ/kg-K, $\varepsilon = 9$. The intake pressure is 0.9 bar and the exhaust pressure 1.15 bar. The atmospheric temperature and pressure are 20°C and 1.0 bar. The fuel is a hydrocarbon with a stoichiometric fuel air ratio of 0.065 and a lower heating value of 46,154 kJ/kg. The fuel air equivalence ratio is 0.95.

If the engine has a cylinder bore of 90 mm, a stroke of 80 mm, four cylinders and is a 4-stroke engine with a friction mean effective pressure (pumping loss included) of 230 kPa at 5000 rpm, find the brake power and the brake specific fuel consumption. ($\eta_i = 0.585$; $imep = 17.5$ bar; $isfc = 133$ g/kWh; $\eta_v = 0.975$, $\eta_f = 0.877$; $BP = 129$ kW; $bsfc = 154$ g/kWh)

Problem 2.5

An ideal Otto engine with a compression ratio of 9.5:1 operates on a stoichiometric mixture of iso-octane (heating value 44649 kJ/kg) and air, with atmospheric temperature of 20°C, atmospheric pressure of 101.325 kPa, intake air temperature of 30°C, intake pressure of 95 kPa and exhaust pressure of 110 kPa. A Fuel Air cycle analysis results in: an indicated efficiency = 0.45394, indicated mean effective pressure = 1430, indicated specific fuel consumption = 177.6; volumetric efficiency = 0.9705, and filling efficiency = 0.8803. Assume that the actual indicated work and efficiency are 80% of that from an ideal engine . For an engine with specifications as in Problem 2.4, find the brake power, the brake mean effective pressure and the brake specific fuel consumption. ($BP = 77.5$ kW; $bmep = 914$ kPa; $bsfc = 278$ g/kWh.)

Problem 2.6

Repeat problems 2.2 and 2.3, with temperature dependent properties. One can use average values for c_p/c_v, air tables, or the EES program (or equivalent) with built-in properties. ($\eta_i = 0.534$, $imep = 1366$ kPa; $\eta_i = 0.543$, $imep=1394$ kPa for $c_v = c_v(T)$)

Problem 2.7

For limited pressure cycles with maximum cylinder pressures of 50, 70 and 90 bar, determine the maximum values of the indicated efficiency and the indicated mean effective pressure as a function of compression ratio, from 3.0 up to the limit of the diesel cycle. Use $q_{in} = 2900 \cdot \frac{\varepsilon - 1}{\varepsilon}$

Problem 2.8

Repeat Problem 2.4 with temperature dependent properties, as in problem 2.8. ($f = 0.03474$, $T_1 =$ 330.1K, $\eta_i = 0.5005$, $imep = 1471$ kPa, $isfc = 155.8$ g/kWh, $bmep = 1241$ kPa, $bsfc = 184.7$ g/kWh).

Problem 2.9

Make a computer simulation of the CFR engine using the heat release function of Equation (2.109). The engine has the following specifications: bore: 82.6 mm; stroke: 114.3 mm; connecting rod length: 254 mm; compression ratio: variable. In order to help with the development and de-bugging of the program, a two-step process is recommended:

Step 1: A program that calculates the pressures and temperatures from the start of the compression stroke ($-180°$) to the end of the expansion stroke ($+180°$). This calculation is to be done with known residual fraction and temperature at the start of compression. Table 2.4 gives simulation results, which can be used to check whether the program calculates correctly and to locate errors.

As a check at an earlier stage of development, try an adiabatic model and compare the results near the end of compression (before combustion starts) to those for adiabatic compression, that is $\gamma = const.$ The result should be quite close. This will confirm that the program calculates correctly without the combustion or heat loss parts. These can them be added later, one at a time to simplify development.

The conditions for this calculation are:

Lower heating value: 44,300 kJ/kg; fuel air ratio: 0.0662; compression ratio: 9.0:1; temperature at $-180°$: 335K; pressure at $-180°$: 95.0 kPa; residual fraction: 0.04; Start of Combustion: 20° BTDC; combustion duration: 50°CA; $\gamma = c_p/c_v = 1.35$; wall temperature: 150°C; speed =1800 RPM.

The results for the first step are: indicated mean effective pressure = 1219.6 kPa, indicated work = 746.97 J, indicated power = 11.20 kW, indicated efficiency = 0.4155, indicated specific fuel consumption = 195.6 g/kW-h, heat loss = -555.2 J.

Step 2. Expand the program to include iteration for the ideal intake and exhaust processes for: intake pressure = 0.95 bar; exhaust pressure = 1.10 bar; intake temperature = 20°C.

The results for this second step are: indicated mean effective pressure = 1285.7 kPa, indicated work = 787.5 J, indicated power = 11.81 kW, maximum pressure = 6790.6 kPa at =14.0° ATDC, indicated efficiency = 0.4198, indicated fuel consumption = 193.6 g/kW-h, heat loss = -558.2 J, start temperature = 319.1K, residual fraction =0.04583

3. When the simulation is operating, an optional extension is to include the Wiebe function, Equation 3.127, which is better suited to the combustion in the CFR engine.

Table 2.4: Results for 1st simulation

Crank Angle	Pressure-kPa	Volume-m^3	Temperature-K	dV/dθ−m^3/°
-180.	95.00	6.890E-04	335.0	0
-179.	95.02	6.890E-04	335.1	-7.231E-08
-178.	95.06	6.889E-04	335.1	-1.446E-07
-177.	95.11	6.887E-04	335.2	-2.169E-07
-176.	95.17	6.885E-04	335.3	-2.892E-07
-21.	1264.88	1.013E-04	655.9	-2.318E-06
-20.	1303.57	9.906E-05	660.8	-2.215E-06
-19.	1348.84	9.690E-05	668.9	-2.110E-06
-18.	1407.54	9.484E-05	683.2	-2.005E-06
-17.	1480.29	9.289E-05	703.7	-1.899E-06
-4.	3700.11	7.747E-05	1467.1	-4.565E-07
-3.	3939.31	7.707E-05	1553.8	-3.426E-07
-2.	4179.82	7.679E-05	1642.6	-2.285E-07
-1.	4419.34	7.662E-05	1732.8	-1.143E-07
0.	4655.49	7.656E-05	1824.1	0.000E-00
1.	4885.92	7.662E-05	1915.8	1.143E-07
2.	5108.34	7.679E-05	2007.5	2.285E-07
3.	5320.54	7.707E-05	2098.7	3.426E-07
4.	5520.45	7.747E-05	2188.8	4.565E-07
12.	6500.07	8.474E-05	2819.0	1.356E-06
13.	6533.15	8.615E-05	2880.5	1.466E-06
14.	6546.01	8.767E-05	2937.1	1.575E-06
15.	6539.16	8.930E-05	2988.6	1.684E-06
16.	6513.28	9.104E-05	3034.7	1.792E-06
17.	6469.19	9.289E-05	3075.2	1.899E-06
18.	6407.82	9.484E-05	3110.1	2.005E-06
19.	6330.21	9.690E-05	3139.1	2.110E-06
20.	6237.48	9.906E-05	3162.1	2.215E-06
26.	5429.32	1.142E-04	3172.4	2.817E-06
27.	5263.63	1.170E-04	3152.8	2.913E-06
28.	5092.05	1.200E-04	3127.1	3.008E-06
29.	4915.59	1.231E-04	3095.6	3.101E-06
30.	4735.24	1.262E-04	3058.3	3.193E-06
178.	347.16	6.889E-04	1223.9	1.446E-07
179.	346.58	6.890E-04	1222.1	7.231E-08
180.	346.05	6.890E-04	1220.3	0

Chapter 3

Combustion

3.1 Stoichiometry

3.1.1 The Fuel Air Ratio.

A very important factor in determining engine performance and emissions is the ratio between the amount of fuel and the amount of air during the combustion process. Some basic definitions and concepts relative for fuel air ratios will be discussed.

In spark ignition engines, the mixture is ideally the same throughout the combustion chamber during the entire engine cycle. The air fuel ratio found by chemical analysis of the mixed exhaust gasses or by measurement of the fuel and airflows to the engine should be the same. In the diesel engine also, measurement of the fuel and airflows should give the same value as a fuel air ratio determined by chemical analysis of the mixed exhaust gasses. However, during the diesel combustion process, an infinite range of fuel air ratios can be found locally throughout the combustion chamber as the mixing and reaction processes occur. The overall fuel air ratio in diesel engines is more related to engine load, but still has a significant effect on emissions.

The following discusses the determination of fuel air ratio from the measured exhaust gas composition. This is important for two reasons:

1. The techniques may be used to determine the mixture ratio in an engine (or other combustion systems) in the case where fuel and airflows are not available.

2. The fuel air ratio determined from an exhaust gas analysis can be used to check for errors in the case where fuel and air flow measurements are available, since fuel air ratios from both methods should agree closely.

There are different parameters used to describe the fuel air ratios in engines and combustion systems. The first is the mass ratio of the fuel and air, the unit of mass by convention:

$$Fuel\,Air\,Ratio\ (FA) \equiv \frac{fuel\ mass}{air\ mass} = \frac{fuel\ massflow}{air\ massflow} \tag{3.1}$$

Similarly, the air fuel ratio is defined:

$$Air\,Fuel\,Ratio\ (AF) \equiv \frac{air\ mass}{fuel\ mass} = \frac{air\ massflow}{fuel\ massflow} \tag{3.2}$$

An important parameter for a given fuel is the *stoichiometric*, or chemically correct fuel air ratio. This is determined from the condition where there is just enough fuel present to completely oxidize the fuel to CO_2 and H_2O, with no oxygen left over in the combustion products. Current engine fuels are hydrocarbon based, while new fuels containing oxygen are starting to appear, so the fuels considered will contain C, H, O, and in some cases sulfur, S.

The general balanced reaction for the stoichiometric reaction of an organic fuel containing oxygen is:

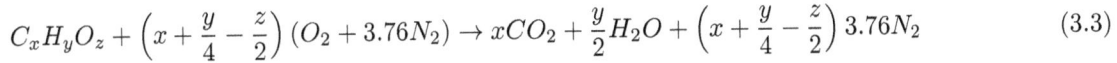

$$C_xH_yO_z + \left(x + \frac{y}{4} - \frac{z}{2}\right)(O_2 + 3.76N_2) \rightarrow xCO_2 + \frac{y}{2}H_2O + \left(x + \frac{y}{4} - \frac{z}{2}\right)3.76N_2 \qquad (3.3)$$

In Reaction (3.3), the nitrogen does not participate in the ideal reaction. But since our source of oxygen, air, consists of about 21% O_2 and 79% N_2 by volume, the nitrogen plays an important role in determining the air flow[1]. It is also a major heat sink for the energy released by combustion and is, therefore, significant in determining flame temperatures. Finally, the nitrogen is not completely inert, a small amount is converted to oxides of nitrogen, which gives rise to serious concerns with exhaust emissions and air pollution.

Using propane, C_3H_8, as an example, the stoichiometric condition can be shown as:

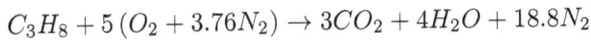

$$C_3H_8 + 5(O_2 + 3.76N_2) \rightarrow 3CO_2 + 4H_2O + 18.8N_2$$

In this case, the stoichiometric fuel air ratio, F/A_s in kg_{fuel}/kg_{air} is calculated as:

$$FA_s = \frac{1\ mole\ C_3H_8 \cdot \frac{44kg\ C_3H_8}{mole\ C_3H_8}}{5 \cdot 4.76\frac{28.95\ kg_{air}}{mole\ air}} = 0.0638kg_f/kg_a$$

The stoichiometric air fuel ratio for propane is then:

$$AF_s = \frac{1}{FA_s} = 15.7kg_a/kg_f$$

In some cases, the composition of the fuel is known on a mass basis. This is often the case when a commercial laboratory conducts an analysis of the fuel, particularly solid fuels. The following shows how to calculate the stoichiometric fuel air ratio from a mass based analysis of the fuel without converting to a mole basis. While it is possible to convert the mass analysis to a molar analysis, if the only consideration is to determine the stoichiometric fuel air ratio, the following method may be simpler.

The complete combustion of a fuel requires a theoretical minimum quantity of air, which can be determined from the fuel mass composition. The required quantities can be calculated as follows.

Carbon: the reaction of carbon and oxygen is:

$$C + O_2 \rightarrow CO_2 \qquad (3.4)$$
$$12kg + 32kg \rightarrow 44kg \qquad (3.5)$$

That is, the combustion of 1 kg carbon requires $32/12 = 2.667$ kg of oxygen

Hydrogen:

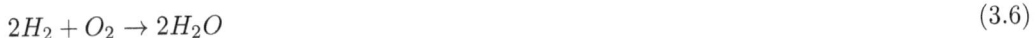

$$2H_2 + O_2 \rightarrow 2H_2O \qquad (3.6)$$
$$4.03kg + 32kg \rightarrow 36.03\ kg \qquad (3.7)$$

That is, the combustion of 1 kg hydrogen requires $32/4.03 = 7.94$ kg of oxygen. Sulfur:

$$S + O_2 \rightarrow SO_2 \qquad (3.8)$$
$$32kg + 32kg \rightarrow 64kg \qquad (3.9)$$

Some fuels, particularly heavy marine fuels, may contain sulfur.

That is, the combustion of 1 kg S requires $32/32 = 1.00$ kg of oxygen.

If c, h, s and o denote weight-% of carbon, hydrogen, sulfur and oxygen in the fuel, the required amount of oxygen is calculated by:

$$(O_2)_{min} = 0.01(2.667 \cdot c + 7.94 \cdot h + s - o)$$

while the necessary amount of air required for complete combustion, AF_s is calculated as:

$$AF_s = (2.667 \cdot c + 7.94 \cdot h + s - o)/23.2 \qquad (3.10)$$

where 23.2 is the mass percent of oxygen in atmospheric air.

---

[1]Actually, there is about 0.9% Argon in the air. It is usually lumped together with the nitrogen as in the proportion given here.

Example 1

Consider a fuel with 86 mass% C and 14 mass% H: Then:

$$AF_s = (2.667 \cdot 86 + 7.94 \cdot 14)/23.2 = 14.68 \; kg \; air/kg \; fuel$$

Example 2

Cetane, $C_{16}H_{34}$. In this case it is necessary to calculate the mass composition first.

$$c = \frac{12 \cdot 16}{12 \cdot 16 + 34 \cdot 1.075} \cdot 100 = 84.9\%, h = 100 - c = 15.1\%$$

Then,

$$AF_s = (2.667 \cdot 84.9 + 7.94 \cdot 15.1)/23.2 = 14.93 \; kg \; air/kg \; cetane$$

As a comparison of methods, for cetane:

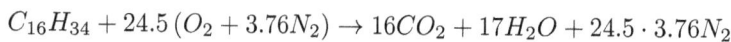

$$C_{16}H_{34} + 24.5 \left(O_2 + 3.76N_2\right) \rightarrow 16CO_2 + 17H_2O + 24.5 \cdot 3.76N_2$$

and:

$$AF_s = \frac{24.5 \cdot 4.76 \cdot 28.95}{16 \cdot 12 + 34 \cdot 1.075} = 14.93 \; kg \; air/kg \; cetane$$

Here air is considered as 1 kmole O_2+ 3.76 kmoles N_2 for a total of 4.76 kmoles with an average molecular weight of 28.95 kg/kmole

Table 2.1 gives the stoichiometric fuel air ratios for a variety of fuel compounds. It is clear that the stoichiometric mixture strength depends on the fuel composition. A fuel air ratio of 0.0639 is stoichiometric for propane, lean for benzene, and rich for methane.

Therefore, when considering different fuels, it is very important to utilize an expression for mixture strength that denotes whether the mixture is *rich* (excess fuel), stoichiometric, or *lean* (excess air), regardless of the fuel in question. There are two commonly used terms.

The first parameter is the *excess air ratio*, normally given the symbol, λ. It is defined in the following way:

$$\lambda \equiv \frac{AF}{AF_s} = \frac{FA_s}{FA} \tag{3.11}$$

For a value of $\lambda > 1$, the mixture is lean, while for $\lambda < 1$ the mixture is rich.

The second parameter is the *fuel air equivalence ratio*, normally denoted as ϕ, and defined as:

$$\phi \equiv \frac{FA}{FA_s} = \frac{AF_s}{AF} = \frac{1}{\lambda} \tag{3.12}$$

The equivalence ratio is often used in the case of very lean mixtures, since it approaches zero for pure air, instead of infinity in the case of the excess air ratio. Both terms are often used in the literature. Typically, in European literature, excess air ratio is more common, while in American literature the fuel air equivalence ratio is more common, but there are many exceptions.

3.1.2 Calculation of Exhaust Gas Composition from the Fuel Air Ratio

In the following, examples will be given of the prediction of the overall composition of the exhaust gasses, based on a given value for the fuel air ratio. This is of primary interest in terms of the estimation of CO emissions from spark ignition engines operating with rich mixtures, and some simplified procedures for estimating the properties of the combustion products at moderate (exhaust) temperatures. Estimation of CO emissions from engines with lean mixtures involves a much more complicated chemical process and will not be discussed here. The overall calculation presented here is not capable of estimating the concentration of HC or NO_x emissions in spark ignition engines. These mechanisms will be discussed in Chapter 6.

Calculation of any of the emissions of CO, unburned hydrocarbons, soot or oxides of nitrogen from diesel engines is a very complicated procedure, and is beyond the scope of this text.

One can calculate the concentrations of the main gaseous components in the exhaust of an engine with reasonable accuracy by using some simple methods. One starts by writing a chemical reaction and solving the equations that arise due to the conservation of all atoms.

Lean and Stoichiometric Mixtures

For lean and stoichiometric mixtures, the calculation is relatively simple. For example, considering octane one can write:

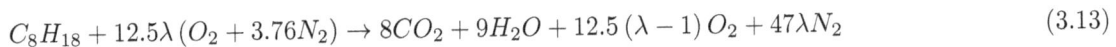

$$C_8H_{18} + 12.5\lambda\,(O_2 + 3.76N_2) \rightarrow 8CO_2 + 9H_2O + 12.5\,(\lambda - 1)\,O_2 + 47\lambda N_2 \qquad (3.13)$$

By giving the value of the excess air ratio, λ, one can calculate values for all molecules and thus find the concentrations of all the main species in the combustion products. The results are often calculated for a dry condition, that is, under the assumption that all the water has been removed from the combustion products before the gas analysis is conducted. This is a normal procedure for certain emissions in the exhaust gasses of engines (CO and CO_2), where the instruments most commonly used for these gasses cannot tolerate condensed water on windows in optical systems.

As an example, the composition of the exhaust gasses from octane can be calculated for an excess air ratio of 1.1:

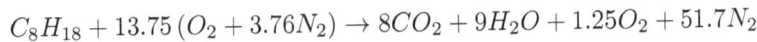

$$C_8H_{18} + 13.75\,(O_2 + 3.76N_2) \rightarrow 8CO_2 + 9H_2O + 1.25O_2 + 51.7N_2$$

This gives the composition, calculated on a molar (volume) basis, shown in Table 3.1:

Table 3.1: Ideal exhaust gas composition for combustion of octane with 110 % theoretical air

Specie	Wet volume %	Dry volume %
H_2O	12.9	-
$CO2$	11.4	13.1
O_2	1.8	2.1
N_2	73.9	84.8

For lean mixtures, one can use the mole numbers from Equation (3.3) to the exhaust gas composition resulting from the combustion of any hydrocarbon, C_xH_y:

On a wet volume basis:

$$\%CO_2 = \frac{100x}{D} \qquad (3.14)$$

$$\%H_2O = \frac{100y}{2D} \qquad (3.15)$$

$$\%O_2 = \frac{100\left(x + \frac{y}{4}\right)(\lambda - 1)}{D} \qquad (3.16)$$

$$\%N_2 = \frac{376\lambda\left(x + \frac{y}{4}\right)}{D} \qquad (3.17)$$

where:

$$D = x + \frac{y}{2} + \left(x + \frac{y}{4}\right)(4.76\lambda - 1) \qquad (3.18)$$

On a dry volume basis:

$$\%CO_2 = \frac{100x}{E} \tag{3.19}$$

$$\%O_2 = \frac{100\left(x + \frac{y}{4}\right)(\lambda - 1)}{E} \tag{3.20}$$

$$\%N_2 = \frac{376\lambda\left(x + \frac{y}{4}\right)}{E} \tag{3.21}$$

where:

$$E = x + \left(x + \frac{y}{4}\right)(4.76\lambda - 1) \tag{3.22}$$

Calculation of the Exhaust Composition of Rich Mixtures

Although there are some additional smaller sources, the appearance of CO in spark ignition engine exhaust is primarily due to operation with rich mixtures. There is not sufficient air to convert all carbon molecules to CO_2, therefore, there is a partial combustion resulting in the presence of CO and CO_2 among the products.

In terms of calculating the exhaust gas composition, the situation with rich mixtures is more complicated than for lean mixtures. In the following, the formation of solid carbon (soot) is neglected, that is, the mixtures are leaner than a value of λ of approximately 0.5 ($\phi < 2$). There are two new gaseous products to consider in the rich combustion gasses in addition to those found with lean mixtures. These new products are CO and H_2, while for ideal combustion, oxygen in rich combustion products can be neglected. In such a case, a general composition of the exhaust gasses can be described by the following, where: α, β, γ, and δ are unknowns, and x and y are known:

$$\phi C_x H_y + \left(x + \frac{y}{4}\right)(O_2 + 3.76N_2) \rightarrow \alpha CO_2 + \beta CO + \gamma H_2O + \delta H_2 + \left(x + \frac{y}{4}\right)3.76N_2 \tag{3.23}$$

In Equation (3.23) the nitrogen balance has already been used. The balances for C, H, and O atoms are then:

$$C: \quad \phi x = \alpha + \beta \tag{3.24}$$

$$H: \quad \phi y = 2\gamma + 2\delta \tag{3.25}$$

$$O: \quad 2\left(x + \frac{y}{4}\right) = 2\alpha + \beta + \gamma \tag{3.26}$$

Since there are 4 unknowns and 3 equations, one additional equation is needed. There are 3 simple alternative assumptions that can be applied:

1. The first is to assume that there is no molecular hydrogen, that is, $\delta = 0$. H_2 is usually in the lowest concentration of the major stable exhaust gas species, and so as a first approximation, it is assumed not to exist. The atomic balances are easy to solve in this case.

2. A second, somewhat better assumption, is that there is a known ratio between the CO and H_2 for example, $\beta/\delta = 2$. This is based on experience and the ratio can be adjusted to match one's experience. This is the additional equation needed to solve the above reaction system.

3. The third assumption is that the water gas equilibrium prevails. This reaction involves the major species in the exhaust gas, as the minor species have very little importance at exhaust gas temperatures (See Figure 2.11.) The water gas reaction is:

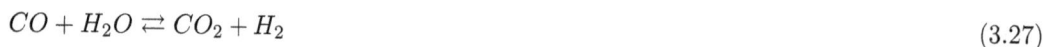

$$CO + H_2O \rightleftarrows CO_2 + H_2 \tag{3.27}$$

The third case is discussed in the following.

$K_p(T)$ is the equilibrium constant for the water gas reaction, and is a function of the temperature (see Section 2.4.1). For the reaction (3.27) it is written as:

$$\frac{p_{CO_2} \cdot p_{H_2}}{p_{CO} \cdot p_{H_2O}} = K_p(T) \tag{3.28}$$

where p_i is the partial pressure of specie, i.

Since the partial pressure of a compound is proportional to its mole fraction, and Reaction (3.27) has the same number of moles of reactants and products, the equilibrium constant can be written:

$$\frac{\alpha\delta}{\beta\gamma} = K_p(T) \tag{3.29}$$

If the value of the equilibrium constant is known, the chemical balances and the equilibrium constant can be solved to obtain the composition of the exhaust gasses as follows, where the terms correspond to the number of moles in Equation (3.23):

$$\beta = \frac{-B + \sqrt{B^2 - 4AC}}{2A} \tag{3.30}$$

$$\delta = \frac{y \cdot K_p \cdot \frac{\beta}{2}}{x + \beta(K_p - 1)} \tag{3.31}$$

$$\alpha = x - \beta \tag{3.32}$$

$$\gamma = \frac{y}{2} - \delta \tag{3.33}$$

where:

$$A = (K_p - 1) \tag{3.34}$$

$$B = 2\left(x + \frac{y}{4}\right)(\lambda - 1)(K_p - 1) + x + K_p\frac{y}{2} \tag{3.35}$$

$$C = 2x\left(x + \frac{y}{4}\right)(\lambda - 1) \tag{3.36}$$

The value of equilibrium constant, K_p, can be calculated with good accuracy using the following equation for T in K.

$$\log_{10} K_p = -1.3543 + 1.28 \cdot 10^3 \cdot T^{-1} + 2.333 \cdot 10^5 \cdot T^{-2} \tag{3.37}$$

There is still a problem remaining in connection with the above solution in that the temperature to be used in determining the equilibrium constant is not known before hand. In the exhaust gasses, it is found that the chemical reactions which maintain equilibrium tend to stop when the temperature falls below a certain level, which "freezes" the composition at this temperature during expansion. In such a case, the composition will correspond to that at the temperature at which the reaction freezes.

The same process occurs in the formation of oxides of nitrogen, although this temperature is considerably higher than that at which the water gas equilibrium freezes. (See the discussion on NO_x formation in Chapter 6). Figure 3.1 shows a comparison between the exhaust gas composition from engine measurements from a single cylinder spark ignition running on a very homogeneous intake mixture of octane and air [20], and the calculated composition using the above method.

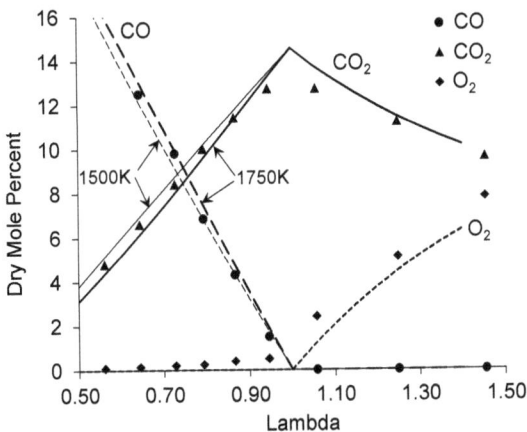

Figure 3.1: Calculated dry exhaust concentrations for water gas equilibrium at two temperatures compared to measurements (symbols).

For rich mixtures, good agreement is obtained by using a freezing temperature of about 1500 - 1750 K. The calculated composition is not strongly dependent on the temperature.

For lean mixtures, a good agreement for CO_2 and O_2 is obtained by using the equations for the lean mixtures in the previous section. Since one, nevertheless, has to assume a freezing temperature, it can also be said that the third method for rich mixtures is also a kind of rule of thumb just like the first two methods for rich mixtures. The difference is that with the third method it is possible to take differing fuels into consideration, since the ratio between the carbon and hydrogen in the fuel is involved in determining the composition and ratio between CO and H_2. The freezing temperature does not depend on the carbon hydrogen ratio to any significant extent.

3.1.3   Calculation of the FA Ratio from the Exhaust Composition

As seen above, the fuel air mixture ratio is an important combustion parameter, and there is a close connection between it and the composition of the products of combustion, as was seen in Section, 3.1.2. The following describes how to reverse the calculation of the previous section and determine the mixture ratio based on measurements of the exhaust gas composition. This is possible in cases where the fuel composition is known, or not known. The method is useful in checking the consistency of experimental measurements, since the fuel air ratio calculated from the exhaust composition should be the same as that from the measured fuel and air flows. Inconsistencies indicate measurement problems.

In addition, there are some applications where the analysis of the composition of the combustion products is the only practical way to determine the fuel air ratio. Examples of this are in-cylinder sampling of combustion products in diesel engines, and the special sampling techniques used to determine the scavenging parameters in 2-stroke diesel engines by determining the fuel air ratio of the gasses trapped in the cylinder during combustion.

Stoichiometric combustion of a fuel can be determined from Reaction (3.3).

For $\lambda < 1$ additional products besides CO_2 and H_2O are formed. Neglecting oxides of nitrogen, the process can be written:

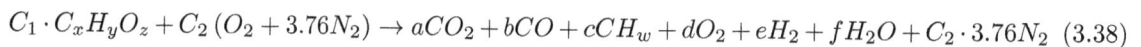

$$C_1 \cdot C_x H_y O_z + C_2 \left(O_2 + 3.76 N_2\right) \rightarrow a CO_2 + b CO + c C H_w + d O_2 + e H_2 + f H_2 O + C_2 \cdot 3.76 N_2 \quad (3.38)$$

The exhaust gas hydrocarbon composition is not known, since even with a single component fuel, some partial reactions occur in the exhaust system. Hence, it is written: CH_w, where in most cases, $w \approx 2$, though it may be higher in the case of natural gas. Unburned hydrocarbon concentrations are normally low, so the value of w does not have a major effect on the calculated air fuel ratio, and fuel composition could also be used.

Since the average molecular weight of air is 28.95, the air fuel ratio, AF, for a given mixture can be written as:

$$AF = \frac{4.76 \cdot 28.95 \cdot C_2}{C_1 \left(12x + y + 16z\right)} \quad (3.39)$$

The stoichiometric air fuel ratio can be written:

$$AF_s = \frac{4.76 \cdot 28.95 \left(x + \frac{y}{4} + \frac{z}{2}\right)}{\left(12x + y + 16z\right)} \quad (3.40)$$

This gives:

$$\lambda = \frac{AF}{AF_s} = \frac{4.76 \cdot 28.95 \cdot C_2}{C_1 \left(12x + y + 16z\right)} \cdot \frac{12x + y + 16z}{4.76 \left(x + \frac{y}{4} + \frac{z}{2}\right) 28.95} = \frac{4 C_2}{4 C_1 x + C_1 y + 2 C_1 y} \quad (3.41)$$

If measurements of CO, CO_2, H_2, HC and O_2 are at hand, the method is the following:

In Equation (3.38) there are 4 unknowns: C_1, C_2, e and f, so 4 equations are needed. Three of the equations are obtained by writing atomic balances for C, H, and O respectively. The fourth equation can be obtained from a value for the water gas constant, as described in the preceding section. In the equation below, the value of the water gas constant is assumed to be 0.286. The value is dependent on the process involved, but as seen in the Section 3.1.2, 0.286 is a reasonable value for the exhaust gasses of a spark ignition engine.

The equations become:

$$xC_1 = a + b + c \quad C - balance \tag{3.42}$$
$$2C_2 = 2a + b + 2d + f \quad O - balance \tag{3.43}$$
$$yC_1 = c \cdot w + 2e + 2f \quad H - balance \tag{3.44}$$
$$K_p = 0.286 = \frac{a \cdot e}{b \cdot f} \quad Water\ Gas\ Equilibrium \tag{3.45}$$

In the above example, the number of measured species is limited to 5. On the other hand the fuel composition is known, and the values for x, y, and w are therefore known.

Example

For a fuel with an unknown composition, but without oxygen, the dry exhaust products are known to have the composition in mole % shown in Table 3.2. Determine the excess air ratio, λ.

Table 3.2: Example exhaust gas composition

Specie	Mole %
CO_2	8.7
CO	8.9
O_2	0.3
H_2	3.7
CH_2	0.6

Then: C Balance: $C_1 x = 8.7 + 8.9 + 0.6 = 18.2$

Equilibrium Constant:

$$K_p = 0.286 = \frac{a \cdot e}{b \cdot f} = \frac{8.7e}{8.9 \cdot f} \Rightarrow f = 3.421 \cdot 3.7 = 12.66$$

O Balance: $2C_2 = 2(8.7) + 8.9 + 2(0.3) + 12.66$

$C_2 = 19.78$

H balance: $yC_1 = 2(0.6) + 2(3.7) + 2(12.66) = 33.92$

Then:

$$\lambda = \frac{4 \cdot C_2}{4 \cdot C_1 x + C_1 y} = \frac{4\,(19.78)}{4 \cdot 18.2 + 33.92} = 0.741$$

Note that the composition of the fuel has not been needed yet, since the values of x and y are combined with the constants C_1 and C_2 in Equation (3.41). To determine the actual fuel air ratio, x and y are needed. If fact, the relative fuel composition can be determined from the constants $C_1 x$ and $C_1 y$:

$$\frac{y}{x} = \frac{C_1 y}{C_1 x} = 1.864$$

Then assuming x = 1 (this is arbitrary), :

$$C_1 = C_1 x = 18.2$$
$$y = 1.864$$

Using Equation (3.39)

$$AF = \frac{4.76 \cdot 28.95 \cdot 19.78}{18.2\,(12 \cdot 1 + 1.864)} = 10.8$$

An alternative method can be used in the case where the hydrogen concentration is not available, as it is not commonly measured in engine testing. The main assumption here is that the sum of the dry concentrations is 100%. Since there is one less known, there must be some indication of the composition of the fuel, at least on a relative molar basis for hydrogen and carbon. For most modern diesel or SI fuels, a hydrogen to carbon ratio of about 1.8 - 1.85 is appropriate.

Assuming x and y, or their ratio, to be known, the balanced chemical reaction can be written:

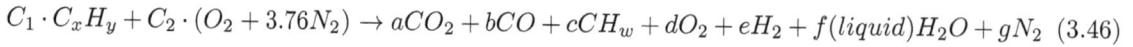

$$C_1 \cdot C_x H_y + C_2 \cdot (O_2 + 3.76 N_2) \rightarrow a CO_2 + b CO + c CH_w + d O_2 + e H_2 + f(liquid) H_2 O + g N_2 \quad (3.46)$$

If all the gas phase products, with the exception of N_2 are included in the composition, the sum of all the dry products must be 100 %:

$$100 = a + b + c + d + e + g \Rightarrow e + g = 100 - (a + b + c + d) \quad (3.47)$$

The nitrogen balance gives:

$$3.76 C_2 = g \quad (3.48)$$

Combining with the condition for the sum of the dry products:

$$C_2 = \frac{100 - (a + b + c + d + e)}{3.76} \quad (3.49)$$

Then the carbon, oxygen and hydrogen balances are:

Carbon balance:

$$C_1 x = a + b + c \quad (3.50)$$

Oxygen balance:

$$2 C_2 = 2a + b + 2d + f \quad (3.51)$$

Hydrogen balance:

$$C_1 y = cw + 2e + 2f \quad (3.52)$$

Which can be solved for $C_1 x$, f, and $C_1 y$. Dividing $C_1 x$ by $C_1 y$, the atomic hydrogen to carbon ratio of the fuel is obtained.

Example

Take the concentrations of all the products in Table 3.2, with the exception of hydrogen. The values of the concentrations will be the same as in the table, but the hydrogen concentration is now included in the non-measured species.

Assume that the hydrogen to carbon ratio in the fuel is y, equivalent to assuming $x = 1$, and that the composition of the dry exhaust products is the same as in Table 3.2

Then the chemical equation can be written:

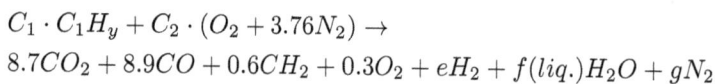

$$C_1 \cdot C_1 H_y + C_2 \cdot (O_2 + 3.76 N_2) \rightarrow$$
$$8.7 CO_2 + 8.9 CO + 0.6 CH_2 + 0.3 O_2 + e H_2 + f(liq.) H_2 O + g N_2$$

For all dry products:

$100 = 8.7 + 8.9 + 0.3 + 3.7 + 0.6 + e + g \Rightarrow e + g = 81.5$

Nitrogen balance: $3.76 C_2 = g$

Carbon Balance: $C_1 = 8.7 + 8.9 + 0.6 = 18.2$

Oxygen Balance: $2 C_2 = 2(8.7) + 8.9 + 2(0.3) + f = 26.9 + f$

Hydrogen Balance: $C_1 y = 2(0.6) + 2(e + f) = 1.2 + 2e + 2f$

Equilibrium constant:

$$f = \frac{8.7}{8.9} \frac{e}{0.286} = 3.42e$$

The preceding 6 equations involve the 6 variables, C_1, C_2, y, e, f, and g. The solution is: $C_1 = 19.1$, $C_2 = 20.33$, $e = 4.142$, $f = 12.87$, $g = 76.46$ and $y = 1.844$.

The air fuel ratio is:

$$AF = \frac{C_2 \cdot 4.76 \cdot 28.95}{C_1 \cdot 12 \cdot 1 + C_1 y} = \frac{20.33 \cdot 4.76 \cdot 28.95}{19.1 \cdot 12 \cdot 1 + 19.1 \cdot 1.844} = 10.6$$

and using Equation (3.41):

$$\lambda = \frac{4C_2}{4C_1 x + C_1 y} = 0.7287$$

The above calculation methods should be accurate enough for a check on the fuel air ratio as measured by intake flows. In lean mixtures, CO and H_2 are often quite low and can be ignored. In that case, the above balances allow FA determination in different manners. With most emissions measurements, carbon measurements are most common and most accurate (CO, CO_2, and unburned hydrocarbons) and should probably be the primary source for the calculation of the fuel air ratio from exhaust gas composition.

3.2   Thermodynamics in Reacting systems

3.2.1   Heat of Formation and the First Law of Thermodynamics

When dealing with changes in thermodynamic properties of non-reacting systems, it is sufficient to define property changes relative to a reference state consisting of a temperature and a pressure. This is not adequate in systems where chemical reactions occur. This can be seen by considering the substances hydrogen and water vapor. They can exist at the same temperature and pressure, but there is a large difference in the amount of energy the molecules contain. For example, if the hydrogen is burned, a lot of energy has to be removed in order to bring it back to the original temperature. This is because there is chemical energy contained in the hydrogen molecule that can be released through an oxidation reaction to make water.

Thus, when calculating energy changes for systems where chemical reactions can occur, it is necessary to define a reference composition as well as a temperature and pressure. Properties are then calculated as changes from this reference condition. When we refer to the internal energy or enthalpy of a substance, what is really meant is the change of the property relative to an arbitrarily (but hopefully conveniently) chosen reference condition (phase, temperature, pressure, and composition).

For reacting systems, it is then necessary to choose a reference chemical composition. In combustion systems, many have chosen the state of combustion products at standard atmospheric conditions as the reference condition, with the idea that these products contain no useful chemical energy. This can be successful for combustion systems but the scientific community has chosen another condition for reference, namely, the state at which a chemical element is likely to be found in nature. The standard temperature and pressure are taken as 25°C and 101.325 kPa = 1 atm. For example, the gasses argon, oxygen as O_2, nitrogen as N_2, hydrogen as H_2, and solid carbon as graphite have been selected as reference conditions, and assigned the enthalpy of zero. Since all chemical compounds are constructed from the elements, this system is very general. In combustion systems it has the minor irritation that sometimes negative energies and enthalpies are encountered. All this means is that the enthalpy of the compound is lower than that of the individual elements of which it is composed.

Then taking enthalpy as a property for any compound, an enthalpy of formation (also commonly called the heat of formation) can be defined as the difference between the enthalpy of a substance and the enthalpy of the elements from which it has been formed, at the same reference temperature and pressure. With water as an example, the enthalpy of formation of water would be the change in enthalpy for the following reaction:

$$H_2 + \frac{1}{2}O_2 \rightarrow H_2O \tag{3.53}$$

where:

$$\Delta h_R = h_{H_2O} - h_{H_2} - \frac{1}{2}h_{O_2} \equiv h^o_{f,H_2O} \tag{3.54}$$

The superscript, o, denotes that the standard state pressure of 1 atm was used. Since the reference temperature is arbitrary, though normally 298.15K, the enthalpy of formation is often written: $h^o_{f,i}(T_{ref})$. It is normally given in molar units, $i.$ $e.$, kJ/kmol., $etc.$

The enthalpy of formation refers only to the chemical portion of a change in the enthalpy. For a given substance, i, relative to the reference state, the enthalpy at any given temperature is then the sum of the change due chemical composition, and the change due to temperature (sensible enthalpy). In the case of an ideal gas, which applies for engine combustion systems, then:

$$h_i(T) = h^o_{f,i}(T_{ref}) + \int_{T_{ref}}^{T} c_{p,i}dT \tag{3.55}$$

This is a most important relationship in reacting systems. Conventionally, the enthalpy of formation is used, and is the most common value tabulated. In principle, an internal energy of formation can be defined, but this is rarely used. Instead, Equation (3.55) is used along with the definition of enthalpy, such that for any substance:

$$u_i(T) = h_i - pv = h^o_{f,i}(T_{ref}) + \int_{T_{ref}}^{T} c_{p,i}dT - pv \tag{3.56}$$

For ideal gasses:

$$u_i(T) = h_i - RT = h^o_{f,i}(T_{ref}) + \int_{T_{ref}}^{T} c_{p,i}dT - R_oT \tag{3.57}$$

$$= h^o_{f,i}(T_{ref}) + \int_{T_{ref}}^{T} c_{v,i}dT - R_oT_{ref} \tag{3.58}$$

R_o is the universal gas constant in molar units. The term: $h^o_{f,i}(T_{ref}) - R_oT_{ref}$ is actually the internal energy of formation of the compound i, so it is not necessary to tabulate it.

For a general reaction, as shown in Reaction (2.94) where species A and B react to form species C and D with the relative number of moles ν_A, ν_B, ν_C, and ν_D:

$$\nu_A A + \nu_B B \Rightarrow \nu_C C + \nu_D D$$

The enthalpy change for the reaction at the reference temperature, $\Delta H_R(T_{ref})$, can be given in terms of the heats of formation, where the values of the enthalpies of formation are given on a molar basis:

$$\Delta H_R(T_{ref}) = \nu_C h^o_{f,C}(T_{ref}) + \nu_D h^o_{f,D}(T_{ref}) - \nu_A h^o_{f,A}(T_{ref}) - \nu_B h^o_{f,B}(T_{ref}) \tag{3.59}$$

For a combustion reaction, the enthalpy change of the reaction is called the $heat$ of $combustion$. Since a combustion reaction is exothermic, the heat of combustion is negative, and so the absolute value of the heat of reaction (3.59) is defined as the $heating$ $value$ of the fuel, when calculated in terms of a kmole or kg of fuel.

An example is shown for methane:

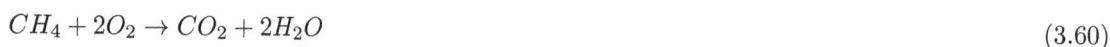

$$CH_4 + 2O_2 \rightarrow CO_2 + 2H_2O \tag{3.60}$$

$$\Delta H_R(298) = h^o_{f,CO_2}(298) + 2h^o_{f,H_2O}(298) - h^o_{f,CH_4}(298) - 2h^o_{f,O_2}(298)$$

In the tables of Section 17.1, the following values of the enthalpies of formation can be found: Note that consistent with the definition of heat of formation, these values are the enthalpies of the compounds at the reference temperature at 25°C. Then:

$$\Delta H_R(298) = (-393,492) + 2 \cdot (-241,815) - (-74,597) - 2 \cdot 0 = -802585 \; kJ/kmole \; CH_4$$

CO_2 $h^o_{f,CO_2}(298) = -393,492 \text{ kJ/kmol}$
H_2O vapor $h^o_{f,H_2O}(298) = -241,815 \text{ kJ/kmole}$
CH_4 $h^o_{f,CH_4}(298) = -74,597 \text{ kJ/kmole}$
O_2 $h^o_{f,O_2}(298) = 0 \text{ kJ/kmole}$

Since methane has a molecular weight of 16.043 kg/kmol, the heating value on a mass basis is equal to the absolute value of the heat of combustion: $|-802585|/16.043 = 50027 \ kJ/kgCH_4$, which is the lower heating value of methane at constant pressure, 25°C, since the water in the products is assumed to be in the vapor phase.

Using the above definition of enthalpy, and the resulting internal energy, the first law of thermodynamics appears the same with or without reactions.

$$\dot{Q}_{CV} - \dot{W}_{CV} = \frac{d(mu)}{dt} + \sum_{out} \dot{m} \left(h + \frac{V^2}{2} \right) - \sum_{in} \dot{m} \left(h + \frac{V^2}{2} \right) \tag{3.61}$$

But in Equation (3.61), it is implicit that the enthalpies include the enthalpies of formation, so the equation is also valid for reacting systems. If there is no chemical reaction, the amount of each specie entering a control volume is the same as that exiting, and the terms involving the enthalpy of formation are the same on each side of Equation (3.61) and, therefore, usually omitted for non-reacting systems.

For a system at steady state, and neglecting kinetic energy of the flows in and out of the control volume, the first law for a reacting system becomes:

$$\dot{Q}_{CV} + \sum_{i=1,N_r} \dot{n}_i \cdot \left(h^o_{f,i}(T_{ref}) + \int_{T_{ref}}^{T_{in}} c_{p,i}dT \right) = \dot{W}_{CV} + \sum_{i=1,N_p} \dot{n}_i \cdot \left(h^o_{f,i}(T_{ref}) + \int_{T_{ref}}^{T_{out}} c_{p,i}dT \right) \tag{3.62}$$

where n_i represents number of moles, N_r is the number of reactants, and N_p is the number of reaction products.

It is assumed that reactants enter and products exit. The equation is written in terms of moles, because tabulations of enthalpies of formation and specific heats and their integrals are most often given in molar units. In many texts, it is a practice to list the integrals of $c_p dT$, and the enthalpies of formation separately, thus, the form above is useful. However, it is simple to add the enthalpies of formation to the integrals of $c_p dT$, and then obtain the enthalpies of compounds relative to the standard reference. This has been done in the tables which are included in Section 17.1. In that case, Equation (3.61) can be used directly. This is also the case with some modern software, for example the program EES, where enthalpies of ideal gasses and gaseous fuels include heats formation. That is to say, that they are calculated with reference to the standard temperature, pressure and composition described earlier in this section.

Equation (3.55) can be used in the calculation of the adiabatic flame temperature. For a flame at constant pressure with no heat transfer, the first law of thermodynamics, Equation (3.61), gives:

$$h_1 = h_2 \tag{3.63}$$

Then using Equation (3.55), where the enthalpies are on a mole basis:

$$\sum_{i=1,N_r} x_i \cdot h_i = \sum_{i=1,N_p} x_i \cdot h_i \tag{3.64}$$

$$\sum_{i=1,N_r} x_i \cdot \left(h^o_{f,i}(T_{ref}) + \int_{T_{ref}}^{T_1} c_{p,i}dT \right) = \sum_{i=1,N_p} x_i \cdot \left(h^o_{f,i}(T_{ref}) + \int_{T_{ref}}^{T_2} c_{p,i}dT \right) \tag{3.65}$$

where N_r is the number of reactants, and N_p is the number of products. In the case of ideal combustion products, all the mole fractions are assumed to be known for the products and reactants from stoichiometric considerations. In the thermodynamic tables in the literature, the heats of formations are listed, along with the integrals of $c_p \cdot dT$. Then in principle, Equation (3.65) can be solved by guessing a flame temperature, T_2 and substituting the integrals of $c_p \cdot dT$ into the equation. When the proper temperature is found, the equation is satisfied. A more detailed discussion is presented in Section 3.4

3.2.2 Entropy

In ideal gas systems, the internal energy and enthalpy are only functions of temperature, but the entropy is a function of two state variables, even when considering an ideal gas. Here the temperature and pressure will be used. Then in reacting systems, it is necessary to include a reaction dependent portion of the entropy, but the third law of thermodynamics tells us that an absolute entropy can be defined at absolute zero. For all the gaseous substances involved in combustion, the absolute entropy is equal to 0 at 0K and 1 atm. By a knowledge of the specific heat of a substance, the entropy change can be calculated from this reference and it is commonly tabulated as a function of temperature for a pressure of 1 std. atmosphere, and called the absolute entropy. If the absolute entropy is only available at the standard temperature of 298K, then the change in entropy from this standard condition can be added in order to determine the absolute entropy of any substance at a given temperature and pressure. Here only ideal gasses are considered, and in molar units:

$$s_i(T,p) = s_i^0(T) - R\ln\frac{p_i}{p^o} = s_i^o(T_{ref}) + \int_{T_{ref}}^{T} c_{p,i}\frac{dT}{T} - R_o\ln\frac{p_i}{p^o} \tag{3.66}$$

Where p_i is the partial pressure of component, i, calculated as the product of the mole fraction of that species and the total pressure of the system, and p^o is 1 atm.

In isentropic calculations for mixtures, Equation (3.66) is used to calculate the entropy of each component for a given temperature and pressure. Assuming the calculation is on a mole basis, the total entropy per mole of mixture is then calculated as:

$$s = \sum_{i=1,N_{components}} x_i \cdot s_i \tag{3.67}$$

It is this total entropy that remains constant for a compression process of a mixture. The molar entropies of the individual components can change, but the weighted sum must remain constant for an isentropic process. Equation (3.67) is used in the calculation of the isentropic processes in the fuel air cycle. On the expansion stroke, the situation is additionally complicated by the fact that the composition (the $x_i's$) are also functions of temperature and pressure. This makes it necessary to perform an iterative process for solution of isentropic compressions and expansions in engines.

3.3 Heating Value

3.3.1 Introduction

An important aspect of combustion is the energy content of the fuel used in the combustion process. The chemical energy released during the combustion is the source of the high temperatures and pressures that are necessary in order that energy conversion devices such as piston engines and gas turbines, *etc.* can function. This energy is the heating value of the fuel, to be examined more closely in the following. The heating value is basically the magnitude of energy per unit fuel quantity that must be removed from a system that has undergone a combustion reaction, to return it to its initial temperature and either pressure or volume. The amount of energy which must be removed from the products is called the *heat of combustion*. It is negative, because of a sign convention which defines heat transfer into a system as positive. The heating value is the absolute value of the heat of combustion.

3.3.2 Energy and Thermodynamics

The calculation of flame temperature involves a conversion of chemical energy to internal or "thermal" energy, so it is appropriate to review some fundamental concepts which will be used in connection with the calculation of flame temperature.

It is well known that when one adds energy to a substance, it's temperature will rise, if there is no phase transformation. When chemical reactions are not considered, the internal energy, U, of a system is the energy it has because of its temperature, also called *sensible energy*. The relationship between the change in internal energy and the change in temperature is called the specific heat or heat capacity, which is given the symbol c_v:

$$c_v = \frac{\partial u}{\partial T}_v \qquad (3.68)$$

where: u is energy per unit mass or mole, and T = temperature.

One of the most fundamental concepts in classical physics in the conservation of energy. This means that energy can be converted from one form to another (internal $\Rightarrow$ kinetic, kinetic $\Rightarrow$ potential, *etc.*), but total amount of energy is always conserved.

We also know that we can add or remove energy from a system in different ways. There are only two of these ways that are of interest in the current discussion: heat transfer and mechanical work. Heat transfer consists of an exchange of energy from one system to another by virtue of a temperature difference. Mechanical work consists of energy transfer by the exertion of a force over a distance. The convention in mechanical engineering thermodynamics is that work is denoted by W and heat transfer is denoted by Q, typically in Joules or kiloJoules. Usually in mechanical engineering, heat transfer to a system is considered as positive, and work extracted (performed on the surroundings) from a system is also considered to be positive.

With these definitions, one can express the conservation of energy by saying that in a process consisting of varying equilibrium states, the change in the energy content of a system during the process is equal to the difference between the energy added and the energy removed. This is the well known first law of thermodynamics, which is expressed mathematically in Equation (3.69), for a closed (constant mass) system where changes in both kinetic and potential energy are negligible.

$$Q - W = \Delta U = U_2 - U_1 \qquad (3.69)$$

Equation (3.69) can be interpreted such that if during a process one adds an amount of energy, Q, by heat transfer and during the same process removes that amount of energy W, by work transfer, the energy content of the system will be changed from U_1 to U_2.

If we return to the heating process for gasses, it is found that there is a difference in the process, depending on whether it occurs with constant pressure or constant volume. We can see this from the ideal gas law:

$$pV = mRT = nR_oT \qquad (3.70)$$

where p = pressure i Pa, V = volume i m^3, m = mass in kg, n = number of kmoles, R = gas constant in J/kg-K (dependent on the molecular weight of the gas) and R_o is the universal gas constant, 8314 J/kmol-K.

If we heat a gas or vapor without restricting its size, there will be a significant expansion, which is not the case with a solid or liquid, where the changes in volume are normally quite small. From Equation (3.70) we can see that if we maintain a constant pressure the volume will increase in direct proportion to the temperature. But as the gas expands, it has to push the surroundings away. In physics, work has been defined as a force exerted through a distance. The pressure represents a force per unit area, and a change in volume corresponds to movement through a distance. It is then clear, that a change in volume involves work.

The work, δW, connected with a change in volume, dV, of a substance with a pressure, p, is:

$$\delta W = p \cdot dV \qquad (3.71)$$

Let us now look at the difference between two different heating processes for an ideal gas. Imagine 2 different situations. The first is that a gas is heated in a closed container which maintains a constant volume, and that an amount of heat is transferred to the substance. Since the volume is constant and the container is closed (dV=0), no work is performed and if no chemical reactions occur,the first law is:

$$Q = U_2 - U_1 = m \int_1^2 c_v \cdot dT \qquad (3.72)$$

The other situation is the case where a gas is heated in a closed container which does not exert any resistance to expansion, such that the pressure is constant during the heating process. In this situation the system performs work on the surroundings. According to Equation (3.71) this work can be written as the product of the constant pressure, p, and the change in volume. Substituting Equation (3.71) in the first law, Equation (3.69), one obtains:

$$Q - p(V_2 - V_1) = U_2 - U_1 \tag{3.73}$$

Since $p_1 = p_2$, one can rewrite Equation (3.73) as:

$$Q = (U_2 + p_2 V_2) - (U_1 + p_1 V_1) \tag{3.74}$$

In thermodynamics a useful property called enthalpy was defined from the results in Equation (3.74) as :

$$H \equiv U + pV \tag{3.75}$$

This definition is very useful, since one often has systems where the processes occur at constant pressure. By using this definition one obtains an equation for constant pressure heat addition which resembles that for constant volume:

$$Q = H_2 - H_1 \tag{3.76}$$

and by defining c_p :

$$c_v = \frac{\partial h}{\partial T}_p \tag{3.77}$$

Where: h is the enthalpy per unit mass, in J/kg or kJ/kg.

One obtains for a non-reacting system:

$$Q = m \int_1^2 c_p \cdot dT \tag{3.78}$$

For solids and liquids, the volume change associated with the heating is normally so small that it can be neglected. For these substances, one most often uses the term c_v for the heat capacity, or often just c.

The heating value and the heat formation of the fuel are closely related, and a heating value obtained from a calorimeter measurement is often used to determine the heat of formation. Consider the case of a reaction in a flow system at constant pressure, where there is heat transfer from the system in order to maintain the same temperature before and after the reaction. Equation (3.62) can be used, and it is assumed that $T_1 = T_2 = T_{ref}$, and that there is no work performed. It is also assumed that the reaction takes place with a stoichiometric or lean mixture, such that the reaction is completed to form the products CO_2 and H_2O for hydrocarbon fuels. Then:

$$\dot{Q}_{CV} = \sum_{i=1,N_p} \dot{n}_i \cdot h_{f,i}^o(T_{ref}) - \sum_{i=1,N_r} \dot{n}_i \cdot h_{f,i}^o(T_{ref}) \tag{3.79}$$

Defining the heating value as the absolute value of the heat of the combustion reaction:

$$H_u(T_{ref}) = \sum_{i=1,N_r} n_i \cdot h_{f,i}^o(T_{ref}) - \sum_{i=1,N_p} n_i \cdot h_{f,i}^o(T_{ref}) \tag{3.80}$$

where the n_i are the number of moles obtained from the stoichiometric reaction. Note that any substances not participating in the reaction will cancel in Equation (3.80). For example, 3.76 moles of N_2 in the products will have the same contribution to the reactants, since the temperature is constant. Thus, the mole numbers from a stoichiometric reaction of one mole of fuel with the appropriate amount of O_2 can be used in Equation (3.80), giving the heating value in typical units units of kJ/mole fuel. Dividing by the fuel molecular weight gives the heating value per unit mass.

Phase Change

As mentioned earlier, the temperature normally increases when energy is added to a system. But in certain situations energy can be added without causing a temperature increase, that is in the case of a change of phase. The most common case is, of course, that of water. There are two phase changes that we all know well: melting and boiling. In these phase changes the temperature is constant while, for example, one goes from 100% water to 100% vapor. After the water phase is completely evaporated, the temperature of the vapor will then increase if more heat addition occurs. Even though the temperature does not change during the evaporation, there is a large amount of energy transferred during the evaporation. The amount of energy needed to evaporate the liquid will depend on the process.

The phase change which is most important in connection with combustion of engine fuels is that of water between the gas and liquid phases. At high temperatures found during combustion, the water is formed as a vapor. During the cooling and expansion processes that follow combustion, it is possible that some or all of the water formed condenses. Then the energy which was absorbed by the evaporation is released to the system. Condensation of the water produced in combustion can then increase the efficiency of some of the accompanying thermodynamic processes, where one can utilize the heat of condensation.

The amount of energy which is relevant in connection with the condensation of water is the latent heat of vaporization of water. For evaporation at constant pressure, for example in an open, flow system at atmospheric pressure, the enthalpy of evaporation of water at 25°C is: $h_{fg} = 2442$ kJ/kg-H_2O. For evaporation/condensation of water in closed systems, one uses the internal energy of vaporization: u_{fg} = 2305 kJ/kg-H_2O.

Due to water condensation, two heating values can be defined: The *higher heating value* and the *lower heating value*. If the water formed during combustion is allowed to condense, the heat of vaporization of that water must also be removed from the system, while if the water remains as a vapor this energy does not have to be removed. With condensation there is then a greater overall amount of energy that must be removed from the system, and the heating value will be larger, and it is, therefore, called the *higher heating value*. If the water remains vaporized, less energy will be removed from the system during the combustion and cooling process, and the *lower heating value* is encountered. For most engine combustion and exhaust systems only the lower heating value is of interest, since the exhaust products typically remain at temperatures above the dew point of the exhaust gasses.

The evaporation of the fuel also has an influence on the flame temperature. If, for example, one sprays a liquid fuel into a combustion chamber, the heat of vaporization of the fuel must be transferred to the fuel from the surrounding air as the phase change occurs. This lowers the temperature before combustion and consequently the temperature after combustion. An example of this is seen just before the start of combustion in the heat release diagram in Figure 3.45.

The effect of fuel vaporization is normally not as large as the effect of the evaporation/condensation of the water in the combustion products, but for some fuels the heat of vaporization is quite large. Methanol, for example has a large heat of vaporization as compared to many hydrocarbon based fuels. Table 3.3 gives a list of the enthalpy of vaporization for several typical fuel components.

Table 3.3: Heats of vaporization for typical fuel components at 25°C or boiling point.

Fuel	Enthalpy of Vaporization in kJ/kg
Methane - CH_4	509
Propane - C_3H_8	425
Benzene C_6H_6	393
n-Heptane - C_7H_{16}	316
n-Octane - C_8H_{18}	300
n-Decane - $C_{10}H_{22}$	277
n-Hexadecane - $C_{16}H_{34}$	358
Methanol - CH_3OH	1103
Ethanol - C_2H_5OH	840

3.3.3 Calorimeters

The following discussion will be limited to fuels consisting of the elements C, H, and O. These elements have the most significant effects on the energy content of standard fuels, and most of the proposed alternatives. Other elements such as N and S as well as ash are normally found in smaller quantities, so they do not have a large effect on the heating value. Bunker fuel, which is used for heavy marine engines has larger amounts of impurities which can affect heating values. Measured heating values for these fuels include the effects of these substances on the heating values, but for simplification, they are not included in the following theoretical portions of the discussion.

For a fuel, the heating value is an expression of how much chemical energy is released when the fuel is converted to the ideal combustion products: CO_2, and H_2O. Even though there may be other elements present after the reaction, for example, O_2 and N_2, they have no significant effect concerning the amount of energy released per kg of fuel burned. This is of course under the condition that there is adequate oxygen present to facilitate the complete oxidation of the fuel. As was the case with the general stoichiometric calculation discussed earlier, no consideration will be taken of reaction of either N_2, or any excess O_2.

This means that the heating value for the fuel is the same whether it is burned in pure oxygen or in air. The temperature resulting from combustion in these different atmospheres will be greatly different, but it is not a reflection of any difference in the heating value of the fuel. The difference arises because air is composed of approximately 79 vol % N_2, which is inert. The heating of the nitrogen consumes a large portion of the chemical energy released in the combustion, and to a large degree determines the final flame temperature. In a combustion process with pure oxygen, there are fewer products to absorb the energy released in the combustion, and the temperature will be much higher. As was shown in Section 2.4.1, there are additional factors such as dissociation and chemical equilibrium which will play an important role in flame temperature determination.

It is a theoretical condition for the determination of the heating value that the fuel is completely converted to combustion products. This means that in practice, the experimental measurements of the heating value are conducted with an excess of oxygen. One must therefore be careful in situations with a rich mixture, since combustion in this case will not release the complete heating value, and a portion of the chemical energy is still bound up in molecules such as CO and H_2. This situation can be handled by the use of heats of formation, but one still needs to know the composition of the combustion products.

To illustrate heating values we can use two different experiments. The first consists of a constant volume container or bomb, which in practice is immersed in an insulated tank of water. The system is shown schematically in Figure 3.2. In the water bath, there is a thermometer which is used to accurately measure the temperature of the water. Such a system is called a constant volume or bomb calorimeter.

In order to determine the heating value of a given fuel, a known quantity of the fuel is placed in the bomb together with an excess of air or oxygen, shown in Figure 3.2. The bomb is placed in the water bath, and when the fuel is evaporated it is ignited, after which a complete combustion process occurs. One then must wait until the temperature throughout the entire system is uniform, that is the combustion products, bomb, and water all have the same temperature.

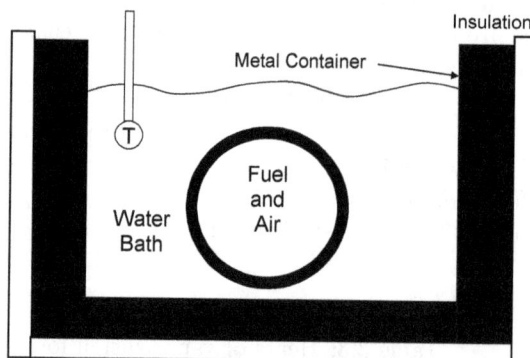

Figure 3.2: Schematic figure of a bomb calorimeter.

When the mass of water and bomb material are significantly large, the temperature increase between the start and the completion of the combustion will be small, and in practice the process can essentially be regarded as isothermal. The process also occurs at constant volume, and by measuring the small temperature increase, once can determine the energy which has been released during the combustion process and transferred to the surrounding materials.

The heating value is, then, the amount of energy which has been released when a fuel is completely burned, and the combustion products have been cooled down to the original temperature of the fuel before combustion. Actually the experimental process is not completely isothermal, but with a knowledge of the composition of the combustion products corrections for small temperature variations can be readily made, and the heating value is valid for a truly isothermal process. It is also possible to make corrections using the first law for reacting systems, Equation (3.62).

By writing the first law of thermodynamics for a system consisting of only the gasses in the bomb, we obtain:

$$\delta Q - \delta W = U_2 - U_1 \tag{3.81}$$

Since the volume is constant and there is no other form of mechanical work, $\delta W = 0$. Therefore:

$$\delta Q = U_{prod,T_2} - U_{react,T_1} \tag{3.82}$$

Obviously the gasses will reach a higher temperature during the combustion, so energy must be removed from the gasses in order to cool them to the original temperature T_1 and so δQ is then less than zero. δQ is called the *heat of combustion*, and it has a negative value, since heat transfer from the system must occur to achieve constant temperature. One is often only interested in the magnitude of the heat of combustion, and so the absolute value of the heat of combustion is used. That is actually what is used as the *heating value*: the absolute value of the heat of combustion. To make the heat of combustion a property, it is given in units of energy per unit mass or mole of fuel.

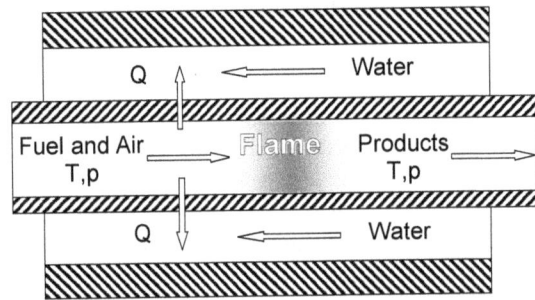

Figure 3.3: Schematic figure of a flow calorimeter.

There are different nomenclatures used for the heating value in the technical literature. In English language literature the terms HHV and LHV are often used with subscript v or p for *Higher Heating Value* or *Lower Heating Value* at constant volume or pressure respectively. In German literature one finds correspondingly H_o and H_u, for higher heating value (Ober Hiezwert) and lower heating value (Unter Heizwert). Other authors use q_c or q_{LHV}. In this text, the German notation will be used.

The second type of experiment to determine heating values can be conducted in an open system as shown in Figure 3.3. The system consists of a pipe or tube with a heat exchanger enclosing it and is called a flow calorimeter. A mixture of fuel and air flows in one end of the tube, and a combustion reaction occurs inside the tube. Any pressure drop in the tube is usually neglected. The temperature of the gasses leaving the tube is determined by the flow rate and temperature of the cooling water, as well as the effectiveness of the heat exchanger.

Similar to the experiment with constant volume, we wish to measure the energy which must be removed from the gasses to achieve a flow of exhaust products out of the tube at the temperature of the inlet gasses. This is possible, in principle, by regulating the temperature and flow of the cooling water. The heat transferred to the cooling water can then be determined by measuring the flow rate and temperature change of the cooling water. When the temperature of the gasses leaving the tube is equal to the temperature of the gasses entering the tube, the heat transferred to cooling water is only due to the chemical energy released by the combustion. By dividing the heat transfer rate (J/s) in this case by the flow rate of the fuel (kg/s), the heating value in J/kg-fuel is obtained. In practice, the temperature of the outlet gasses is measured, and a correction is made, since the outlet gasses are not usually at the inlet temperature. The correcting is done by using the first law for reacting systems (Equation (3.62)) to determine the heat of formation, which can then be converted to a heating value.

Since the system is open, one can write the following:

$$\dot{Q} = \dot{m} \left(h_{prod,T_2} - h_{react,T_1} \right) \tag{3.83}$$

where $\dot{m}$ is the total mass flow through the tube and h denotes the enthalpy per unit mass. As in the case for constant volume, if one divides rate of heat transfer, $\dot{Q}$ by the mass flow of the fuel, the heat of combustion is also obtained per unit mass. The heat of combustion in this case is also negative, and the heating value is the absolute value of the heat of combustion.

In principle, this experiment could also have been conducted in a closed system, where the pressure was maintained constant by placing a weight on top of a piston in a cylinder with a perfect, frictionless seal. In practice such a process would be extremely difficult to conduct, but one can write the 1st law for such a process anyway. Since $\delta W = p \cdot \Delta V$ for constant pressure:

$$\delta Q = H_{prod,T_2} - H_{react,T_1} \tag{3.84}$$

The difference between Equations (3.83) and (3.84) is that Equation (3.83) is based on an open system where one uses the mass flow (kg/s), while Equation (3.84) is based on a closed system where there is now a defined mass, m. When based on one kilogram of fuel, the results (heating value) are identical.

Example 1

1 gram of evaporated octane, C_8H_{18}, is mixed with air which is saturated with water vapor, and placed in an iron bomb, which weighs 2 kg. The bomb is placed in a perfectly insulated water bath, which contains 6 liters of water. The temperature of the entire system is initially 25°C. Then the fuel air mixture is ignited. When the system comes to thermal equilibrium, the temperature is measured to be 26.86°C.

a. Calculate the heating value. b. Is it the higher or the lower heating value?

Solution: The mass of air may be assumed to be unimportant, since the density of the air is about 1/1000th than that of the water or steel. The combustion energy released is transferred to the bomb and the water, where it can be observed through an increase in the temperature. The first law of thermodynamics for the system consisting of the bomb and the water gives:

$$\delta Q = \Delta U_{iron\ and\ water} \tag{3.85}$$
$$= \Delta U_{iron} + \Delta U_{water} \tag{3.86}$$
$$= (m_{iron}c_{v,iron} + m_{water}c_{v,water})\,\Delta T \tag{3.87}$$

In handbooks one can find: $c_{v,iron} = 0,47$ kJ/kg-K $c_{v,water} = 4,18$ kJ/kg-K

$$\delta Q = (m_{iron}c_{v,iron} + m_{water}c_{v,water})\,\Delta T$$
$$\delta Q = (2 \cdot 0.47 + 6 \cdot 4.18)\,1.86$$
$$= 48.4 kJ \tag{3.88}$$

Q for the system consisting of water and iron is positive, because heat is added to the system. For the system consisting of the combustion gasses, Q is negative. Because the temperature change is very small, the heating value obtained can be assumed to be the value at 25°C.

The heating value is found by calculation of the energy on the basis of 1 kg fuel:

$$H_o = \frac{48.4}{0.001} = 48400 kJ/kg$$

b. Since the air was originally saturated with water vapor, it cannot hold more water vapor, and the amount of water that was produced during the combustion condenses before the final temperature is reached. Therefore, the measurement gives the higher heating value for constant volume.

Example 2

A flow calorimeter is water cooled and has an air flow of 1.0 g/s. Methane is added to the air such that the excess air ratio, $\lambda = 2.00$. Complete combustion occurs inside of the calorimeter. The inlet state of the gasses is the same as the exit state: 25°C and 1.0 bar. The entrance and exit temperatures of the cooling water are 20°C and 24°C, and the flow rate of the cooling water is = 363 liter/hour. Under the assumption that the water generated during combustion remains evaporated, calculate the heating value for methane at constant pressure.

Solution:

For stoichiometric conditions:
$$CH_4 + 2 \cdot (O_2 + 3.76 \cdot N_2) \rightarrow CO_2 + 2H_2O + 7.52N_2$$
For $\lambda = 2.0$:

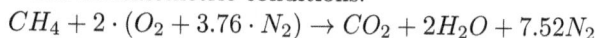

$$CH_4 + 4 \cdot (O_2 + 3.76 \cdot N_2) \rightarrow CO_2 + 2H_2O + 15.042N_2 + 2O_2$$

The Fuel Air ratio, FA:

$$FA = \frac{12 + 4}{4 \cdot 4.76 \cdot 29.85} = 0.029$$

The fuel mass flow:

$$\dot{m}_f = FA \cdot \dot{m}_a = 0.029 \cdot 1.00 = 0.029 g/s$$

The heat transfer to the water:

$$\dot{Q} = \dot{m}_{H_2O} \cdot c_{p,H_2O} \cdot \Delta T_{H2O}$$
$$= \frac{363l/h \cdot 1kg/l \cdot 4.18kJ/kg - K \cdot (24 - 20)}{3600s/h} = 1.45kW$$

The heating value is found by calculating the heat transfer based of 1 kg of fuel:

$$H_{u,p} = \frac{-\dot{Q}}{\dot{m}_f} = \frac{1.45kJ/s}{.029 \cdot 10^{-3}kg/s} = 50500kJ/kg \tag{3.89}$$

It is possible to define several heating values. The four most common are the higher and lower heating values with constant pressure and volume respectively, all for a constant temperature. If one of the heating values is known as well as the composition of the fuel, all the others can be calculated by use of the fuel composition and the evaporation properties of water.

For an engine viewed as an overall system, with intake and exhaust flows at atmospheric pressure, the lower heating value at constant pressure is normally used for the thermal efficiency. This is because it is assumed that the water remains vaporized when leaving the exhaust pipe, and the system seen from this viewpoint is an open flow system.

Consider for example octane from Example 1, where the higher heating value for constant volume was calculated. We can convert to constant pressure conditions by making the following calculation : For constant volume:

$$Q_{v,T} = U_{prod,T} - U_{react,T} \tag{3.90}$$

While for constant pressure:

$$Q_{p,T} = H_{prod,T} - H_{react,T} \tag{3.91}$$

Subtracting Equation (3.90) from Equation (3.91):

$$Q_{p,T} - Q_{v,T} = (H_{prod,T} - U_{prod,T}) - (H_{react,T} - U_{react,T}) \tag{3.92}$$

But from the definition of enthalpy, Equation (3.76) and the ideal gas law:

$$H = U - pV = U - n_{gasses}R_oT \tag{3.93}$$

Then we can rewrite Equation (3.93):

$$Q_{p,T} - Q_{v,T} = (n_{prod} - n_{react})_{gas} R_oT \tag{3.94}$$

Where n_{prod} = kmol of combustion products in the gas phase, and n_{react} = kmol reactants (fuel air mixture) in the gas phase.

Example 3

On the basis of the results in Example 1, calculate the higher heating value of octane with constant pressure.

Solution: For octane the stoichiometric equation is:

$$C_8H_{18} + 12.5\left(O_2 + 3.76N_2\right) \rightarrow 8CO_2 + 9H_2O + 47N_2$$

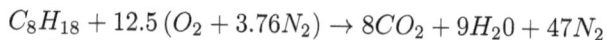

From Example 1 we have, that $H_{o,v} = 48400$ kJ/kg-octane at $25°$ C. When we calculate n_{prod}-n_{react}, we must consider whether we are dealing with the higher or lower heating value. In this example, we have the higher heating value, which means that the water in the products is condensed and so it is assumed that its volume is negligible compared to that of the gasses. There is typically a factor of about 1000 difference between the specific volume of a liquid and its vapor.

Then we have:

$$
\begin{aligned}
Q_{p,T} &= Q_{v,T} + \left(n_{prod} - n_{react}\right)_{gas} R_o T \\
&= -48400\frac{kJ}{kg} + \frac{(8-1-12.5)\,kmol}{kmol C_8H_{18}} \cdot \frac{8.314\,kJ}{kmol-K} 298.15K \frac{1}{114.23}\frac{kmol}{kg C_8H_{18}} \\
&= H_{o,p} = 48520\frac{kJ}{kg C_8 H_{18}}
\end{aligned}
\tag{3.95}
$$

Notice that the water is not included in the number of moles of gaseous products, since it is condensed and N_2 not included in δn, since it doesn't react.

The difference between the higher heating value and the lower heating value is simply the heat of vaporization of the amount of water which is produced per kg fuel.

Example 4

From the results in Example 1, calculate the lower heating value of octane for constant pressure.

If we convert the higher heating value of octane for constant pressure to the lower, we use the following procedure:

$$H_{o,p} = H_{u,p} + n_{H_2O} h_{fg,H_2O} \tag{3.96}$$

Where n_{H_2O} is the number of kmol water produced per kilogram fuel. Then:

$$
\begin{aligned}
H_{u,p} &= \frac{48400 \cdot kJ}{kg C_8 H_{18}} - \frac{9 \cdot kmol H_2 0}{kmol C_8 H_{18}}\frac{18\,kmol H_2 O}{kmol C_8 H_{18}}\frac{2442 \cdot kJ}{kg \cdot H_2 O}\frac{kmol C8 H_{18}}{114.23 kg} \\
&= 44900\frac{kJ}{kg C_8 H_{18}}
\end{aligned}
$$

For the heating values at constant volume the procedure is the same, except that one uses u_{fg} instead of h_{fg}. Heating values for selected hydrocarbons can be found in Table 3.4.

The heating values are related to the heats of formation and can be calculated from them and the stoichiometric chemical reaction. This can be seen from Equation (3.84):

$$
\begin{aligned}
H_{u,p}\left(T_1\right) &= H_{react,T_1} - H_{prod,T_1} \\
&= \sum_{i=1}^{react} n_i \left(h_{f,i}^o + \int_{T_{ref}}^{T_1} c_{p,i} dT\right) - \sum_{i=1}^{prod} n_i \left(h_{f,i}^o + \int_{T_{ref}}^{T_1} c_{p,i} dT\right)
\end{aligned}
\tag{3.97}
$$

For $T_1 = T_{ref} = 298.15K$ for example,

$$H_{u,p}\left(T_{ref}\right) = \sum_{i=1}^{react} n_i \cdot h_{f,i}^o - \sum_{i=1}^{prod} n_i \cdot h_{f,i}^o \tag{3.98}$$

As an example, consider octane, C_8H_{18}. The stoichiometric equation is:

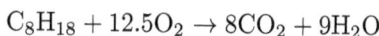

$$C_8H_{18} + 12.5O_2 \rightarrow 8CO_2 + 9H_2O$$

Then from Equation (3.98) using the values found in the table of thermodynamic properties in Sec-

Table 3.4: Heating values for constant pressure combustion for different hydrocarbons, unless noted, fuel assumed evaporated before.

Hydrocarbon	Chemical	Mole mass formula kg/kmol	Higher Heating Value constant pressure kJ/kg	Lower Heating Value constant pressure kJ/kg
Paraffins	-	-	-	-
Methane	CH_4	16.043	55,496	50,026
Ethane	C_2H_6	30.070	51,899	47,511
Propane	C_3H_8	44.097	50,323	46,333
n-Butane	C_4H_{10}	58.124	49,503	45,719
n-Pentane	C_5H_{12}	72.151	49,004	45,346
n-Hexane	C_6H_{14}	86.178	48,676	45,103
n-Heptane	C_7H_{16}	100.205	48,433	44,922
n-Heptane -liq	C_7H_{16}	100.205	48,069	44,557
n-Octane	C_8H_{18}	114.232	48,249	44,784
n-Octane -liq	C_8H_{18}	114.232	47,886	44,421
i-Octane	C_8H_{18}	114.232	48,116	44,650
i-Octane -liq	C_8H_{18}	114.232	47,808	44,343
Decane	$C_{10}H_{22}$	128.259	48,000	44,598
Dodecane	$C_{12}H_{26}$	143.286	47,824	44,467
Olefins				
Ethylene	C_2H_4	28.053	50,303	47,167
Propylene	C_3H_6	42.081	48,905	45,769
1-Butene	C_4H_8	56.108	48,421	45,284
1-Pentene	C_5H_{10}	70.135	48,126	44,990
1-Hexene	C_6H_{12}	84.160	47,929	44,794
1-Heptene	C_7H_{14}	98.189	47,790	44,655
1-Octene	C_8H_{16}	112.216	47,685	44,549
Nonene	C_9H_{18}	126.243	47,611	44,475
Decene	$C_{10}H_{20}$	140.270	47,546	44,410
Aromatics				
Benzene	C_6H_6	78.114	42,264	40,574
Methylbenzene	C_7H_8	92.141	42,847	40,938
Ethylbenzene	C_8H_{10}	106.168	43,387	41,316
Propylbenzene	C_9H_{12}	120.195	43,799	41,603
Butylbenzene	$C_{10}H_{14}$	134.222	44,122	41,828

tion 17.1:

$$H_{u,p}(T_1) = H_{react,T_1} - H_{prod,T_1}$$
$$= h^o_{f,C_8H_{18}} + 12.5h^o_{f,O_2} - 8h^o_{f,CO_2} - 9h^o_{f,H_2O}$$
$$= -208741 + 12.5 \cdot 0 - 8 \cdot (-393492) - 9 \cdot (-241815)$$
$$= 5115530 \frac{kJ}{kmolC_8H_{18}} \frac{kmolC_8H_{18}}{114.23kg} = 44783 \frac{kJ}{kgC_8H_{18}}$$

3.4 Flame Temperature

3.4.1 Introduction

In the previous sections, some simple thermodynamic concepts and a description of heating values were presented. This section expands on this and shows how to calculate flame temperatures. The flame temperature has a major influence on the emissions of NO_x from various combustion systems, and so determination and control of the flame temperature is important.

In the previous chapter, the first law of thermodynamics was discussed for systems at constant pressure or volume. The heating values were based on combustion processes in which the temperatures before and after combustion are equal.

Opposed to the heating value process, the flame temperature is the result of an adiabatic process, in which none of the energy released from the combustion is removed by heat transfer. The temperature arising from this situation is called the *adiabatic flame temperature*. As would be expected from the discussion of heating values, there will be different flame temperatures, depending on whether the process occurs at constant volume or constant pressure.

3.4.2 Constant Volume Flames

As a starting point, consider a constant volume container. As we did in the previous section, we will start with a mixture of fuel and air at a known condition and ignite the fuel. Contrary to the assumption made for the heating value calculation used in the previous chapter, let us now assume that the container is perfectly insulated, such that there can be no energy loss from the gasses through heat transfer. The constant volume precludes mechanical work. In this situation, the first law for the gasses in the bomb is:

$$\delta Q - \delta W = \Delta U = U_{prod,T_2} - U_{react,T_1} = 0 \tag{3.99}$$

Equation (3.99) shows that there is no exchange of energy between the interior of the bomb and the surroundings, but there is a conversion of the chemical energy in the fuel and the internal or thermal energy of the combustion products. What is sought is T_2, the temperature after the combustion has occurred, that is, the adiabatic flame temperature, in this instance for constant volume.

Equation (3.99) can be rewritten:

$$U_{prod,T_2} = U_{react,T_1} = \sum_{i=1}^{N_{prod}} n_i \cdot u_i(T_2) = \sum_{i=1}^{N_{react}} n_i \cdot u_i(T_1) \tag{3.100}$$

In systems without chemical reactions, one can calculate a change in internal energy by integrating the specific heat over a given temperature interval. We can see from Equations (3.99) and (3.100) that this is not the case for the flame temperature calculation. We know that for an adiabatic combustion process at constant volume, we have a constant energy, but also a large increase in the temperature. It is necessary to separate the change in internal energy due to the chemical reaction from the change due to the temperature. This is one of the reasons for defining the heating value as a change in energy (or enthalpy) for a constant temperature, such that the only energy changes involved are due to chemical reactions.

From the definition of the heat of combustion and the heating values at constant volume (see Equation (3.82) in the previous section) it was found that:

$$\delta Q = -H_{u,v,T} = U_{prod,T} - U_{react,T} \tag{3.101}$$

If it is assumed that the temperature, T, in Equation (3.101) is equal to T_1 and combining Equations (3.99) and (3.101), we obtain:

$$H_{u,v}(T_1) = U_{prod,T_2} - U_{prod,T_1} \tag{3.102}$$

In Equation (3.102), the heating value is written as a function of the temperature, in this case T_1, as a reminder that the heating value is a (weak) function of the temperature, and that we have assumed that the heating value at the start temperature, T_1, is known.

It can be seen from Equation (3.102) that the determination of the flame temperature, T_2, consists of finding a value of T_2 such that Equation (3.102) is satisfied. In Equation (3.102) there is only one chemical composition, which is that of the combustion products. Therefore, we can calculate the change in the internal energy of the combustion products without a change in composition, and use of the integral of $c_v dT$ between T_1 and T_2 is acceptable. It is assumed the composition is fixed, *i.e.* equilibrium effects are neglected in this case.

This can be more easily seen if we rewrite Equation (3.102):

$$H_{u,v}(T_1) = U_{prod,T_2} - U_{prod,T_1} = \sum_{i=1}^{n_{prod}} n_i \int_{T_1}^{T_2} c_{v,i} dT \tag{3.103}$$

where: n_i = number of kmol of product i per kmol fuel, $c_{v,i}$ = specific heat for product i, n_{prod} = number of species in the combustion products.

Equation (3.103) contains a heating value (see Table 3.4), stoichiometric coefficients, specific heats, a known temperature, T_1 and an unknown temperature, T_2 and can therefore be solved for the unknown temperature, T_2. In order to demonstrate the principle, and the importance of some variables, a simplified case will be shown first. Assuming the specific heats are constant, Equation (3.103) can be written:

$$T_2 = T_1 + \frac{H_{u,v}(T_1)}{\sum_{i=1}^{n_{prod}} n_i c_{v,i}} \tag{3.104}$$

Actually, the specific heats are not constant over the temperature range T_1 to T_2 for typical combustion systems. This makes the calculation slightly more complicated, but still straight forward. The procedure is shown in the following.

Equation (3.103) can be written:

$$H_{u,v}(T_1) = \sum_{i=1}^{n_{prod}} n_i \int_{T_{ref}}^{T_2} c_{v,i} dT - \sum_{i=1}^{n_{prod}} n_i \int_{T_{ref}}^{T_1} c_{v,i} dT \tag{3.105}$$

Since the specific heats are known functions of the temperature, it is easy to construct tables which correspond to the integrals found in Equation (3.105). This is done by selecting a reference temperature and calculating the integrals from T_{ref} to T for a range of T which can be expected to be observed in the combustion products. These results are given in the form of tables in Section 17.1 for compounds found the ideal combustion products for carbon and hydrogen based fuels. The table can be replaced by a computer program or subroutine, which calculates the values of the integrals, using input of mole fractions and temperatures.

In order to solve Equation (3.105) it is necessary to guess a flame temperature, T_2, look up (calculate) the integrals, calculate the right hand side of Equation (3.105), and see if it is satisfied. If not, a new value of T_2 is selected and the process continued until Equation (3.105) is satisfied. The temperature at which this occurs is the adiabatic flame temperature. With a computer program to calculate the properties, agreement can be achieved to any desired degree of accuracy. If tables are used, interpolation between the two bracketing temperatures is used.

The procedures are illustrated in the following examples:

Example 1

Calculate the adiabatic flame temperature for a mixture of propane and air with an excess air ratio, λ of 1.5. The combustion occurs at constant volume from a temperature of $25°C$ and a pressure of 1 bar.

Assume that the specific heats are constant and equal to the values at 25°C. Calculate the pressure after combustion.

Solution: The first step is to write the stoichiometric relationship for propane:

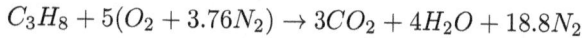

$C_3H_8 + 5(O_2 + 3.76N_2) \rightarrow 3CO_2 + 4H_2O + 18.8N_2$

For $\lambda = 1.5$,

$C_3H_8 + 7.5(O_2 + 3.76N_2) \rightarrow 3CO_2 + 4H_2O + 2.5O_2 + 28.2N_2$

Equation (3.104) is used, since it is assumed that the specific heats are constant. The specific heat values are found in the tables in Section 17.1 and listed in Table 3.5 for 25°C. An improved approximation would be to use an average temperature.

Table 3.5: The number of moles and the specific heats for the flame temperature example with constant volume and specific heats

Specie	n_i - kmol/kmol	$c_{v,i}$ - kJ/kmol
CO_2	3.0	28.82
H_2O	4.0	25.27
O_2	2.5	21.06
N_2	28.2	20.81

Heating values are found in Table 3.4. In the calculation of the flame temperature, the lower heating value must be used, since the water formed in the combustion process will always be in the gas phase after the combustion. Since the initial temperature = 25°C, the heating values for Table 3.4 can be used. But since these are values for constant pressure, they must be converted to constant volume.

Using Equation (3.95):

$$Q_{v,T} = -H_{u,v} = Q_{p,T} - (n_{prod} - n_{react})_{gas} R_o T$$
$$= -H_{u,p} - (n_{prod} - n_{react})_{gas} R_o T$$

It must be remembered that the heat of combustion is negative. This is important when converting the heats of combustion by the use of Equation (3.95).

Inserting the values from the tables and the stoichiometric equation (Nitrogen is not included in the mole change, since it does not react):

$$Q_{v,T} = -46335 \frac{kJ}{kgC_3H_8} 44.1 \frac{kgC_3H_8}{kmolC_3H_8}$$
$$- (3 + 4 + 2.5 - 1 - 7.5) \frac{kmol}{kmolC_3H_8} 8.314 \frac{kJ}{kmol - K} 298.15K$$
$$= 2,045,852 \frac{kJ}{kmolC_3H_8} \frac{1}{44.1} \frac{kmolC_3H_8}{kgC_3H_8}$$
$$= 46391 \frac{kJ}{kgC_3H_8}$$

From Equation (3.104):

$$T_2 = 298.15 + \frac{2045852 \frac{kJ}{kmolC_3H_8}}{(3 \cdot 28.82 + 4 \cdot 25.27 + 2.5 \cdot 21.06 + 28.2 \cdot 20.81) \frac{kmol}{kmolC_3H_8} \frac{kJ}{kmol-K}}$$
$$= 2772K$$

The pressure is found through the use of the ideal gas law. Since the volume is constant:

$$V = \frac{n_1 R_o T_1}{p_1} = \frac{n_2 R_o T_2}{p_2} \tag{3.106}$$

and:

$$p_2 = p_1 \frac{n_2 T_2}{n_1 T_1} = 1bar \frac{(3 + 4 + 2.5 + 28.2)}{(1 + 7.5 \cdot 4.76)} \frac{2772}{298.15}$$
$$= 9.53 \ bar$$

It can be observed that it is largely N_2 that determines the flame temperature. The denominator in calculation of T_2 has the value of 829 kJ/kmol C_3H_8, of which the contribution of the N_2 alone is 580 kJ/kmol C_3H_8. This means that 70 % of the energy released is used to heat the N_2. The N_2 itself does not participate in the reaction, but is present because the fuel is burned in air which contains 79 volume % N_2.

It can also be seen that the extra air helps to lower the flame temperature. The extra O_2 (2.5 kmol/kmol C_3H_8) and associated N_2 (9.4 kmol/kmol C3H8) contributed with 30 % of the value of the denominator, and reduces the temperature relative to a stoichiometric combustion. (See Problem 1).

As will be discussed more thoroughly later, high flame temperatures contribute to the production of nitric oxide (NO) in combustion systems. Already on the basis of a simplified calculation of the flame temperature, we can see possibilities for the reduction of the flame temperature and thereby the reduction of the production of NO.

One method is to use lean mixtures in the combustion systems. The method is used in some practical situations, and in some systems, lean combustion is necessary to keep the temperature low enough that some mechanical components of the system are not destroyed. The gas turbine engine is a classic example of this. But one must be aware that if the mixture is only a little lean, that is if λ is about $1.1 \lesssim \lambda \lesssim 1.3$, the flame temperature will still be high, and the extra air present will actually more than compensate for a slightly lower temperature with respect to the production of NO. The emissions of NO are actually highest for mixture strengths in this region.

Another related method used to reduce the flame temperature is to use an inert gas to dilute the fuel air mixture. Two examples of this are water injection and exhaust gas recirculation (EGR). These substances can be added to the intake air in a combustion system. Neither of these substances contain chemical energy, and can absorb the chemical energy liberated during the combustion and reduce the flame temperature. From Equation (3.104) it is clear that it is advantageous to use a substance with a large specific heat. CO_2 and H_2O are, for example much more effective than He or Ar [21].

Recirculation of exhaust gasses is a very common method for the reduction of NO_x emissions from both spark ignition and diesel engines, and a recirculation of up to 30 % of the gasses is not uncommon.

In Example 1 we assumed that the specific heats were constant with values at 25°C. From the result (2772 K) and the values in Section 17.1, it is obvious that this is not a very good assumption. In the temperature range here, the value of c_v increases by between 40-85 % depending on the compound. In addition, the value of c_v is non-linear, which also makes it difficult to choose an appropriate average value of c_v. By using Equation (3.102) we can make a much more accurate calculation of the flame temperature. The method is simple enough to make the use of an average value of c_v unnecessary, and is shown in Example 2:

Example 2

Solve Example 1, but use specific heats which are functions of the temperature.

Solution: By looking at Equation (3.102) it can be seen that the heating value is known, and that the integrals from T_{ref} to T_1 are also known. In this case, $T_{ref} = T_1 = 298.15$ K, and the integrals for T_1 are all equal to zero.

To be able to solve Equation (3.105) we must:

1. Guess a flame temperature. T_2

2. Look up the integrals in Section 17.1.

3. Calculate the right hand side of Equation (3.105) and compare with the left hand side.

4. Repeat 1 - 3 until Equation (3.105) is satisfied or T_2 is between 2 adjacent temperatures in the tables of Section 17.1.

5. Interpolate between these 2 temperatures.

Guess T_2: We know that 2727 K is too high, so we start with a guess of 2000 K. Then we have the values shown in the following table:

Specie	n_i - kmol/kmolC$_3$H$_8$	$u(T_2) - u(T_{ref})$ kJ/kmol
CO$_2$	3.0	77284
H$_2$O	4.0	58891
O$_2$	2.5	45049
N$_2$	28.2	41965
$\sum n_i \cdot u_i$	-	1763987

Since this value is lower than $H_{u,v} = 2.045852 \cdot 10^6$ we must try again with a higher temperature.

The results for some different temperatures are shown in the following table. When we find two adjacent temperatures between which the temperature lies, we can interpolate and find T_2.

Specie	n_i - kmol/kmol C$_3$H$_8$	$u(2200) - u(298)$ kJ/kmol	$u(2300) - u(298)$ kJ/kmol
CO$_2$	3.0	87741	93005
H$_2$O	4.0	67717	72227
O$_2$	2.5	50990	53996
N$_2$	28.2	47517	50315
$\sum n_i \cdot u_i$	-	2,002,363	2,122,698

By interpolation between 2200 and 2300 K we find that the temperature and pressure are:

$$T_2 = 2236K$$

$$p_2 = p_1 \frac{n_2 T_2}{n_1 T_1} = 1bar \frac{(3+4+2.5+28.2)}{(1+7.5 \cdot 4.76)} \frac{2240}{298.15}$$
$$= 7.70 \ bar$$

A comparison with the result from Example 1 shows that the flame temperature is 536 K lower when the correct values for the specific heats are used.

Example 3

The above method is shown since the thermodynamic data are often tabulated in the form of the heats of formation and the sensible energy and enthalpy. If the enthalpies and internal energies are tabulated including the chemical reference state, as in Section 17.1, then the problem is even simpler. In this case, Equation (3.100) can be used directly.

An example is shown here to illustrate both the use of the method, and the effect of EGR on the flame temperture. Consider stoichiometric propane and the use of 20 molar % EGR from a temperature of 700K and pressure of 2000 kPa[2]. To find the temperature reduction, it is first necessary to find the temperature without EGR. For stoichiometric propane, the reaction equation with air is:

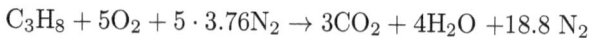

$$C_3H_8 + 5O_2 + 5 \cdot 3.76N_2 \rightarrow 3CO_2 + 4H_2O + 18.8 \ N_2$$

If 20% of the exhaust products is recirculated:

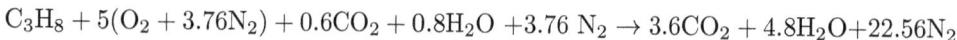

$$C_3H_8 + 5(O_2 + 3.76N_2) + 0.6CO_2 + 0.8H_2O + 3.76 \ N_2 \rightarrow 3.6CO_2 + 4.8H_2O + 22.56N_2$$

For no recirculation, the internal energy of the reactants is:

Specie	n_i - kmol/kmol-C$_3$H$_8$	$u_i(700)$ kJ/kmol
C$_3$H$_8$	1.0	-61662
O$_2$	5	6680
N$_2$	18.8	6116
$\sum n_i \cdot u_i$	-	86719

And for the products and the bracketing temperatures:

Specie	n_i - kmol/kmol C$_3$H$_8$	$u(3100)$ kJ/kmol	$u(3200)$ kJ/kmol
CO$_2$	3.0	-260236	-254835
H$_2$O	4.0	-134238	-129339
N$_2$	18.8	70641	73522
$\sum n_i \cdot u_i$	-	10391	100353

[2]In engines, EGR is usually on a mass basis. Molar (volume) % is used here to simplify the calculation.

By interpolation, $T_2 = 3185$K, and $p_2 = 9467$ kPa.

For 20% recirculation, the internal energy of the reactants is:

Specie	n_i - kmol/kmol-C_3H_8	$u_i(700)$ kJ/kmol
C_3H_8	1.0	-61662
O_2	5	6680
N_2	22.56	6116
CO_2	0.6	-381557
H_2O	0.8	-233444
$\sum n_i \cdot u_i$	-	-305974

And for the products and the bracketing temperatures:

Specie	n_i - kmol/kmol C_3H_8	$u(2800)$ kJ/kmol	$u(2900)$ kJ/kmol
CO_2	3.6	-276369	-271004
H_2O	4.8	-148736	-143938
N_2	22.56	62038	648990
$\sum n_i \cdot u_i$	-	-309284	-202395

By interpolation, $T_2 = 2803$K, and $p_2 = 8276$ kPa.

The use of exhaust gas recirculation in this example gives a temperature reduction of 382K. This gives the potential for a significant reduction in NO_x emissions.

The use of tables in this form is convenient when the initial temperature is not equal to the reference temperature, a common situation in engines. It should be noted, however, that the tables are for gaseous fuel, and for water as vapor. Conversion to other conditions require the use of the principles described previously, but will not be discussed here.

3.4.3 Constant Pressure Flames

In the preceding, we have only been concerned with constant volume flames. There are of course, many combustion processes which occur at constant pressure. In fact, most practical combustion systems operate with constant pressure flames in open systems. At any instant during SI engine premixed flame propagation the flame burns to give to local constant pressure adiabatic flame temperature at those conditions. The burned gasses formed are subject to compression as cylinder pressure rises and falls. In the special case where the volume remains constant during combustion, and the burned gasses not mixed, there would be a temperature gradient across the combustion chamber. The final temperature of the burned gasses, if mixed, would correspond to the adiabatic flame temperature at constant volume. This same temperature (mass averaged temperature) would be found by using the unmixed final pressure and the ideal gas law.

For a closed system in which combustion occurs at constant pressure, the system performs a quantity of work:

$$W = \int p\,dV = p_2 V_2 - p_1 V_1 = n_2 R_o T_2 - n_1 R_o T_1 \tag{3.107}$$

From Equations (3.99) and (3.107) as well as the definition of enthalpy ($H \equiv U + pV$) the first law of thermodynamics for an adiabatic, constant pressure combustion system is obtained.

$$H_{prod,T_2} = H_{react,T_1} = \sum_{i=1}^{N_{prod}} n_i \cdot h_i(T_2) = \sum_{i=1}^{N_{react}} n_i \cdot h_i(T_1) \tag{3.108}$$

Equation (3.108) is valid for a closed, adiabatic system with constant pressure, and for an open adiabatic system in steady-state without mechanical work. In the latter case, the enthalpy is usually written in terms of the amount of mass (or moles) flowing through the system.

From the definition of the heating values we have that:

$$Q_{p,T} = -H_{u,p}(T) = H_{prod,T} - H_{react,T} \tag{3.109}$$

Equation (3.109) differs from Equation (3.82) in that the heating value is for a constant pressure combustion and so the enthalpy is involved instead of the internal energy. As was the case with the development of Equation (3.102), Equation (3.108) can be combined with the definition of the heating value in Equation (3.109). Again an equation is obtained which contains only the heating value of the relevant fuel and a change in the enthalpy of the combustion products between the initial temperature and the flame temperature, T_2:

$$H_{u,p}(T_1) = \sum_{i=1}^{n_{prod}} n_i \int_{T_{ref}}^{T_2} c_{p,i} dT - \sum_{i=1}^{n_{prod}} n_i \int_{T_{ref}}^{T_1} c_{p,i} dT \tag{3.110}$$

Equation (3.110) corresponds to Equation (3.105), but Equation (3.110) is valid for a constant pressure combustion process instead of a constant volume process.

The method of solving Equation (3.110) is the same as shown in Example 2. Example 4 demonstrates the difference in combustion temperature for constant pressure combustion as compared to constant volume.

Example 4

Calculate the adiabatic flame temperature for a mixture of propane and air for an excess air factor, λ of 1.5. The combustion occurs with constant pressure from an initial temperature of 25°C with a pressure of 1 bar.

Since the initial temperature is 25°C, the lower heating value from Table 3.4 in the previous chapter can be used with Equation (3.110). It is more convenient to convert it to mole units by multiplying by the molecular weight of C_3H_8.

$$H_{u,p} = 44.097 \frac{kg C_3 H_8}{kmol\ C_3 H_8} \cdot 46333 \frac{kJ}{kg C_3 H_8} = 2043146 kJ/kmol\ C_3 H_8$$

The results for the bracketing temperatures with 1 kmole of fuel are:

Specie	n_i - kmol	$h(1800) - h(298)$ kJ/kmol	$h(1900) - h(298)$ kJ/kmol
CO_2	3.0	79429	85415
H_2O	4.0	62843	67902
O_2	2.5	51690	55433
N_2	28.2	48976	52545
$\sum$	-	2,000,007	2,148,205

By interpolation: $T_2 = 1829 K$

The result shows that the flame temperature for a constant pressure combustion process is considerably lower than for a constant volume process. In Examples 2 and 4 the difference is 407 K. The difference is dependent on λ and the initial temperature, but for the same λ and initial temperature, the flame temperature for constant pressure is always lower than for constant volume. The reason is that the hot gasses formed during combustion will expand if they are not restrained by a container. The expansion requires that the gasses do work on the surroundings, and this work reduces the amount of energy remaining in the gasses after combustion and expansion. This is related to the difference in the efficiencies of the Otto and Diesel cycles as discussed in Section 2.1.

From the discussions up to this point concerning NO_x formation, one would anticipate that combustion with constant pressure will give lower NO_x emissions. This is correct if everything else is the same. But in some situations, especially in combustion engines, combustion which approaches constant pressure yields a lower efficiency than combustion which occurs closer to constant volume. This means that a compromise between low NO_x emissions and high efficiency is often necessary,

In most cases in an engine, the gasses are compressed before combustion, and so the temperature at the start of combustion is not normally equal to the temperature for which the heating value was determined, or is available. The heating value is a function of temperature, and so in this case, and it must be determined for the temperature at the start of combustion, though the heating value is not a strong function of temperature. The procedure for calculating the flame temperature for the case where the initial temperature is not equal to that of the heating value is shown in the following.

The correct procedure is to use either Equation (3.100) or Equation (3.108). If the internal energy and entropy are available as calculated from both the reference temperature and composition, then these equations are solved as above. If only heats of formations and sensible internal energies and enthalpies available, then Equations (3.57) and (3.55) must be used in addition. Procedures are the same. That is, the final temperature is assumed until the equation balances, as above.

3.4.4 Equilibrium Flame Temperature

From Section 2.4, combustion products such as CO_2 and H_2O dissociate at high temperatures, the energy required for this reaction reducing the flame temperature. The solution for the temperature when chemical equilibrium in the products is considered is much more complicated than the above. It requires a calculation of the equilibrium composition of the combustion products as a function of temperature, pressure and mixture ratio. This must be combined with the first law of thermodynamics for the reacting system as in Equation (3.99), where the composition of the products depends on the temperature and pressure. The reactant composition is fixed. The solution involves an iterative process, where the initial energy of the reactants is calculated as in the above examples, but the temperature and pressure of the combustion products must be estimated and the equilibrium composition of the products determined before their energy can be determined.

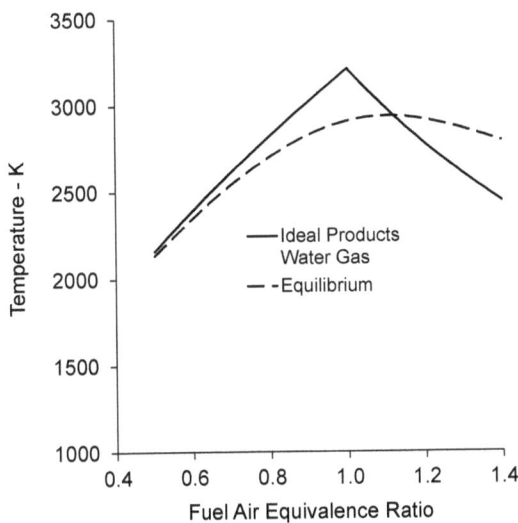

Figure 3.4: Constant volume flame temperature as a function of fuel-air equivalence ratio for n-octane for full chemical equilibrium and ideal combustion products. For rich mixtures, the ideal products are the water gas equilibrium products at 1750K.

The equilibrium flame temperature calculation requires determination of the equilibrium composition in combination with the energy balance as shown in previous sections. The initial energy is calculated, a flame temperature estimated and the energy of the products at that temperature compared to the initial value. The final temperature is then determined by iteration, as with the known composition. The complication here is that the composition is now a function of the flame temperature.

Figure 3.4 shows the flame temperature for n-octane as a function of equivalence ratio for two different assumptions. The first is that of ideal products of combustion for lean mixtures combined with the water gas assumption for rich products at 1750K. The water gas constant can also be calculated at the combustion temperature, but the difference is at most a few degrees, and the calculation more difficult. The second assumption is that of full chemical equilibrium. Both calculations use temperature dependent specific heats. For lean mixtures, the difference is small until the mixture becomes richer than about $\phi \approx 0.70$. At stoichiometric conditions, the effect of chemical equilibrium is to lower the temperature about 300K. Pressures are reduced similarly. While the effect on performance calculations is not major, such a temperature difference will have a major impact on the calculation of NO emissions, and in this case, equilibrium conditions must be used, as shown in Section 6.2.

For rich conditions, the equilibrium flame temperature remains nearly constant for the region of $1.0 < \phi < 1.4$, with a maximum occurring at $\phi \approx 1.15$. The water gas assumption results in a flame temperature that decreases immediately as soon as $\phi > 1.0$. At combustion temperatures, the complete chemical equilibrium is much more complicated than the water gas reaction, though the latter give a good approximation of exhaust composition.

The shape of the equilibrium flame temperature as a function of equivalence ratio is very important for SI engine performance. Since maximum temperatures and pressures are obtained with a slightly rich mixture, maximum engine output is also obtained here, both because the pressures are higher, and also a secondary effect of slightly higher flame speeds.

3.5 Spark Ignition Combustion

3.5.1 Flame Speed in Spark Ignition Engines

The flame in a spark ignition engine is characterized by the fact that it is a flame "wave" that propagates through a homogeneous mixture of fuel and air, a so-called *pre-mixed* flame. The speed at which the wave moves relative to the unburned mixture is called the *flame speed*, nearly always turbulent for engine conditions. To understand how the engine variables affect combustion rate, it is of interest to look at a classic pre-mixed flame. Various theories have been presented through the years, and today it is possible to calculate the behavior of laminar premixed flames with the inclusion of detailed chemistry, and multi-component diffusion. Here, however, one of the older theories will be mentioned, since the physical processes are basically those occurring in an actual, detailed flame, and the overall dominating mechanisms are the focus of interest. The case shown is that of a laminar premixed flame, with a single chemical reaction of the fuel, A (highly idealized) $A \to B$ with a reaction rate as a function of temperature:

$$Rate = [A]^n Z \exp\left(-\frac{E}{RT}\right) \tag{3.111}$$

where E, Z and n are constants, R is the universal gas constant, and $[A]$ is the concentration of species A, typically in moles per cc.

E varies widely from reaction to reaction, but an order of magnitude estimate is that $E/R \approx 20000$ for an overall combustion reaction. In this case, a temperature increase from 1000K to 1500K will give an increase in the reaction rate of a factor of 1800 according to (3.111). So temperature is a very important factor.

Assuming the reaction above, and simple bimolecular diffusion of A and B, a simplified function for the flame speed in a laminar flow situation can be obtained [9]:

$$S_l \propto \sqrt{\kappa \cdot \exp\frac{E}{RT_f}} \tag{3.112}$$

Where T_f is the adiabatic flame temperature and κ is the thermal conductivity.

There are two basic parameters of importance in equation (3.112) the first being the adiabatic flame temperature, and the second being the thermal conductivity. For simple situations, the Lewis number ($Le = \frac{\kappa}{c_p \rho D}$) is equal to one, then the thermal conductivity and mass diffusion coefficient, D are proportional. That is, it is assumed that the transport of mass and energy occur by the related mechanisms.

In a simple flame model then, there are two dominant factors: The flame temperature, and the mass/energy transport rate. Given this, it is expected that the flame temperature plays an important role in determining the flame speed. Flame speeds and their dependence on temperature, pressure and equivalence ratio can be found in the book by Turns [8]. This is shown to be the case in Figure 3.5, which is a plot of the laminar flame speed at atmospheric temperature and pressure and adiabatic equilibrium flame temperature of methane as functions of the fuel-air equivalence ratio. There is obviously a very close relationship.

How does the flame travel? First, the local gasses must be brought to a high enough temperature that the reactions proceed quickly. This is typically done with a spark, where the local temperatures are very high. Now there is a region of high temperature a very short distance from the unreacted gasses. Consequently, a heat transfer occurs to the unreacted gasses. This heats up the unreacted gasses

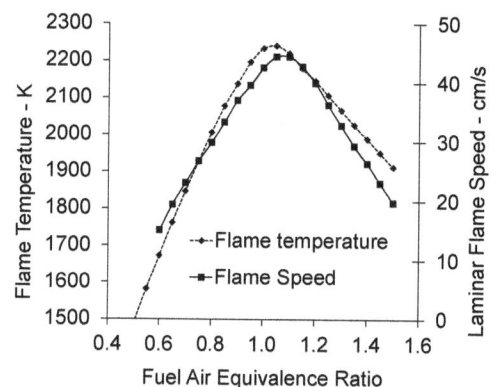

Figure 3.5: Laminar flame speed [8] and adiabatic flame temperature for methane air mixtures at a pressure of 100 kPa and an initial temperature of 298K.

close to the reaction zone, and as expected from Equation (3.111), the combustion reaction there occurs more rapidly. This reaction increases the chemical energy conversion rate, which further increases the temperature, and the reaction accelerates until the reactants are consumed and the adiabatic flame temperature is reached. The reaction wave propagates through the unburned mixture in this manner. In a real flame, in addition to the energy transfer through heat transfer, there is a mass transfer of reactive species from the flame zone into the unburned mixture, and these species also help to increase the reaction rate. These transport mechanisms are related at the molecular level, and similar physical processes are involved.

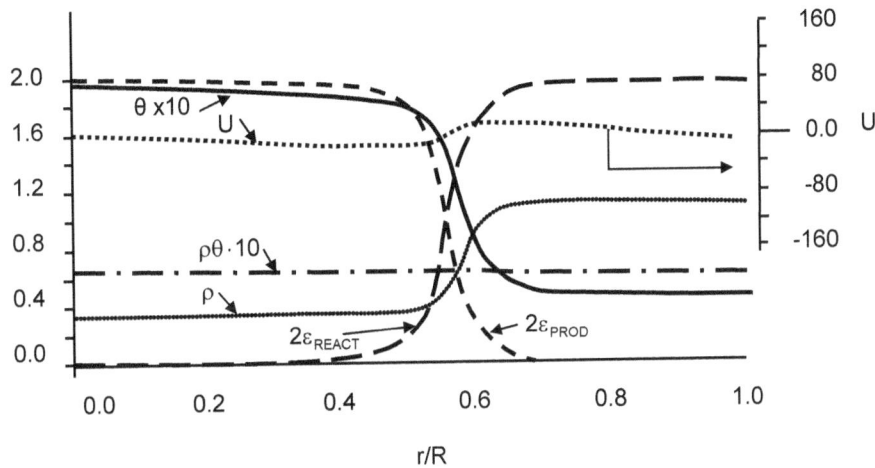

Figure 3.6: The structure of a theoretical premixed turbulent flame in a cylindrical container, showing non-dimensional temperature θ, density, ρ, velocity, U, pressure $(\rho\theta)$ and mole fraction, ε.

Figure 3.6 shows a sketch of the structure of a premixed flame propagating from the left to the right through a mixture, from the calculation of a turbulent flame. The turbulence is very much simplified and is simulated by increasing the value of the transport properties, but the essence of the structure of the flame is valid. In actual flames, the details of the combustion zone are quite complicated, and the surface of the flame zone very irregular. The flame zone here can be thought of as an average of these detailed processes.

To the left are the burned gasses at the adiabatic flame temperature. To the right are the unburned gasses, fuel and air. There is a large temperature gradient, from the flame temperature down to the unburned temperature. Similarly, there is a large gradient in the concentration of the gasses. The pre-mixed flame is driven by these gradients, as heat and reactive combustion products diffuse into the unburned gasses, causing them to react.

As was seen previously, the combustion energy per unit air mass is similar for most hydrocarbon and organic fuels. This gives comparable flame temperatures, and consequently, similar laminar flame speeds for most fuels encountered in engines. Table 3.6 shows the laminar flame speed for several different hydrocarbons for unburned mixtures at room temperature and pressure. As would be expected from Equation (3.112) and Figure 3.5, the two are closely related. In the simple terms of the thermal flame model, the flame temperature "drives" the flame, if it is high, the flame speed is high.

A noticeable exception in Table 3.6 is hydrogen, which has a flame speed about 5 times greater than the organic compounds. It should be recalled that (3.112) indicated that temperature *and* transport properties play a role in the flame speed. Hydrogen is a very light molecule, and its diffusion is much more rapid than the organic compounds, which gives the high flame speed.

The maximum flame speed at atmospheric temperature and pressure for typical fuel hydrocarbons is about 40 cm/s. Let's apply this speed to an engine, and see what happens. Take an engine with an 8.0 cm bore, and assume that the flame starts in the middle of the combustion chamber and spreads radially. Then the combustion time will be $\frac{4cm}{40cm/s} = 0.1s$. If the engine speed is taken to be 1800 rpm, then in

Table 3.6: A list of the maximum laminar flame speed at atmospheric temperature and pressure and the adiabatic flame temperature for a stoichiometric mixture of gaseous fuel and air[10], [8], author's calculations.

Compound	Flame speed - cm/s	Flame Temperature - K
Methane - CH_4	40	2226
Ethylene - C_2H_4	67	2369
Ethane - C_2H_6	43	2260
Propane - C_3H_8	44	2270
n-Hexane - C_6H_{14}	39	2273
Benzene - C_6H_6	40	2342
iso-Octane - C_8H_{18}	35	2275
Methanol - CH_3OH	48	2245
Hydrogen - H_2	260	2390

terms of engine crank angle the time it takes to burn the fuel is:

$$\delta\theta = 0.1s \cdot \frac{1800 rev/min \cdot 360°/rev}{60 s/min} = 1080° = 3 \; revolutions!$$

This seems totally unreasonable, the engine cannot run if it takes this long for the flame to pass. What are the reasons that a spark ignition engine actually can run, and at high speeds? First of all, engines run on a compression process where the unburned fuel and air are compressed before combustion. From Section 2.4 a temperature of 700 - 800 K at the time of combustion would be reasonable, with corresponding pressures on the order of 2000 kPa. This is considerably above the room temperature of Figure 3.5, and according to the correlations of [8] gives a substantial increase in flame speed. At 700 K, the laminar flame speed at stoichiometric mixtures increases to about 1.3 - 1.5 m/s. This gives a decrease in the combustion duration of about a factor of 4 from the previous example, and shortens the combustion duration to about 270 - 300 ° of crank angle rotation. This is better, but still too long.

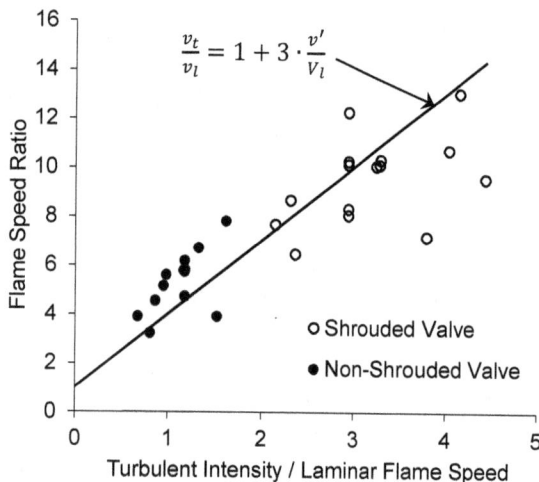

Figure 3.7: The ratio of the turbulent flame speed, v_t, to the laminar flame speed, v_l, in a CFR engine as a function of the turbulence intensity in the cylinder before combustion. Data of [22] modified with flame speeds from [8].

The other mechanism of importance is that of mass and temperature transport. Note that in the previous discussion, the flame speed was referred to as the laminar flame speed. In an engine, there is a lot of turbulence in the cylinder, primarily generated on the intake stroke. As is well known from basic heat and mass transfer, turbulence increases mass and energy transfer rates substantially, and thus we experience a further increase in the flame speed in an actual engine due to the turbulence, since this increases the speed of the transfer of mass and energy across the flame front. According to Lancaster, et. al. [13], the turbulent flame speeds for an engine running at 1800 rpm can be from 5 - 15 times that of the laminar flame speed, depending on engine design and other conditions. Assuming a value of 7, then the combustion duration is about 40 - 50 degrees, a more reasonable value.

The point of all of this is that there are two primary factors that determine the combustion rate in a spark ignition engine. The first is the flame temperature, determined primarily by the mixture ratio. The second is the turbulence level. It has been shown that it is the turbulence parameter called the intensity that is the significant variable [13], [23]. The turbulent intensity is basically the standard deviation of the local velocity with respect to the average velocity at that time in the cycle, where the average is obtained over a large number of cycles. Measurements of Lancaster [22], given in Figure 3.7, show that the turbulent flame speed, v_t, is linearly proportional to the turbulence intensity, v'. In these experiments, turbulence was modified at similar engine conditions by modifying the intake valve. In a shrouded valve, about half of the flow area is restricted, increasing the gas velocity through the valve and thus increasing turbulence. The relation between intensity and flame speed has been demonstrated to hold true for different engines and fuels [24], where they developed a correlation similar to the equation shown in Figure 3.7. The value of the coefficient is dependent on the value of the laminar flame speed at engine conditions, and there is some uncertainty as to the effect of pressure on flame speeds.

Now an interesting fact is that the turbulence intensity is linearly proportional to the engine speed for a given engine. This is shown in Figure 3.8, and means that the turbulent flame speed in a spark ignition engine, "automatically" increases in proportion to the engine speed! If the process was precisely linear, then the combustion duration in crank angle degrees for a given engine would be constant, no matter what the speed. This is not strictly true, as the very first part of the combustion deviates from this, as the flame is smaller than the structure of the turbulence, but by and large, the combustion duration in terms of crank angle degrees is nearly constant at all speeds for a given engine. This is shown later in Figure 3.42, where there is a factor of 3 difference in the speeds, yet the combustion duration in crank angle degree is constant. This implies a three-fold increase in the combustion rate as the speed increases, which is in agreement with the above discussion of turbulence intensity and flame speed. Therefore, we usually show pressure time diagrams in terms of engine crank angle rather than seconds, as they are nearly independent of engine speed.

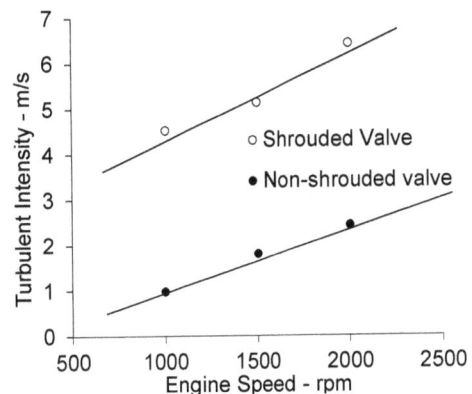

Figure 3.8: The turbulence intensity in the cylinder of a single cylinder engine during combustion [22].

The simple functions mentioned previously for SI heat release models are thus well suited to the simulation of "standard" engines. It is also possible to determine the relationship between the combustion parameters and the laminar flame speed and cylinder turbulence [25]. The method of using the simple trigonometric or Wiebe functions to be shown in Section 3.8.4 has the disadvantage that there is no direct connection between the combustion parameters and the engine geometry. If the engine does not vary significantly from standard design practice, that is normally not a problem.

If a geometrically based calculation is the goal, it is possible to use a procedure developed by Lancaster et. al., [13], the so called quasi-dimensional model referred to previously in Section 2.5. The method is based on the assumption that the flame in a spark ignition engine is spherical in shape and spreads uniformly outwards in all directions from the spark plug. For an average process, this is a good assumption, if there isn't too much air swirl in the cylinder [23]. Figure 3.9 shows a depiction of the shape of the flame for throughout the combustion for a spark ignition engines. The measurements were obtained using high speed photography of the flames, and the outer circle near the outside represents the limit of visibility of the window in the combustion chamber of the engine. Although the flame has some irregularities, the nature of the flame propagation can clearly be seen and the pictures indicate that the assumption of a spherical flame is reasonable. In cases where a large amount of swirling motion is introduced, this assumption is not as valid.

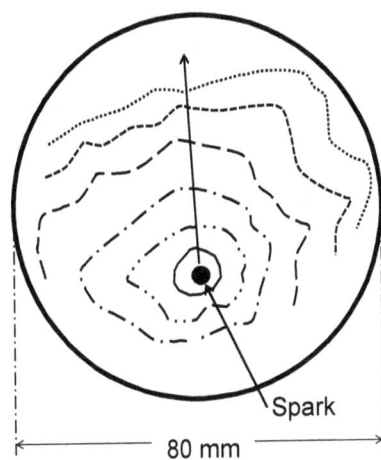

Figure 3.9: Position of the flame in a spark ignition engine as a function of time. Redrawn from the data of [23].

The calculation procedure in this case is based on a heat release calculation as discussed in Reference [19]. The combustion chamber is divided into 2 zones, a burned and an unburned, and the heat release analysis results in the determination of the volume and thermodynamic state of each of these zones throughout the entire combustion process, and results in a determination of the mass burning rate of the fuel, $\dot{m}_f$. Then it is possible to calculate the actual flame velocity using Equation (3.113):

$$\dot{m}_u = \rho_u \cdot v_f \cdot A_f \qquad (3.113)$$

where: $\dot{m}_u$ = the burning rate of the unburned gasses, ρ_u = density of the unburned gasses, v_f = flame speed, and A_f = flame area.

The unburned gasses consist of fuel, air, and residual gasses, and so:

$$\dot{m}_u = \frac{d(m_{f,u} + m_{a,u} + m_{r,u})}{dt} \qquad (3.114)$$

and the burning rate of the fuel itself is then:

$$\dot{m}_f = \dot{m}_u \frac{(1-f)FA}{1+FA} \qquad (3.115)$$

Since the volume of the 2 zones is known from the heat release calculations, Equation (3.113) indicates that it is necessary to know the flame area as a function of the volume of the burned gasses. Therefore, it is necessary to assume a specified shape of the flame. Even though one assumes that the flame spreads as a sphere, the calculation of the flame area can be complicated, since the spherical flame meets an irregular geometry when it contacts the walls of the combustion chamber, the shape of which is dependent on the speed of the flame as well as the timing of the combustion process. For example, in the case of an early flame, the flame can contact the piston before it contacts the cylinder walls, while the reverse can be true in the case of very late ignition timing. In addition, the geometry of the combustion chamber in the clearance volume is often irregular, and can not be readily described with mathematical functions. This normally means that the geometry must be read into computer simulations in the form of timing dependent tables.

3.5.2 Influence of Engine Parameters on Combustion Speed.

In simplest manner there are two parameters that can describe the combustion process in the spark ignition engine. They are the time for the early development of combustion and the duration of combustion. In this section, the dependence of these parameters on various engine parameters is shown. The results are shown for typical engines, or in some cases for single cylinder research engines. Care must be taken when comparing results for different engines, or attempting to simulate engines with different combustion chambers. For example, an engine with dual spark plugs, or other arrangements to increase combustion rate, will typically have shorter delay times and combustion durations than engines with one spark plug per cylinder. The CFR research engine can exhibit long delays and combustion durations, due to the design of the combustion chamber which gives long flame travel distances. This engine often uses a special, shrouded, intake valve where a portion of the intake port is blocked by a protrusion on the valve. This increases the intake velocity through the intake port, increasing the turbulence intensity and flame speed, as indicated by Figure 3.8. While the absolute magnitudes will vary from engine to engine, the trends exhibited by changing the specific engine parameters such as speed, fuel air ratio, etc. will be similar, regardless of engine design.

Fuel Air Ratio

Figure 3.10A shows the dependency of the spark ignition engine combustion parameters on fuel air ratio for a CFR engine with a shrouded valve operating at 1200 rpm, compression ratio of 7.0:1 and MBT timing. For reference, Figure 3.10A also shows the laminar flame speed of iso-octane calculated from the correlations of Turns, [8], for conditions typical of those for combustion in an SI engine. It can be seen that the most rapid combustion occurs for mixtures that are slightly rich. The figure uses the fuel/air equivalence ratio to describe the mixture strength. In the case of Figure 3.10, the chemically correct fuel air ratio is about 0.068, The maximum power of this particular spark ignition engine is found at a rich mixture which corresponds closely to the equivalence ratio where the maximum flame speed is observed.

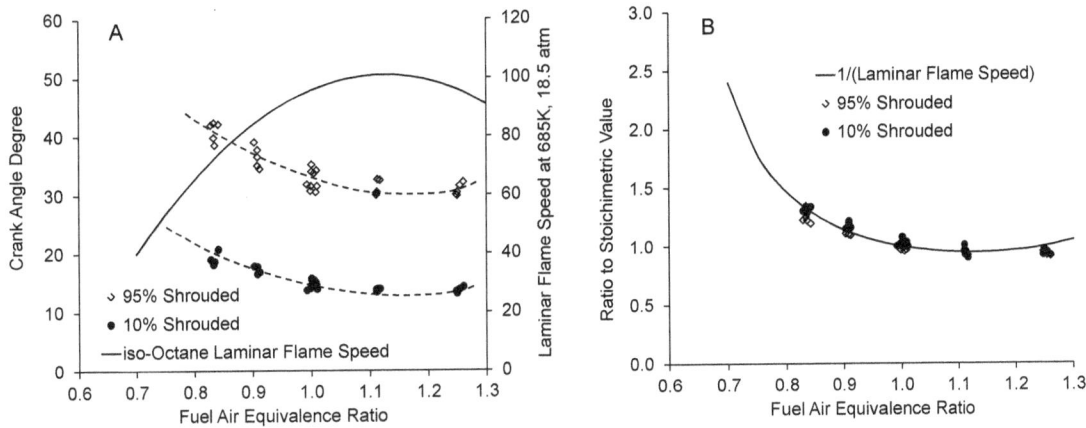

Figure 3.10: Flame speed parameters as a function of fuel air ratio for a CFR engine with a shrouded valve. A: Time for 10 % and 95% of combustion, and the laminar flame speed of iso-octane at typical engine conditions, flame speed from correlations of [8]. B: Combustion durations, and reciprocal laminar flame speed divided by the value at $\phi = 1.0$.

The equivalence ratio in this case is about 1.20 ($FA \approx 0.087$). In Figure 3.10, the values shown to define combustion duration are for 10% fuel combustion, and 95% fuel combustion.

From Figure 3.10 it can be seen that the combustion is slow for very lean mixtures, both in terms of a longer ignition delay and combustion duration. For example, at an equivalence ratio of 0.8, ($FA = 0.054$) about 22° are required for the first 10% of combustion, and about 43° are needed to complete combustion in this engine. For the maximum combustion speed, which occurs at an equivalence ratio of about 1.2 ($\lambda \approx 0.833$) the ignition delay is about 13° and the total duration about 33°. It can be shown that these changes are due to variations in the laminar flame speed. Figure 3.10A shows that the shortest durations occur at the highest laminar flame speed.

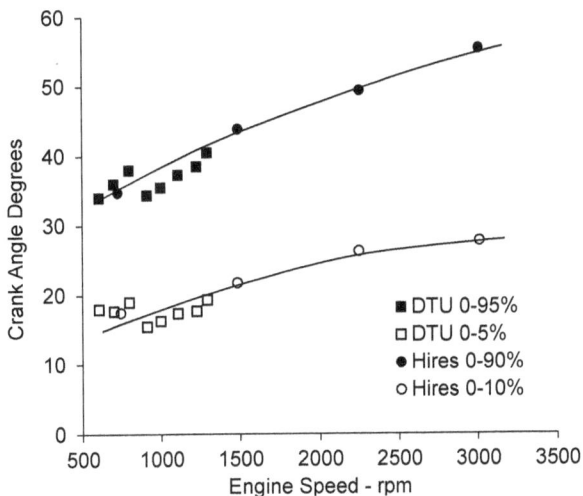

The relation between laminar flame speed and combustion rates is shown in Figure 3.10B. Here, the combustion durations have been arbitrarily normalized by the values for a stoichiometric mixture. The curve also shows the same normalization for the reciprocal of the laminar flame speed. The latter should represent a combustion time. The results indicate that the major cause of combustion variations with fuel air ratio is the variation of laminar flame speed with fuel air ratio. For an engine at a given speed and throttle position, the turbulence should be unaffected by the changes in fuel mixture ratio.

The curves in Figure 3.10 show that the flame speed drops rapidly at lean mixtures, and that combustion durations become quite long. Though not shown, this also happens at very rich mixtures. In practice, the flame can become so slow that it is actually quenched during the expansion stroke [26]. This happens at mixtures leaner than $FA \approx 0.055$ or richer than

Figure 3.11: Flame speed parameters as a function of engine speed for a CFR engine with a shrouded valve, author's values and those of [25].

$FA \approx 0.115$. So there is a mixture which is at the limit of stable operation/misfire for both rich and lean mixtures. These mixture ratios are called the *Lean Flammability Limit*, and the *Rich Flammability Limit*. The values are somewhat dependent on engine design and operating conditions, but the point is that there is a rich FA and a lean FA beyond which operation is not possible, or at best unreliable. This is characteristic for an SI engine, where the overall fuel air ratio is the same as the fuel ratio where combustion occurs. This is not the case with a diesel engine.

Figure 3.10 illustrates an important problem in connection with operation at leaner mixtures, that is, the combustion is very slow at these mixture ratios. This counteracts the effects of lean mixtures on efficiency shown in Figure 2.13, as the slow flame speed will reduce efficiency, because the combustion will extend too long into the expansion stroke, even before the lean flammability limit is reached. This reduces the effective expansion of the combustion energy released in the cylinder, which is what the comparison of the Otto and Diesel ideal cycles showed. Therefore, it is important to find methods to increase the combustion rate for engines operating with lean mixtures. As discussed previously, increasing in-cylinder turbulence is an obvious option. Today, the most common engines operating with lean mixtures are stationary natural gas engines, used for pumping and co-generation units. They are operated lean for high efficiency and in order to reduce emissions, especially NO_x. Spark ignition engines in road transport nearly all operate at stoichiometric mixtures due to emission control with the three-way catalyst.

Engine Speed

Figure 3.11 shows the combustion parameters as functions of engine speed for the CFR engine. It can be seen that the initial delay increases with engine speed when expressed in terms of degrees. The combustion duration in crank angle degrees is nearly constant over the speed range shown. To obtain optimum performance in an SI engine over a speed range, the spark timing must be adjusted to compensate for speed. Figure 3.11 indicates that the compensation is due mainly to the increase in the number of crank angles required for the initial part of combustion.

Figure 3.12 shows results at higher speeds for the same engine type [22], but operating on propane with and without a shrouded valve. The parameter for the start of combustion was defined as the first noticeable increase in heat release after spark, and is considerably shorter than the 10% shown in the previous picture. The turbulence increase due to the higher velocity of the air through the intake valve gives an increase in the combustion rate, both with regard to the start, where the flame is developing, and throughout the entire combustion

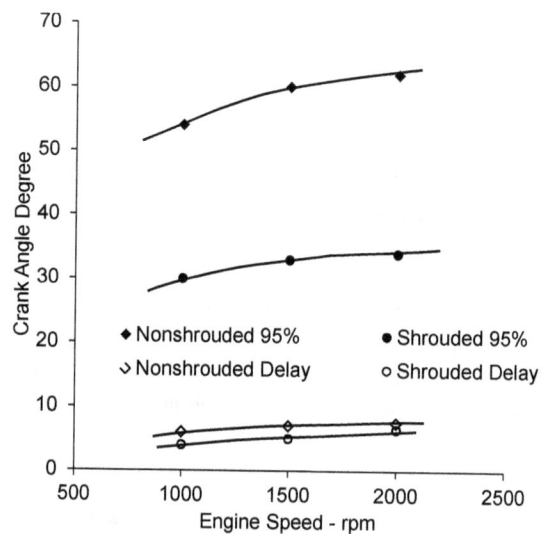

Figure 3.12: Flame speed parameters as a function of engine speed for a CFR engine with and without a shrouded valve [22].

period. These results are for the variation in turbulence intensity shown in Figure 3.8. Note that a linear increase in flame speed with increasing speed would result in a constant duration in crank angle over the speed increase. Thus the roughly constant duration in crank angle degrees of the main portion of the combustion is consistent with the increase in turbulence. The first portion of the combustion has a generally duration in terms of crank angle degrees with increasing speed, since it occurs where the flame is small, and its size is less than that of the turbulence scale.

Ignition Timing

When the fuel air mixture is ignited at different times in the engine cycle, different combustion rates can be expected due to the variation in cylinder temperature and pressure during the flame propagation. This can be seen in Figure 3.13, where the dependency of the combustion parameters on ignition timing is shown for a CFR engine operating at 1500 rpm on standard gasoline. The fastest combustion (shortest delays and duration) is achieved by ignition of the mixture at about 25° BTDC. Here the pressure and temperature increase caused by the combustion in addition to the compression gives the highest unburned temperatures and pressures, hence the fastest laminar flame speeds. This ignition timing is also close to that for maximum power. The speed of the flame depends on the temperature and to a lesser degree, the pressure when the flame consumes the fuel and air.

If the ignition is too early, the temperatures and pressures are generally lower, since highest temperatures and pressure due to compression are obtained at about TDC. Similarly, if the ignition is too late, the pressure and temperature of the unburned mixture will be falling as the ignition occurs, and the combustion rate will be slower than for earlier timing. Figure 3.43 shows similar results.

Intake Pressure (Load)

In the spark ignition engine, the output of the engine is regulated by changing the pressure of the incoming charge of premixed fuel and air, that is the intake manifold pressure. Figure 3.14 shows the dependence of the combustion parameters on the intake pressure. The pressure shown here is that at BDC at the end of the intake stroke. Since the CFR

Figure 3.13: Flame speed parameters as a function of spark timing for a CFR engine.

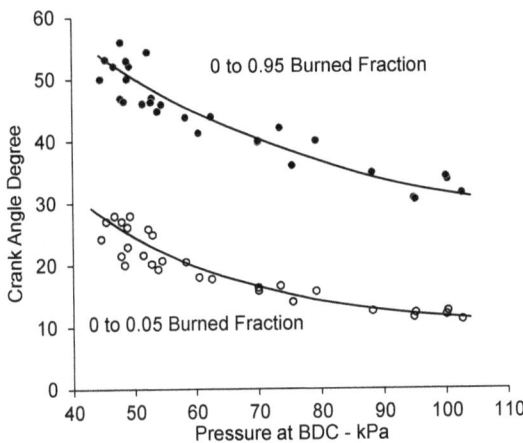

is a single cylinder engine, is has strong pressure oscillations in the intake pipe that make it difficult to define a meaningful intake manifold pressure. The pressure at BDC on intake is closely related to the intake manifold pressure in engines with uniform manifold pressures. It can be seen that the combustion speed increases (duration decreases) with increasing manifold pressure. The increase in the speed of the initial flame development in the most important factor, the duration of the main burning (difference between 0.05 and 0.095) decreasing only slightly as the pressure increases. This is due to a lower residual fraction at higher loads, since the higher pressure charge of fuel and air occupies a larger percentage of the cylinder mass as the pressure (and load) is increased. This causes an increase in the flame temperature, which translates into a higher flame speed. There is also a pressure dependence for the laminar flame speed, and an increase in pressure will increase the flame speed, though this dependence is not as strong as that of the temperature.

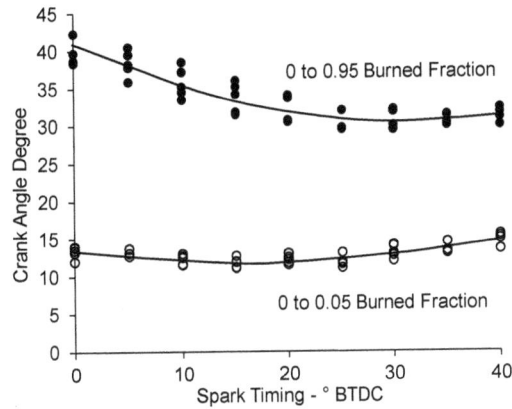

Figure 3.14: Flame speed parameters as a function of pressure at BDC for a CFR engine.

the flame speed parameters in SI engines.

The results presented here are typical of trends of spark ignition engines. It should be noted that the results are for the CFR engine with a predominantly cylindrical combustion chamber. The general tendencies, however, should be be the same regardless of the engine type. That is to say that the combustion rate (per second) will increase with speed for all spark ignition engines, *etc.* For detailed calculations, experiments on specific engines should be used to find the proper values of the combustion parameters. However, with the simple simulation methods that can be used for predicting general performance and fuel consumption trends, the absolute values of combustion parameters are not critical. In the case of predicting exhaust emissions, the errors in the basic modeling procedure (number of zones, heat transfer rates, *etc.*), can be as large or larger than errors in the combustion parameters. Reference [25] gives results for other types of engines, as well as presenting a model for predicting

3.6 CI Combustion

3.6.1 Introduction

The diesel combustion process in an internal combustion engine is a very complex process for many years not well fully understood. However, recent developments in laser imaging techniques have enabled new experiments to shed light on the subject and modified some of the previous concepts. This chapter will describe the diesel combustion process in general, using the results of these new methods to give a better physical picture of the combustion. The mathematical prediction of the details of the diesel combustion process is still a formidable task, though recent developments in CFD techniques have made computer modeling of CI combustion much more realistic.

The diesel, or compression ignition, combustion process is due to the work of Dr. Rudolph Diesel in the late 19th century. Diesel's goal was to develop a constant temperature combustion process, with Carnot's cycle being the model. To do this, control of combustion rate was necessary, and the intention was to do this by achieving cylinder conditions that allow rapid ignition of the fuel and then controlling the fuel introduction rate to keep the temperature constant. This necessitated a much higher compression ratio than the previous Otto cycle. As seen in Section 2.1, this in turn gave a much higher thermal efficiency. When it later could be shown that this combustion was closer to constant pressure than constant temperature as he had claimed, Diesel lost a law suit on patent rights to the engine. Diesel's real contribution was the compression ignition process, and the necessary high compression ratio still gives the highest efficiency for piston engines today.

3.6.2 Diesel Engine Configurations

Diesel engines have achieved great popularity because of their high efficiency. Diesel engines run with a variable fuel air ratio, and the lack of intake throttling also helps to increase the efficiency, particularly at part load, in addition to the high compression ratio. However, diesel combustion is a slower process than spark ignition combustion for mainly two reasons. The fuel is injected into the combustion chamber at the end of the compression stroke and thus evaporation and mixing of fuel and air must take place during the combustion process itself - limiting the combustion speed. In addition, typical diesel fuel consists of much heavier compounds than are found in petrol. The heavy components affect the process in two opposing ways. The large molecules are more difficult to evaporate than petrol, but when they are evaporated, they are much easier to ignite than petrol components. The latter is, of course, a requirement for a good diesel fuel.

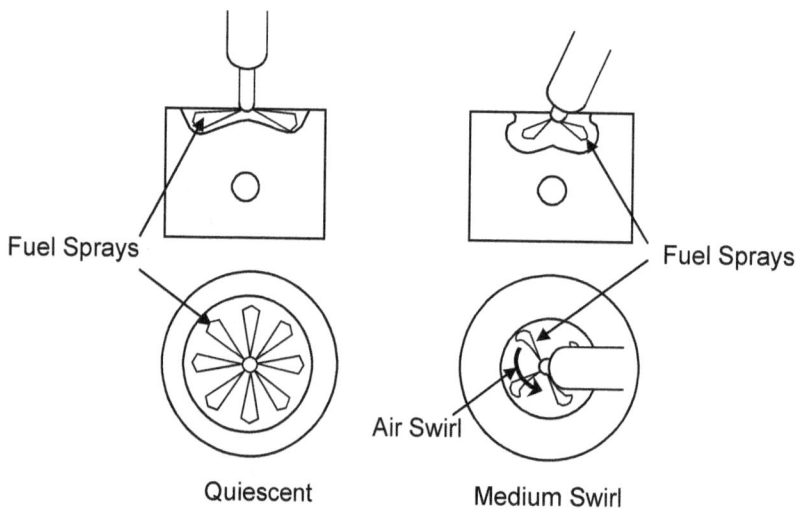

Figure 3.15: Side and top views of two different Direct Injection diesel combustion systems: (a) Quiescent chamber with a multihole injector, (b) Medium swirl, bowl-in-piston chamber with a multihole injector and swirl.

Diesel fuel burns at it's best in large slow engines (Trucks, ships, stationary power generation units) where the time available in the combustion stroke is long enough for the reactions to take place without any special aids. These engine types use the process called direct injection (DI), since the fuel is injected directly into the cylinder. Because of the high efficiency, it is desirable to also apply diesel technology to passenger cars and portable power generation units. These engines have much higher rated speeds than the larger engines, leaving very little time for the combustion to take place. The main challenge in diesel engine combustion, is to get the fuel and air mixed in the time available for combustion. Combustion can normally be considered to be air limited, in that it is easy to inject more fuel than can be effectively burned, without reaching the stoichiometric fuel air ratio. Two of the most common arrangements for road vehicle engines are shown in Figure 3.15.

In these arrangement there is only one combustion chamber, and the principle is called direct injection (DI), since the fuel is injected directly into the combustion chamber. Configuration (a) is called a quiescent chamber, since there is very little rotary motion of the air. The distribution of the fuel is determined by the number of nozzle holes, in this case 8, and air motion for the completion of combustion is created by the kinetic energy of the injected fuel. This system is most commonly used on larger engines used for heavy duty vehicles. Configuration (b) is called a medium swirl engine, as there is a rotary motion of the air in the cylinder. Here, there are only 4 nozzle holes, and the air motion serves to help initial evaporation of the fuel, and spread the fuel throughout the chamber. The swirl speed and the number of holes are determined in combination with each other. Too high a swirl or too many holes will cause sprays to overlap each other and result in a deterioration of combustion (see Figure 5.14). To successfully use the diesel combustion process in smaller, high speed engines, the combustion/mixing rate has to be increased. The air motion is this engine is complicated, but has three main elements.

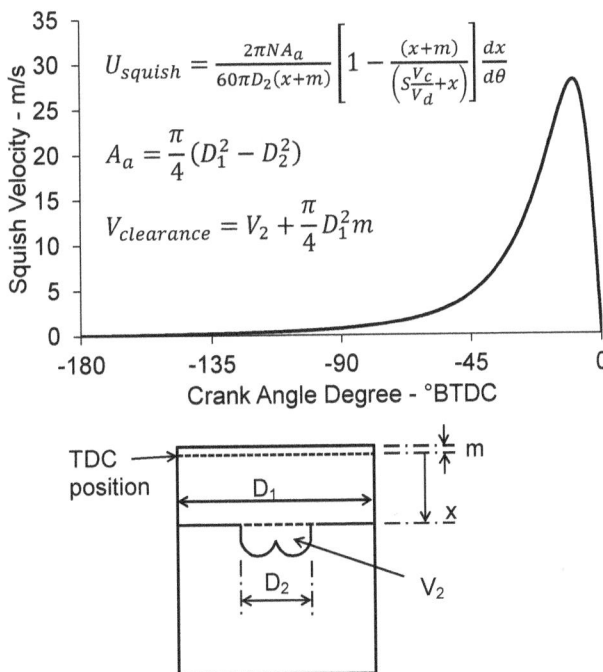

The equations shown in the figure:

$$U_{squish} = \frac{2\pi N A_a}{60\pi D_2 (x+m)} \left[1 - \frac{(x+m)}{\left(S\frac{V_c}{V_d}+x\right)} \right] \frac{dx}{d\theta}$$

$$A_a = \frac{\pi}{4}(D_1^2 - D_2^2)$$

$$V_{clearance} = V_2 + \frac{\pi}{4}D_1^2 m$$

Figure 3.16: Calculation of a simple estimate of the squish velocity, U_{Squish}, in a diesel engine with $B = 90$ mm, $S=85$ mm, rod length $= 180$ mm, $m=1$ mm, $N = 2500$ rpm, compression ratio $= 18$:1, $dx/d\theta$ in m/radian, $D_2 = D_1/2$. Piston motion from Equations (13.13) and (13.14).

The principal air motion is introduced into the cylinder by forming the intake port such that a strong rotational flow called swirl is created around the cylinder axis. Swirl is established by the rotation created by the intake flow on the intake stroke, and is increased during the compression stroke as the air is pushed into the bowl of the piston. Conservation of angular momentum increases the swirl speed in the bowl by the time of combustion to the desired level.

As indicated in Figure 3.15, there is little clearance between the outer, flat region of the piston and the cylinder head as the piston approaches TDC. The air is forced out of this region to create an inward air motion as the piston approaches TDC. This motion is known as squish, and has a maximum value shortly before TDC. A simple estimate of the squish velocity can be made by assuming that the cylinder is initially filled with mass at BDC, and then a portion of it is displaced from that annular volume above the outer part of the piston into the bowl, with a uniform density, which is related to the total volume. The velocity is that across the cylinder formed between the rim of the bowl and the cylinder, and is related to the relative masses in these two regions. An example is shown in Figure 3.16. The squish velocity is quite small until just before TDC, and in the example shown, reaches a maximum value 10 degrees BTDC. can be seen in Figure 3.15. This air motion is present before and during fuel injection, and helps to promote the breakup and mixing of the fuel spray.

During combustion, the spray is injected into the bowl, at a high velocity, and it often impacts on the piston wall. The geometry of the bowl and the lip on the edge of the bowl help keep the burning spray in the bowl where velocities are higher and combustion can be made more complete. This is a fairly recent development, called the reentrant design and is common on especially smaller engines. The crater or bowl in the piston is formed such that the burning fuel and air jet can be directed back towards the middle of the combustion chamber where combustion is most effective. All of the above types of air motion are shown schematically in Figure 3.17

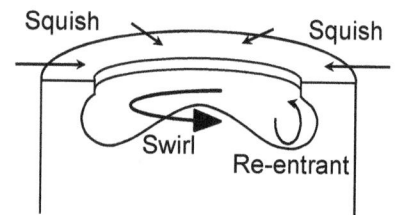

Figure 3.17: Different types of air motion in the medium swirl direct injection diesel engine.

In the smallest engines in the past, the measures above were not enough to ensure sufficient mixing and good combustion, especially at high speeds. Hence, an auxiliary (pre)chamber was added adjacent to the cylinder. The prechamber is connected to the main chamber in the cylinder by a small passage. During the compression, air is forced through this passage from the cylinder into the prechamber creating a very strong swirling flow in the auxiliary chamber. Fuel is injected into the pre-chamber, usually at low pressures and with a single hole nozzle to reduce liquid impingement on the walls, and as the mixture ignites, the burning mixture is pressed out into the cylinder where it mixes with the remainder of the air, and the rest of the combustion takes place. This principle is called indirect injection (IDI).

Figure 3.18 shows a typical IDI system. As the burning fuel and air leave the orifice between the two chambers, it impacts on the piston, the top of which has a recessed area, to impart an additional rotary motion to the fuel and air to promote further mixing in the main combustion chamber. Due to flow losses for the flow between the chambers, and increased heat loss due to the high velocity of the gas impacting the piston, this engine has lower efficiency than a DI engine. In addition, the combustion must be delayed to prevent pushing the burning fuel and air back into the prechamber, where there is no air left to burn. This means the combustion must be delayed, and the engine operates closer to the Diesel cycle than does the DI engine. The engine also requires a high compression ratio to achieve ignition, due to heat losses for the flow passing through the orifice, and the poorer mixing of the spray due to low injection pressures. Compression ratios of 21-22 are common, but still the engine has a lower indicated thermal efficiency than the DI engine.

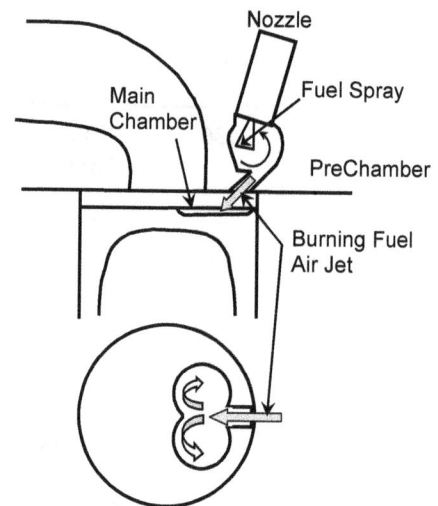

Figure 3.18: A schematic diagram of an IDI combustion system, the pre-chamber here is called a swirl chamber.

For many years, this was the dominant design for small high speed diesel engines, but recent progress in engine design has almost rendered the IDI engine obsolete, and since around the year 2000 in automobiles it has largely been replaced by a new type of DI engine that is turbocharged and uses an injection system capable of very high pressures (on the order of 50-150 MPa compared to 8-10 MPa in a typical IDI engine). In addition to the high pressure, electronic control of the combustion process is now achieved with computer controlled injection.

The system used in modern DI engines is the common rail system, see Figure 5.5. All the nozzles are connected to the same fuel line (the common rail), where a constant pressure is maintained. The pressure can be controlled independently of the engine load and rotational speed, since the pump and metering process have been separated, in contrast to earlier systems. Nozzle operation is electronically controlled by opening and closing a small solenoid valve near the nozzle. The newest injection nozzles in common rail systems use a piezo-electric mechanism to open the nozzle needle itself.

This is in contrast to earlier purely mechanical systems, where fuel pressure increase at the nozzle is generated by a positive displacement mechanical pump connected to the engine crankshaft. Injection in the conventional system starts when the pressure increases over the threshold set for the nozzle needle. The pressure during injection with this older system is determined by the nozzle hole size and the volumetric displacement of fuel by the pump, and is dependent on engine speed.

The new electronically controlled system allows very precise control of the injection rate and the amount of fuel injected. The high level of control combined with very high swirl rates provides sufficient mixing to eliminate the need for an auxiliary chamber. An important additional feature with the electronically controlled common rail system is the use of what is called pilot injection. In this concept, a small amount of fuel is injected into the cylinder air several degrees before the main fuel charge. This fuel can start to react, which causes an increase in the cylinder temperature above that of pure compression. So when the main fuel is injected, the ignition delay is shorter than without pilot injection, improving combustion characteristics and reducing engine noise. The effect of these features on the combustion process is shown in Section 3.8.3. Fuel injection is described in more detail in Section 5.4.

The final combustion configuration to be shown for diesel engines is that typical of large marine diesel engines, and is shown in Figure 3.19. These engines typically operate on the two-stroke cycle, are heavily turbocharged and run at low speed. Bore sizes over one meter and strokes on the order of 3 meters can be found in the largest. They typically have an exhaust valve in the cylinder head and piston controlled intake ports in the bottom of the cylinder, the latter being used to create a swirl motion for the combustion. Because of the slow engine speed, there is much time for the combustion, reducing the demand for a large number of fuel sprays. In addition, due to the high degree of turbocharging, these engines typically run with a smaller maximum fuel air ratio than

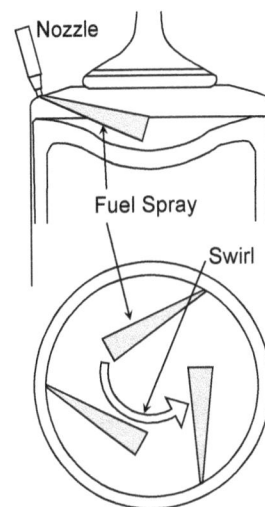

Figure 3.19: A schematic diagram of an direct injection combustion system for a large, two-stroke marine diesel engine.

that common on smaller engines, so there is an excess of air available for completing the combustion. Combustion processes are limited by maximum cylinder pressure, and these engines have brake thermal efficiencies over 50%.

3.6.3 Combustion Process

The combustion process in a diesel engine can be characterized by four phases caused by different physical and chemical processes. Portions of these processes take place concurrently and are in that way not completely time-wise separated. Nevertheless, the concepts may still serve as guidance when determining the nature of the combustion. The four phases are:

- Ignition delay period
- Premixed combustion phase
- Mixing controlled combustion phase
- Late combustion phase

Figure 3.20 shows a typical heat release rate diagram marking the different classical combustion phases in a direct injection engine without intake charge compression. The engine here is a naturally aspirated DI engine, where the premixed combustion phase is quite pronounced. The use of pilot injection and turbocharging has reduced the significance of the first phase of combustion by shortening the time between

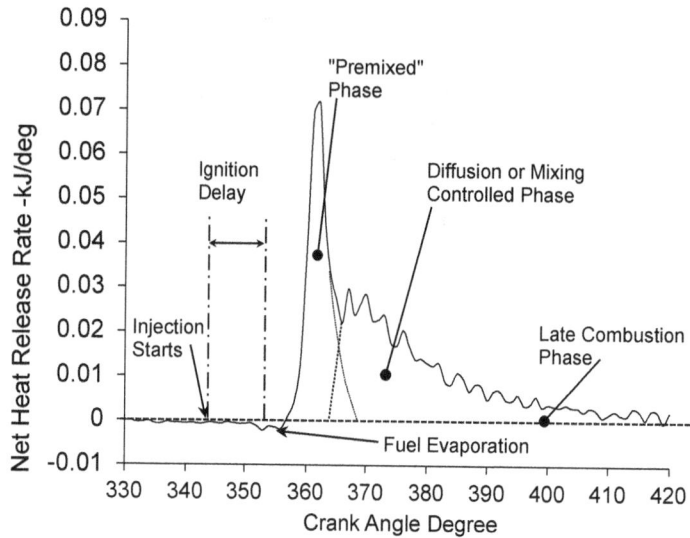

Figure 3.20: Typical DI engine heat release rate diagram, non-turbocharged engine.

the start of injection and the start of ignition, reducing the ignition delay, and the importance of the premixed combustion phase, see for example Figure 3.47.

A systematic description of the events taking place during the combustion process is given in the following. The bulk of the data upon which the description is based are from the model presented by Dec [27], [28], [29]. The concepts were derived from experiments in a heavy-duty engine with no swirl.

Injection Timing

Fuel injection begins near the end of the compression stroke. The optimum time for this depends on the engine type and load condition. The object is to start injection late enough, so that the pressure increase from the combustion will not attempt to slow down the piston movement towards TDC, but early enough to make sure that combustion doesn't continue too long into the expansion stroke. This is important, since mixing rates decrease as the cylinder volume increases, and there may be difficulties in achieving complete combustion if mixing isn't sufficient. High soot and carbon monoxide emissions are often symptoms of this problem. The start of injection (SOI) is usually 20-5° BTDC in direct injection engines.

Depending on engine load, different amounts of fuel have to be injected. That is done by either varying the injection pressure or the injection duration, or a combination of the two. The change of duration with respect to load for a typical engine is shown in Figure 3.48. In systems prior to the common rail, the injection pressure could not readily be independently varied. For the engine from which the data were obtained, the amount of fuel was only controlled by varying the duration of injection. The pump type used there employs a positive displacement of a variable fuel volume in a piston cylinder arrangement. It is typically called a jerk pump system.

Spray Formation and Ignition Delay

Fuel is injected into the cylinder through one or more nozzle holes by applying a very high pressure difference across the nozzle (typically in the range 10 - 150 MPa). The nozzle diameter plays a major role in determining the pressure, where in conventional pump systems that are mechanically connected to the engine crankshaft/camshaft, it depends on the volumetric displacement rate of the fuel. The amount of fuel is determined by the load, and the maximum injection duration is limited by the requirement of adequate time for good mixing and complete combustion at the end of the process. Then, as is discussed in the Section 5.4, the flow rate through the nozzle is determined by the ratio of the mass of fuel injected to the injection time required. The nozzle hole size is then chosen to give the required injection pressure (pressure drop across the nozzle), in accordance with Equation (5.13). Once injected, the fuel forms a spray that can be characterized as a transient jet of evaporating liquid. A series of photographs of a typical fuel jet can be seen in Figure 3.21.

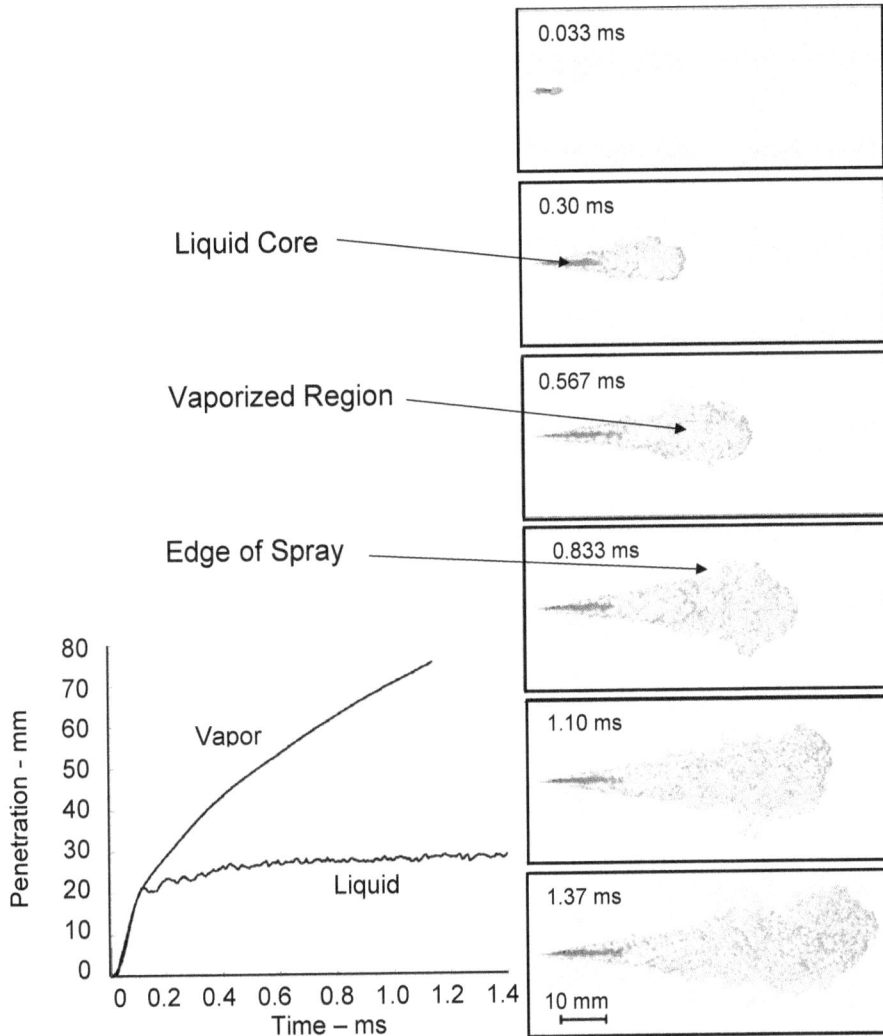

Figure 3.21: Temporal development of an evaporating spray injected into nitrogen/combustion products at 3,4 MPa and 670 K. The pictures are combined Schleiren and scattered light photographs. The curves show the penetration of the tip of the spray plume and the end of the liquid region of the spray. Times is after the time after the start of injection. Figure courtesy of Fredrik Westlye.

There are two regions shown in Figure 3.21, which shown superimposed Schleiren and scattered light photographs of a spray from a common rail injector into an inert environment. The first region is the liquid region, which is seen by the scattered light from the droplets. It is found at the exit of the nozzle, and consists mainly of a vast number of small fuel droplets, formed by the high velocity exit of the fuel. There is a small conical shaped region at the very exit of the nozzle (not visible here) where the liquid is still in the form of a single stream, but essentially all the liquid shown in this figure is in the form of droplets. There is a distribution of sizes, the mean value of which dependents on injection parameters, but typical sizes are on the order of a few microns. As seen in the figure, the liquid region reaches a steady state length of about 20 mm. This means that after a short time, the rate at which fuel is injected into the chamber is equal to the rate at which it evaporates.

The second region is the vaporized fuel region. This is shown by the Schlieren images, which shows density gradients in the gas phase. In the injection chamber of the photographs, there was no air motion, and so all the subsequent motion of the spray has been due to the kinetic energy of the liquid fuel leaving the nozzle. As the fuel leaves the nozzle and enters the chamber, it draws hot air along with it (entrainment), and the vaporized fuel and entrained air form the vaporized region of the spray. As the injection continues and the jet progresses across the combustion chamber, more air is entrained in the jet and the droplets evaporate, creating a combustible mixture of fuel vapor and air. The mixture in this area is very fuel rich with an equivalence ratio of 2-4 [27]. Sprays are often characterized by an spray angle and penetration length, though it can be seen from the pictures, that an average angle must be used, since portions of the vaporized region of the spray break off, and move in a direction perpendicular to the axis of the spray. The penetration of the tip of the spray is typically proportional to the square root of the time since the start of injection.

The injection chamber temperature and pressure are typical for diesel combustion, but there is no combustion in these pictures, since fuel was injected into an inert environment. In an actual engine, chemical reactions occur within the vaporized region of the spray, which lead to ignition, and the situation changes. Additionally, the presence of swirl will change the shape of the spray and in medium swirl engines, it will often impinge on the wall of the piston bowl.

A temporal sequence of schematics showing the evolution of the diesel spray in the first period after start of injection is shown in Figure 3.22. The system described is very similar to that of Figure 3.21. The results in this figure are based on a number of studies of diesel combustion, in which very advanced laser and optical technologies have been used to determine detailed properties during the combustion process.

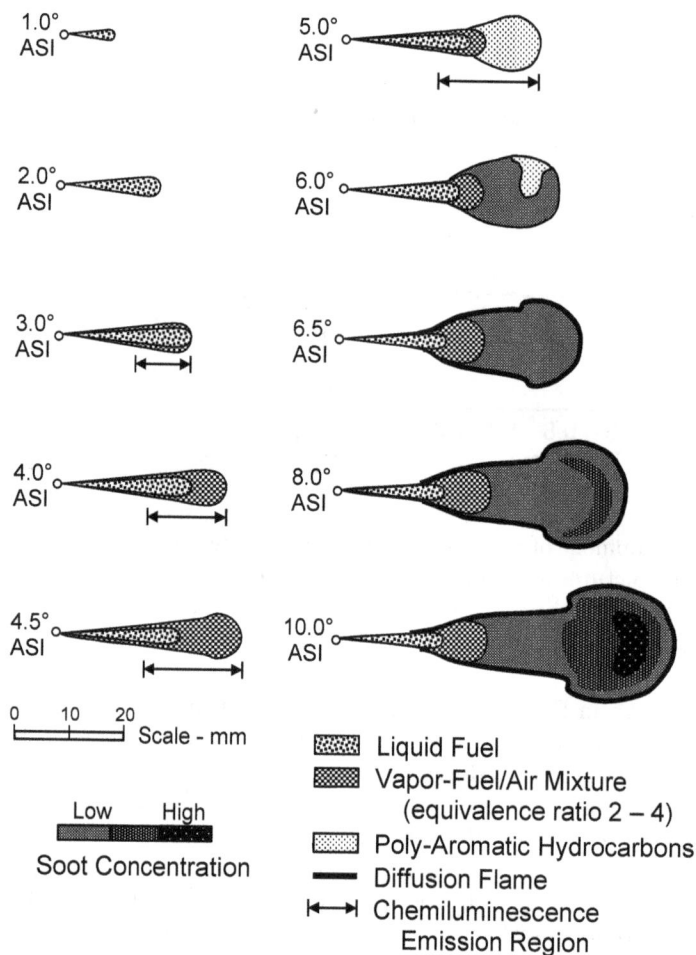

Figure 3.22: A temporal sequence of schematics showing how DI diesel combustion evolves from start of injection through the early part of the mixing controlled burn. The temporal and spatial scales depict combustion at 1200 rpm in a heavy-duty engine. ASI denotes after start of injection. From From John Dec, used by permission.

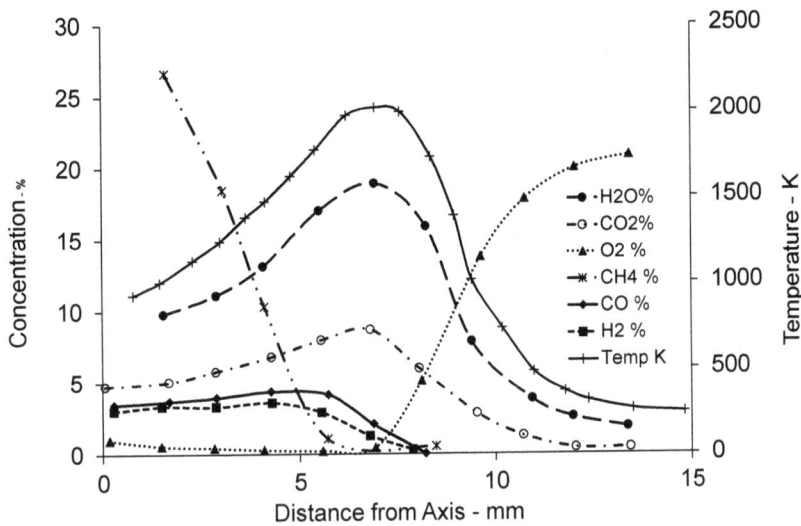

Figure 3.23: The structure of a laminar methane diffusion flame. Adapted from [30].

When the temperature of the fuel vapor is high enough, chemical reactions start taking place. These reactions give off light in certain wavelengths, and their location is shown as the chemiluminescence region in Figure 3.22. These reactions are primarily thought to be thermal decomposition of the large fuel molecules. While they do produce some energy conversion, they are not considered to be the start of combustion, but rather a preparatory phase and as such part of the ignition delay period. After a while, the fuel/air mixture ignites almost simultaneously at several locations inside the spray.

Premixed Combustion Phase

This phase is the short period during which the fuel air mixture prepared during the ignition delay period burns, causing a very rapid rise in heat release rate (see Figure 3.20). As seen in Figure 3.22, the mixture inside the spray is very rich. The combustion in this phase is, therefore, not complete, but rather it produces a wide range of intermediate products. Specifically a large amount of soot is formed in the core of the spray. The soot is distributed fairy equally over the full volume of the mixture indicating that combustion takes place at many location in the spray nearly simultaneously. The soot formation is indicated in the last four schematics in Figure 3.22. As the available oxygen in the mixture inside the spray is used up, the heat release rate decreases rapidly again to a much lower level. Toward the end of the premixed phase, temperatures at the periphery of the jet reach a level that allows a diffusion flame to form between the products and the surrounding air (shown in Figure 3.22 starting at 6.5° ASI).

Mixing Controlled Phase

This period is characterized by much lower heat release rates than in the premixed phase, and the heat release is primarily generated by the diffusion flame sheet that now surrounds the jet. Figure 3.23 shows a simple laminar diffusion flame for methane. It is shown to illustrate the general nature of a diffusion flame, and is a classical diffusion flame, in which the fuel and oxidizer do not coexist. It is assumed that the chemical reactions between the fuel and oxygen are very fast compared to the rate at which the reactants are transported (diffuse) to the flame zone, and so the rate of mixing determines the speed at which this part of the combustion proceeds. Fuel and oxidizer are assumed to react instantaneously upon contact, a common assumption in diffusion flame modeling, and the physical mixing process totally dominates the combustion rate. The flame front is characterized by the point at which a stoichiometric mixture of fuel and air (actually their combustion products) is found. Thus, even though a diesel may be operating at an overall lean mixture, the temperature where the reaction between fuel and air occurs during mixing controlled combustion is essentially the adiabatic flame temperature for constant pressure combustion at the prevailing pressure at the time of combustion. This has a large impact on the formation of NO in diesel engines.

These same general features can be seen by sampling the combustion zone in a swirling flow inside a DI diesel engine as the flame passes a fixed location, in Figure 3.24. Initially, the sample is from inside the spray, before ignition. There is a large amount of oxygen with the fuel until ignition. After ignition (at about 7 degrees BTDC), the oxygen falls to nearly zero, and there are rich combustion products like CO and soot, consistent with the rich ignition inside the spray, as shown in Figure 3.22. As the spray moves past the sampling point, the CO_2 concentration rises to about 10%, the temperature rises to near the stoichiometric adiabatic flame temperature and the O_2 is near zero, similar to the simple diffusion flame. Note that the NO_x is formed in the region near the flame with high temperatures and adequate O_2.

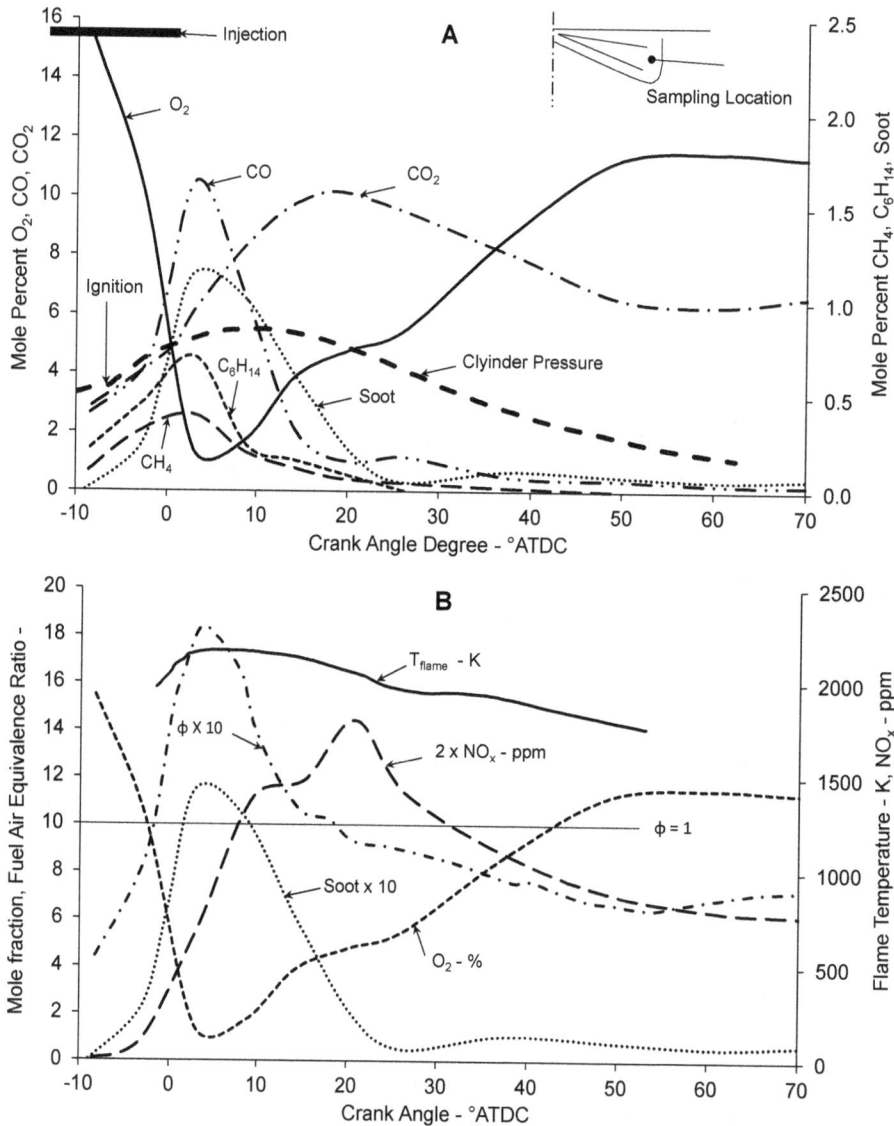

Figure 3.24: The composition of the combustion chamber gasses in a direct injection diesel engine as the flame passes the sampling point. Adapted from [31].

This period of the combustion process is characterized by much lower heat release rates, than in the pre-mixed phase, and the heat release is primarily generated by the diffusion flame sheet that now surrounds the jet or jets. The part of the jet downstream from the nozzle now contains partially combusted fuel in a near homogenous mixture. It has been shown [2], [3] that the soot particles in this region increase in size as they travel towards the flame sheet. This is thought to happen since the higher temperatures near the flame sheet (1700-1800K) promote further decomposition of the remaining carbon hydrates leaving more free carbon to form soot. That the majority of the soot particles burn in the diffusion flame is also confirmed in Figure 3.24.

The concentration of soot particles is highest in the area well downstream of the liquid core at the end of the spray, but there are particles within the entire spray. This is shown in Figure 3.25, which is the situation for mixing controlled combustion but before the end of injection. The soot oxidation zone is shown here as surrounding the entire jet, in a diffusion flame as discussed above. Also shown is the location of the production of NO, on the lean side of the diffusion flame, as indicated in the discussion above.

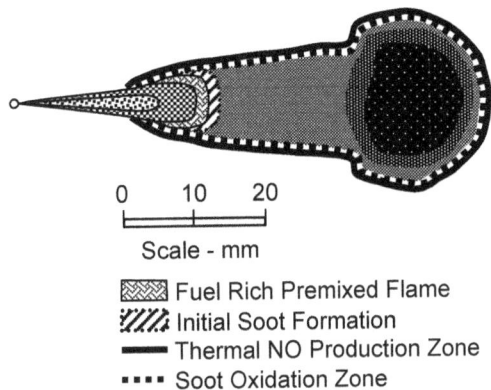

Figure 3.25: A schematic continuation of the combusting spray well into the mixing controlled burn, but before end of injection. Also shown are the soot production and NO formation zones. From John Dec, used by permission.

Late Combustion Phase

While the combustion process is fairly well documented for the period from start of injection through onset of combustion until the end of injection, the situation from end of injection until the combustion is over, either upon flame extinction or when the exhaust valve opens has not yet been subject to investigation with laser imaging technology, so there is not as detailed a basis for interpretation as in the earlier stages. However, the recent experiments show as well that diesel combustion is a much more organized process than was previously thought, so an attempt will be made.

Once injection stops the fuel just injected will continue to move towards the premixed flame. However, since the injection pressure at the end of injection is lower than during the main portion of the injection and since no more fuel is pushing from behind, less air will be dragged into the spray and hence the mixing rate will be lower. This can be seen in the simple heat release model of Equation (3.131), where the basic shape of the heat release rate curves changes when fuel is no longer injected.

Furthermore, due to the lower speed of the fuel, the premixed flame may retract even further upstream creating a situation, where the mixture combusted has an even higher equivalence ratio or maybe even contains fuel droplets. In any case, the premixed combustion will form significantly more soot towards the end.

As the fuel rich flame retracts, the flow of products into the plume decreases and stops when the last of the fuel vapor has been consumed. The diffusion flame continues to "eat away" at the periphery and the plume size will decrease. Eventually, when all the products have passed through the flame sheet, combustion is over.

3.6.4 Droplet Evaporation

As mentioned in the discussion of Figure 3.20, the fuel jet leaving the nozzle quickly breaks up into small droplet, that are drawn into the spray and evaporate into the surrounding gasses. Though the actual situation in the spray is very complicated, due to a wide range of temperatures, gas velocities and gas composition, a short presentation of the basics of droplet evaporation will be shown here, to give an idea of how the process works. The trends are of interest, absolute values should not be regarded as accurate.

Complete presentations of droplet vaporization can be found in classical combustion texts, for example [8], from which the following originates. A simple theoretical analysis of a droplet evaporating into quiescent air says that the evaporation rate, $\dot{m}_f$ is given by:

$$\dot{m}_f = \frac{4\pi \cdot k_g \cdot r_d}{c_{p,g}} \ln\left(B_q + 1\right) \tag{3.116}$$

where:

$$B_q = \frac{c_{p,g}\left(T_\infty - T_{boil}\right)}{h_{fg}} \tag{3.117}$$

and k_g and $c_{p,g}$ are the gas phase thermal conductivity and specific heat, r_d is the droplet radius, T_∞ and T_{boil} are the surround temperature and boiling point of the droplet, and h_{fg} the latent heat of vaporization of the fluid. Equation (3.116) gives rise to a classic relationship that the square of the droplet diameter decreases linearly with time.

$$\frac{dD^2}{dt} = -\frac{8 \cdot k_g}{\rho_l c_{p,g}} \ln\left(B_q + 1\right) = constant \tag{3.118}$$

where: ρ_l is the density of the liquid.

For droplets in a spray, there are the additional effects of convection. In that case, Equation (3.116) can be written:

$$\dot{m}_f = \frac{2\pi \cdot k_g \cdot r_d \cdot Nu}{c_{p,g}} \ln\left(B_q + 1\right) \tag{3.119}$$

where Nu is the Nusselt number of the drop, and can be determined from heat transfer correlations for convection over spheres. Turns recommends the following [8]:

$$Nu_D = 2 + \frac{0.555 Re_D^{1/2} Pr^{1/3}}{1 + \dfrac{1.232}{Re_D Pr^{4/3}}} \tag{3.120}$$

where: Re_D is the Reynolds number based on the diameter and Pr is the Prandtl number of the gas.

Along with the evaporation, there are aerodynamic forces, or drag involved, changing the velocity of the droplet.

$$F_d = m_d \frac{dv_d}{dt} = m_d v_d \frac{dv_d}{dx} = C_d \frac{\pi D^2}{4} \frac{\rho_g v_{rel}^2}{2} \tag{3.121}$$

where v_d and v_{rel} are the spatial velocity of the drop, and the velocity of the drop relative to the gas motion, C_D is the drag coefficient for a sphere, and ρ_g is the gas density.

Turns recommends the following correlation for drag on a sphere [8]:

$$C_d \approx \frac{24}{Re_{D,rel}} + \frac{6}{1 + \sqrt{Re_{D,rel}}} + 0.4 \tag{3.122}$$

where $Re_{D,rel}$ is the Reynolds number of the drop based on the relative velocity between the drop and the gas. Using $v_d = \frac{dx}{dt}$, Equations (3.118) and (3.121) can be integrated to solve for the position and size of a drop as it evaporates. In using Equation (3.121) care must be taken with the sign, to make sure the force is in the proper direction.

To show the behavior of an evaporating drop in a situation similar to that of diesel injection, the above model was solved, using octane fuel droplets injected at a speed of 500 m/s into nearly quiescent air at a conditions at the end of compression with a compression ratio of 15:1. The results are shown in Figure 3.26. First of all, the velocity of the droplets falls very rapidly due to the high aerodynamic drag,

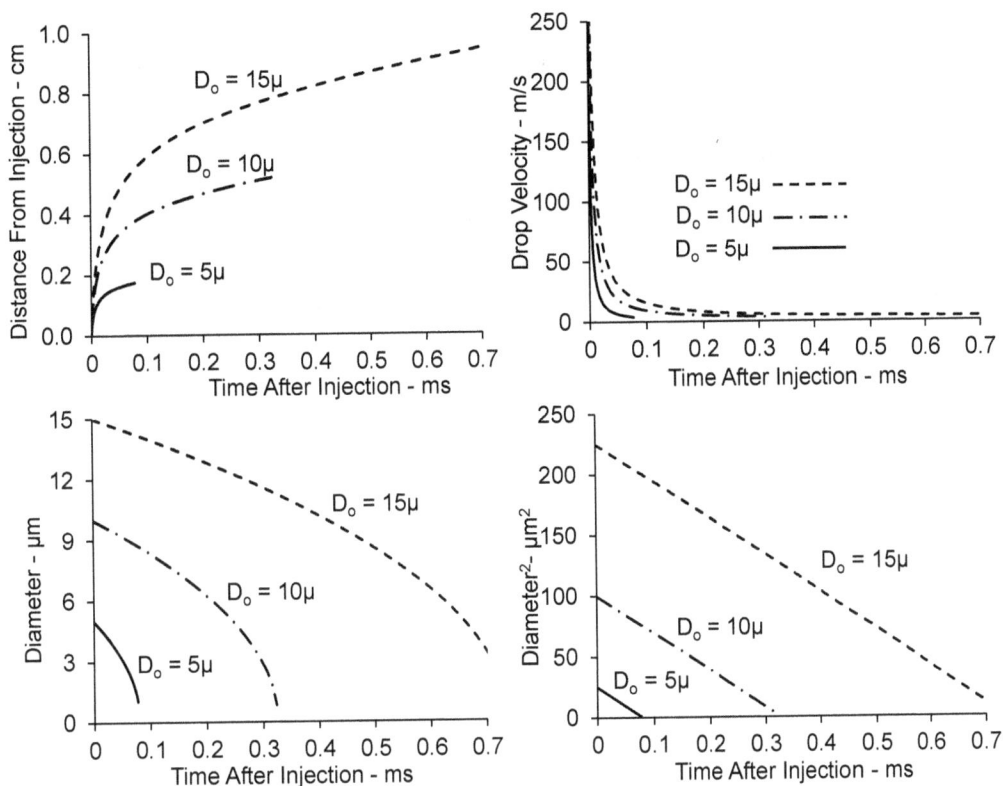

Figure 3.26: Simulated behavior of octane drops injected into quiescent air at 868K, 42 bar. Drop injection velocity = 500 m/s.

with the smaller drops slowing fastest. The velocity is within a few microseconds already reduced to a 10th of the injection velocity. The penetration distances are also strongly dependent on the size, with the largest drops penetrating farthest into the chamber. The penetration distances for the $5\mu m$ drops is on the order of 20mm, similar to that seen in the injection photographs in Figure 3.15 with times less than a millisecond. Given all the uncertainties and theoretical simplifications, these results indicate that this model should give reasonable trends. A version of this model is often used to model sprays in CFD simulation, where the different parameters in the model are typically adjusted to give better agreement with experiments.

3.6.5 Autoignition

The second major process occurring in diesel combustion is the chemical reactions leading up to ignition. The fuel air mixtures are compressed and at high temperatures and pressures, the reactions lead to a sudden increase in temperature and pressure. Since there is no external source of ignition, the process is called autoignition. The ignition starts locally, and spreads to the rest of the flame, leading to the last stage of combustion, the diffusion based combustion of the spray. Seen from a chemical point of view, the process is the same as that leading to knock in the SI engine, where it is autoignition of the end gasses near the end of combustion that gives rise to the rapid pressure rise called knock. The process is the same in both engines, but the demands on the fuel the opposite: In diesel engines, it is desirable that the fuel ignites readily, so that combustion can be better controlled, while in the spark ignition engine, the desirable fuel is resistant to autoignition, so that it is consumed by the propagating flame at a controlled rate, instead of igniting instantaneously and causing damage from the high rate of pressure rise.

This causes some fundamental differences in the possibilities for improving engine performance. The major one is that the diesel engine is extremely well suited to turbocharging. The higher cylinder temperatures and pressures, shorten the ignition delay, providing for a quicker, smoother ignition phase of combustion. This can be seen from the heat release curves of a naturally aspirated engine, Figure 3.45 with the very rapid premixed phase compared to that of a turbocharged engine, Figure 3.47, where the premixed phase of combustion is nearly absent.

The chemical reactions for the ignition of organic compounds is extremely complex, involving hundreds of individual reactions between molecules and free radicals. Modern computer methods have made it possible to solve such complicated reaction systems, but for many purposes a simpler model is needed. Such a model was developed in the 1970's [32] and forms the basis for ignition models today, including CFD calculations [33].

The model consists of series of generic reactions, that is reactive species are lumped into one specie that has the same effect. The reactive specie corresponds to the free radicals, which attack stable molecules, and eventually form the products of reaction, giving off chemical energy in the process. The specie RH represents the fuel, R^* is a generic reactive fuel fragment (The superscript * indicates a reactive substance or free radical, which can remove a carbon or hydrogen atom from another organic molecule.)

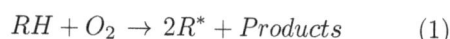

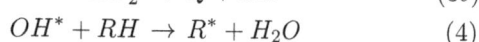

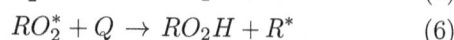

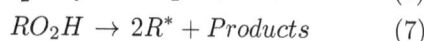

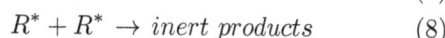

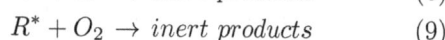

$$RH + O_2 \rightarrow 2R^* + Products \qquad (1)$$
$$R^* + O_2 \rightarrow RO_2^* \qquad (2)$$
$$RO_2^* \rightarrow Products + OH^* \qquad (3a)$$
$$RO_2^* \rightarrow Q + OH^* \qquad (3b)$$
$$OH^* + RH \rightarrow R^* + H_2O \qquad (4)$$
$$RO_2^* + RH \rightarrow RO_2H + R^* \qquad (5)$$
$$RO_2^* + Q \rightarrow RO_2H + R^* \qquad (6)$$
$$RO_2H \rightarrow 2R^* + Products \qquad (7)$$
$$R^* + R^* \rightarrow inert\ products \qquad (8)$$
$$R^* + O_2 \rightarrow inert\ products \qquad (9)$$

Reaction 1 is a basic attack on the fuel which gets the reaction started. The reactive substances formed begin to attack the fuel and the semi-stable intermediates formed. Reactions 2 to 4 are balanced with regard to the production of reactive species, in that there is one reactive specie on each side of the reaction. This is called propagation, and the reaction slowly increases due to the release of energy from forming the reaction products (CO_2 and H_2O) Reaction 5 is also a propagation reaction, but the product RO_2H is not stable and decomposes in Reaction 7 to give 2 reactive species instead of one. This is called a branching reaction, and gives a geometric increase in the number of reactive species. If the fate of the RO_2^* specie changes from Reaction 5 to Reaction 6, the reaction can change character and accelerate. Many of the the reactions in this mechanism are strongly temperature dependent, and a reaction that may not be important at the start of the ignition process can wind up being the dominant reaction later on, if it is more temperature sensitive than the other.

Figure 3.27: Calculated ignition of a stoichiometric mixture of n-heptane and air as a function of initial temperature at constant volume, 20 bar pressure.

This gives rise to the possibility that the way the fuel oxidizes changes through the ignition process. At low temperatures one reaction route can dominate, while at higher temperatures, the other route can dominate. At intermediate temperatures, both routes can be active, and what is called two stage ignition occurs. That is, there is a delay time before the first marked temperature rise. This is the first

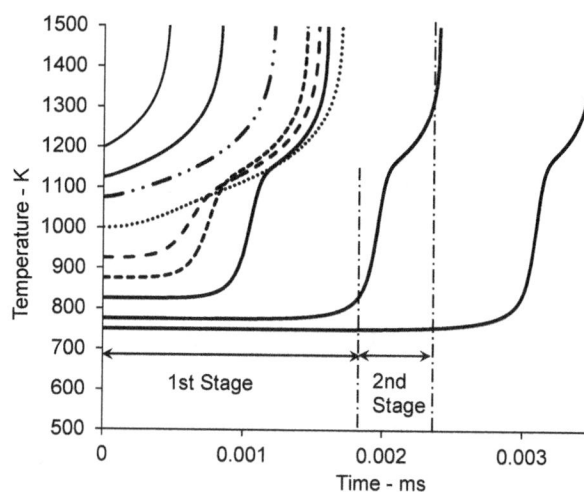

stage. It is followed by another region of slow temperature rise until the appearance of a second marked temperature rise, which is the end of the second stage of ignition.

An example of this is shown in Figure 3.27, using the above mechanism with the parameters reported for n-heptane ignition [33]. For the lower initial temperatures, the two stage ignition process is clearly seen. As the temperature rises, different reactions in the mechanism become important, and the first stage becomes shorter and merges into the entire process. For an initial temperature of 1000K, the total delay is actually longer the delay for 800K. This is contrary to the normal idea that reactions increase with temperature, and is called the negative temperature dependence, and was first clearly explained and demonstrated by Halstead, *et. al* [32].

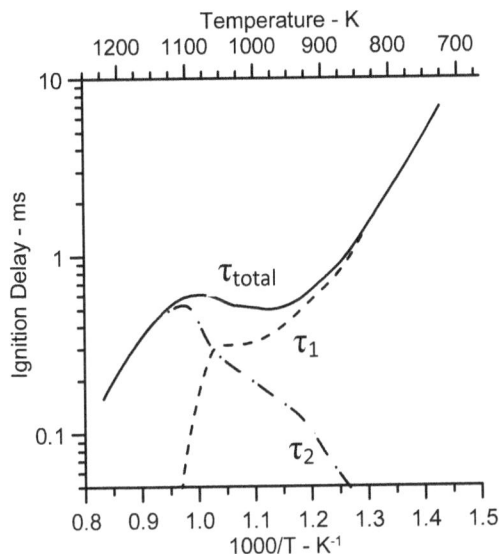

Figure 3.28: Calculated ignition delays of a stoichiometric mixture of n-heptane and air as a function of initial temperature at constant volume, 40 bar pressure.

The change in the mechanism for varying temperature can be seen in Figure 3.28, where the total ignition delay is shown, as well as the two first ignition delays. In chemical kinetic calculations, it is usually convenient to plot a semi-log curve, with the ordinate being the reciprocal temperature. That the mechanism changes is apparent from these curves, as the first stage process dominates at low temperatures, but is replaced by the second stage process. In the regions where the two processes are active at the same time, the negative temperature dependence region can occur.

The situation is complicated, as Reactions 3b, and 7 are very temperature dependent, Reactions 3a, 6 are moderately temperature, and the others have little temperature dependence. That means that as the temperature changes through the energy release from the initial reaction, the reactions which are important change (mechanism change). At the very start, Reaction 1 produces some reactive species, and their growth rate is controlled by a balance between the branching reaction, Reaction 7, and the termination processes, Reactions, 8 and 9. The temperature rises as the reactive species increase and this in turn speeds up all reactions, some more than others. Eventually, the temperature begins to rise rapidly, indicating ignition, and the end of the first stage. But under proper conditions, the increase in temperature and the number of reactive species increases the relative importance of the termination reactions, and the rate of formation of the reactive species becomes negative and the temperature rise slows down drastically. This is the start of the second stage of ignition.

While this has been occurring, the species RO_2H has been increasing, since Reaction 6, which forms it, has become faster due to the temperature rise. This forms an alternate source for the branching specie through Reaction 7, which also has a higher rate because of the temperature rise, and the reactive species increases in concentration again. This gives rise to another temperature rise, which speeds up all the reactions until a second sharp temperature rise occurs, marking the end of the second ignition phase. After this, the model is no longer valid, but is has proven very useful for calculating diesel ignition and knock in SI engine, the latter being its original goal [32].

As Figure 3.28 indicates, the relative importance of the two stages changes with temperature, as it also does with other variables such as pressure, equivalence ratio and fuel composition. The effect of pressure on n-heptane ignition at stoichiometric conditions and constant volume is shown in 3.29. Shown in the figures are the first, second and total ignition delays, where applicable, and an empirical equation to model the effect of pressure, called "Fit" in the figures. This is calculated by referring the total ignition delays at 2000 kPa with a simple, exponential function:

$$\tau(p) = \tau(2000) \cdot (p/2000)^n \tag{3.123}$$

where $n = -0.6$ for a temperature of 700K, and -1.4 for the remaining temperatures. There is single stage ignition at 700K and 1100K, while there is two stage ignition at the intermediate temperature. In

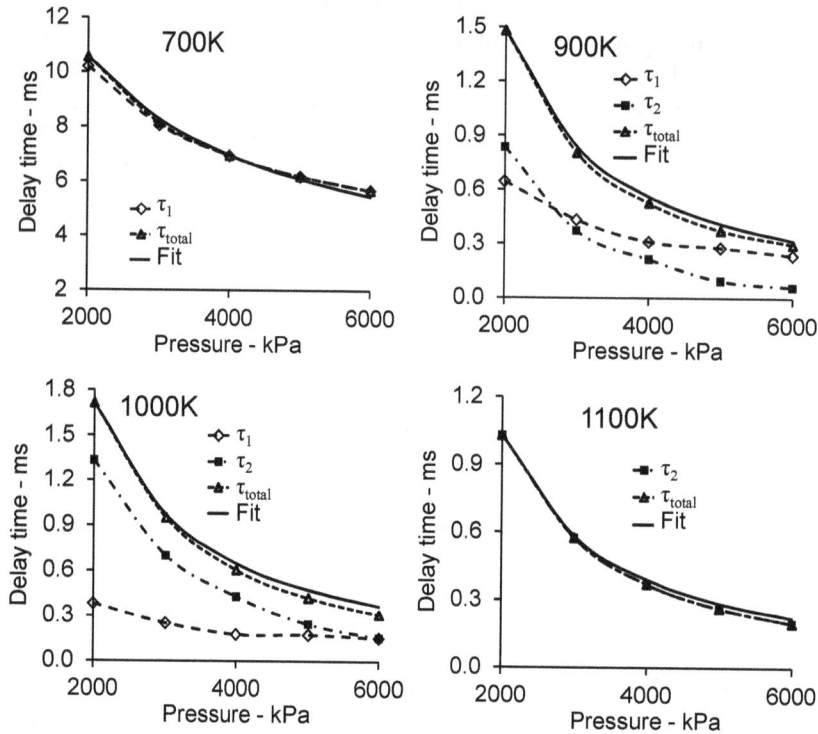

Figure 3.29: Calculated ignition of a stoichiometric mixture of n-heptane and air as a function of pressure and temperature at constant volume.

all cases, an increase in pressure shortens the ignition delay.

The final parameter changed with the model is the fuel air equivalence ratio. Calculations were performed for n-heptane with a variation of fuel air ratio from rich to lean at a pressure of 40 bar in a constant volume container. The results, show in Figure 3.30 indicate that as the mixture is made richer, the ignition delay shortens, consistent with a knock tendencies with spark ignition engines [4].

The fact that both temperature and pressure increase cause more rapid ignition has significant implications for for spark and compression ignition engines. In the former, it means the high pressures and temperatures promote autoignition of the end gasses, leading to knock. This limits important performance and economy factors such as high compression ratio, and the amount of pressure charging that can be accepted. Fuel composition is important, and new combustion methods (for example in-

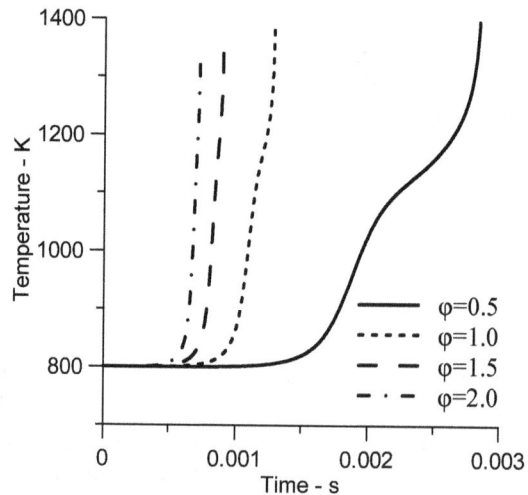

Figure 3.30: Calculated ignition of a stoichiometric mixture of n-heptane and air as a function of fuel air equivalence ratio, φ, at a pressure of 40 bar at constant volume.

cylinder injection) reduce the time the end gasses are exposed to high temperature and pressure, as well as cooling the cylinder gasses through fuel evaporation. Diesel engines, on the other hand, thrive on a fuel that is easily ignitable, and high compression ratios and pressure charging improve the combustion process by shortening the ignition delay, major factor in the diesel engines good fuel economy. Details of the model shown here, solution methods and parameters for a variety of fuels are available in the literature, for example in References [32], [34], [33] [35].

Application to other Types of Engines

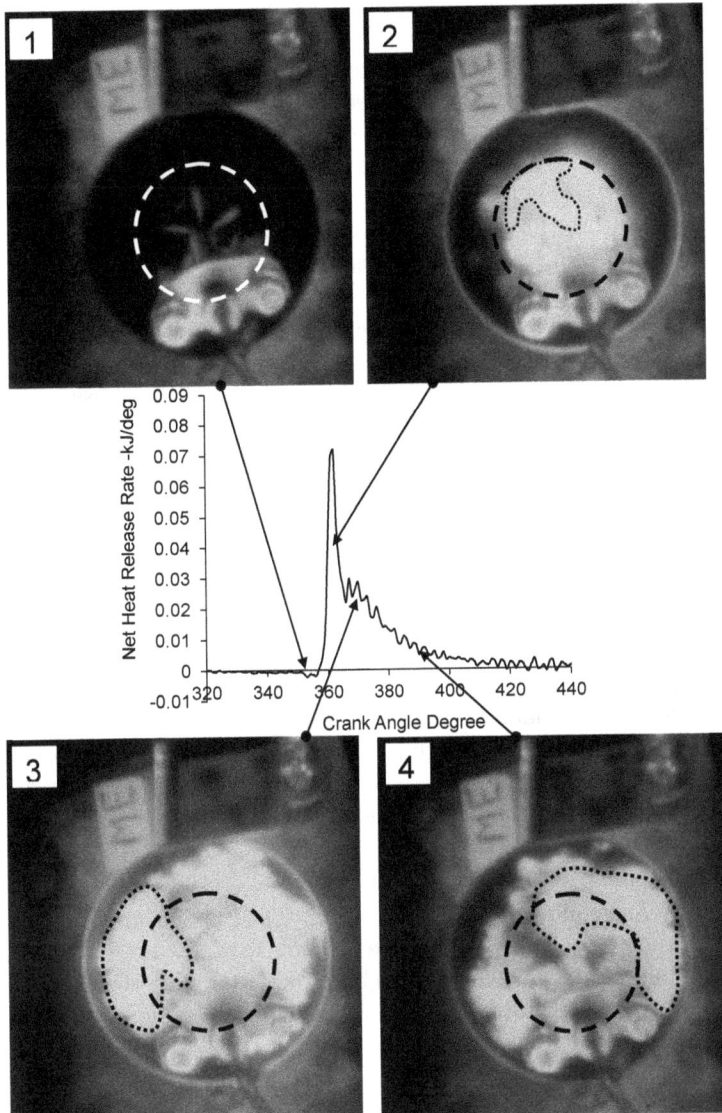

Figure 3.31: High speed photographs of the injection and combustion in a naturally aspirated direct injection engine with low injection pressure. Engine speed approx. 1200 rpm, full load. The engine is a two stroke engine, the view is from the top looking down, and the swirl is counterclockwise. Dashed line is the rim of the piston crater

The results presented for the laser diagnostics and the ensuing model are based on a system representative of a fairly large heavy-duty size engine. The cylinder is large enough that neither liquid nor intermediate products impinge on the cylinder walls. There is essentially no air motion within the cylinder. The engines used in passenger cars have much smaller bores and even smaller piston bowls. This makes wall impingement of the spray virtually inevitable, but is difficult to say to which extent impingement occurs in small engines. The large bore experiments show less liquid penetration than it was traditionally expected; also, they show that all the liquid evaporates before combustion, so there are no droplets burning. Note that in Figure 3.21, the liquid fuel extends only about 20 mm into the spray. It was previously thought that there was liquid impingement throughout the combustion process, but for this engine at least, that does not appear to be the case. Other studies have shown that liquid exists for a similar distance. For larger engines, then, if the spray impinges on the walls, it is likely to be a spray of vaporized fuel and not liquid.

While impingement as such, in view of this, may occur to a lesser extent than what was previously thought, data are lacking to explain what happens if impingement does occur. Combustion pictures do show that wall impingement of liquid fuel does occur before combustion if the ignition delay is long enough. It is generally expected that if indeed wall impingement occurs, impinged vapor will experience lower mixing rates and hence lower burning rates. Impinged liquid will stick to the wall and either burn directly from the wall or continue to evaporate. In both cases, the result will be much lower burn rates and probably incomplete combustion. Swirl is introduced as well to help avoid wall impingement, by blowing the spray into the areas of the cylinder with air as seen in 3.17. The swirl may help to burn liquid fuel from the walls of the piston, through convective heat and mass transfer. Figure 3.31 shows some of these processes using an older direct injection engine, on which was mounted a clear cylinder head. The engine is a two stroke engine with both intake and exhaust ports, thus the only obstruction to view is the injection nozzle and line. The engine bore is about 110 cm, and the bowl in the piston about half that. There are 4 holes in the injection nozzle. Floodlights were used to make the spray visible, and the injection system is a conventional jerk pump mechanical system. Injection pressures are not known, but are expected to be much lower than current common rail pressures.

View 1 shows the injection process, about 5 CA degrees after the start. The visible spray as can be seen has already reached the walls of the piston bowl, and conventional correlations for spray penetration indicate that this is nearly unavoidable with engines of this size. With higher injection pressures, the penetration rate will be even higher. The placement in the cycle with regard to the combustion rate is indicated by the arrows to the heat release diagram. View 1 is before combustion, and the effects of the fuel evaporation can be seen in that a negative heat release is observed.

View 2 show combustion during the premixed phase near the transition to the diffusion flame. The outline of the flame from the topmost injection spray has been outlined with a dotted line, to make it more visible. The regions are not sharply defined, and these areas are a rough indication of the location of the combustion. It can be seen that the spray has impinged on the wall, and is being deflected along the walls. The piston bowl has vertical walls, and does not use the reentrant design. View 3 shows combustion of this region in the diffusion combustion region. It has rotated to the left, and the combustion has now expanded into the region above the outer reaches of the piston, in a reverse squish type of motion. The visible (diffusion) flame is an envelope surrounding the burning jet.

View 4 shows combustion near the end of the diffusion period. The plume shown is the later stages of combustion of the plume from the nozzle hole clockwise to the hole of views 1-3, indicating the effect of the swirl. The portion of the plume in the bowl appears to have rotated farther than the portion over the outer reaches of the piston. This is thought to be due to the higher swirl in the bowl, since conservation of angular momentum says that it must rotate slower.

While the pictures here do not give a detailed analysis of the process that advanced laser diagnostics and the like can, they, along with the results of Figure 3.24, indicate that the qualitative picture of combustion determined from the advanced studies for quiescent chambers has a high degree of applicability to the medium swirl engine.

3.6.6 Fuel Effects

The quality of the diesel fuel plays a role in the combustion process. Especially the ignition process is important. One way to show this and other features of diesel combustion is to use a very different fuel for comparison. Such a fuel is dimethyl ether(DME), which has recently been shown to be a very good diesel fuel, since it operates without generating any soot at all, and has very good ignition characteristics. A comparison of DME with more conventional diesel fuel is shown in Table 3.7.

DME, synthesized from natural gas or other carbonaceous materials, is physically very much like liquefied petroleum gas (LPG), but due to its low ignition temperature and high vapor pressure, ignites

Table 3.7: Properties of typical diesel fuel and DME

Property	Diesel	DME
Chemical Formula	$C_nH_{1.8n}$	CH_3OCH_3
Molecular Weight	> 120	46
Oxygen Content - mass %	0	34.8
Stoichiometric Air fuel ratio - kg/kg	14.8	9.1
Lower Heating Value - kJ/kg	42,500	28,800
Liquid Density -g/ml @ 15° C	0.8 - 0.85	0.668
Boiling Point - ° C	> 150	-24.9
Viscosity -kg/m-s @ 25° C kg/m-s	2 - 4	0.122
Vapor Pressure @ 25° C - bar	< 0.01	5.1
Ignition Temperature - ° C	450 - 500	235
Cetane Number	40 - 55	$\gg 55$

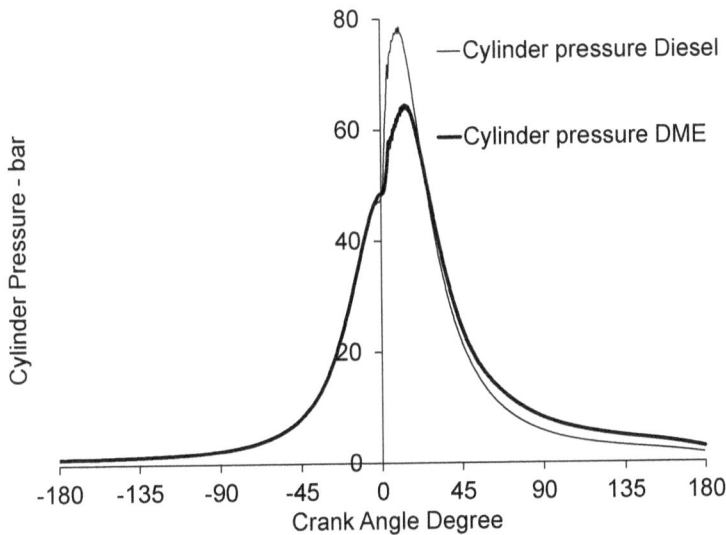

Figure 3.32: Pressure time diagrams for a 2 cylinder, 1 liter 4-stroke NA diesel engine operating on DME and diesel fuel at a brake mean effective pressure of 600 kPa.

very readily in a diesel engine, as shown by its high cetane number. On the other hand, because it is an oxygenated substance, its low stoichiometric air fuel ratio, combined with a lower heating value below that of diesel, and a lower liquid density demand a larger volume to be injected to maintain the same energy input to an engine. Some experiments have been performed comparing DME to diesel fuel, using the same fuel injection pump. These tests should indicate some of the differences in the diesel combustion process due to fuel differences.

The tests were made on a 2 cylinder, naturally aspirated, direct injection diesel engine. Figure 3.32 shows a comparison of the pressure time diagrams for diesel and DME operation at a brake mean effective pressure of 600 kPa.From the pressure time diagrams, it can be seen that the DME has a much lower rate of pressure rise at the start of combustion, and that the peak cylinder pressures are lower. This results in lower noise. In general higher cetane fuels will result in lower noise from a diesel engine.

Figure 3.33A shows the injection line pressure for both fuels in a traditional jerk pump-line-nozzle-system. The fuel injection nozzle opens at a pressure of 100 bar for DME and 230 bar for the diesel. The opening pressure is pre-determined by adjusting the needle spring in the nozzle. Looking at Figure 3.33A, it can be seen that with the injection pump in the same condition for both fuels, the pressure rise occurs earlier for the diesel fuel and is much faster than for the DME. The DME is more compressible than the diesel fuel, and therefore, the pressure rise that occurs when the pump begins to compress the fuel is lower for the DME, and the fuel initially is compressed rather than injected.

The consequences of DME's higher compressibility can also be seen in the injection pressure data.

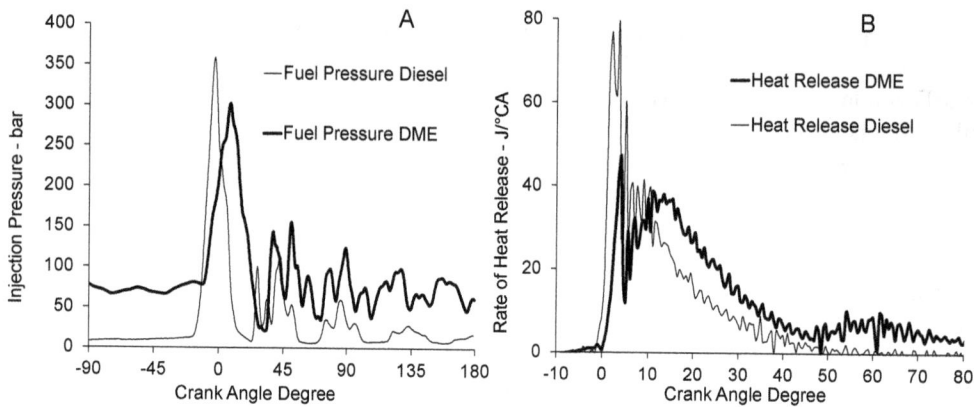

Figure 3.33: Injection pressure - time (A) and Heat Release - time (B) diagrams for a 2 cylinder, 1 liter 4-stroke NA diesel engine operating on DME and diesel fuel at a brake mean effective pressure of 600 kPa.

For both fuels, pressure oscillations arise in the line between the high pressure plunger and the injection nozzle. They arise due to the rapid start and stop of the fuel during the short injection period. The DME pressure oscillations are larger than for diesel, because more energy is stored in the compression of DME. Similarly, the residual pressure (the average pressure in the line when the nozzle is closed and the plunger not moving,) is also higher for the same reason. Pressure oscillations arise in all high pressure fuel systems and need to be taken into consideration in system design.

This has the potential to give some problems, which can also be seen in Figure 3.33. In the injection system used, the injection nozzle uses a spring, to hold a needle closed and prevent injection. If the injection pressure is above the opening pressure of that needle, fuel is injected. The high residual pressure and the large pressure oscillations with DME cause the injection pressure to exceed the opening pressure of 100 bar for a short period around a crank angle position of 45° after top dead center. This means that some fuel could be injected very late in the combustion process. This would not be good for fuel consumption or possibly exhaust emissions, since there is very little pressure in the fuel system to promote a spray and good combustion.

A comparison of the combustion for the two fuels is shown in Figure 3.33B. The injection of DME is slightly later than the diesel fuel, due to the higher compressibility, and possible some leakage, since the DME has a much lower viscosity than the diesel fuel. But the first phase of the ignition of the DME is not as violent at that of the diesel, in agreement with the pressure data in Figure 3.32.

It can be seen that more of the DME burns in the diffusion flame mode, after the ignition of the fuel than does the diesel fuel. An additional factor is that the DME combustion period is longer than that of the diesel. This is due in part to the fact that a greater volume of DME must be injected to inject the same amount of energy, due to DME's lower heating value and density. Since the piston of the fuel injection pump was not changed, the increase in fuel must come through a longer injection period, seen in Figure 3.33A.

Also noticeable in Figure 3.33B is an increase in the heat release rate shortly after 50° ATDC for the DME combustion. This is most likely due to the late injection of fuel (after injection) mentioned above. It was also found that the engine on DME had higher emissions of carbon monoxide, symptomatic of an inadequate diffusion type of combustion. After injections can also occur with diesel fuel and give increases soot/particulate and CO emissions, though they are not common with a properly developed combustion system. They comprise an example of phenomena to be avoided in practice.

Other oxygenated fuels have been tested in engines. In particular, vegetable oil based fuels have been found to give good combustion, though less than spectacular emissions benefits. These fuels have the potential to improve on greenhouse gas emissions, but the energy required to produce them remains a problem in the total balance of CO_2 emissions.

3.6.7 Conclusions

The general features of the combustion process in diesel engines have been discussed here. The topic is a very complicated one, and in combustion research recent years has involved major developments like the increased use of computational fluid mechanics to study the flow processes involved in combustion and fuel chamber. Modern diagnostic methods based on laser technology have also greatly improved our knowledge of the important elements of diesel combustion. Reference [29] is a good example of our greatly improved knowledge due to these diagnostic tools. Since it has the highest efficiency of all engines today, it will be relied upon in an increasing degree to help the problems of the emission of CO_2 as a greenhouse gas. While doing this, the local environment cannot be ignored, and great challenges are presented for future engines to operate at very low NO_x and particulate levels at the same time. All of the understanding acquired to date, and more, will be necessary to solve this challenge if the small diesel engine is to survive in the urban environment.

3.7 Cylinder Pressure Measurements

When discussing combustion in engines, it is important to start with $p-V$ or $p-t$ ($p-\theta$) diagrams, the so-called indicator diagram. This is because a large amount of information concerning the combustion process is available from this diagram. The measurement of cylinder pressure in combustion engines has become a standard procedure, though one must be careful when making these measurements [36], [37]. Pressure measurements are discussed in Section 1.2.2.

The $p-V$ diagrams usually constructed from pressure-crank angle measurements, from which one calculates the cylinder volume using the engine geometry, such as Equation (2.106), and reference timing markers. By far the most common geometry is the slider crank mechanism, but it is possible to use other geometries, for example the Wankel engine, and swash plate engines, *etc.* The $p-V$ diagrams from these engines resemble each other to a large extent. As will be seen later in the determination of friction, it is very important to have an accurate knowledge of piston position, since even small errors in position are magnified when determining indicated work and especially engine friction.

When making pressure measurements, it is important to have representative measurements. It is found that consecutive pressure time or volume diagrams are not identical, even though the engine is operating at a steady condition. Figure 3.34 shows some typical pressure-time diagrams for a spark ignition engine.

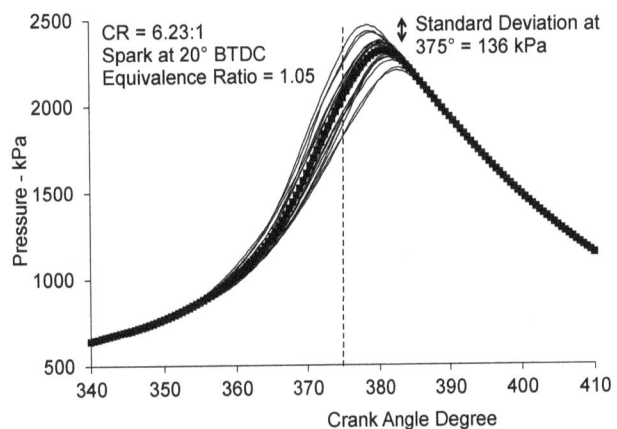

Figure 3.34: Typical pressure time measurements for 20 consecutive cycles of a CFR spark ignition engine at 1200 rpm, showing the standard deviation of cylinder pressure at 15° After TDC. The square data points show the average pressure for the 20 cycles.

The pressure traces are different for each cycle, even though the engine is operating under steady state conditions, that is constant speed and load. The differences in the pressure diagrams are due to a phenomenon called cyclic variation, which is a subject that has been discussed for a long time in connection with spark ignition engines. Studies in SI engines have shown the variations to be due to random fluctuations of the local flow velocity at the spark plug at the time of spark [38]. In some cycles, the first stage of the flame is blown from the spark plug away from the center of the combustion chamber, where the flame is slow because it is blocked by the combustion chamber wall. This corresponds to the slowly rising curves in Figure 3.34. In other cycles, the initial flame is blown towards the center of the combustion chamber, where there is a large flame area, and the combustion is rapid. This corresponds to the fastest rising curves in Figure 3.34. Notice in Figure 3.34 that the pressures are identical before the time when the spark is ignited (340°, or 20° Before Top Dead Center-BTDC).

Given the cycle-to-cycle variations experienced, it is common to collect a number of consecutive cycles and then average them to obtain an average rate of combustion when doing combustion analysis. Combustion rates (heat release rates) for individual cycles can also be obtained if desired, as will be shown later in this section. It is common to collect from 100 to 300 consecutive cycles for averaging. CI engines tend to have smaller cycle-to-cycle variations than SI engines, but cyclic variation should not be ignored for accurate work in CI engines.

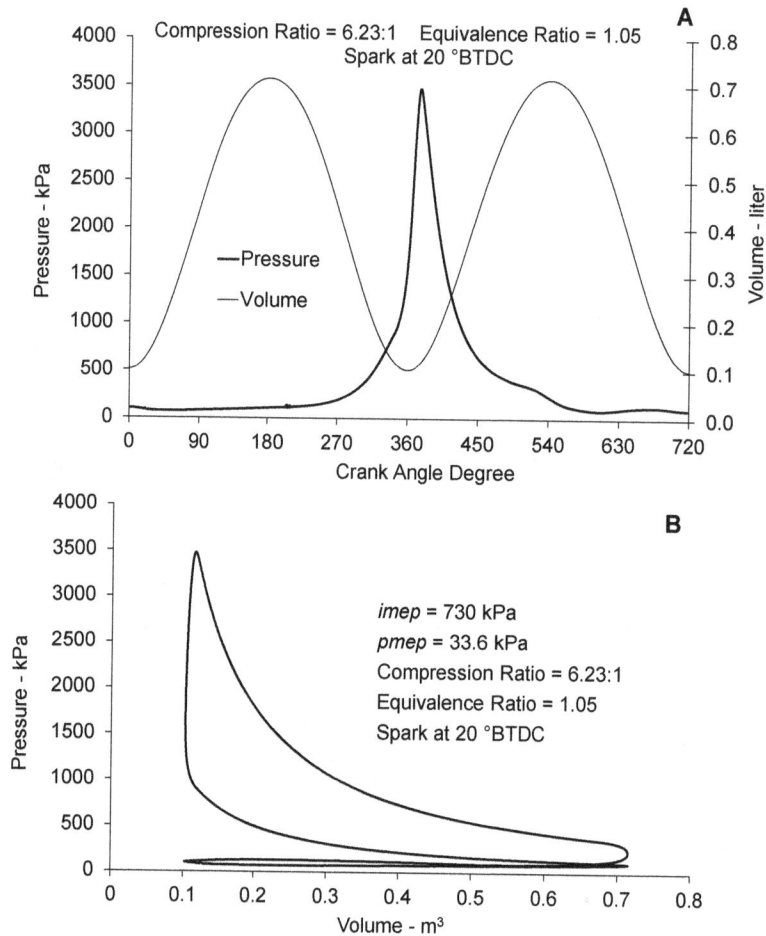

Figure 3.35: Basic data for volume and cylinder pressure time measurements for one cycle of a CFR single cylinder spark ignition engine operating at 1200 rpm with wide open throttle.

Figure 3.35A shows the cylinder pressure and volume of a CFR engine as a function of crank angle. The pressure data used here is the average cylinder pressure from the data shown in Figure 3.34. This data can be replotted as a $p - V$ diagram, and the results are shown in Figure 3.35B. The indicated mean effective pressure has been determined from integrating the $p - V$ diagram over the compression and expansion processes. Similarly, the magnitude of the pumping mean effective pressure ($pmep$) has been determined by the area of the pumping loop at the bottom of the $p - V$ diagram.

Figure 3.36 shows a pressure time curve for a non-turbocharged, direct injection diesel engine and its corresponding pressure volume curve. Of particular interest is that there is a sharp pressure rise at the start of the combustion process in the diesel engine. This is caused by the autoignition of the fuel a short time after the injection starts. It is a kind of explosion, similar to that which occurs in a knocking spark ignition engine. However, here it is an integral part of the combustion process, and occurs at the start of combustion, not at the end as in a knocking spark ignition engine. It gives rise to the characteristic diesel noise, and diesel engines must be built to withstand this.

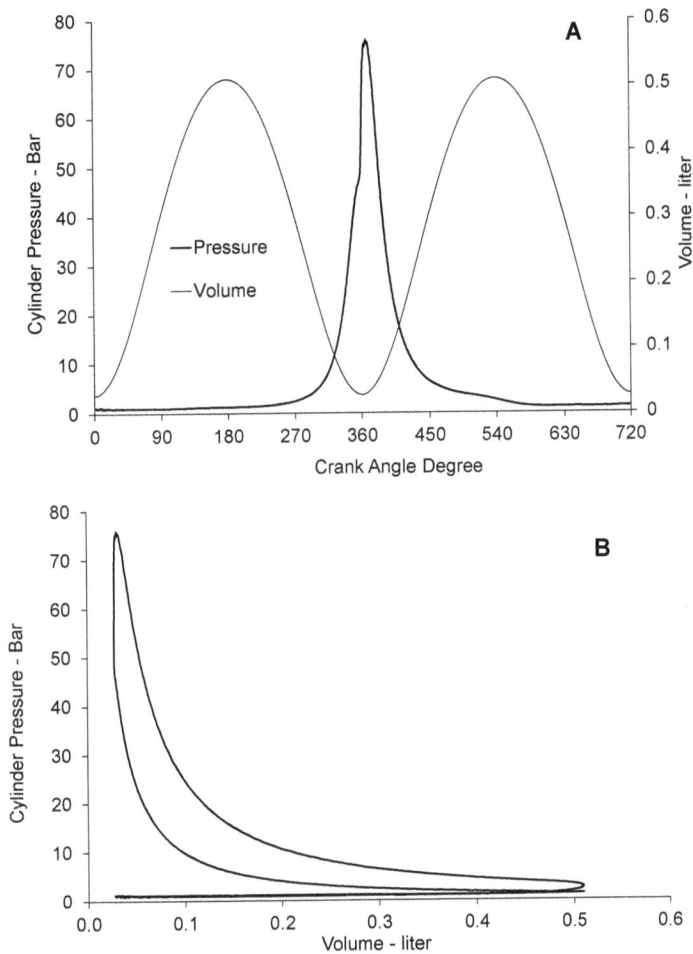

Figure 3.36: Basic data for volume and cylinder pressure time measurements for one cycle of a naturally aspirated direct injection diesel engine: 2-cyl with 0.483l/cyl, 2355 rpm, bmep= 655 kPa, imep=832 kPa.

A significant amount of information is available directly from the indicator diagrams themselves. For example, Figure 3.37A shows spark ignition engine p-V diagrams for 3 different ignition timings. At a timing of 20° BTDC, the maximum output is obtained, which is MBT timing. The diagram resembles those for the ideal processes, in which the combustion is said to occur with constant volume. This is nearly true in an engine that operates with maximum torque, since the movement of the piston near top dead center is very small.

Figure 3.37A also shows a $p-V$ diagram for a spark ignition engine where the spark timing is advanced about 12 degrees in relation to the optimum timing. That is, the ignition occurs too early. The figure shows that the pressure is much higher during the later portion of the compression stroke, resulting in an increase in the compression work, and a much higher compression pressure. The increase in compression work causes the output of the engine to decrease. Another disadvantage of the very early timing is that the very high cylinder pressure increases the tendency of the engine to knock, described above.

The opposite timing situation is shown in the third curve in Figure 3.37A, where the spark timing is very late, in this case, at top dead center. The combustion occurs late and the pressure rise from combustion is nearly balanced by the expansion as the piston proceeds downwards. This approaches constant pressure combustion, which was shown in Section 2.1.3 to be less efficient than the constant volume process.

When the mixture burns so late, full advantage cannot be taken of the complete expansion stroke of the engine, and thus power is reduced over that when the combustion is completed nearer top dead

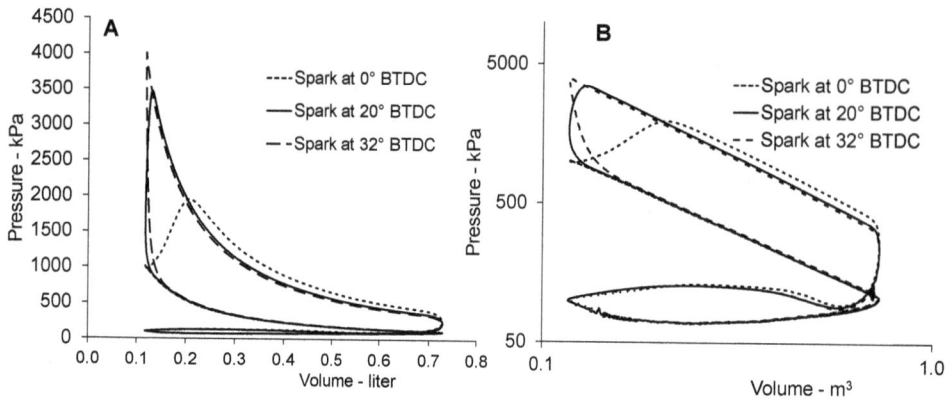

Figure 3.37: A: Pressure-volume diagram for a spark ignition engine for different spark timing, 20° BTDC, the optimum for this engine and condition, advanced timing 32° BTDC, and retarded timing, 0° BTDC. B: Logarithmic diagrams of the same data. The engine is operating at 1200 rpm with wide open throttle, with a fuel air equivalence ratio of 1.05 and a compression ratio of 6.23:1.

center. Late combustion is characterized by high exhaust temperature, since the less complete expansion results in a larger amount of energy in the exhaust gasses leaving the cylinder.

The curves on the compression and expansion strokes of the piston are close to being polytropic, that is, they are quite close to being linear when the logarithm of the pressure is plotted as a function of the logarithm of the volume. Examples of this are shown in Figure 3.37B for the same data. This characteristic is often useful when examining engine pressure data, as if the logarithmic $p - V$ plot is not linear on the compression and expansion strokes of the piston, there is most likely some kind of error in the basic engine data that has been recorded.

Since the combustion process is equivalent to a heat addition process, it is not isentropic, and during combustion, the $\ln p$ - $\ln V$ curves should not follow the linear relationship seen on the compression and expansion processes. This is quite clear from the curves in Figure 3.37B. The start, end and duration of the combustion process can clearly be seen from the logarithmic curves, whereas they are basically indiscernible from the linear p-V diagrams. This information is an immediate advantage to plotting pressure volume information on a logarithmic scale. In addition, errors in pressure measurements are easier to see.

For a spark ignition engine, the pressure-crank angle curves are normally smooth, (see Figure 3.34) as the flame propagates uniformly across the combustion chamber. In abnormal circumstances, some of the mixture explodes before the flame reaches it and the pressure rises sharply. This process is known as *knock*. The pressure-time history of a knocking cycle is shown in Figure 3.38. The chemical process is the same as that for autoignition in the diesel engine described in Section 3.6.5, but with significantly different reaction rates. Note that the sharp pressure rise comes at the end of the combustion process. The pressure rise, which occurs in a portion of the combustion gasses, gives rise to the noise as well as creating shock waves that propagate back and forth in the cylinder. This process substantially increases the rate of heat transfer through the boundary layer at the combustion walls. This in turn increases the thermal load on engine components and if the engine is not designed to withstand this extra load, it can be destroyed, often in a very short time.

A final set of curves is shown in Figure 3.39, which shows $p-V$ diagrams with varying load for an SI engine and a non naturally aspirated, direct injection DI engine. These curves show the fundamental difference in operation between

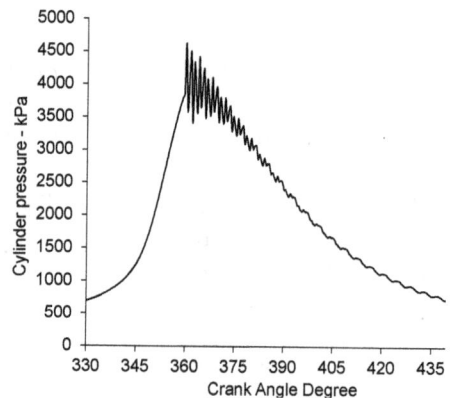

Figure 3.38: A pressure time diagram for a spark ignition under knocking conditions.

these two types, as discussed in previous sections. The SI engine shows the characteristic behavior of being regulated by changing the intake pressure (throttling) to regulate the amount of fuel and premixed air. This gives a change in the pumping loop, and lower pressures throughout the cycle as the load is reduced. The CI engine on the other hand, is non-throttled, resulting in a constant intake pressure and pressure-time history on the compression stroke. The load variation is manifested by the increase in the cylinder pressure at the end of combustion, and the higher cylinder pressure throughout the expansion stroke when the load is increased. For a turbocharged CI engine, the entire pressure curve would increase, as both the intake and exhaust pressures increase with load.

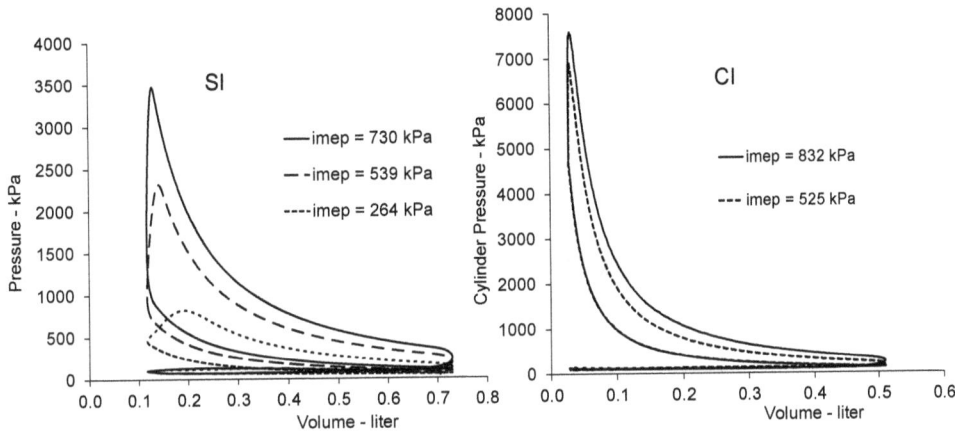

Figure 3.39: Pressure-volume diagrams for a CFR spark ignition engine (SI) for different 3 different loads. The engine is operating at 1200 rpm with a fuel air equivalence ratio of 1.05 and a compression ratio of 6.23:1. The non-turbocharged DI diesel engine (CI) is operating with two loads at a 2300 rpm, with a compression ratio of 18:1.

3.8 Heat Release Analysis

3.8.1 Calculation of Heat Release Rate

In order to understand the combustion process more directly, it is necessary to obtain a direct expression for the combustion rate. This is done with the so-called heat release calculation. Heat release is the term very often used to describe combustion rate. The expression comes from early experiments with the measurement of combustion rates in engines, where the effects of combustion were compared to those of an equivalent rate of heat "released" by the combustion process to the cylinder gasses. The term apparent heat release is often used to emphasize that it is not really heat transfer, but equivalent effects that are calculated. Here, the term heat release will be used for simplicity.

The basic idea of the process is discussed in Section 2.7.1. Equation (2.104) can used with experimental pressure time/crank angle curves as input, as well as a knowledge of the cylinder volume as a function of the time/crank angle. For more complicated analysis, the cylinder can be divided into 2 or more zones and chemical equilibrium of the combustion products is considered, but the procedure is basically the same [19]. Equilibrium effects are most important for spark ignition engines, as they operate with richer mixtures, with higher temperatures of the combustion products than diesel engine.

Equation (2.103) is often used to calculate the net apparent combustion rate, that is the rate at which energy apparently has been added to the combustion chamber to give the measured pressure rise. The net heat release rate is the combined result of the energy conversion by combustion and the heat loss to the walls of the combustion chamber. It is often used because the heat transfer process is difficult to calculate accurately. In particular, heat transfer in the diesel engine is very difficult to model, due to the combination of radiation and convection heat transfer. The effects of energy release from combustion and heat transfer to the cylinder walls are, therefore, combined in this calculation. Though not the true combustion rate, the results are useful for showing combustion rates.

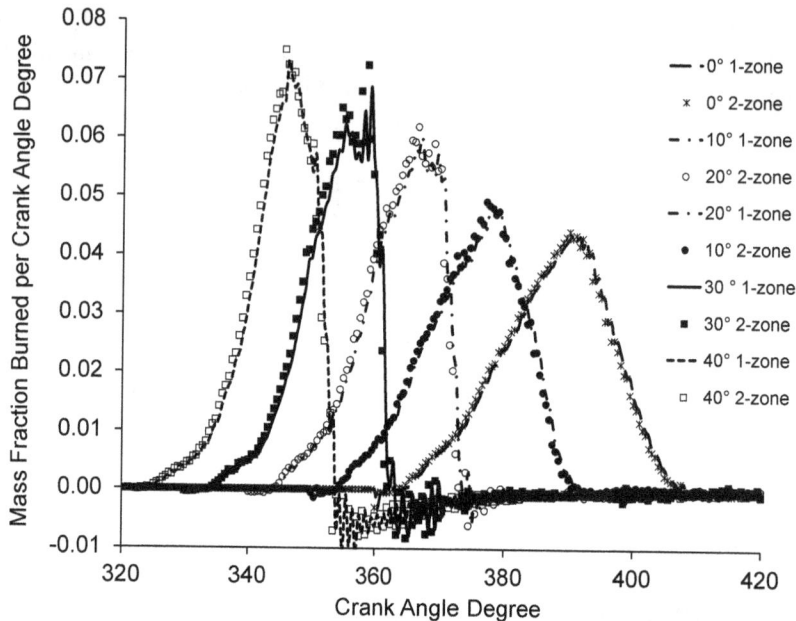

Figure 3.40: Instantaneous rate of heat release as function of crank angle degree for a CFR spark ignition engine operating at full load at 1000 rpm using 1-zone and 2-zone combustion models. For the single zone model, a constant specific heat ratio of 1.26 was used. The spark timings are degrees BTDC.

Figure 3.40 shows the net apparent heat release in a CFR engine as a function of spark timing using two different calculational methods. The first is a very simple single zone model using Equation (2.104). The second method uses the two-zone model of Equation (2.136). It can be seen that the basic shape and duration of the combustion rate are the same irrespective of the methods used. In fact, by choosing and appropriate value of the specific heat ratio, the results become nearly identical. In terms of determining the timing, rate and duration of combustion, then, a simple model is quite adequate. There are often oscillations in the heat release rate curves, especially for individual cycles. With the exception of knocking conditions, these normally arise from numerical differentiation of experimental pressure data, but can reflect pressure oscillations in the cylinder or pressure sampling system. If such oscillations occur, they affect the validity of the heat release calculation, though the normal combustion parameters usually can be observed.

3.8.2   Heat Release from Spark Ignition Engines

Figure 3.41 shows typical results for the rate of heat release as a function of the crank angle degree for a spark ignition engine. It can be seen that the combustion is smooth, without sharp increases or decreases in the rate. This is typical for a spark ignition engine that is not knocking. Empirical functions that resemble the curves of Figure 3.41 are used in simulation programs for SI engine combustion. While these empirical curves will be useful, they are not really connected to the physical processes controlling combustion in spark ignition engines. Also shown in Figure 3.41 is the cumulative heat release as a function of crank angle degree for a spark ignition engine. This results from an integration of rate of heat release throughout the combustion.

$$Q_{total} = \int_{\theta_{ignition}}^{\theta_{end}} \dot{Q}_{HR} d\theta \tag{3.124}$$

The value of Q_{total} in Equation (3.124) will ideally result in the total amount of energy burned, which in the case of lean or stoichiometric mixtures is equal to the mass of fuel burned times the lower heating value of the fuel. This requires a correction for the heat loss to the cylinder walls, typically with the Woschni relationship for SI engines. Since the heat transfer calculation is uncertain, the result cannot be expected to be precise, but it can be used as a check on the calculation, where the integrated heat release should be slightly less than the chemical energy available in the amount of fuel burned. For rich mixtures, this check cannot be used, since the lack of oxygen limits the amount of energy release by combustion to a value lover than the heating value of the fuel. The general shapes of the curves are little affected by the heat loss. An estimate of the magnitude of the effect of heat transfer can be made by using the simple simulation program of Section 2.5, where the heat transfer is included.

Figure 3.38 showed a pressure time diagram for a spark ignition engine that is knocking. The rate of pressure rise is very high with knocking, and it is not possible to measure pressure accurately under knocking conditions. The pressure rise during knocking is so fast that pressure gradients exist in the combustion chamber as shock waves propagate back and forth, and the assumptions of the heat release calculation are no longer valid. Analysis of the combustion rate during knocking is normally not of interest, since it is a condition that the engine designer tries very much to avoid.

Figure 3.42 shows the calculated rate of heat release for a spark ignition engine as a function of crank angle degree for 3 different engine speeds [40]. It is of interest to note that the shape and duration of the curves as a func-

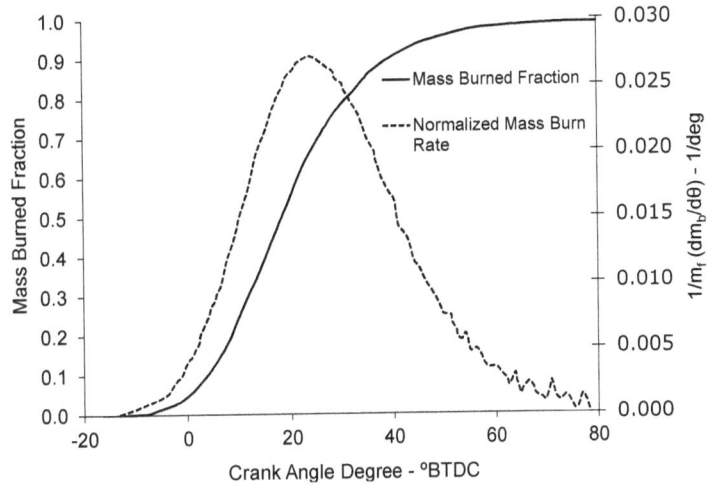

Figure 3.41: Instantaneous rate of heat release (dotted) and cumulative heat release (solid) as functions of crank angle degree for a spark ignition engine. Adapted from [39].

tion of crank angle degree is very similar, even though the engine speed varies by a factor of three. Rotation through one crank angle degree at 1000 rpm takes 0.167 ms, while at 3000 it takes only 0.0556 ms. This implies that the flame speed in m/s actually increases nearly linearly with engine speed. This is very important in the operation of engines at varying speeds. This is primarily caused by an increase in combustion chamber turbulence when the engine speed increases, permits spark ignition engine operation over a tenfold variation in speed. Because of this, the use of crank angle degrees instead of time in seconds as the cycle time variable is common in engine work.

Another set of spark ignition engine heat release curves is shown in Figure 3.43. In this case, the heat release in shown as a function of ignition timing. Although the curves are displaced in time, due to the start of combustion at different times, the basic shape of the curves is similar in all cases, and similar to that of the previous figure. This will be used later in the presentation of a simplified empirical model for the heat release rate in spark ignition engines.

Also to be seen in Figure 3.43 is the fact that the heat release rate first becomes noticeable at a significantly later time than that of the spark. For example, in the case of the spark timing of 30° BTDC (330°), the heat release curve begins to rise above the x-axis at about 338°, which is about 8° after the spark. During this time interval, the flame is very small, but growing. It takes about 15° in this case for the flame to be large enough to release sufficient energy to be detectable in the measurement system. The time between the spark and the first observable

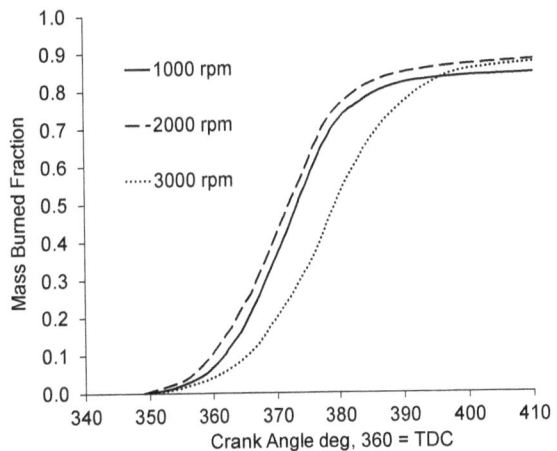

Figure 3.42: Cumulative heat release for a spark ignition engine as a function of engine speed. Adapted from [40].

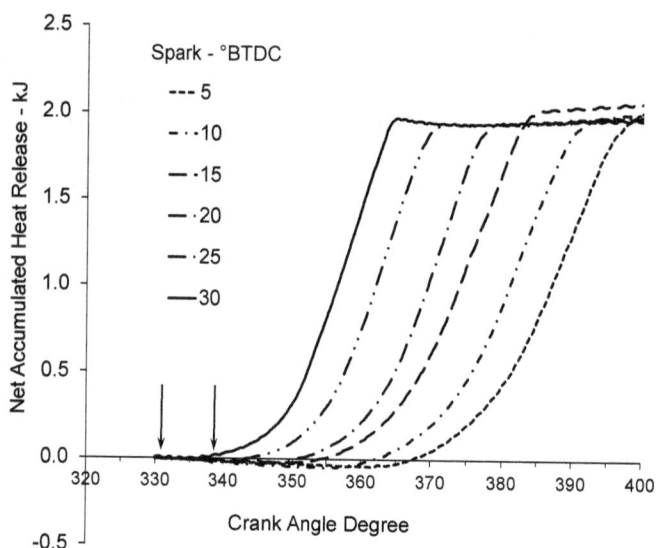

Figure 3.43: The net cumulative heat release for a spark ignition engine as a function of ignition timing. The arrow on the left indicates spark timing at 30° BTDC, and the arrow on the right the start of detectable heat release at that timing. CFR engine operating at 1200 rpm, at full load, stoichiometric mixture.

heat release is often called an ignition delay. It is not truly a delay, since the combustion is actually proceeding during this time. It is important, however, to understand the effect on the combustion process, especially when simulating the effects of spark timing. The effects of engine parameters on this "delay" are given in Section 3.5.2.

Figure 3.44 shows the cumulative heat release for an SI engine as a function of fuel air equivalence ratio. The burning rate shows very little effect of fuel air ratio when the mixture is stoichiometric or richer. The curves retain the basic shapes, but the combustion duration is longer with the mixtures leaner than stoichiometric. The delay between the time of the spark firing and the first measurable heat release is also longer when the mixture is made leaner.

Figure 3.44 also shows the effect of mixture strength on the amount of energy available from combustion. As the mixture becomes leaner than stoichiometric, the final value of the accumulated heat release, seen at crank angles after 390°, decreases. This is because there is less fuel for the same amount of air in the cylinder as the mixture is leaned, since the engine is operating at full load (approximately atmospheric pressure in the intake). As the mixture is made richer, more fuel added to the cylinder, and therefore the potential chemical energy in the cylinder increases. However, once the mixture is richer than stoichiometric, there is no longer any more oxygen to completely burn all the fuel. As a result, the total amount of heat release is oxygen limited and remains approximately the same for stoichiometric and richer mixtures. As will be shown later in Section 3.1.2, the exhaust gas will show an increase in the amounts of H_2 and CO as the mixture becomes richer than stoichiometric, and they contain the chemical energy that has not been released during the combustion. The fact that the energy released for rich mixtures is nearly constant at the stoichiometric value, is a useful piece of information for simple engine simulations.

An important result from this section is that the general shape of the cumulative heat release curve as a function of crank angle does not change much as engine variables such as fuel air ratio, speed and ignition timing change. This is consistent with the discussion in Section 3.5 which descri ed the effect of increasing turbulence with increasing engine speed on the turbulent flame speed. This behavior is very useful for modeling purposes.

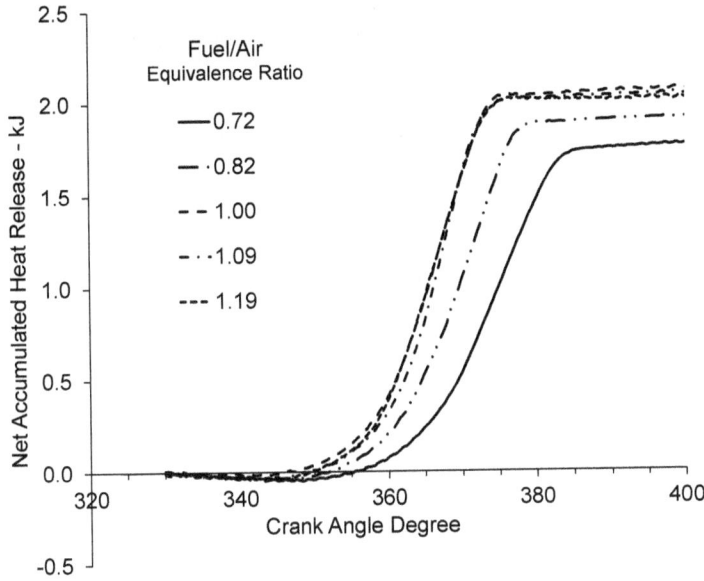

Figure 3.44: The net cumulative heat release for a spark ignition engine as a function of fuel air equivalence ratio. CFR engine operating at 1200 rpm, at full load, spark timing at 20° BTDC.

3.8.3   Heat Release from Diesel Engines

Direct Injection (DI) Engines

As might be expected from the nature of the combustion process, diesel engine heat release diagrams look quite a bit different than those for spark ignition engines. A typical heat release diagram for a naturally aspirated direct injection diesel engine is shown in Figure 3.45. The $p - \theta$ and $p - V$ diagrams

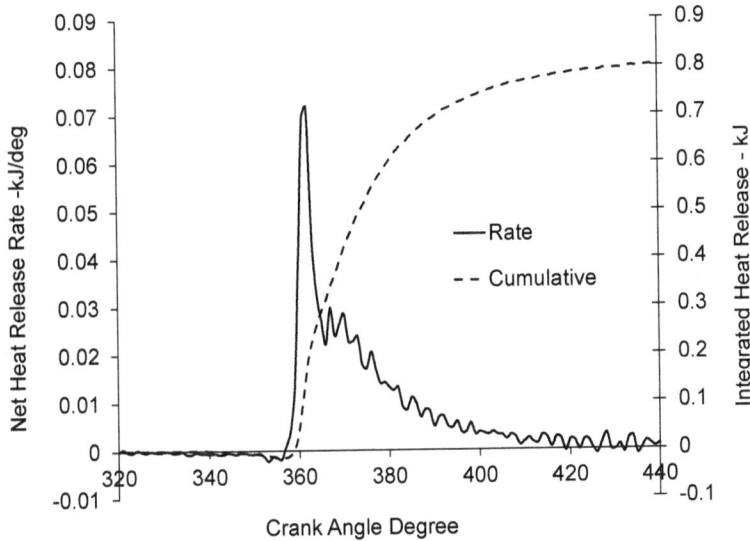

Figure 3.45: A net heat release - crank angle diagram for a naturally aspirated diesel engine. 2 cyl with 0.483l/cyl, 2355 rpm, bmep= 655 kPa, imep=832 kPa.

for this condition can be seen in Figure 3.36. Notice the very high rate of heat release at the start of the process and then the later, slower combustion towards the end. The first part, often called the *"premixed"* combustion phase, is due to the rapid combustion of the first portion of the fuel that has been injected, vaporized and mixed with air before the chemical reactions leading to combustion have progressed to the state of ignition. One can actually see a "negative heat release" just before the sharp initial rise in the heat release rate. This is due to the energy absorbed from the air in connection with the evaporation of the first portion of fuel injected. When the ignition finally occurs, there can be a large portion of fuel ready to burn, and it usually burns very quickly. This gives rise to a rapid increase in pressure, and is responsible for the noise commonly associated with diesel combustion.

When the first phase is completed, the combustion slows down, as it is now the physical mixing of the remaining air in the combustion chamber with the remaining fuel or combustion products that determines the combustion rate. This second phase is often called the *"diffusion"* phase of diesel combustion.

The two phases overlap, but they can be distinguished from each other on the cumulative heat release diagram, also shown in Figure 3.45. The curve shows that when approximately 30% of the fuel has burned, there is a marked change in the slope. This gives an approximate indication of the amount of fuel burned in the first stage of the combustion. The combustion rate at the end of the process is slow and the actual end of combustion is difficult to define precisely.

There are different forms of the heat release analysis (see for example References [19] and [41]). They differ mainly in the description of the thermodynamic properties. Equation (2.103) is in a simplified form, in that the thermodynamic properties are simply represented by a specific heat ratio. This can be a function of temperature and equivalence ratio, which improves accuracy. In a more exact analysis, one should consider the effects of chemical equilibrium on the combustion process. However, in a diesel engine, the overall mixture ratios are lean, and the effects of equilibrium are not large. Thus, a simple model as in Equation (2.103), can be expected to give a good indication of the combustion rate.

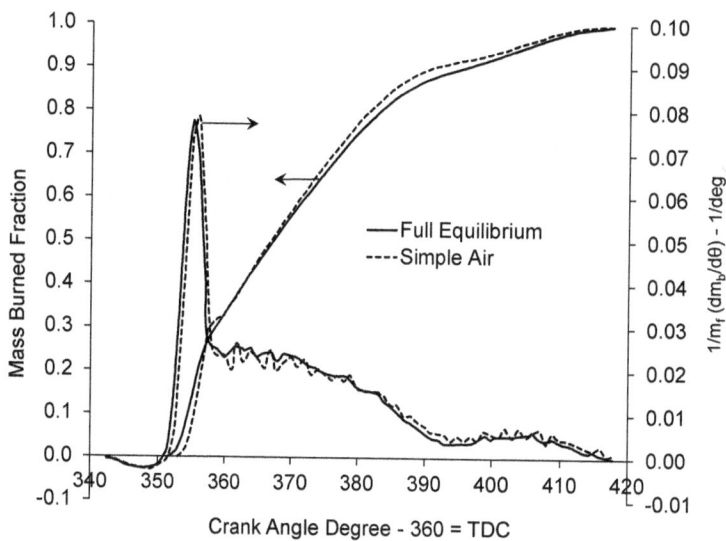

Figure 3.46: Normalized burning rate (a) and burned mass fraction for heat release calculation with a simple procedure using air with variable specific heat [42], and the full equilibrium procedure of Krieger and Borman [19]. The engine is a 3.08 liter, naturally aspirated direct injection engine operating at 1800 rpm

Figure 3.46 shows a comparison of the heat release rate for a diesel engine calculated with simple properties and temperature dependent specific heat with the rate of heat release calculated assuming full chemical equilibrium. The results indicate that there is not much difference between the two methods if a temperature dependent value of c_p is used. Notice again, the two phases of combustion, as well as a late combustion, indicating a secondary injection (see Section3.6.6). The two initial phases are especially pronounced in the case of non-turbocharged engines, as is the case in Figure 3.46.

For diesel engines then, the effects of chemical equilibrium have been shown not to be of major importance when using a single zone combustion model. This is because of the lean mixtures in diesel engine, which cause temperatures lower than those required for significant dissociation to occur. This is due to the excess air in the diesel combustion, such that the effects of chemical equilibrium are not as important as they are in spark ignition engines. Another difference in heat release analysis methods is the assumption of the distribution of the gasses in the combustion chamber. For spark ignition engines, it is common to divide the combustion chamber into at least two zones, a burned zone and an unburned zone. This is natural, since the combustion of the homogeneous mixtures results in two volumes: a burned and an unburned. The effects of thermodynamic properties are more important in spark ignition engine heat release analysis, due to the higher temperature in the burned zone. The zones are not as sharply defined in diesel engines and there is more mixing than in spark ignition engines, so single zone models are more appropriate.

The previous diesel engine heat release results have all been for naturally aspirated engines, that is, engines without a turbocharger. In modern usage, most diesel engines are equipped with turbochargers to increase the specific power of the engine as well as the engine efficiency (through a better mechanical efficiency). Turbocharging gives higher intake pressures and temperatures, which result in higher temperatures and pressures at the time of fuel injection. These hasten the ignition of the fuel, and the first premixed peak in a turbocharged DI engine will be less pronounced than that in a naturally aspirated engine. At the same time, recent emission control regulations have led to higher injection pressures (> 1000 bar) which reduce the particulate emissions caused by the later injection timing which is used to reduce NO_x emissions. The increased intake charge pressure and temperature, higher injection pressures, and later timing (such that the cylinder pressure and temperature are higher) all contribute to a shorter ignition delay. This means that less fuel has been vaporized and prepared to burn when ignition occurs, resulting in a less violent ignition.

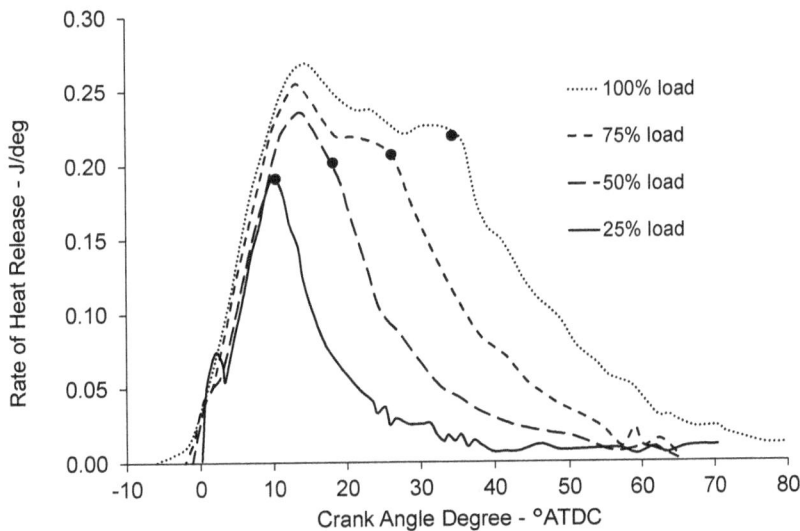

Figure 3.47: Rate of heat release for different loads in a turbocharged direct injection diesel engine, with a displacement volume of 2 liters per cylinder. Dots indicate end of injection. Data from [43].

Some examples of heat release in a modern turbocharged diesel engine are shown in Figure 3.47. The injection timing in all cases was 3° BTDC. This injection is rather late, partly in order to reduce NO_x emissions and partly to prevent the combustion from occurring too early in the cycle and thus avoid exceeding maximum pressure limitations for the engine.

First of all, Figure 3.47 shows that the ignition delay is quite small. Because of this, the first phase of combustion is not nearly so prominent as it is for the naturally aspirated engines shown previously. The second thing of interest is that as the load increases, the first phase of combustion becomes less and less prominent, and at the highest load, cannot be distinguished from the rest of the combustion. The start of heat release also occurs slightly earlier as the load increases. With a turbocharged engine, the increased load causes higher intake pressure and temperature, which in turn increase the temperature and pressure at the time of injection, which reduces the ignition delay. The combustion rate in turbocharged engines is primarily controlled by the rate at which the fuel and air can be mixed in the combustion chamber, and the chemical reactions play a less important role than with naturally aspirated engines.

Also to be observed in Figure 3.47 is the increase in duration of combustion corresponding to the increase in load. The engine operates with a unit injector, and at higher loads, the increased amount of fuel is achieved by a longer stroke of the injection pump. The initial portions of the heat release diagrams are almost identical, with only small differences arising from the very first portion of the heat release. The fuel goes into the cylinder at the same rate, regardless of the load, so the start of combustion is the same at for all loads. Increasing fuel amounts are provided by lengthening the period of injection. The end of injection is shown by the dots in Figure 3.47. This occurs at 10° ATDC for the *bmep* of 5 bar (25% load), and at 33° ATDC for the *bmep* of 22 bar (100% load). It can be seen that there is a marked change in the shape of the heat release curve after the injection stops. After the end of injection, the heat release rate falls rapidly. For the higher loads, there is a period where the heat release rate is approximately constant, indicating that the spray is stable, and there is a balance between input fuel and combustion.

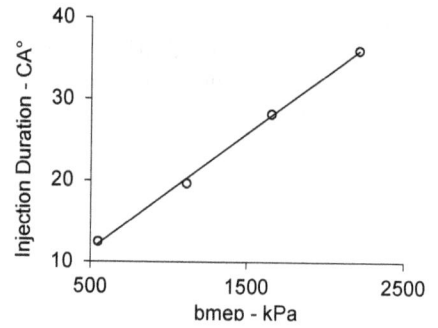

Figure 3.48: Injection duration for the engine of Figure 3.47 as a function of brake mean effective pressure.

The injection duration as a function of loading is shown in Figure 3.48. A straight line has been drawn through the results to indicate that the relationship is nearly linear. A small upward curvature might be expected due to the higher load and longer duration giving combustion further on in the expansion stroke, with a resulting efficiency loss, which would mean a little more fuel needed per increment in mean effective pressure. Such loss here is small. For a shorter duration, the curve would not be linear, since the opening and closing times of the injector would play a more significant role. From Figure 3.48, it appears that at the *bmep* of 5 bar, the nozzle is fully open.

Figure 3.49 shows the pressure volume diagram and the injection rate and rate of heat release curves for a large marine diesel engine. The heat release curve is quite smooth, and an initial phase of combustion

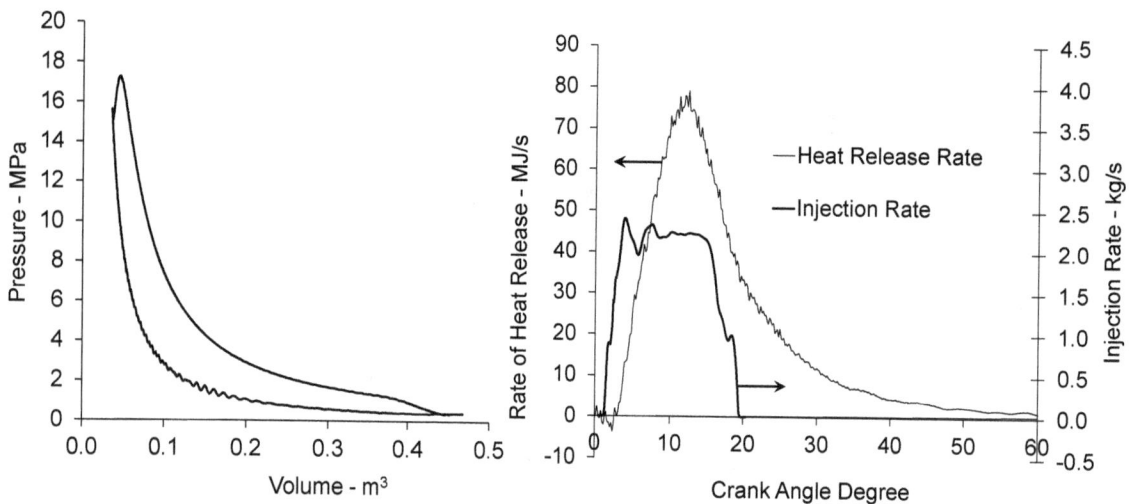

Figure 3.49: Pressure volume diagram (left) and injection and heat release rates (right) for a large 2-stroke marine diesel engine. The engine has a bore of 0.50 m, a stroke of 2.2 m and produces 1.79 MW per cylinder at a speed of 123 rpm.

as found in smaller engines is not to be seen, though there is a slight "negative" heat release before the start of combustion due to the energy required for fuel evaporation after injection. The ignition delay in the premixed phase determined by chemical reactions, which do not "scale" with engine speed. Thus at this slow engine speed, an ignition delay of a few milliseconds is not very significant. This type of engine normally operates on very heavy fuel, with different characteristics than normal diesel fuel. The engine is highly turbocharged, and injection of fuel occurs after TDC in order to limit cylinder pressure, as was discussed in Section 2.1.4. The high temperatures and pressures at injection also contribute to a relatively short ignition delay.

A major development affecting the nature of the heat release curve in CI engines today is the use of electronics. Electronically controlled injections systems are much more flexible than their mechanically based predecessors, and combined with common rail technology have substantially altered CI combustion. With rapid electronically controlled injectors, it is possible to have multiple injections for a combustion process. One example of this is the use of pilot injection. In this process, a small preliminary, or pilot, amount of fuel is injected a considerable time before the main combustion is desired. This mixture reacts and can ignite, but because it is a small amount of fuel, the effects of its combustion on the pressure diagram and engine noise are small.

But since the pilot fuel reacts, it serves to provide additional heating for cylinder gasses into which the main portion of the fuel to be injected. This reduces the delay between the injection of the main portion of the fuel and its ignition. This in turn reduces the large pressure rise normally experienced in a diesel engine. This helps to reduce the noise and the emissions of NO_x.

As a final example, for direct injection diesel engines, Figure 3.50 shows the net rate of heat release for a modern, common rail direct injection diesel engine with electronic injection control and pilot injection. From Figure 3.50 it is immediately apparent that there is pilot injection. This can be seen by the small amount of heat release starting at about 10° BTDC. There is not a major amount of heat release, but its effect on the rest of the process can be seen by the lack of the very sharp premixed" ignition spike at the start of the main combustion. This is due to a combination of the temperature and pressure rise from the pilot injection, and the overall cycle temperatures and pressures resulting from the turbocharging. Both of these things makes the ignition of the main fuel quicker, resulting in a lower rate of combustion at the start, controlled more by the mixing of the fuel and air than by rapid chemical reactions.

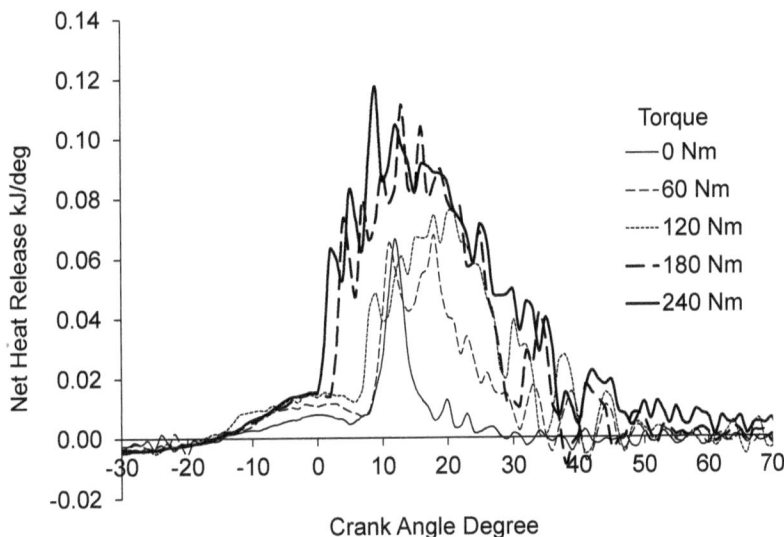

Figure 3.50: Heat release rate as a function of load for a modern, common rail 2 liter direct injection diesel engine operating at a speed of 2500 rpm.

For the lower loads, the ignition of the main portion of the fuel is maintained at about TDC. The start of injection becomes earlier as the load increases, while the end of combustion remains relatively constant at about 40° ATDC, indicative of an advancing timing. With electronic control of the rail pressure[3], the injection rate can be increased with the load to keep the combustion duration from becoming too long. The combustion duration does increase with load, but not as much as with the turbocharged engine with a positive displacement fuel pump (Figure 3.47). Limiting the combustion duration helps to reduce particulate emissions, as combustion of the last portion of fuel becomes less effective with the lower temperatures later on the expansion stroke. In the case of the engine shown, the rail pressure varied from about 50 MPa at idle to about 130 MPa at high loads. The maximum duration of injection was

---

[3]Fuel injection pressure at the inlet to the electronically controlled nozzle

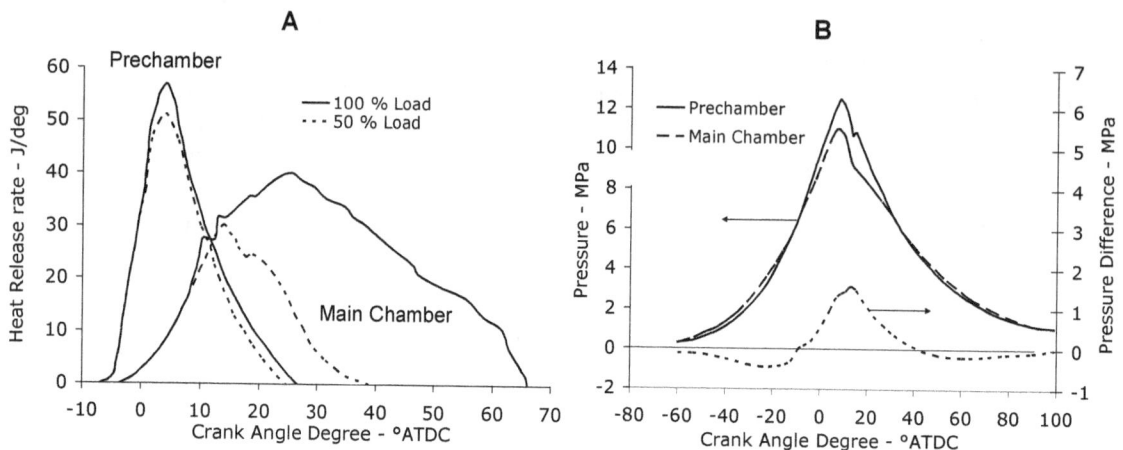

Figure 3.51: Heat release rate (A) and pressure time (B) diagrams for a swirl chamber IDI diesel engine operating at 3600 rpm. The pressure difference is across the connecting orifice. Adapted from [44].

maintained at 30 - 35°. Because of the pilot injection, the main portion of the combustion looks like the diffusion controlled part of the combustion in older engines. The data indicate the extreme flexibility available in modern diesel engines, with independent control of injection time, number of injection pulses, and injection pressure and duration. his flexibility is needed if ever stricter emissions standards are to be met.

Indirect Injection (IDI) Engines

Figure 3.18 shows the combustion configuration of an indirect injection, or IDI, diesel engine, in this case a swirl chamber engine. In this type of engine, fuel is injected into a constant volume (pre)chamber in the cylinder head, into which air is compressed on the compression stroke. Combustion raises the pressure in the prechamber and the burning fuel air mixture is forced into the main combustion chamber for completion. The energy from the initial combustion is used to promote the mixing of the last portions of fuel and air to burn.

The IDI diesel engine has been used for many years, especially in applications requiring a larger range of engine operating speeds. The most common applications, therefore, have been smaller vehicles, such as passenger cars, delivery vans and the like. IDI engines typically have lower smoke emissions than their DI counterparts especially at higher speeds. Typically, the IDI smoke is low until a certain load is reached, after which the smoke increases sharply. With the DI engine, smoke typically increases more or less continually with load increase. In recent years, the advance of common-rail diesel injection systems is changing this picture, in that the common rail injection system allows injection pressure control independent of engine speed. Therefore, it is now possible to optimize the combustion system in a smaller DI diesel engine over the wide speed range required for light duty vehicle operation. Therefore, the IDI engine is rapidly being replaced by more efficient DI engines in vehicle applications, even down to very small engines.

Figure 3.51 shows the heat release rates (A) and pressures in the two chambers of a swirl chamber IDI engine, as well as their difference (B). On the compression stroke, the pressure in the main chamber is greater than that in the prechamber, as the air is forced into the prechamber. The fuel is injected near top dead center and when combustion starts, the pressure in the prechamber chamber becomes greater than the main chamber pressure, causing a strong flow out of the prechamber. The overall equivalence ratio in the prechamber during this period is quite rich. The combustion rate in the main chamber is determined by the mixing rate between the jet outlet flow and the air.

The heat release analysis for an indirect engine is more complicated than in the direct injection engine, and Equation (2.103) is no longer valid, since there is flow between the chambers. There are two heat release curves: one for the prechamber, where the ignition process occurs, and one for the main chamber where the final combustion process occcurs. Analysis methods for calculating IDI engine heat release can be found in the literature, and will not be discussed here [7].

The combustion is typically placed later in the cycle for an IDI engine than a DI diesel engine. The combustion in the prechamber starts close to top dead center. This is necessary to prevent too much back flow into the prechamber near the end of the compression stroke, and to facilitate a strong flow from the prechamber into the main chamber to complete the main combustion once the combustion starts. If the fuel is injected too early, the compression will tend to keep the burning fuel and air in the prechamber. The rapid rate of heat release associated with ignition will be observed in the prechamber as it is here that the ignition occurs. The combustion rate will slow down quickly in the prechamber under moderate to heavy loads, because there will be a lack of air, since the prechamber contains only a fraction of the total cylinder air charge. Figure 3.51 shows the rapid rate of heat release in the prechamber, giving rise to a noisy engine operation, just as in a non-turbocharged DI engine. Note that the combustion rate in the prechamber is nearly the same regardless of load. The first stage of combustion occurs during injection and is basically limited by the amount of air in the prechamber available for combustion, and so not very dependent on load. At the larger load, more fuel is injected, but there is little air remaining for combustion in the prechamber, and the combustion is only completed when the prechamber gasses flow out and mix with the air in the prechamber.

The differences with respect to load can be seen in the main chamber. Since the heat release in the prechamber is nearly the same, the pressure rise there will not be very dependent on load, and the emerging jet will have about the same strength, as indicated by the first portion of the combustion in the main chamber. The main difference with respect to the engine load then, is the later stages of combustion, which will lengthen with load as was shown with previous engines (see for example Figure 3.47 and 3.50). A larger portion of the heat release period in the IDI engine will be characterized by the slower, mixing controlled combustion as the load increases, since the jet leaving the pre-chamber controls the mixing rate and thereby the combustion rate. There is more fuel to be burned, and it takes a longer time to complete the combustion at high load. Because of the late start of combustion the second, mixing controlled, stage of combustion is also late and occurs further down on the expansion stroke. For this reason, the prechamber engine is that which comes closest to approximating the constant pressure diesel cycle. This is disadvantageous for the efficiency, but that is partly compensated for by the high compression ratio of the IDI engines.

Due to pressure and heat transfer losses in the prechamber and connecting nozzle, IDI engines typically have a higher compression ratio (21-23) than DI engines (15-19). These losses cause the efficiency of an IDI diesel engine to be lower than those of the IDI engine, in spite of the higher compression ratio. The heat loss is also high, since the jet issuing from the prechamber impacts the piston in the main chamber, causing locally high heat transfer coefficients. This also causes a greater thermal loading on the piston.

Comparing Figure 3.51 to Figure 3.45, it is seen that both types of combustion systems exhibit the rapid phase of combustion followed by the diffusion controlled phase. In the IDI engine, the first phase is confined to the prechamber, and the diffusion controlled phase to the main chamber.

3.8.4 Simple Heat Release Models

The models presented here are intended for use in engine simulations, such as that in Section 2.7, where it is desired to predict engine operation and conditions for working parts of engines. Some examples are development of air exchange systems, thermal loading of parts, mechanical stresses and performance of practical systems involving engines. Since the models are empirical, they cannot be used for combustion system development, more advanced analytical and experimental methods are needed for that.

Spark Ignition Engines

After the heat release calculation has been performed, it is possible to study the combustion process in detail. One simple way is to look at the overall parameters that describe the combustion. There are the time between the spark and the first significant heat release, and the duration of combustion.

There are different ways to describe the combustion rate. The first is to regard the heat release curve as a mathematical function with only a few parameters. One of the main reasons for this is to be able to use simple mathematical functions to describe the combustion rate in engine simulation models. That this can reasonably be done can be seen in Figure 3.43, where the heat release rate for a spark ignition engine was shown for 6 different ignition timings.

Even though the curves are displaced in time, it can easily be seen that the shape of the curves is similar for all the timings. The same general shape is observed in spite of variations in the other engine variables. Therefore, it is normal to use at least two parameters to describe such a curve. The first parameter is typically the combustion duration, that is, the amount of time elapsed between the first detectable effects of combustion and its completion. Since the curves have low slopes near the start and end, it is very difficult to accurately define when the combustion actually starts or ends. Therefore, when describing experimental data, it is common to use defined percentages of heat release to establish the shape of the curve. Values such as 5 or 10 % of the fuel burned are often used to define the start, and 95 or 90 % of combustion to define the completion of combustion.

The other parameter is the start of combustion. It can also be seen from Figure 3.43 that the heat release curves first begin to rise several degrees after the spark is fired, as mentioned previously. So in practice it is necessary to compensate for the time elapsed between the spark itself and the arbitrarily defined percentage for the start of combustion as mentioned in the preceding section. This is not a true ignition delay as encountered with diesel engines, as the flame is actually in progress from the instant of the spark. However, the burning mass in this initial combustion phase is so small that it cannot be detected within the accuracy of the pressure measurements available.

Normally the start of combustion and its duration are sufficient to describe the progress of SI combustion to a sufficient degree of accuracy for simulation purposes with zero-dimensional models. It has been found that the exact shape of the combustion chamber does not have major influence on the performance of the engine as long as the combustion is placed near the optimum timing. This can be verified by using the simple simulation with Wiebe functions of different shapes, Equation (3.127). While the MBT timings for the different shapes will not be the same, the indicated efficiency at MBT timing will not vary much, unless the combustion duration is also lengthened significantly (See section 2.7).

There are two functions that typically are used for the combustion curve for spark ignition engine simulations. These functions are strictly empirical and they are used because they are simple, yet give a good reproduction of the heat release rate for non-dimensional simulations. The first is a simple trigonometric function, shown in Equation (3.125), and used in the simple simulation [12].

$$y = \frac{1}{2}\left[1 - \cos\left(\pi\frac{\theta - \theta_o}{\Delta\theta_c}\right)\right] \tag{3.125}$$

where y denotes the fraction of the fuel burned. ($y = 0$ before combustion, and $y = 1$ after combustion). Equation (3.125) can be written in terms of the combustion rate by differentiating.

$$\frac{dy}{d\theta} = \frac{\pi}{2\Delta\theta_c}\sin\left(\pi\frac{\theta - \theta_o}{\Delta\theta_c}\right) \tag{3.126}$$

In Equations (3.125) and (3.126), θ_o denotes the time for the start of actual heat release. That is, the crank angle where the first detectable combustion occurs. This is the time of spark plus the flame build up delay mentioned earlier in this section. For example, if the spark is at 340° and the ignition delay 15°, then $\theta_o = 355°$. $\Delta\theta_c$ denotes the duration of the combustion. Under these assumptions, Equation (3.125) can describe the combustion rate with only two parameters. It is, of course, necessary to know how these parameters vary with engine conditions in order to perform accurate simulations over a range of engine conditions.

A second model often used in connection with the simulation of the course of combustion in engines is the so-called Wiebe function [45]. The function is shown in Equation (3.127) in a form suitable for use in the simulation of the combustion in spark ignition engines.

$$y = 1 - \exp\left[-a\left(\phi\right)^{(b+1)}\right] \tag{3.127}$$

where: y = mass burned fraction, and $\phi = \frac{\theta - \theta_o}{\Delta\theta_c}$

In addition to the two parameters used in Equations (3.125) and (3.126) to determine the timing and duration of combustion, there are two additional parameters in the Wiebe function that control its shape. The first is the so-called efficiency parameter, a, and the second the shape factor, b. The efficiency parameter determines the value of the function when $\theta = \theta_o + \Delta\theta_c$. For use with spark ignition engines, the use of $a = 5$ and $b = 2.2$ is recommended for the combustion function in a conventional, well developed engine. The parameter b has a large impact on the shape of the functions and other values of the shape parameter should be used with care.

Equation (3.127) can also be differentiated to convert to a combustion rate, shown in the following equation:

$$\frac{dy}{d\theta} = \frac{a(b+1)\phi^b}{\Delta\theta_c} \exp\left[-a(\phi)^{(b+1)}\right] \tag{3.128}$$

Curves from Equations (3.127) and (3.128) are shown in Figure 3.52. Also shown are curves for the trigonometric function used in Equations (3.125) and (3.126). The abscissa corresponds to the variable, ϕ, in Equation (3.127). For intermediate values of the shape parameter, the Wiebe function and the trigonometric function are similar. The Wiebe function has the advantage that it has greater flexibility and can simulate variations in combustion rates, especially if the rate profile is not symmetric. On the other hand, there is the additional shape parameter, b, which also needs to be determined as a function of the engine's operating condition. It should be emphasized, that these curves are basically empirical curves that resemble experimental results. There is no real theoretical justification for either curve.

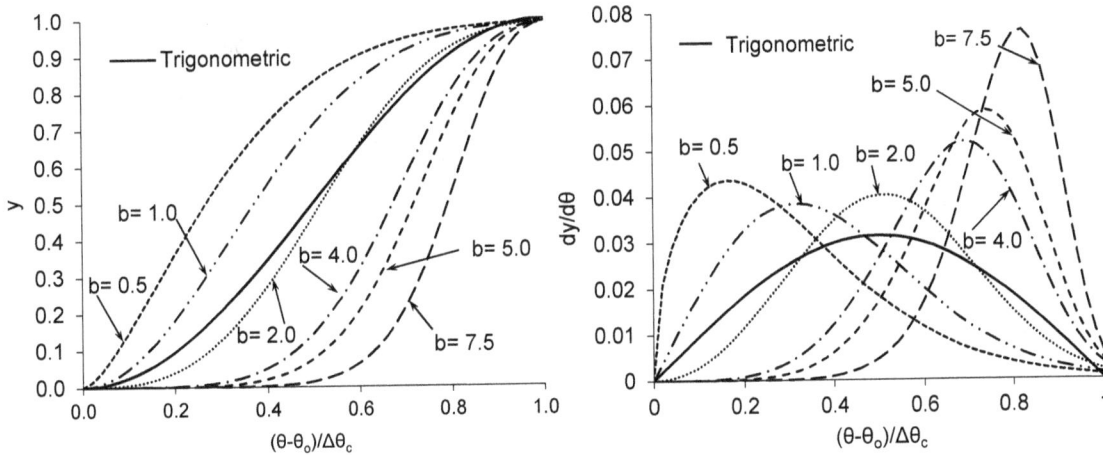

Figure 3.52: The Wiebe function and its derivative used for simulation combustion rates in SI engines compared to the trigonometric model. For the curves, $a = 5$.

It should also be remembered, that Equations (3.125) to (3.128) are only valid during the time of combustion, that is:

$$y = 0, \frac{dy}{d\theta} = 0 \quad for \quad \theta \leq \theta_o \quad and \tag{3.129}$$

$$y = 1, \frac{dy}{d\theta} = 0 \quad for \quad \theta \geq \theta_o + \Delta\theta_c \tag{3.130}$$

The values of the 10 and 95% burn points can be substituted into Equation (3.125) or (3.127) to determine the values of the parameters θ_o and $\Delta\theta_c$ in the trigonometric model. An example is given below:

Consider the case where 10 % of the fuel is burned $42°$ after the spark, and 90 % is burned $69°$ after the spark. Assume that the combustion rate follows the trigonometric function of Equation (3.125). Then for the two conditions:

$$0.1 = \frac{1}{2}\left(1 - \cos\pi\frac{42 - \theta_o}{\Delta\theta_c}\right)$$

and

$$0.9 = \frac{1}{2}\left(1 - \cos\pi\frac{69 - \theta_o}{\Delta\theta_c}\right)$$

This results in two simultaneous equations for the parameters θ_o and $\Delta\theta_c$:

$$\frac{\Delta\theta_c}{\pi}\cos^{-1}\left(1 - 2\cdot 0.9\right) = 69 - \theta_o$$

$$\frac{\Delta\theta_c}{\pi}\cos^{-1}\left(1 - 2\cdot 0.1\right) = 42 - \theta_o$$

Solving, one obtains: $\theta_o = 32.6°$ and $\Delta\theta_c = 45.7°$. In this case, the ignition delay, between the spark and the first observance of heat release, is $42 - 32.6 = 9.4°$. The parameters estimated here should only be used for the trigonometric model, Equation (2.109). Due to the exponential shape of the Weibe function and its dependence on the shape factor, different values of θ_o and $\Delta\theta_c$ will be obtained. This is shown in Figure 3.53, for both the Weibe and trigonometric functions. For the trigonometric function, $\theta_o = 352°$, and $\Delta\theta_c = 22°$ For the Weibe function, a=4, b= 3.8, $\theta_o = 340.5°$, and $\Delta\theta_c = 33°$ The parameters have been determined such that 50% of the fuel is burned at $362.2°$, the same as that of the experimental data. The Weibe function better captures the non-symmetric nature of the combustion and the slow start. When used in simulation calculations, the two functions give close to the same results.

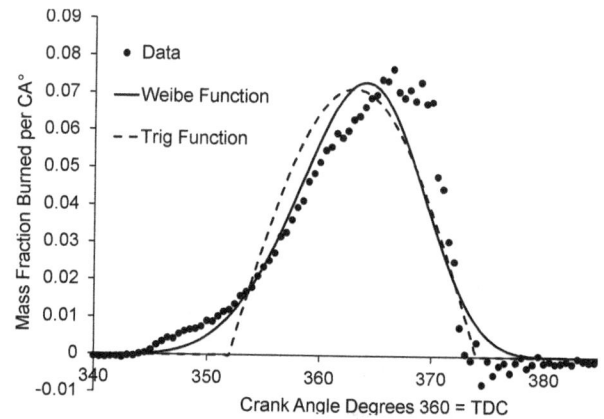

Figure 3.53: Comparison of Weibe and trigonometric heat release functions with CFR engine experimental data for a spark timing of 340° at 1000 rpm, stoichiometric combustion.

Compression Ignition Engines

The diesel combustion process is more complicated than that in the spark ignition engine, and in some cases, more complicated models are needed. However, there is one case where a very simple model can be used. That is in the case of a diesel engine with a very short ignition delay, as compared to the remainder of combustion [46]. The model has only one empirical parameter, but is based on a knowledge of the injection rate of fuel to the cylinder.

$$\dot{Q}_{HR} = k\left(Q_{fuel} - Q_{HR}\right) \qquad (3.131)$$

where: Q_{fuel} is the accumulated energy injected with the fuel, $\dot{Q}_{HR}$ is the apparent rate of heat release from combustion and Q_{HR} is the accumulated heat release.

The model is a first order, inhomogeneous differential equation through the injection period, and a homogeneous differential equation throughout the remainder of combustion. It

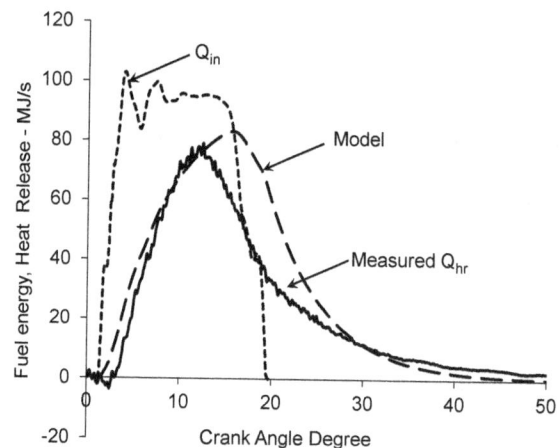

Figure 3.54: The comparison of the predicted rate of heat release from the model of Equation (3.131) with the measured heat release from the large marine diesel of Figure 8.5. The constant k in Equation (3.131) is equal to 120 s^{-1}. The rate of injected energy is also shown.

is not an explicit function, but can easily be integrated numerically throughout the combustion period when solving Equation 2.105. A simple model of the injection rate can often be used, for example, a constant injection rate over a given time interval. An example of the use of this heat release function is shown

in Figure 3.54, where it was applied using experimental values of the fuel injection rate ($\dot{Q}_{in} = \dot{m}_f \cdot H_u$). This engine is a large 2-stroke direct injection marine engine.

The model gives a reasonable description of the trends in the heat release, though there are some differences in the values. This is not unexpected, since there is only one parameter in the model. The experimental results show that there is no significant ignition delay, and the start of the combustion is quite smooth, so the model can be used. The large marine engines with very high cylinder pressures and low engine speed give this situation.

This special case does not apply to smaller engines operating at higher speeds. A portion of the ignition delay is determined by the chemical reactions leading up to the ignition of the vaporized fuel. These reactions are not significantly affected by the speed of the engine, and so there is a larger build up of evaporated fuel before ignition in the smaller high speed engine than in the large slow speed engine. This gives rise to a significant ignition delay, and a sharp rate of heat release at the start of combustion, as seen in Figure 3.45.

The more complicated nature of combustion means that a more complicated model with more parameters must be used. First of all, some prediction of the ignition delay must be used. This is commonly some variation of an empirical equation of the type:

$$\tau = A \cdot p_{O_2}^b \exp\left(\frac{E}{T}\right) \tag{3.132}$$

where A, b and E are empirical parameters, and p_{O_2} is the partial pressure of the O_2 in the combustion chamber at the time of injection. The parameters need to be determined for a given engine, and depend on fuel type, injector characteristics, and the flow situation in the engine. This model has some serious limitations at lower temperatures, but is usually accepted at diesel ignition conditions. (See Section 3.6.5).

When the combustion starts, one way to describe the combustion rate is through the use of Wiebe functions. Two are used, one to describe the first ("premixed") phase of combustion and the second to describe the second ("diffusion") phase of combustion. As seen in the previous section, the Wiebe functions are quite flexible, allowing a variety of shapes with the adjustment of just one parameter. For a diesel engine, the mass burned fraction of the total amount of injected fuel can be written:

$$x = f_p \cdot x_p + (1 - f_p) \cdot x_d \tag{3.133}$$

where f_p is the total fraction of the fuel burned during the first, or "premixed" phase of combustion, and x_p and x_d are fractions of fuel burned in the respective

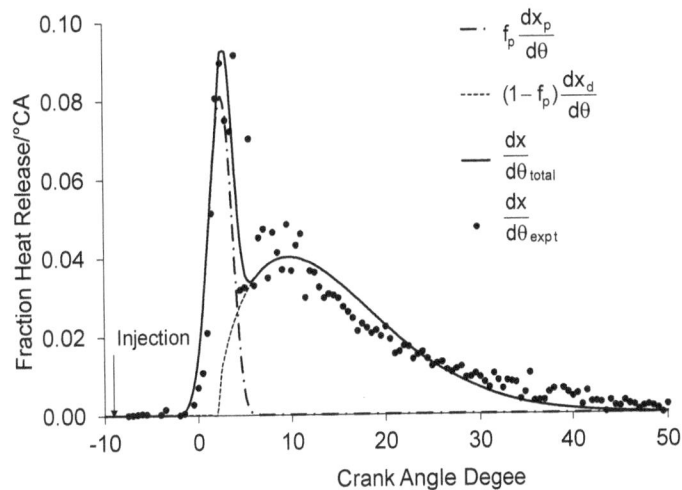

Figure 3.55: The comparison of the predicted rate of heat release from the model of Equation (3.134) with the measured heat release for small naturally aspirated direct injection diesel engine. The heat release rate has been normalized by the total heat release.

"premixed" and "diffusive" phases of combustion. The basic Wiebe function in Equations (3.127) and (3.128) are used for each phase of combustion, with separate parameters a, b, θ_o, and $\Delta\theta_c$ for each phase.

$$\frac{dx}{d\theta} = f_p \frac{dx_p}{d\theta} + (1 - f_p) \frac{dx_d}{d\theta} \tag{3.134}$$

From this, the rate of heat release can be calculated by multiplying Equation (3.134) by the mass of fuel injected to the cylinder:

$$\dot{Q}_{HR} = m_a \cdot FA \cdot H_u (1 - f) \frac{dx}{d\theta} \tag{3.135}$$

where m_a is the air mass in the cylinder, and FA the overall fuel air ratio.

Two examples are given to show the flexibility of the Wiebe function when used for modeling diesel engine heat release. The first is that of a classic, naturally aspirated direct injection diesel engine, with a cylinder volume of 500 cm^3 operating at a bmep of 638 kPa. Figure 3.55 shows that a very good representation of the heat release curve can be obtained by adjusting the empirical parameters.

For the calculation of Figure 3.55, the model parameters used for the different phases of the combustion process are given in Table 3.8. It should be noted that this procedure is simple curve fitting, there is no theoretical justification for the equation. Since, however, the curve gives a good representation of the heat release rate, there is the potential of fitting the parameters as a function of engine variables in connection with experiments, so that a more general, empirical description of an engine combustion system could be developed. Note also, that the ignition delay period is not described with these functions. It is likely that the parameters for the first, or premixed phase of combustion, depend on the length of the ignition delay and the fuel injection rate.

The flexibility of the Wiebe function is further shown in Figure 3.56, where the heat release rate for the common rail, direct injection engine with pilot injection of Figure 3.50 has been fit to Equation (3.134). The result here also gives a very good description of the heat release rate, and should be suitable for general simulation purposes. Note that the curve for the second, diffusion phase appears similar to that predicted by Equation (3.131). The pilot injection basically eliminates the ignition delay for the second portion of fuel, satisfying the conditions for that heat release model. Again, the problem remains of obtaining the values of the parameters for the model and determining their dependency on the relevant engine operating variables and conditions.

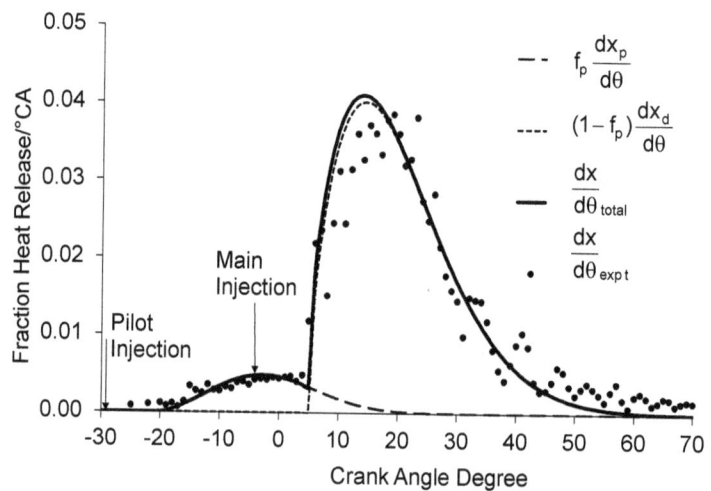

Figure 3.56: The comparison of the predicted rate of heat release from the model of Equation (3.134) with the measured heat release for a turbocharged direct injection diesel engine with pilot injection. The heat release rate has been normalized by the total heat release.

Table 3.8: Parameters used in the heat release model of Equation (3.134).

| Parameter | Figure 3.55 | | Figure 3.56 | |
	Premixed	Diffusion	Pilot	Diffusion
a	5	10	5	10
b	4	0.6	1.5	0.6
θ_o	4	8	-20	5
$\Delta\theta_c$	8	65	40	72

3.9   Problems

Problem 3.1

Calculate the flame temperature for a stoichiometric mixture of propane and air for constant volume combustion from 298 K, 1 bar. Compare the result with the result for $\lambda = 1.5$ in Example 2.

Problem 3.2

Compare the flame temperature for a fuel with a low hydrogen to carbon ratio, benzene, and a fuel with a high hydrogen to carbon ratio, methane for constant pressure combustion from 298 K and 1 bar. For both fuels use an excess air factor of 1.2.

Problem 3.3

Calculate the flame temperature for a stoichiometric mixture of ethane and air for constant volume combustion from 600 K and 10 bar. Calculate the pressure after the combustion.

Problem 3.4

a. Calculate the adiabatic flame temperature at constant volume for a stoichiometric mixture of iso-octane C_8H_{18}, and air, for an initial temperature and pressure of 700K, 1800 kPa. Use ideal combustion products with no dissociation, but with temperature dependent specific heats. ($T_f = 3214$K)

   b. Compare the answer in a. with the equilibrium flame temperature from the program "aft.exe". (2901K)

   c. Compare the answer in b. with the equilibrium flame temperature for methane, CH4, and benzene, C6H6 under the same conditions. (2840.8K, 2972.3.8K)

   d. Find the equilibrium flame temperature and the equilibrium concentration of NO for octane as a function of Lambda for the same start conditions as a.

Problem 3.5

A motor operates on a gas mixture that consists of 20% CH_4, 50% H_2, 30% CO_2 (vol.-%).
   Find:
   a. The heating value of the gas mixture; b. The stoichiometric mixture ratio; c. The energy content for a mixture with an excess air ratio of = 1.1 in kJ/kg, d. The exhaust gas composition for question c. on a mole basis
   (Lower heating value = 15990 kJ/Kg; AF_{th} = 5.1;E = 2419.5 kJ/Kg. Mole fractions: CO_2 = 0.1204, H_2O = 0.2167, N_2 = 0.6473, O_2 = 0.01565)

Problem 3.6

For octane, calculate the lower heating value with constant pressure: a. starting with the higher heating value for constant pressure. b. starting with the higher heating value for constant volume. (Note that the result for a. = Result for b.)

Problem 3.7

Verify the difference in the values in Table 3.4 for benzene. Calculate the heating values for benzene for constant volume.

Problem 3.8

Calculate the flame temperature for a stoichiometric mixture of propane and air for constant volume combustion from 298 K, 1 bar using ideal reaction products and variable specific heats. Compare the

result with the result for $\lambda = 1.5$ in Example 2. Repeat the calculation for the equilibrium flame temperature.

Problem 3.9

Compare the flame temperature for a fuel with a low hydrogen to carbon ratio, benzene, and a fuel with a high hydrogen to carbon ratio, methane for constant pressure combustion from 298 K and 1 bar using ideal reaction products and variable specific heats. For both fuels use an excess air ratio of 1.2. Repeat the calculation for the equilibrium flame temperature.

Problem 3.10

Calculate the flame temperature for a stoichiometric mixture of ethane and air for constant volume combustion from 600 K and 10 bar using ideal reaction products and variable specific heats. Calculate the pressure after the combustion. Repeat the calculation for the equilibrium flame temperature.

Problem 3.11

Methane (CH_4) is burned with an excess air ratio, $\lambda = 0.9$. Determine the coefficients of the chemical reaction equation using the assumption: a. there is no H_2 present, b. $CO/H_2 = 2$ and c. the water gas equilibrium constant, $K_p = 0.286$. (a: 0.6, 0.4, 2, 0; b: 0.733, 0.267, 1.867, 0.133: c: 0.7631, 0.2369, 1.837, 0.1631)

Problem 3.12

A hydrocarbon fuel burns with air. The dry exhaust products have the following molar composition: $CO_2 = 10$, $CO = 7$, $O_2 = 1$, $H_2 = 4$, and $CH_2 = 1$. Determine the air fuel ratio, the excess air ratio, and the relative (hydrogen/carbon ratio) of the fuel. ($AF = 11.29$, $\lambda = 0.773$, fuel $= CH_{1.884}$)

Problem 3.13

A direct injection diesel engine has a bore of 120 mm, stroke of 140 mm, compression ratio of 16:1 and operates at 1800 rpm. The diameter of the crater in the piston bowl is half the bore, and the clearance between the piston top and the cylinder head at TDC is 1 mm. Calculate the volume of the crater in the piston, and estimate the squish velocity as a function of crank angle. ($V_{crater} = 94.2$ cm^3, $v_{squish} = 43.2$ m/s at 10° BTDC).

Problem 3.14

The equilibrium flame temperature has a modest dependency on the fuel type for hydrocarbon fuels. As shown in Table 2.1, the product of the stoichiometric fuel air ratio and the heating value for benzene is higher than that of methane. This is connected to the relative energy released when converting carbon or hydrogen to CO_2 and H_2O, and is shown in terms of the hydrogen to carbon ratio of the fuel, Figure Figure 3.57A. This has some effect on the NO formation as shown in Figure 3.57B.

1. To see this effect, compare the constant volume flame temperatures for methane and benzene without dissociation to those of this figure. Use a stoichiometric mixture, burning at constant volume from a temperature of 700 K and a pressure of 1500 kPa.

2. From the equilibrium flame temperature calculation it was found that the mole fractions of H_2O in the combustion products are 0.07183 and 0.1766 for the benzene and methane flames respectively. For CO_2, the mole fractions are 0.1202 and 0.07454 for benzene and methane. Calculate the relative amounts of H_2O and CO_2 that have dissociated, relative to complete combustion (no dissociation). (3353, 3404, 0.109, 0.0709, 0.255, 0.216) (flame temperature, H_2O, and CO_2 for benzene and methane respectively)

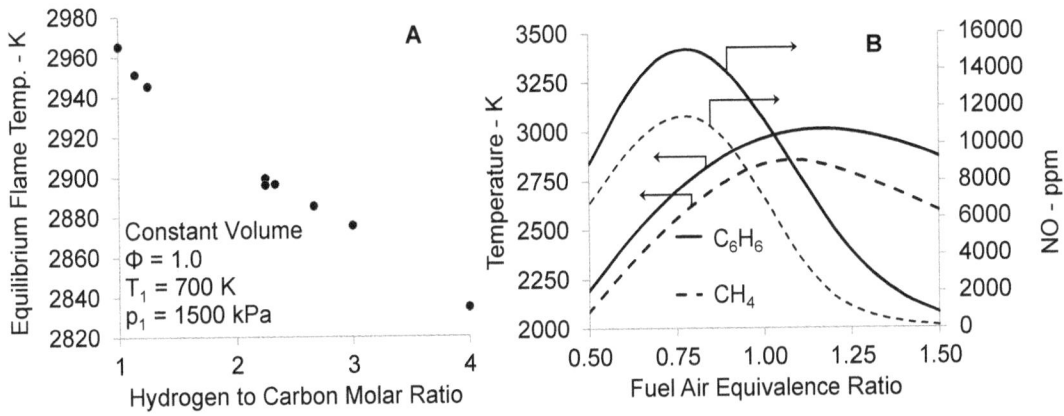

Figure 3.57: A. Equilibrium flame temperature as a function of hydrogen to carbon ratio of the fuel at stoichiometric fuel air ratio and B. Equilibrium flame temperature and equilibrium NO concentration as a function of fuel air equivalence ratio for methane (C/H=4) and benzene (C/H=1). For constant volume combustion from 700K and 1500 kPa.

Chapter 4

Fuels

4.1 Introduction

As might be expected from the different types of combustion, there are quite different characteristics required of the fuels for either a diesel or a spark ignition engine. In the case of the traditional spark ignition engine, there are the following basic requirements:

1. The fuel must be volatile enough to evaporate easily and form a homogeneous mixture at the time of combustion

2. The fuel must be resistant to the tendency to ignite (explode) when mixed with air and exposed to high temperatures and pressures. That is, the fuel must be knock resistant

There are many secondary requirements, but these are the essential ones. Without satisfying them, performance is not acceptable or even impossible.

For diesel fuels, the prime requirement is that the fuel will ignite rapidly when exposed to compression pressure and temperature in engines. Because high pressure injection is used, the vaporization characteristics of diesel fuels are not nearly as important those of SI engine fuels. Diesel engines can operate on fuels that are gaseous at room temperature or with fuels that are so heavy they need to be heated to flow. The fuels that are used in practice for SI and CI engines are quite different in nature.

The current source for the vast majority of fuels is crude oil, which varies in properties from the various sources around the world. There is a range of production techniques and processes used to convert the crude oil to the final fuel products. These products are made to conform to standards agreed upon by various authorities, engine manufacturers, and refiners. The production methods for vehicle fuels will not be discussed here, but can be found in various books on engines and combustion in general.

Common for most of the fuels in use in engines today is that they are hydrocarbon based. Only recently have limited amounts of organic compounds containing oxygen been introduced into the fuel market. These components include ethanol and MTBE (Methyl tertiary butyl ether) in SI fuels, and vegetable oil or derivatives thereof in CI fuel. The addition of these components has had little effect on the combustion characteristics of engines, but gives some effects on exhaust emissions and knocking tendencies. While there are many additives in fuel, the amounts are normally so small that they are ignored when exhaust products, fuel air ratios and the like are considered. These additives do, however, have significant effects on engine cleanliness and durability.

4.2 Fuel Specifications

4.2.1 Ignition Quality and Knock

Diesel engines and spark ignition engines have diametrically opposite ignition requirements. A good fuel in a diesel engine should readily ignite when it is mixed with air and subjected to the high temperatures and pressures found at the end of the compression stroke. If the fuel doesn't ignite, the engine doesn't operate.

In the spark ignition engine, the ignition is controlled by the spark, and it is intended that the premixed flame spreads uniformly away from the spark and eventually consumes all the fuel. If the unburned fuel air mixture does not have sufficient resistance to the high temperatures due to compression near the end of the combustion process, the unburned gasses autoignite and the very rapid combustion gives rise to knock and subsequent problems. This is the opposite of what is needed in the diesel engine.

A schematic diagram of the SI combustion process is shown in Figure 4.1. Cylinder pressures during knocking combustion were shown in Figure 3.38. There is basically a competition between the propagating flame and the chemical reactions leading up to autoignition of the end gasses. If the end gasses autoignite before the flame arrives, knock occurs.

A very simple way to look at autoignition and knock is through the idea of an ignition delay dominated by chemistry. The concept is that a fuel-air mixture subject to elevated temperature and pressure will spontaneously ignite after a certain length of time (ms). This general concept is correct, but the mathematical description of the process is often simplified, in that the form that the ignition delay, τ, is assumed to be a function of the temperature and pressure of the form:

$$\tau = k_1 p^{k_2} \exp\left(\frac{k_3}{T}\right) \qquad (4.1)$$

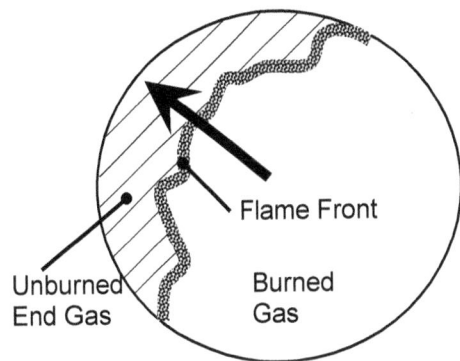

Figure 4.1: A schematic diagram of the combustion in an SI engine. The flame progresses away from the combustion source, compressing the unburned gasses (end gasses) in the process.

Where the k's are empirical constants. k_1 varies considerably, while $-.5 \gtrsim k_2 \gtrsim -1.5, k_3 \approx 3000$ for T in K for typical SI fuels. The ignition chemistry is actually much more complicated than this, as shown in Section 3.6, but Equation (4.1) is an approximation that provides general tendencies.

Some approximate values of the constants in Equation (4.1) indicate that the ignition delay for autoignition is reduced when the temperature and pressure of the fuel air mixture increase. For a temperature increase of 700 K to 900 K, the ignition delay is reduced by about a factor of three. For a 3 fold increase in pressure, corresponding to full load vs. idle, the ignition delay decreases by a factor on the order of 2 - 4. The implication for SI knock is that higher temperatures and pressure of what is called the *end gas*, that is, the last portion of the fuel air mixture to burn, increases the tendency to knock. That is because the shorter ignition delay makes it more likely for the end gas mixture to autoignite before the flame arrives.

The situation is reversed in the diesel engines. Here, knock occurs at the very start of combustion, as seen in Section 3.6. The very high rate of heat release associated with this autoignition gives rise to the characteristic diesel engine noise, and is an inherent element of that kind of combustion. As seen above, the chemical delay is strongly affected by the temperature and pressure to which the fuel air mixture is exposed. This is what basically makes the diesel engine so well suited to turbocharging and the spark ignition engine much less suited. Figure 3.47 showed this quite clearly. At the high cylinder pressures encountered with the turbocharged engine, the rapid initial part of ignition was very much diminished. There are two advantages here - less noise, and a lower peak cylinder pressure.

In the consideration of engine knock there is a very simple rule, with not many exceptions: What is good for an SI engine is not good for a CI engine[1]. Therefore there are two fuel standards for measuring ignition quality each related to a specific kind of engine. The idea is to name a "good" fuel and a "bad" fuel for comparison purposes.

The first specification of this type is the *octane number*, used for rating SI engine fuels for knock resistance. The basis of the procedure is to define two reference fuels, which are given a value of 0 and 100. The scale is then defined by the percentage of the "good" reference fuel in a mixture of the

---

[1]Interestingly, the same CFR engine is used to test each fuel type, though with a different cylinder head.

two reference fuels, which gives the same tendency to knock as the fuel being tested. The fuels used to define SI engine knock characteristics are n-heptane, a normal paraffin with the composition C_7H_{16}, which is very susceptible to autoignition, and many years ago was defined to have a value of zero on the ignition quality scale. On the other hand, the fuel 2,2,4 trimethyl pentane, was found to be resistant to autoignition and given the value 100. This fuel, with the overall chemical formula of C_8H_{18} is one of the many isomers of octane. In engine work it is popularly called iso-octane, though this name is not very descriptive to a chemist. If a fuel knocks with the same intensity as a volumetric mixture of 85% iso-octane and 15% n-heptane, then it is said to have an octane rating, or *octane number* of 85. Because these two chemicals are the basis for the definition of the octane scales, they are called Primary Reference Fuels, often denoted by the abbreviation, PRF

Unfortunately, all fuels have differing knock sensitivity in response to different operating, so the octane number is not absolute value for a fuel, but is related to a given test. The tests are performed in an engine, designed decades ago to knock and to withstand knocking, called a CFR (Cooperative Fuel Research) engine. It has long been a standard engine for fuel testing and research. There are two main octane rating tests, the Research test and the Motor test, which result in the Research Octane Number (RON) and the Motor Octane Number (MON). These numbers are usually different for the same fuel. The RON is normally higher than the MON, and the difference between them is an indication of sensitivity of fuel knock resistance to engine conditions. In Europe, the RON is normally posted at the filling station, while American practice is the average of the two values. The octane scale was defined decades ago, and there are fuels which have octane numbers over 100. Methods have been developed to determine octane ratings for these fuels by adding tetra ethyl lead to iso-octane. The two tests are briefly outlined in Table 4.1.

Table 4.1: CFR engine conditions for the Research and Motor octane test methods.

Variable	Research Method - RON	Motor Method - MON
Speed - rpm	600	900
Intake pressure	atmospheric	atmospheric
Intake temperature - ° C	≈ 107	150
Spark	13° BTDC	$f(CR)$ 19 - 26 ° BTDC
Compression Ratio	adjust to standard knock	adjust to standard knock

A major factor affecting knock is fuel composition. It has long been known that certain fuel structures have a higher resistance to autoignition than others. Figure 4.2 (from 1948) shows the maximum allowable compression ratio in a test engine for fuels as a function of fuel structure for alkanes, that is, hydrocarbons containing only single carbon-carbon bonds.

Small molecules have good autoignition resistance, with methane showing the highest resistance to autoignition. Long, straight chain molecules have low autoignition resistance, and branched chain molecules have better autoignition resistance than straight chain molecules for a given carbon number. The two SI primary reference fuels are denoted by the boxes in Figure 4.2. Note that the structure is very important. The straight chain n-octane has an octane number below 0! Note also that there are fuels which would be expected to have octane numbers above 100.

Figure 4.3 shows additional fuel structure effects on knock resistance. The inclusion of unsaturated (double or triple) carbon-carbon bonds reduces knock resistance for a given carbon number. For a given carbon number, a cyclic structure stabilizes the molecule and gives it more knock resistance. For example, cyclo-hexane has a better knock resistance than n-hexane. Aromatic structures have good knock resistance. Benzene has one of the highest knock resistances for a pure hydrocarbon. Unfortunately, it is also a carcinogenic substance, and its concentration in SI engine fuel is limited by regulations in Europe and North America. Aromatic content of SI fuel has increased as lead has been removed from fuel. In addition, some oxygenated compounds have good knock resistance. Methanol has an RON of 106 and an MON of 92, ethanol 107 and 89 respectively. Oxygenated compounds can be beneficial when blended with regular fuel. This is described by a value called the octane blending value, which is basically the value of the octane number of a hydrocarbon fuel that would have the same benefit on octane rating as the oxygenate. That is, if the benefit to the octane number is the same as adding a fuel with an octane number of 110, then the oxygenate would have a blending octane number of 110. Table 4.2 shows octane blending values for some oxygenated compounds. MTBE (methyl-tertiary butyl ether) is formed by the

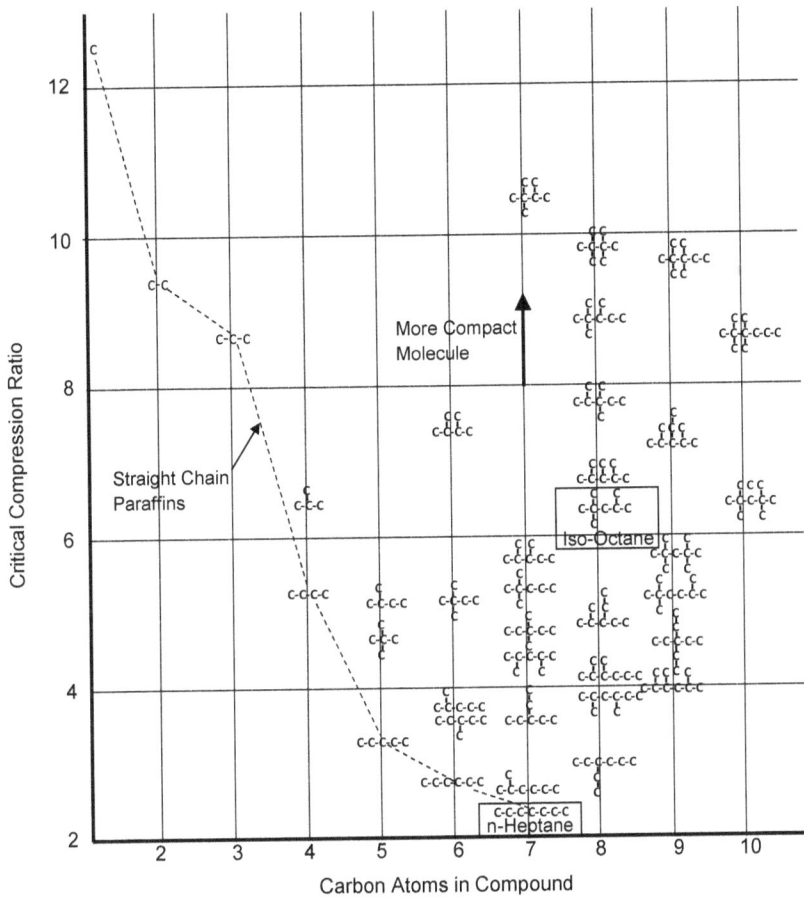

Figure 4.2: The effect of molecule structure on the knock resistance of alkanes. Adapted from [47].

reaction of butane and methanol. It has been used for octane improvement after the removal of tetra ethyl lead, but in the case of spills, has been a source of ground water pollution, and its use is being discontinued. ETBE (ethyl-tertiary butyl ether) is a similar compound.

The tendency to knock in an SI engine is influenced by anything that affects the ignition delay of the end gases relative to the flame propagation time. Some of the engine operating and design factors affecting knock are:

- *Temperature of the end gas* - High end gas temperatures can be caused by higher compression ratio, advanced spark timing, and heat transfer effects. One way to reduce knock is to locate the spark plug such that the end gasses are in a cooled location, that is away from the exhaust valve, and in a location where there is a large surface to volume area so that they are cooled, called a quench area. For this reason, spark plugs are normally located near the exhaust valve, and the warmest region of the combustion chamber burned first. For pressure charged engines, intake charge cooling is very important.

- *Pressure of the end gas* - As above, high compression ratio and advanced spark timing promote

Table 4.2: Octane blending values for some oxygenated compounds that can be used as gasoline additives.

Substance	Blending RON	Blending MON
Methanol	127-136	99-104
Ethanol	120-135	100-106
ETBE	110-119	95-104

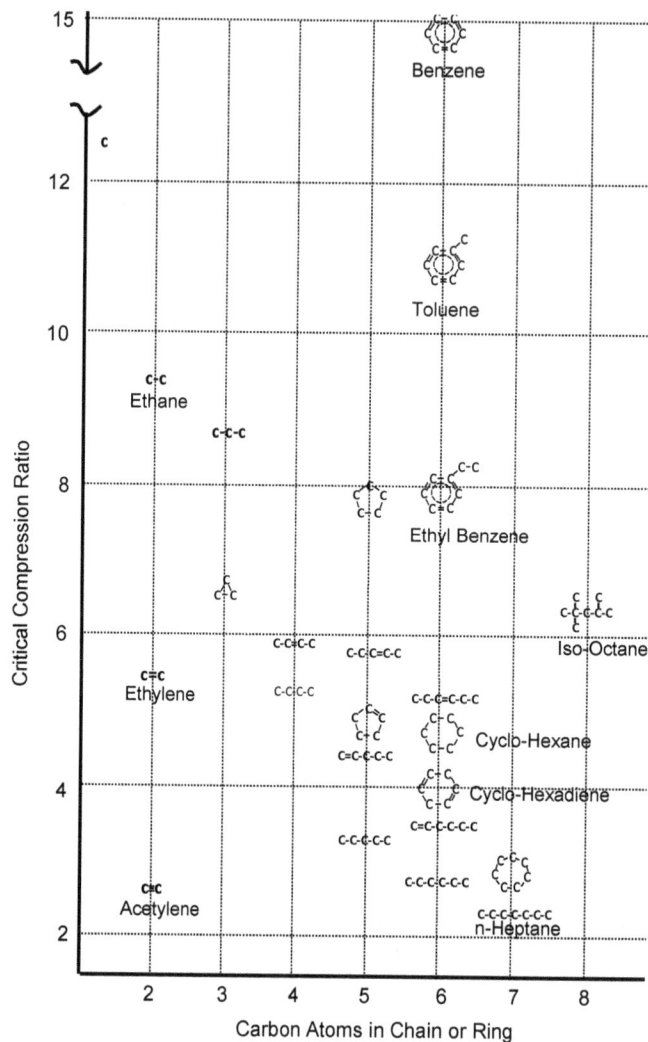

Figure 4.3: The effect of molecule structure on the knock resistance of non-saturated and cyclic molecules. Adapted from [47].

knock in SI engines. In addition, the load has a strong effect, since the pressure is directly proportional to the load in an SI engine. Thus knocking is more likely at wide open throttle conditions.

- *Flame travel time* - Knock is prevented by the flame consuming the end gasses before they can autoignite, so an increase in flame travel speed reduces knocking tendency. The chemical reactions causing knock are nearly independent of the turbulence level in the engine, since the end gasses have a homogeneous composition. Thus, in contrast to the flame propagation speed, the ignition delay in seconds is not affected significantly by engine speed (there may be some second order heat transfer effects). That means that knock tendency is worst at slow engine speed, since the flame speed measured in m/s is slower, and there is more time (in ms) for the end gasses to reach ignition conditions before the flame arrives than at higher engine speed. Flame travel time is also affected by the length of flame travel. A spark plug located close to the center of mass of the change is beneficial to preventing knock. Multiple spark plugs are also beneficial, as is increasing the level of turbulence in the engine to increase the flame speed.

- *Volume of the end gas* - This does not actually have a significant effect on the tendency to knock, but is very important if an engine should happen to knock. The smaller the mass of end gas ignition, the lower the intensity of the knock, simply because less energy is released. This is an additional advantage of the quench area mentioned in the first item. The quench area usually is very thin, so the volume there is usually small.

Figure 4.4 shows the effects of different combustion chamber geometry on the octane number required to prevent knock for a test engine. A high octane requirement indicates a greater tendency to knock. This figure illustrates the effects of some of the factors mentioned in the above list. The most effective combustion chambers are those where the greatest part of the combustion gasses are centered near the spark plug, and where quench areas are significant for the end gasses. This is seen by the low octane requirement for the compact combustion chamber show at the bottom of the figure. The design requiring an octane number of 96 is similar to that of the CFR engine, which was designed to knock, due to the long flame travel and lack of quench areas

Octane Requirement

Figure 4.4: The effect of combustion chamber shape on the octane requirement for a spark ignition engine, adapted from [48].

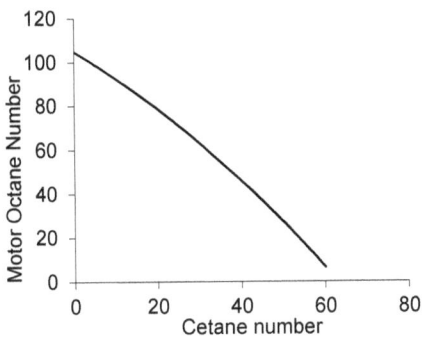

The other specification related to fuel ignition quality is for diesel fuel ignition quality and is the cetane number. The cetane test is conducted on the CFR engine, on which a pre-chamber diesel cylinder head has been mounted, instead of the SI cylinder head. The engine is operated at 900 rpm, with an intake temperature of $66°C$ and atmospheric intake pressure. The compression ratio is adjusted until a standard ignition delay of 13 CA° is achieved. The same general idea as the octane scale is used for knock testing of diesel fuels, though obviously different reference fuels are used. The original fuels, that is primary reference fuels, are an aromatic compound, alpha-methyl naphthalene $C_{11}H_{10}$, with a value of zero and a straight chain paraffin, n-cetane (n-hexadecane) $C_{16}H_{34}$, with a rating of 100. The cetane number is defined at the volume percent of n-cetane in a mixture of these two reference fuels, that gives an ignition delay of 13 CA° for the compression ratio of the test fuel. More recently, heptamethylnonane, an isomer of cetane is used to replace alpha-methyl naphthalene. It is given an cetane number of 15, and the cetane number is then the percent cetane in a mixture of heptamethylnonane and cetane plus 0.15 times the volume percent of heptamethylnonane.

Figure 4.2 and 4.3 were developed with knock in SI engines in mind but can also be used for general tendencies for CI engines, but with opposite tendencies. As a point of reference, n-heptane has a cetane number of 56. There is a general relationship between octane number and cetane number [49], and an empirical correlation is shown in Figure 4.5. The results are indicative of general trends, and show a

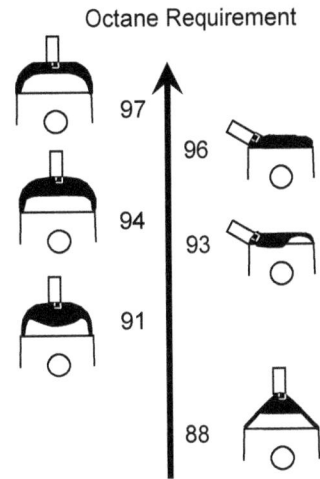

$$MON = 104.7 - 1.135CN - 0.00837CN^2$$

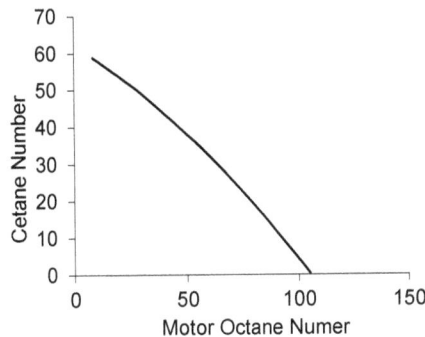

$$CN = 62.0 - 0.383MON - 0.00198MON^2$$

Figure 4.5: An empirical relationship between cetane number and octane number for some typical engine fuels

high cetane number corresponds to a low octane number.

The autoignition is an integral part of the start of the diesel combustion process, and it is desired that the fuel ignite rapidly after injection. Thus, those conditions that are undesirable from the point of view of preventing knock in SI engines are desirable for CI engines. In particular, high temperatures and pressures are beneficial to diesel operation, and result in shortened ignition delay, less noise and lower cylinder pressure. The latter through its effect on cylinder temperature, can be used to reduce NO_x emissions. The beneficial effect of high temperature and pressure makes turbocharging a natural application for diesel engines, while it is the increase in the tendency to knock that limits the use of pressure charging in spark ignition engines.

In terms of fuel structure, diesel fuel is generally different than SI fuel. Instead of small, branched hydrocarbons and aromatic compounds, typical of gasoline, diesel fuel contains a larger amount of heavier, long chain compounds, which have low ignition temperatures. They have lower vapor pressures, but the force of the diesel fuel injection spray and the high cylinder temperature when injected help to overcome this disadvantage. As seen from Figure 4.2 and 4.3, the longer chain hydrocarbons have low autoignition resistance, and are therefore well suited as diesel fuels.

4.3 Fuel Volatility

It is necessary for all liquid fuels in engines that they evaporate before they burn, so it is important to be able to specify the volatility characteristics of a fuel. The SI and CI engines have different requirements in this respect, since the CI engine promotes vaporization through the injection and atomization process, which forms an immense number of drops with a resulting large surface area moving through the air in the cylinder at high velocity. SI engines do not use such powerful injection, though in recent years, injection systems have become predominant in automobile engines, and they help vaporization. Typical pressure differences are a few atmospheres in SI engines, while modern diesel engine operate with injection pressures of 1000-2000 bar. GDI injection pressures are intermediate between the SI engine and the diesel, the fuel being standard gasoline.

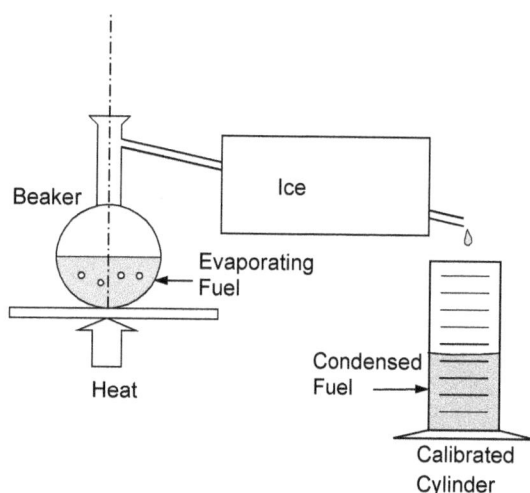

Figure 4.6: The ASTM distillation test procedure for determining the volatility characteristics of engine fuels.

The test method for fuel volatility is a simple distillation of the fuel. The apparatus used is shown in Figure 4.6. The test result is a curve of vapor temperature as a function of the percent of fuel evaporated from the test flask. Since the fuel evaporates into fuel vapors, higher temperatures are encountered than would be the case for evaporation into air. This difference can be accounted for by some empirical procedures, but for fuel specification purposes, the ASTM curve, defined through some of the evaporation points, is adequate. Points such as the 10%, 50% and 90% evaporation temperature are commonly used.

Some results for sample spark ignition and diesel fuels are shown in Figure 4.7. These results indicate a large difference in temperatures during evaporation between the two fuels. For the SI fuel, basically all of the fuel is evaporated at temperatures lower than about 200 °C, while for the diesel fuel at this temperature, only about 10 % has evaporated below this temperature. This is primarily an indication of the requirement of the SI fuel to evaporate readily at low temperatures, especially in consideration of cold starting of engines. Values for the temperatures of the fuel volatility test for different European fuel classifications are shown later in Figure 4.13.

The CI engine is much less sensitive to volatility, and can operate on much less volatile fuels, though operation on high volatility fuels is not a problem. Large marine diesel engines operate successfully on bunker fuel, which is very heavy and has a very low volatility. The high boiling portion of the fuel is important with respect to particulate emissions. If the final boiling point (FBP) is too high, the fuel has an increased tendency to form soot and particulate emissions.

Gasoline (petrol) then typically consists of fairly light hydrocarbons, with a carbon number under 10 or so. The volatility of hydrocarbons depends primarily on the number of carbon atoms in the molecule.

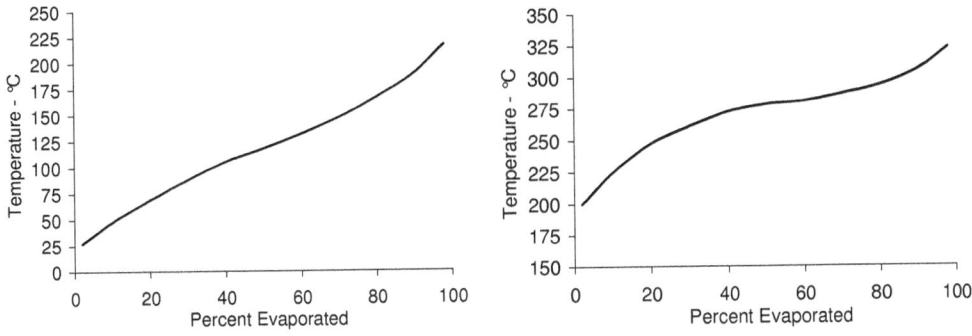

Figure 4.7: Typical ASTM distillation results for an SI engine fuel (left) and a CI engine fuel (right).

Table 4.3: Molecular weight and boiling points at atmospheric pressure for selected hydrocarbons and organic compounds

Fuel	Mole Weight -kg/kmole	Boiling Point - °C
Methane, CH_4	16.04	-164
Propylene, C_3H_6	42.08	-47.4
Propane, C_3H_8	44.096	-42.1
n-Butane, C_4H_{10}	58.123	−0.5
Butene-1,C_4H_8	56.104	17.2
n-Pentane, C_5H_{12}	72.15	36.1
n-Hexane, C_6H_{14}	86.177	69
Cyclohexane, C_6H_{12}	84.156	80.6
Hexene-1,C_6H_{12}	63.3	80.1
n-Heptane, C_7H_{16}	100.203	98.4
Octane,C_8H_{18}	114.23	125.7
iso-Octane, C_8H_{18}	114.23	117.9
Hexadecane, $C_{16}H_{34}$	226.43	286.7
Benzene,C_6H_6	78.114	80.1
Ethyl Benzene,C_8H_{10}	106.16	136.1
Toluene, C_7H_8	92.141	110.6
p-Xylene, C_8H_{10}	106.169	138.4
Methanol, CH_3OH	32.04	65
Ethanol, C_2H_5OH	46.06	77.8
Dimethyl Ether, CH_3OCH_3	46.06	-24.7

Table 4.3 shows values of boiling points for a range of fuel components.

Estimates of the effects of fuel composition on fuel vapor pressure can be obtained with the following tools. One can assume that each fuel component behaves as an ideal gas in a single phase mixture:

$$p_i \cdot V = \frac{m_i}{MW_i} \cdot R_u \cdot T \tag{4.2}$$

where p_i is the partial pressure of component i, m_i is the mass of the component, and R_u is the universal gas constant.

In a mixture of gasses, it is assumed that Dalton's law is valid, that is, the total pressure is the sum of the partial pressure of the components:

$$p = \sum_{i=1}^{N} p_i \tag{4.3}$$

and

$$p_i = \frac{n_i}{\sum_{i=1}^{N} n_i} \cdot p = x_i \cdot p \tag{4.4}$$

Table 4.4: Constants for the calculation of vapor pressure for some selected liquid hydrocarbons and alcohols and water. The constants are for use in Equation (4.8) where the temperature is in K, and the vapor pressure in kPa.

Substance	A	B
C_3H_8	-2408	15.07
$n-C_5H_{12}$	-3441	15.70
$n-C_6H_{14}$	-3865	15.84
$n-C_8H_{18}$	-4572	15.94
$i-C_8H_{18}$	-4056	15.51
$n-C_9H_{20}$	-5076	16.57
$n-C_{10}H_{22}$	-5389	16.63
$n-C_{11}H_{24}$	-5682	16.70
$n-C_{12}H_{26}$	-6030	16.87
$n-C_{14}H_{30}$	-6712	17.34
$n-C_{16}H_{34}$	-7222	17.41
CH_3OH	-4602	18.22
C_2H_5OH	-4699	17.76
H_2O	-5320	18.87

That is, the partial pressure for a component i is equal to the mole fraction x_i of the component in the gas phase times the total pressure of the system.

For two phase systems, it is assumed that Raoult's law applies, that is, in the case of a liquid mixture in equilibrium with a corresponding gas mixture:

$$p_i = x_{i,l} \cdot p_{v,i} \tag{4.5}$$

where $x_{i,l}$ is the mole fraction of the component in the liquid phase, and $p_{v,i}$ is the saturated vapor pressure of component i. Raoult's law is an acceptable approximation for hydrocarbon mixtures.

The Clausius-Claperon vapor pressure is used to calculate the change in the saturated vapor pressure with respect to temperature:

$$\frac{d \ln p_{v,i}}{dT} = \frac{h_{fg}}{RT^2} \tag{4.6}$$

If the heat of vaporization is approximately constant, Equation (4.6) can be integrated to give:

$$\ln \frac{p_{v,2}}{p_{v,1}} = -\frac{h_{fg}}{R} \cdot \left(\frac{1}{T_2} - \frac{1}{T_1} \right) \tag{4.7}$$

This leads to a function that can describe the variation of the vapor pressure for a single specie as a function of the temperature:

$$\ln p_v = \frac{A}{T} + B \tag{4.8}$$

where A and B are empirical constants. Values for some fuel components are given in Table 4.4

Evaporation in the manifold on an SI engine is dependent on the volatility of the fuel. As will be discussed in the Section 5.6.3, fuel in the intake manifold of a manifold injected SI engine is divided into two phases, liquid on the manifold walls and vapor in the intake air. The process with multi-component fuel is complicated, so the principles will be illustrated with a single component fuel. It is also assumed that the evaporation is in a state of equilibrium. This will not be the case in an operating engine, but the intent here is to show tendencies.

Consider a pure fuel. In the liquid phase, its mole fraction is 1, and if equilibrium occurs, the partial pressure of the fuel in the air will be equal to its saturated vapor pressure. If the temperature is above the boiling point, of course, then all the fuel is evaporated. For the two phase case, assuming a manifold volume, V, for fuel vapor in the fuel air mix:

$$p_f \cdot V = \frac{m_f}{MW_f} \cdot R_o T \tag{4.9}$$

where the subscript f denotes fuel vapor.

Similarly for the air:

$$p_a \cdot V = \frac{m_a}{MW_a} \cdot R_o T \qquad (4.10)$$

where the subscript a denotes air. From this:

$$\frac{p_f}{p_a} = \frac{MW_a}{MW_f} \cdot \frac{m_f}{m_a} \qquad (4.11)$$

And the air fuel ratio in the vapor phase is:

$$\frac{m_a}{m_f} = \frac{MW_a}{MW_f} \cdot \frac{p_a}{p_f} = \frac{MW_a}{MW_f} \cdot \frac{p_{tot} - p_f}{p_f} \qquad (4.12)$$

Note that m_f is the mass of the fuel vapor. If not all of the fuel is vaporized, p_f is the vapor pressure, which means that the fuel air ratio in the vapor phase will vary with temperature, since the vapor pressure is a function of temperature. The vapor phase mixture ratio is important, because this is what is available at the spark for ignition. If it is too lean the engine will not start. The problem of most interest is that of cold start at low temperatures, where fuel volatility is low, and a substantial portion of fuel may be left in the manifold, due to low air velocity at engine cranking speed. The mixture of vapors entering the cylinder could be too lean to ignite.

Example

Consider a mixture of air and fuel in a container with liquid fuel in equilibrium with the mixture. The temperature is 10°C, the total pressure 1 bar, and the fuel is n-octane. Calculate the air fuel ratio of the vapor phase.

Since there is liquid fuel present (note the the temperature is below the boiling point of 125.6 °C from Table 4.3), the air is assumed saturated with fuel vapor. From Table 4.4, the vapor pressure can be calculated to be:

$$p_f = \exp\left(\frac{A}{T} + B\right) = \exp\left(\frac{-4572}{283.15} + 15.94\right) = 0.813 kPa$$

Then from Equation (4.12) the air fuel ratio in the vapor phase is:

$$\frac{m_a}{m_f} = \frac{28.95}{114.23} \cdot \frac{100 - 0.813}{0.813} = 30.9$$

This mixture is probably too lean to ignite.

The pressure plays a role in the process, not because it affects the fuel vaporization under these ideal assumptions, but because it determines the amount of air present. In the above example, if the pressure is lowered to 50 kPa, then the fuel air ratio becomes:

$$\frac{m_a}{m_f} = \frac{28.95}{114.23} \cdot \frac{50 - 0.813}{0.813} = 15.3$$

This mixture is very close to stoichiometric. Looked at in a reverse situation, with a two phase system, and equilibrium between the fuel liquid and vapor and the air, increasing the total pressure will result in a leaner vapor mixture ratio, as long as all the fuel is not evaporated.

This is part of the reason for temporary enrichment of the fuel air mixture upon acceleration of SI engines. This was the function of the accelerator pump in older vehicles, but is now part of the electronic injection control strategy.

The effects of fuel temperature on evaporation and vapor phase fuel air ratio are shown in Figure 4.8. If a stoichiometric mixture of fuel and air is supplied to the system and the temperature is below 23.2°C, the vapor air mixture will be leaner than stoichiometric and some unevaporated fuel will remain. If the temperature is raised to 23.2°C, the vapor air ratio will be stoichiometric and all of the fuel evaporated.

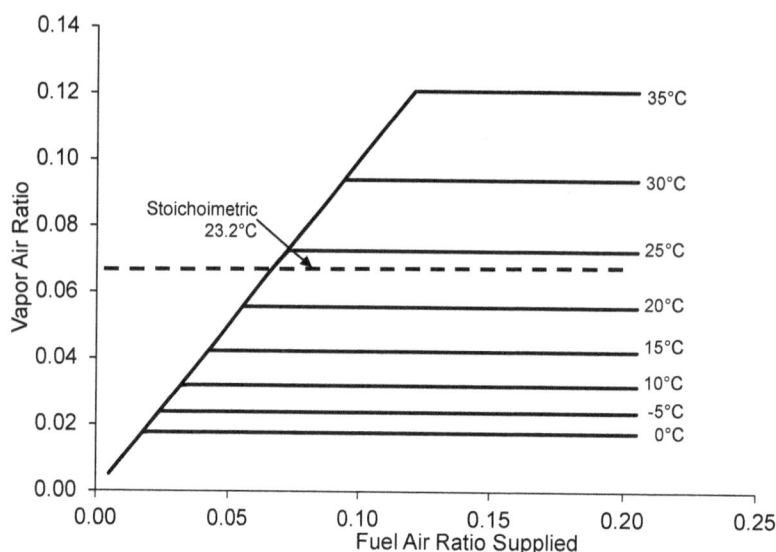

Figure 4.8: Equilibrium vapor phase fuel air ratio for n-octane in air as a function of fuel air ratio supplied at a pressure of 101 kPa. .

If the temperature is maintained at 23.2°C and more fuel added, the vapor air mixture will remain at stoichiometric and the added fuel remain as liquid.

For a single fuel component, a certain temperature is needed to obtain a combustible mixture at start, if fuel is injected into the manifold. In practice, the situation is alleviated somewhat by the fact that some liquid is drawn into the cylinder even on starting, and that the compression will help to evaporate at least some of it. Though if the liquid is deposited on very cold surfaces at engine start-up, it may not evaporate, and a fuel film on a spark plug can prevent the engine from starting.

Actual SI fuels contain many components, and the volatility of the fuel is controlled to accommodate for varying atmospheric conditions. The results of Figure 4.8 will be significantly different for an actual commercial fuel, since as the fuel evaporates, the remaining liquid changes composition, as the higher boiling components evaporate first. The evaporation process can be studied in a test called the equilibrium air distillation (EAD), where fuel and air are mixed and flow into a temperature controlled system. The flows of liquid fuel in and out are used to determine the amount of vaporization and, therefore, also the vapor phase fuel air ratio. These results can be estimated using empirical correlations based on the fuel distillation curves, and are only guidelines, as the fuel air process in a real engine does not achieve equilibrium. Nonetheless, the results can illustrate the problem of cold starting of engines.

Figure 4.9 shows typical results for EAD distillation of a multi-component gasoline. Note that the curves are not as sharply defined as they were for a single component with a well defined vapor pressure, as seen in Figure 4.8. This is due to the mixture of components and some non-ideal behavior of the liquid mixture. At low temperatures, a very rich mixture must be supplied in order for the fuel air ratio in the vapor phase to approach the lean limit, in this case for a temperature of -10°C, about 1.6 times the stoichiometric amount of fuel must be added, and about 5 times to achieve a stoichiometric vapor air ratio. At a temperature of 20°C, only about 0.8 times the stoichiometric amount of fuel must be added to achieve the lean limit, and 1.9 time to achieve a stoichiometric vapor air ratio. These results are only indicative, but in practice a rich fuel mixture is supplied to the engine a short time during start and warm up to ensure stable operation. Remember that the engine starts at low intake pressure, and when the throttle is opened the mixture will become leaner. The pressure effect described for the single component fuel is also valid for the multi-component fuel. In practice, these effects are accounted for by regulating the volatility of the fuel as discussed above, and by experimental determination of the fuel air mixture ratio requirements during cold start and warm up. An effective strategy is programmed into the engine management system in modern engines. Since EAD performance can be estimated on the basis of fuel distillation curves, [4], volatility specifications based on distillation curves alone are adequate.

Added to the problems of mixture control with start up and acceleration is the need to operate the

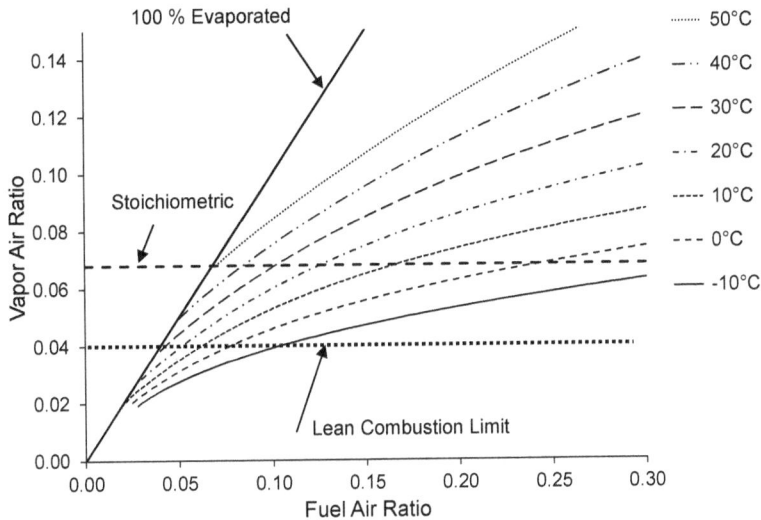

Figure 4.9: Typical equilibrium air distillation results for a spark ignition engine fuel at atmospheric pressure.

engine at stoichiometric mixtures as soon as possible after starting, in order to have good efficiency of the exhaust gas catalyst. The major amount of hydrocarbon and carbon monoxide emissions from modern SI engines equipped with 3-way catalysts originate from the cold start of an engine. Once a spark ignition engine is warm and operating at a stoichiometric fuel air ratio with a warm catalyst, the emissions are very low.

There is another commonly used test method for fuel volatility in gasoline, and that is the Reid Vapor Pressure, abbreviated RVP. In this simple test, a cooled chamber of fuel is connected to a chamber of air with a volume 4 times that of the fuel. The two chambers are connected, and immersed in a water bath at a specified temperature, usually 37.8°C. When the system is stabilized, the pressure is read from a pressure gauge, the value being the Reid vapor pressure. It is useful in determining the tendency of the fuel to form vapor in the fuel lines of a vehicle.

In fuel systems, fuel enters the air in the form of sprays that rapidly break up into droplets, and it is the droplet form of the fuel that evaporates. This is an area that has been studied for many years and is still of research interest. Droplet behavior in sprays, turbulent flow, and super critical conditions can be quite complicated. A very simple introduction was presented in Section 3.6.4, in order to present some very elementary concepts. The same assumptions are used for SI engines with manifold or in-cylinder injection.

Simple droplet theory can be used to calculate a simple form of evaporation rate. For a spherical droplet in a convective environment, the evaporation rate, $\dot{m}_f$ can be shown to be an alternate form of Equation (3.119) [8]:

$$\dot{m}_f = 2\pi\rho \mathbf{D} r N u_d \ln\left(1 + \frac{c_p(T_\infty - T_s)}{h_{fg}}\right) \tag{4.13}$$

where T_∞ is the ambient temperature T_s, is the boiling temperature of the fuel h_{fg} is the heat of vaporization of the fuel, ρ the density of the surrounding gas, r, d the droplet radius and diameter, and $\mathbf{D}$ is the diffusion coefficient.

The theory assumes that the droplet temperature is its boiling point throughout the evaporation. Equation (4.13) leads to a well known form of behavior for drops called the d^2 law, which implies that for an evaporating droplet, the square of the diameter decreases linearly with time. The fluid properties appear in Equation (4.13) , but in such a way that the logarithmic term varies little for most typical fuel components, including alcohols. Therefore, it the the transport properties, here represented by the diffusion coefficient, that play a major role in determining differences in evaporation rates. The main influence here is the velocity of the droplet relative to the surrounding air. Increasing droplet velocity greatly increases the evaporation rate, as seen in Section 3.6.4.

4.4   Fuel Specifications

In addition to volatility and ignition characteristics, there are a variety of properties of engine fuels that need to be regulated. They encompass a variety of areas such as fuel system corrosion, lubrication of fuel pumps, deposit formation in engines and storage stability. Several of these properties and related test procedures are briefly discussed in the following. Detailed descriptions and specifications for the relevant test procedures can be purchased from national standards organizations, ISO, or the ASTM in the United States.

- *Cetane Index* - In lieu of a cetane test of the fuel, the cetane index can be used to estimate the cetane number of hydrocarbon based diesel fuel. It is an empirical relationship based on the density of the fuel and its vaporization characteristics.

- *Total Contamination* - This is a measure of the non-soluble contaminants in fuel. Fuel is tested for contamination by a filtration method under specified conditions. Contaminants in engine fuels can result in premature plugging of fuel system filters.

- *Carbon Residue* - This test gives an indication of the fuel to coke or char in engines. A sample of the fuel is placed in a container called a coking bulb, which is heated in a metal furnace to a high temperature for a certain length of time. The material remaining is called carbon residue.

- *Ash* - This is a measure of non-combustible residue from the fuel. It is tested by burning a given amount of fuel, then heating the remainder at a high temperature in an oven. This temperature is higher than that of the carbon residue, so all the carbon is oxidized. The material remaining is called ash. It can lead to deposits in the engine combustion chamber and in the exhaust system.

- *Flash Point* - This is a measure of the ignitability of a fuel in a container exposed to air. The flash point is the highest temperature of the fuel at which it will not ignite when a flame is passed over its surface. At this temperature, a flammable air vapor mixture is produced. The flash point is related to handling safety of the fuel.

- *Gum* - Certain components in fuel can oxidize and form thick liquids or solids that can be deposited at different locations in engines, particularly in spark ignition engines, where the fuel is present in the intake system. These oxidized substances are called gum, which often have a high content of organic acids, leave deposits in manifolds and on valve train components in and can also leave deposits on pistons and rings. This can eventually lead to sticking of valves and rings, with poorer operation, even possibly engine failure as a result. Gum can be present in the fuel as manufactured, or can be created in the fuel over a period of time during fuel storage. To determine the amount of gum present in fuel, a certain volume is evaporated from a glass dish by passing hot air over it for a specified length of time. The fuel remaining is designated gum, and is typically specified in terms of mg of gum per 100 ml of fuel. The tendency of the fuel to form gum is tested in a glass lined bomb, in which a specified volume of fuel is placed and then the bomb charged with pressurized oxygen and placed in boiling water. The time required for the pressure to drop at a specified rate is a measure of the stability of the fuel, the longer the time, the more stable the fuel.

- *Copper Corrosion* - Many elements of engine fuel systems are made using copper or copper alloys. Certain elements in fuels can attack the copper, causing corrosion, with leakage as a potential result. To determine the corrosive nature of a fuel, a fuel sample is placed in a container with a polished copper strip at a given temperature for a given time. At the completion of the test, the condition of the copper strip is compared to that of standards from testing organizations, and given a numerical rating.

- *Pour Point* - This is a parameter used to indicate the utility of diesel fuel at low ambient temperatures. It is an indication of at which temperature the fuel can no longer be poured. In the test for pour point, a fuel sample in a transparent container is heated to a defined temperature, then cooled at a specific rate. At 3°C intervals, it is checked to see if there is movement of the fuel when tipping the container. The lowest temperature for which the fuel does not move is called the pour point.

- *Cloud Point* - This is a measure of the ability to use a diesel fuel at low temperatures. In the test, a sample of the fuel is cooled at a certain rate, and visually examined to determine when wax crystals begin to appear as a cloud in the fuel. The temperature at which this occurs is called the cloud point. Wax crystals can plug components in the fuel injection system, blocking fuel flow and causing engines to stop, or prevent them from starting in cold weather.

- *Cold Filter Plugging Point (CFPP)-* This is a measure of the the ability to use a given diesel fuel at low ambient temperatures, and was developed for use in Europe, and is related to the cloud point, as it is concerned with wax crystals or solids plugging filters in engines, and stopping the flow of fuel. The American version of the test is called the Low-Temperature Flow Test. In these tests, fuel samples are gradually cooled, and the checked to see if a specified volume can flow through a standard wire mesh within a certain period of time.

- *Viscosity* - In diesel injection systems, the fuel is pumped to high pressure, and there are small clearances in the injection systems. In the high pressure pumps and at injection nozzles, it is the fuel itself which must lubricate the moving parts, and prevent wear and system deterioration. Viscosity is important for preventing wear when the lubrication is in the hydrodynamic regime. Viscosity also plays an important role in fuel leakage through small clearances. Fuel spray and droplet formation and evaporation in the combustion system are also affected by fuel viscosity, but the test procedures and standards have been developed to insure satisfactory operation and durability of the fuel injection systems.

  Viscosity is typically measured with a viscometer, which is a calibrated glass capillary tube. The viscosity is determined from the time it takes fuel to flow through a calibrated volume of the viscometer.

- *Lubricity* - In addition to viscosity, another characteristic of fuel which is related to fuel system wear and durability is the lubricity of the fuel. Lubricity is the ability of a fluid to protect surfaces from wear when they are subject to lubrication in the boundary lubrication regime. This occurs in cases where there is periodic motion and the relative motion between the surfaces is slow or stopped, for example at TDC and BDC. The lubricating film disappears and can no longer provide lubrication. In modern fuel, lubricity is assured by special additives, which apparently attach them selves to the surfaces to be lubricate, and prevent actual surface to surface contact.

  Lubricity is measured by the High Frequency Reciprocating Rig. In this test procedure, a metal ball is pressed against a metal plate with a specified force and they are submersed in the fuel to be tested. The ball is reciprocated against the plate with a given stroke and frequency for a certain length of time. The wear scar on the ball is then measured, and its mean diameter in microns is used as the measure of lubricity.

4.4.1 Diesel Fuel Standards

There are many different requirements for diesel fuels. These are established through cooperation of oil companies, engine manufacturers, and public authorities. Figure 4.10 shows specifications for diesel fuel in the European Union from 2010. Similar standards are found in other parts of the world. Properties are specified relative to many areas. In addition to the parameters mentioned above, other important specifications are seen in Figure 4.10. Fuel sulphur is a double-edged sword. A certain amount of sulphur is very helpful in the lubrication of fuel pumps and components. Unfortunately, it plays an important role in the formation of diesel particulate emissions, and is very incompatible with emission control systems, especially catalysts. Consequently the European sulphur standard in Figure 4.10 has since been lowered to under 10 mg/kg (10 ppm mass). Because of this low amount of sulphur and its detrimental effect on the fuel lubricity, it has been necessary to add lubricity improving additives to diesel fuel, to prevent fuel system failure. This has resulted in the inclusion of a lubricity standard, as shown in Figure 4.10.

All Fuel Grades		Test Method
Cetane Number, mimimum	51	EN ISO 5165 EN ISO 15195
Cetane Index, minimum	46	EN ISO 4264
Density at 15°C – kg/m^3	820-845	EN ISO 3675 EN ISO 12185
Polycyclic aromatic hydrocarbons, mass %, maximum	8.0	EN 12916
Sulphur Content – mg/kg, maximum	10.0	EN ISO 20846 EN ISO 20884
Flash Point - °C, minimum	55	EN ISO 2719
Carbon Residue – mass %, maximum	0.30	EN ISO 10370
Ash – mass %, maximum	0.01	EN ISO 6245
Water - mg/kg, maximum	200	EN ISO 12937
Total Contamination – mg/kg, maximum	24	EN 12662
Copper Corrosion, 3 hrs at 50°C, rating	Class 1	EN ISO 2160
Fatty acid methyl ester (FAME) – volume %, maximum	7.0	EN 14078
Oxidation Stability – g/m^3, maximum – h, minimum	25 20	EN ISO 12205 EN 15751
Lubricity, corrected wear scar diameter at 50°C, μm, max.	460	EN ISO 12156-1
Viscosity at 40°C - mm^2/s	2.00 – 4.50	EN ISO 3104
Distillation		EN ISO 3405
Maximum % volume recovered at 250°C:	65	
Minimum % volume recovered at 350°C:	85	
95 % volume recovered at - °C	360	

Figure 4.10: European specifications for diesel fuel (2010).

Requirements of polycyclic aromatic hydrocarbons are included, as several of these compounds are carcinogenic. Additionally, the presence of these compounds in the fuel has a detrimental effect on the diesel particulate emissions.

A final specification mentioned is that of fatty acid methyl ester. Vegetable oils are compounds of this type, and with the interest in using renewable resources to fuel engines, vegetable oil methyl esters have been added to standard diesel fuels in some areas. Vegetable oil methyl esters are good fuels for diesel engines, but have different stoichiometric air requirements, and in high concentrations can attack some of the plastics or polymers used in fuel systems and seals.

Due to a large variation in climatic conditions from the Arctic to the Mediterranean, European fuel standards have been implemented for varying climates. For diesel fuel, the main consideration is that of cold weather starting. Several classes of fuels are developed, and individual countries can decide which of these classes are to be applied to the conditions of that country. The particular parameters of interest are the CFPP and the cloud point, as these relate to the plugging filters at low temperature. The fuel must obviously be able to flow through the filters at low temperatures. A lower viscosity is also an advantage at low temperatures, as this enables better fuel flow.

4.4.2 SI Fuel Standards

There are also many different specifications for spark ignition engine fuels. They are also established through cooperation of oil companies, engine manufacturers, and public authorities. Figure 4.12 shows specifications for spark ignition engine fuel in the European Union from 2008.

Probably the most important specification for SI engine fuel is the octane number. Operation of an engine on a fuel with an inadequate octane rating can rapidly lead to the mechanical failure of the engine. Thus it is important that the fuels on the market satisfy the requirements that were established with the design and development of the engine. Using an engine with an octane rating higher than required

Limits for Temperate Climates							Test Method
Cold filter plugging point	Grade						Test Method
	A	B	C	D	E	F	
CFPP -°C, max	+5	0	-5	-10	-15	-20	EN 116

Limits for Arctic or Severe Winter Climates						Test Method
Property	Class					Test Method
	0	1	2	3	4	
CFPP -°C, max	-20	-26	-32	-38	-44	EN 116
Cloud Point -°C, max	-10	-16	-22	-28	-34	EN 23015
Density at 15°C – kg/m³	800 - 845	800 - 845	800 – 840	800 - 840	800 - 840	EN ISO 3675 EN ISO 12185
Viscosity at 40°C - mm²/s	1.50 – 4.00	1.50 – 4.00	1.50 – 4.00	1.40 – 4.00	1.20 – 4.00	EN ISO 3104
Cetane Number, mimimum	49.0	49.0	48.0	47.0	47.0	EN ISO 5165 EN ISO 15195
Cetane Index, minimum	46.0	46.0	46.0	43.0	43.0	EN ISO 4264
Distillation Maximum % volume recovered at 180°C:	10	10	10	10	10	EN ISO 3405
Minimum % volume recovered at 340°C:	95	95	95	95	95	

Figure 4.11: European specifications for diesel fuel (2010) for low temperature conditions.

does not normally provide any additional economy or performance. Some modern engines are equipped with knock sensors, and can adjust the ignition timing to best advantage while avoiding knock for the fuel currently being used in the engine. In such a system octane rating can affect engine economy and performance.

Sulphur content in SI fuels been reduced, again due to the sulphur sensitivity of the emission control systems required for latest emission standards. Another specification is the concentration of lead. Formerly, lead was added to gasoline in the form of tetra-ethyl lead as a very effective means to reduce knock[2]. Tetra-ethyl lead is a poison, and so were the lead particles that were deposited along travel routes in significant quantities. Lead was not removed from gasoline, however, until standards for gaseous pollutant emission reduction resulted in the need for catalytic converters in the exhaust. Lead deposits block active sites on catalysts, and lead was removed in order to enable the use of exhaust gas catalysts, which made it possible to meet more strict emissions standards. Lead is no longer found in gasoline today in the industrialized countries of the world.

Fuel volatility is very important in SI engines, since fuel evaporation at temperatures close to atmospheric values is required. It is also important with respect to formation of vapor pockets (vapor lock) in fuel lines, which can lead to engine stoppage, and plays an important role in determining the evaporative emissions for gasoline powered vehicles. Due to variations in local climate and weather, different volatility specifications apply depending on the country and time of years. Common European classes have been developed to accommodate the variations found within the continent. It is up to each local country to decide which of these volatility classes should be used there. The volatility classifications are shown in Figure 4.13.

---

[2]One of very few "miracle" additives ever to be effective in engines.

Property	Units	Limits		Test Method
		Min.	Max.	
Research Octane Number RON		95.0		EN ISO 5164
Motor Octane Number MON		85.0		EN ISO 5164
Lead content	mg/l		5.0	EN 237
Sulphur content	mg/kg		10	EN ISO 20846 EN ISO 20884
Oxidation stability	Minutes	360		EN ISO 7536
Gum content	mg/100ml		5	EN ISO 6246

Hydrocarbon type content
  -olefins
  -aromatics
	Vol %		18.0
			35.0
				EN 14517
				EN 15553

Benzene content	Vol %		1.00	EN 238 EN 12177 EN 14517
Oxygen content	Mass %		2.7	EN 1601 EN 12177 EN 14517

Oxygenates content
 	
  -methanol
  -ethanol
  -iso-propyl alcohol
  -iso-butyl alcohol
  -tert-butyl alcohol
  -ethers (5 or more C atoms)
  -other oxygenates
	Vol %		
			
			3.0
			5.0
			10.0
			10.0
			7.0
			15.0
			10.0	

Figure 4.12: European specifications for 95 octane gasoline (2008).

Property	Units	Limits						Test method
		Class A	Class B	Class C/C1	Class D/D1	Class E/E1	Class F/F1	
Reid Vapor Pressure	kPa, min kPa, max	45.0 60.0	45.0 70.0	50.0 80.0	60.0 90.0	65.0 95.0	70.0 100.0	EN13016-1
% Evap at 70°C, E70	Vol %, min Vol %, max	20.0 48.0	20.0 48.0	22.0 50.0	22.0 50.0	22.0 50.0	22.0 50.0	EN ISO 3405
% Evap at 100°C, E100	Vol %, min Vol %, max	46.0 71.0	46.0 71.0	46.0 71.0	46.0 71.0	46.0 71.0	46.0 71.0	EN ISO 3405
% Evap at 150°C, E150	Vol %, min	75.0	75.0	75.0	75.0	75.0	75.0	EN ISO 3405
Final Boiling point (FBP)	°C, max	210	210	210	210	210	210	EN ISO 3405
Distillation residue	Vol %, max	2	2	2	2	2	2	EN ISO 3405
Vapor Lock Index (VLI) (10VP +7 E70)	index, max	-	-	C -	D -	E -	F -	
Vapor Lock Index (VLI) (10VP +7 E70)	index, max	-	-	C1 1050	D1 1150	E1 1200	F1 1250	

Figure 4.13: European specifications for volatility of gasoline (2008). VP denotes Reid vapor pressure, and E70 denotes the percentage of fuel evaporated at 70°C for the standard distillation test.

4.5 Alternative Fuels

Most engine fuels today consist of hydrocarbons obtained from crude oil and processed to suit the needs discussed above. For various reasons, alternatives to the current type of fuel have been discussed and investigated for many years, though as of this writing have yet to achieve major usage in vehicles. There are two fundamental reasons for using different fuels than petroleum based fuels:

- Fuel supply issues. The first petroleum reserves used have been the cheapest to exploit. As these are depleted and demand increases, more expensive fuel sources are used. Any petroleum reservoir has an optimum production rate, and there is a finite number of these reservoirs, whatever that number is. With increasing demand for fuel, prices increase, and supplies may become uncertain. Thus, there are both short and long term advantages to having different fuel sources available.

  Other petroleum sources, such as oil sands, can be come economically attractive as oil prices increase, and so become part of the supply system. A high oil price also makes alternatives to petroleum fuel more competitive, as they have traditionally been higher priced that petroleum based fuels, and often give high CO_2 emissions during production.

- Environmental Issues. Petroleum fuels have different degrees of emissions problems. On the local scale, one of the serious remaining problems is that of particulate emissions from diesel engines, a fundamental part of the combustion process of hydrocarbon fuels in diesel engines. Particulate emissions have proven to be much more difficult to control than gaseous emissions. Some alternative fuels eliminate this problems, but are not used currently for reasons discussed later. At the time of writing, the more important issue is considered to be that of global warming, and the contribution of combustion of petroleum based fuels to the buildup of CO_2 in the atmosphere is receiving intense interest. This has led to an revival in the interest in renewable fuels, where carbon is recycled, or carbon free fuels are used, in which there is no carbon in the combustion process.

There are many alternatives that have been proposed. Some suitable alternative fuels for spark ignition engines are shown in Table 4.5. Some of them are discussed in the following.

4.5.1 Natural Gas/LPG

To many, natural gas is considered an alternative fuel, though its production today is often connected with the production of crude oil. It has mainly been considered an alternative fuel where issues of local environment have been involved, as natural gas from oil and gas reservoirs takes CO_2 out of "storage" and puts it into the atmosphere, the same as oil. However, the use of natural gas (methane) from renewable sources would contribute to lower CO_2 emissions. Natural gas has been used in engines for many years, a primary application being large, spark ignition engines used to drive compressors on natural gas pipe lines. The availability of fuel here is obvious.

The property of natural gas which affects its operation as a vehicle fuel are the fact that it normally exists as a gas, which complicates storage and any form of high pressure injection. Since it is composed primarily of methane, and has very high resistance to auto ignition, (See Figure 4.2)it is primarily used in spark ignition engines. Because of good knock resistance, natural gas powered engines can be operated at high compression ratios to improve efficiency. Compression ratios on the order of 12:1 are common for

Table 4.5: Some properties of alternative fuels suitable for SI engines. LPG consists of propane, n-butane and i-butane in varying proportions

Compound	Formula	Mol. Wt.	FA_s	H_u	Boiling Point	RON/MON
Natural Gas	$\approx C_1H_{3.73}$	≈ 18	≈ 0.059	≈ 49.4	$\approx -163°C$	$\approx 115-120$
Propane	C_3H_8	44.09	0.0637	46.3	$-42°C$	112/97
n-Butane	C_4H_{10}	58.12	0.0656	45.7	$0\ °C$	94/90
i-Butane	C_4H_{10}	58.12	0.0656	45.6	$-12°C$	102/98
Methanol	CH_3OH	32.04	0.154	19.9	$65°C$	106/92
Ethanol	C_2H_5OH	46.07	0.111	26.8	$78°C$	107/89

larger natural gas engines. Spark ignition engines still have part load pumping losses, so SI natural gas engines are not as efficient as CI engines in vehicle applications, though they approach them at higher loads.

The main use of natural gas in vehicles has been natural gas powered SI engines for city busses. The prime reason here is local emissions problems. This type of engine does not have a significant emission of particles, and the homogeneous charge and operation near stoichiometric conditions makes the application of 3-way or oxidation catalysts effective, thus natural gas engines have lower emissions than diesel engines. Since they are SI engines, however, the natural gas busses have higher fuel consumption than diesel powered busses.

LPG (Liquified Petroleum Gas) is similar to natural gas, it is a gas a atmospheric temperature and pressure, and has a high resistance to autoignition, though not as high as natural gas. It has been used for many years in vehicle applications. There was a time when LPG was economically attractive for vehicle use, and it has been used for indoor vehicles where CO emissions can be a problem. LPG powered SI engines can successfully operate with quite lean mixtures, which gives low CO emissions. LPG is produced as a by-product from oil production, so there is a question as to whether it is a true alternative fuel, but it is at least mentioned here. LPG is known to have a widely varying composition, depending on the source and other factors. The fact that LPG must be kept under light pressure (about 4-5 atm) is not a major problem with SI engines.

4.5.2   Alcohols

There are two alcohols that are primarily considered as alternative fuels, in that there is the possibility of the production of the large quantities of these compounds needed for a transport fuel, without being extremely expensive. They are methanol, (wood alcohol) and ethanol (grain alcohol or spirits). Some relevant properties for use as SI engine fuels are listed in Table 4.5. That these two compounds are well suited as SI engine fuels can be seen by the octane numbers, which are higher than those of conventional fuels today. A quick estimate (product of stoichiometric fuel air ratio and heating value) shows that without any other changes than the fuel, an SI engine operating at stoichiometric conditions will deliver roughly the same power regardless of the fuel, gasoline or alcohol. More fuel will be needed when operating on alcohols, due to the lower heating values. If the compression ratio is increased for the higher octane alcohols, the comparison would be more in their favor on an efficiency basis.

$$bmep = \eta_v \cdot \rho_{int} \cdot FA_s \cdot H_u \cdot \eta_i - fmep$$
$$bmep_{Gasoline} = 0.9 \cdot 1.15 \cdot 0.068 \cdot 44500 \cdot 0.36 - 250 = 877kPa$$
$$bmep_{Ethanol} = 0.9 \cdot 1.15 \cdot 0.111 \cdot 26800 \cdot 0.36 - 250 = 858kPa$$
$$bmep_{Methanol} = 0.9 \cdot 1.15 \cdot 0.154 \cdot 19900 \cdot 0.36 - 250 = 892kPa$$

These fuels are not found in any significant quantities in nature, which means that they have to be made from other sources. Today, methanol is made from natural gas, in a process where the gas is decomposed to a mixture of synthesis gas, composed of hydrogen, carbon dioxide, carbon monoxide and water vapor. The synthesis gas is reformed over a catalyst to form methanol. Since the initial feed stock is decomposed, it is possible to make methanol from a variety of sources, including coal and biomass. Methanol can then be made from biological fuels, and can be seen as a renewable resource, a favorable characteristic with respect to CO_2 emissions. Made from coal, on the other hand, it would contribute to the addition of CO_2 to the atmosphere. Reasons why biological based methanol is not used as an engine fuel today are largely economic. The collection of biomass contributes to this, as well as initial investments in manufacturing facilities.

Ethanol has been manufactured by a fermentation and distillation process from various kinds of fruit and grain. In the large quantities needed for a fuel additive, it is produced by the fermentation and distillation of corn/grain. Research work is underway to produce ethanol in greater yields through other biological processes, including production from the cellulose material from plants, with the goal of reducing the energy requirements for ethanol production, and increasing the supply of raw materials.

From an engine point of view, methanol is quite an acceptable SI fuel, though some changes are needed relative to gasoline. Methanol is more aggressive towards materials, so some components of the fuel system need to be changed. Methanol has a higher heat of vaporization than gasoline (1100 kJ/kg compared to about 350 kJ/kg for gasoline), and about twice the mass of methanol per kg of air is needed due to the high stoichiometric fuel air ratio. A lot of energy is then needed to accomplish the evaporation of the fuel, which must be taken into consideration in the design of the system.

The example above shows that a methanol engine can provide slightly more power than when operated on gasoline, since the product of the stoichiometric fuel air ratio and the heating value is slightly higher than for gasoline. (See Table 2.1). Depending on how the fuel is evaporated, the larger heat of vaporization may result in a lower charge temperature at the start of compression. In addition, the higher octane number should allow a higher compression ratio, further increasing a power advantage. The indicated thermal efficiency of a methanol engine is very close to that of a gasoline powered engine at the same compression ratio.

The emissions of a methanol powered engine will be of a magnitude similar to a gasoline powered SI engine, with some exceptions. The NO_x emissions may be slightly lower if charge temperatures are lower. A significant problem with methanol engines in the past has been the tendency to produce large amounts of formaldehyde. This is because formaldehyde is the first product to be formed in the oxidation of methanol, and most of the unburned "hydrocarbons" are really methanol. Oxidation catalysts can effectively reduce formaldehyde emissions, though they can be a problem particularly with cold starting during the short time when catalysts are not yet effective.

An ethanol engine would produce marginally less power than a gasoline engine according to the above example, though as is the case with methanol, the engine could accept a higher compression ratio than conventional gasoline, winning back the loss in power [50]. The emissions are similar in magnitude to gasoline engines. An ethanol engine does not produce nearly as much formaldehyde as a methanol engine.

From a technical point of view, there are no major obstacles to operating engines on either methanol or ethanol, and prototypes have been built, and limited numbers of vehicles run in practice. The major replacement of gasoline with these fuels has not occurred due to the higher prices of alcohols, infrastructure changes and the general inertia of the transportation manufacturing and fuel supply system. If alcohols can be produced from biomass at competitive prices, they are good candidates for replacing petroleum fuels in a sustainable fashion.

It should be added, that ethanol is currently used in substantial quantities as a fuel additive in the United States and Europe. Fuels contain up to 10% ethanol and are sold extensively. The original reason for adding ethanol to gasoline was to reduce CO emissions in certain urban areas. Adding ethanol to gasoline requires a richer mixture, if stoichiometric conditions are to be maintained. If this is not done, particularly in the case where the cold start engine control strategy does not account for this, the engine operates with a leaner mixture than with gasoline. It was seen in Section 3.1 that this reduces CO emissions in the exhaust. Emphasis on ethanol at the time of writing has shifted toward replacing and reducing dependency on petroleum fuels.

Recently, E85 vehicles have been produced that operate on an 85% ethanol - 15% gasoline mixture. E85 is available in many American cities, particularly those in the corn producing states of the Middle West.

A large amount of ethanol is used for vehicles in Brazil, partly because of the large amount of sugar cane as a natural resource. The addition of ethanol to gasoline in Europe is increasing, with Sweden having the highest usage at the time of writing.

Alcohols can, and have been used in diesel engines. It is difficult, but can be accomplished through special engine design, and the use of ignition improving additives, though the latter are expensive. Pure ethanol does not form particles in diesel combustion, but smooth operation requires ignition improvers in an amount that can produce enough particles to require a particulate filter to satisfy future emissions standards.

4.5.3 Vegetable Oil Based Fuels

Since the first oil crises of the 1970's, vegetable oils and some of their derivatives have been tested as diesel fuel substitutes [51]. The most common vegetable oils mentioned are rape seed oil in Europe and soybean oil in North America. The oils are pressed from the seeds of these plants, and other plants such as peanuts, sunflowers, and palms. The oils as such will burn successfully in a diesel engine, though they have the disadvantages of having a rather high viscosity of approximately 32 cs at 40°C, and in

Table 4.6: Some properties of alternative fuels suitable for CI engines. The oils and esters decompose above 300°C, and DME boils at -24.5°C, requiring a pressurized container.

Compound	Formula	FA_s	H_u - MJ/kg	Cetane
Soybean Oil	$C_{18}H_{33}O_{1.9}{}^a$	0.079	39.3	38
Soybean Oil Methyl Ester	$C_{18}H_{33.6}O_{1.7}^a$	0.0831	39.8	51
Rapeseed Oil	$C_{18}H_{32}O_{1.9}^a$	0.0807	36.7	44
Rapeseed Oil Methyl Ester	$C_{18}H_{32}O_2^a$	0.0813	37.1	53
DiMethyl Ether	CH_3OCH_3	0.111	28.9	> 70

^{a}normalized to C_{18}

unmodified engines may form carbon deposits around the injection nozzle. They also give satisfactory combustion when blended with diesel oil. Some properties vegetable oils are given in Table 4.6 along with those of Dimethyl Ether, to be discussed in the following section. The vegetable oil properties are only indicative, they vary according to type of oil and reported values in the literature vary considerably.

The flow properties of the vegetable oils can be improved by reacting them with methanol in the presence of a NaOH catalyst to form a vegetable oil methyl ester, with glycerine as a byproduct. Viscosities of methyl esters are quite close to those of diesel oils, typical values being about 4 cs at 40°C compared to the 2-4 cs of diesel fuel. The combustion of methyl esters is very similar to that of diesel fuel, and can be accomplished with basically no modifications to the combustion chamber [52]. Methyl esters are good solvents and care must be taken with polymers in the fuel system, as they may be dissolved.

In terms of exhaust emissions, vegetable oil fuels have similar emissions of NO_x, CO and unburned hydrocarbons, though the latter have a different composition than from standard diesel fuel, and are more likely to leave deposits due to the large amount of oxygenated compounds present. With respect to particulate emissions, the vegetable oil fuels also produce particles in amounts close to those of diesel fuel. The composition of the particles is different because the basic fuel is different.

From an operational point of view, vegetable oil fuels can be blended into diesel oil without major problems, at least in moderate amounts, and "bio-diesel" is commercially available at many European locations at the time of writing. Problems with the fuels are primarily related to production, as there is a question as to the overall CO_2 balance of the fuels when the energy required to cultivate the land, and the natural gas converted to fertilizer is considered.

4.5.4 DiMethylEther

Dimethyl ether (DME) is made in a manner that is very similar to methanol production. In fact, for the current major use of DME today as an aerosol propellant, it is made from methanol. For fuel production, DME can be made the sole product of the process. Since the process is like methanol, DME can be made from natural gas, which is cheapest at the present, coal and biomass, so it also has the potential to be produced from renewable resources.

DME is the simplest ether, and has many properties that are advantageous compared to the other ethers. It is stable in storage, non-poisonous, rapidly degradable in the atmosphere, and because it is stored under modest pressure, like LPG, it is not evaporated to the atmosphere. The pressurized storage is a disadvantage compared to diesel fuel which as a liquid is easily stored and handled. Some DME properties are shown in Table 4.6.

DME was shown to be an excellent diesel fuel in the mid 1990's, evidenced by its high cetane number [53]. Diesel engines operating on DME make no smoke at all, so particulate emissions are very low, and only originate from the engine's lubricating oil. As a result, a diesel engine may be adjusted solely on the basis on NO_x control without regard to smoke. Some of the very first DME engines tested exceeded the EU emissions standards for 2007 and 2008, over 10 years before they were put into effect [54].

There are some reasons why DME has not yet become a major participant in the diesel fuel market though, not the least of which is the lack of production facilities. It requires time and a lot of money to build production facilities of the size needed to supply transport fuels, and since DME has not been available, engine manufacturers have had to work on emission solutions based on diesel fuel. Fuel production of DME has been started in China, using surplus methanol production facilities.

Though it has excellent combustion characteristics, there are problem areas for DME in diesel engines. The first is that it is a gas at standard temperature and pressure, this requires some special treatment, though common rail systems perform well with DME. Due to its lack of soot formation, high injection pressures have proven unnecessary, and DME works well with injection pressures on as low as 300 bar, much below modern common rail diesel systems, which have pressures in excess of 1500 bar. Because it has a very low viscosity, it is subject to leakage, and part tolerances and design need special care with DME. In addition, DME has a low lubricity, and DME fuel injection systems are subject to high rates of wear as compared to diesel fuel systems. A long term DME injection system has not been demonstrated at the time of writing, though promising systems are under development and progress is being made.

DME has an advantage over many fuels, in that it can be used in a variety of applications. It is a fine substitute for LPG in cooking and heating, and is currently produced as a fuel for these purposes in China on a limited, but expanding scale. It is an excellent turbine fuel, useable for electrical generation, and has many applications in the chemical industry. The first major DME plants will be dedicated to making DME to produce proplyene for the chemical industry. The vision of DME promoters is a multi-source, multi-use fuel/chemical. Given the suitability of DME's closely related chemical, methanol, for SI engines, a process of making fuels for all kinds of transport engines from gasification of bio-fuels is an interesting prospect in the effort to reduce global CO_2 emissions. One very attractive source of bio-DME is black liquor from paper production. A demonstration project is currently underway in Sweden using on-road diesel vehicle powered by DME made from black liquor. Because bio-mass material has already been collected, one major cost of bio-fuel usage has been eliminated.

A very thorough presentation of all aspects of DME applications, production and economics is available in the DME handbook [55].

4.5.5 Hydrogen

In the discussion of global warming and CO_2 emission reduction, hydrogen has been discussed a great deal, primarily as a fuel for fuel cells. Hydrogen does not exist naturally in a form or quantity suitable for use as a fuel, and must be produced. This is a problem shared by many alternative fuels. Hydrolysis of water is named most often as a production method, and if this can be produced by renewable power, hydrogen would not contribute to CO_2 emissions. Other methods include decomposition of organic compounds, for example methanol or natural gas.

Hydrogen has the disadvantage of being difficult to store in large quantities as a fuel. It is not trivial to compress hydrogen, and even at high pressures, has a low energy/volume ratio than hydrocarbons and other alternative fuels. It can be stored advantageously as a hydrate with metals, and new proposals store hydrogen as ammonia in a solid phase with simple salts. This gives a large hydrogen density, but the hydrogen must be obtained again by the decomposition of the ammonia.

Hydrogen is usually mentioned as a fuel cell fuel, but it can also be used as an engine fuel, hence its inclusion in this chapter. Historically, one of the very first engines proposed[3] was to operate on hydrogen. The method of operation was to burn hydrogen at atmospheric pressure, and then operate the engine on the vacuum created when the water produced in the combustion gasses condensed!

Hydrogen has been tested as a spark ignition fuel, it has a high octane number, but very low ignition energy, and as such is very susceptible to backfiring in the intake manifold, when used as a premixed fuel. As an SI fuel, hydrogen has a very fast flame speed. Table 3.6 shows that it has a laminar flame speed about 5 to 6 times that of hydrocarbons. Engine testing with spark ignition of homogeneous charge hydrogen has sometimes shown an MBT spark timing after TDC [56]. The flame speed can be slowed considerably by using large amounts of exhaust gas recirculation. Hydrogen forms NO_x in the same way as hydrocarbons in combustion engines, and levels are similar to or above those of hydrocarbon fuels. Another advantages of hydrogen in an SI engine is that it has very wide flammability limits. This means that it is possible to operate a hydrogen powered SI engine by mixture strength alone, so that throttling is not needed, except for idle and possibly very low loads, reducing pumping losses.

The heating value of hydrogen is 120900, but its stoichiometric fuel air ratio is 0.029. The product of these two is 3500 kJ/kg-air, higher than the corresponding value for carbon based fuels. Using the same

---

[3]Proposed by a Reverend Cecil in England in 1830

parameters as for gasoline and alcohols, a first estimate of the power of a hydrogen powered SI engine is:

$$bmep = \eta_v \cdot \rho_{int} \cdot FA \cdot H_u \cdot \eta_i - fmep$$
$$bmep_{hydrogen} = 0.9 \cdot 1.15 \cdot 0.0290 \cdot 120900 \cdot 0.36 - 250 = 1056 kPa$$

A disadvantage of hydrogen when used in premixed combustion is its high specific volume. At a stoichiometric mixture, hydrogen occupies 22.8% of the volume of intake gasses, displacing oxygen and resulting in lower power on that basis. The above power could be obtained by in-cylinder injection, but if normal induction of a fuel air mixture is used, the displacement of oxygen would reduce the power to about the same level as that of gasoline and alcohols.

At the time of writing, it appears that a form of in-cylinder injection would be the best choice for a hydrogen powered spark ignition engine. There has not been a lot of engine work with hydrogen, but if it becomes a common fuel, there are potential applications in transport engines.

4.6 Problems

Problem 4.1

There is a mixture of n-octane (C_8H_{18}) and air in a container. The stoichiometric air/fuel ratio is 15.13 kg air/kg octane.

1. At a total pressure of l atm and a temperature of $10°C$, only 50% of the fuel is evaporated. Calculate the air/fuel ratio and the excess air ratio for the vapor phase and the overall excess air ratio. Table 4.4 can be used in the solution.

2. Calculate the lowest temperature (for a total pressure of l atm) at which a combustible air/vapor-mixture can occur under equilibrium conditions. The lean flammability limit is at an excess air ratio of 1.60.

3. Calculate the highest total pressure, at which a combustible air/vapor-mixture can occur at equilibrium, when the temperature of the mixture is $10°C$. ($\lambda_v = 2.078$; $\lambda_{tot} = 1.039$; $T_{min} = 14.6°C$; $p_{max} = 778.2kPa$)

Problem 4.2

A gasoline mixture consists of 5% n-butane (C_4H_{10}), 90% iso-octane (C_8H_{18}) and 5% n-hexane (C_6H_{14}). The following data for the three compounds are found:

Compound	Mol. wt.	Boiling pt.	Vapor pressure at $37.9°C$
n-butane	58.12	$-0.5°C$	355.7 kPa
n-hexane	86.17	$68.7°C$	34.2 kPa
iso-octane	114.22	$99.2°C$	11.8 kPa

The mixture composition is given in weight-%. Atmospheric pressure is 1.01325 bar.

1. Calculate the mixture vapor pressure at $20°C$

2. Calculate the mixture stoichiometric air/fuel-ratio.

($p_{v-mix} = 24.1$ kPa bar; $AF_s = 15.11$)

Chapter 5

Fuel Systems

5.1 Fundamental Flow Principles

In the following, the physical principles and fundamental equations needed for fuel system calculations are described.

Air Flow

Air flow in engines is normally treated by using isentropic, compressible flow as the standard, and modifying results with empirical correction factors (flow coefficiencts). These equations are used for flow through a venturi (carburetor) and are normally also applied to the flow through the intake throttle and valves. A sketch of a general system is shown in Figure 5.1

For this system, the first law for adiabatic flow and no work transfer becomes:

$$h_1 + \frac{v_1^2}{2} = h_2 + \frac{v_2^2}{2} \tag{5.1}$$

Normally, a reference condition is chosen at a point where the velocity is small (point 1 here), and the velocity at the exit is expressed as a function of the pressure difference. If the flow is not choked,

$$v_2 = \sqrt{2c_p T_1 \left[1 - \left(\frac{p_2}{p_1} \right)^{\frac{\gamma-1}{\gamma}} \right]} \tag{5.2}$$

Then using $\dot{m} = \rho v A$, the isentropic relations for ideal gas, and $c_p = \frac{\gamma R}{\gamma - 1}$, and a flow coefficient, C_a, which is the empirical ratio of the actual flow to the ideal flow:

$$\dot{m}_a = \frac{p_1 \cdot A \cdot C_a}{\sqrt{T_1}} \sqrt{\frac{2\gamma}{R(\gamma-1)}} \sqrt{\left(\frac{p_2}{p_1} \right)^{\frac{2}{\gamma}} - \left(\frac{p_2}{p_1} \right)^{\frac{\gamma+1}{\gamma}}} \tag{5.3}$$

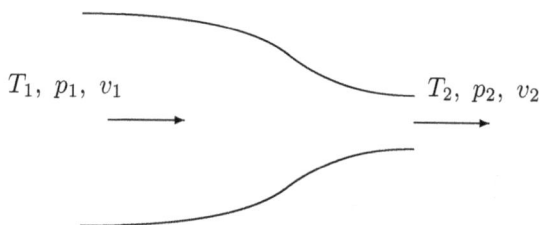

$T_1, \ p_1, \ v_1$ $\qquad\qquad$ $T_2, \ p_2, \ v_2$

Figure 5.1: General system for isentropic flow

The flow is choked when the Mach number is equal to unity at the minimum area, which gives the following pressure ratio:

$$\frac{p_2^*}{p_1} = \left(\frac{2}{\gamma + 1}\right)^{\left(\frac{\gamma}{\gamma-1}\right)} \tag{5.4}$$

which is equal to 0.5283 for $\gamma = 1.4$ Here, p_2^* denotes the critical throat pressure.

In this case, the flow is no longer dependent on the conditions downstream of the minimum area. Substituting Equation (5.4) into Equation (5.3), the choked air flow rate is:

$$\dot{m}_a^* = \frac{p_1 \cdot A \cdot C_a}{\sqrt{T_1}} \sqrt{\frac{\gamma}{R} \left(\frac{2}{\gamma + 1}\right)^{\frac{\gamma+1}{\gamma-1}}} \tag{5.5}$$

For the case of $\gamma = 1.4$, the square root term of Equation (5.5) is equal to 0.532.

The most common usage of the above equations is for flow through a venturi, across a throttle plate or reduction valve, the flow through a poppet valve, and the flow through intake and exhaust ports in a 2-stroke engine. They can also be used for the flow of gaseous fuels, such as in a natural gas, vaporized LPG and the like by changing the relevant thermodynamic parameters.

Fuel Flow

Flow of liquid fuel is simpler than of air, since compressibility can be neglected. In this case, Bernoulli's equation is applicable. Using the conceptual sketch of Figure 5.1:

$$v_2 = \sqrt{\frac{2}{\rho_f} (p_1 - p_2)} \tag{5.6}$$

Then using $\dot{m} = \rho v A$ and a flow coefficient, C_f, which is the empirical ratio of the actual flow to the ideal flow:

$$\dot{m}_f = A \cdot C_f \sqrt{2\rho_f (p_1 - p_2)} \tag{5.7}$$

5.2   Calculations with Fuel Injection Engines

Only fuel injected engines are considered here, since carburetors have basically disappeared in all but the smallest spark ignition engines. Carburetors are described in detail in many previous engine texts. In the case of fuel injection engines, the fuel system is physically separate from the air system. In SI engines, there is a need for an estimate or measurement (or both) of the air flow in the engine, which is used to determine the fuel flow, normally by the computer in the engine management system. The airflow is normally regulated with a throttle, and the amount of fuel added is regulated in relation to that, in order to obtain a target air fuel ratio. The fuel is sprayed into the intake system, as will be discussed later. In practice, these systems work with the opening and closing of a solenoid valve, and the output of the engine management system is a pulse length (opening time) for the fuel injector. The timing of the fuel pulse is often determined by reference to a given crankshaft position through some sort of sensor. Timing of manifold injection in SI engines is not normally critical to performance, though some timings have minor advantages.

In a diesel engine, the quantity of fuel is important, as it determines the engine power. The fuel air ratio is determined as a result of the engine operation rather than an input parameter, especially in the case of turbocharged engines. Since with injection engines the fuel quantity must be determined for each injection process, it is best to look at the process in terms of each injection, or using control system terminology, on an *event basis*. The event basis is used for both jerk pump type and common rail type systems. The physical means by which diesel injection systems function are described in Section 5.4.

An additional complication with the diesel engine compared to the spark ignition engine is that the timing of the fuel injection pulse is very important. Thus, additional facilities are needed in the control system for accurate measurement of piston position, and adjustment of the precise time of the fuel injection pulses.

For any kind of injection engine, the basic equation involved is Bernoulli's equation, converted to mass flow in Equation (5.7). However, care must be taken in determining the mass flow. In injection systems, the flow is periodic. That is, Equation (5.7) can only be applied during the time when the fuel flow is occurring, and the mass flow here should *not* be confused with the average fuel flow rate as given by the product of the air flow rate and the fuel air ratio measured in the exhaust gasses. In gasoline injection and common rail diesel systems, the fuel pressure is basically constant, as is the manifold or cylinder pressure. Thus to a good first approximation, the instantaneous mass flow through the injection system when fuel is flowing is the mass of fuel injected divided by the injection duration:

$$\dot{m}_f = \frac{m_f}{\delta t_{inj}} = \frac{m_a \cdot FA}{\delta t_{inj}} \tag{5.8}$$

For a 4-stroke engine, the air mass is:

$$m_a = \eta_v \cdot \frac{p_{in}}{RT_{in}} \cdot V_d \tag{5.9}$$

For a 2-stroke engine, the air mass is:

$$m_a = \eta_s \cdot \frac{p_s}{RT_s} \cdot V_d \frac{\varepsilon}{\varepsilon - 1} \tag{5.10}$$

Where the subscript s refers to the scavenging reference condition as discussed in Chapter 8. The displacement volume is for one cylinder in the case of diesel engines, GDI engines, and port injected SI engines where each injection is for one cylinder. It is conceivable that the injection process in an SI engine is designed to supply fuel to more than one cylinder at a time. In that case, the appropriate displacement volume must be used in Equation (5.9) or (5.10)

It is usually more convenient to refer to the injection time in terms of crank angle degrees rather than time in seconds. This is especially important in a jerk pump system, where the injection is controlled by the plunger position, and the plunger is directly connected to the camshaft and crankshaft. The injection duration, δt_{inj} in seconds is related to the duration in crank angle degrees $\delta\theta_{inj}$ in the following manner:

$$\delta t_{inj} = \delta\theta_{inj} \frac{rev}{360°} \frac{min}{N \ rev} \frac{60s}{min} = \frac{\delta\theta_{inj}}{6N} \tag{5.11}$$

In a jerk pump system, Bernoulli's equation still applies, but the flow is not constant. In this system, the motion of the camshaft driven plunger determines the rate at which the fuel moves through the system. Since the camshaft velocity is not necessarily constant, the volume flow rate of fuel can be estimated throughout the injection process:

$$\dot{v}_f = A_p \cdot v_p \tag{5.12}$$

where v_p is the instantaneous velocity of the piston in the injection pump, and A_p is its cross section area. Figure 5.2 shows the volumetric displacement of the fuel in a unit injector system, where pressure pulsations are not large. The displacement can be seen to be related to a camshaft type of displacement. At the end of the process, the flow drops suddenly, as a relief valve opens to limit the amount of fuel delivered to the desired quantity. If the fuel is assumed to be incompressible, then the instantaneous fuel mass flow rate through the injection system is:

$$\dot{m}_f = \rho_f \cdot A_p \cdot v_p = A \cdot C_f \sqrt{2\rho_f (p_1 - p_2)} \tag{5.13}$$

If Equation (5.8) is applied to a jerk pump system using the total amount of fuel injected, the average injection pressure can be determined. Combining Equations (5.7) and (5.8) and rearranging, one can solve for the injection pressure:

$$\Delta p_{inj} = \frac{1}{2\rho_f} \left(\frac{24m_f \cdot N}{\pi \cdot \delta\theta_{inj} \cdot C_f} \cdot \frac{1}{n_h} \right)^2 \frac{1}{d_h^4} = \frac{1}{2\rho_f} \left(\frac{4m_f}{\pi \cdot \delta t_{inj} \cdot C_f} \cdot \frac{1}{n_h} \right)^2 \frac{1}{d_h^4} \tag{5.14}$$

where n_h, is the number of holes in the injection nozzle tip.

These basic equations can be applied to both spark ignition and diesel engines, with appropriate choice of parameters and variables.

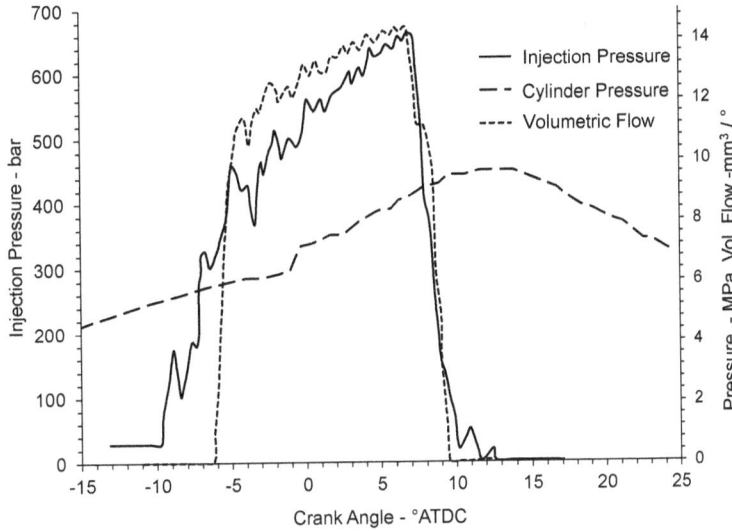

Figure 5.2: Fuel flow, cylinder pressure and injection pressure in a unit injector for a 2-liter per cylinder diesel engine, adapted from [57].

Example

Consider a 4 stroke, DI diesel engine with a bore of 120 mm and a stroke of 130 mm. The engine operates at 1700 rpm, with an intake pressure of 200 kPa (absolute) and intake temperature of 45° C, with a volumetric efficiency of 0.91. The fuel air ratio is 0.039. The engine is equipped with a 4-hole injection nozzle, and an injection duration of 32 CA° is desired. The fuel has a density of 0.837 g/cm^3, and the injection system is a common rail type with an injection pressure difference of 1000 bar above atmospheric. If the flow coefficient of the nozzle is 0.822, calculate the hole size in the nozzle.

Solution:

Find the instantaneous fuel flow rate through the nozzle when it is open and apply Bernoulli's equation. The fuel mass is:

$$
\begin{aligned}
m_f &= m_a \cdot FA = \eta_v \cdot \rho_{in} \cdot V_d \cdot FA \\
&= \eta_v \cdot \frac{p_{in}}{RT_{in}} \cdot \frac{\pi}{4} B^2 S \cdot FA \\
&= 0.91 \cdot \frac{200}{0.287 \cdot (45 + 273)} \cdot \frac{\pi}{4} \cdot 0.12^2 \cdot 0.13 \cdot 0.039 \\
&= 1.143 \cdot 10^{-4} kg
\end{aligned}
$$

To calculate the fuel flow rate, the injection duration in seconds is needed:

$$
\delta t_{inj} = \frac{\delta \theta_{inj}}{6N} = \frac{32}{6 \cdot 1700} = 3.131 \cdot 10^{-3} \ s
$$

and the mass flow rate for the open nozzle is:

$$
\dot{m}_f = \frac{m_f}{\delta t_{inj}} = \frac{1.143 \cdot 10^{-4} kg}{3.131 \cdot 10^{-3}} = 3.643 \cdot 10^{-2} \ kg/s
$$

Equation (5.13) can be rearranged to solve for the hole diameter:

$$d = \sqrt{\frac{4\dot{m}_f}{\pi C_f n_h \sqrt{2\rho_f \Delta p}}}$$

$$= \sqrt{\frac{4 \cdot 3.643 \cdot 10^{-2}}{\pi \cdot 0.822 \cdot 4\sqrt{2 \cdot 837 \cdot 1000 \cdot 10^5}}}$$

$$= 1.857 \cdot 10^{-4} \ m = 0.1857 \ mm$$

When using Equation (5.13) it is a good idea to put everything into basic SI units, kg/s, kg/m^3, m and Pa.

5.3 Introduction to Fuel Systems

There are two classic types of engine combustion systems used today: Spark Ignition (SI) and Compression Ignition (CI). These are described in more detail in other sections, but the most fundamental difference between the two is the fact the spark ignition engines in essence operate with a homogeneous combustion process, while compression ignition engines operate with a heterogeneous combustion process. Though, like many classifications, there are exceptions. For example, the compression ignition process may be modified in the future, as at the time of writing, there is considerable work underway to develop what is called the "*H*omogeneous *C*harge *C*ompression *I*gnition" or HCCI engine. Another intermediate system, which has recently appeared in passenger cars engines, is called "*G*asoline *D*irect *I*njection" and involves the ignition of a heterogeneous fuel air charge with a spark plug, this is discussed in detail in Chapter 16.

The fuel system for any kind of combustion system has two main functions:

1. To deliver the proper amount of fuel to the engine (metering)

2. To assist in the proper mixing of the fuel and the combustion air

In addition, for diesel engines and engines with cylinder injection, the fuel system must provide fuel injection at the proper time in the engine cycle.

In the spark ignition engine, the mixing function is normally not so critical, as the fuel and air usually are mixed outside the engine, and a long (in engine time scales!) time is available for fuel evaporation and mixing. Fuels for SI engines are also produced with adequate volatility to ensure the evaporation. In the standard diesel engine the time for mixing and combustion is very short, as the fuel is injected just before combustion. The differences in the fuel preparation have a lot to do with the design parameters of the fuel system, not the least the pressures at which it operates.

Historically, spark ignition engine fuel systems were based on carburetors to meter the fuel and air. This system was dominant on all SI engines until about 1980, when vehicle emissions standards became so strict that computer control of engine operation to satisfy the needs of exhaust gas catalysts became a necessity. Electronically controlled fuel injection is now dominant on spark ignition engine powered vehicles. A few smaller vehicle engines still use carburetors and they can still found on small engine powered tools, predominantly 2-stroke engines.

5.4 Diesel Injection Systems

The fundamental elements of a diesel injection system are the pump used to create the high fuel pressure, a nozzle in the combustion chamber for creating and directing the fuel spray, and a connection between the two. Diesel injection systems vary in design, the major factors being the location of the high pressure pump element and the method of controlling injection.

Some common systems are:

Pump-Pipe-Nozzle This is the oldest type of injection system, and consists of a central, engine driven pump and lines connecting the pump to the nozzles, which are placed in the individual cylinders.

Many pump designs are used, all of which use a tight fitting plunger in a cylinder to press the fuel through the system, the high pressure arising by the relation between the nozzle area and the rate of flow through it, Equation (5.14). Pumps for pump-pipe-nozzle systems can be divided into two types:

- Individual pump system - where there is one pump element (piston/cylinder) for each cylinder.

- Distributer pump - where there is one piston/cylinder combination for the entire engine. The pump includes a mechanism for distributing the fuel to the proper cylinder at the proper time.

In the pump-pipe-nozzle system, care must be take for line lengths to be the same for all cylinders, to ensure proper timing for all cylinders at all engine speeds. (See Section 5.4.1)

A check valve, called the delivery valve is located at the outlet of the high pressure pump in a pump-pipe-nozzle system. This spring-loaded valve opens when the pump forces fuel out of its cylinder, and closes when the flow stops. When the injection stops, the delivery valve moves slightly, allowing a faster stop of injection, and helps to prevent after injection. Since it can only move a small distance, there is still high pressure fuel in the injection line when the next injection starts, which gives a more rapid start of injection. This is called the residual pressure.

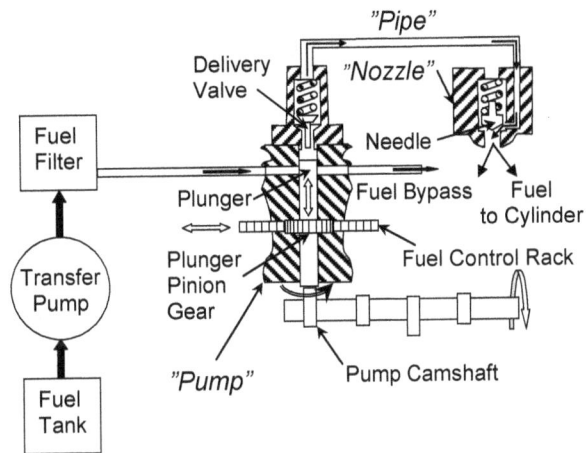

Figure 5.3: An example of the pump-pipe-nozzle system, with an individual pump element for each cylinder.

Unit Injector

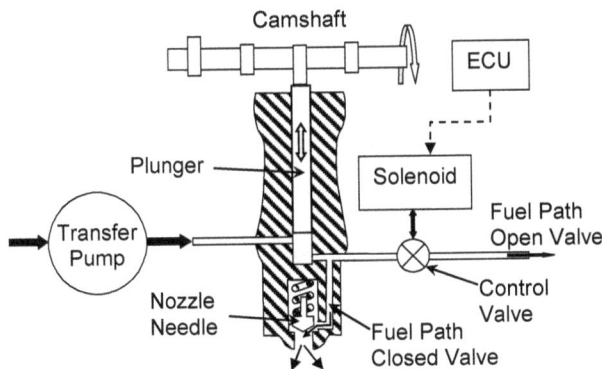

Figure 5.4: An electronically controlled unit injector. The start and end of injection controlled by the solenoid valve.

In this system, the pump element is enclosed in the same structure as the nozzle, and both are inserted into the cylinder head in a common unit. The pump is driven by a camshaft. There is a pump element for each cylinder. This system eliminates the connecting line between the pump and the nozzle, the distance between the two being on the order of a few centimeters with a unit injector. The construction of an engine with unit injectors is more complicated, because the unit injector is larger than the standard nozzle, and because of the extra complexity of the camshaft system to drive both the valves and the unit injector. Figure 5.4 shows a drawing of an electronically controlled unit injector. The short distance between the nozzle tip and the pump piston greatly reduces effects of pressure pulse travel time between the pump and the nozzle, and also reduces the pressure loss between the pump and the nozzle.

Both of the above systems can be adapted to computer control. The pump-pipe-nozzle and unit injectors utilize electronically controlled pressure relief valves to start and end the injection process. If these valves are open, the high pressure piston presses the fuel into a by-pass instead of through the injector into the cylinder. When injection is to start, the by-pass valve is closed, the fuel must flow through the nozzle to the cylinder until the relief valve is opened again, such that the injection stops. These valves are typically placed near the outlet of the high pressure piston in the pump-pipe-nozzle system.

Common Rail Whereas the previous two systems create a high pressure pulse of fuel when it is needed for combustion, the common rail system uses a high pressure fuel reservoir, called the rail, from which fuel is "tapped" when needed. The latter function is controlled by an electronic valve of some kind, which accomplishes the metering and timing functions. The same principle is used on modern SI injection systems, but with much lower pressures.

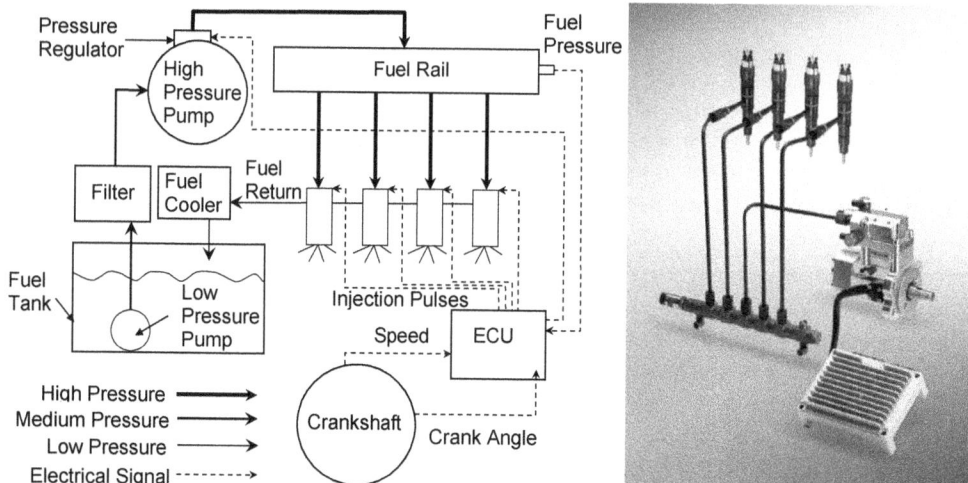

Figure 5.5: The common-rail fuel injection system. The high pressure pump supplies fuel to a high pressure manifold (common rail) from which fuel is sent to the injectors by electronically controlled solenoid valves. Photo courtesy of Motorpal a.s., used by permission.

The first common rail systems use an electronic valve to control injection timing and duration. In this case, the valve is placed between the high pressure rail and the injection nozzle. When injection is to start this valve is opened, and a high pressure pulse travels to the injector, which opens and allows the fuel to be injected. At the end of injection, the valve is closed, the high pressure source is no longer in communication with the injection nozzle, and the injection stops. Recently, the switching function has been built into the injector itself, where fast acting piezo-electric switches control the fuel in a similar, but faster manner.

A major advantage of the common rail system, is that the injection pressure can be controlled independently of engine operating conditions. In the pump-pipe-nozzle and unit injector systems, the pressure is related to the amount of fuel that flows through the injector, and to the time required for the pump to displace a given quantity of fuel. Equation (5.14) indicates that there will be a substantial increase in injection pressure as engine speed increases, for a constant fuel mass and injection duration in crank angle degrees. The flexibility of the common rail system is responsible for the ability to meet more stringent emissions standards, and plays a key role in the recent application of small direct

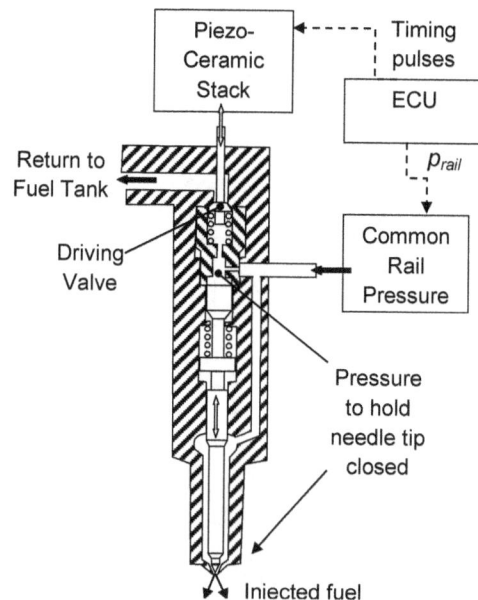

Figure 5.6: A typical piezo-electric fuel injector. Common rail pressure on one end of the needle holds the nozzle closed, until the driving valve opens, relieving that pressure and enabling injection.

injection engines to light duty vehicles. A common rail system is shown in Figure 5.5.

A simplified drawing of the piezo-electric injector is shown in Figure 5.6. The rail pressure, normally quite high, is used to hold the injector needle closed to prevent injection. When injection is desired, an electronic signal is sent to a stack of several hundred thin piezo-electric crystals, which very rapidly expand, and open a so-called driving valve. The opening of this valve reduces the pressure on the back of the needle and the high pressure at the opposite end of the needle, by the tip, forces the needle to open, enabling injection. Because of the high pressures involved, and the rapid response of the piezo-stack, the whole process is very rapid, enabling multiple injections, as was seen in the heat release diagrams of Figure 3.50.

Diesel Pumps To create the high pressure required for diesel combustion, diesel engines have traditionally used what is called a "jerk pump" concept. The principle is very similar to spraying water with a syringe. A piston that fits tightly in a cylinder is filled with a given amount of fluid (metering), sealed, and then the piston is moved rapidly towards the small hole at one end of the cylinder. The small outlet hole (nozzle) in the cylinder causes a large pressure drop, and the liquid is forced out of the nozzle and into the combustion chamber at a high velocity. In the diesel engine, this high velocity fuel creates the fuel spray, where the fuel is broken up into countless small drops, that evaporate, mix and burn with the cylinder air. Most of the variations in former diesel fuel injection systems have been concerned with the mechanism used to control the amount of fuel injected by a plunger stroke, but the basic physics are the same.

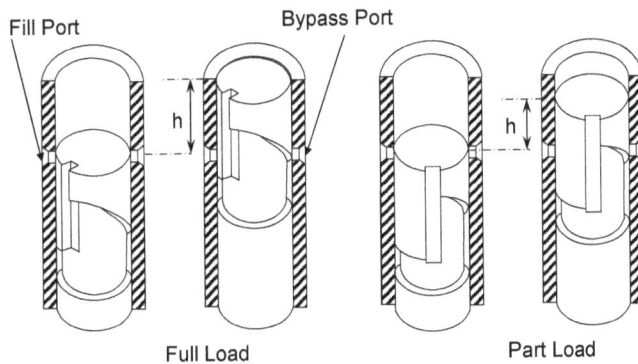

Figure 5.7: The scroll pump used in diesel injection systems. Fuel injection starts when the top of the top of piston covers the bypass port and ends when the scroll uncovers the bypass port. The effective pump stroke is denoted by h.

The piston, which fits very tightly in the cylinder, is driven up and down by a camshaft in the pump. In addition, there is a gear mounted to the piston, and a rack which meshes with the gear, as seen in Figure 5.3. When the rack is moved in a linear motion, the piston rotates, and the effective stroke of the piston is changed. If the slot in the pump is in line with the bypass port there is no injection. The effective stroke is denoted by h in Figure 5.7. There are variations of the geometry, but this principle is often used in individual pump systems.

When the ports are closed, the fuel is forced out of the cylinder through a high pressure line to the injection nozzle. An example of a nozzle is shown in Figure 5.8. The nozzle is mounted in the cylinder head, with the tip extending into the combustion chamber. The inlet is connected to the pump by a high pressure line. The tip of the nozzle contains the hole(s) through which the fuel finally passes into the combustion chamber upon injection. When the pump begins to deliver fuel, a high pressure pulse is transmitted to the inlet of the nozzle. The pressure continues

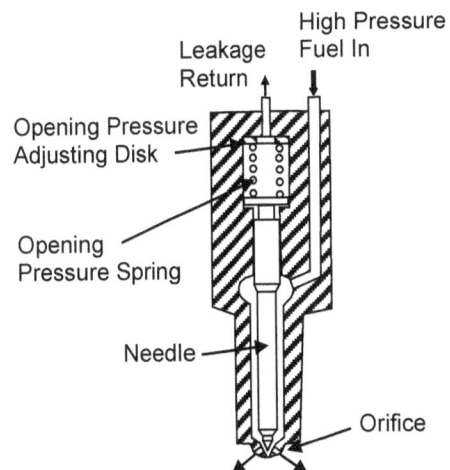

Figure 5.8: A schematic diagram of a diesel fuel injection nozzle.

to rise as the pump moves, since initially, the nozzle is closed. At some point, the pressure acting on the tip of the needle is greater that the spring pre-load on the needle, and the needle opens. Fuel is then pressed out through the nozzle tip and into the combustion chamber.

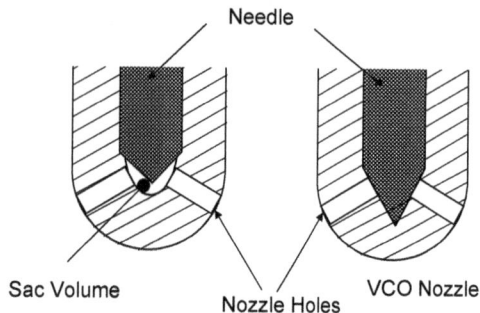

Figure 5.9: Two nozzle tips, a traditional design with a sac volume and the newer Valve Closes Orifice design.

The holes are in the tip of the nozzle, in a part that can be changed to adjust orifice size with minimum time and cost. In recent years, a new kind of nozzle tip has been used, the Valve Closes Orifice Nozzle (VCO). In previous nozzle designs, there was a little (sac) volume between the tip of the needle and the nozzles holes. This design made sealing between the needle and the injector body easier. After it was discovered that after the end of injection, the fuel left in the sac volume boiled out into the combustion late on the expansion stroke and was a major contributor to hydrocarbon emissions, the VCO nozzle has been developed and taken in use. In this design, there is no sac volume, and the needle tip closes the holes (orifices) completely, reducing hydrocarbon emissions greatly. The two designs are shown schematically in Figure 5.9.

The spring pre-load determines only the opening and closing pressures of the nozzle. The pressure during injection itself when the nozzle is open depends on the size and number of nozzle holes, and the velocity of the flow, primarily related to the instantaneous piston velocity in the pump. This is the relationship shown in Equation (5.14). It is undesirable to have injection start immediately upon the first motion of the pump, as the spray into the combustion chamber would be weak, with little mixing. The spring in the nozzle prevents it from opening until an adequate pressure is reached. At the end of injection, the nozzle is closed while there still is an advantageous pressure difference across the nozzle. This prevents large, slow drops from entering the combustion chamber and creating excess smoke. Should there be inadequate pressure during injection, the nozzle can close and possibly open again, there can be multiple injections. Since these are not really controllable, they should be avoided. In the systems with cam driven high pressure plungers, the pressure during injection is speed dependent, since at higher speeds, the same volume of fuel pressed through the orifices in a shorter time will give a higher pressure. This makes combustion control more difficult, and is one of the reason why mechanical jerk pumps are being replaced by common rail systems, especially in smaller engines that must operate over a wide speed range.

In the unit injector in Figure 5.4, the pump displaces fuel when acted upon by the cam, but as long as the control valve is open, the fuel spills into the by-pass, and not into the cylinder. When the control valve is closed, injection starts and continues until the control valve is opened again. Thus, there is control of the timing of the injection process. The pressure is determined by the rate of fuel displacement and the nozzle area, according to Equation (5.14)

Another type of pump that has been used on diesel engines is the distributer pump. This is a variation of the pump-pipe-nozzle concept. The idea here is to use one high pressure element to provide high pressure to all cylinders. In addition to providing high pressure fuel, the distributer pump must insure the direction of high pressure fuel to the proper cylinder at the proper time. One example of a distributer pump for lighter duty engines is shown in Figure 5.10.

This system uses a rotating plug, operating inside a cylinder in which a pair of opposing pistons is mounted. They are pressed towards each other as they rotate inside a ring cam. There is a connecting passage from the fuel source outside the pump, through filling holes in the cylinder, and leading into the chamber between the two pistons. That passage is open when the rotor is in a position such that the pistons are farthest away from each other. This allows a metered amount of fuel to enter the volume between the two pistons. As the plunger rotates further, the inlet connection is blocked off, and the pistons are pushed towards each other by the cam at their bases. The fuel charge is determined by a variable restriction in the body of the pump. At the same time, an outlet line (shown at the far end of the plug from the pistons) is opened between the chamber above the pistons and one of the fuel injection lines to a cylinder by the holes at the end of the rotating plug. Thus, when the fuel begins to be compressed, there is a pathway toward an injection nozzle, and one pump can deliver fuel to several cylinders.

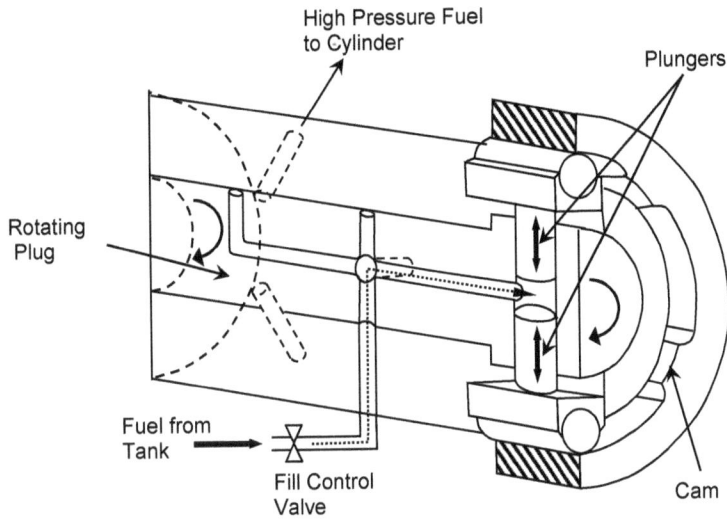

Figure 5.10: A distributer pump for light duty diesel engines.

Other versions of distributer pumps are used, some of them using variations on the variable stroke concept described above for the individual pump. Common to all distributer pumps is a rotary output system, where the injection line to the proper injector is opened when the pump piston begins its compression movement. Distributer pumps can be more complicated than individual pumps, but the total number of parts is less than for an engine with 4 to 8 individual pumps, and the distributer pump is often cheaper. However, one piston, or set of pistons delivers all the fuel to the engine, so the wear on that one cylinder is greater than would be experienced with an individual pump system.

In the common rail system, the fuel pump is not related to engine events, and pressure is controlled with a conventional type of pressure regulator. The pump is still based on a piston cylinder arrangement, but does not have fuel metering functions, and is simpler in design than the jerk pump. Different arrangements can be used, in-line cylinders or rotary pumps with swash plates have been used.

5.4.1 Fuel Compressibility

At low pressures, diesel fuel and gasoline can be assumed to be incompressible. They do have a finite compressibility, though, which can be significant at higher pressures in diesel engines. There are some things to consider in connection with fuel compressibility. One of the most important is that the speed of sound is related to the fuel compressibility, which has implications concerning timing in diesel engines with a pump-line-nozzle injection system. The other is the fuel volumes concerned, in that compressibility affects the amount of fuel in a given volume, and if fuel is compressed instead of delivered, adjustments in fuel systems need to be made.

First, compressibility of the fuel must be defined. The term used to do that is called the bulk modulus of the fuel. It is the ratio of an applied pressure to the relative change in the volume of a fuel:

$$K = \frac{\Delta p}{\frac{\delta v}{v}} \tag{5.15}$$

It can be shown that the speed of sound in a fluid is related to the bulk modulus:

$$V_s = \sqrt{\frac{K}{\rho}} \tag{5.16}$$

The compressibility and speed of sound of typical fuel components is shown in Table 5.1. The fuels become less compressible at higher pressures. At 1000 atm, there is about a 5% change in the volume of the liquids due to the effects of compressibility. This has little effect on sizing of the volumes of fuel systems, even for high pressure.

Table 5.1: Density, Bulk modulus and speed of sound for some typical fuels. [58], [59]

Substance	Density kg/m3	Bulk modulus @1 atm - MPa	Vs m/s	Bulk modulus @1000 atm - MPa
Benzene	874	1067	1105	2025
Docecane	755	1126	1221	2025
Ethanol	785	889	1064	2025
n-Hexane	655	633	983	1580
Petroleum Oil	840	1448	1313	1920
Soybean Oils	885	1650	1366	2550
m-Xylene	862	1267	1212	-
Di-methyl Ether	668	400	774	-

The sonic speed may be of importance in injection systems with fuel lines, especially larger engines. Taking a sonic velocity of 1300 m/s as being typical for a petroleum oil, this means that at 1800 rpm, the number of crank angles required for a pressure pulse to travel 1 meter is:

$$t = \frac{L}{v_s} = \frac{1m}{1300m/s} = 7.7 \cdot 10^{-4} s \frac{1800 \; rev}{min} \cdot \frac{min}{60s} \cdot \frac{360deg}{rev} = 8.3 \; deg$$

This can be quite important for timing, and is one reason why engine with pump-line-nozzle systems have injection lines of the same length. This also results in equal pressure drop between the pump and injector for all cylinders.

Compressibility is also important with some alternate fuels. DME for example, has a high compressibility and low speed of sound. As seen in the discussion of diesel combustion, this can lead to high residual pressures, pressure oscillations and after injection. The effects can be seen in Figure 5.11, which shows injection line pressure oscillations for diesel fuel and DME in the same injection testing equipment.

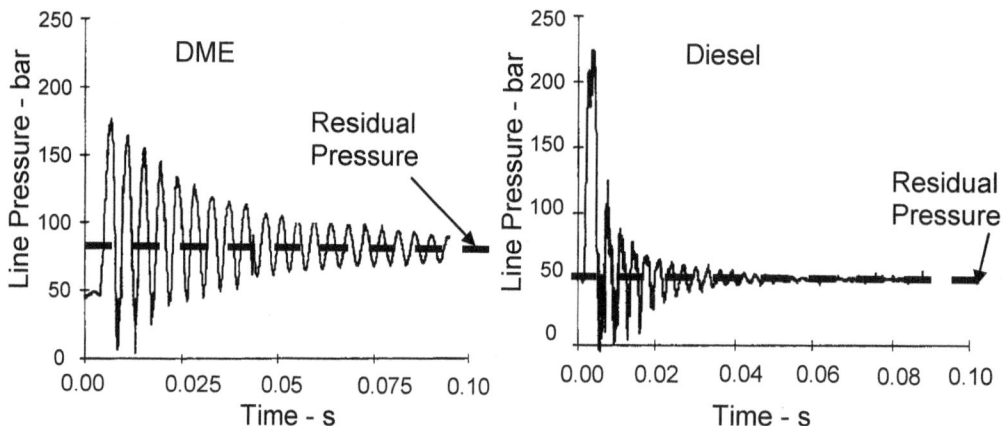

Figure 5.11: Pressure pulses in an injection test rig operating on DME (left) and diesel fuel (right) with a valve opening pressure of 130 bar.

The consequences of the higher compressibility can be seen as a later opening time (the first peak), a lower peak pressure (injection occurs during the first peak), a higher oscillation of the pressure waves after completion of injection, and a higher residual pressure in the system after the end of the process for DME. The latter is a result of the extra work done on the fuel due to its higher compressibility. Extra work will give higher fuel temperatures. That the compressibility is higher for DME (lower bulk modulus) can be seen by the lower frequency of the pressure waves for the DME system. Engines with similar injection systems have been found to operate successfully on DME, but the results show that for an optimum design, considerations should be taken of greater fuel compressibility.

CI Engine Considerations

Looking at the first formulation of Equation (5.14) one can see a fundamental problem with the jerk pump system. If a given mass of fuel needs to be injected, this corresponds in principle to a given volume of fuel, which is intended to be injected over a given number of crank angle degrees. In this case, the injection pressure is proportional to the square of the engine speed. That is, if the injection pressure is appropriate to give good injection at a high engine speed, for example rated speed, then when the speed is decreased, the injection pressure will decrease with a second order dependence on the engine speed. This will give a slow speed of the injected jet at low engine speed, and in all probability, a poorer combustion, with likelihood for higher smoke and particulate emissions

With the second formulation in Equation (5.14), the idea of using a computer controlled pulse length to regulate the injection can be seen in the case of a common rail system. If the rail pressure is maintained constant, then the same injection time in milliseconds can be used to deliver the same fuel mass, regardless of the speed, and the injection spray will have the same strength, giving a better combustion over a wider speed range.

The prechamber engine was used extensively on small high speed diesel engines until a few years ago. A sketch of the combustion chamber is shown in Figure 5.12. The version shown is also called a swirl chamber engine, because of the high degree of rotary motion of the air in the prechamber. In this engine, the powerful flow in and out of the prechamber has a strong effect on combustion, and reduces sensitivity to the injection process. Thus, prechamber engines could operate at higher speeds than DI engines with jerk pumps. Injection nozzles in prechamber engines were typically of the single hole type, as the powerful air motion in the prechamber was adequate to mix the fuel and air. The recent appearance of small, high-speed DI diesel engines is connected with the development of the common rail system and better control of the injection/combustion process over a wider speed range. This is important for fuel economy, since the DI engine has a lower fuel consumption due to the elimination of flow and heat transfer losses associated with the flow in an out of the prechamber. Chapter 12 on engine maps shows some typical results for these two engine types.

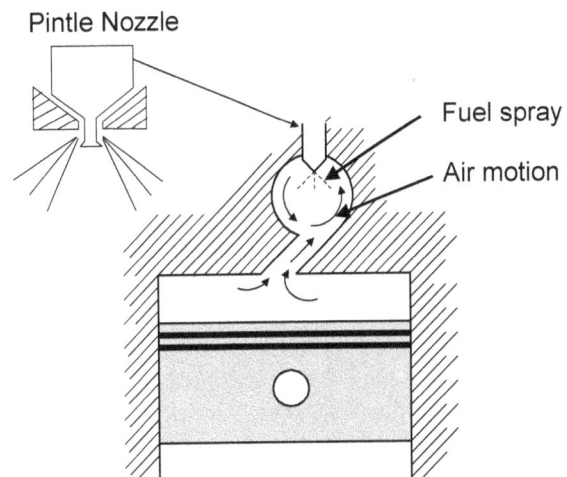

Figure 5.12: A sketch of the combustion system in a swirl chamber diesel engine. Because of the strong air motion in the prechamber, a single hole nozzle is adequate

The importance of other factors can be seen in Equation (5.14). The nozzle hole diameter is the most significant factor in determining injection pressures, and the injection pressure drop is inversely proportional to the fourth power of the nozzle hole size for a given injection rate. The number of holes is important as well, but this is to a large extent determined by the type of combustion system. In the prechamber engine, one hole is sufficient, as the flow into the prechamber helps to break up the spray, and the flow out of the prechamber helps to mix the burning fuel and air. In addition, a pintle nozzle was also used. In this type of nozzle, a pintle plugs the middle of the fuel orifice, and a taper on the tip of the pintle leads to a conical spray, which enhances mixing. A sketch of the pintle nozzle is shown in Figure 5.12.

For direct injection engines, several nozzle holes are needed in order to distribute the fuel into the air in the combustion chamber. The number of nozzle holes depends primarily on the amount of rotary air motion (swirl) existing in the combustion chamber at the time of injection. The more swirl, the fewer holes needed. Medium swirl DI engines typically use between 3 and 6 holes in the injection nozzle. A sketch of the combustion system and the position of fuel sprays for this type of engine are shown in Figure 5.13. Larger engines tend to have less swirl, and therefore have more nozzle holes than medium

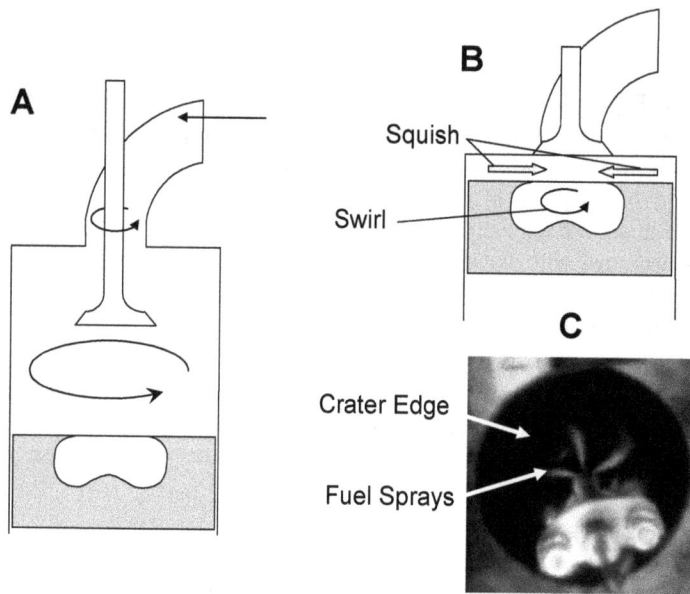

Figure 5.13: A sketch of the combustion system in a medium swirl DI diesel engine. A. shows the air swirl set up on the intake stroke. B. shows the air motion resulting from forcing the air to the piston crater, and C. shows the patterns of the fuel sprays from the 4-hole nozzle, swirl is counter clockwise.

swirl engines. Diesel engines with no swirl (*Quiescent*) typically have between 8 and 11 holes in the tip of the nozzle. Once the number of holes is decided, and the injection duration determined (there is not so much leeway here) the final injection pressure is determined by the hole size.

In order to make an estimate of the rotation speed necessary, a simple calculation can be made. Assume that there are a number of equally spaced holes in the nozzle of a direct injection engine. Figure 5.14 shows the sprays from two of them. The edges of the sprays are separated by the angle β, and the sprays have a width of δ. In order to spread the fuel throughout the combustion chamber without hitting the adjacent spray, the maximum angle of rotation of the air during injection of the fuel, β, must equal the angle between the two adjacent sprays less the angle describing the width of the spray, δ. That is:

$$\beta = \frac{360}{n_{holes}} - \delta \qquad (5.17)$$

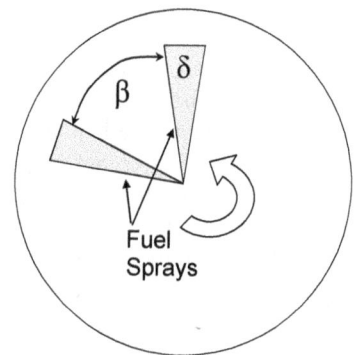

Figure 5.14: A sketch of two fuel sprays in a medium swirl, direct injection diesel engine.

The required rotational speed of the air then can be estimated as the angle through which the air rotates and the time that is available for the rotation. The latter is determined by the maximum duration of the fuel injection.

$$\omega_{spray} = \frac{\beta}{\Delta t_{inj}} \frac{deg}{s} = \frac{\beta \cdot 6N}{\Delta\theta_{inj}} \qquad (5.18)$$

Converting degrees per second to revolutions per minute:

$$N_{swirl} = \frac{\beta}{\Delta\theta_{inj}} N_{engine} \qquad (5.19)$$

The fewer holes, the larger the angle β and the higher the rotational speed required. There are limits, of course, and eventually when there are 8 or more holes, the engine is designed to be rotation free, a quiescent chamber.

Example

To estimate the relation between the rotational speed of the air and the rotational speed of the engine, some typical values are used. Assume that the nozzle has 4 holes, and that the maximum engine speed is 2500 rpm, and that the width of the spray is 20°. Further, assume that the maximum duration of the fuel injection is 30 CA°. Then,

$$\beta = \frac{360}{4} - 20 = 70°$$

and

$$N_{swirl} = \frac{70}{30} \cdot 2500 = 5830 \; rpm$$

This value is a first estimate, and can be used as a starting point for future designs. Modern modelling techniques making use of CFD programs can model the process in much greater detail, but this estimate can be used to obtain an appreciation for the order of magnitude involved. The rotation of the air is achieved by the design of the intake port, a common method being to shape the intake port by the intake valve in a form of spiral that imparts rotation to the air as it comes into the cylinder.

It must also be remembered that the rotating air at the end of the intake process is compressed into a smaller diameter in the piston bowl near the end of the compression. According to well known physical principles, this increases the speed of rotation of the air by conservation of angular momentum. In a perfect system, this would give an increase in rotation proportional to the square of the ratios of the cylinder and bowl diameters. In practice, frictional losses reduce this dependency to an approximately linear function of the diameter ratio.

Timing Considerations

Note that Equations (5.9) and (5.10) make no direct reference to engine speed. Assuming a constant rail pressure, the amount of fuel is determined by the length of injection. Hole sizes and rail pressure are selected to give injection durations that result in good fuel consumption and emissions. For a constant rail pressure and constant amount of fuel injected, the injection duration in crank angle degrees decreases as the engine speed decreases. With the common rail system, one can adjust the rail pressure to give essentially any injection duration desired. An example of this is shown in the heat release diagrams in Figure 3.50, where the injection rail pressure was increased at high loads to maintain an acceptable injection duration.

5.4.2   Fuel Injection Examples:

Example 1

Consider a 4-stroke gasoline engine with injection in each intake port. The bore and stroke are 85 mm and 80 mm respectively. The engine operates with a manifold pressure of 65 kPa, and a volumetric efficiency of 0.825. The manifold temperature is 30°. Since the engine uses a 3-way catalyst for emission control, the engine must operate at a stoichiometric mixture ratio, which for the fuel at hand is 0.0685. The fuel pressure regulator holds the fuel pressure constant at 2.5 bar above the intake manifold pressure. Assume that it has been found desirable to limit the maximum injection pulse duration to 170 CA° at the maximum engine speed of 6000 rpm. If the flow coefficient of the nozzle is 0.81, determine the flow area of the injection nozzle. The fuel density is 0.735 g/cm^3. Since there is separate injection for each cylinder, the calculation uses the displacement volume of one cylinder. The mass of fuel to be injected is:

$$m_f = m_a \cdot FA_s = \eta_v \frac{p_{in}}{RT_{in}} \cdot V_d \cdot FA_s$$

$$= 0.825 \frac{65}{0.287 \cdot 303} \cdot \frac{\pi}{4} 0.085^2 \cdot 0.08 \cdot 0.0685$$

$$= 1.92 \cdot 10^{-5} kg/inj$$

The time for injection is determined from duration at the maximum speed:

$$\delta t_{inj} = \frac{\delta \theta_{inj}}{6N} = \frac{170}{6 \cdot 6000} = 4.73 \cdot 10^{-3} s$$

Then the fuel flow when the nozzle is open is:

$$\dot{m}_f = \frac{m_f}{\delta t_{inj}} = \frac{1.92 \cdot 10^{-5}}{4.73 \cdot 10^{-3}} = 4.06 \cdot 10^{-3} kg/s$$

From Bernoulli's equation:

$$A_f = \frac{\dot{m}_f}{C_f \sqrt{2 \rho_f \Delta p_{inj}}} = \frac{4.06 \cdot 10^{-3}}{0.81 \sqrt{2 \cdot 735 \cdot 2.5 \cdot 10^5}} = 2.62 \cdot 10^{-7} m^2 = 0.262 \ mm^2$$

Example 2

For a four cylinder SI engine it is desired to use an injection system that only injects fuel while the intake values are open. A multi-point system is to be used that injects fuel onto each intake valve. The maximum engine speed is 6000 rpm, and the intake valve is open from 50° BTDC to 50° ABDC. What does the timing diagram look like when the time is given in both seconds and crank angle degrees? What happens with full load when the speed is 3000 rpm?

At 6000 rev/min there are:

$$\frac{6000 rev}{min} \cdot \frac{360°}{rev} \cdot \frac{min}{60 \ s} = 36000°/s$$

The injection period is $50 + 180 + 50 = 280$ ° which corresponds to $280/36000 = 7.78$ ms. Figure 5.15 shows the pulse duration in both crank angle degrees and seconds.

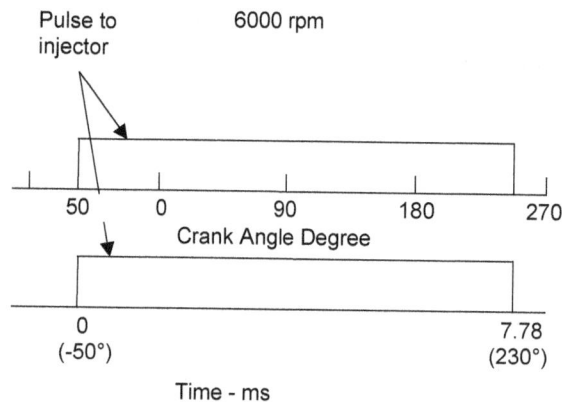

Figure 5.15: The injection pulse for Example 2 in time and crank angle degrees at 6000 rpm.

If the pulse length is constant at 7.78 ms, at 3000 rev/min the same mass of fuel will be injected into the cylinder as at 6000 rpm, since the nozzle pressure ratio and flow area are constant. The injection period is constant in seconds and the only difference is that when the engine runs slower, the injection will take place over a smaller number of crank angle degrees, and occur half as frequently. The timing diagram in this case is shown in Figure 5.16.

If the volumetric efficiency is not changed between the two speeds, the air and fuel masses inducted per intake remain unchanged. In principle then, when the conditions at maximum speed have been satisfied, one only needs to be concerned with variation in injection quantity with respect to load variations at lower speeds, though small variations in volumetric efficiency will occur.

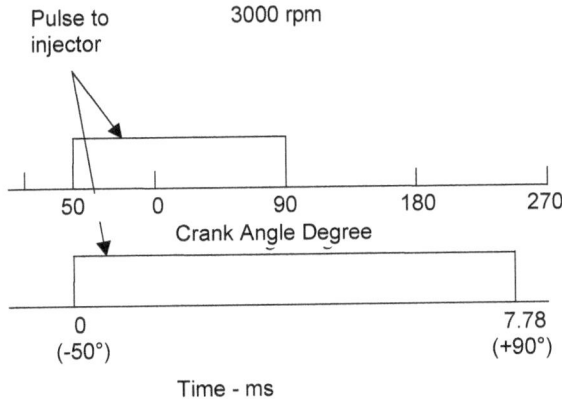

Figure 5.16: The injection pulse for Example 2 in time and crank angle degrees at 3000 rpm.

Example 3

Diesel fuel injection. Consider the engine from the sample calculation in Chapter 1: A 4-cylinder, 4-stroke direct injection diesel engine (Common Rail) with turbocharger and intercooler.

Bore: 88 mm, Stroke 84 mm, compression ratio 19:1. The nozzle has 6 holes, each with a diameter of 0.169 mm. It is estimated that the flow coefficient for the nozzle is about 0.82.

For operation at an engine speed of 1500 rpm, and bmep of 1000 kPa, the bsfc is 215 g/kW-h. The common rail pressure is 800 bar, and it is estimated that the cylinder pressure at the time of injection is about 60 bar. For this condition, estimate the duration of the fuel injection in crank angles degrees, neglecting compressibility effects. Assume that diesel fuel has a density of 0.835 g/cm^3. Note: This is an estimate of the engine operating with all injection occurring in a single pulse. In the actual engine a small amount of fuel is injected earlier in the cycle (pilot injection).

It is necessary to find the amount of fuel injected per injection. Since the pressures, area and flow coefficient are known, the duration can be calculated from applying Bernoulli's equation when the injector is open. In an actual system, the pressure will change somewhat during the flow process, but that is neglected here.

The average fuel flow to the engine is calculated from Equation (1.20), but first the power must be calculated. The displacement volume was calculated in the section on sample calculations to be 2.15 liter:

Then the power is calculated from the speed and *bmep*

$$BP = \frac{bmep \cdot V_d \cdot N}{120}$$

$$BP = \frac{1000 \cdot 0.00204 \cdot 1500}{120} = 25.5kW$$

The average fuel flow to the engine is then:

$$\dot{m}_f = BP \cdot bsfc = 25.5kW \cdot 215\frac{g}{kWh} \cdot \frac{1kg}{1000g} \cdot \frac{1h}{3600s} = 1.525 \cdot 10^{-3}\frac{kg}{s}$$

From the average fuel flow, the fuel per injection can be calculated:

$$m_f = \frac{2 \cdot 60 \cdot \dot{m}_f}{n_{cyl} \cdot N} = \frac{2 \cdot 60 \cdot 1.525 \cdot 10^{-3}}{4 \cdot 1500} = 3.05 \cdot 10^{-5}kg$$

Rewriting Equation (5.14), one can solve for the injection duration:

$$\delta\theta_{inj} = \frac{24 \cdot N \cdot m_f}{\pi \cdot C_f \cdot n_h \cdot d_h^2\sqrt{2\rho_f \cdot \Delta p}}$$

$$= \frac{24 \cdot 1500 \cdot 3.05 \cdot 10^{-5}}{\pi \cdot 0.82 \cdot 6 \cdot 0.000169^2\sqrt{2 \cdot 835 \cdot (800 - 60)\,10^5}} = 7.08°$$

5.5 Spark Ignition Systems

5.5.1 Mixture Requirements.

Historically, there have been several requirements for the fuel air ratio of spark ignition engines. They will be mentioned here, though for vehicle spark ignition engines, the use of the 3-way catalyst has drastically reduced, and in some engines, eliminated the possibility for mixture ratio adjustment, as only stoichiometric operation gives effective catalyst operation.

The fundamental requirement for a spark ignition engine is that the mixture can be ignited. If the fuel air mixture is leaner than λ of about 1.5, the engine will be unstable, or misfire. Similarly, if it is richer than λ of about 0.5, operation the engine will misfire. There is little incentive to operate that rich, but such conditions might be encountered under cold start where fuel enrichment is used in marginal conditions. With stoichiometric operation there is no problem with ignitable mixtures.

A classic curve for engine operation as a function of Fuel Air ratio is shown in Figure 5.17. It shows that the maximum power is obtained

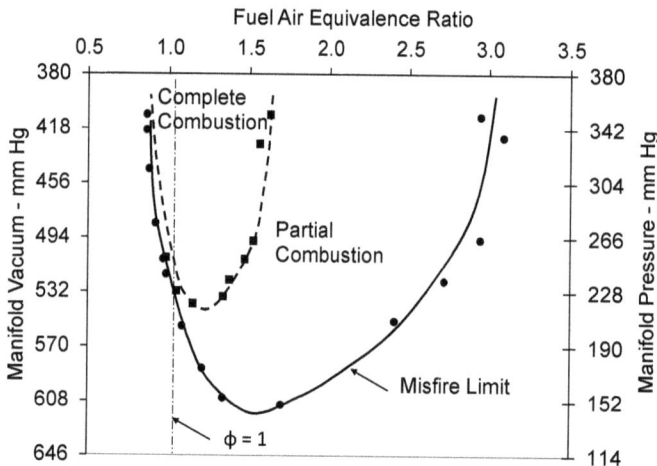

Figure 5.17: The bsfc and bmep of an SI engine as a function of fuel air ratio for full throttle operation adapted from [60].

with a slightly rich mixture, in line with the predictions of Fuel Air Equilibrium Cycles. On the other hand, maximum efficiency is obtained with a lean mixture, also in keeping with the predictions of the Fuel Air Equilibrium Cycles shown in Chapter 2. The $bsfc$ starts to rise for leaner mixtures. This is due to the slower flame speed with the leaner mixtures as discussed in Section 3.5.1, since the Fuel Air Equilibrium cycle predicts good efficiency for leaner mixtures with constant volume combustion. Ignition limits for this engine are ends of the shown curves.

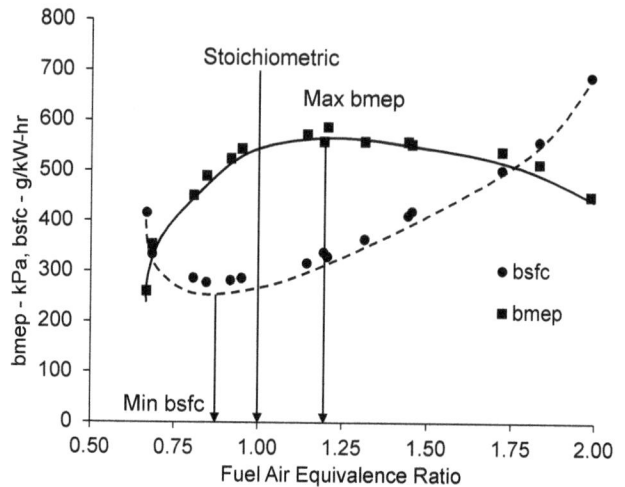

Figure 5.18: The effect of intake manifold pressure and fuel air ratio on stable operation of an SI engine at idle (500 rpm, TDC spark, compression ratio = 6.7.). Adapted from [61].

In normal practice with 3-way catalysts and stoichiometric operation, neither max power or min $bsfc$ will be reached, though Figure 5.17 shows the penalties in fuel consumption and power not to be too important for normal operation. Some vehicles do operate with rich mixtures at full load to produce slightly more power if these conditions are not encountered on emissions testing procedures.

Another condition is stable operation at idle. It is desired that the engine is stable for the purpose of driver comfort and satisfaction in a vehicle, and that the idle speed be as low as possible in order to use as little fuel as possible when idling. A problem with idle conditions, as can easily be shown from ideal engine calculations, is the increasing residual gas fraction as the load is decreased. This reduces the flame speed, and flammability limits. Results

shown in Figure 5.18 show that the lowest manifold pressure (highest vacuum) for idle operation is found at a rich mixtures of about FA=0.10. This, however, is not compatible with the stoichiometric operation of a 3-way catalyst, and one of the compromises that has been made for stoichiometric operation for emission control is a higher idle manifold pressure that gives a higher idle speed. Historically, this means the idle speeds for SI vehicle engines have increased from pre-control values of about 600 rpm to current engine of around 1000 rpm, with a slight fuel penalty in overall driving. Idle speeds have also increased, due to increase load on the engine from increased use of accessories in modern vehicles.

The all-dominating requirement for vehicle SI engines, though, is stoichiometric operation at all conditions for normal driving to ensure optimum exhaust gas catalyst operation. That was the primary driving force for the introduction of electronically controlled fuel injection in SI engines. The flexibility of the computer control has mitigated some of the disadvantages that might have arisen from non-optimal operation in accordance with the above figure. In addition, cold start operation has vastly improved over that of carbureted engines, and due to durability requirements for emission control systems, engine maintenance has been greatly reduced.

5.5.2 Carburetors

Though carburetors on vehicles are rapidly becoming a thing of the past, a very cursory summary will be given here, as there are still a few applications where they are not eliminated from consideration by strict emissions standards. In a carburetor, the air flows through a passageway with smaller area than the rest of the intake channel, called the venturi. From Bernoulli's principle, the reduction in area increases the velocity, which decreases the pressure in the venturi and draws fuel from a reservoir through a fuel duct in which there is an orifice called the fuel jet. The size of the venturi is chosen such that the maximum air flow required by the engine is less than the critical air flow through the venturi, where the Mach number would reach 1.0. In some engines the venturi size is chosen to operate with choked flow to limit the maximum air flow and thereby limit the power output of the engine.

The fuel air ratio is determined by the size of the venturi in Equation (5.3) and by the size of the fuel jet in Equation (5.7). The pressure drop in the venturi is determined by the air flow through the engine through Equation (5.3). The fuel air ratio is calculated by the ratio of Equation (5.7) to Equation (5.3). Since the flow dependence on pressure ratio is different for compressible and incompressible flows, the fuel air ratio is dependent on the flow rate in a simple carburetor. Through the years, many systems were developed to compensate for this, and they can be found in older books on internal combustion engines such as [4] and [7].

The above is based on the assumption that the flow is steady, whidh is a reasonable assumption for multi-cylinder engine. In the remaining applications where carburetors are used, single cylinder engines are very common, and the flow is not steady. Carburetor sizing here is better performed by assuming that the engine is a 3 or 4 cylinder engine to calculate the flow, as the average flow in this case is closer to the maximum flow.

5.5.3 Electronically Controlled Fuel Injection

The advent of strict emission controls on spark ignition engines, and later diesel engines has changed the fuel systems over the past twenty years. The need to precisely control engine conditions and the advent of microelectronics has led to computer controlled fuel systems on most road vehicles today. This has resulted in fuel systems for spark ignition and diesel engines that more and more resemble each other, though there are still variations in the diesel fuel systems. What is meant here is the similarity between the electronic fuel injection system for SI engines and the electronically controlled common rail system for modern CI engines.

The physics of operation of these two systems is the same, the basic difference is the pressure in the fuel systems. The principle of operation here is no more complicated than turning a water faucet on and off. For a given fuel orifice and pressure, the amount of fuel delivered by a fuel system is determined by how long a valve is opened. This is very well suited to digital systems. Mathematically, the same general equations are used for gasoline and diesel injection systems. They were demonstrated in Section 5.4.2.

Since gasoline is more volatile than diesel fuel, and the normal wish is for a homogeneous, stoichiometric mixture in the combustion chamber, it is sufficient to inject the fuel into the intake manifold, with larger nozzles than the diesel engine and with much lower pressures. The turbulence on the intake stroke and the increasing temperatures during compression will normally be sufficient to assure complete vaporization and mixing.

In gasoline injection, the requirements for the length of injection are much less strict than for the diesel engine. Since the mixture at the time of combustion should be homogeneous, the injection duration has very little effect on combustion provided it is long enough to give adequate vaporization and mixing. The only real requirement is that the maximum injection period is short enough that at the highest speed and load, there is time to close the injector between pulses. If the pulse width is too long, the injector would be open continuously, and adjustment of the fuel rate no longer possible. Other than that, one is in principle free to choose a timing and duration of injection that gives acceptable performance, and is convenient to facilitate in the control system software.

In single-point systems (that is one injector to supply all cylinders), it is common to inject fuel for each individual intake process. For example, the injection pulses can be synchronized with the ignition pulses. If this is the case, then the speed-pulse length requirements are stricter than for multi-point systems (one injector per cylinder), since there would only be $(360/n_{cyl})$ crank angle degrees available for each pulse, less a small amount of opening and closing time. Note that in Equation (5.9) or (5.10) the engine speed does not appear explicitly. The variation in fuel quantity is only with respect to load. By triggering the pulse at a fixed time (s) per cycle, the speed variation is automatically included, provided that the pulse width at maximum speed is sufficiently short to avoid pulse overlap, and that any necessary compensation has been made for changes in the volumetric efficiency with engine speed.

5.6 Control of Spark Ignition Engines

5.6.1 Introduction to Gasoline Injection

Requirements for emissions from vehicle engines continue to be tightened. At the same time, the desire for better fuel consumption and higher power demand a more accurate control of the fuel air mixture. Especially the appearance of the 3-way catalyst and cheaper microprocessors and other electric components has made electronically controlled fuel injection the dominant form of fuel system for gasoline powered vehicles. The main initial reason for the use of electronically controlled injection systems was the appearance of the 3-way catalysts for emission reduction. It is a must that the fuel air ratio is always maintained at a value very close to the stoichiometric value. The Gasoline Direct Injection (GDI) is somewhat of a hybrid between standard gasoline and diesel engines. Its operation is discussed separately in Chapter 16 .

5.6.2 Basic Principles of SI Injection Systems

A typical SI injector is shown in Figure 5.19. Basically, there are two general types of spark ignition injection systems, that is the single point system and the multi-point system. The GDI (Gasoline Direct Injection) engine falls into the latter category, but the injection is directly into the cylinder and the system has more in common with diesel engines than spark ignition and these injectors operate in a pressure regime intermediate between that of SI and CI engines. The single point system is also called "Throttle Body Injection" or "Central Fuel Injection" (TBI or CFI). In this case, one or two solenoid operated injectors are placed directly above the throttle plate, and in essence replace the venturi and fuel nozzle of the carburetor. Typical systems use one injector to supply fuel to 3 or 4 cylinders. For example, on a V-8 engine, two injectors were mounted above the throttle plate. This was the most common type of system used in initial electronically controlled injection systems.

A newer development is multi-point injection. Multi-point injection system is most commonly called "Electronic Fuel Injec-

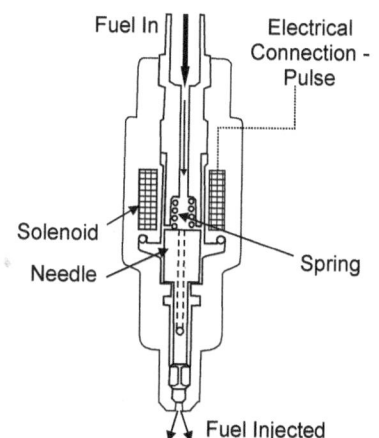

Figure 5.19: Cutaway view of a fuel injector for SI engine systems.

tion" (EFI), though both CFI and EFI in reality are electronic fuel injection. In the following, these systems will be discussed. The EFI system is become the predominant system in modern SI engines.

Not shown, the throttle body unit is typically more compact, since it is designed to sit on top of the throttle plate of the engine. The EFI unit is longer because it has to extend through the manifold into the port of the engine, and provisions must be made for sealing, since it is exposed to sub-atmospheric pressure in the intake manifold. Both use a needle to seal the orifice opening, where the needle is moved by energizing and de-energizing the coil of the solenoid in the injector. Pressurized fuel is supplied by a fuel pump from the tank at a pressure in gasoline injection systems much lower than that encountered in diesel engines.

Single Point Injection

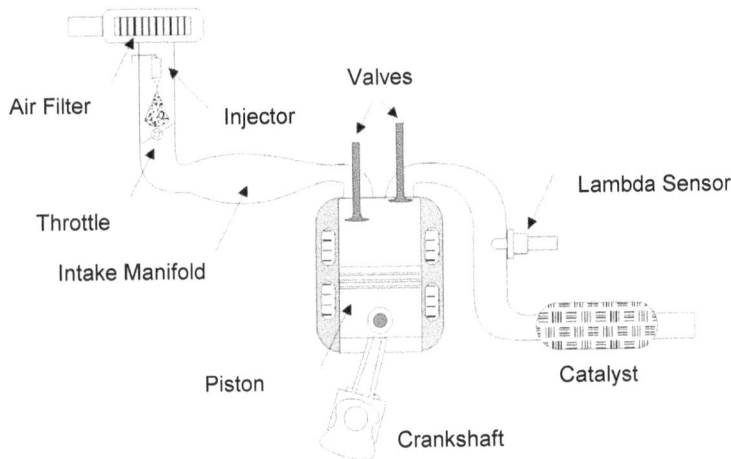

Figure 5.20: A CFI-system. The fuel is injected just before the throttle plate. The throttle position determines the quantity of air that enters the intake manifolds. The exhaust gasses are routed through the 3-way catalyst to reduce emissions. A lambda-sensor measures the Air Fuel ratio in the exhaust which is maintained at stoichiometric to ensure effective catalyst operation.

A system with single point injection based on a centrally placed injector for gasoline injection, is shown in Figure 5.20. The injector is placed just in front of the throttle such that it sprays fuel onto it. The most important parameters are the time of injection and its duration. It should be added that because of the location of the injector, the timing is not so critical for this system. It is a relatively cheap system, because of fewer components.

Figure 5.21 shows the events for the air exchange process, the injection and the spark from the spark plugs for 2 revolutions or 720° crank angle degrees for a CFI-engine with 4 cylinders. The basic strategy shown here is to inject fuel at at time corresponding to each intake stroke.

The main advantage of the single-point system is that it is simpler and cheaper. For a 4-cylinder engine, only one fuel injector is needed, and the construction of an intake manifold is simpler. On the other hand, the fuel must travel a longer distance from the injector to the intake valve. This gives a delay in the transit of the fuel, as a portion of the fuel finds its way to the manifold walls as liquid and must either be transported into the cylinder as a liquid, or evaporate and be transported along with the air and fuel from subsequent injections. It is important to compensate for this delay in modern engines with 3-way catalysts if low emission operation is to be achieved successfully.

Another problem with single point injection is that of the equal distribution of fuel to the different cylinders. The fuel film on the manifold will flow differently than the air flow. This was also a problem with carbureted engines, as there are different flow paths for the different cylinders. As emissions and knock tendencies are strongly dependent on fuel air ratio, it is advantageous to have all cylinders operating at nearly the same fuel air ratio, even though the overall fuel air ratio (as "seen" by the catalyst) may be stoichiometric in an engine with cylinder to cylinder variation in the mixture ratio. As emissions requirements become stricter, single-point systems are becoming less common.

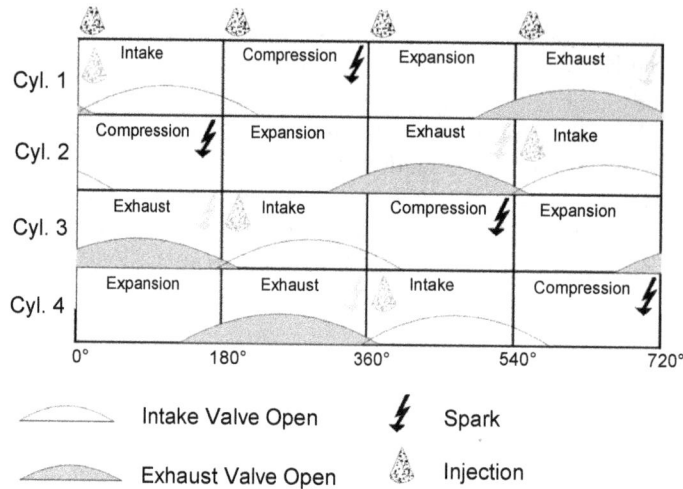

Figure 5.21: Timings for the air exchange process, the fuel injection and the spark for a centrally located injector. The injection symbols over the diagram show that there is only one injector. The grey symbols show for which cylinder the injection is intended. The firing order shown here is 1-3-4-2, but it can also be 1-4-3-2.

Multi-Point Injection

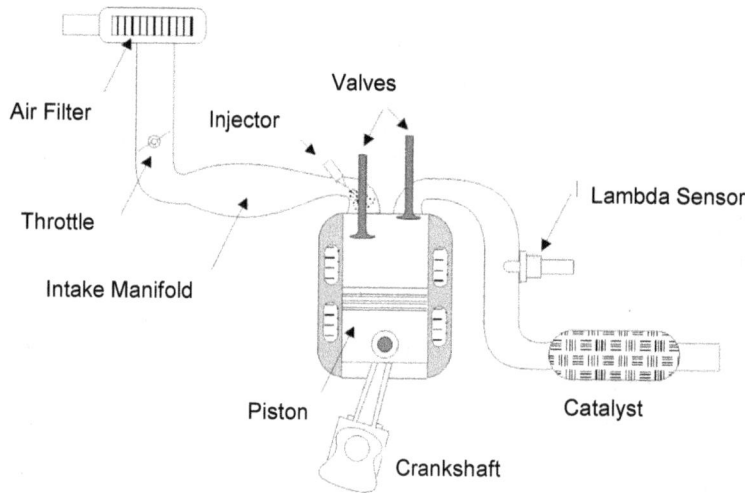

Figure 5.22: An EFI, (SEFI), system. The fuel is injected in the area near the intake valve. The throttle determines the amount of air flow to the intake manifold. A 3-way catalyst limits the emissions in the exhaust gasses. A lambda-sensor measures the air fuel ratio in the exhaust. Only one cylinder is shown for simplicity.

In this case, one injector is used per cylinder, and the injectors are typically placed close to the intake valves in the intake port/head. It has become common practice to spray the fuel directly on the head of the intake valve, as this has been found to improve vaporization and mixing. A multi-point injection system normally consists of the same number of injectors as cylinders, as shown in Figure 5.22. In some applications, a single injector near the throttle has been used for supplying extra fuel for cold starting.

Each injector is placed such that the fuel is injected into the individual intake port, and commonly sprayed directly onto the base of the intake valve, from where the fuel is drawn into the cylinder when the valve opens. The injection can also occur in the air flow while the intake valve is open. It is not uncommon for the fuel pulse to be sent to more than one cylinder at a time, as this gives a slightly simpler control system. Relative to the CFI system, this system shows a little improvement of power and torque, due to a more uniform fuel distribution, and a slightly lower temperature at the start of compression. This is due to the fuel evaporation in the port. The engine also reacts more quickly to changes, since the fuel is injected closer to the intake valve. The more uniform fuel distribution means that a slightly higher compression ratio can be used, since it is not as necessary to compensate for one cylinder that may have a higher knock tendency than the others. Due to the number of injectors, this system is more expensive than the CFI-system. The timing of events in an EFI system is shown in Figure 5.23.

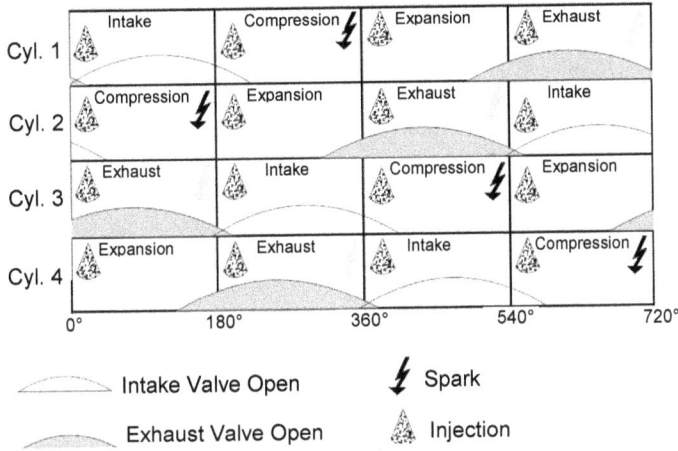

Figure 5.23: The times of the air exchange processes, fuel injection and spark for an EFI-system with 4 port-injectors. The time of injection is every 180 crank angle degrees, or for each "event" as it is also called. It could also be every 360 crank angle degrees or something similar. The firing order is 1-3-4-2.

A variation on the EFI-system is the SEFI (Sequential Electronic Fuel Injection) where the physical placement of the injectors is the same as in the EFI-system. The only difference is in the implementation of the engine control software, such that the injection timing for each cylinder is controlled individually to assure that fuel is only injected on closed valves. In relation to the EFI-system this gives a little improvement in the power and torque. The improvement arises because the injection timing can be advanced or retarded depending on the speed and load of the engine. The system is more complicated to implement due to the expanded use of software and hardware, and is the most expensive of the systems shown here. Figure 5.24 shows the order of events for a SEFI-system.

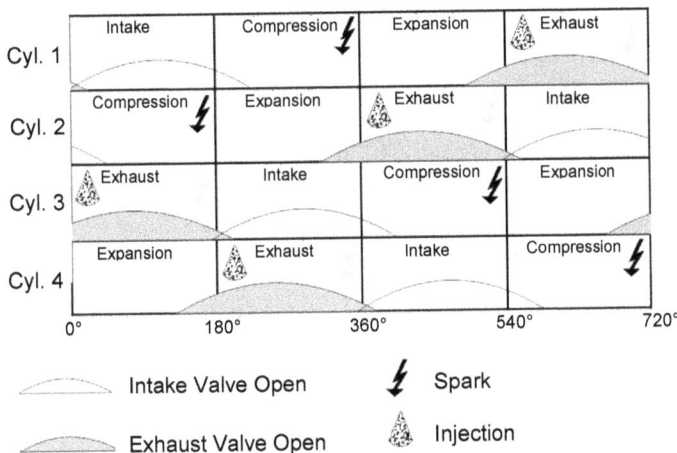

Figure 5.24: Timing for the air exchange process, fuel injection and spark for a SEFI-system. The injection system has 4 port-injectors. The injection is timed to take place on closed intake valves. The firing order is 1-3-4-2.

For both the EFI and SEFI it is necessary to use a higher injection pressure than with the CFI-system in order to avoid the formation of fuel vapor pockets in the fuel line, since the injector in the EFI/SEFI system is located in a warmer place than the CFI-system.

The multi-point injection system reduces the delay time and evaporation problems in the intake manifold, as well as improves cylinder to cylinder fuel distribution. The disadvantage is that the system has more components and thus costs more. Since the injectors are placed downstream from the throttle valve, the injection takes place in the manifold, where the pressure changes as the load is varied. Thus compensation must be made in the control system for the manifold pressure in Equation (5.9). The injectors are sealed with o-rings to prevent leakage of air into the manifold.

5.6.3 Mean Value Models

One of the major developments in engine technology in the past 30 years has been that of electronic engine control. This was initially set in motion by increasingly strict emission standards that required precise fuel air control in SI engines. First generation control systems were based on detailed steady state engine mapping, "look up" tables and simple control techniques. From the start, control of transient engine operation was the most challenging aspect of engine control. In order to improve transient control and produce more robust control systems, model based control systems were developed. These are based on engine models designed to predict transient engine performance on a time scale of 3-5 revolutions. These models were then included in what are called observer based control systems, and were given the name Mean Value Engine Models, (MVEM) [62]. On this time scale, detailed models on a crank angle basis are not needed, and the demands on the on-board computer capacity reduced. MVEM's can be developed to give an accuracy of a few percent over the entire engine operating range. They are based on well-known engine principles and relationships.[1]

The initial object of the SI engine MVEM was to give an accurate prediction of the fuel air ratio under transient operation, in order for the 3-way catalyst to operate optimally. As such, emission predictions were not a goal of the MVEM's, as the catalyst reduces tailpipe emissions to acceptable levels. Emission predictions, with the possible exception of CO, are problematic, even with full cycle models, and need to be based on experimental data and empirical fits if they are to be included in MVEM's.

One advantage of an observer based control system is that it allows the use of cheaper and simpler sensors in the engine. The models can accurately predict engine performance under transient conditions, often more accurately than they can be measured. With the observer based system, one can use simple sensors under steady state operation to ensure predictions are correct. This is done with the use of a control technique called a Kalman filter [63]. When the engine operates under highly transient conditions, the performance can be predicted using the MVEM rather than the sensors. Even with more expensive sensors, the observer control system can be used the improve accuracy, and compensate for deficiencies in sensors, for example the location of a mass air flow sensor within the intake system. In the following, a basic MVEM model for a non-pressure charged SI engine is presented. Models for CI engines and pressure charged engines have also been developed. The turbocharged diesel engine calculation in Section 11.12 is a form of MVEM, though the turbocharger model shown is complicated, and simpler methods can be used for control system MVEM's with turbochcarging [64], [65].

There are three systems that need to be modeled in the SI MVEM: the crankshaft dynamics, or engine speed, the mass flow of fuel, and the flow of the air, or the intake manifold filling dynamics.

The Crankshaft Dynamics

To determine the speed variations, an energy balance on the crankshaft balances the input energy (indicated power), output energy (brake, or load power), loss (friction) and inertial forces (acceleration/deceleration).

$$\frac{d}{dt}\left[\frac{I}{2}\left(\frac{2\pi N}{60}\right)^2\right] = \frac{d}{dt}\left[\frac{I\omega^2}{2}\right] = (FP_m + FP_p + BP) + \dot{m}_f H_u \eta_i (t - \tau_d) \tag{5.20}$$

where:

$$I = (I_{eng} + I_{load}) \tag{5.21}$$

---

[1]Professor Hendricks has made a MVEM for an SI engine available on his website, including Matlab files and a more thorough explanation: http://www.iau.dtu.dk/~eh/

I = moment of inertia of the engine

The engine load, BP, depends on the vehicle characteristics and for road vehicles, the concepts of Chapters 14 and 15 can be used. The time delay, τ_d, shows that there is a delay between the start of a fuel flow step input and the change in engine speed, this is approximately:

$$\tau_d = \frac{2\pi}{n_{cyl}N/60} = \frac{60}{2n_{cyl}N} \tag{5.22}$$

The friction power here is given as the sum of the mechanical friction FP_m and the pumping work, FP_p. Correlations as indicated in Chapter 9 are adequate for calculating full load engine friction in the MVEM. They can be adjusted slightly to fit specific engines, but retain the same functional dependence on variables. A simple model can also be developed for the friction over the entire load range using a simple relationships. The basis is the full load correlation, as given in Chapter 9. To modify it for part load conditions, two ideas are used. The first is related to the simple air exchange model, and assumes that the cylinder pumping is proportional to the difference between the intake and exhaust pressures. The second includes flow losses through other intake components such as air filer, intake pipes and such. In accordance with general flow characteristics, it is assumed that there is a loss proportional to the square of the mass flow rate.

Figure 5.25: Motoring friction of a 1.8 liter SI engine fit to Equation (5.23) as a function of engine speed and manifold pressure.

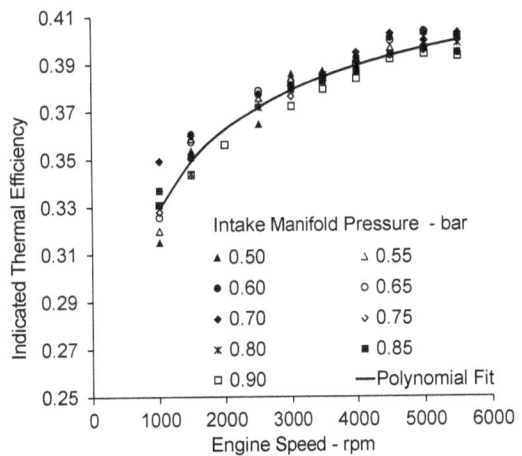

Given this, a model of the following form of Equation (5.23) was fit to the motoring data of the 1.8 liter engine shown in Chapter 12. The results are shown in Figure 5.25.

$$fmep = K_1 + K_2 \cdot (N/1000) + K_3 \cdot (N/1000)^2 + K_4 \cdot (p_{exh} - p_{int}) + K_5 \cdot \dot{m}_{air}^2 \tag{5.23}$$

where the K_i's are empirical constants. For the engine shown, with speed in rpm, pressure in kPa and mass flow in g/s, the constants are: $K_1 = 50$, $K_2 = 12$, $K_3 = 2$, $K_4 = 0.9$ and $K_5 = 0.014$.

$\dot{m}_f$ is the flow of fuel (liquid and vapor) entering the engine at the intake port. The relationship between the fuel and air flow is determined by the desired fuel air ratio. In engines with three-way catalysts, this is the stoichiometric fuel air ratio. This is determined by the fuel input to the manifold, and the behavior of the fuel in the intake manifold, as described later.

It was seen in Chapter 2 that fuel air ratio, heat transfer, compression ratio and ignition timing are major factors in determining the indicated thermal efficiency, η_i. For a given engine, the compression ratio is constant, and one can assume optimum ignition timing. For most modern SI engines, the fuel air ratio of modern engines is stoichiometric, and only changes at full load. Relative trends of thermal efficiency with respect to fuel air ratio can be used, if needed, from the ideal fuel air ratio cycles of Section 2.4. The main factor left influencing η_i is then the heat transfer. Chapter 10 shows that the fraction of energy lost to the coolant is a polynomial function of engine speed, so this type of correlation is an obvious choice. Figure 5.26 shows the indicated efficiency of 1.1 liter SI engine as a function of speed for the entire operating range for stoichiometric

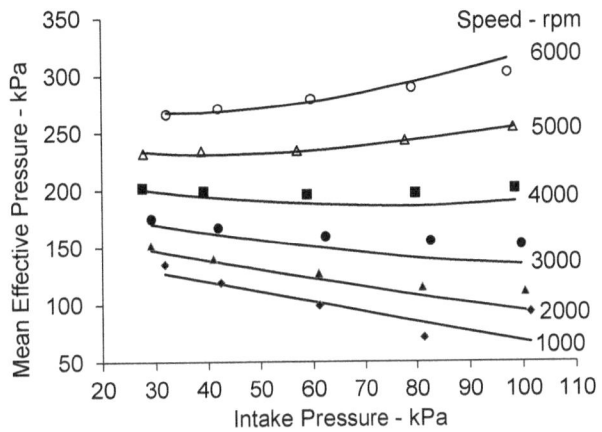

Figure 5.26: Indicated thermal efficiency for a 1.1 liter SI engine as a function of engine speed and manifold pressure.

operation. There is a small effect of manifold pressure, but the entire range can be closely fit with one curve, the curve here being:

$$\eta_i = 0.506 \left[1 - 0.3502 \left(\frac{N}{1000} \right)^{-0.3} \right]$$

(5.24)

Note that the curve is simple and has few parameters. This insures that the model is stable over the entire engine operating range, and not subject to problems encountered with high order polynomial fits. A low order (2) polynomial can be used to correct the pressure dependence [62].

The thermal efficiencies shown are for operation with stoichiometric fuel air ratio at optimum timing. Empirical results shown in Figure 5.27 can be used to estimate relative changes if desired. These can be fit with simple correlations, ensuring stable predictions over the entire engine operating range [62]. The Fuel Cycle gives a realistic trend, but a lean mixtures, engine sensitive lean flames begin to affect the results. The timing curve is related to Figure 2.18, but differs somewhat due to the effect of pressure and temperature during combustion, which affects the flame speed, so combustion duration is a function of timing.

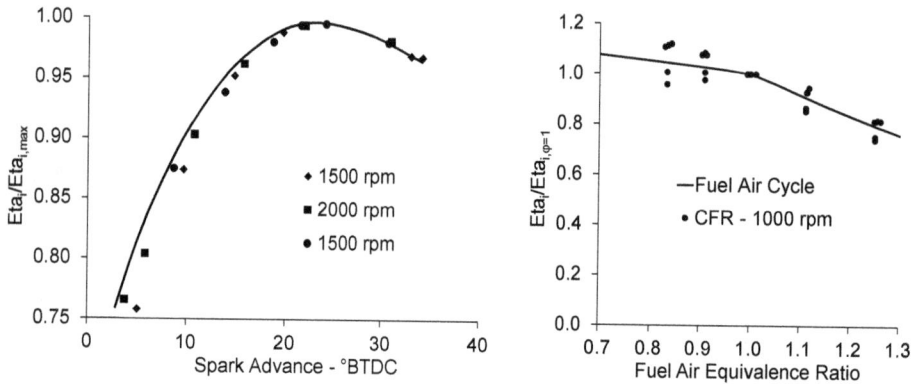

Figure 5.27: Correction of indicated thermal efficiency for changes from stoichiometric operation at constant spark timing and optimal spark timing at stoichiometric fuel air ratio.

The Dynamics of Fuel Injection

The operating condition of an engine is determined by the amounts of air and fuel than have been inducted into the engine, and it is important with modern engines that this ratio in the exhaust is kept very close to the stoichiometric mixture ratio, or $\lambda = 1$, so the catalyst works properly. The mixture ratio of the exhaust gasses is that of the cylinder gasses after during combustion. A precise knowledge of instantaneous air consumption as well as the fuel flow into the cylinder is an absolute necessity. That the fuel flow into the cylinder is not necessarily the same as that injected in the manifold will be illustrated in the following. In some SI engines, fuel is injected directly into the cylinder, and in this case, the manifold fuel

Figure 5.28: The CFI-system. The injector is placed upstream of the throttle. A portion of the injected quantity of fuel is deposited on the throttle and the manifold walls and the rest continues on as vapor.

phenomena are no longer relevant. A knowledge of the air flow is, though, crucial to maintaining the desired fuel air ratio.

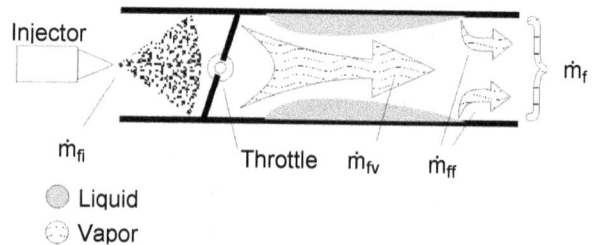

Figure 5.28 shows a drawing of the intake manifold processes occurring when the fuel is injected in a CFI-injection system. The speed of the fuel spray when it leaves the injector is great enough to break the spray up into small drops. At the same time, the interaction with the air near the throttle will increase the breakup into small drops, the smallest of which will be accelerated by the air flow. Some fuel will also impact on the throttle plate, a disadvantage with the CFI concept. When the fuel enters the manifold, it consists of both vapor and small fuel drops.

Since the density of the drops is larger than the surrounding air, they will not be able to follow the air flow completely when the air changes direction, which means that some of the drops will be deposited on the manifold walls. Depending on the temperature of the manifold wall, the deposited fuel will evaporate with a delay that can be characterized by a time constant.

For an EFI-system, the picture of the injection process is a bit different. In this case, the fuel is injected in the area around the intake valves. Figure 5.29 shows injection in an EFI-system. Some of the injected fuel quantity will evaporate almost instantaneously, and if the intake valve is open, it will be immediately drawn into the cylinder with the intake air. Another portion will not evaporate, but will be deposited on the valve and the walls of the intake port. This portion of the fuel will evaporate with a delay characterized by another time constant, which will be substantially smaller than for the CFI-system.

A simplified model for the fuel flow for both the CFI- and EFI-systems can be written with two algebraic and one differential equations [66]:

$$\dot{m}_{fv} = (1 - X)\,\dot{m}_{fi} \tag{5.25}$$

$$\frac{d\dot{m}_{ff}}{dt} = \frac{1}{\tau_f}\left(-\dot{m}_{ff} + X\dot{m}_{fi}\right) \tag{5.26}$$

$$\dot{m}_f = \dot{m}_{fv} + \dot{m}_{ff} \tag{5.27}$$

The injected fuel quantity, $\dot{m}_{fi}$, is the input value to the model, it is determined by the flow rate and pulse width of the nozzle, while $\dot{m}_f$ is the output value, that is the flow of the fuel in both the liquid and vapor phase that enters the cylinder and is available for combustion. The liquid fuel drawn into the cylinder evaporates on the intake process, with the final evaporation possibly occurring as late as the compression stroke. For the CFI system, X is that portion of the fuel that contacts the throttle plate and manifold walls. τ_f is the time constant determining how quickly the fuel evaporates from the walls. For an EFI system, X expresses how much of the fuel contacts the walls around the intake valve, and τ_f is the time constant for how fast the fuel evaporates from the port walls and intake valve. The variable $\dot{m}_{fv}$ is therefore the amount of immediately evaporated fuel. Then $\dot{m}_{ff}$ is the flow rate of the fuel in the film. Experimentally, τ_f has be found to vary be-

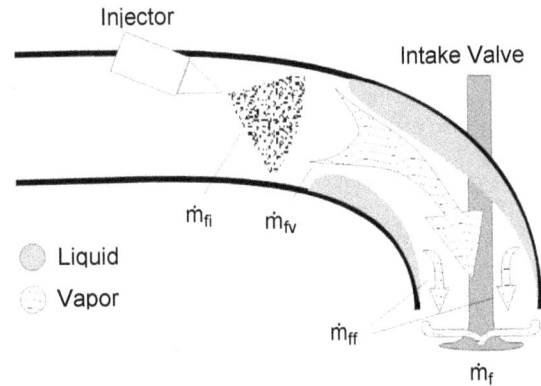

Figure 5.29: The EFI-system. The injector is placed near the intake port, such that the fuel is injected directly on or near the intake valve. Some of the fuel is deposited on the wall, and the rest evaporates quickly and is drawn into the engine.

tween 0.5 and 1.5 s for a CFI-system, and 0.2-0.4 s for an EFI-system. Even though the model is quite simple, in comparison to the detailed physical process, it has been shown to give satisfactory results in engine control system development [67].

For a CFI-system, X depends on the throttle angle, α

$$X(\alpha) = 0, \quad 0 \le \alpha \le \alpha_0 \tag{5.28}$$

$$X(\alpha) = a_0\left(1 - \cos\left(\alpha_1\right) - b_0\right) \quad \alpha_0 \le \alpha \le \alpha_1 \tag{5.29}$$

$$X(\alpha) = X_1, \quad \alpha_1 \le \alpha \le 90° \tag{5.30}$$

where: $a_0 \approx 1$, α_0 is the angle when the throttle is closed, $b_0 \approx \frac{\alpha_0^2}{2!}$ and $\alpha_1 \approx 60°$, for a typical engine,

$X_1 \approx 0.5 - 0.6$. The dependence of X on $1 - cos(\alpha)$ can be explained by seeing that the open area of the throttle has the same dependence on the throttle angle as X. It is reasonable that the open area determines the amount of fuel that can contact the manifold wall. When a certain limiting opening angle of $\alpha_1 \approx 60°$ is reached, the dependence will disappear. This is partly due to the increased air flow, and partly because the further opening does not expose more wall area, seen from the injector.

In an EFI system, X will not depend on the throttle angle, since the fuel is injected downstream from the throttle in the intake port and valve area. Then for an EFI system, it is found that X is engine design dependent and depends on the manifold pressure and the engine speed, that is:

$$X = X(p_{man}, N) \tag{5.31}$$

Equations (5.25) through (5.27) form the basis of the mean value fuel flow model, given the above behavior of X and τ. For engines with in-cylinder fuel injection only, the fuel dynamic calculation is not needed and the mean value model simplified. However, the indicated efficiency of this engine may not be as well behaved as that of the "homogeneous charge engine, especially if operated in the GDI mode.

The Dynamics of Air Induction

In order to determine the correct amount of fuel to inject, it is necessary to determine the instantaneous air flow. The air flow through the intake valves is determined by the engine's displacement and speed. The mass flow to the intake valve of the engine is:

$$\dot{m}_{ap} = \frac{\eta_v p_{in} V_d N}{120 R T_{in}} \tag{5.32}$$

This relationship is used in what is called the speed density control system, as the mass flow is primarily due to changes in the intake density and engine speed. The speed has an influence on the volumetric efficiency, but it is not nearly as strong as the density and the speed *per se*. The port air flow is then calculated on the basis of $\rho(p_{in}, T_{in})$ and the speed, hence the name *speed-density*. In the mean value model, a the volumetric efficiency can normally be described by a fairly low order polynomial.

As an alternative to modeling the volumetric efficiency, it has been found that the product of the volumetric efficiency and the intake pressure are linear functions of the intake pressure at a given speed:

$$\eta_v \cdot p_i = s_1(N) \cdot p_i + y_i(N) \tag{5.33}$$

where $s_1(N)$ and $y_i(N)$ are weak functions of the engine speed and dependent on manifold design [68]. This formulation can then be used with Equation (5.32).

To a reasonable approximation, the air flow past the throttle can be approximated by compressible flow through a nozzle [62], though a more exact expression has been developed [68]. The throttle is considered to be the nozzle, and the area is a function of the throttle position. The flow through the throttle is calculated as:

$$\dot{m}_{at} = \dot{m}_{at1} \beta_1(\alpha) \beta_2(p_{man}) + \dot{m}_{at0} \tag{5.34}$$

where: $\dot{m}_{at1} = C_f p_o \frac{\pi}{4} \sqrt{\frac{2\gamma}{(\gamma - 1) R T_o}}$ and $\dot{m}_{at0}$ is the air flow through the throttle in it's most closed position. p_o and T_o are the temperature and pressure upstream of the throttle, C_f the flow coefficient and β_1 is given by:

$$\beta_1(\alpha) = 0, \quad 0 \leq \alpha \leq \alpha_0 \tag{5.35}$$

$$\beta_1(\alpha) = 1 - cos(\alpha) - \frac{\alpha_0^2}{2!}, \quad \alpha_0 \leq \alpha \leq \alpha_1 \tag{5.36}$$

$$\beta_1(\alpha) = A_N \quad \alpha_1 \leq \alpha \leq 90° \tag{5.37}$$

where: $\alpha_1 \approx cos^{-1}\left[\frac{d}{D} cos(\alpha_o)\right]$ $A_N = \left(1 - \frac{4d}{\pi D}\right)$

α_0 (in radians) is the angle of the throttle relative to vertical, (assuming the manifold is horizontal) when it is completely closed and α_1 is the angle where the opening area is the maximum, this is before the maximum opening of the throttle [69]. This is because when a given opening angle is reached, the throttle shaft makes up a significant portion of the throttling effect. A_N is the normalized area at the angle α_2, d is the diameter of the throttle shaft, and D is the diameter of the intake manifold where the throttle is located. β_2 is found from a form of the isentropic flow equations:

$$\beta_2(p_{man}) = \sqrt{p_r^{\frac{2}{\gamma}} - p_r^{\frac{\gamma+1}{\gamma}}} \quad if \quad p_r \geq \left(\frac{2}{\gamma+1}\right)^{\left(\frac{2}{\gamma+1}\right)} \tag{5.38}$$

$$\beta_2(p_{man}) = \sqrt{\left(\frac{\gamma-1}{2\gamma}\frac{2}{\gamma+1}\right)^{\frac{\gamma+1}{\gamma-1}}} \quad if \quad p_r < \left(\frac{2}{\gamma+1}\right)^{\left(\frac{2}{\gamma+1}\right)} \tag{5.39}$$

where: γ is the specific heat ratio and p_r is the ratio between the intake manifold pressure and the pressure at the inlet to the throttle, $p_r = p_{man}/p_o$. In a naturally aspirated engine the latter is the ambient pressure less whatever pressure drop there may be due to the pipes and air filter.

In order to calculate the air flow into the cylinder, the manifold pressure must be known, as seen in Equation (5.32). The manifold pressure is primarily determined by the difference in the inlet flow through the carburetor, and the flow drawn into the port. This is done by writing the mass conservation for the manifold mass, m:

$$\frac{dm}{dt} = \dot{m}_{at} - \dot{m}_{ap} \tag{5.40}$$

An additional condition is obtained by differentiating the ideal gas law $PV_{man} = mRT_{man}$, applied to the mass in the manifold and using (5.40):

$$\frac{dp_{man}}{dt} = \frac{RT_{man}}{V_{man}}\left(\dot{m}_{at} - \dot{m}_{ap}\right) + \frac{p_{man}}{T_{man}}\frac{dT_{man}}{dt} \tag{5.41}$$

To determine the temperature change, energy conservation is applied to the manifold with heat transfer to the manifold assumed positive:

$$\frac{dU}{dt} = \dot{m}h_{out} - \dot{m}h_{in} = \frac{d\left(mc_v T_{man}\right)}{dt} = \dot{m}_{at}c_p T_o - \dot{m}_{ap}c_p T_{man} \tag{5.42}$$

Combining with mass conservation:

$$\frac{dT_{man}}{dt} = \frac{RT_{man}^2}{pV_{man}}\left[\left(\frac{T_o}{T_{man}}\gamma - 1\right)\dot{m}_{at} - (\gamma - 1)\dot{m}_{ap} - \frac{\dot{q}}{c_v T_{man}}\right] \tag{5.43}$$

Combining with Equation (5.41), an equation for the derivative of the manifold pressure is obtained:

$$\frac{dp_{man}}{dt} = \frac{\gamma RT_{man}}{V_{man}}\left[\frac{T_o}{T_{man}}\dot{m}_{at} - \dot{m}_{ap} - \frac{\dot{q}}{c_p T_{man}}\right] \tag{5.44}$$

When the engine is operating under transient conditions, equations (5.43) and (5.44), should be solved simultaneously to determine the temperature and pressure of the intake manifold. The temperature change associated with sudden opening or closing of the throttle is normally very fast, and upon filling or emptying, the temperature is close to that given by assuming that the manifold temperature is constant. In that case, the pressure change is simply described by:

$$\frac{dp_{man}}{dt} = \frac{RT_{man}}{V_{man}}\left(\dot{m}_{at} - \dot{m}_{ap}\right) \tag{5.45}$$

Though not strictly correct, this result has been used in many systems with fair success. However, as the emissions requirements become stricter, and even better air fuel control is needed, the change in manifold temperature should be taken into consideration, as it is not a major difficulty in a modern engine computer unit.

From a qualitative point of view, Equation (5.45) makes a very important point. That is there are different air flows into and out of the manifold when the engine is not operating at steady state. This

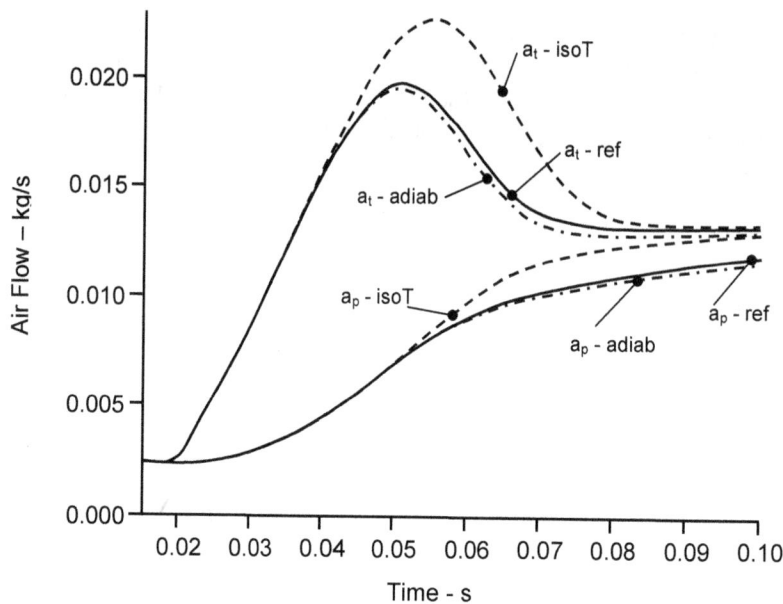

Figure 5.30: The transient air flow into and out of an intake manifold, calculated with three different assumptions. a_t refers to throttle air, a_p refers to port air, *isoT* is constant manifold temperature, *adiab* refers to adiabatic conditions, and *ref* includes temperature change and heat transfer to the manifold walls [65].

means, that even if an accurate air flow meter is used to measure engine air flow, the measured flow is not the flow going into the cylinder unless the engine is at steady state. Therefore, the above equations should be used to calculate the proper flow into the cylinder in transient operation, in order to maintain the best fuel air ratio control. Figure 5.30 shows the difference in air flows for three different conditions: 1. The reference condition, including heat transfer and variable manifold temperature, 2. an adiabatic condition, including variable manifold temperature and no heat transfer, and 3. an iso-thermal case, *i.e.*, constant manifold temperature. There is a substantial difference in the calculation of the flows, especially when the manifold temperature is assumed constant. A peak error of about 30 % in the throttle flow is encountered briefly for the isothermal case. The error in the port flow, which is more important for proper fuel air ratio control, is on the order of 10 % during the transient. This translates into a similar error in the transient fuel ratio, enough to prevent optimum operation of a three-way catalyst during the transient.

5.6.4 Control Strategies

This section will show the principles in the most common gasoline fuel injection systems for spark ignition engines, as well as how the intake air flow is determined in the individual systems.

Speed-Throttle

A speed-throttle is a cheap and simple control system, as it only needs to contain three sensors:

1. A throttle angle sensor.

2. An engine speed sensor.

3. A manifold temperature sensor.

Because simplicity and low price are prime objectives, this type of system is most commonly encountered with the CFI injector. Figure 5.31 shows the basic elements of a speed-throttle system. The air flow is determined by "calibrating" the throttle, that is, the flow to the engine is determined as a function

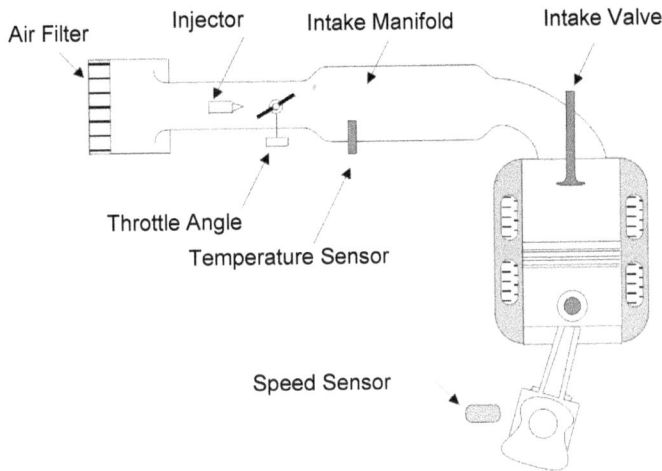

Figure 5.31: A speed-throttle CFI system. The throttle determines the air flow to the engine, and the throttle angle is measured with a sensor. The engine speed is determined by a sensor mounted in proximity to the crankshaft.

of throttle angle and engine speed. These values are programmed into a look-up table in the engine management system for later use.

The main problem with the system is at low engine speeds. Here, the air flow rate is low, and since the difference in throttle position to give the variation in manifold pressure between full throttle operation and that for idle is small, the flow regulation requires a only small area change. Said another way, at low engine speeds, after only a small opening of the throttle, it becomes ineffective, and further opening of the throttle geometrically has no effect on the engine flow. Thus while the throttle angle excursion to regulate the air flow at maximum speeds is from fully open to fully closed, at low speeds, only a small motion of the throttle gives large changes in air flow. This gives a requirement for very accurate measurements of throttle position at low speeds, as well as a high sensitivity to error, especially at low engine speeds and flows. Since much vehicle operation is at fairly low speed and load, this is a major drawback, and the system has lost favor through the years [66].

Speed-Density

Relative to a speed-throttle system, a speed-density system is expanded to include a manifold pressure sensor. The system must contain the following sensors to work:

1. A manifold pressure sensor.

2. A throttle position sensor.

3. An engine speed sensor.

4. A manifold temperature sensor.

Figure 5.32 shows a schematic diagram of the necessary elements of a speed-density system. The throttle position sensor is mainly used to detect changes in throttle position in connection with acceleration or deceleration, and not the actual position. In more advanced control systems, engine models, called observers, are built into the engine control system [70]. The use of observers makes it possible to calculate the filling process of the intake manifold in real time (See Section 5.6.3). This is important in fuel air regulation, as it takes some time to fill and empty the manifold. The throttle position is important with observers, since it determines the flow area limiting the flow of air into the intake manifold. In observer systems, the rate of change of throttle position is also used to change from measurement of manifold pressure to a calculation, as this is more accurate under rapid acceleration. Figure 5.32 shows a CFI system, an EFI or SEFI system would look the same, except the injectors would be located near the intake valve.

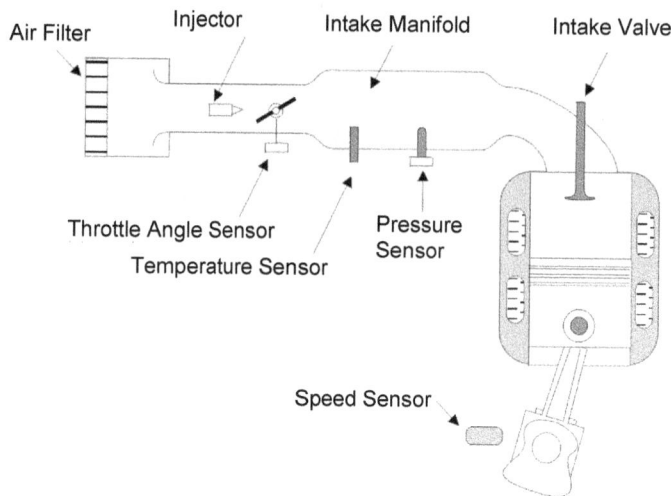

Figure 5.32: A speed-density CFI system. The injector sprays fuel upstream of the throttle. In a multi-point injection system, the fuel is injected in close to the intake valve of each cylinder. A throttle position sensor is used to detect changes in throttle position, pressure and temperature sensors measure manifold pressure and temperature, and the engine speed is measured from a speed sensor.

The air flow is determined by the basic speed density relationship from Equation (5.32). The target fuel flow rate is then determined through the desired air fuel ratio:

$$\dot{m}_{fi} = \frac{\dot{m}_{ap}(N, p_{in})}{\lambda_{target} \cdot AF_s} \tag{5.46}$$

Except for the volumetric efficiency, all the variables in Equation (5.32) are determined by measurement. The volumetric efficiency is basically only dependent on the intake pressure and the engine speed, and this needs to be determined by a mapping of the engine. The volumetric efficiency is then programmed into a table or function in the engine management system. Note that the fuel flow here includes the combination of the liquid film off the walls of the manifold and the vaporized fuel coming in with the air, as calculated from the fuel film model, Equations (5.25) to (5.27).

MAF-System

The MAF-system is an abbreviation for Manifold Air Flow, and in this system a flow measuring device is used to measure the flow directly. Typically a hot-wire sensor upstream of the throttle is used, but other types of flow sensors have been used previously, including vortex shedding frequency instruments, and mechanical vanes, whose motion is calibrated as a function of air flow. Compared to speed-throttle and speed-density, a MAF-sensor is typically an expensive system since the sensor includes a hot-wire sensor and electronics to compensate for measurement problems at low speeds and with pulsating flow. Therefore, initially it was only seen with EFI, SEFI injection systems, in "top of the line" systems. It is increasingly becoming more common. In principle, the system only needs one sensor, that is, a Manifold Air Flow sensor. Typically, MAF systems also include an engine speed sensor, and manifold temperature and pressure sensors.

Though in principle, the MAF sensor measures flow directly, there are some problems with the system. The first is that the air flow is measured a substantial distance away from the intake port. This means that compensation must be still made for filling and emptying of the intake manifold when determining the instantaneous flow into the cylinder, as it is the air flow at the port that determines the amount of fuel needed during acceleration and deceleration.

A second problem, especially with hot wire sensors, is that they are not directionally sensitive. That is, in a pulsating flow backwards flow is registered just as much as forward flow, as the system cannot determine the flow direction. If there are pulsations, there will be errors in the flow. Compensations are made, but this is difficult and engine dependent. Careful design and software, speed-density systems

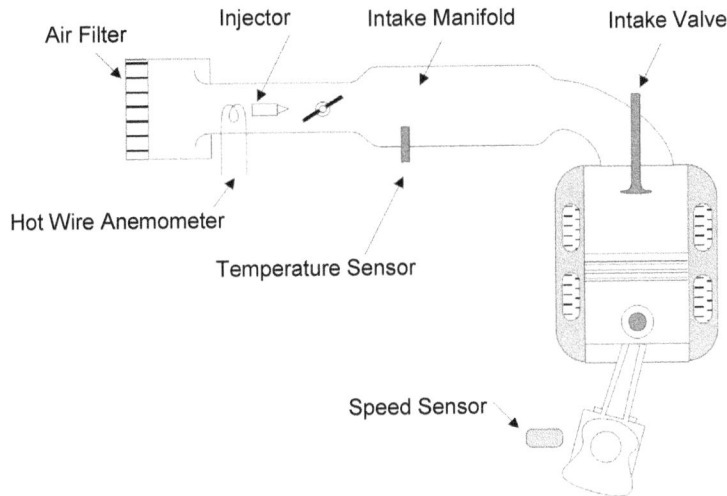

Figure 5.33: An illustration of a MAF-system. The magnitude of the air flow is measured directly with a hot wire placed upstream of the throttle. This system is an EFI-system with intake port injection.

can give equivalent performance to MAF systems, though the latter have achieved substantial use at the time of writing.

5.7 Problems

Problem 5.1

A 4-stroke diesel engine has a cylinder diameter of 150 mm and a stroke of 175 mm. The engine has a turbocharger and operates such that the intake pressure is 2.5 bar at maximum load and intake temperature is 428K. The volumetric efficiency based on intake conditions is 91%. The engine operates on diesel oil, which has a density of 0.85 g/cm3. It is desired that the engine operate with a fuel air ratio of 0.041 at maximum load, which occurs at 1750 rpm. Calculate the volume of fuel that is injected per pump stroke in mm^3. ($V_f = 276$ mm^3/stroke)

Problem 5.2

A 4-cylinder spark ignition engine has a displacement volume of 1.3 liter, and operates with a maximum engine speed of 6000 rpm. At this speed, the engine has a volumetric efficiency of 0.70 based intake conditions of 20°C and 1.0 bar. The engine uses Central Fuel Injection, that is, 1 nozzle supplies fuel to all cylinders. The injection strategy is that fuel is injected once for every intake stroke, which corresponds to 4 injections for 2 revolutions. It is necessary to maintain a minimum closing time of 1.0 ms between injections to insure adequate regulation. The engine is equipped with a 3-way catalyst, so the excess air ration will always be 1.00. The pressure regulator functions such that there is always a pressure difference of 1.0 bar between the gasoline in the nozzle and the manifold pressure. The flow coefficient can be assumed to be 0.70 for the nozzle. The fuel air ratio is 0.0669 and the fuel specific gravity is 0.74. Calculate flow area of the nozzle. Above which engine speed will it be impossible to maintain the desired excess air ratio of 1.0? ($A_d = 0.531$ mm^2; $N = 7500$ RPM)

Problem 5.3

The engine in problem 5.1 is equipped with a nozzle that has 4 holes with diameter of 0.28 mm. The fuel is injected over a period of 25 crank angle degrees. The engine operates at 1800 rpm. The nozzle has a flow coefficient of 0.80. Determine the pressure difference over the nozzle. ($\Delta p = 1559$ bar)

Problem 5.4

A 4-stroke diesel engine has a cylinder diameter of 145 mm and a stroke of 155 mm. At maximum load it has a fuel/air ratio of 0.045. The volumetric efficiency is 81% at 1 bar and 20°C when the engine operates at 1800 rpm. The fuel is to be injected over a duration of 35 crank angle degrees with a pressure of 200 bar. During injection, the average cylinder pressure is 45 bar. The nozzle has 4 holes with a flow coefficient of 0.75 and the fuel has a density of 0.826 g/cm^3.

Find: a) Volume of the injection pump in mm^3, b) Size of the holes in the nozzle. ($V_f = 134.3$ mm^3; $d_d = 0.301$ mm)

Problem 5.5

The engine in problem 1.1, Chapter 1, has "common rail" injection, with a pressure difference of 1200 bar. With full load and a speed of 2500 rpm, what nozzle hole diameter is necessary with a 4-hole nozzle for an injection duration of 25 crank angle degrees? The flow coefficient of the nozzle = 0.80. ($d_h = 0.165$ mm)

Problem 5.6

The curve in Figure 5.2 shows measurements of the volumetric flow of fuel through the nozzle, injection pressure, and cylinder pressure as functions of time in crank angle degrees, measured during an experiment with a 2-liter, 1-cylinder diesel engine with direct injection. It can be seen that the fuel volume flow is not constant for the entire injection process. This is because the piston in the fuel pump moves with a non-constant speed.

The integrated volume flow for the entire injection process is 179 mm^3. The nozzle has 8 holes with a diameter of 0.24 mm, and the engine operates at 1350 rpm. From experience, the flow coefficient can be estimated at 0.80. The fuel has a density of 0.845 g/cm3.

a. Calculate the average value of the pressure difference over the nozzle for the entire injection process, based on the given injection parameters

b. Use the measurements to calculate the instantaneous value of the flow coefficient for the injection conditions at Top Dead Center from the injection conditions at Top Dead Center. (a: 539 bar; b. 0.843)

Problem 5.7

Consider a turbocharged 6-cylinder, 4-stroke diesel engine with direct injection. Cylinder diameter = 110 mm, stroke = 120 mm, compression ratio = 17:1. The engine operates with a speed of 1500 rpm and a brake mean effective pressure of 1500 kPa with a fuel air ratio of 0.035. Intake pressure and temperature are 258 kPa and 330K, and the volumetric efficiency = 0.938. The injection nozzle has 5 holes, with diameter of 0.14 mm. The fuel specific gravity is 0.84. The system is of the common rail type, where the injection pressure difference is held constant at 1100 bar and the duration is regulated via pulse length for a solenoid. Find the injection duration in crank angle degrees, if the nozzle flow coefficient = 0.80. (34.8°)

Problem 5.8

A 4 cylinder SI engine has a bore of 85 mm and stroke of 80 mm, and operates at stoichiometric conditions on a fuel with a stoichiometric $FA = 0.068$, $H_u = 44MJ/kg$. An approximation for the volumetric efficiency as a function of speed and manifold pressure in kPa is (typical for non-tuned manifolds):

$$\eta_v = 0.5223 + 0.08253 \cdot (N/1000) - 0.01382 \cdot (N/1000)^2 + 0.002363 \cdot p_{in} \tag{5.47}$$

The engine operates at MBT timing, has friction according to Equation (5.23), and the indicated efficiency of Equation (5.24).

1. At full load, assume a constant intake pressure of 95 kPa, an intake temperature of 310K, and exhaust pressure of 105 kPa. Calculate and plot the full load curves of brake power, torque, and brake specific fuel consumption as a function of speed from 1000 to 6000 rpm.

2. For a speed of 2500 rpm, plot the part load curve of brake specific fuel consumption as a function of brake mean effective pressure for intake pressures from 30 to 95 kPa.

3. Assume that for manifold pressure between 89 and 95 kPa, the fuel air equivalence ratio increases from stoichiometric to a fuel air equivalence ratio of 1.2. The Fuel Air cycle curve in Figure 5.27 can be described with the following curve fit:

$$\frac{\eta_i}{\eta_{i,\phi=1}} = 1.1197 + 0.0653\phi - 0.1838\phi^2 \qquad \phi < 1.0 \tag{5.48}$$

$$\frac{\eta_i}{\eta_{i,\phi=1}} = 2.3148 - 1.7023\phi + 0.3912\phi^2 \qquad \phi \geq 1.0 \tag{5.49}$$

Repeat question 2.

Chapter 6

Exhaust Emissions

Due to extensive use in automobiles, vehicle engines have been major contributors to urban air pollution. One of the major driving forces in engine development since the 1970's has been reduction of emissions from engines. Through improved engine design, electronic engine control, and catalyst technology, emissions from spark ignition engines from automobiles have been greatly reduced. Diesel engine emissions have also been greatly reduced, though at a somewhat slower rate. This chapter discusses the different emissions from engines, the reasons for their appearance in engine exhaust, and some control technologies.

The deleterious effects of various air pollutants are described in various texts on air pollution, and will only be briefly covered here. The exhaust gas air pollutants from engine of concern in the local environment are:

- *Carbon Monoxide*, the molecule CO, which is poisonous. In concentrations over about 500 parts per million CO is fatal, and at lower concentrations (on the order of 50-100 ppm) it can lead to headache and reduced mental activity [71].

- *Unburned Hydrocarbons* (UHC, or HC) are unburned or partially burned fuel components that are emitted in the engine exhaust. In order to include oxygenated substances, a more general term, *Volatile Organic Compounds* or *VOC* has recently become more widely used. Hydrocarbons belong to this classification. Some hydrocarbons have toxic effects, though in general they do not have high toxicities. One compound of concern is benzene, which is a known carcinogen. The most serious air pollution problems from these compounds arise from their reaction with oxides of nitrogen in sunlight to form photochemical smog, which consists of a number of complicated organic substances irritating to eyes and respiratory systems, as well as nitrogen dioxide and ozone.

- *Oxides of Nitrogen* refers to a mixture of NO and NO_2 commonly denoted as NO_x, since their relative concentrations vary in different situations. The primary oxide formed in combustion processes is nitric oxide, NO. NO oxidizes to form nitrogen dioxide, NO_2, in oxygen-containing environments. NO has lower toxicity than NO_2, which is the anhydride of nitric acid. Nitrous oxide, N_2O is formed in improperly designed catalytic converters, but is not normally measured and not included with the NO_x results from engines. It is a very strong greenhouse gas, though so far, limited attention has been paid to its emissions from vehicles.

- *Particulate Matter (PM)*, is commonly referred to as particulates. PM is solid material, or material in the exhaust that condenses at environmental conditions. PM from engines usually contains significant amounts of harmful aromatic compounds (PolyAromatic Hydrocarbons or PAH), many of which are carcinogens. In addition, there is evidence that particles themselves, regardless of composition, cause adverse health effects when absorbed in the body.

In addition to these pollutants, CO_2 has become of great concern in the discussion of global warming. Since it is a direct product of fuel oxidation, it is directly related to fuel consumption, and can be reduced only by reducing fuel consumption or changing to a renewable or carbon free fuel.

6.1 SI Engine Emissions

6.1.1 CO Emissions From SI Engines

The main source of CO emissions in SI engines is rich operation. This was seen in the calculation of the composition of rich combustion mixtures in Section 3.1. SI engines formerly were operated with rich mixtures at idle in order to give stable operation and low idle speed, see Figure 5.18. CO concentrations of several percent or more were often observed, as is still the case for some small SI utility engines. At high loads, some SI engines operate at rich mixtures to increase the maximum power, which increases CO emissions at these conditions. For most operation today, vehicle SI engines operate with stoichiometric mixtures, where CO concentrations are low. The CO that is present is effectively reduced by the 3-way catalyst, to be discussed later. In addition to rich mixtures, some CO is formed by the partial oxidation of the unburned hydrocarbons in the exhaust. The oxidation of hydrocarbons proceeds through the following overall mechanism:

$$HC \Rightarrow CO \Rightarrow CO_2 \tag{6.1}$$

If temperatures in the exhaust fall below about 1000K, the oxidation reactions will become too slow for the complete oxidation of the carbon not accomplished and as a result, CO will appear in the exhaust. Also, in the equilibrium products of combustion, the CO can be frozen in a similar manner to the NO, as will be discussed below. The resulting levels of CO from this process are normally small.

6.1.2 Unburned Hydrocarbons From SI Engines

Hydrocarbons (HC) emitted from combustion processes arise to a large extent because a portion of the hydrocarbon based fuel escapes the combustion or is only partially oxidized. Spark ignition engines have much higher HC emissions than diesels, typically about 10 to 100 times. This section focusses on the mechanisms responsible for the emission of unburned hydrocarbons from spark ignition engines.

Experimental measurements of hydrocarbons in exhaust gasses are made with an instrument called the *Flame Ionization Detector* (FID). In this instrument, a small amount of the exhaust gasses are passed through a flame of hydrogen and oxygen. The presence of the carbon atoms from the hydrocarbons in the exhaust gasses gives rise to the formation of ions in the flame. CO and CO_2 do not form ions in the H_2 flame. By exerting a voltage across the H_2 flame, a current due to the presence of the ions is obtained. This current is very linearly proportional to the amount of hydrocarbons present in the exhaust gas.

The FID does not, however, distinguish between different types of hydrocarbons. That is, 1 atom of hexane, C_6H_{14} will give essentially the same response as 6 atoms of methane, CH_4 or 2 atoms of propylene, C_3H_6. (There are some minor differences in the responses of these atoms, of only a percent or two, which is normally neglected). The composition of hydrocarbons reported from an FID is then reported as equivalent to a given hydrocarbon molecule, or number of carbon atoms in a molecule. For example, all of the following concentrations give the same response on an FID:

$$6000 \ ppm CH_4/CH_x = 2000 \ ppm C_3H_8 = 1500 \ ppm C_4H_8$$
$$= 1000 ppm C_6H_{12} = 500 \ ppm C_{12}H_{24}$$

where: ppm denotes parts per million volume (1 ppm = a mole fraction of 10^{-6})

The form CH_x in the above list is probably the most common, and one normally assumes a molar hydrogen to carbon ratio of 1.8 to 2 for most gasoline and diesel fuels. The ratio of carbon and hydrogen molecules in the exhaust gas hydrocarbons is related to, but not necessarily equal to that of the ratio in the fuel. This is due to partial oxidation reactions in the exhaust gasses, which change the composition of the emitted hydrocarbons from that of the original fuel. A precise determination of the carbon to hydrogen ratio of the exhaust gas hydrocarbon requires a detailed analysis of all the specific hydrocarbons present. This is very time consuming, and normally, all of the hydrocarbons are measured together, and given an average composition as described above and emission regulations are written on this basis.

There are special situations, for example in some American emissions standards, where reference is made to non-methane hydrocarbons. This is related to the low activity of methane in the process of photo-chemical smog formation. In this case, special accessories are attached to the FID, where the methane is separated from the gasses before they are sent to the FID.

HC Formation Mechanism

Originally, the common assumption was that hydrocarbons in the exhaust gas of spark ignition engines are caused by the quenching of the flame at the cold walls of the surface of the combustion chamber. That is, when a flame approaches a colder wall, it loses energy (and reactive species) to the cold wall, and the flame is extinguished, or quenched. Later results indicated that the vast majority of hydrocarbons resulting from wall quenching of the overall surfaces of an engine combustion chamber would be quickly mixed with the hot combustion products and oxidized to CO_2 and H_2O. At the current time, other mechanisms, some of which still deal with flame quenching, are believed to be mainly responsible to the presence of hydrocarbons in spark ignition engine exhaust gasses. The following mechanisms have been mentioned as possible:

1. *Flame quenching* at the cylinder walls, mentioned above.

2. *Crevices* in the combustion chamber, into which the flame cannot propagate.

3. *Absorption* of HC in the oil film and deposits on combustion chamber surfaces and subsequent release later in the cycle.

4. *Bulk quenching*[1], which occurs with rapid expansion and a slow flame under marginal combustion conditions, and the flame is extinguished due to rapid cooling of the cylinder gasses. This usually is found with mixtures near the lean combustion limit [26].

Just after the flame has traversed the combustion chamber and has been quenched at the walls, the bulk gas temperature is still very high and therefore an oxidation of the remaining unburned hydrocarbons from the fuel can occur until the combustion product temperature reaches a certain limit, which is on the order of 1000K. The unburned hydrocarbons originating with mechanisms 1 - 3 will, to a certain extent, be oxidized when mixed with combustion products, particularly with lean mixtures due to the oxygen present. For the wall quench mechanism it is expected that this oxidation will be very extensive, while the delay implicit in the release of the hydrocarbons from the crevices, oil film, and deposits will lead to the situation where the temperature of the combustion products can be too low for significant oxidation of this portion of unburned hydrocarbons in the combustion products.

The oxidation rate of hydrocarbons in combustion products is strongly dependent on the temperature and, to a lesser degree, the pressure. The rate of oxidation can be expressed by Equation (6.2):

$$\frac{[HC]}{dt} = Z[HC]^b[O_2]^b exp\left(-\frac{E_a}{R_c T}\right) \tag{6.2}$$

where square brackets, [], indicate concentration $gmol/cm^3$, calculated as the product of the mole fraction of the given specie and the total molar density in $gmol/cm^3$, R_c is the universal gas constant = 1.986 cal/gmol-K, and Z, a, b, and E_a are empirical constants [72]. As an example, the oxidation rates of 100 ppm n-butane have been calculated at temperatures and pressure for conditions after combustion, with 2% O_2, and are shown in Figure 6.1. Using the data of [72], $Z = 2.26 \cdot 10^{14}, a = 0.6449, b, = 1.064$ and $E_a = 42270$ cal/mol. The units of the reaction rate are $gmol/s\text{-}cm^3$. The curve on the left shows the extreme importance of the temperature. At 900K, the reaction is barely started after 50 ms, while at 1100K, the reaction is completed in less than 20 ms. The temperature of 1000 K is somewhat critical, in in some applications, it is assumed that as long as the temperature is above this HC reactions are infinitely fast, and below it, the reaction speed is zero.

Several models have been developed to calculate the thickness of the quenching layer and therefore the contribution of the wall flame quenching mechanism to the HC emission. Reference [73] gives, for example, the following expression for the amount of carbon which is due to wall quenching in the boundary layer of a wall with an area, A in m^2:

$$\frac{m_c}{A} = 3.3 \cdot \frac{g_u \cdot k_u}{S_u^o \cdot c_p} \tag{6.3}$$

where: m_c is the mass of carbon in kg, g_u is the mass fraction of C in the unburned mixture, k_u er thermal conductivity of the unburned mixture in W/m-s, S_u^o is the laminar flame speed in m/s and c_p is the mean specific heat of the unburned mixture at constant pressure in J/kg-K.

---

[1]Bulk gasses refer to the main portion of the gasses in the cylinder, away from the influence of the cylinder walls.

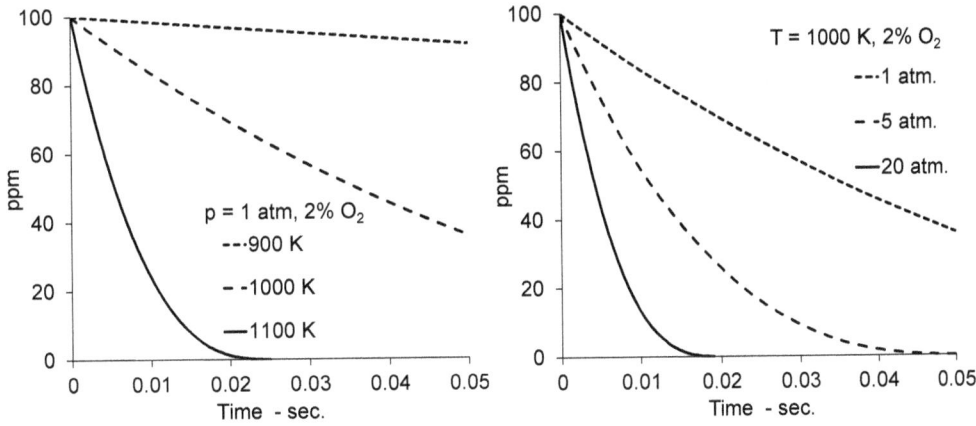

Figure 6.1: Concentration vs. time for the reaction of n-butane in combustion products.

In experimental investigations of the importance of crevices in HC emissions, combustion bombs are often used, since this enables the close control of the experimental conditions. Such a bomb with crevices can be theoretically described as a main chamber with an auxiliary chamber with the same volume as the total crevice volume. (see Figure 6.2) The crevices are all the locations where the distance between two adjacent walls is so small that the flame cannot propagate into the crevice.

In spark ignition engines, the most important crevices are thought to be around the top land clearance of the piston, and the crevice produced between the head and block, due to the cylinder head gasket. There may also be crevices associated with the installation of the spark plug.

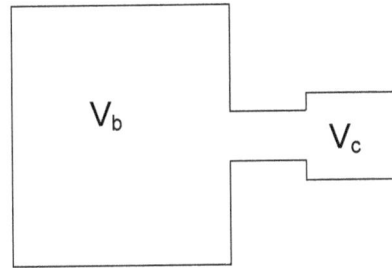

Figure 6.2: Combustion chamber with crevice volumes which are theoretically set equivalent to an auxiliary chamber with a volume V_c.

Depending on the combustion conditions, the flame will not propagate into a region where the distance between the combustion chamber walls is less than about 1 mm.

In order to calculate the contribution of crevices to the HC emission in a combustion chamber, let it be assumed that the pressure during combustion is the same throughout the main and auxiliary chambers. The number of moles of C, as C_1, which are stored in the auxiliary chamber can be written as:

$$N_c = f_u \frac{p_p \cdot V_c}{R_o \cdot T_c} \tag{6.4}$$

N_c is the number of moles of C in crevices, f_u is the mole fraction of C in the unburned mixture, p_p is the maximum pressure after combustion, V_c is the total crevice volume, R_o is the molar gas constant and T_c is the temperature of the unburned mixture in the crevice. Since the mixture in the crevice does not burn, it can be no hotter than the temperature on the unburned gasses when the flame approached the crevice. The temperature of the crevice gasses would be between that temperature and that of the surrounding surface, probably closer to the latter, since the crevice likely has a large surface area to volume ratio. Equation (6.4) gives the maximum value which N_c can attain, since the maximum pressure p_p is used in the expression and it is assumed that gasses are not burned before they are pressed into the crevices.

With respect to the contribution of the hydrocarbons in the oil film, that which determines how much HC "hides" in the oil film during combustion is determined by the conditions at the boundary between the combustion gasses in the cylinder and the oil film. The relation between the concentration of unburned hydrocarbon in the gas phase (cylinder gasses) and the concentration in the oil film can be described by Henry's law:

$$p_b = H * y_b \tag{6.5}$$

where: p_b = partial pressure of unburned fuel in the gas phase (combustion chamber), H* = Henry's constant and y_b = concentration of fuel dissolved in liquid phase (oil film)

A large value of Henry's constant means that the fuel is less likely to be dissolved in the oil film, which can be seen by rearranging in the following way:

$$y_b = \frac{p_b}{H*} \tag{6.6}$$

In recent years, methods to calculate the role of lubricating oil film in hydrocarbon emissions have been developed. The calculational procedures involve the solution of partial differential equations for the movement of the fuel through the oil film [74].

Example

During an experiment with a combustion bomb where mixture of propane and air was used, a HC concentration of 38 $ppmC_1$ was measured in the expanded and cooled combustion products when a mixture with $\lambda = 1, T_1 = 325$ K and a pressure of 5 bar was ignited [75]. The total crevice volume in the bomb is 60 mm^3. Evaluate which relevant effects that could have contributed to the formation of unburned hydrocarbons (HC).

Adiabatic combustion at constant volume gives a maximum pressure, p_{max}, of 46 bar, and a combustion temperature of 2875 K. For the calculation, $R_o = 8.3143$ J/(gmole-K), $T_c = T_B = 325$ K, $k_u = 0.0262$ W/(m-K), $S_u^o, = 0.4$ m/s, and $c_p = 1100$ J/(kg-K).

1. Crevice contribution, assuming crevices filled with unburned gas at p_{max}, and T_B:

Use Equation (6.4), where f_u is found from:

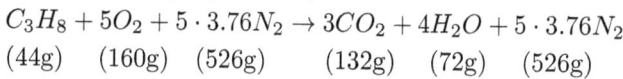

$$C_3H_8 + 5O_2 + 5 \cdot 3.76N_2 \rightarrow 3CO_2 + 4H_2O + 5 \cdot 3.76N_2$$
$$(44g) \quad (160g) \quad (526g) \quad\quad (132g) \quad (72g) \quad (526g)$$

$$f_u = \frac{3}{5 \cdot 4.76 + 3 \cdot 1} = 0.112$$

and therefore the number of moles of C in the crevice volume is:

$$N_c = \frac{f_u \cdot p_{max} \cdot V_c}{RT_B} = \frac{0.112 \cdot 46 \cdot 10^5 \cdot 60 \cdot 10^{-9}}{8.314 \cdot 325}$$
$$= 11.4 \cdot 10^{-6} gmoleC = 1.37 \cdot 10^{-4} gC = 1.60 \cdot 10^{-4} gCH_2$$

2. Contribution from wall quenching:

$$\frac{m_c}{A} = 3.3 \cdot \frac{g_u \cdot k_u}{S_u^o \cdot c_p}$$
$$g_u = \frac{3 \cdot 12}{44 + 526 + 160} = 0.0493$$
$$A = 4\pi r^2 = 4\pi \cdot 0.0756^2 = 0.0726 m^2$$

$$m_c = 0.0.0726 \cdot 3.3 \frac{0.0493 \cdot 0.0262}{0.4 \cdot 1100} \cdot 1000$$
$$= 7.03 \cdot 10^{-4} g \ C_1 = 5.85 \cdot 10^{-5} gmoleC$$

The volume of the bomb is:

$$V_u = \pi \frac{d^3}{6} = \pi \frac{0.152^3}{6} = 1.85 \cdot 10^{-3} m^3$$

The number of moles in the bomb before combustion is:

$$n_1 = \frac{p_1 V}{R_o T} = \frac{5 \cdot 10^5 \cdot 1.85 \cdot 10^{-3}}{8.314 \cdot 325} = 0.0.340$$

Due to the reaction, there is an increase in the number of moles from 24.8 moles to 25.8 moles, so the number of moles in the bomb after the reaction is:

$$n_2 = 0.340 \frac{25.8}{24.8} = 0.354$$

The maximum volumetric concentration of carbon due to the crevice is then:

$$ppmC = \frac{N_{crevice}}{n_2} 10^6 = \frac{1.14 \cdot 10^{-5}}{0.354} 10^6 = 32.3 ppm$$

The maximum volumetric concentration of carbon due to the wall quenching is then:

$$ppmC = \frac{N_{crevice}}{n_2} 10^6 = \frac{5.85 \cdot 10^{-5}}{0.354} 10^6 = 165.5 ppm$$

And the total maximum concentration:

$$ppmC = ppmC_{crevice} + ppmC_{quench} = 32.3 + 165.5 = 197.8 ppmC$$

The estimated formation is quite a bit larger than the measured quantity. Reference [75] found that about 95 % of the quench gas oxidized, likely because it is located close to the high temperature combustion products. Using that value and the assumption that none of the crevice gasses oxidized, a estimate of about 41 ppm is obtained. Whether the estimates of the relative oxidation of the two portions is correct, the above indicates that oxidation of the hydrocarbons left in the combustion chamber after combustion is very important.

6.2 NO_x Formation in Combustion Systems

In the combustion systems encountered in internal combustion engines, gas turbines, and other gas and light oil units, a significant amount of the oxides of nitrogen is formed. In exhaust gasses, the oxides of nitrogen are primarily composed of NO and a smaller portion of NO_2. In this section, only the formation of NO_x from atmospheric air at high temperature will be discussed. The formation of NO_x from nitrogen in the fuel will not be discussed, as this is more important in combustion of coal and possibly very heavy fuel oil. The oxides of nitrogen formed by the high temperature gas reactions after combustion are often called "thermal NO_x". The basic process is the same in SI and CI engines, though the local values of temperature, pressure and mixture ratio may vary.

The NO_x formation mechanism to be discussed will be in terms of homogeneous mixtures. But the basic concepts will also apply to diesel combustion. In those engines, there is a variation of fuel mixture ratios throughout the time and space variation of the combustion process, making it difficult to use simple mathematical models. The NO_x is formed in a diffusion flame, near stoichiometric conditions. However, the basic concepts are relevant, and the general principles apply to NO_x formation in CI engines as well.

The only oxide of nitrogen formed in significant amounts during the combustion process is nitric oxide, NO, but some of the NO formed is converted to NO_2 later in the exhaust system. Only the formation of NO will be discussed here. In reporting emissions measurements, however, NO_x is always calculated as if it is NO_2, that is, with a molecular weight of 46 kg/kmol.

NO formation is a well-understood process. At high temperatures oxygen and nitrogen react to form NO according to the following overall Reaction: 6.7:

$$N_2 + O_2 \rightleftharpoons 2NO \tag{6.7}$$

The equilibrium constant for Reaction 6.7 is written:

$$K_p = \frac{p_{NO}^2}{p_{N_2} \cdot p_{O_2}} = \frac{x_{NO}^2}{x_{N_2} \cdot x_{O_2}} \tag{6.8}$$

The value of the equilibrium constant K_p for Reaction 6.7 is shown in Table 6.1 as a function of temperature. The equilibrium concentration of NO is also shown for an atmosphere consisting of 98 % N_2 and 2 % O_2 at the given temperature.

Table 6.1: Equilibrium constant for Reaction 6.7, and equilibrium concentration of NO for an initial gas composition of 98% N_2 and 2 % O_2.

Temperature - K	K_p	x_{NO}
1500	$1.07 \cdot 10^{-5}$	$9.06 \cdot 10^{-4}$
2000	$4.00 \cdot 10^{-4}$	$5.24 \cdot 10^{-3}$
2500	$3.52 \cdot 10^{-3}$	$1.35 \cdot 10^{-2}$
3000	0.0149	$2.25 \cdot 10^{-2}$
3500	0.0416	$2.93 \cdot 10^{-2}$

Since K_p increases greatly with the temperature, Equation (6.7) shows that the most NO in relation to N_2 and O_2 is produced with the highest temperatures. On the other hand, it is expected that the equilibrium concentration of NO will be very small for temperatures < 1000 K.

In other words, if there is enough oxygen and nitrogen present, and if the temperature is high enough, chemical equilibrium says that a significant amount of NO will be produced. The flame temperature depends on the mixture ratio and combustion process and to a lesser extent the fuel as can be seen with sample calculations using the program "AFT.EXE". In combustion engines, it can normally be said that the temperature must be higher than about 1800 K for NO formation to be important. This temperature occurs in engines for only a short time after the combustion has occurred. This means that NO formation only takes place in combustion products, and only lasts a short time after combustion, as will be explained later.

Normally, it is assumed that after combustion occurs all the products achieve equilibrium immediately, with the exception of the NO. The oxygen which is needed for NO formation must therefore be found in the equilibrium combustion products. When one is concerned with the products of rich combustion, there is no oxygen in the ideal products. For combustion in air, there is always a large excess of nitrogen, and it is only the O_2 which limits the formation of the NO. As was shown in Chapter 3, on account of dissociation and chemical equilibrium there are some molecules containing oxygen, H_2O and CO_2 for example, which dissociate and thereby produce O_2 even in rich combustion products as well as with stoichiometric and lean mixtures. When this happens, there is adequate oxygen in the hot post-combustion gasses to be able to form NO, even in rich mixtures. Section 2.4 also shows that there are many reactions which occur among the combustion products.

When all of these reactions are considered and a calculation of chemical equilibrium is made, the composition of the combustion products is obtained. Figures 17.1 - 17.6 in Section 17.2 show the composition of the equilibrium products of combustion from octane (C_8H_{18}) as function of the temperature, pressure and fuel/air equivalence ratio, ϕ ($\phi = 1/\lambda$, where λ denotes the excess air ratio). These figures show that there is a large difference between the compositions at high and low temperatures. At the lower temperatures, the composition is similar to that of ideal products.

From an equilibrium point of view, NO formation then requires at least two conditions: 1: There must be sufficient oxygen present. 2: The temperature must be high enough. These two conditions can be used to explain the dependence of NO emissions on the fuel air ratio in gasoline engines. For low excess air ratio, there is not much oxygen in the equilibrium combustion products. NO emissions are therefore low for rich mixtures. When the excess air ratio is increased, the amount of oxygen increases in the equilibrium combustion products,

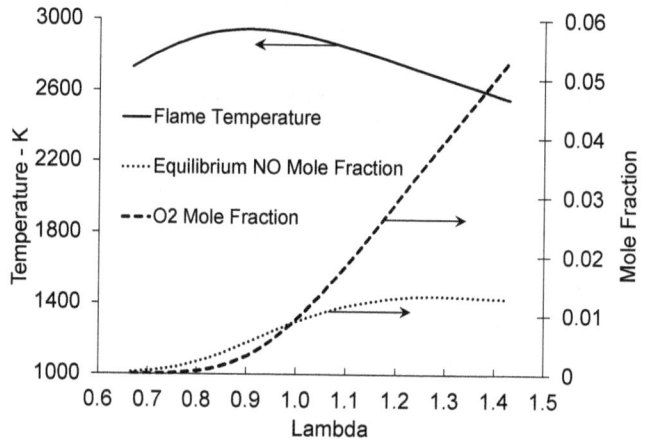

Figure 6.3: Equilibrium flame temperature, and equilibrium concentrations of NO and O_2 for constant volume combustion of iso-octane and air as a function of excess air ratio.

which results in a larger NO concentration. The combustion temperature for $\lambda < 1.0$ is not strongly dependent on the mixture ratio, as opposed to lean mixtures. (see Figure 3.4)

Figure 6.3 shows equilibrium flame temperature and concentrations of NO and O_2 for constant volume combustion of octane from an initial condition of 700 K and 1800 kPa. These conditions are similar to those encountered before ignition in a spark ignition engine. Since the combustion in an engine does not occur at exactly constant volume and is not adiabatic, the temperatures and pressures after combustion in an actual engine will actually be a bit lower. The tendencies with respect to changes in the mixture ratio will, however, be the same.

Figure 6.3 should be seen in the light of Reaction 6.7. The equilibrium formation of NO is driven by high temperature and high oxygen concentration. Figure 6.3 shows that the flame temperature is high for rich mixtures, but since the oxygen concentration after combustion is low, there is little NO formed. As the mixture becomes leaner, the flame temperature increases slightly toward the stoichiometric condition, and the oxygen concentration increases. At the maximum temperature, corresponding to a lambda value of about one, the NO concentration is not at its maximum since the oxygen concentration is low. Passing over towards leaner mixtures, it is found that the flame temperature starts to fall, lowering one of the driving forces for NO formation, but the oxygen concentration increases, raising the other driving force. The result of these two opposing trends is a maximum in the NO concentration for a lambda value of about 1.25. This is due to the rapidly rising oxygen concentration. The temperature remains high enough to give high NO concentrations until the mixture is leaner than $\lambda \simeq 1.25$. After this, the falling temperature becomes dominant, and the NO value falls as the mixture becomes leaner.

But as the equilibrium NO concentration is strongly dependent on temperature, it is to be expected that at exhaust conditions, there should be a lot less NO than at the peak temperatures and pressure. It is then of interest to compare the ideal NO concentrations at peak temperature to those at exhaust conditions, and finally to compare them to typical engine emissions.

Figure 6.4 shows the difference between the equilibrium concentration of NO at the flame temperature and at 1000 K and 1 atm. The latter condition is intended to illustrate the exhaust condition of a spark ignition engine. The y-axis in Figure 6.4 is logarithmic, and spans a range of concentrations from $10^{-2} \Rightarrow 10^{-12}$, a range of 10 orders of magnitude. From equilibrium calculations, one then expects that the NO concentration at 1000 K (equilibrium in the exhaust gas) will always be

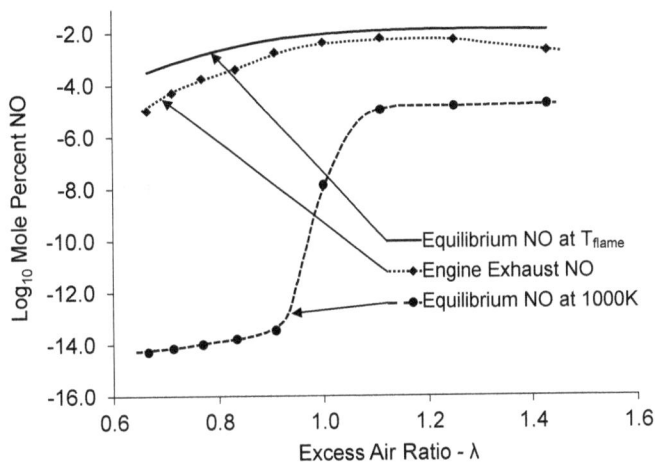

Figure 6.4: Equilibrium concentrations of NO at peak temperature and pressure, at exhaust conditions (1000K, 1 atm) compared to typical NO_x emissions from a spark ignition engine operating at full load.

under 10 ppm. In reality, the exhaust temperature in engines is usually lower than 1000 K, so this could be considered a high estimate.

Figure 6.4 also shows the measured concentrations of NO_x in the exhaust of a typical spark ignition engine. The concentrations of NO_x in the exhaust gasses are within a factor of 2 of the maximum NO concentrations at the equilibrium flame temperature. They are also at least 600 times greater than the equilibrium concentrations at 1000 K and 1 atm. It can then be seen that the NO emissions are closely related to the highest temperatures in the combustion process, and are much higher than equilibrium concentrations for the exhaust gas conditions. This requires an explanation.

The explanation lies in the speed of the reactions which lead to the maintenance of equilibrium in the gasses after combustion. These reactions happen on a time scale that means that equilibrium conditions for NO are only achieved a small amount of the time, if at all in combustion engines.

This can be illustrated with some measurements. Figure 6.5 shows measurements and calculations of the NO shortly after the passage of the flame in a closed constant volume combustion chamber burning with hydrogen [76]. The measurements were obtained by measuring the radiation given off by the NO molecules. It can be seen that the NO concentration increases exponentially shortly after the flame has passed the measurement location. This indicates that the NO concentration is increasing toward its equilibrium value in a short period of time after the flame passage. Other results show that all other components achieve equilibrium much more rapidly than the NO after the flame has past. Only NO differs from the other components with its relatively slow formation speed. This allows us to separate the NO reaction from all the other reactions occurring after combustion.

But these relatively slow NO reactions can also prevent equilibrium conditions from being satisfied when the temperature falls. That is, they limit the conversion of NO back to N_2 and O_2 according to Reaction 6.7. This can be seen in some calculations for a spark ignition engine, as shown in Figure 6.6. The calculations were for two different ignition timings: 1. where the start of heat release (trigonometric heat release function) location where the combustion occurs early, (30° BTDC) and the products of combustion are compressed to an even higher temperature as the flame travels throughout the combustion chamber, and 2: at timing where combustion occurs late (BTDC) and the pressure in the engine is decreasing, due to the movement of the piston away from the cylinder head and resulting expansion of the cylinder gasses.

Figure 6.5: The NO concentration as a function of time in the period after the flame in a constant volume bomb.

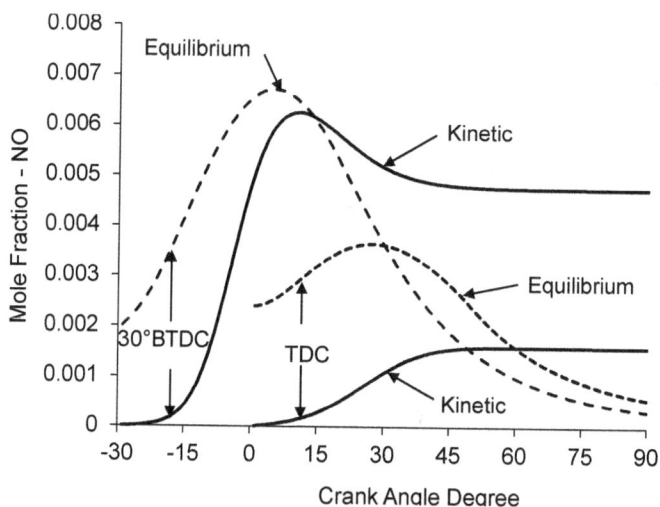

Figure 6.6: A comparison of the calculated NO concentrations at equilibrium conditions and from reaction kinetics controlled conditions for two timings for a spark ignition engine. Stoichiometric combustion, 1800 rpm, 8:.1:1 compression ratio.

For each timing, two different NO curves are indicated. The dashed lines represent the equilibrium concentration for the burned temperature throughout the combustion process. That is, the NO concentration if the conditions of thermodynamic equilibrium were always satisfied. The solid line indicates the calculated development of the NO concentration from the rates of the reactions determining the NO concentration, described in detail later in this section.

Initially, the measured NO concentration is lower than the equilibrium value. For the early combustion timing, after a slow start, there is a period where the measured NO concentration is close to the equilibrium concentration and begins to follow its decrease. But then the reaction rate slows down, the concentration no longer follows equilibrium, and eventually the reaction stops, leaving a concentration somewhere in the vicinity of the maximum equilibrium value. For the later timing, the temperature is low, and the reaction rates so slow that the kinetically determined NO concentration never follows the equilibrium curve, and the reaction has stopped before equilibrium has been achieved, leaving a

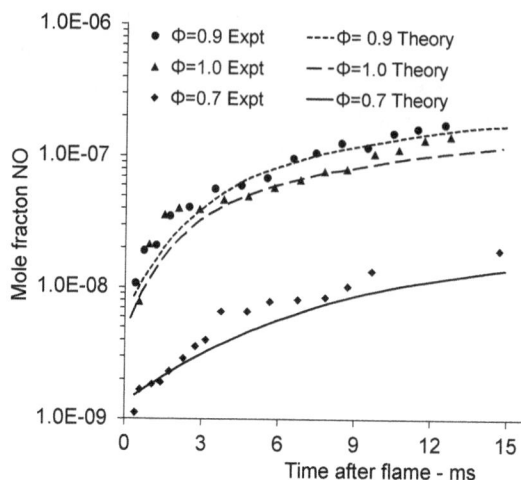

concentration of about half the maximum equilibrium value.

These results explain why the NO concentration in the exhaust gas is very close to the concentrations expected with the maximum cylinder temperatures. It is only at the highest temperatures that the NO forming reactions are fast enough to approach equilibrium conditions, and thereby form large amounts of NO. But as soon as the temperature falls, the reactions that reduce the NO are not fast enough to reduce the NO to the concentrations expected with equilibrium at the lower temperature.

This is a process where the NO concentration is "frozen" at high levels, since the reactions which are needed to maintain equilibrium conditions basically stop shortly after combustion. This is because the expansion is much too rapid and the temperature falls so much that the reaction rates become very slow. In order to maintain equilibrium during the expansion to exhaust conditions, the expansion would have to occur on a time scale much longer than on a scale of seconds. This is completely unrealistic to expect in an engine.

For lean mixtures there is significantly more oxygen present in the combustion products. Therefore, the NO emissions initially increase dramatically when the excess air ratio becomes greater than 1.0. But then when the excess air ratio becomes larger than 1, the flame temperature begins to fall, as was seen in the chapter on flames temperatures. Therefore, the NO emissions begin to decrease when the excess air ratio is increased beyond about 1.2. The maximum emission of NO from a gasoline engine occurs at an excess air ratio of about 1.1 (see Figure 6.4). As a first approximation, one obtains a reasonable estimate of NO emissions by calculating the equilibrium concentration right after combustion.

Since the NO formation process is determined by reaction kinetics, it is not sufficient to use equilibrium calculations alone for calculation of the NO formation process. Fortunately, the relevant chemistry is relatively simple. In many cases, it is sufficient to use only two elementary reactions, which constitute the so-called Zeldovich mechanism, named after the Russian scientist. The first version of the reaction is the so-called simple Zeldovich mechanism:

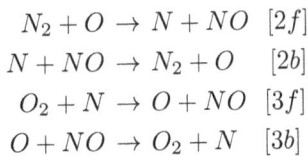

$$N_2 + O \rightarrow N + NO \quad [2f]$$
$$N + NO \rightarrow N_2 + O \quad [2b]$$
$$O_2 + N \rightarrow O + NO \quad [3f]$$
$$O + NO \rightarrow O_2 + N \quad [3b]$$

Especially for rich mixtures, it has been found advantageous to extend the mechanism to what is called the extended Zeldovich mechanism by addition one more reaction:

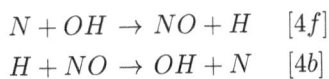

$$N + OH \rightarrow NO + H \quad [4f]$$
$$H + NO \rightarrow OH + N \quad [4b]$$

For T in K, the values for the reaction rate constants in $cm^3/gmol$-s are:

$$k_{2f} = 7 \cdot 10^{13} \exp(-37800/T) \tag{6.9}$$

$$k_{2b} = 1.55 \cdot 10^{13} \tag{6.10}$$

$$k_{3f} = 1.33 \cdot 10^{10} \cdot T \cdot \exp(-3600/T) \tag{6.11}$$

$$k_{3b} = 3.2 \cdot 10^9 \cdot T \cdot \exp(-19700/T) \tag{6.12}$$

$$k_{4f} = 7.1 \cdot 10^{13} \exp -450/T) \tag{6.13}$$

$$k_{4b} = 1.7 \cdot 10^{14} \cdot \exp(-24560/T) \tag{6.14}$$

For pressure in atmospheres, temperature in K and concentration in $gmol/cm^3$, the molar density can be calculated by using $R_o = 82.05$ atm-cm^3/gmol-K. The concentration of each substance is then its mole fraction times the total molar density.

Experiments have shown that in combustion systems, there is no NO formed before the flame occurs. Figure 6.7 shows the calculation of the development of the products of combustion just before and just after a fast combustion at high temperature and pressure. The calculation is for the compression ignition

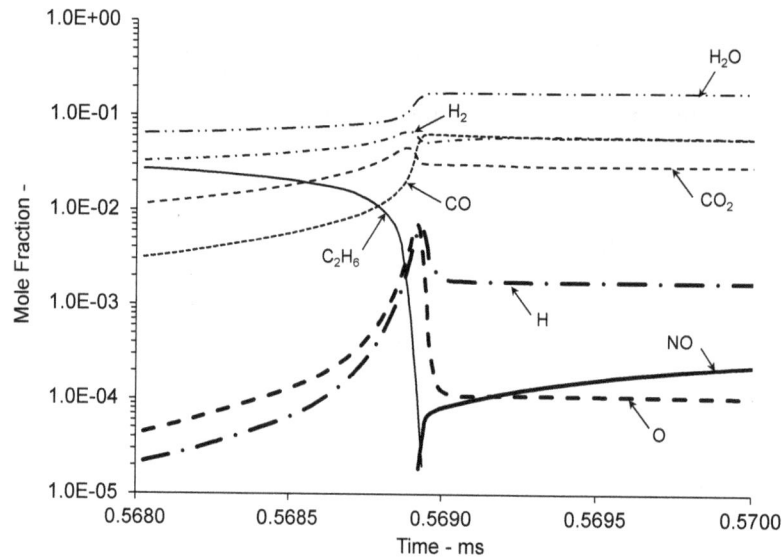

Figure 6.7: Calculated composition of the combustion gasses immediately before and after homogeneous charge compression ignition at high pressure and temperature. The pressure and temperature after the combustion are 190 atm and 3000 K, and the fuel air equivalence ratio is 1.40 and the engine speed 2000 rpm.

of a homogeneous rich mixture of ethane, air and residual gasses for a compression ratio of 21:1 at an engine speed of 2000 rpm. The point of ignition is at 6.8 degrees ATDC. The calculation is based on a detailed chemical reaction mechanism, including 130 elementary (molecular level) chemical reactions for 30 species. Except for the NO, all other combustion products attain equilibrium concentrations very rapidly, and remain essentially constant. The formation of the NO occurs on a time scale which is at least 10 times slower than all the other species. A small amount of the NO is formed in the combustion process itself (called prompt NO), but it is normally a small amount of the total. For leaner mixtures, it is even less important that for the rich mixture shown here.

In the calculation of the NO concentration one can, therefore, assume that all other components than NO attain their equilibrium concentrations immediately after the combustion. The NO concentration history is then calculated by using reaction kinetics to describe the speed of reactions (2), (3) and (4) in which the concentrations of all components other than NO are determined by an equilibrium calculation.

In most combustion systems, the temperature, and possibly the pressure, change rapidly, and in order to calculate the NO time history, one must be able to calculate the equilibrium concentration for all conditions. Since this is too complicated for teaching purposes, we will start with a simplified calculation method, which includes all of the essential elements in the more detailed calculation.

One assumes the following process: There is an instantaneous combustion, in which all gasses have the same state. It is also assumed that the pressure and temperature after the combustion are held constant, and that all species except the NO are at their equilibrium concentration. These can be obtained from the curves in Figures 17.1 - 17.6 in the Appendix, or through the use of a program that can calculate equilibrium compositions for example [9] or [77].

Since we know that reactions 2f, 2b, 3f, and 3b, 4f, and 4b are elementary reactions (involve actual molecular collisions), using chemical kinetic theory we can write:

$$\frac{d[NO]}{dt} =$$

$$k_{2f}[N_2][O] - k_{2b}[NO][N] + k_{3f}[O_2][N] - k_{3b}[NO][O] + k_{4f}[OH][N] - k_{4b}[NO][H] \qquad (6.15)$$

It is assumed that at time t = 0 there is a known concentration of NO which is called $[NO]_o$

If the pressure and temperature are maintained constant, all the equilibrium concentrations in Equation (6.15) are also constant. Since the rate constants for reactions $k_{2f}, k_{2b}, k_{3f}, k_{3b}, k_{4f}, k_{4b}$ are only functions of temperature, they are also constant.

Since the concentration of a species, i, in gmol/cm^3 can be found from:

$$[A_i] = x_i \frac{p}{R_o T} \qquad (6.16)$$

where x_i is the mole fraction. Equation (6.15) becomes:

$$\frac{dx_{NO}}{dt} = A - \frac{x_{NO}}{\tau} \qquad (6.17)$$

where:

$$A = \frac{p}{R_o T} \left[k_{2f} x_{N_2,e} x_{O,e} + k_{3f} x_{O_2,e} x_{N,e} + k_{4f} x_{OH_e} x_{N,e} \right]$$

$$\frac{1}{\tau} = \frac{p}{R_o T} \left[k_{2b} x_{N,e} + k_{3b} x_{O,e} + k_{4b} x_{H,e} \right]$$

and $x_{NO} = x_{NO,o}$ at t = 0.

All mole fractions except NO are given the subscript e (equilibrium) to indicate that they are the equilibrium values for the given pressure and temperature.

Equation (6.17) can now be integrated:

$$x_{NO} = x_{NO,o} \exp\left(-\frac{t}{\tau}\right) + x_{NO,\infty} \left[1 - \exp\left(-\frac{t}{\tau}\right)\right] \qquad (6.18)$$

where:

$$x_{NO,\infty} = A \cdot \tau = \frac{k_{2f} x_{N_2,e} x_{O,e} + k_{3f} x_{O_2,e} x_{N,e} + k_{4f} x_{OH_e} x_{N,e}}{k_{2b} x_{N,e} + k_{3b} x_{O,e} + k_{4b} x_{H,e}} \qquad (6.19)$$

and:

$$\tau = \frac{R_o T}{p} \frac{1}{k_{2b} x_{N,e} + k_{3b} x_{O,e} + k_{4b} x_{H,e}} \qquad (6.20)$$

The values in the rate constants are given in Equations (6.9) through (6.14)

Let us now examine some special cases of Equation (6.18). The first is for $t \to \infty$. In this case we obtain $x_{NO} \to x_{NO,\infty}$. That is, $x_{NO,\infty}$ is the concentration of NO which is reached after infinitely long time. It will be shown that $x_{NO,\infty}$ from the above is very close to the full equilibrium concentration of NO for the given pressure and temperature.

The second case is when $x_{NO,o} = 0$. This corresponds to the instant just after combustion occurs, if there is no NO formation in the flame. In this case:

$$\frac{x_{NO}}{x_{NO,\infty}} = 1 - \exp\left(-\frac{t}{\tau}\right) \qquad (6.21)$$

We see that the concentration increases with time as a first order system. This resembles Figures 6.5 and 6.7. The time constant can be calculated from the concentrations in Figures 17.1 - 17.6 in the 17.2 and the reaction rate constants for the relevant reactions.

Figure 6.8 shows τ as a function of pressure temperature, and mixture ratio. Initially, we note that τ is very strongly dependent on the temperature. A decrease in the temperature from 2500 K to 2000 K can reduce τ by a factor of about 1000. It can also be seen that τ increases when the pressure decreases and that τ is the lowest with lean mixtures for a given pressure and temperature. For a temperature of 1500 K, $\tau > 20$ min. This make is essentially impossible to cool combustion products slowly enough to maintain equilibrium NO concentrations in practical systems. The time constant falls for two reasons: firstly, the temperature reduces the values of the rate constants, and secondly the there is a decrease concentration of the reactants need to obtain equilibrium (O, H, and N) appearing the the denominator of τ.

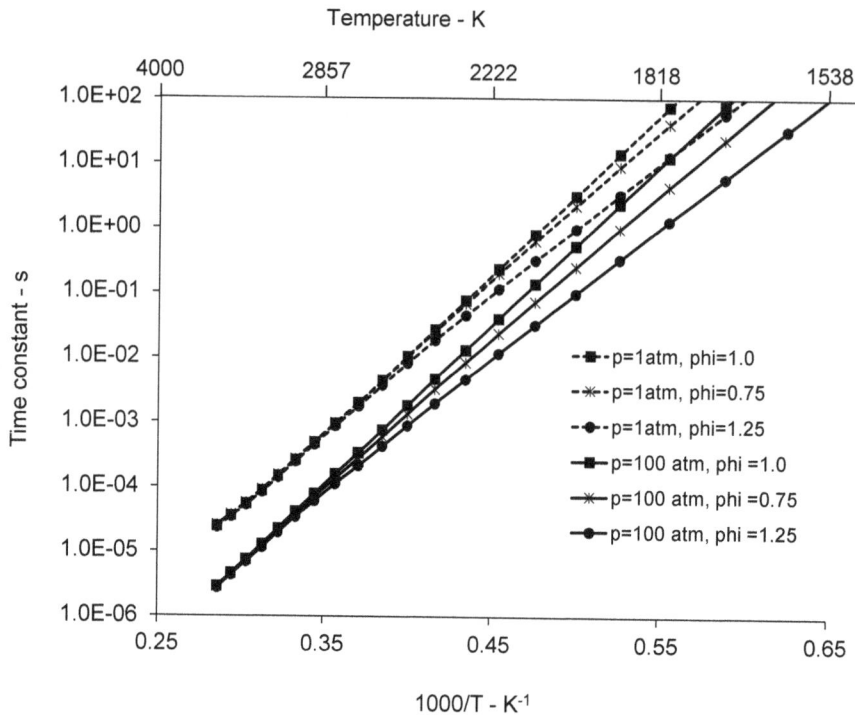

Figure 6.8: Calculated values of the time constant for the NO reactions as a function of temperature, pressure and equivalence, iso-octane was the fuel used to form the combustion products.

Example

Start with a system in which an instantaneous combustion occurs with the result that the time constant $\tau = 4$ ms and $x_{NO,\infty} = 0.002$ (2000ppm). Assume that the system is held at constant pressure and temperature for 5 ms after which the pressure and temperature increase instantaneously such that now $\tau = 2$ ms and $x_{NO,\infty} = 0.004$ (4000ppm) and are constant in the next 5 ms. Finally, assume that the pressure and temperature decrease instantaneously such that $\tau = 8$ ms and $x_{NO,\infty} = 0.001$ (1000ppm). Calculate the NO concentration after 15 ms.

Use Equation (6.18). For the first 5 ms $x_{NO,o} = 0$. Then we have:

$$x_{NO} = 0.002 \left[1 - \exp\left(-\frac{5}{4}\right)\right] = 0.00143$$

For the next 5 ms $x_{NO,o} = 0.00143$ and $x_{NO,\infty} = 0.004$ and we have:

$$x_{NO} = 0.00143 \exp\left(-\frac{5}{2}\right) + 0.004 \left[1 - \exp\left(-\frac{5}{2}\right)\right] = .00379$$

Notice that the time starts over at zero seconds, since the initial conditions have changed. Finally:

$$x_{NO} = 0.00379 \exp\left(-\frac{5}{8}\right) + 0.001 \left[1 - \exp\left(-\frac{5}{8}\right)\right] = .00249$$

The process is depicted in Figure 6.9. In the first 5 ms it is seen that the concentration increases from zero toward the final value of 2000 ppm, but reaches only 70 % of this value. In the next 5 ms, the NO concentration almost reaches the equilibrium value, since the time constant is smaller. Finally, in the third stage of the example, the time constant has become larger and the reaction proceeds more slowly, and after a total of 15 ms has passed, the concentration is still a long way from the low equilibrium concentration. With a further decrease in the temperature, the time constant will become so large that the reaction will basically stop. The general process can be seen to be the same as shown in Figures 6.6 and 6.7.

Even though the three steps shown in the above example are a rough approximation, one can use the concept of a series of small time steps with constant temperature, pressure and composition to describe a process with rapidly changing temperature and pressure. If the time steps are sufficiently small, the approximation will be quite good. Thus, the method can be used in connection with computer simulation of engines, and with a time step of one crank angle degree gives good results.

The calculation of the NO concentration in a practical combustion system requires that the pressure, temperature, and mixture strength throughout the entire system are known as a function of time for the period during which the NO reactions are active. This subject is beyond the scope of this text. Such calculations are often more complicated than the calculation of the NO chemistry itself. But computer programs are available which have been developed for this purpose for hydrocarbon fuels (see for example Reference [9]). With an understanding of the fundamental factors and processes which determine the development of the NO concentration, one is in any event able to make a qualitative evaluation of the effects of changes in the combustion system on NO emissions.

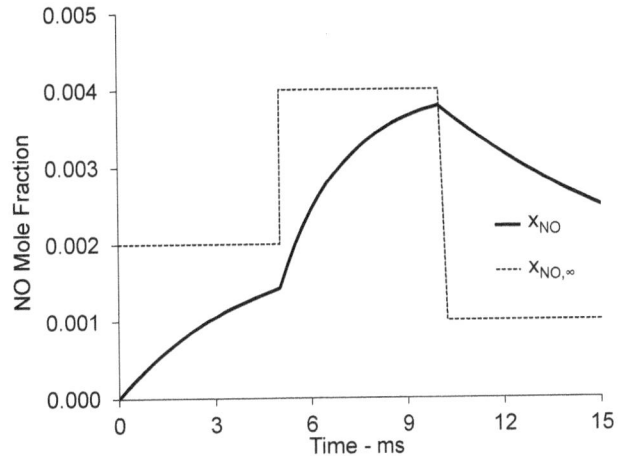

Figure 6.9: NO concentration as a function of time for the simple model in Example 1.

Since the formation of NO occurs in the short period of time just after the combustion, it is clear that if one wishes to limit the formation of NO, one needs to control the conditions in that period of time. For the first, the temperature must be held as low as possible. For the second, one can attempt to reduce the temperature of the combustion products as quickly as possible after the combustion in order to stop the NO reaction system before the concentration reaches its maximum value. This can limit the production of NO which, once formed, is difficult to remove again.

A method to reduce the NO emissions is simply to reduce the temperature during the entire course of the combustion. One of the most commonly used techniques for this is *exhaust gas recirculation*, EGR. The exhaust gasses do not contain any chemical energy, and therefore absorb a portion of the chemical energy released by the combustion and reduce the temperature (see Section 3.4). One can also inject water into the intake gasses of the combustion system, but this often involves a series of practical problems.

Lean combustion itself is actually a form of dilution of the combustion gasses. The excess air does not contain chemical energy either, and the flame temperature decreases directly when the mixture is made more lean. (see Section 3.4). But one must be sure that the mixture ratio is sufficiently lean that the reduced temperature is not counteracted by the increased oxygen in the combustion products. In practice, the mixture should be such that approximately, $1.25 > \lambda$ in order to achieve a significant reduction in the NO emission. It must be noted that λ in this case refers to the local mixture ratio in the combustion gasses during the time when the NO reactions are active.

There are also chemical and catalytic methods available for the reduction of NO_x emissions from exhaust gasses. It should be clear that the demands on these after-treatment techniques are reduced if the concentration of NO_x in the exhaust gasses is lowered by a well designed combustion system.

Example 2

Combustion of a stoichiometric mixture of octane and air occurs in a chamber with a constant volume without heat loss. The temperature and pressure after combustion are 2800K and 100 atm. Calculate the final NO concentration achievable after infinite time, and how much time is required to reach 80% of this value.

Since all components except for NO can be assumed to be in chemical equilibrium, their concentrations can be read from the figures in the Appendix Section 17.2 for $T = 2800\,K, p = 100\,atm$, and $\phi = 1/\lambda = 1$.

$x_{N_2,e} = 0.713$, $x_{O2,e} = 6.76 \cdot 10^{-3}$, $x_{O,e} = 4.47 \cdot 10^{-4}$ $x_{N,e} = 2.97 \cdot 10^{-7}$, $x_{H,e} = 4.90 \cdot 10^{-4}$, $x_{OH,e} = 5.13 \cdot 10^{-3}$

The reaction rate constants in cc/gmol-sec can be calculated to be: $k_{2f} = 9.60 \cdot 10^7$, $k_{2b} = 1.55 \cdot 10^{13}$, $k_{3f} = 1.03 \cdot 10^{13}$, $k_{3b} = 7.88 \cdot 10^9$, $k_{4f} = 6.05 \cdot 10^{13}$, $k_{4b} = 2.64 \cdot 10^{10}$

Then using Equation (6.19):

$$k_{2f} x_{N_2,e} x_{O,e} = 9.60 \cdot 10^7 \cdot 0.715 \cdot 4.47 \cdot 10^{-4} = 3.06 \cdot 10^4$$

$$k_{3f} x_{O_2,e} x_{N,e} = 1.03 \cdot 10^{13} \cdot 6.76 \cdot 10^{-3} \cdot 2.97 \cdot 10^{-7} = 2.07 \cdot 10^4$$

$$k_{4f} x_{OH_e} x_{N,e} = 6.05 \cdot 10^{13} \cdot 5.13 \cdot 10^{-3} \cdot 2.97 \cdot 10^{-7} = 9.21 \cdot 10^4$$

$$k_{2b} x_{N,e} = 1.55 \cdot 10^{13} \cdot 2.97 \cdot 10^{-7} = 4.61 \cdot 10^6$$

$$k_{3b} x_{O,e} = 7.88 \cdot 10^9 \cdot 4.47 \cdot 10^{-4} = 3.52 \cdot 10^6$$

$$k_{4b} x_{H,e} = 2.64 \cdot 10^{10} \cdot 4.90 \cdot 10^{-4} = 1.29 \cdot 10^7$$

$$x_{NO,\infty} = \frac{3.05 \cdot 10^4 + 2.07 \cdot 10^4 + 9.21 \cdot 10^4}{4.61 \cdot 10^6 + 3.52 \cdot 10^6 + 1.29 \cdot 10^7} = 0.00682$$

One can compare this value with the equilibrium concentration of NO from Figure 17.6 of about $6.5 \cdot 10^{-3}$ and see that there is good agreement.

The time constant τ is found from Equation (6.20):

$$\tau = \frac{(82.05 \cdot 2800)}{100} \frac{1}{4.61 \cdot 10^6 + 3.52 \cdot 10^6 + 1.29 \cdot 10^7} = 1.64 \cdot 10^{-4} \ s$$

Equation (6.21) is used to determine the time, t, such that:

$$\frac{x_{NO}}{x_{NO,\infty}} = 0.80 = 1 - \exp\left(-\frac{t}{\tau}\right)$$

$$\exp\left(-\frac{t}{\tau}\right) = 0.2$$

$$t = (-\tau) \cdot \ln(.2) = \left(-9.30 \cdot 10^{-3}\right) \cdot (-1.61) = 1.50 \cdot 10^{-2} s$$

6.3   SI Emission Control

Early emission control for SI engines used engine modifications to operate under lower emission conditions. Spark retard arrangements were used for lower NO_x and HC emissions. Retarding the spark reduces combustion temperatures and increases the exhaust temperature (see the comparison of the Otto and Diesel cycles in Chapter 2). The lower combustion temperature reduces NO formation, and the higher exhaust temperature promotes oxidation of the hydrocarbons in the exhaust gasses. These steps obviously had a detrimental effect of fuel economy. Exhaust gas recirculation was also used to reduce NO_x emissions, and changes in fuel air ratios and idle speeds were used to reduce CO and HC emissions, particularly at idle. Eventually emissions requirements became so strict that engine modifications alone were not adequate. Consequently, after-treatment of the exhaust gasses was used. The first of these strategies involved the use of an oxidation catalyst to reduce CO and HC emissions. NO_x requirements could be met by other means and oxidation catalysts operate effectively as long as there is some O_2 in the exhaust.

Figure 6.10: Typical performance of a noble metal catalyst in SI engine exhaust as a function of fuel air equivalence ratio. Adapted from [78].

Eventually NO_x emissions standards became strict enough that after treatment was also needed for them. The control technique made use of basically the same equipment as the oxidation catalyst, operated in a slightly different manner. Figure 6.10 shows the performance of a catalyst in the exhaust of an SI engine as a function of the fuel air equivalence ratio. When the mixture is lean, the oxidation occurs in the following general reactions:

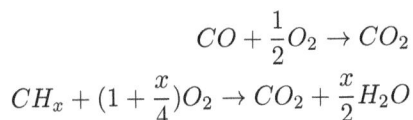

$$CO + \frac{1}{2}O_2 \to CO_2$$
$$CH_x + (1 + \frac{x}{4})O_2 \to CO_2 + \frac{x}{2}H_2O$$

Over 90 % of the HC and CO can be removed in this way.

When the mixture is rich, the catalyst promotes different reactions, since the exhaust now presents a low O_2 concentration, which gives a reducing environment. In this case, the NO_x is reduced through the following reactions:

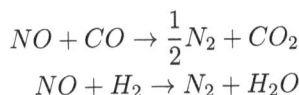

$$NO + CO \to \frac{1}{2}N_2 + CO_2$$
$$NO + H_2 \to N_2 + H_2O$$

The key to the success of the catalyst in SI engines can be seen in the operating region very close to the stoichiometric mixture ratio. Here, there is a significant reduction of all three emissions components at the same time, hence the name *three way catalyst*. This refers as much to an operating mode as to a piece of equipment. The fact that it can be used in practice is due to the principle of combustion with a homogeneous mixture ratio. Control of the intake mixture ratio at its stoichiometric value enables the three way operation. The three way catalyst cannot be used in a diesel engine, since operation there is always with a lean mixture ratio.

As Figure 6.10 shows, the allowable operating region for three way operation is not large, though operating with an oscillating mixture around the stoichiometric average widens the permissible range. Lambda must be kept constant to within a very few % of 1. This is only possible through the use of

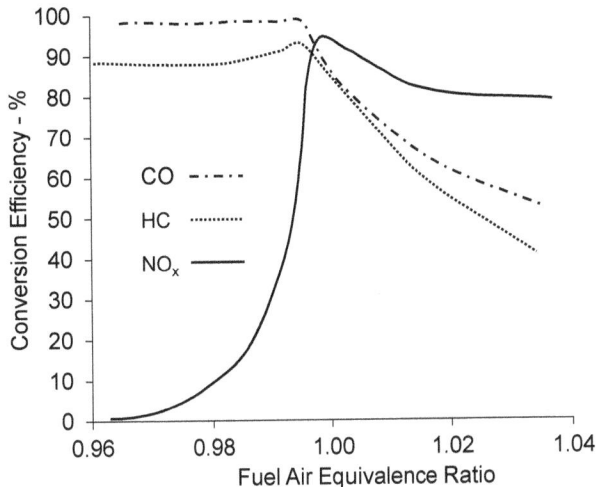

electronically controlled fuel injection and computerized control, with feedback control of the exhaust air fuel ratio. Systems for this are presented in Section 5.5.

The catalytic material consists of a mixture of precious metals, platinum, palladium and rhodium. Many other substances will perform as catalysts, but the precious metals have been found the best in terms of low temperature performance and durability. The catalytic materials are laid upon a substrate made of cordierite, which is normally extruded into a fine honeycomb structure. Substrates of coils of corrugated stainless steel have also been used, specially in applications where the catalyst is mounted close to the engine for better cold starting performance. A thin layer, called a wash coat, of highly porous Al_2O_3 is deposited on top of the substrate. The precious metals are in the pores of the wash coat, and as the exhaust gasses pass through the honeycomb structure, they diffuse to the surface where the reactions occur.

The use of the three-way catalyst has produced additional benefits. One is the removal of poisonous lead substances from fuel. The lead deposits in the exhaust rapidly block the pores of the wash coat and render the catalyst ineffective. Due to the precise control of fuel air ratio needed, significant improvement in engine control and the widespread use of electronic fuel injection have also occurred. Improved durability, better cold starting, and optimized engine tuning have resulted.

The catalyst becomes active or "lights off" at a temperature of about 250 - 300 °C. It is normally not the catalyst temperature that determines when the emission control is effective, but rather the engine temperature, which determines when the rich operation needed for good cold start can be terminated. When the engine is warm and the engine operating at stoichiometric conditions, the exhaust emissions are very low. For emissions testing, most of the pollutants are emitted in the first few minutes of testing while the engine is warming up and operating at non-stoichiometric conditions.

The emissions mechanisms mentioned for HC, CO and NO_x in the previous sections are valid for determining the input conditions to the catalyst, and low emission design from the start can result in a smaller catalyst or a longer durability. But the use of the catalyst has also allowed engines to be operated under higher "engine out" emission conditions, with an improvement in fuel economy compared to emission reduction with engine design and operating changes alone. The 3-way catalyst allows SI engine operation at emissions levels that are not attainable with engine modification alone, and has become a dominant technology for SI vehicle engines today.

6.4 CI Engine Emissions

6.4.1 CO and HC

Since diesel engines always operate with a lean fuel air mixture, CO and HC emissions are normally very low, with the possible exception of idle operation, where an adjustment in the fuel injection timing can often be enough to secure adequate emission reduction. CO emission from CI engines comes from conditions where the temperature is too low to complete oxidation. This includes very light loads, where the fuel is often "overmixed", so that combustion temperatures are low, and injection conditions where combustion occurs late on the expansion stroke, where oxygen concentrations are low as well as the temperature. Unintended after-injection of fuel can often give rise to higher CO emissions.

Since there is no premixed fuel and air in the diesel engine combustion system, there is no fuel stored in crevices, little wall quenching, and in general very low hydrocarbon emissions. It was found in early studies of emission control, that a significant portion of HC emissions in older engines came from the small amount of fuel in the sac volume in the nozzle. This is the small volume at the tip of the nozzle between the nozzle plunger and the combustion chamber shown in Figure 5.8. Reducing this volume and closing off injector hole has reduced this emission significantly. The latter is accomplished with the use of a Valve Closes Orifice (VCO) nozzle as shown in Figure 5.9. HC emissions are normally not considered much of a problem with CI engines. At most, to date, an oxidation catalyst might be used, which also reduces the amount of volatile substances adsorbed on the particulate matter, reducing particulate emissions as well.

6.4.2 Particulate Matter

The emissions of great importance for CI engines are those of NO_x and particulate matter, PM. Particulate emissions are what is commonly perceived as black diesel smoke, but are more complicated than that. With respect to emissions legislation, particulate emissions are defined as the mass that is deposited on a filter through which a volume of diluted exhaust gas has been passed. The particulate matter deposited on the filters has several components. The basis of the particulate matter is small carbon spheres formed during combustion, with diameters on the order of 50 nm, that agglomerate to form longer chains, or string-like structures, with a characteristic length on the order of $0.1 \rightsquigarrow 1.0 \mu m$. This structure can be

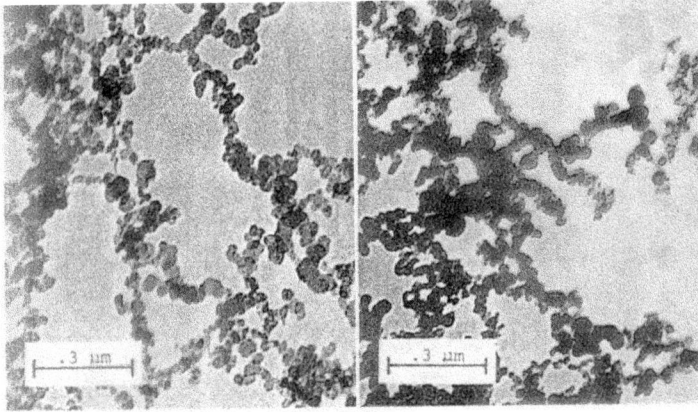

Figure 6.11: Electron microscope pictures of diesel particles. The picture on the left is of particles taken directly from the exhaust pipe, the picture on the right shows particles after the exhaust gas has been diluted with air. Photo courtesy Prof John H. Johnson, used by permission.

seen in electron microscope pictures of the particles in Figure 6.11. In the sampling procedures for engine emissions measurements, the exhaust gas is diluted with air before the measurements are made, in order to simulate the processes in the atmosphere. Since the particles have a large amount of active carbon, they absorb various components of the exhaust gas in the dilution process. The photograph on the right side of Figure 6.11 shows particles after they have been diluted, indicating a difference in the physical characteristics after dilution. In the dilution process, some of the vapor phase hydrocarbons are adsorbed on the particles.

The particulate matter consists, then, of more than just carbon. Vapor phase hydrocarbons, from both the fuel and the lubricating oil are found on the particles, there is an amount of sulfur, normally as sulfates, there is a small amount of metals from engine wear and fuel additives, and in the laboratory there is even some water. Measurement techniques require that the filters and samples be conditioned with air with a defined humidity range before weighing, in order to stabilize the amount of water on the filter samples.

When analyzing particles in detail, an extraction process is used, where a portion of the particulate matter is analyzed. This is called the *Soluble Organic Fraction*, or SOF. It contains a wide range of substances, including unburned fuel, lubricating oil and poly-aromatic hydrocarbons, or PAH. While detailed information regarding the composition of particles is very valuable in determining how they are formed and how to reduce them, the current measurement methods for legal certifications are based on only the mass of particles. Particle size has been raised as a measurement issue, as the smaller the particle, the deeper the penetration into the human respiratory system. There are several methods for measuring particulate matter size distribution, but none has been included in legislation as this is being written.

6.4.3 NO_x and Particulate Emissions in CI Engines

While these two emissions have different sources and are formed by different processes, it is relevant to discuss them together, since reduction of one normally leads to an increase of the other when dealing with hydrocarbon fuels in CI engines. The source of the NO_x and particulate emissions has already been indicated in Figures 3.22 and 3.25. The basis of particulate matter is the soot or solid carbon formed inside of the diffusion flame indicated in the figures, due to the combustion of the very rich mixture. The chemical composition of the reacting gasses inside the spray is very complicated. Some equilibrium calculations have been performed to determine the equilibrium concentration of solid carbon in reaction products of very rich mixtures. The results of these calculations are shown in Figure 6.12 for a H/C ratio of 2.0 for the fuel and a pressure of 100 atm.

According to the results shown in Figure 3.22, the combustion inside the spray is at an equivalence ratio of between 2 and 4 [28]. The equilibrium results show that there is a tendency to form solid carbon at these conditions, and that tendency increases as the mixture is richer. Raising the temperature helps to reduce the tendency to form soot. The adiabatic flame temperature for rich mixtures is not strongly dependent on mixture ratio, and is on the order of 2500 K in diesel engine combustion.

It was already seen that NO formation is very low at the equivalence ratios found inside the spray region. Also shown in the Section 3.6 was the structure of the diffusion flame, showing NO formation in a region near the flame containing O_2, with the flame itself occurring at a stoichiometric mixture ratio. That is, the fuel (in this case the rich products from the reaction inside the spray) and the oxidizer (the air outside

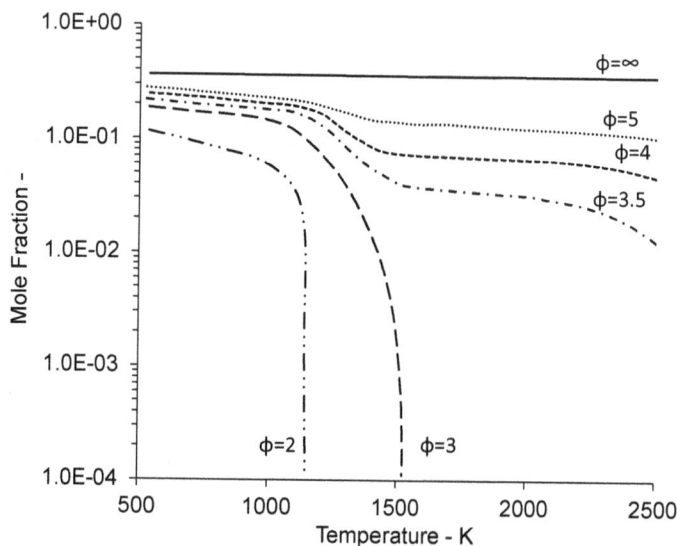

Figure 6.12: The equilibrium formation of solid carbon in the combustion products of a fuel with a H/C ratio of 2.0 at a pressure of 100 atm, as a function of equivalence ratio and temperature adapted from [79].

the spray) do not exist at the same time and place. The rich combustion products inside the spray include the soot formed in the first part of the combustion. The reactions between the fuel and air are very fast compared to the time it takes to mix the fuel and air, the classic diffusion flame assumption.

So the NO is formed at nearly the same location where the soot is burned up. To burn soot, a high temperature is desired. But that is precisely what is NOT needed to limit NO formation. Thus in CI combustion at the very fundamental level, a conflict arises between the two pollutants of primary concern - particulate matter and NO_x. This so-called NO_x-particulate trade off has been known for a long time, and has been a difficult problem to deal with in diesel engines. That it is fundamental can also be illustrated with some results for particulate and NO_x emissions in which the flame temperature was varied by different means, such as enriching the air with oxygen, EGR and other methods. The results are given in Figure 6.13 as a function of the reciprocal of the flame temperature. The flame temperatures range from about 2220 to 2900 K in these graphs. The two curves have directly opposite slopes, indicating the nature of the problem. If NO_x control is desired, the simplest method is to retard the combustion timing. This gives combustion later in the cycle, and as with the SI engine, lower NO_x emissions. But the lower temperatures slows down the combustion of the soot, and based on equilibrium considerations shown in Figure in 6.12, have the tendency to increase the amount of soot formed inside the spray.

An example of the effects of lowering cylinder temperatures in a diesel engine by retarding timing is shown in Figure 6.14. At the condition where the timing is most retarded, the injection starts 6° after top dead center. While NO_x reductions are possible, there are many disadvantages. All the other emissions increase. The increase in particulate emission is the most important, and its increase is in agreement with the general model described above. As expected from fundamental cycle analysis, the mean effective pressure decreases and specific fuel consumption increases as the timing is retarded for a constant fuel injection quantity.

As suggested by this and our experience with cycles, there is a correlation between the NO_x emission

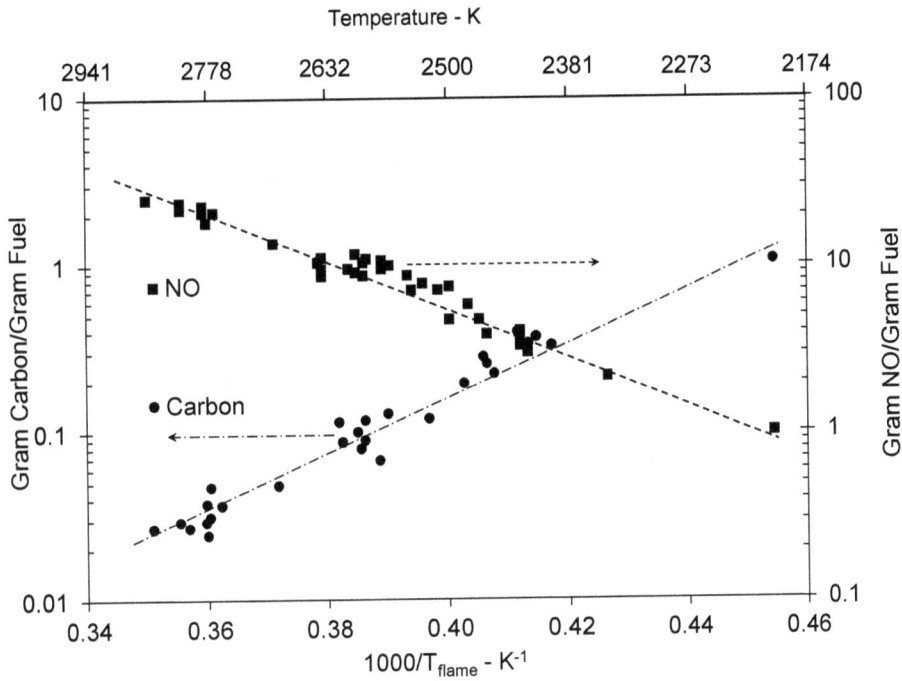

Figure 6.13: The variation of NO and soot emissions from diesel engines as a function of reciprocal flame temperature adapted from [80].

and the peak cylinder pressure for a given engine, with the NO_x emission increasing as the pressure rises, since this increases the temperature at which the fuel burns. This is shown for a naturally aspirated engine in Figure 6.15. The experiments covered a range of nozzle sizes and timings, as well as two different fuels, diesel fuel and soybean oil methyl ester.

The emission value plotted is gram NO_x per MJ energy released by combustion. This is used because the fuels have two different heating values and fuel air ratios. For diesel fuel, the stoichiometric fuel air ratio is 0.06869 , for the methyl ester it is 0.07951. The corresponding heating values were measured at 42.78 MJ/kg and 37.26 MJ/kg. The product of the stoichiometric fuel air ratio and heating values for the two fuels are nearly the same, at 2.940 MJ/kg-air for diesel and 2.962 MJ/kg-air. Then the two fuels should have the same adiabatic flame temperature, and since NO_x is formed after the fuel is burned, then the NO_x emission per energy release should be the same regardless molecular structure of the fuel.

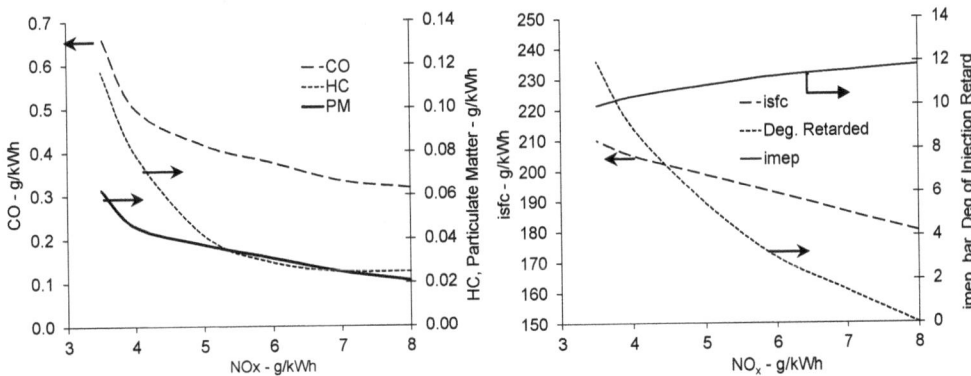

Figure 6.14: Emissions and performance of a heavy duty diesel as a function of NO_x emission level as the injection timing is retarded at constant speed. From the data of Reference [81].

This idea is consistent with the results shown in Figure 6.15, where the NO$_x$ emissions depend only on the maximum pressure for two chemically quite different fuels. This indicates that NO$_x$ emissions are due to combustion conditions and not fuel chemistry and the fuel itself should not have much effect on NO$_x$ emissions. Fuel effects would be indirect, that is the fuel effect on NO$_x$ emissions would only be due the the fuel's effect on the combustion rate. But there can be a big NO$_x$ advantage in using a fuel that does not form particles. Since DME, for example, does not form carbon particles inside the spray, it is possible to change combustion conditions to reduce NO$_x$ without the risk of forming particles. Thus the particulate emission problem is eliminated by using a fuel containing sufficient oxygen, though there are other practical problems to consider.

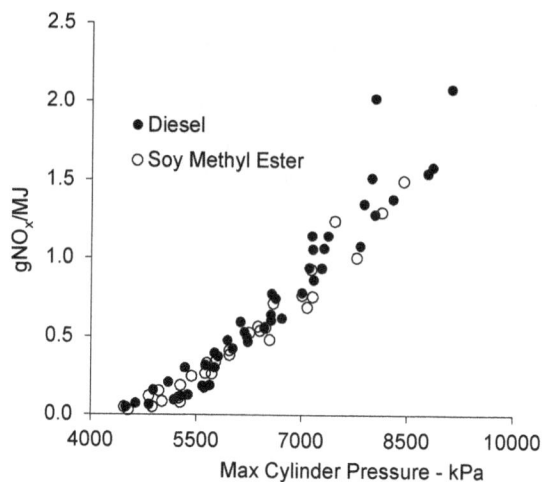

Figure 6.15: The NO$_x$ emission from a naturally aspirated DI diesel engine as a function of maximum cylinder pressure for diesel fuel and soybean oil methyl ester. Author's data, see also [52].

6.5 CI Emission Control

Some of the methods for emission control have been suggested in the previous material in connection with showing the effects of engine variables. Changing injection timing is one such possibility.

One way to improve the overall combustion process from the point of view of emissions is to increase the intensity of the injection process, that is, increase the injection pressure. There are several benefits to this. First of all, the higher spray velocity will break the fuel into smaller drops. This helps them to evaporate quicker, and so when the ignition occurs after a shorter delay because of the higher injection pressure, the mixture may not be as rich, leading to lower soot, based on equilibrium tendencies. The higher velocity of the spray will also pull more air into the spray, giving less rich conditions, and less tendency to form soot in the first place. In addition, the shorter delay can lead to lower rates of pressure rise at the start of combustion. This in itself can help to reduce NO$_x$ emissions, as it will result in lower temperature and pressure throughout the entire cycle, as seen in Figure 6.15. This allows more leeway to reduce particulate emissions. In recent years, injection pressures have increased greatly to help reduce emissions. Injection pressures on the order of 1500-2000 bar are not uncommon.

Pilot injection also can be used. This is easily accomplished through common rail system, where only a solenoid movement is require to give injection, and was shown in Figure 3.50. The small amount of energy released when the pilot fuel burned does not create significant NO, and the higher temperature into which the main portion of the fuel is injected reduces the intensity of the initial combustion, giving lower cycle temperatures and lower NO$_x$ emissions, again in accordance with Figure 6.15. Increasing the injection pressure with load, possible with the flexibility of the common rail system, keeps the injection duration from becoming too long, and preventing the combustion from being delayed too late on the expansion stroke to burn the last particles.

Detailed design of the combustion chamber, with respect to nozzle geometry, swirl, detailed piston crater design and such all have an impact on emissions. This discussion would go beyond the basic purpose of this text to concentrate on fundamental principles. There is a number of text books which go deeper into the details of the diesel combustion process, as well as a very large number of technical papers in the literature. It is intended that the discussion here will give a fundamental basis for understanding and evaluating these works on detailed combustion development.

6.5.1 EGR

Exhaust gas recirculation is used in SI engine to reduce flame temperature and NO_x emissions, and it can also be used in diesel engines. Figure 6.16 shows the emissions from two different technology levels for direct injection engines, a naturally aspirated engine, A, and a turbocharged engine with common rail system, intake throttle, pilot injection and exhaust gas catalyst B. In both cases, the NO_x emission is reduced by over 50% when an EGR in excess of 15% is used. In the naturally aspirated engine, the HC and CO emissions increase significantly with the EGR. This extra emission can be eliminated and actually reduced through the use of an intake throttle and a catalyst. The use of the throttle increases the fuel air ratio, giving a higher exhaust temperature, making the catalyst more effective. Note that by increasing the *overall* fuel air ratio, the flame is not affected that much, since diffusion flames occur at *locally* stoichiometric conditions. With EGR, the mixture is still stoichiometric at the flame, but the combustion gasses are diluted with the more inert exhaust products, lowering the local flame temperature.

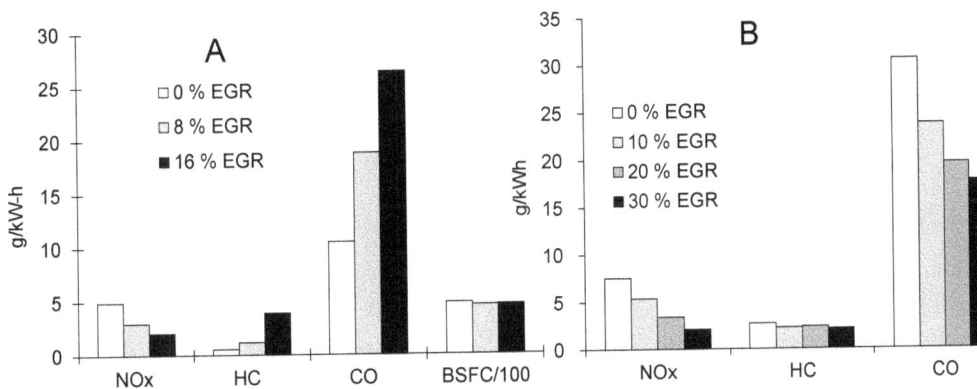

Figure 6.16: The effect of EGR on low load gas phase emissions from: A. A naturally aspirated DI engine operating on dimethyl ether without exhaust gas catalyst, B. A turbocharged, Common rail DI engine, with intake throttling and exhaust gas catalyst operating on diesel fuel.

So it is possible to lower the flame temperature and increase the exhaust temperature at the same time, though at the price of a small increase in pumping work.

The effectiveness of the catalyst can be seen by the results in Figure 6.16B for the engine with common rail injection and electronic injection control. In this case, the CO and HC emissions do not increase with the increase in EGR. When the EGR was increased for this engine, the intake pressure was lowered, and the exhaust temperature was found to increase. Particle emissions are known to increase as EGR increases.

6.5.2 Exhaust Gas Catalysts

Because diesel engines always operate with a lean mixture, the use of a three-way catalyst is ruled out. The most natural catalyst application is that of oxidation, since there is always oxygen present in diesel exhaust. Unfortunately, the substances that can be easily oxidized, HC and CO, normally do not present that great a problem, though modern diesel engines on small vehicles use catalysts to reduce the emissions of these substances. One area where there is some benefit in this regard is the oxidation of heavier hydrocarbons in the vapor phase. If these substances can be oxidized before they reach particles and become absorbed there, that will serve to reduce the mass of particulate emissions, and many catalysts are used on diesel

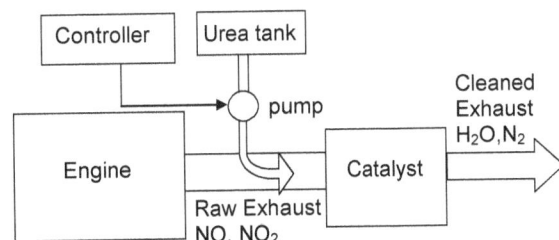

Figure 6.17: A schematic diagram of the selective catalytic reduction system used to reduce NO_x emissions in diesel exhaust.

engines for this purpose today. An added benefit, though not a legal requirement, is a reduction of odor by oxidizing some of the compounds that give rise to this problems. Oxidation catalysts, which are passively mounted in the exhaust gas system, have little effect on NO_x emissions. There can be a slight increase in the amount of NO_2 as some NO is oxidized.

There are applications where the use of catalysts in diesel exhaust will be significant. One of these is a process called *Selective Catalytic Reduction*, or SCR. A schematic diagram of an SCR system is shown in Figure 6.17. The idea here is to add a reducing agent to the diesel exhaust that can react with the NO_x in the presence of a suitable catalyst. Different substances can be used for this, including hydrocarbons such as unburned fuels, but the most effective agent is ammonia. The current systems use urea, which decomposes to form ammonia at temperatures above $200°$. The catalysts require a minimum operating temperature in order to be effective, and light load operation can be a problem because of the low exhaust temperatures that accompany the very lean mixtures. In steady state operation, SCR systems can reduce nearly all the NO_x emissions at an adequate load, but in transient operation on vehicles, lower results are expected, as there will be operation at very low load some of the time, and care must be taken under transient operation not to supply too much urea, otherwise unreacted ammonia could be emitted. Very low sulphur fuel is required for a system of this kind. EURO 5 emissions level diesel engines can require SCR technology, and urea solutions can now be purchased at many filling stations in Europe to satisfy this need.

6.5.3 Diesel Particle Filters

Even with the highest quality hydrocarbon diesel fuels, it appears that the diesel combustion process will still produce too many particles to meet future emissions standards. The use of alternate fuels containing over about 30% oxygen by mass can eliminate the particles formed in combustion, but the current engine and fuel industries prefer to work with hydrocarbon fuels and not to work for alternative fuels.

It is, therefore, likely that there will be the need for a filtration method to remove particulate matter from diesel exhaust. Filtration methods have been established for quite some time. A device used to physically filter the particles is called a *diesel particle filter* or DPF . There are different filter configurations, but the most common one is that called the wall flow filter. The filter is a structure consisting of a large number of small channels with walls made of a material with a porous structure. The channels have a large length to width ratio, and alternate channels are closed at alternate ends. The extruded ceramic structure is very similar to that of a catalyst, with the difference that alternate ends are closed, whereas all the channels are open with a catalytic converter. The principle of the filtration and a cutaway cross section are shown in Figure 6.18.

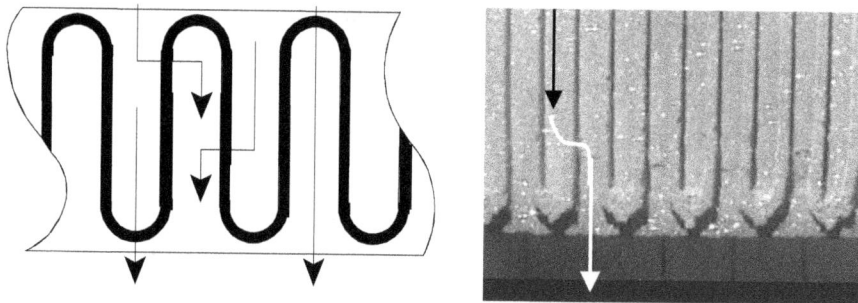

Figure 6.18: A schematic diagram of the principle of a wall flow diesel particle filter (left) and a cutaway photograph of the filter structure showing the closure of alternate flow channels and the flow path through the filter (right).

Particle laden flow enters the half of the channels that are open at the inlet end. Since these channels are blocked at the outlet end, the flow must pass through the walls in order to get to the other half of the channels that are blocked off at the inlet but free at the outlet. If the pore size is adequate, on the order of $20 \rightsquigarrow 40 \mu m$, the particles will be deposited on the walls of the filter. In filtration, it is only the very first particles that are stopped by the filter wall itself. A layer of particles is soon built up, called the filter cake, which does the filtration for the remainder of the process until the filter is cleaned. Pressure drop in the filters is important as the filter becomes loaded with particles. It can be calculated through Darcy's law:

$$\Delta p = \frac{v \cdot \mu \cdot t}{\alpha} \tag{6.22}$$

Where: Δp is the pressure drop, v the face velocity, that is the linear flow velocity into the surface of the wall, μ the gas viscosity, t the thickness of the wall, and α a property of the filtering material called the permeability. For sintered silicon carbide filters, the permeability is on the order of $2.1 \cdot 10^{-12} \ m^2 = 2.1$ Darcy, and that of cordierite about 0.65 Darcy.

The pressure drop in the particles is more difficult to estimate, but is very important as the collection process goes on. The problem is that the thickness of the particle layer is not known as the particles collect. One can assume a uniform layer of particles and estimate an equivalent thickness, δ_p:

$$\delta_p = \frac{m_p}{A_f \cdot \rho_p} \tag{6.23}$$

where: m_p is the mass of particles collected, A_f the effective surface area of the filter, and ρ_p the bulk density of the layer of particles on the filter. The surface gas velocity can be calculated from the continuity equation:

$$v = \frac{\dot{m}_{ex} \cdot R \cdot T_{ex}}{A_f \cdot p_{ex}} \tag{6.24}$$

Combining the above gives the pressure drop across the particle layer:

$$\Delta p_p = \frac{\dot{m}_{ex} \cdot R \cdot T_{ex} \cdot m_p}{\mu \cdot A_f^2 \cdot p_{ex}} \cdot \frac{1}{\rho_p \cdot \alpha_p} \tag{6.25}$$

The total pressure drop across a filter is the sum of the contributions from the filter material itself, Equation (6.22) and the particle layer Equation (6.25).

A combination of properties appears in Equation (6.25): the product of the permeability and the density. Measurements indicate that this product varies roughly linearly from about $0.7 \cdot 10^{-12}$ kg/m for low pressure drops across a filter to about $0.5 \cdot 10^{-12}$ kg/m with a large pressure drop of about 50 kPa [82]. The value appears to vary with the filter substrate material, but the above values can be used for a first estimate, and measurements used to adjust the values for specific cases. This calculation can be used to estimate how long an engine can be operated before the filter must be cleaned, or regenerated. This time is highly dependent on the emission level, and flow through the filter, that is the power and speed of the engine, but is on the order of magnitude of hours.

Clearly then, means must be provided to clean the filter to keep engine back pressure to an acceptable limit. This is the real problem with a DPF, as there are several acceptable filtration concepts. In principle, the particles can simply be burned off by supplying high temperature ex-

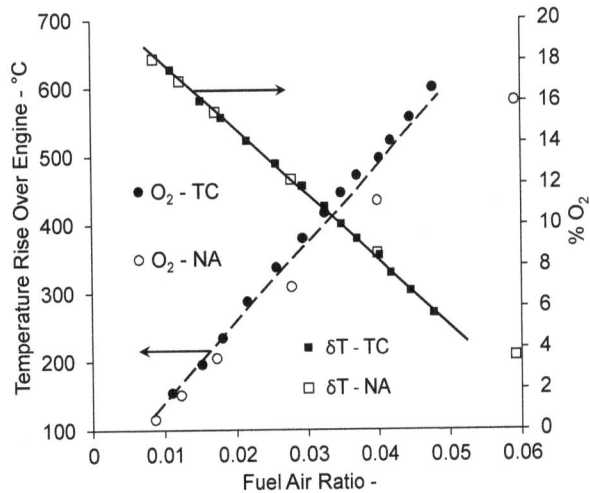

Figure 6.19: Exhaust temperature and oxygen concentration as a function of fuel air ratio for a turbocharged and naturally aspirated DI diesel engines. 2000 rpm, no EGR.

haust gas containing oxygen. But to burn the particulate matter, which consists of a large amount of solid carbon, the temperature needs to be on the order of $500 \leadsto 600°C$, and a diesel engine operates with a variable fuel air ratio. This means that there is a varying temperature and oxygen content in the exhaust. To get a high temperature in the exhaust, more fuel must be burned, that is the fuel air ratio must increase. But this reduces the oxygen content in the exhaust. Figure 6.19 shows this very clearly for an older naturally aspirated CI engine and a newer DI engine with turbocharging, common rail injection and EGR at light load. The temperatures will be affected by engine speed and design, but the same trend will be observed for all diesel engines.

To obtain an adequate temperature to burn the particulate matter off the filter, the engine must operate at close to full load, at lighter loads the particulate matter accumulates. This problem can be even more difficult in turbocharged engines, where because of the amount of air available, full load fuel air ratios are often lower than for naturally aspirated engines. It turns out that the limiting factor is the temperature. For diesel engines, the 3 to 5 % oxygen at full load is adequate to burn the particulate matter.

Since most vehicle engines do not operate at full load enough of the time to keep filters clean, other methods must be found. An immediate choice seems to be a catalytic coating on the trap, but that is not so effective, as only the particles at the interface between the trap wall and the filter cake are exposed to the catalyst. A more effective way to accomplish this is through the use of a fuel additive which leaves deposits intimately dispersed throughout the filter cake. These deposits, metal oxides, can reduce the oxidation temperature to the vicinity of $250 \leadsto 350°C$. Very small amounts of additives are used, and they leave deposits, but the technology has been shown to be applicable to some transport engines. The active metals used are either based on cerium or iron. Copper is very effective, but has not been used for fear of making dioxin.

One method that can be used with modern electronic controls is the periodic adjustment of the air fuel ratio to raise the temperature in accordance with Figure 6.19. An intake throttle can be closed when needed for a short time to restrict air flow, while maintaining fuel flow for power. This raises the exhaust temperature to where particles can burn, or at least ignite, and can be done when needed by monitoring the pressure drop across the filter. The method results in a modest fuel economy penalty.

A catalytic method is what is called the Continuously Regenerating Trap, or CRT. It makes use of the interesting characteristic that NO_2 acts as an oxidant for the particulate matter. Since the relative amount of NO_2 in the NO_x mixture is normally small, a catalyst upstream of a filter is used to convert the NO portion of the NO_x to NO_2, which then oxidizes the particles. There are temperature limitations with this system, as the exhaust gas must be hot enough for the catalyst to oxidize the NO. This temperature is in the vicinity of 300°C. The oxidation of the particulate matter with NO_2 is also temperature sensitive. The catalyst requires very low sulphur content of the fuel, on the order of 10 ppm. An area of concern is the generation of an increased amount of NO_2, since it is more harmful than NO, and an excess of NO_2 is needed to complete the oxidation of the particles. This application has seen fairly extensive use, and is particularly adaptable to engines that operate continuously with moderate to heavy loading. As with other regeneration methods, light load is problematic.

A final issue that should be raised concerning particle filters is the choice of material. For spark ignition engines, the material of choice for catalysts has been cordierite, a ceramic consisting of a mixture of three metal oxides in the proportions: $2MgO_2\ Al_2O_3\ 5SiO_2$. To make the structures for catalysts and DPF's the oxides are mixed in a paste, extruded and fired. Cordierite has excellent thermal expansion properties, and is the dominant material in SI engine catalysts. For DPF's there is a problem connected with unintended regeneration. It is possible for the particles to start to burn with an engine operating at a high power condition, and then experience a sudden drop in engine load and air flow. The regeneration process will continue, and without the flow to remove the energy liberated by combustion, trap temperatures on the order of 1400°C are encountered. This will melt a cordierite trap, and special measures are needed to avoid the catastrophic situation.

In an effort to make a DPF with a higher temperature resistance, silicon carbide (SiC) was introduced as a trap material, as it can withstand temperatures in excess of 1600°C [83]. SiC has less desirable thermal expansion than cordierite, but that can be accommodated with appropriate design. The filter consists of an extruded, fired monolith, similar to that of the cordierite, but as it is made of SiC particles, it has a more uniform pore structure. Figure 6.20 shows the structure of a Cordierite filter and an SiC. Because of the more uniform pore structure, SiC filters exhibit somewhat lower pressure drops under particle collection. SiC filters generally weigh more and cost more than cordierite filters. First introduced in 1993, SiC filters are now used on production vehicles for DPF's.

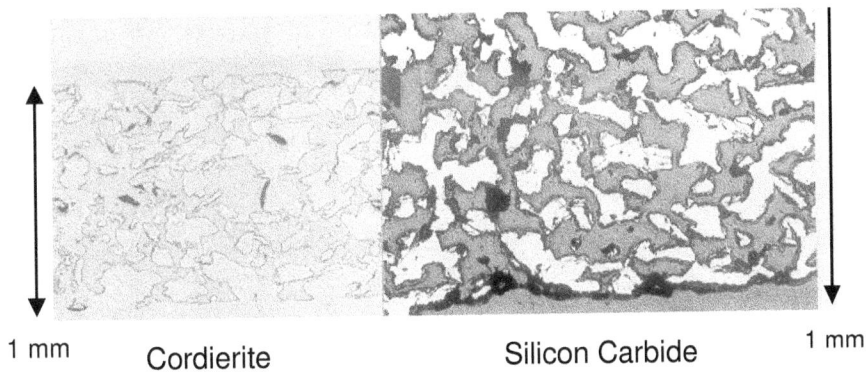

Figure 6.20: The structure of two materials for wall flow diesel particulate filters

Increased use of DPF's is anticipated with increasingly stringent emission requirements. Improvements in materials, designs and operating strategies will undoubtedly occur.

6.6 Emissions Test Procedures

For legal purposes, it is not enough to specify emission limits, it is also necessary to specify under which operating conditions and with which measurement techniques the emissions are to be determined. The first of these considerations is the test measurement equipment. For light duty vehicles, it is desired to measure the emissions under controlled, realistic driving conditions. This is done through the use of a chassis dynamometer, which is capable of producing a load of the same magnitude and speed dependence as a real vehicle. This can be done with a water brake, which as seen previously has the proper speed-load dependence, or an electric dynamometer, which can be programmed to give the proper behavior. In addition, inertial discs, or electrically simulated inertial load are used to simulate the energy required for acceleration and braking of a vehicle operating on the road.

6.6.1 Light Duty Vehicles

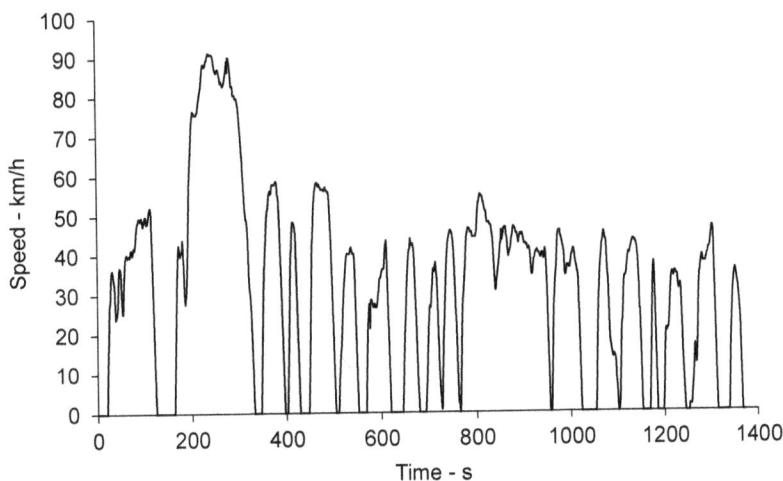

Figure 6.21: The FTP driving cycle used in the American emission test procedure for light duty vehicles.

With this equipment, the vehicle is operated on a specified driving cycle, which specifies the speed time profile through which the vehicle operates. There are currently two cycles used. The first cycle developed was the American FTP (Federal Test Procedure) cycle, based on speed time measurements of vehicles operating in traffic in Los Angeles. The cycle lasts 1374 seconds, covers 12 km, and has a maximum speed of 91.2 km/h, and is shown in Figure 6.21.

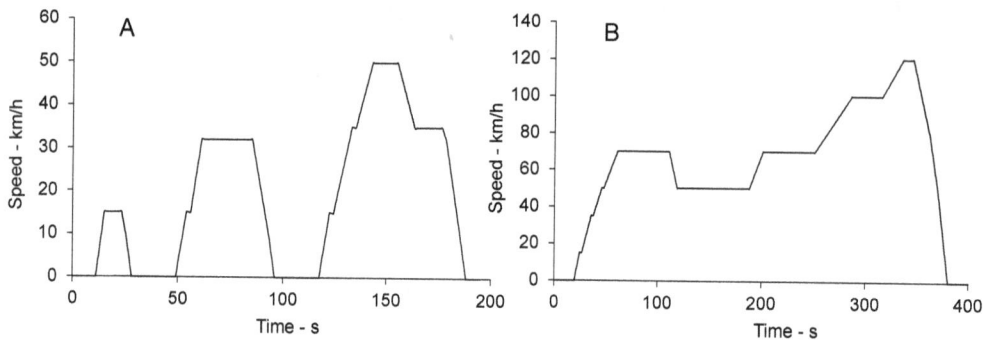

Figure 6.22: The driving cycle used in the European test procedure for light duty vehicles. The low speed portion, A, is repeated 4 times, followed by the high speed portion, B.

The European driving cycle consists of two portions. The first, lower speed portion called the Urban Driving Cycle simulates urban driving, and has been used since the start of European emissions regulations. It lasts 195 seconds, covers 1 km, and has a maximum speed of 50 km/h. In the early 1990's the cycle was expanded to include a higher speed portion, to simulate highway and motorway operation. It lasts 400 seconds, covers 7 km, and has a maximum speed of 120 km/h. The total procedure consists of 4 repetitions of the low speed portion from a cold (25°C) engine start, concluded with the high-speed portion.

The emission standards for light duty vehicles are given in gram/km, and so a method is needed to determine the total mass emitted over the above cycles. Since the engine operates over a range of speeds and loads, it is not sufficient to just measure exhaust concentrations, since the flow and emissions concentrations change independently. Rather than measure the instantaneous flow and emissions out of the engine, a method has been developed to give the same result, but requires no direct engine attachment or modification. The method is called the CVS (Constant Volume Sampler) method, and utilizes a device that maintains a constant flow. Either a positive displacement blower operating at constant speed or a venturi with choked flow can be used. The device is fed with a combination of atmospheric air and engine exhaust. When the engine produces a large flow, there will be less flow of air into the device, since the total flow is constant. This procedure produces a variable dilution of the engine exhaust. At the outlet of the constant flow device, a portion of the exhaust gases is withdrawn at a constant rate and stored in an inert plastic bag until the end of the test, at which time it is analyzed. A schematic diagram of the CVS system is shown in Figure 6.23.

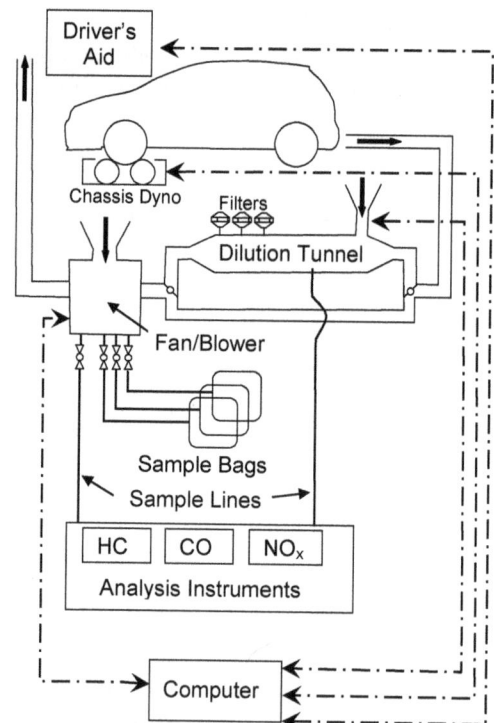

Figure 6.23: The instrumentation used for determining mass emissions from light duty vehicles, the CVS system.

Since the percentage of the exhaust gas in the mixture of air and exhaust leaving the CVS unit is proportional to the engine flow, the variable dilution "weights" the flow of exhaust, and the concentration of the pollutant in the mixture is determined by both the exhaust concentration and the amount of engine flow at this concentration. Thus, if the sampling flow is constant, the concentration of the pollutant in the bag at the end of the test is the same as that which would be measured from the total amount of diluted exhaust gas. The total grams of pollutant emitted are then calculated by multiplying the concentration in the sampling bag by the total amount of flow. The latter is known, since the constant flow device is calibrated and the flow held constant during the measuring period, the duration of which is known.

The European emissions standards for passenger cars are listed in Table 6.2 for passenger cars and in Table for diesel powered passenger cars. There has been a very great reduction in emissions from all vehicles since the first standards were issued in 1971.

6.6.2 Heavy Duty Vehicles

The emissions testing procedure for heavy duty vehicles is considerably different that that for light duty vehicles. A given heavy duty engine may be used in a variety of applications, with vehicles of differing sizes, transmissions, operating speeds and such. Some of these vehicle configurations are produced in very limited numbers, making it impractical to test the entire vehicle. This is opposed to the light duty vehicle, where vehicles are mass produced with fewer engine-chassis-drivetrain combinations. Given these conditions, the engines for heavy duty vehicles are tested separately, without regard to the application in which they will be used. The idea is that a clean engine will be clean in any application if the test procedure is relevant.

Initial heavy duty engine testing was performed on a steady state operating cycle, where the engine is operated at 5 loads each at rated speed and an intermediate speed, as well as three repeated idle measurements for a total of 13 engine test modes. Similar cycles were used in the United States and Europe, with different weighting factors for the respective modes. The emission rates in g/hr in each mode and the power in each mode are multiplied by a weighting factor, and the weighted flow rate divided by the weighted power gives a specific emission in g/kW-h on that cycle. The emissions in these units must be lower than legislated standards for the engine to be approved for use. There are different standards and time schedules for implementation of road vehicle and non-road engines.

It has been decided that a steady state test is not entirely representative of actual engine operation, and so transient test cycles have been produced in North America and Europe for heavy duty engines. Both have been developed by statistical analysis of operating mode distribution for operation of heavy duty road vehicles in an urban environment, on rural roads and on motorways. The engine cycles consist of long strings of operating points, in which the engine speed and torque are specified on a second-by-second basis. This includes negative torque to simulate the operation of the engine under vehicle braking. The measurement techniques requires a large dynamometer capable of both absorbing and delivering power to and from the engine, and with the ability to switch rapidly between the two conditions. Combined with the large outputs of heavy duty engines, this makes for an expensive device.

The European transient cycle from EU Directive 1999/96/EU for heavy duty engines is shown in Figure 6.24. The cycle is run in one unit for 1800 seconds, but is shown in the three segments which represent driving conditions in urban traffic (A), conventional highways (B) and motorways (C). The cycle is prescribed in terms of speed and load variations. The load varies between 100% of maximum torque and motoring, which is shown as -10% in Figure 6.24, while the speed varies between idle and maximum rated speed. The distribution of speeds and loads for the entire cycle is shown in 6.24D. Two

Table 6.2: Historical development of European emissions standards for light duty gasoline vehicles. Emissions in g/km on the European Driving Cycle. Emissions standards in g/km on prevailing ECE/EU test.

Standard	Year	CO	HC	NO_x
ECE 15	1971	39.7	$HC < 3.0$	-
ECE 15-04	1984	19.8	$HC + NO_x < 6.32$	-
91/441/	1993	2.72	$HC + NO_x < 0.97$	
EURO 2	1996	2.2	$HC + NO_x < 0.5$	
EURO 3	1997	2.3	0.2	0.15
EURO 4	2005	1	0.1	0.08
EURO 5	2008	1	0.075	0.06

Table 6.3: Historical development of European emissions standards for light duty diesel vehicles. Emissions in g/km on the European Driving Cycle. Emissions standards in g/km on prevailing ECE/EU test.

Standard	Year	CO	HC	NO$_x$	PM
ECE 15	1971	39.7	HC < 3.0	-	-
ECE 15-04	1984	19.8	HC $+$ NO$_x$ < 6.32	-	-
91/441/	1993	2.72	HC $+$ NO$_x$ < 2.72		1.14
94/12/ IDI diesel	1997	1	HC $+$ NO$_x$ < 0.7		0.08
94/12/ DI diesel	1997	1	HC $+$ NO$_x$ < 0.9		0.1
EURO 3	2000	0.64	HC $+$ NO$_x$ < 0.56	0.5	0.05
EURO 4	2005	0.50	HC $+$ NO$_x$ < 0.30	0.25	0.025
EURO 5	2008	0.50	HC $+$ NO$_x$ < 0.25	0.20	0.005

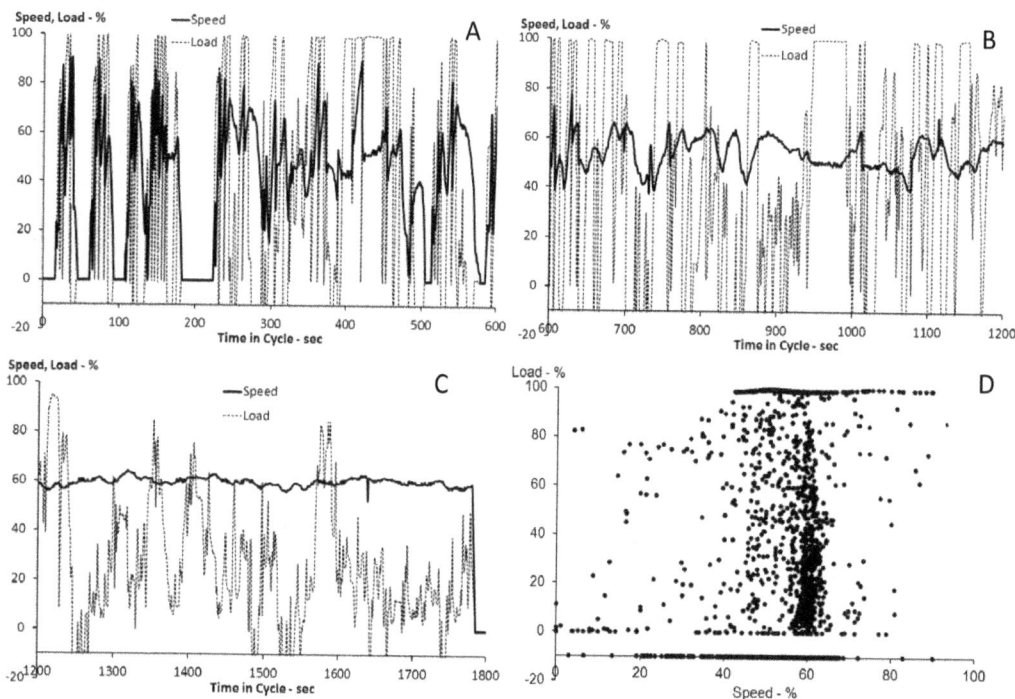

Figure 6.24: The transient heavy duty test cycle used for engine emissions testing in Europe.

variations are apparent in the T-shaped graph. Speed varies at nearly 100% load, which corresponds to acceleration through the gears up to a desired driving speed. At an intermediate speed, the load varies from motoring to 100%, which corresponds to load variations (for example up and down hills) while driving at a constant vehicle speed.

In addition to the transient cycle, European testing procedures include a steady state cycle, in which one of the engine test speeds can be freely selected between limits. This is to prevent specialized tuning of the engine to give good emissions performance at a known test speed, and to provide for emissions testing over a wider range of engine operating conditions. The emissions standards for heavy duty engines for road vehicle application are given in Table 6.4. Emissions standards were implemented later for heavy duty engines, but large reductions have been achieved to date, with further reductions on the way. The EURO 5 standards are expected to involve particle filters and some NO$_x$ after treatment for diesel fuel at the time of writing.

Table 6.4: Historical development of European emissions standards for heavy duty vehicles. Emissions in g/km on the European Driving Cycle. Emissions standards in g/kWh on prevailing ECE/EU test.

Standard	Year	CO	HC	NO_x	PM
EURO 1 < 85kW	1992	4.5	1.1	8.0	0.5
EURO 1 > 85kW	1992	4.5	1.1	8.0	0.36
EURO 2	1995	4.0	1.1	7.0	0.15
EURO 3	2000	2.1	0.66	5.0	0.1/0.13-small high speed engines
EURO 4	2005.	1.5	0.46	3.5	0.02
EURO 5	2008.	1.5	0.46	2.0	0.02

6.7 Problems

Problem 6.1

In order to calculate the NO formation tendency for a petrol engine, it can be assumed that the combustion occurs instantaneously, and that the combustion products are held at constant temperature and pressure for a number of milliseconds, after which an instantaneous expansion occurs to a condition where the NO reactions are stopped. Such a model can be used to evaluate the effects of a reduction in the temperature of the combustion products.

For the state of the gasses, assume that the combustion products for a mixture of octane and air have a temperature of 2500K, a pressure of 50 atm, and a fuel/air equivalence ratio ϕ of 0.75. Through the use of exhaust gas recirculation, the temperature is reduced to 2200K, while the pressure and mixture ratio can be assumed to be unchanged.

Assume that there are 5 milliseconds between the combustion and expansion, and calculate the reduction of the NO concentration obtained the lower temperature.

Problem 6.2

A diesel motor operates such that the NO_x emission corresponds to the following process: 1. Combustion with a stoichiometric mixture, giving a final pressure and temperature of 100 bar and 2300 K. 2. The combustion products are held at this condition for 8.2 milliseconds. 3. The pressure and temperature decrease instantaneously to levels where the NO_x reactions become infinitely slow. The engine produces 173 kW and has an exhaust gas volume of 0.188 Nm3/s. NO_x has a density of 2.055 kg/Nm^3. Calculate the emission of NO_x from the engine in gram/kW-hour.

Problem 6.3

A 4-cylinder, 4-stroke spark ignition engine has a cylinder diameter, B = 70.61 mm, a stroke S = 81.28 mm and operates at 2010 rpm. The engine has a volumetric efficiency of 0.787 with an intake pressure of 96.75 kPa and intake temperature of 35.6C. The torque is 79.55Nm. On basis of fuel and air flow measurements; an air/fuel ratio of 14.56 is found.

Exhaust measurements (on a dry volume basis) are: CO_2 = 14.17 %, CO = 1.06%, O_2 = 0.721%, UHC (as $CH_{1.85}$) = 736 ppm, and NO_x = 2932 ppm.

Calculate brake specific emissions of CO, UHC, and NOx in g/kW-h. (46.8; 1.61; 21.25). If it is assumed that the UHC consists of unburned fuel, $CH_{1.85}$ with a heating value of 44.5 MJ/kg, what portion of the ingested fuel is emitted as UHC? (0.59%). How much energy is found in the exhaust gasses in the form of UHC and CO? (0.332 kW, 2.20 kW). Calculate the air/fuel ratio from the exhaust gas composition. (14.45)

Chapter 7

Air Exchange in 4-Stroke Engines

7.1 Filling Efficiency

The amount of work performed during each cycle, and thereby the engine power produced at a given speed, depends on the amount of fuel burned in that cycle. Since combustion requires a much larger amount of air than that of the fuel on either a mass or volume basis, the amount of air drawn into the cylinder is very important in determining the power output of the engine. It is, therefore, important to carefully look at the intake and exhaust processes and their impact on the overall air exchange process.

For a 4-stroke engine the ideal intake flow of air can be written:

$$\dot{m}_{a,id} = \frac{\rho_{a,in} \cdot V_d \cdot N}{120} \tag{7.1}$$

In order to describe and compare the effectiveness with which an engine performs the air exchange process, it is practical to define a reference value for the air content of the engine cylinder at the start of compression.

One common reference value for air capacity is that corresponding to the condition where the ingested air fills the cylinder displacement with standard, atmospheric conditions. One can use the given atmospheric conditions, or refer to a "standard" atmosphere, as 760 mm Hg and 20°C, where the density of a standard atmosphere is $\rho_{st} = 1.20 kg/m^3$.

Using this reference, it is possible to define a filling efficiency for the engine as:

$$\eta_f = \frac{m_a}{V_d \cdot \rho_{st}} = \frac{\dot{m}_a \cdot 120}{\rho_{st} \cdot V_d \cdot N} \tag{7.2}$$

where: m_a is the actual mass of air drawn into the cylinder per intake stroke, and $V_d \cdot \rho_{st}$ is the reference mass, that is the ideal mass in the cylinder if it is full of air at the reference density.

In the description of the efficiency of the air charging process, different efficiencies can arise. The main difference between these is the choice of the reference density. When reading the literature, it is unfortunately not always obvious which reference density was selected. It is recommended that when using filling efficiencies for the air charging process in reports *etc.* it be made clear which reference condition was chosen for the ideal air charge.

For non-pressure charged engine, η_f will be less than one. This is partly because there is a pressure drop through the intake system and partly because the air entering the cylinder is heated by the intake passages and valves, lowering its density. Additionally, the mixing with the warm residual gas raises the temperature of the cylinder gasses before the valve closes. This does not affect the filling efficiency though, since it gives rise to a corresponding cooling of the residual gasses, and a corresponding reduction in their volume. When pressure charging is used, η_f will most often be greater than one.

The significance of the filling efficiency on the engine power is found in the following, by expressing the power in terms of the energy fed into the engine:

$$IP = \dot{m}_f \cdot H_u \cdot \eta_i \qquad (7.3)$$

$$= \dot{m}_a \cdot FA \cdot H_u \cdot \eta_i \qquad (7.4)$$

$$= imep \cdot V_d \cdot \frac{N}{60 \cdot 2} \qquad (7.5)$$

Using the definition of the fuel air equivalence ratio and Equation (7.5), it is found that:

$$imep = \eta_f \cdot \rho_{st} \cdot \phi \cdot FA_s \cdot H_u \cdot \eta_i \qquad (7.6)$$

As shown in Table 2.1, the term $FA_s \cdot H_u$ is approximately 3000 kJ/kg for most fuels, and using the standard atmospheric density, Equation (7.6) becomes:

$$imep \simeq 3600 \cdot \eta_f \cdot \phi \cdot \eta_i \qquad (7.7)$$

With the definition of the mechanical efficiency:

$$bmep \simeq 3600 \cdot \eta_f \cdot \phi \cdot \eta_i \cdot \eta_m \qquad (7.8)$$

for *imep* and *bmep* in kPa.

The advantage of pressure charging an engine is immediately apparent from Equation (7.6). Notice that it is applicable to any type of engine. The specific values will depend on the type of engine and operating condition, but it can quickly be seen that for an SI engine without pressure charging and with stoichiometric operation that the bmep cannot be expected to be much over 1000 kPa. ($\eta_f \simeq 0.8$, $\eta_m \simeq 0.85$, $\eta_i \simeq 0.4$). Diesel engines always operate with fuel air equivalence ratios < 1, so the maximum bmep of a non-pressure charged diesel engine will be lower, on the order of 750 - 800 kPa. Naturally, specific engines will differ, due to design changes, but normal engines of the above types will exhibit maximum bmep's in the vicinities mentioned here. The specific outputs of pressure-charged engines will depend on the amount of pressure-charging that the engine allows. See Chapter 12 for further examples.

Since it is possible for the filling efficiency to be greater than one, and in the case of highly pressure-charged engines greater than two, that efficiency does not seem to be very appropriate. Therefore, in this section a different efficiency will be used for the charging of 4-stroke engines. It will be called the volumetric efficiency, η_v.

This efficiency will be based on the intake conditions of the engine rather than atmospheric conditions, and the reference density then becomes the intake manifold density. The means of regulating the intake pressure, such as throttling or pressure charging, are taken into consideration in the density and not the efficiency. It is therefore expected that the volumetric efficiency should be somewhat lower than one for most operating conditions and types of engines.

Using this reference, it is possible to define the volumetric efficiency for the engine as:

$$\eta_v \equiv \frac{m_a}{V_d \cdot \rho_{in}} \qquad (7.9)$$

or, using flows:

$$\eta_v = \frac{\dot{m}_a}{V_d \cdot \rho_{in} \cdot \frac{N}{120}} \qquad (7.10)$$

As seen previously, the power produced by an engine is directly proportional to its air capacity. For a 4-stroke engine:

$$BP = \eta_v \rho_{in} \frac{V_d \cdot N}{60 \cdot 2} FA \cdot H_u \eta_i - FP \qquad (7.11)$$

If one considers a non-pressure charged spark ignition engine at full throttle, or a non-pressure charged diesel engine, the intake pressure will be close to atmospheric pressure, and the values of the filling efficiency and the volumetric efficiency will be close to each other. In many cases in the literature, the volumetric efficiency is based on the actual atmospheric density encountered during testing, but this definition will not be used here. Suffice it to say, that this usage is often found in the literature, and the reader should be aware that it can be found.

The relationship between the filling efficiency and the volumetric efficiency can be seen in Figure 7.1, which shows the intake and exhaust pressures, and the filling and volumetric efficiencies as a function of brake mean effective pressure for a conventional spark ignition engine. The filling efficiency and the intake pressure follow the same trend with load. As shown previously, the intake pressure change is responsible for the load change. That the curve is not precisely linear is an indication of changing thermal efficiency. To a large extent, the volumetric efficiency is constant, though there is a decrease at the lowest intake pressures. This will be explained later, but is related to the increasing difference between the intake and exhaust pressure. At the highest load, the intake pressure is nearly atmospheric, and the filling and volumetric efficiencies are nearly equal as would be expected from their definitions.

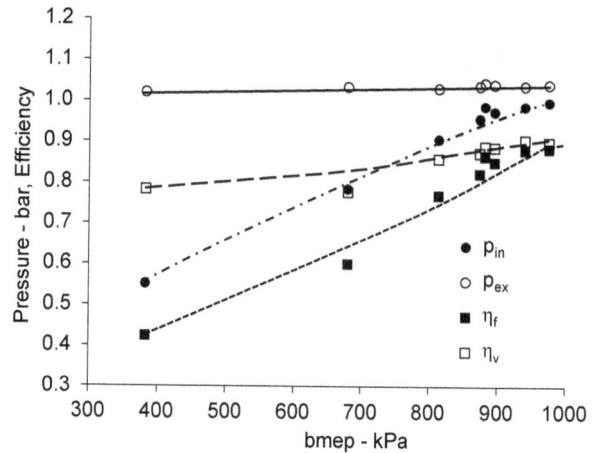

Figure 7.1: A comparison of the volumetric and filling efficiencies of a 1.1 liter 4 cylinder spark ignition engine operating at 2500 rpm as a function of *bmep*.

One way to look at this is to say that seen from the point of view of the atmosphere, the engine is less effective in terms of being filled with air as the load decreases. However, this is due to the throttling of the intake air, and when seen in terms of the intake air available at the intake port the engine has approximately the same effectiveness with respect to being filled with the air available in the intake manifold at any load (volumetric efficiency).

7.1.1 Effects of Engine Variables on Volumetric Efficiency

A simple analysis can lead to some relevant trends of the volumetric efficiency. This is basically the simple air exchange process from Section 2.2 looked at in a slightly different manner. Starting with Equation (2.76) and considering a total amount of heat transfer on the intake process, Q, then:

$$Q - \int_{TDC}^{BDC} p\frac{dV}{dt}dt = \int_{TDC}^{BDC} \frac{d(mu)}{dt}dt - \int_{TDC}^{BDC} \dot{m}_{in}h_{in}dt \tag{7.12}$$

Consider an intake process where the exhaust valve closes at TDC, then the intake valve opens also at TDC. There will be some flow back into the intake manifold when the intake valve opens, but it will be returned to the cylinder on the intake stroke. If it is assumed that this mass does not mix with the charge air, or undergo a heat transfer or pressure change, then the integral of the product of the mass flow times the enthalpy going out will be equal to the same value coming in, and therefore it will cancel out in Equation (7.12), and does not need to appear in the version to be used. Then integrating Equation (7.12), and assuming that the cylinder pressure is equal to the intake until the process stops at BDC:

$$Q - p_{in}\left(V_{BDC} - V_{TDC}\right) = \Delta\left(mu\right) - m_{in}c_pT_{in} \tag{7.13}$$

The following substitutions can be made:

$$\Delta\left(mu\right) = \Delta\left(mc_vT\right) = \Delta\left(\frac{mRT}{\gamma - 1}\right) = \frac{\Delta\left(pV\right)}{\gamma - 1} \tag{7.14}$$

$$m_{in} = \eta_v \cdot \rho_{in} \cdot V_d = \eta_v \frac{p_{in}}{RT_{in}}V_d \tag{7.15}$$

$$\Delta\left(pV\right) = p_{in}V_{BDC} - p_{ex}V_{TDC} \tag{7.16}$$

Equation (7.13) can then be solved for the volumetric efficiency of this idealized process:

$$\eta_v = 1 - \frac{\frac{p_{ex}}{p_{in}} - 1}{\gamma\left(\varepsilon - 1\right)} - \frac{\gamma - 1}{\gamma}\frac{Q}{p_{in}V_d} \tag{7.17}$$

Using an average heat transfer coefficient, $\bar{h}$, and assuming that the heat transfer is through an effective area, A and is due to the temperature difference between the engine coolant temperature, and the intake temperature,

$$\eta_v = 1 - \frac{\frac{p_{ex}}{p_{in}} - 1}{\gamma \left(\varepsilon - 1\right)} - \frac{\gamma - 1}{\gamma} \frac{\bar{h} A \left(T_{cool} - T_{in}\right)}{p_{in} V_d} \tag{7.18}$$

Intake Temperature

The above simple analysis shows that heat transfer to the gasses on the intake process reduces the volumetric efficiency if the coolant is warmer than the intake air. This is because the gasses at the end of the intake process are warmer than the intake gas, and thus have a lower density at the intake pressure. If the intake temperature is increased, Equation (7.18) indicates that the volumetric efficiency will increase, since the engine coolant temperature is typically around 100°C. At first thought, this seems contradictory, since we associate lower air mass with higher intake temperatures. The explanation is that the higher temperature intake gasses will be heated to a lesser degree, since the temperature difference to the engine surfaces is lower. The engine will, though, still have lower power when the intake air temperature is increased due to its lower density.

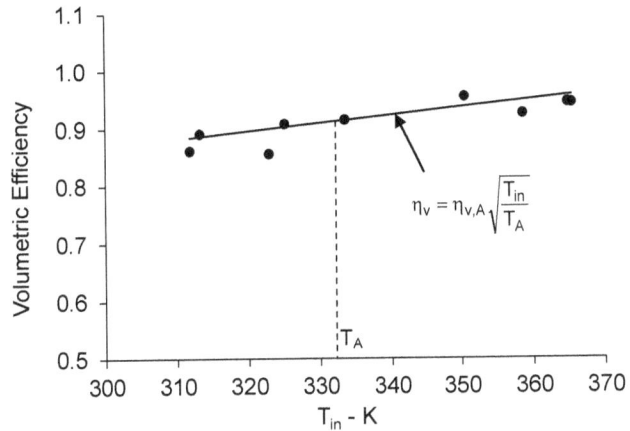

Figure 7.2: Volumetric efficiency as a function of intake temperature for a modern 1.9 turbochaged CI engine with charge air cooling at a constant speed of 2000 rpm.

The decrease in power will not be quite as bad as expected, since the volumetric efficiency will increase somewhat to compensate for it.

The actual amount by which the volumetric efficiency changes will depend on the specific engine. For typical engines, it is common to use the approximate relationship that:

$$\eta_v \propto \sqrt{T_{in}} \tag{7.19}$$

Figure 7.2 shows the variation of volumetric efficiency for a turbocharged direct injection diesel engine as the load is varied at constant speed. The ratio between the intake and exhaust pressure is nearly constant. The variation is close to that of Equation 7.19.

For the relationship in Equation (7.19), the relative change in indicated mean effective pressure for a change in intake temperature alone is:

$$
\begin{aligned}
\frac{imep_2}{imep_1} &= \frac{(\eta_v \cdot \rho_{in} \cdot FA \cdot H_u \cdot \eta_i)_2}{(\eta_v \cdot \rho_{in} \cdot FA \cdot H_u \cdot \eta_i)_1} \\
&= \frac{\eta_{v,2} \cdot T_{in,1}}{\eta_{v,1} \cdot T_{in,2}} \\
&= \sqrt{\frac{T_{in,1}}{T_{in,2}}}
\end{aligned}
\tag{7.20}
$$

If for a given engine it is known that the dependence on volumetric efficiency involves a different exponent than 0.5, the change for Equation (7.20) should be obvious.

Intake and Exhaust Pressure

Equation (7.18) also indicates that the ratio of the intake and exhaust pressures affects the volumetric efficiency. Figure 7.1 suggests this is the case, as the volumetric efficiency is lowest at the lowest intake pressure, in accordance with the trends predicted by Equation (7.18). That this is the case over a larger area is shown by Figure 7.3, which shows a plot of the change in volumetric efficiency as a function of the ratio of the exhaust pressure to the intake pressure. That is, the load decreases as this ratio increases. The volumetric efficiency has been normalized by dividing by the volumetric efficiency at full load at each engine speed, so the effect of speed variation has been reduced. The theoretical line shown is that from the first 2 terms of Equation (7.18), with $\gamma = 1.4$, and using the engine compression ratio of 9.5:1. The tendencies agree reasonably well.

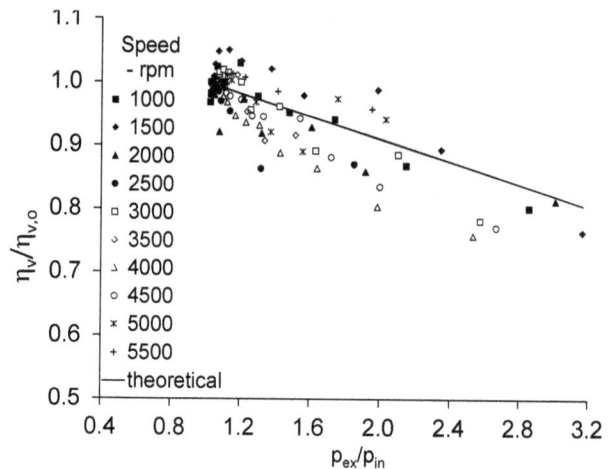

Figure 7.3: Relative changes in the volumetric efficiency of a 1.1 liter SI engine with exhaust to intake pressure ratio. As the load decreases, the ratio of exhaust pressure (roughly constant) to intake pressure (decreasing) increases.

The effect of the ratio of intake to exhaust pressures is normally only important in spark ignition engines operating at part load. Diesel engines operate with roughly equal intake and exhaust pressures, as a naturally aspirated engine operates with unthrottled intake, and turbocharged engines usually have intake and exhaust pressure close to each other. Though in later DI diesel engines for smaller vehicles, intake throttling is used in connection with EGR for emission control, so the variation might have some significance.

7.2 Displacement of Intake Air

The volumetric efficiency is defined on the air flow alone. In an operating engine, the intake gasses consist mainly of air, but they can contain other substances as well. This includes atmospheric humidity, vaporized fuel and possibly **E**xhaust **G**ases **R**ecirculation, (EGR), to the intake in order to reduce NO_x emissions. These gasses occupy space, and reduce the amount of air that can be drawn into the engine. If not cooled, they can also affect the intake temperature, the effects of which are seen in the previous section. This has an effect on engine performance, as the presence of gasses other than air reduces the amount of fuel that can be burned. The capacity of the engine to draw in gasses is not changed by the composition of these gases. That is, the volumetric efficiency *per se* is not affected. To establish the effects of air displacement in the intake manifold, the method from Taylor and Taylor [6] will be used, and expanded to include the effects of EGR.

The basic concept used is that of Dalton's Law, which states that the total pressure of a gas mixture is the sum of the partial pressures of all the gaseous components. In the case of the intake manifold, the total intake pressure can then be written

$$p_{in} = \sum p_i = p_{a,in} + p_f + p_w + p_{EGR} \tag{7.21}$$

The partial pressures are related to the mole fractions:

$$p_i = x_i \cdot p_{tot} \tag{7.22}$$

For an engine, $p_{tot} = p_{in}$, where p_{in} is the intake pressure as measured with a manometer.

The diluents referred to here, *i.e.* fuel, water, and EGR are normally measured in mass terms through the air fuel ratio, absolute humidity and percentage EGR. The definition of the percentage EGR used here is the mass flow rate of the recirculated exhaust gas divided by the total intake mass flow to the engine, including the EGR, that is the total flow at the intake valve.

$$\%EGR \equiv \frac{100 \cdot \dot{m}_{EGR}}{\dot{m}_{tot}} \qquad (7.23)$$

$$= \frac{100 \cdot \dot{m}_{EGR}}{\dot{m}_a + \dot{m}_{f,i} + \dot{m}_w + \dot{m}_{EGR}} \qquad (7.24)$$

Here, the subscript f,i refers to the amount of fuel evaporated in the intake manifold. The fuel vapor is regarded as an ideal gas and the volume of the non-vaporized fuel is assumed to be negligible.

According to Equation (7.22), the partial pressure of the air is determined by the mole fraction of the air in the intake gasses. The partial pressure of the air is used along with the temperature of the intake air to determine the density of the intake air, and thereby the ideal air mass flow according to Equation (7.1). As mentioned above, the parameters describing the other intake gasses are on a mass basis, and must be converted to a mole basis to find the mole fraction of the air.

$$x_a = \frac{n_a}{n_a + n_{f,i} + n_w + n_{EGR}} \qquad (7.25)$$

$$= \frac{m_a}{m_a + m_{f,i} \cdot \frac{M_a}{M_f} + m_w \frac{M_a}{M_w} + m_{EGR} \frac{M_a}{M_{EGR}}} \qquad (7.26)$$

$$= \frac{1}{1 + FA_i \cdot \frac{M_a}{M_f} + h \frac{M_a}{M_w} + \frac{m_{EGR}}{m_a} \frac{M_a}{M_{EGR}}} \qquad (7.27)$$

$$= \frac{1}{1 + FA_i \frac{M_a}{M_f} + h \frac{M_a}{M_w} + \frac{\%EGR}{100 - \%EGR} \frac{M_a}{M_{EGR}}} \qquad (7.28)$$

where h is the absolute humidity of the intake air in kg water vapor per kg dry air, and M_i's are molecular weights.

Then the ideal air flow to the engine is:

$$\dot{m}_{a,id} = x_a \cdot \frac{p_{in}}{RT_{in}} \cdot V_d \cdot \frac{N}{120} \qquad (7.29)$$

Using the value of the mole fraction of the air in the intake manifold, given by Equations (7.27) or (7.28) the power can then be written:

$$BP = \frac{\eta_v p_{in}}{RT_{in}} \cdot x_a \cdot \frac{V_d N}{120} \cdot H_u \cdot FA \cdot \eta_i - FP \qquad (7.30)$$

Similarly, the brake mean effective pressure is:

$$bmep = \frac{\eta_v p_{in}}{RT_{in}} \cdot x_a \cdot H_u \cdot FA \cdot \eta_i - fmep \qquad (7.31)$$

Note that the dilution effects only influence indicated work and are considered separately from the heating and cooling effects of the various substances. The most significant effect of the latter is probably the mixing with EGR gasses, which may not have been cooled to the intake temperature. The impact of the various components with regard to temperature is implicitly included in the intake manifold temperature. In the above, it is assumed that this value is known. For the mixing of gasses with different temperatures, a first law analysis of the intake gasses must be performed in order to establish the value of the intake temperature. That analysis is not considered here.

In many practical situations, Equation (7.30) is not even considered. To see where that would be satisfactory, consider some of the individual terms related to intake charge dilution. The first is that of the effect of the evaporated fuel in the intake system. Clearly, this term has no relevance to a diesel engine or gasoline direct injection engine, as the fuel is injected into the closed cylinder and not the intake. Therefore, $FA_i = 0$.

Consider a typical gasoline, with a stoichiometric fuel/air ratio of about 0.068. Gasoline is a mixture of many compounds so the molecular weight is in practice not normally known. Let us assume that the average molecular weight is approximately that of heptane, C_7H_{16}, which is $100 kg/kmol$. If all the fuel evaporates in the intake manifold, then $FA_i = FA_s$ and the fuel evaporation term becomes:

$$FA_i \frac{M_a}{M_f} = 0.068 \frac{28.9}{100} = 0.0197$$

Even if all the fuel is evaporated, then the volumetric effect is less than 2%. In actual engines, the fuel is not fully evaporated in the intake manifold and the effect even smaller, and thus normally not considered. This is especially the case with modern fuel injected gasoline engines with injection in the intake port. There are cases, however, where this is not true, one of the more important being a natural gas engine. Here, using the properties of methane at stoichiometric conditions:

$$FA_i \frac{M_a}{M_f} = 0.0581 \frac{28.9}{16} = 0.105$$

the situation is quite different, and a power reduction on the order of 10% from gasoline values can be expected for a natural gas SI engine operating with a homogeneous intake charge at stoichiometric conditions. Engines operating on LPG are between these two cases, the final effect being determined by the actual composition of the LPG. For SI engines with hydrogen or other gassifier products in the intake manifold, this term is even more important.

For diesel engines, there is no fuel in the intake manifold, and $FA_i = 0$ and there is no correction needed.

The effect of water can be illustrated by taking a warm, humid day. Assume that the dew point is 30°C, and the pressure 1 atm.. This gives an absolute humidity of about 0.0274 kg-water per kg-dry air. Then the correction term is:

$$h \frac{M_a}{M_w} = 0.0274 \frac{28.9}{18} = 0.044$$

The water correction would for most purposes be under 5%. It is generally considered to be important enough that standard engine test procedures (DIN/SAE) correct performance data to a standard humidity, in addition to temperature and pressure as shown in the following section.

The EGR correction is, of course, dependent on the amount of EGR used. The exhaust gas is approximately 75% nitrogen, and so its molecular weight is close to that of air, so the ratio of molecular weights of air and EGR is close to 1. Similarly, as seen above, the effects of the evaporated fuel and humidity are normally small, so the effects are due only to the amount of EGR used. In a gasoline engine using EGR, a power reduction will result only in the case where the throttle is wide open, that is $p_{in} = p_{atm}$ for a non-pressure charged engine. At part load, the effect of EGR on the intake process will be an increase in intake pressure with an increase in EGR, assuming a constant power. This will have a small effect on the engine friction/pumping loss, and well as some effects on the combustion itself. In SI engines, the combustion rate is normally decreased by the EGR, which has an opposing influence to the lower pumping loss.

7.3 Engine Power Correction

The above indicates that changes in atmospheric temperature, pressure and humidity will affect engine power during testing. In order to allow meaningful comparisons of engines tested under differing conditions, standards organizations have developed empirically based correction factors for engine power.

For spark ignition engines, the correction is of the form:

$$BP_{corr} = BP_{test} \frac{p_{ref}}{p_{test} - p_{H_2O}} \left[\frac{T_{test}}{T_{ref}} \right]^{0.5} \tag{7.32}$$

where p_{H_2O} is the partial pressure of the water in the atmosphere, and the subscripts refer to testing and corrected conditions. For the organizations EEC (European Union), ISO (International Standards Organization), JIS (Japan Industrial Standards) and SAE (Society of Automotive Engineers), the reference values are $p_{ref} = 99 kPa$ and $T_{ref} = 298K$. For the German (DIN) standards, $p_{ref} = 101.3 kPa$

and $T_{ref} = 293$K. Note that it is important to state the standards when referring to power, as the same engine test results will give a higher DIN power than the other standards.

For diesel engines, the DIN procedure is the same as in Equation (7.32) for the same reference conditions. The other organizations use a more complicated procedure. For temperatures i Kelvin the correction is of the form:

$$BP_{corr} = BP_{test}\alpha^{\beta} \tag{7.33}$$

For naturally aspirated engines, or engines with mechanically driven superchargers:

$$\alpha = \frac{p_{ref}}{p_{test} - p_{H_2O}} \left[\frac{T_{test}}{T_{ref}}\right]^{0.7} \tag{7.34}$$

For turbocharged engines, with or without charge air cooling:

$$\alpha = \left[\frac{p_{ref}}{p_{test} - p_{H_2O}}\right]^{0.7} \left[\frac{T_{test}}{T_{ref}}\right]^{1.5} \tag{7.35}$$

The exponent, β, is a function of a combination of engine parameters, here called K_β:

$$K_\beta = \frac{120\dot{m}_f}{V_dN}\frac{p_{atm}}{p_{in}} \tag{7.36}$$

$$\beta = 0.036K_\beta - 1.14 \qquad 40 \leq K_\beta \leq 65 \tag{7.37}$$

$$\beta = 0.3 \qquad K_\beta < 40 \tag{7.38}$$

$$\beta = 1.2 \qquad K_\beta > 65 \tag{7.39}$$

where $\dot{m}_f$ is in g/s, the speed in rpm, and the displacement volume in liters. The reference conditions are the same for both CI and SI engines.

Note that theoretically, this form of correction is not strictly correct, since Equation (1.31) says that it is the indicated power that is affected by atmospheric conditions, and that the mechanical efficiency is not truly constant. If corrections are small, though, the implicit assumption of constant mechanical efficiency is acceptable.

7.4   Effect of Volumetric Efficiency on SI Torque

The volumetric efficiency plays a very important role in determining the torque characteristics of a naturally aspirated SI engine. This can be seen simply through the following. It was seen in Equation (1.14) that the torque of an engine is proportional to the brake mean effective pressure, thus tendencies in *bmep* curves will be repeated in the torque curves. It was also seen earlier that the *bmep* of an engine can be simply expressed in terms of a few fundamental quantities:

$$bmep = \eta_v \cdot \rho_{in} \cdot FA \cdot H_u \cdot \eta_i \cdot \eta_m \tag{7.40}$$

Of interest in the following is the shape and magnitude of the *bmep*/torque curve as a function of speed at full load (WOT). This is important since it gives the maximum torque that can be transferred to the wheels of a vehicle, and is a fundamental part in determining the wheel tractive force and hence the acceleration characteristics of a vehicle, to be discussed in Chapter 15.

In Equation (7.40), the heating value of the fuel is constant, and since we are considering full load power, the intake manifold pressure should be close to atmospheric at all speeds, though it will drop a small amount at the highest speeds. For the current discussion, let it be assumed that changes in intake pressure with engine speed are negligible. The fuel air ratio is controlled by the engine management system, a computer with software with feedback provided by a lambda sensor. If the engine is operated with an operative 3-way catalyst at all loads, then the stoichiometric fuel air ratio is used in Equation (7.40), and is also constant. If not prohibited by emission restrictions, a vehicle will typically operate with an approximately constant fuel air ratio with a lambda value of about 0.8-0.9 at full load conditions to provide maximum power. In either case, it can be assumed that the fuel air ratio at wide open throttle is constant. That is sufficient, since it is only variational trends that are of interest here.

With the above assumptions, there are three efficiencies left to determine the shape of the engine torque curve: the indicated thermal efficiency, the mechanical efficiency and the volumetric efficiency. Since the discussion concerns full throttle performance, there is little difference if the filling efficiency is used instead of the volumetric efficiency.

Figure 7.4 shows the relevant efficiencies as a function of engine speed at wide open throttle. The engine operates at a constant air fuel ratio of 12.7 for maximum power.

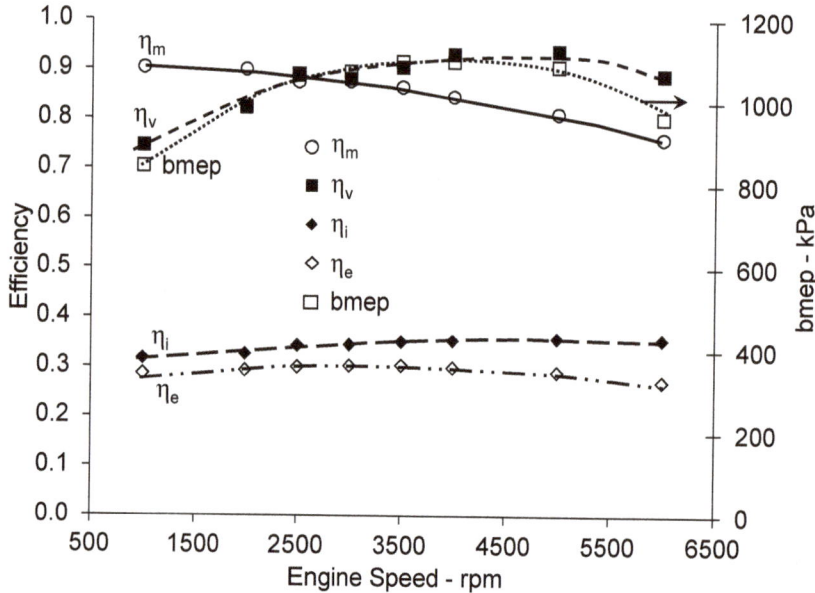

Figure 7.4: Mechanical, indicated, brake and volumetric efficiency and *bmep* as functions of engine speed at full load for a modern 1.8 liter SI engine.

First, the mechanical efficiency falls as the speed increases, primarily since the *fmep* increases nearly linearly with engine speed, while changes in *imep* are small, so a larger portion of the engine power is consumed in friction as the speed increases. The indicated thermal efficiency, on the other hand, increases as the engine speed is increased because the portion of fuel energy that is lost to the coolant during the combustion cycle decreases as the engine speed increases. This is because the heat transfer coefficient is proportional to the engine mass flow raised to a power less than one (Section 10.3).

Then the effects of speed on these efficiencies nearly balance out, as shown by their product, the brake thermal efficiency, η_e, which is nearly constant up to a speed of about 4500 rpm, after which it falls, as friction effects become more important. From Equation (7.40) then, we expect that the *bmep* and the torque should be nearly proportional to the volumetric efficiency. Figure 7.4 shows this to be the case, with some divergence at the high speeds.

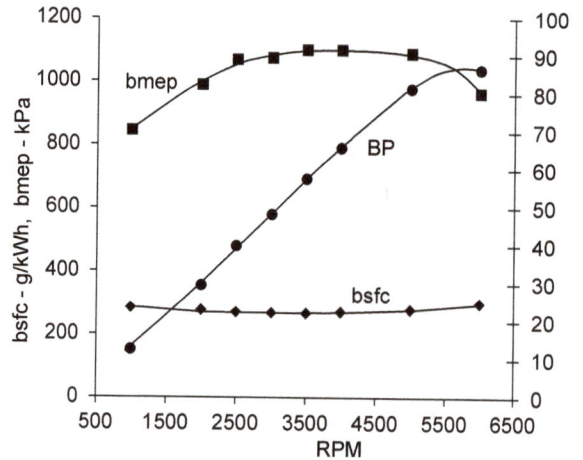

Figure 7.5: Brake power, mean effective pressure and specific fuel consumption as a function of engine speed at full load for a 1.8 liter SI engine.

The *fmep* is primarily determined by engine design choices, and will have a nearly linear increase with speed. Most engines have similar basic designs and *fmep* (see Section 9.7), so options to affect the torque curve are limited. Similarly, for an optimized engine operating at near MBT timing, the effects of heat transfer are determined primarily by physical phenomena that occur beyond one's control. To determine the torque curve for a naturally aspirated engine, then the primary design parameter available is the cam timing. By adjusting the cam timing, it is possible to affect the nature of the torque curve, and importantly, the engine speed at which the maximum torque is obtained. This is very important for vehicle performance characteristics, which will be shown later. In pressure charged engines, the behavior of the turbocharger/supercharger and its control are more responsible for the shape of the torque curve, since the intake density in Equation (7.40) can no longer be assumed constant.

The performance curves for the engine in Figure 7.4 are shown in Figure 7.5, where *bmep* is plotted instead of the torque. The engine is intended for vehicle use. It is common for passenger car SI engines to have maximum torque at about half the maximum speed. In the case of an industrial engine, it could be preferable to have the maximum torque at a different speed. How the torque curve is affected by the valve timing is shown in the following.

The power curve as a function of speed is determined by the torque curve, as the power is proportional to the product of the speed and the torque. The power appears nearly linear with respect to speed except at the highest speed, where the curve falls off due to the quadratic dependence of the friction power on engine speed. Often, the volumetric efficiency also drops at the highest engine speeds due to high fluid friction, and it is often the case that the maximum engine power occurs at a speed lower than the maximum engine speed. Mechanical limitations normally determine the maximum engine speed.

7.5   Cam Timing and Volumetric Efficiency

In Section 2.2, the air exchange process was considered to be ideal. The valves were assumed to open precisely at the top and bottom dead center positions, and the cylinder and manifold pressures were assumed to equalize instantaneously. The only variation was in the intake valve opening, which in a certain case was delayed until the cylinder pressure and the intake manifold pressure were equal.

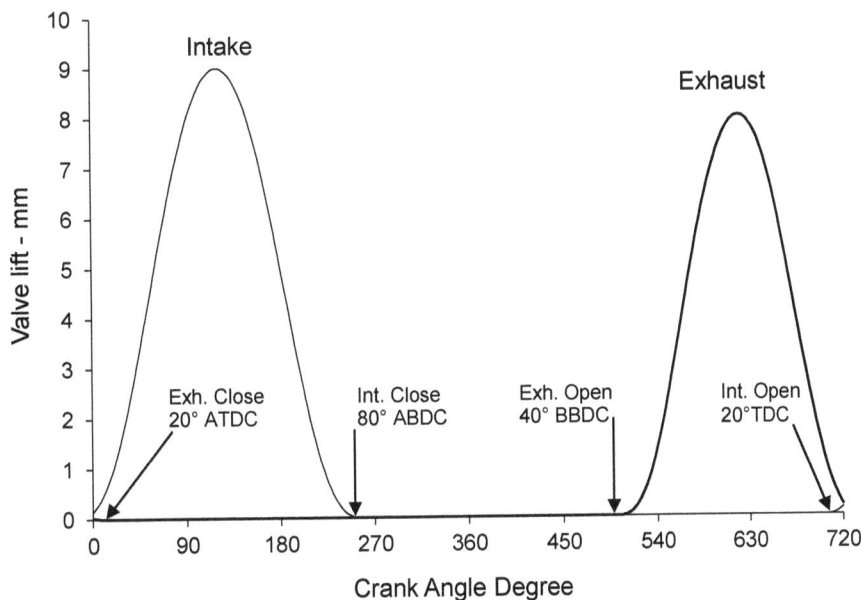

Figure 7.6: An example of valve profiles for a spark ignition engine with a bore of 80 mm.

Ideally, the intake valve would be timed to be fully open at TDC and close at BDC, while the exhaust would open fully at BDC and close at TDC, the instant before the intake valve opens. Physically, such processes cannot occur, since the discontinuity in the motion gives infinite accelerations and infinite forces. In most operating engines, the poppet valves are opened by cams, and the cam shapes are carefully designed to limit accelerations such that the forces encountered in moving the valves are held to an acceptable level. While the detailed shape of the cam is extremely important for force determination, it is much less important in terms of the engine performance, where opening and closing times and maximum lift play a much larger role.

Figure 7.6 shows an example of valve lift profiles for an SI engine. The slow lift at the opening, and the slow closing are evident, their purpose being to reduce the physical loading on the camshaft. As a result, the valves are opened before the ideal opening times, and closed after. This is because opening is defined as the time when the valve area is no longer zero. If the opening and closing were precisely at TDC and BDC, then the valves would not be open sufficiently at the time when the instantaneous piston velocity is high, and a large flow is needed.

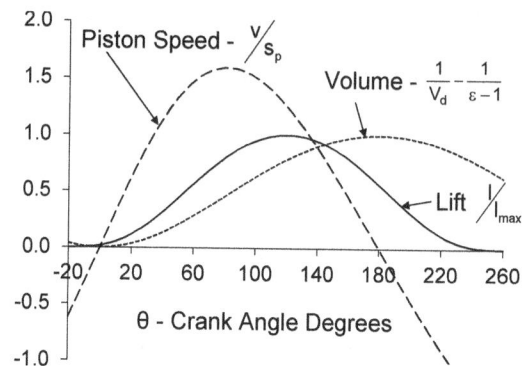

Figure 7.7: Normalized cylinder volume, intake valve lift and instantaneous piston speed as a function of crank angle for the camshaft of Figure 7.6.

The slow opening and closing of the valves when the piston is near top or bottom dead center is not a serious restriction, as shown in Figure 7.7, which shows the connection between piston and valve motion. Shown are normalized values of cylinder volume, instantaneous piston speed, and valve lift as a function of crank angle for the intake valve of an SI engine. The cylinder volume shown is the total cylinder volume less the clearance volume, divided by the displacement volume, the values ranging from 0 to 1. The instantaneous piston speed has been normalized by the mean piston speed, and the valve lift is the fraction of its maximum lift. The maximum valve lift occurs about the time that the piston speed is greatest. Since the volumetric flow ideally is the product of the piston area times the instantaneous piston velocity, the valve has the highest opening area when most needed. Notice that the valve is held open until after bottom dead center. This will be discussed in more detail later, but note that it allows for flow into the cylinder, even though the piston has reversed direction from the intake stroke.

For situations where the pressure difference between cylinder and manifold is small, the instantaneous piston motion can be considered to drive the flow, that is, without any pressure losses, the instantaneous flow in the intake port is proportional to the instantaneous piston speed. Thus it is advantageous to have a large opening area when the piston speed is highest. That this is the case is shown more clearly in Figure 7.8, which shows the normalized valve lift plotted as a function of the normalized instantaneous piston speed. This again shows that when the piston speed is highest, the valve lift, which is proportional to the valve flow area, is also highest. Thus the geometries of the valve and piston are well suited to each other, which is not surprising, since they are both based on periodic motion, closely resembling simple trigonometric functions.

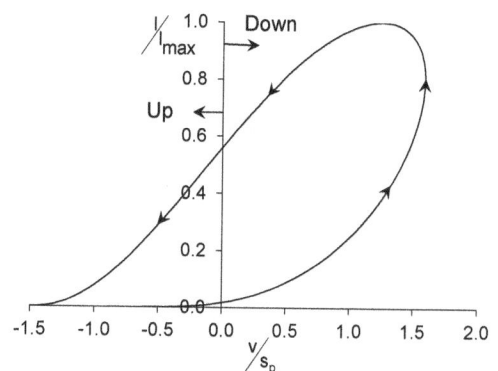

Figure 7.8: Normalized valve lift as a function of normalized instantaneous piston speed for the intake valve of the camshaft of Figure 7.6.

The area to the left of the abscissa represents that portion of time when the piston is moving towards the valve, and thus in a situation with no pressure loss, would be forcing gasses out of the cylinder. If the filling "lags", there could still be a positive pressure gradient and the cylinder continue to fill, in spite of the motion of the piston.

7.5.1 Valve Timing Effects

In the following, considerations are made with respect to valve timings. Cylinder pressure diagrams were obtained using a cycle simulation program for an SI engine with the characteristics shown in Table 7.1. The flow through the valves was determined using a flow area equal to the product of the valve

Table 7.1: Engine parameters for valve timing effects simulation.

Bore - mm	73.96
Stroke - mm	75.48
Cylinders -	1
Displacement - liter	0.324
Connecting Rod Length - mm	150.0
Intake Valve diameter - mm	31.5
Exhaust Valve diameter - mm	31.5
Intake Valve lift - mm	9.0
Exhaust Valve lift - mm	8.0
Compression Ratio	9.3
Intake Manifold Pressure - kPa	101.6
Exhaust Manifold Pressure - kPa	102.1
Reference Timing	
Intake Opens:	40° BTDC
Intake Closes:	50° BTDC
Exhaust Opens:	80° BTDC
Exhaust Closes:	40° BTDC

head perimeter and the instantaneous valve lift. Isentropic flow relations were used, modified by flow coefficients as functions of instantaneous valve lift from [7]. Using the method of [11], the back flow into the intake manifold during overlap was assumed to be perfectly mixed with the intake manifold gasses. Since the objective was to show effects of valve timings on air exchange process, all calculations were performed with constant intake and exhaust pressures. This is not the case in an actual engine, where speed related pressure drop and pressure pulsations (see Section 7.6) must be taken into account. The valve profiles were obtained using a polynomial profile giving profiles similar to those used in production engines [84]. The polynomials are given in Equations (7.41) and (7.42). When changing one of the valve timings, all the others were kept at the reference values shown in Table 7.1.

$$\frac{l}{l_{max}} = 6.09755 \cdot \beta_u^3 - 20.7804 \cdot \beta_u^5 + 26.73155 \cdot \beta_u^6 - 13.60965 \cdot \beta_u^7 + 2.56095 \cdot \beta_u^8 \qquad (7.41)$$

$$\frac{l}{l_{max}} = 1 - 2.63415 \cdot \beta_d^2 + 2.78055 \cdot \beta_d^5 + 3.1705 \cdot \beta_d^6 - 6.87795 \cdot \beta_d^7 + 2.56095 \cdot \beta_d^8 \qquad (7.42)$$

where: l is the valve lift and β_u is the fraction of the duration of the upward portion of the cam, and β_d is the fraction of the duration of the downward portion of the cam $0 \leq \beta_i \leq 1$ for each portion of the profile.

Intake Valve Timing

There is an advantage in not opening the intake valve too soon, since that would increase the overlap flow and increase cylinder residual fraction. The overlap period is the time when both intake and exhaust valves are open at the end of the exhaust and start of the intake. On the other hand, waiting too long to open the intake valve will result in low pressures on the intake stroke, and the valve flow area may not be adequate to allow enough air to enter the engine to keep the cylinder pressure near the intake manifold pressure throughout the intake stroke. This causes more pumping work and can result in a lower cylinder pressure and air charge at the end of the intake stroke. The latter will have an adverse effect on the maximum power output of an engine.

The factors in the compromise are shown in Figure 7.9. At low speed, there is not much effect even for the late opening, though there is a slight increase in the pumping loss while the valve lift reaches a value which allows flow to keep up with piston motion. At high speed, on the other hand, there is a major increase in the pumping loss when the intake opens at top dead center instead of 80° before. At the latter timing, the valve is open adequately throughout the intake stroke to allow flow to keep up with piston motion.

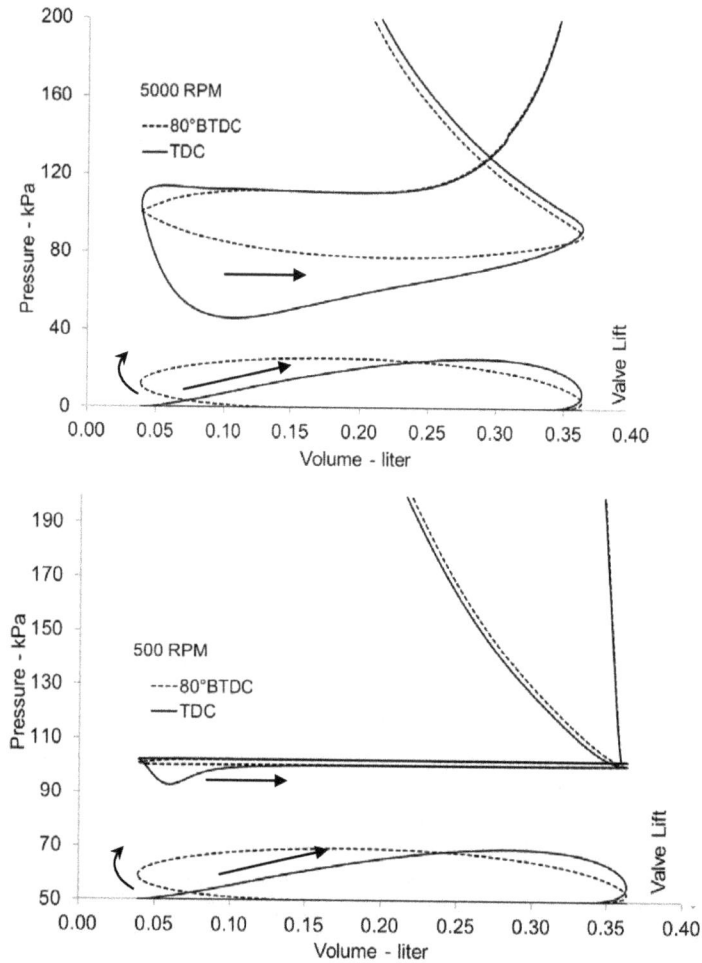

Figure 7.9: The effect of intake valve opening on air exchange processes at 500 and 5000 rpm for intake valve openings of Top Dead Center and 80° BTDC.

The next event, and one of the more important, is the closing of the intake valve. Closing too early results in greater flow restriction near the end of the intake stroke, which results in lower cylinder pressure on the compression process, and less power. If the cylinder pressure at BDC is lower than the intake manifold pressure, then there is no need to close the intake valve, as intake charge can still enter the cylinder, even though the piston may be starting its travel away from BDC. Ideally, the intake valve should not be closed until the cylinder pressure is equal to the intake manifold pressure. There is also an effect due to the momentum of the intake air column, which has been set in motion by the intake process. The reduction in speed of the air column at the intake valve will increase the pressure of the intake charge at the valves, allowing more flow to enter the engine, though this is not included in the simulations.

Thus, the intake valve is held open after BDC, in principle as long as there is the potential for additional flow into the cylinder. The optimum timing depends on the engine speed, as it does for the other openings and closing. If an optimum is achieved at a given engine speed, then at lower speeds and fixed valve timing, the valve will be open too long. In this case, fresh charge that has been drawn into the cylinder will actually be pressed be back into the intake manifold, reducing the trapped mass, and giving lower pressures throughout the rest of the cycle after the intake valve has closed.

The situation is shown in Figure 7.10 at the engine speeds of 500 and 5000 rpm. At 500 rpm it is clear that for closing at 80° ABDC, flow is out of the cylinder at the beginning of the compression stroke, since the pressure does not increase until the valve is almost closed. There is a very small pressure drop for opening at TDC, but the cylinder pressure is much higher on the compression stroke, indicating better

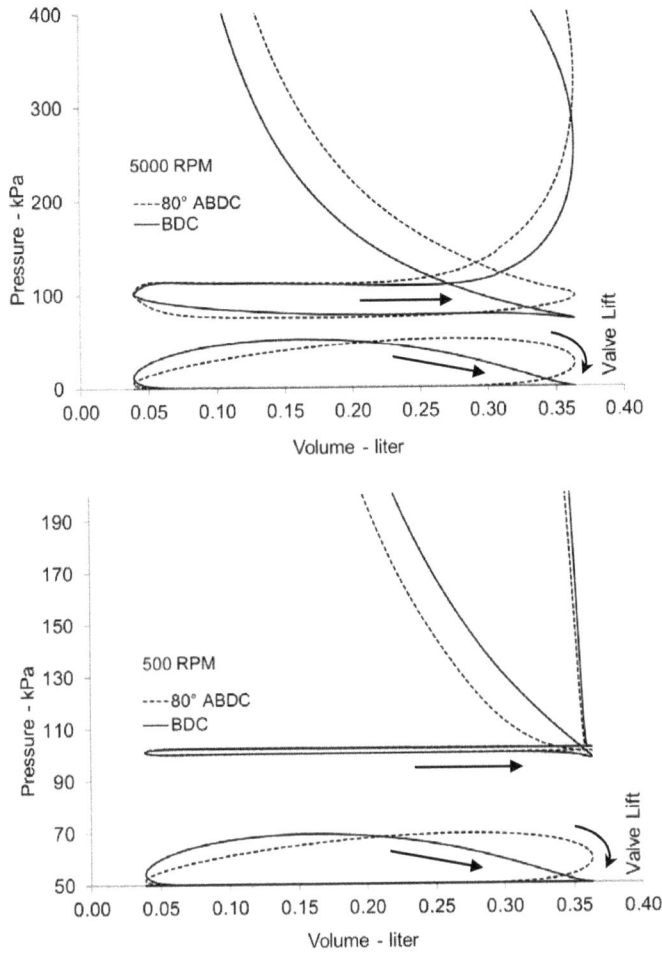

Figure 7.10: The effect of intake valve closing on air exchange processes at 500 and 5000 rpm for intake valve closings of Bottom Dead Center and 80° ABDC.

charging. At 5000 rpm the situation is reversed, and the cylinder pressure at the start of compression is much higher for the opening of 80° ABDC. An increase in cylinder pressure at the very end of the intake stroke is in contrast with the pressure drop seen for the TDC timing. Pumping loss is little affected by the timing at either speed, it is the last portion of the intake charging that is critical.

The later the intake valve closing, the better the charging at high speeds, but the worse the charging at low speeds. The intake valve closing is an important variable determining the speed at which the maximum volumetric efficiency is obtained. In accordance with Equation (7.40) this will have a significant impact on the speed at which the maximum bmep is achieved. Typically, earlier openings and later closing of valves give peak torque at higher speed. Though the peak of the torque occurs at higher speeds, normally the magnitude of the peak torque decreases at higher speeds because of higher gas flow velocities and greater pressure drops.

Exhaust Valve Timing

Ideally, the exhaust valve opens at bottom dead center, in order for the expansion to be as complete as possible, and the greatest amount of work extracted from the combustion products. There is another factor that must be taken into consideration, and that is that quite often the flow through the exhaust valve is choked, since pressure ratios between the cylinder and the exhaust manifold of greater than two are commonly encountered. This restricts the flow, and limits the amount of exhaust gasses that can

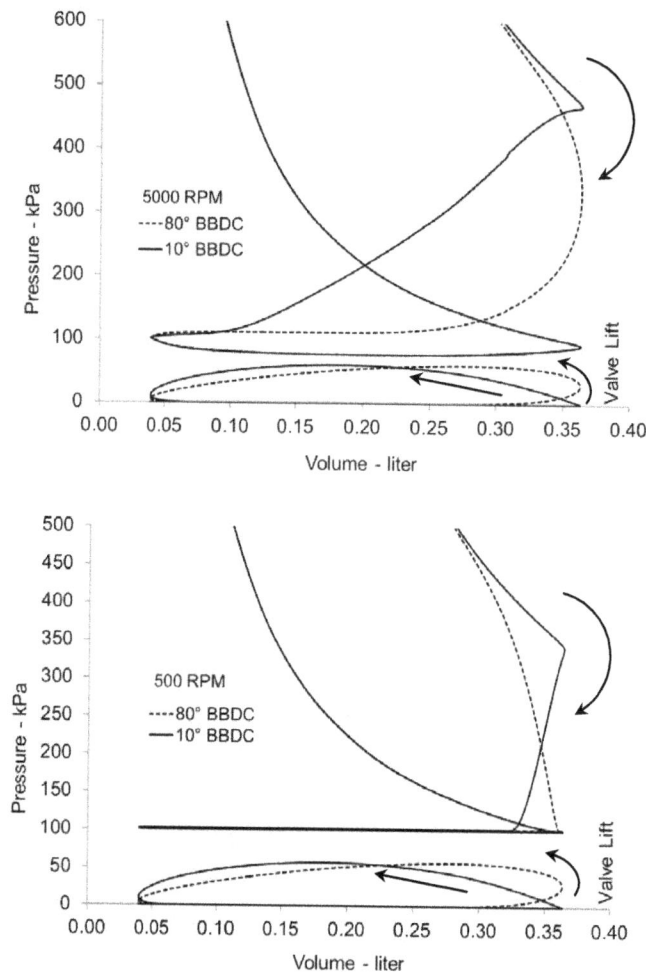

Figure 7.11: The effects of exhaust valve opening on air exchange processes at 500 and 5000 rpm for exhaust valve openings of 10° BBDC and 80° BBDC.

flow out in the first part of the process, since the area is small due to the low valve lift. If the exhaust valve opens too late, the pressure will not drop close to exhaust pressure while the piston is near bottom dead center, and will still be high as the piston moves away from bottom dead center towards top dead center.

This results in cylinder pressures considerably above exhaust pressures, particularly at the start of the exhaust stroke. In addition, the amount of exhaust that flows out of the cylinder during the blowdown process (see Section 2.2.1) is small. This means that later on in the exhaust stroke, more gasses must be displaced from the cylinder as the piston moves toward the exhaust valve and pushes them out. This in turn increases the pressure on the exhaust stroke, resulting in more pumping work and a larger residual gas mass when the exhaust process is ended.

In order to improve this situation, the exhaust valve is opened before bottom dead center, so that more gasses can leave the cylinder during the blowdown process, and give lower pumping losses. The price paid for this is an earlier drop in the cylinder pressure before the piston reaches bottom dead center, resulting in less expansion work. As expected, there is an optimum timing, where the loss from the reduced expansion work is balanced by the reduction in pumping work. The situation is shown in Figure 7.11. The optimum timing is dependent on the engine speed: the higher the speed, the earlier the optimum opening of the exhaust valve.

At 500 rpm, opening the exhaust at 10° BBDC allows the nearly complete expansion of the cylinder gasses, while the pressure drops quickly to the exhaust pressure when the valve opens, though there is an increase in the pumping work. For 80° BBDC, the opening is clearly too early, and there is a significant loss in expansion work, though this is balanced to some extent by less pumping work. At 5000 rpm, opening at 10° BBDC gives a very large pressure throughout the exhaust stroke, as the valve area is too small for much mass to leave the cylinder, and the choked flow continues until past the middle of the exhaust stroke where $p_{cyl} \approx 2p_{exh}$. This leads to a very large pumping loss. At 80° BBDC, there appears to be a good compromise between minimum expansion loss, and low pumping work.

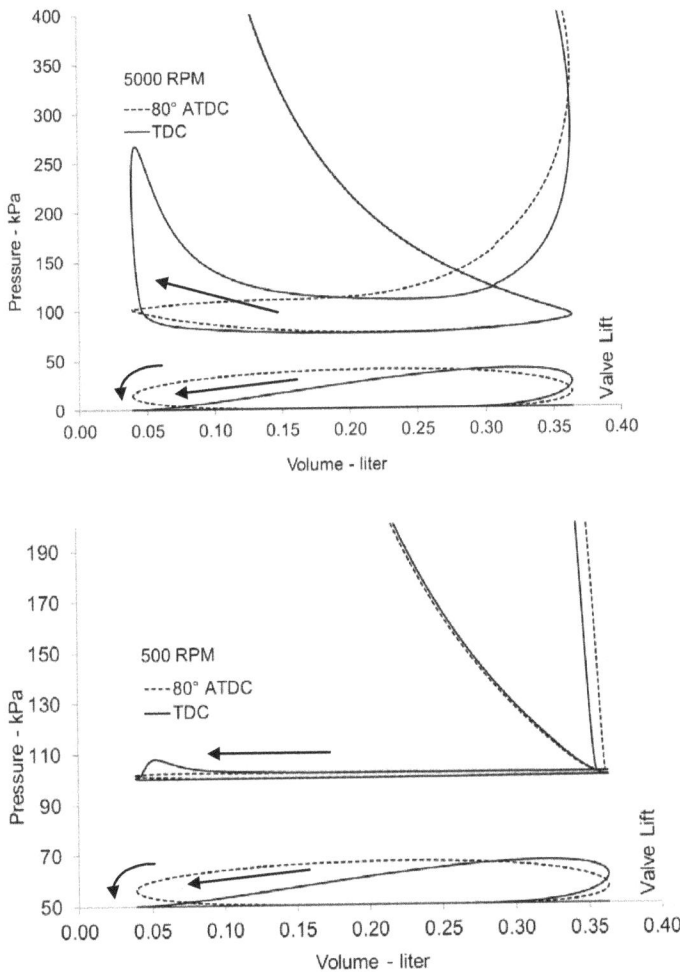

Figure 7.12: The effects of exhaust valve closing on air exchange processes at 500 and 5000 rpm for exhaust valve closings of Bottom Dead Center and 80° ABDC.

Similar considerations can be made at the end of the exhaust stroke. Here, if the exhaust valve is fully closed at top dead center at the end of the exhaust stroke, the flow near the end of the stroke will be restricted as the valve closes due to the small lift, and the pressure at the end of the exhaust stroke will rise, causing more pumping work and a higher residual mass. To compensate for this, the exhaust valve is closed some time after TDC, to keep the pumping work at a lower level. The situation is shown in Figure 7.12.

At 500 rpm there is not much effect, though there is a slight increase in pumping work when the valve closes at TDC. However, at 5000 rpm, the early closing has a dramatic effect, as the cylinder pressure becomes very high at the end of the exhaust stroke. Not only is the pumping work greatly increased with this timing, but the high residual pressure increases residual backflow during the overlap period, increasing intake charge dilution. For closing at 80° ATDC, there is no increase in cylinder pressure when the piston nears TDC, and the pumping work is kept low.

There is a disadvantage to this, as it is at this time that the intake valve is opening, and for a naturally aspirated engine, the intake pressure at this time is lower than the exhaust pressure, and there can be a flow of exhaust gasses back into the intake manifold, displacing the some of the air in the intake charge. This period time is called the *valve overlap period*, since the end of the exhaust process overlaps the start of the intake process.

For a turbocharged engine, the compressor and turbine are often matched such that the intake pressure is slightly higher than the exhaust pressure. Then the flow situation during overlap is reversed, and intake gasses can flow through the engine directly into the exhaust without being trapped in the cylinder. For an SI engine, this increases air pollution and fuel consumption. However, for the turbocharged diesel engine this can be advantageous, since the intake charge contains only air, which does not contribute to air pollution. Through flow of air, then, causes no increase in emissions or fuel consumption, and can prove to be slightly beneficial, since the intake air is cooler that the residual gasses, and an overlap flow can help to cool the exhaust valve, one of the cylinder components most subject to thermal loading.

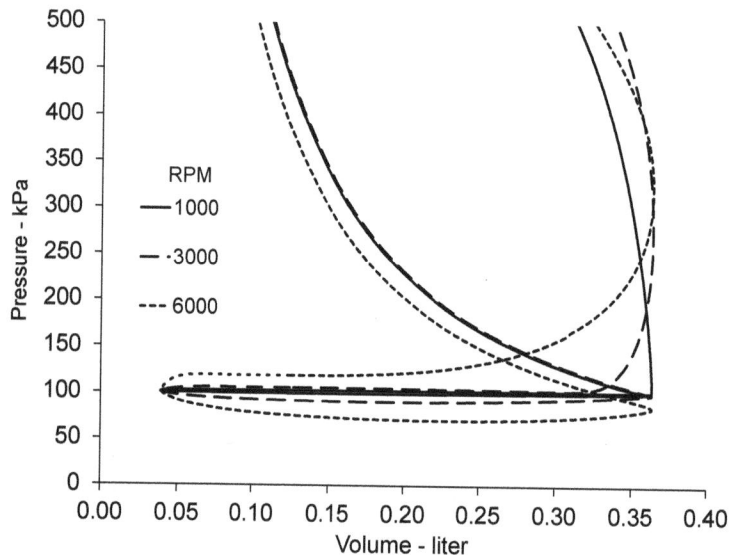

Figure 7.13: Simulated pressure volume diagrams for the reference engine of Table 7.1 at 3 operating speeds.

Some of these processes can be seen in simulation results. The following results were obtained with a computer simulation that includes the flow through the intake and exhaust valves as a function of crank angle. The simulation assumes that the pressures in the intake and exhaust manifolds are constant throughout the entire engine cycle. The engine is a 2.8 liter V-6 engine with a bore of 89 mm and a stroke of 76 mm. The exhaust valve opens at 64 degrees BBDC and closes at 60 degrees ATDC. The intake valve opens 26 degrees BTDC and closes 85 degrees ABDC, giving an overlap of 86 degrees. The simulations were made at full throttle conditions with constant manifold pressure, and neglecting the effects of air column inertia. Figure 7.13 shows the simulated $p - V$ diagram for the engine speeds on an expanded scale in order to focus on the air exchange process. Of particular interest is the relative changes in the pressures on the intake and exhaust strokes as the engine speed changes. It can be seen that the pumping losses become much higher at higher engine speed.

The simulated flows through the intake and exhaust valves as a function of time for the engine above are shown in Figure 7.14. First of all, note the difference in the nature of the intake and exhaust flows. The basic shape of the intake flow is similar in shape to that of the instantaneous piston speed as a function of crank angle. This is because the pressure differences across the intake valve are small, and the instantaneous piston displacement is the driving factor for the flow. At the highest speed, the flows are slightly smaller (on a flow per crank angle basis), due to an increased pressure drop across the valve at high flow velocities. It can also be seen at the highest speed that the flow into the cylinder continues after bottom dead center, since the cylinder pressure is still lower than the intake manifold pressure. There is a flow reversal at the very end of the intake process as the piston pushes cylinder gasses back out of the cylinder. This occurs at all speeds in the simulation, though at the highest speed the effects of the momentum of the intake air column will reduce or likely eliminate this.

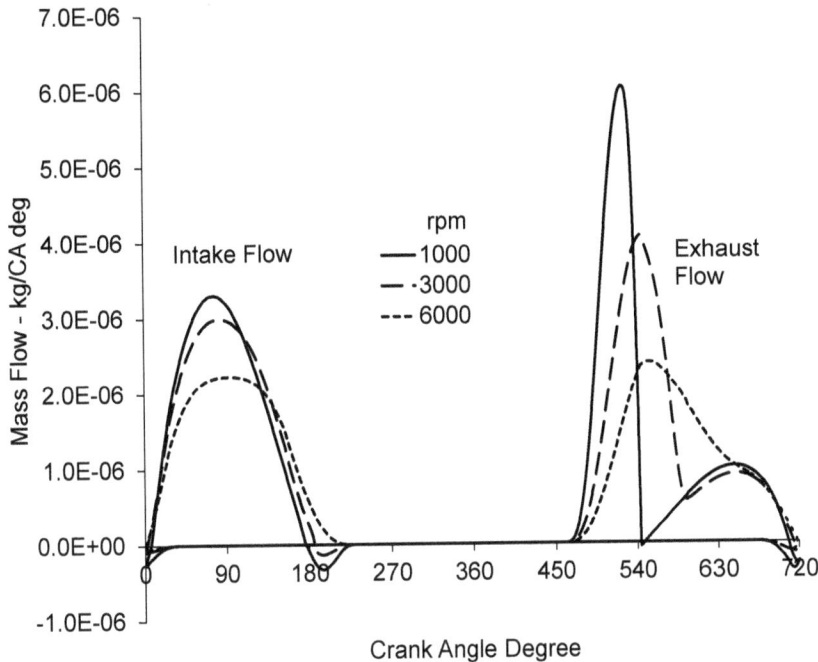

Figure 7.14: Simulated intake and exhaust flows for the reference engine of Table 7.1 at 3 operating speeds.

For exhaust gas, however, there are two distinct phases to the flow. The first phase is very rapid, and takes place at the time when the piston is near its BDC position and moving only little. This is the blowdown process, driven by the pressure difference between the cylinder and the exhaust manifold. At the lowest speed, the flow goes nearly to zero at the end of this phase as the cylinder almost pressure drops to the exhaust manifold pressure. Then the second phase starts, and shows a shape related to the instantaneous piston speed, similar to that of the intake flow. The cylinder pressure decreases rapidly at the end of expansion for the lowest speed, and the cylinder pressure throughout the exhaust stroke remains low.

At the higher speeds, the cylinder pressure does not drop to the exhaust manifold pressure at the end of the blowdown process. The flow is choked so the flow per unit area depends solely on the cylinder conditions, Equation (5.5), not on instantaneous piston velocity as would in the case of positive displacement. At the crank angle of maximum flow at 1000 rpm, the exhaust mass flow rate per unit area is 172 g/m^{2}s, at 3000 it is 191g/m^{2}s, and at 6000 it is 185 g/m^{2}s. This shows the effect of choking, the flow is essentially independent of the engine speed, and depends on the cylinder pressure and temperature alone. The extra mass in the cylinder and the shorter time (in seconds) to move it out of the cylinder is responsible for the high cylinder pressure on the exhaust stroke at higher speed, and its adverse effects on the pumping loss. Especially at the highest speed, the advantages of an earlier exhaust valve can be seen. But in that case, the expansion work loss at the lowest speed, already apparent here, would be worse. The advantages of variable valve timing should begin to be apparent.

Also seen in Figure 7.14 is the backflow at the end of the exhaust stroke. This is the overlap period, and there is flow from the exhaust manifold back towards the cylinder. At light loads, there will be a significant flow of residual gasses into the intake manifold. This is not so prominent in this case, since the intake manifold pressure is near atmospheric.

A summary of valve timing effects is shown in Figures 7.15 and 7.16. Figure 7.15 shows how changing the individual valve effects influences the torque of the engine at a low and high speed. The effect of valve timing on the power and *bmep* values as a function of speed is shown in Figure 7.16. The timings not changed were kept at the reference values of Table 7.1 While all the events have an influence, the intake closing and exhaust opening can be seen to have the most effect, especially at high speeds.

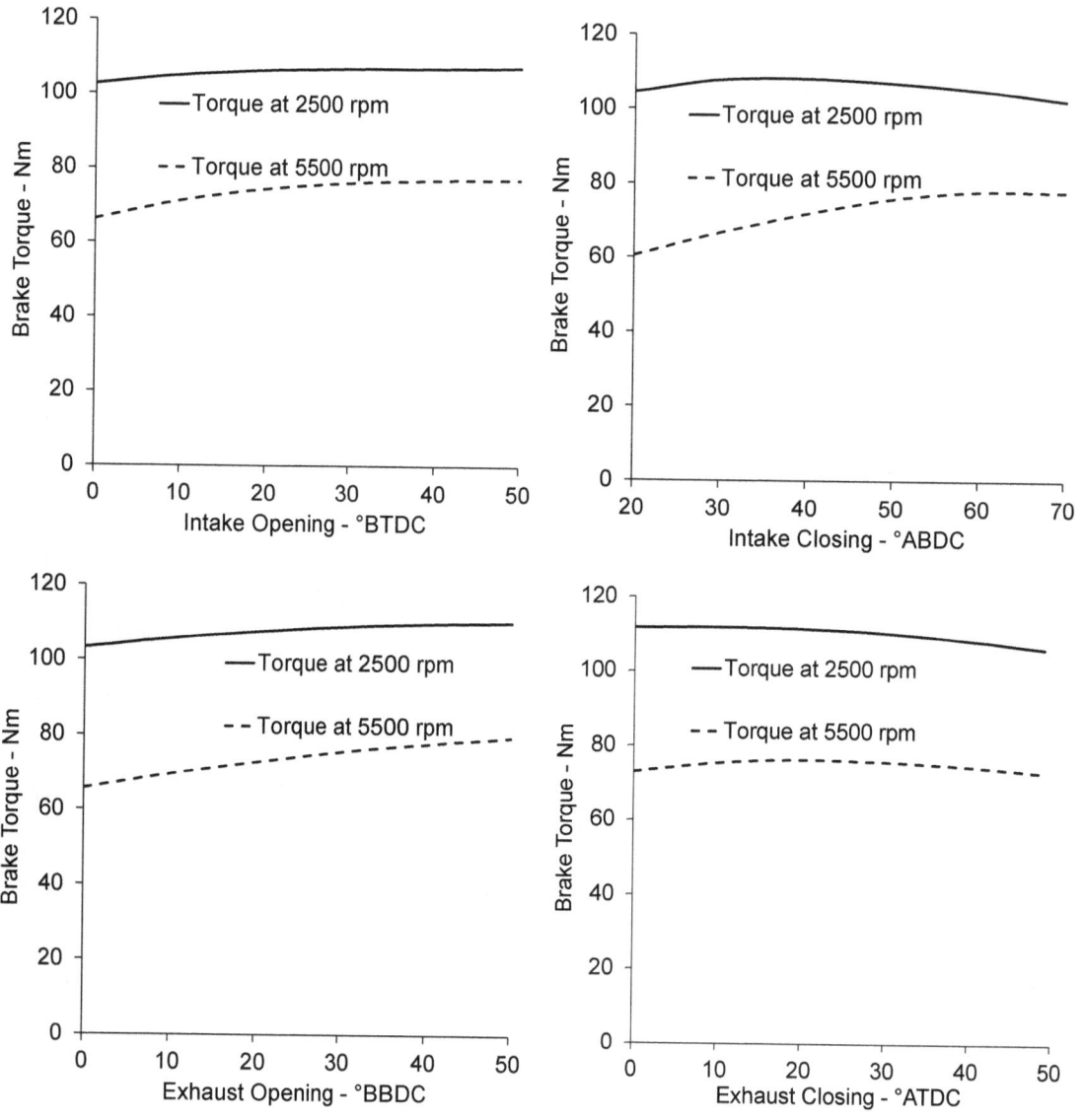

Figure 7.15: Effect of valve timings on engine torque at two engine speeds, for the reference engine of Table 7.1.

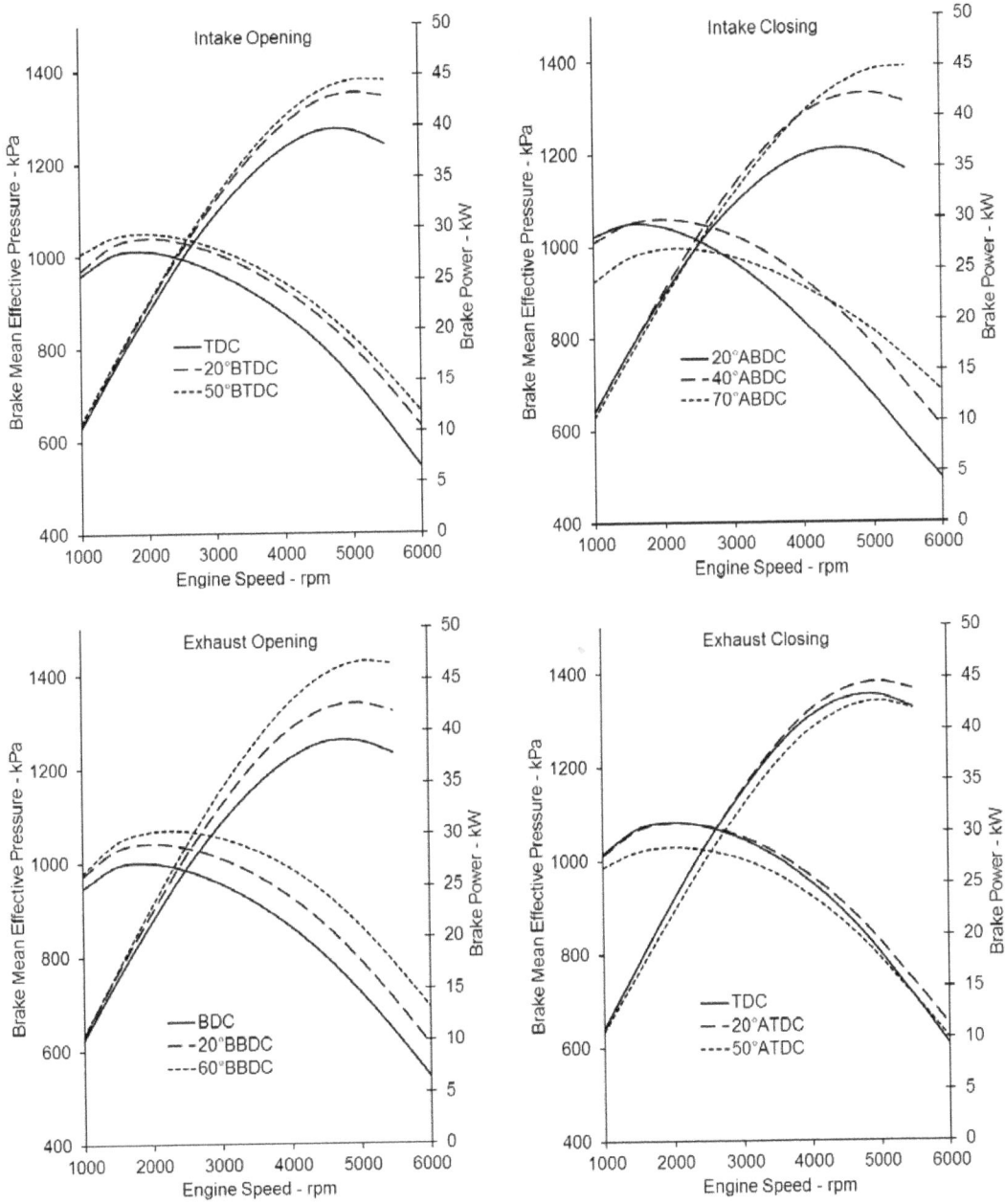

Figure 7.16: Effect of valve timings on engine power and *bmep* for the reference engine of Table 7.1.

Valves

Flow through the valves is treated as isentropic flow, with the rates adjusted by empirical flow coefficients. The flow can be calculated using the gas flow equations of Section 5.1. It is possible for the intake flow to be choked, and as seen above, the exhaust flow is often choked at the beginning of the blowdown process. The area of the valve plays an important role in the flow. On the intake stroke, where the volumetric flow is determined by the piston motion, the area determines the pressure drop, and the time required for pressure equalization between the manifold and the cylinder. For the blowdown process, it is the valve area that determines how quickly the pressure can fall to the exhaust pressure.

There are two factors determining the flow area: The diameter of the valve and the height of the valve above the seat or *valve lift*. To a reasonable first approximation, the valve flow area can be approximated by the surface of a cylinder whose height is the valve lift, and whose diameter is the valve diameter. The minimum flow area actually has a slight dependence on the valve seat geometry, but that is not included here, as the flow is modeled through the use of an empirical flow coefficient anyhow. The situation is shown schematically in Figure 7.17. There is an advantage in making the valve flow area as large as possible, since it has a direct influence on the pressure drop. In isentropic flow, it is the minimum area that restricts the flow. Therefore, it is of interest to see the effect of maximum valve lift on minimum flow area. In Figure 7.17 there are two areas to consider. The first is the cylindrical area at the open valve, A_v. The second is the flow area in the port, A_p, where it is assumed that the port diameter is close to

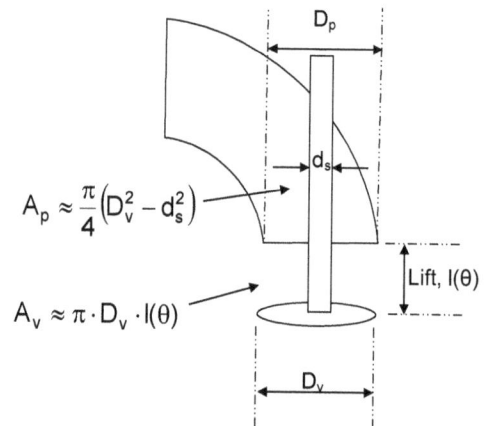

Figure 7.17: A simplified depiction of the relevant flow areas for a valve.

the diameter of the valve. The effective flow area in the port is the cross section of the port minus the area taken up by the valve stem, as shown in Figure 7.17. To find the maximum useful valve lift, these areas are set equal to each other, since at higher valve lift, the constant port area will be the limiting area instead of the valve. Then, assuming that the stem diameter is about 0.2 times the valve diameter:

$$\pi \cdot D_v \cdot l = \frac{\pi}{4}\left(D_v^2 - d_s^2\right) = \frac{\pi}{4}\left(D_v^2 - 0.04 D_v^2\right) \approx \frac{\pi}{4}D_v^2 \tag{7.43}$$

Which gives:

$$\frac{l_{max}}{D_v} \approx 0.25 \tag{7.44}$$

This indicates that once the valve lift is greater than about 25 % of the valve diameter, the effective flow area will not increase with a further increase in valve lift. There are simplifications in the calculation, but this value is not atypical for operating engines. There is a mechanical disadvantage to a greater valve lift than required for flow purposes, since it will increase velocities and accelerations of valve train components, which is not beneficial to wear. There may be some advantages to a greater valve lift than that indicated here, since the flow is proportional to the product of a flow coefficient and the area. It has been shown that with higher valve lifts, the flow through an intake valve becomes straighter, and there is some improvement in the flow from a better flow coefficient if the lift is greater that 25% of the valve diameter.

Since there is a limit to the area increase due to valve lift, another way to improve the breathing capability of an engine is to increase the total valve area. There is a limit on the diameter of an intake and exhaust valve for a given bore size, but a larger portion of the cylinder head area is available for flow if multiple valves are used. This can be seen from simple geometric considerations, as illustrated in Figure 7.18. This example illustrates the situation for a flat cylinder head, such as would be found in a DI diesel engine. The maximum possible valve area, where the valves extend to the edge of the cylinder and are allowed to contact each other is shown for the case of 2 and 4 valves. The center of the 4 valves can be shown by a simple geometric calculation to be a distance of $r_c = \dfrac{\sqrt{2} \cdot B/2}{1 + \sqrt{2}}$ from the center of the

cylinder, and the valve radius equal to $r_v = \dfrac{B/2}{1 + \sqrt{2}}$.

The valve flow areas will be proportional to the circumferences of the circles in Figure 7.18. For the maximum possible valve diameters, the effective flow area (valve circumference) ratio between 4 valves and two valves is 1.657, giving a major increase in the valve flow area. As an example of the more realistic case, where there is clearance between the valves and the edges of the cylinder, if the radius of the valves was decreased by 5% of the cylinder diameter, the effective flow area is still increased by a factor of 1.571 when 4 valves are used instead of 2, so a 50% increase in flow area is not unreasonable. In an actual engine, the intake valves are typically larger than the exhaust valves, so the geometry will be different than in Figure 7.18, but increases of the same magnitude in relative valve flow areas are achievable.

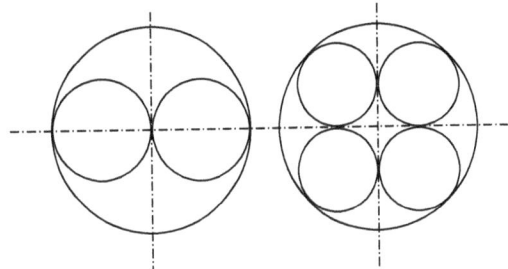

Figure 7.18: Maximum possible valve area for a flat cylinder head with 2 and 4 valves.

Earlier results for the reference engine of Table 7.1 showed a significant decrease in torque for at higher speeds for all valve timings. An obvious way to improve this is to increase the valve area, to improve flow characteristics at these speeds. A simulation was performed at the reference valve timing, but with 4 valves instead of two. An intake valve diameter of 25.91 mm, and exhaust valve diameter of 22.25 mm were used, The respective valve lifts were reduced to 25% of the diamteters, that is 6.5 and 5.6 mm for the intake and exhaust respectively. The increases in flow areas were 35.3 and 25.8% for the intake and exhaust respectively.

Results from this simulation are shown in Figure 7.19. The results show that there is a 7.9% improvement in the volumetric efficiency at 6000 rpm, which gives a 16.5% improvement in the brake torque and power of the engine. The torque and power improvement is greater than that of the volumetric efficiency, since all of the extra work goes into brake work, the friction remaining constant, and thus the mechanical efficiency increases. The extra valve area decreases the pressure drop across the valves and results in higher pressure at the end of the intake stroke, and lower pressure at the end of the exhaust stroke. It can also be seen that at the lowest speeds, there is a slightly lower power with the larger valve area. Under these conditions, there is less restriction for flow out of the cylinder at the end of the in-

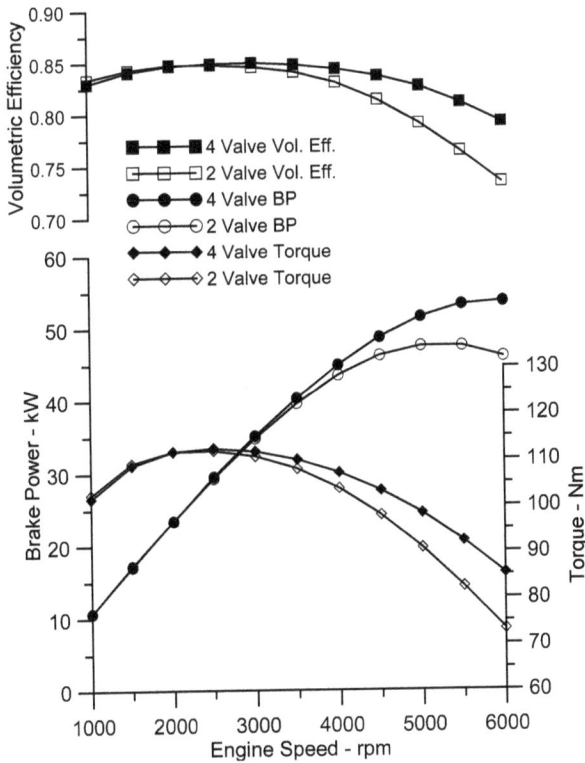

Figure 7.19: Simulated effect of increasing from 2 valves to 4 valves on the performance of the engine in Table 7.1.

take process, as shown in Figure 7.14. This process would be improved by closing the intake valve earlier, but this would reduce the benefit at the highest speed. The valve timing must then be optimized for the new area. This points out the advantages of variable valve timing, in that the overall torque curve can be made flatter if the valve timing is changed as a function of speed. Recent developments in engine technology are making this a more practical option.

The use of variable cam profiles, or variable valve timing is a topic that has long been of interest in engine development. Until recently, devices to accomplish this have been too expensive or too unreliable to be put into practice. However, in recent years developments have been made to achieve some of these goals.

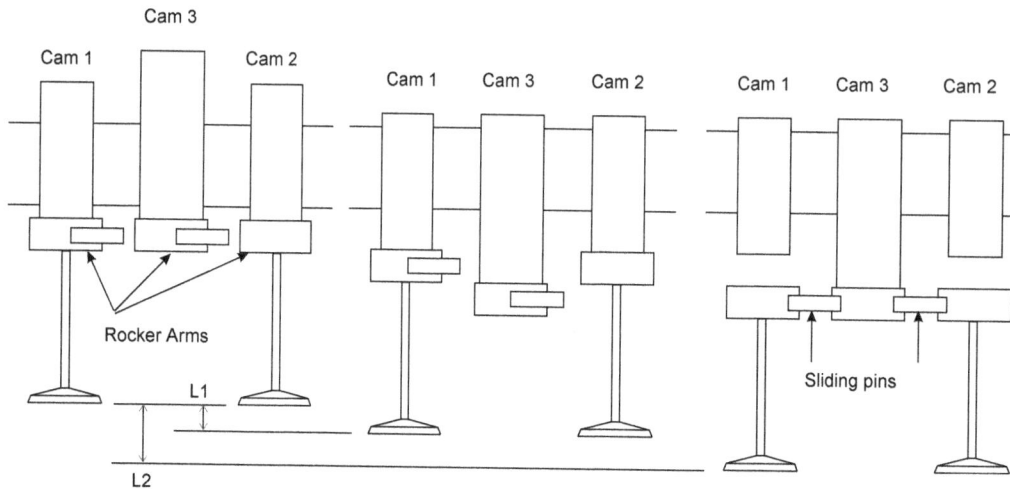

Figure 7.20: A mechanism to achieve variable valve timing in engines.

One of the methods is to have two valves operated by 3 cams. The system is shown in Figure 7.20 and is called the VTEC system, developed by Honda. Cams 1 and 2 on the camshaft have identical profiles, while cam 3 has a different profile. All three cams activate the respective rocker arms on the rocker arm shaft at all times. The rocker arm shaft is located behind the camshaft, and is not shown here, only the portion of the rocker arm which activates the valve. When the sliding pins are moved such that there is no connection between rocker arms 1 and 2 and rocker arm 3, the latter has no function and simply moves up and down. In order to change the cam profile, the sliding pins are moved such that all rocker arms are connected, Cam 3, which has a greater lift and possibly duration, then becomes the decisive factor, and the force on rocker arm 3 is carried to the other two rocker arms, and the valve opens in accordance with the profile of cam 3. Cams 1 and 2 then do not come in contact with the respective rocker arms.

This system does not allow infinite control of the valve timing, but can provide significant improvement in the torque curve. It is possible to have long opening times for high speed operation, while at the same time operating with a more modest profile at low speed, which gives better torque since the intake valve is not held open too long.

A second technique to accomplish variable valve time is cam phasing. A camshaft is driven by a chain, or toothed belt, but instead of a standard gear, there is a pulley with an internal mechanism that allows the cam shaft to rotate inside the pulley. Thus the cam phasing can be changed to different positions. This can be advantageous, though as was seen earlier in the chapter, it is common to extend the duration of the openings of the valve for high speed operation. That is not accomplished with cam phasing, but if the phasing is adjusted in accordance with the most important opening or closing event, improvements in engine operation can be achieved.

7.6 Pressure Pulsations in Manifolds

7.6.1 Basic Concepts

One of the most important properties of a gas in connection with wave motion is the speed of sound. It can be shown that for an ideal gas, the speed of sound is dependent on the specific heat ratio of the gas, its molecular weight, and the local temperature. The speed of sound, a, is given by:

$$a = \sqrt{\frac{\gamma R_o T}{M}} \tag{7.45}$$

In isentropic processes, the speed of sound can also be written as a function of the pressure or density change, where the subscript, $_o$, refers to reference conditions of the undisturbed atmosphere. At $25°$, $a_o = 346$ m/s.

$$\frac{a}{a_o} = \left(\frac{\rho}{\rho_o}\right)^{\frac{\gamma-1}{2}} = \left(\frac{p}{p_o}\right)^{\frac{\gamma-1}{2\gamma}} \tag{7.46}$$

In acoustic waves, the air is still, and the pressure rise from the wave is quite small, therefore, acoustic waves propagate with a velocity equal to the acoustic velocity. A small disturbance can be shown to propagate in both directions with a velocity, du, that is related to the corresponding density change, $d\rho$:

$$du = \pm\frac{d\rho}{\rho} \tag{7.47}$$

Assuming that the pressure and density change in an isentropic manner, Equation (7.47) can be integrated over a change in density, to give the local gas velocity:

$$u = \pm\frac{2a_o}{\gamma-1}\left[\left(\frac{\rho}{\rho_o}\right)^{\frac{\gamma-1}{2}} - 1\right] = \pm\frac{2a_o}{\gamma-1}\left[\left(\frac{p}{p_o}\right)^{\frac{\gamma-1}{2\gamma}} - 1\right] = \pm\frac{2}{\gamma-1}(a - a_o) \tag{7.48}$$

From this, it can be seen that in waves of finite amplitude, the acoustic velocity changes within the wave:

$$a = a_o \pm \frac{\gamma-1}{2}u \tag{7.49}$$

The propagation speed of the finite amplitude wave, u_p is then given by the sum of the local acoustic velocity and the local gas velocity:

$$u_p = a \pm u = a_o \pm \frac{\gamma+1}{2}u = a_o\left(1 \pm \frac{\gamma+1}{\gamma-1}\left[\left(\frac{\rho}{\rho_o}\right)^{\frac{\gamma-1}{2}} - 1\right]\right) \tag{7.50}$$

In engine systems, the waves caused by the pulsating flow involve significant velocity and pressure variations. Therefore, the propagation of waves depends on the flow in which they occur. The speed of sound determines the speed at which a disturbance moves through the gas, relative to the local gas velocity. In quiescent air, a disturbance moves away from its source at the speed of sound alone. If the gas is in motion, however, the disturbance moves away at the relative velocity. In the case of Figure 7.21 for example, the local gas velocity is given by u, where u is positive towards the right, the disturbance moves at the propagation velocity $u_p = a + u$, and towards the left at the velocity $u_p = a - u$. It is important in engine systems to remember that pressure waves move relative to the local gas velocity. This is important since the local velocity in manifolds and pipes changes during the course of an engine cycle, with the driving forces being the drawing of air into the cylinder on the intake stroke and the expulsion of exhaust gasses during the blowdown process and during the exhaust stroke.

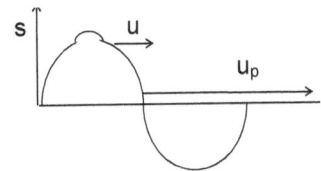

Figure 7.21: Simple representation of a disturbance superimposed on a propagating wave.

A detailed mathematical treatment of the motion of waves will not be given here. Two methods can be used, the method or characteristics [85], or the direct numerical integration of the fundamental conservation equations [86]. Application of the method of characteristics in an engine context is used in the books by Blair [87] and [14]. Some simple relations will be presented here, related to the description and basic calculation of wave properties.

The wave can be described in terms of the pressure amplitude ratio, X, a function of the local and atmospheric properties which is defined as:

$$X \equiv \left(\frac{p}{p_o}\right)^{\left(\frac{\gamma-1}{2\gamma}\right)} \tag{7.51}$$

By using the isentropic relationships and the relation for the speed of sound as a function of temperature, it can be shown that:

$$X = \left(\frac{a}{a_o}\right) = \left(\frac{\rho}{\rho_o}\right)^{\frac{\gamma-1}{2}} = \left(\frac{T}{T_o}\right)^{\frac{1}{2}} \tag{7.52}$$

In order to get a feel for some of the numbers and magnitudes involved, a simple example is shown in Figure 7.22. The wave shown here is called a compression wave since the pressure rises above atmospheric when the disturbance passes the point. It corresponds to a situation where a chamber at the left end of the pipe contains air at a pressure above that in the pipe, and is suddenly opened. In the example, the wave is propagating into gasses which are not moving.

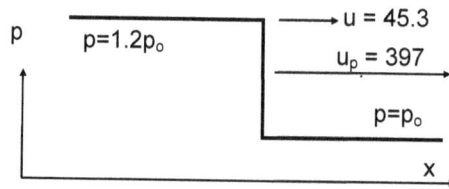

Figure 7.22: A simple compression wave.

Consider a compression wave in a pipe for which the pressure ratio is 1.2, that is, when the wave passes a location, the static pressure increases to 1.2 times its undisturbed value of 1.00 bar. The undisturbed temperature is 20°C (293.15K). It is assumed that the specific heat ratio =1.4. When the pressure increases isentropically, the local temperature will increase, and the speed of sound will increase from its undisturbed value of 343.2 m/s to 352.3 m/s. Using Equation (7.48) with the positive sign, it is possible to calculate the local gas velocity as 45.3 m/s, it is positive, meaning that it is moving in the same direction as the wave. This means that the compression wave is pushing gasses along with the wave. Until the pressure wave arrives, the gasses in the pipe are not moving. After it arrives, there is an outward flow with the velocity given above. This corresponds to the condition where an increase in pressure from the left hand side of a pipe pushes the gases out of the pipe to the right. There is a delay between the increase in pressure on the left, and the time at which the gasses start to be pushed out of the pipe. In a real case, when the pressure wave reaches the other end of the pipe, it will be reflected, and will have an influence on the condition in the remainder of the pipe, so the condition above only corresponds to the start of an exhaust process.

The velocity at which the wave propagates can be calculated from Equation (7.50) with the positive sign, and it is found to be 397 m/s, which is the sum of the local gas velocity and the speed of sound as the wave passes. The flow rate at the time the wave passes can be calculated from the continuity equation, with the density being determined from the isentropic relationships.

An expansion wave is a wave in which the pressure falls below the atmospheric pressure. An expansion wave also moving from left to right is shown in Figure 7.23. It corresponds to the case where there is a chamber at the left end of the pipe, with a pressure below that in the pipe, and the chamber is suddenly opened.

Consider a case where the static pressure falls to 0.8 bar when the wave passes. The pressure amplitude factor is now less than 1.0. The other properties at the time of the wave passage can be calculated using the same equations as in the previous case. $u_p = 278.5$ m/s and $u = -53.9$ m/s. Note that in this case the velocity is negative, which means that as the wave passes, the gas moves in a direction opposite to the wave motion, in this case from a region of higher pressure in front of the wave to a region of lower pressure behind the wave. The wave propagation velocity of 278.5 m/s is lower than the sonic velocity of the undisturbed

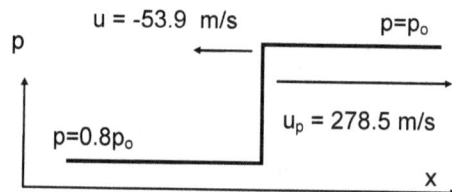

Figure 7.23: A simple expansion wave.

mixture. As was the case with the compression wave, the wave does not move at the sonic velocity. That is only the case for infinitesimally small disturbances.

7.6.2 Superposition

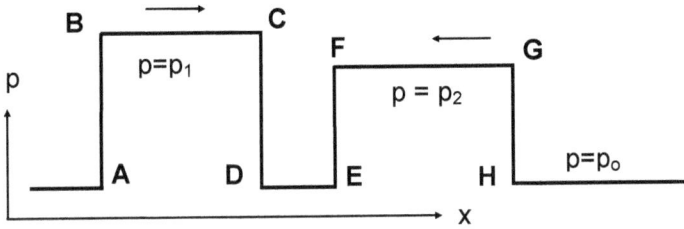

A. Two compression wave moving in opposite directions, about to meet.

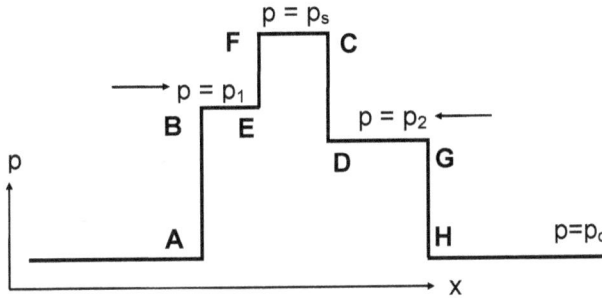

B. The compression waves passing each other.

Figure 7.24: The intersection (superposition) of two opposite moving compression waves. A. Before intersection. B. During the superposition.

In the simple example above, a single disturbance passed a point. However, this is the exception, rather than the rule, since waves propagating through pipes are reflected at the ends, and disturbances occur over a finite time interval, for example throughout the intake or exhaust process of an engine. Therefore, it is important to examine the processes that occur when two waves meet. A simple case is shown in Figure 7.24. Here there are two compression waves moving in opposite directions. Figure 7.24A shows the situation before the two waves meet. As time progresses, the two waves will intersect, as shown in Figure 7.24B. The interacting wave can be treated as a wave propagating in each direction, and the properties determined (See Blair [87], for example). Two relations describing conditions during the meeting of the waves, can be calculated. The first involves the pressure amplitude during the time of intersection, or superposition X_s:

$$X_s = X_1 + X_2 - 1 \tag{7.53}$$

The second involves the local gas velocity during the time of superposition, u_s:

$$u_s = \frac{2a_o}{\gamma - 1}(X_1 - 1) - \frac{2a_o}{\gamma - 1}(X_2 - 1) \tag{7.54}$$

When two waves meet, what is observed in measurements is the net result. If two compression waves of the same magnitude and opposite directions meet, Equations (7.53) and (7.54) say that one will measure a high pressure (X) and zero velocity.

The superposition of waves in a one-dimensional system then complicates the interpretation of the pressure time signal, which is often only measured at a single point. This is shown by the following example. Consider two waves, a compression wave with the properties of the simple compression wave discussed above, meeting the expansion wave of the previous example moving in the opposite direction. As before for the compression wave, $\frac{p_1}{p_o} = 1.2$, $X_1 = 1.0264$, $u_p = 397$ m/s and $u = 45.3$ m/s. Similarly for the expansion wave, $\frac{p_2}{p_o} = 0.80$, $X_2 = 0.9686$, $u_p = 278.5$ m/s and $u = -53.9$ m/s relative to the wave, that is $+53.9$ m/s in laboratory coordinates. By the principle of superposition, $X_s = X_1 + X_2 - 1 = 0.995$. From the definition of X, $\frac{p_s}{p_1} = 0.9655$. Then as the two fairly strong waves interact at this position, an observer looking only at the pressure trace as the waves pass would only observe a very small pressure difference relative to the atmosphere.

7.6.3 Wave Reflections

Closed End

Reflections occur when waves encounter different forms of changes in flow conditions such as an open or closed pipe end or a change in the flow area. These processes will be described briefly here, and can be used to implement boundary conditions for different solution techniques for the relevant differential equations. The closed end of a pipe is characterized by the fact that the velocity of the gas must be equal to zero. This results in the kinetic energy of the gas in the wave being converted to pressure as the wave comes to a halt. The resulting pressure sends a compression wave back in the opposite direction.

The characteristics of this wave can be shown through the following. Consider a compression wave with a pressure ratio of 1.2 approaching a closed end of a pipe from the left as shown in Figure 7.25. At the end of the pipe, both the local gas velocity u and the wave propagation velocity u_p are equal to zero. Figure 7.25 indicates that there is a wave, X_i coming in that will be reflected, X_r. At the instant of reflection, the two waves interact, and the equations of superposition can be used to determine the conditions.

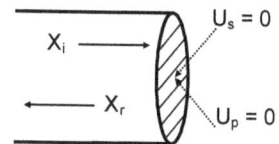

Figure 7.25: A pressure wave approaching a closed end of a pipe.

As an example, consider the compression wave from the previous example. The situation around the time of the reflection is shown in Figure 7.26. The incident wave has a pressure ratio of 1.2, wave propagation velocity of 397 m/s and a gas velocity of 45.27 m/s. At the instant of reflection, Equation (7.53) gives $X_s = 1.0528$, from which it can be determined that the pressure ratio is 1.434. After reflection, the wave moves away from the wall with the same velocity as the incident velocity.

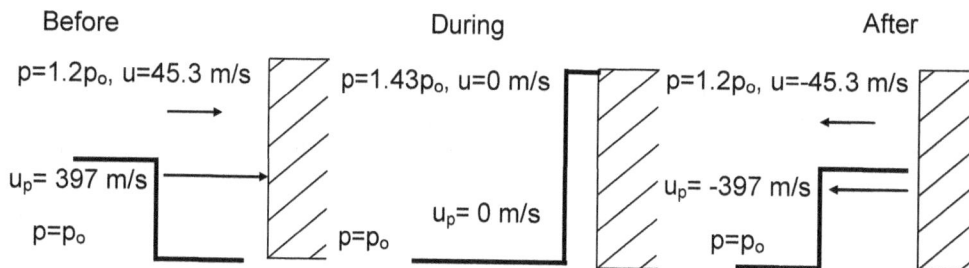

Figure 7.26: Reflection of a compression wave at a closed end of a pipe.

At a closed end, a wave is reflected as the same kind of wave with the same magnitude, that is: (Compression $\Rightarrow$ Compression and Expansion $\Rightarrow$ Expansion).

Open End

The situation at the open end of a pipe is more complex than that at the closed end, since the flow can be either inward or outward. The open end is connected to a pressure reservoir, where the pressure is assumed to be constant. This could be the atmosphere, or a chamber held at a given pressure. In the case of a compression wave approaching the open end, we have seen that the flow is in the same direction as the propagating wave, and so the flow is outward at the open end. The flow is also outward in the case of an expansion wave moving away from the open end, since the gas velocity is opposite to the wave propagation velocity for an expansion wave. For the opposite cases, a compression wave moving away from the end or an expansion wave toward the end, the flow at the end of the pipe is inward, and depends on the properties of the fluid outside the pipe. The case of the incident compression wave is the simplest and will be considered first. The basic assumption made in the case of outward flowing waves is that the pressure at the outlet section of the pipe is equal to that of the reservoir. In terms of the pressure amplitude ratio, it is set equal to one at the pipe outlet. Figure 7.27 shows the reflection process for a wave at the open end.

The reservoir properties are denoted with the subscript o. These could refer to atmospheric values, but can also refer to a plenum in an engine intake system. It is a location where the pressure can be taken as constant during the reflection of the wave at the end of the pipe time step. For example the cylinder of an engine could be considered a reservoir, since the pressure changes are normally slow with respect to wave phenomena. It is assumed that the pressure at the outlet plane is the reservoir pressure. At the instant of reflection, there are two waves, the incident and the reflected. For an incident compression wave at the open end with flow out of the pipe, the reflected wave will be an expansion wave. For the same compression wave as in the previous example reflecting from an open end of a pipe, with a pressure $= p_o$ outside of the pipe. The principle

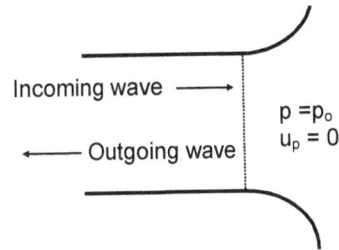

Figure 7.27: Incident and reflected waves at the open end of a pipe.

of superposition can be applied with $x_s = 1$. The pressure amplitude ratio of the reflected wave can be calculated from Equation (7.53) to be $x_r = x_s + 1 - x_i = 1 + 1 - 1.0264$. Then from Equation (7.51), the pressure ratio is$= 0.8292$. The gas velocity of the reflected wave can be calculated to be -45.29m/s using Equation (7.48). The negative value means that the velocity flows in the opposite direction of the wave propagation. Since the reflected wave is moving to the left, the flow is outwards, even after the reflection. The wave propagation velocity after the reflection can be found to be 288.8 m/s (from right to left) from Equation (7.50).

Now consider an expansion wave reaching the open end of a pipe. We know that an expansion wave moving to the right in Figure 7.27 has a local gas velocity in the opposite direction of the wave propagation. Therefore, the flow will be into the pipe when the wave reaches the open end of the pipe. The flow into the pipe will be a function of the pressure difference between the reservoir and the pressure at the end of the pipe at the time of reflection. For the expansion wave previously discussed, with a pressure ratio of 0.8 approaching the open end of the pipe from the left, the pressure ratio of the reflected wave can be found to be 1.241 using Equations (7.51) and (7.53). This means that the reflected wave is now a compression wave. So at the open end of a pipe, a compression wave is converted to an expansion and vice versa. The gas velocity at the wave is 40.8 m/s and the wave propagation velocity is 392.1 m/s. Since the local gas velocity is positive, it is traveling in the same direction as the wave, that is from right to left, which is consistent with flow into the pipe.

At an open end, a wave is reflected as the opposite kind of wave: (Compression $\Rightarrow$ Expansion and Expansion $\Rightarrow$ Compression).

Area Change

A common occurrence in manifold systems is a change in the cross section area of the pipe. This is especially encountered in 2-stroke motorcycle engines, in order to make maximum benefit of the pressure pulsations existing in the exhaust pipe. It could be said that the conditions discussed up to now have been special cases of this, as the area change from a finite value to 0 is the closed end, while an increase to an infinitely large area corresponds to the conditions at an open. Given these limitations, it

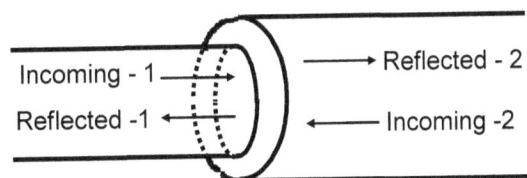

Figure 7.28: Sudden change in the cross section area of the pipe.

is not unreasonable to expect that the actual conditions with finite area changes correspond to something in between these two limits, and the results of area changes can be either an expansion or a contraction, as shown in Figure 7.28.

Figure 7.28, shows incident waves from both the left and the right. Even if the gas on the right hand side is quiescent, a wave approaching from the left will encounter gas there with a pressure ratio of =1.0. The reflected wave on the right, will actually be the transmitted portion of the incoming wave at 1.

As an example, consider the compression wave of the previous examples entering a sudden expansion with an area ratio of 2. The pressure amplitude ratio of the quiescent gas is 1.0, as shown in Figure 7.29.

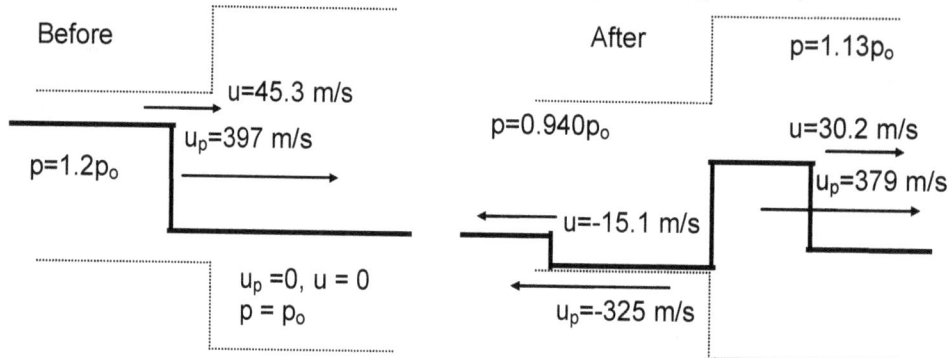

Figure 7.29: Compression wave encountering a sudden expansion.

Though they are not presented here, calculations will show that the reflected wave to the left has a pressure ratio of 0.940. See for example Blair's book [87]. The forward propagating "reflection", has a pressure ratio of 1.130. The wave can be seen as being divided, with an expansion wave being reflected with a pressure magnitude factor smaller than that for a completely open end, where the pressure ratio was 0.8292, and a compression wave continuing to propagate to the right with a pressure ratio of 1.13, a smaller magnitude than the original (1.2). A second example is the case of the same area enlargement, but for an incident expansion wave with pressure ratio of 0.8. In this case the reflected wave has a pressure ratio of 1.073 and the wave continuing in a forward direction has a pressure amplitude ratio of 0.8265. Again the wave is divided with the reflected portion as a compression wave, and the part continuing being an expansion wave like the incident wave but with a smaller magnitude. That is, the area expansion acts like a less effective open ended pipe than for a complete expansion to a reservoir.

A contraction behaves in a similar manner. That is, it functions as a partially but not completely closed end of the pipe. This can be shown by a couple of numerical examples. First consider a contraction in the pipe where the area ratio = 0.5, and the same compression wave is incident from the left. As before, the pressure ratio for the left side incoming wave is 1.2 and that for the right side = 1.0 due to the quiescent gas. Calculations show that pressure ratio of the reflected wave on the left = 1.057, and that of the "reflected" wave to the right = 1.274. In this case, both the reflected and the transmitted waves are compression waves, the reflected wave is weaker than the incident because of the partial reflection, while the transmitted wave is stronger, this is due to the smaller area and the acceleration of the flow. The final case is the contraction with an incident expansion wave, with a pressure ratio of 0.8. The reflected wave is also an expansion wave, with the pressure ratio = 0.9288, which is not as strong a wave as the incident wave, and the transmitted wave with a pressure ratio of 0.741. As before, the reflected wave is weaker, but the transmitted wave is stronger.

Example

To investigate the actions of waves in the air exchange process of engines, a simplified simulation developed by Professor Gordon Blair of Queen's University Belfast was used [87]. The program simulates the filling process of a constant volume cylinder through a port with a time dependent area. The flow into the cylinder is regulated by a port, which opens in a linear fashion with respect to time up to a maximum value, then closes in a reverse linear fashion. Between the atmosphere and the intake port, there is an intake pipe of constant area, the length of which can be chosen as a parameter in the calculation. The process is calculated on a cyclic manner from an initial condition, at an equivalent speed of a two-stroke engine. The flow in the pipe is calculated by a method called the method of characteristics, which calculates the wave motion in the pipe, including the reflection at the closed end when the port is closed, and reflection at the open ends of the pipe. The open ends are the end open to the atmosphere and the intake port, during the period of time when it is open to the cylinder. Results of the calculation include the mass and pressure in the cylinder throughout the cycle, and the velocity and pressure profiles along the intake pipe as a function of time. Since waves propagate in two directions, the calculation also gives values of pressure pulses moving in both directions.

The conditions for the calculation are: cylinder volume = 600 cubic cm, intake pipe diameter = 30 mm, initial cylinder temperature = 50°C, intake port diameter = 27 mm, intake port opening duration = 180° CA, atmospheric temperature = 25°C, equivalent engine speed = 6000 rpm, intake pipe length = 100 - 1500 mm

The calculations start at a time when there is no motion in the intake pipe and the pressure in the cylinder is assumed to be 0.6 bar. Figure 7.30A shows the conditions for a pipe length of 250 mm at an elapsed time of 0.86 ms, which is shortly after the intake port has opened. At the top left, the static pressure along the pipe is shown, and it can be seen that the pipe pressure near the cylinder drops to a low level as flow starts to enter the pipe. The curve below shows the Mach number of the gas along the length of the intake pipe. The height above the axis is proportional to the magnitude, and the arrows indicate that the flow is into the cylinder. To the upper right in the figure, an expansion wave created by the low cylinder pressure can be seen propagating away from the cylinder, and it has just reached the end of the pipe. As mentioned earlier, expansion waves are reflected as compression waves at open ends, and in the curve below this, an inward moving compression wave can be seen to start towards the cylinder. Since the original expansion propagated into quiescent air in the pipe, the speed at which it reaches the end should be the sonic velocity. This is confirmed with a simple calculation, where the expansion reaches the end of the pipe at about 26 crank angle degrees, which corresponds to a time of $L/V = .250/345 = .725ms = 0.725 \cdot 10^{-3} \cdot 6 \cdot 6000 = 26°CA$, consistent with arrival slightly before the graph in Figure 7.30A.

The bottom curve in Figure 7.30A shows the time histories of the pressures in the cylinder and the intake port. The cylinder pressure has already increased to a value of 0.63 bar as air flows in from the pipe, and the port pressure has fallen to a value near that of the cylinder. The cylinder mass ratio, shown in the table at the lower right, is the ratio of the mass in the cylinder to that of the cylinder filled with atmospheric air. At the start of the calculation, with the temperature of 50°C, the mass ratio was 0.554. The height of the white block at the left of the Mach number plot is proportional to the cylinder mass, and the distance between the two small black blocks at the left hand entrance to the pipe is proportional to the open area of the intake port.

The reflected compression wave will propagate towards the cylinder and result in a high pressure at the intake port, affecting the flow process there, and consequently the mass in the cylinder. The situation at the arrival of this reflected compression wave is shown in Figure 7.30B, which is for a time of 1.64 ms after the opening of the intake port. It can be seen that the reflected compression wave has arrived at the intake port, and that both the port pressure and the cylinder pressure are beginning to increase. There is a strong flow throughout the entire length of the pipe, filling the cylinder with air (Mach number plot). The cylinder pressure at this time has increased to 0.71 bar, and the cylinder mass ratio has increased to a value of 0.626. There is still an outward expansion pulse, as there is still a low pressure at the intake end (open port) of the pipe. The superposition of the two pulses gives the total pressure ratio in the pipe, shown in the upper left graph.

The compression wave arriving at the open port will then be partially reflected as a compression wave. This compression wave will then travel to the open end of the pipe, where it will again be reflected, this time as an expansion wave. Each time the wave is reflected at the open end, it will lose a little of its intensity, due to the flow conditions there. In this case, it is assumed that the open end of the pipe has a bell mouth structure, which results in slightly lower losses than a sharp edged pipe.

This wave, which was established by the sudden flow of air into the cylinder due to a pressure difference of 0.4 bar, will then propagate back and forth along the pipe throughout the cycle, reflecting at the boundaries as was described previously. The phasing of these reflections is determined by the speed of the "engine" and the length of the intake pipe. They determine the conditions at the cylinder inlet and, therefore, the rate of flow into and (potentially) out of the cylinder.

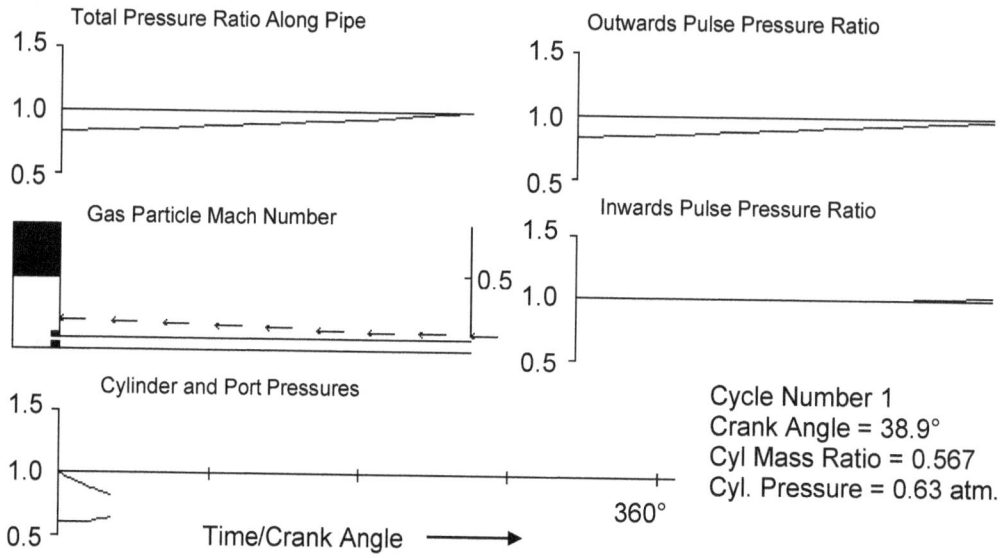

A

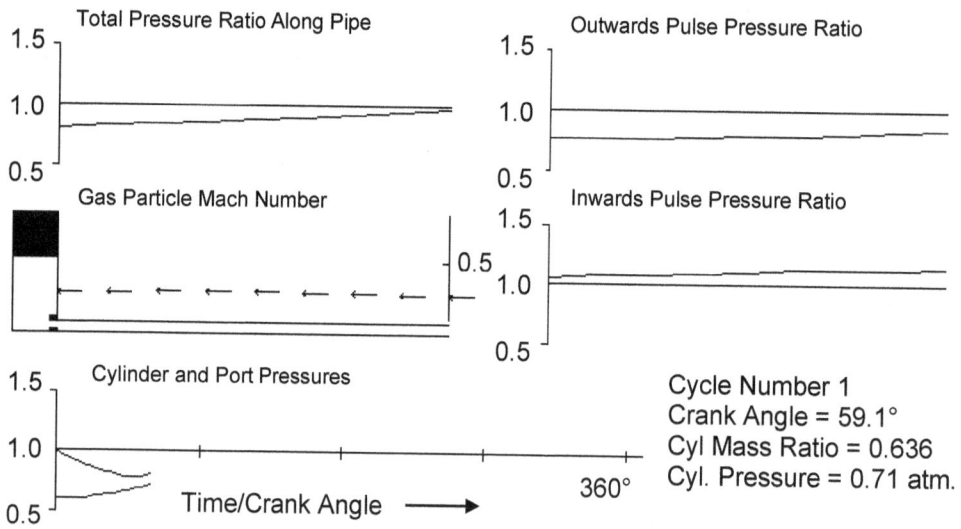

B

Figure 7.30: Calculation of the filling of a constant volume cylinder, 0.86 ms (A) and 1.64 ms (B) after opening of the pipe on the first cycle.

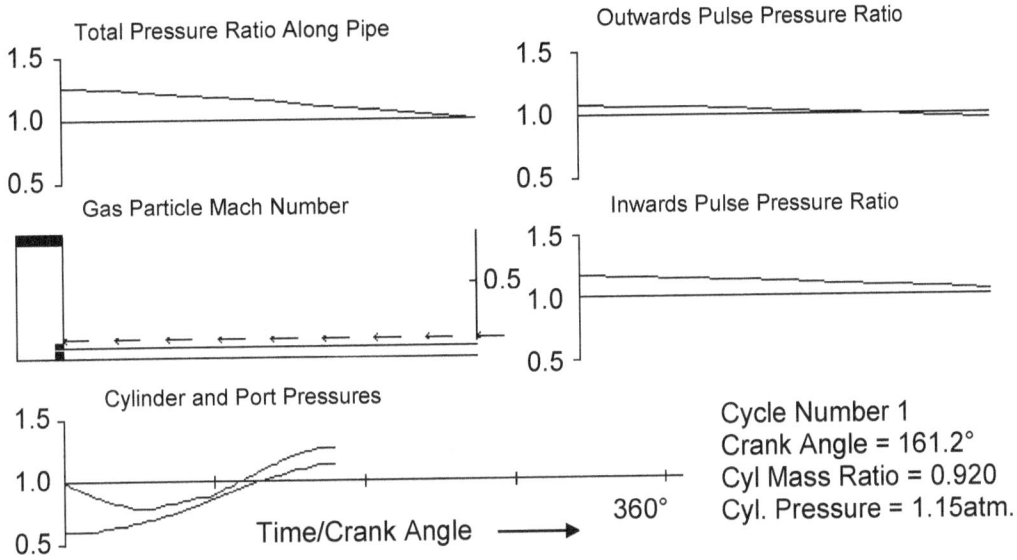

A

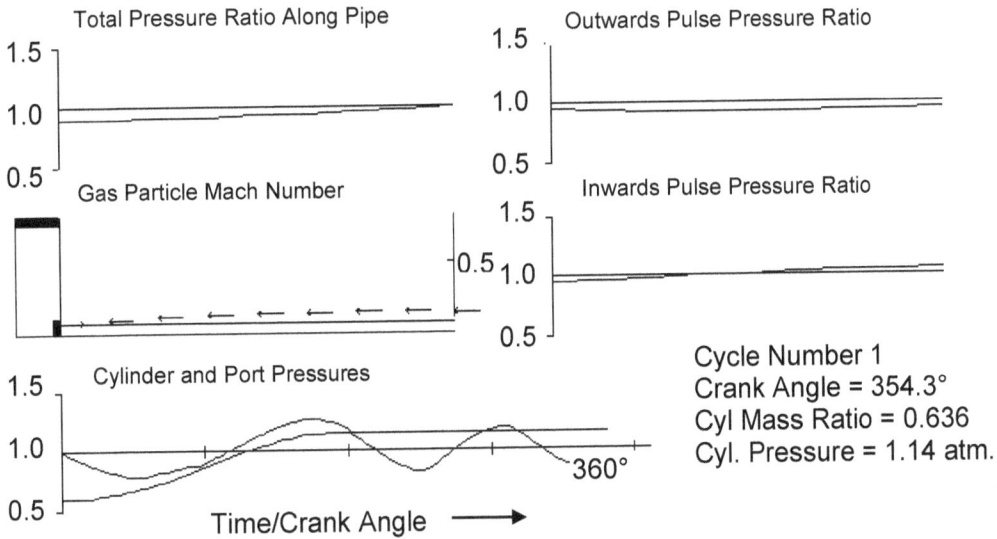

B

Figure 7.31: Calculation of the filling of a constant volume cylinder, 4.48 ms and 9.5 ms after opening of the pipe on the first cycle.

The next time of interest in the above calculation is near the time of intake port closing. This is shown in Figure 7.31A for a time of 4.48 ms after opening of the intake port. Note that the intake port is nearly closed now, and that the velocities are low.The phasing of the reflecting waves is such that there is a high pressure at the intake port and in the cylinder at this time, and the cylinder pressure is 1.14 bar, considerably above atmospheric. The cylinder mass ratio is also high, at a value of 0.92, so the filling process has been efficient. Had the pressure at the time of intake port closing been low, the situation would have been different, and there would have been much less mass in the system. After the intake port closes, the pressure wave will continue to oscillate, but can no longer affect the cylinder mass. The state at the end of the first cycle is shown in Figure 7.31B, at a time of 9.5 ms. The pressure oscillations continue throughout the entire cycle, with the closed intake port acting as a closed end for the remainder of the process.

The above process is for the first cycle after the port opens with a quiescent gas in the intake pipe. In this case the port pressure is close to atmospheric pressure at the end of the first cycle, so the pressure difference when the port opens on the following cycle will be similar to that when it opened at the start of the first cycle with quiescent air in the pipe. That this is the case is only a matter of the resonance of the wave in the pipe. By changing the length of the pipe, the situation could arise where the pressure in the port at the opening of the intake port could be considerably above or below atmospheric, and the flow processes affected correspondingly. Thus, the example shown is useful for demonstrating the movement and effect of the pressure wave established with the initial flow into the cylinder, but in an operating engine, the port flow for a given cycle will be dependent on the gas dynamic situation of the preceding cycle, and it is necessary to perform an iteration on the processes in the intake and exhaust to obtain correct results.

A situation where the wave motion established in a initial cycle has a significant effect on the air exchange process is shown in Figure 7.32. The length of the intake pipe has been increased to 400 mm, otherwise the conditions are the same as in the previous cycle. The upper set of curves in Figure 7.32 shows the cylinder pressure, port pressure, cylinder mass, and Mach number at the port during the first cycle, where the air in the pipe is stationary.

First, there are fewer waves in the cycles than in the previous example, which is due to the longer pipe length and the longer time required for a wave to reach an end of the pipe and be reflected. At the end of the first cycle, there is a high pressure in the port, which assists in the charging of the cylinder on the second cycle. The effect is moderated somewhat by the fact that the port is not open very much. By the time the port is open significantly, the pressure has fallen due to an expansion wave created by the first cycle, and in the middle of the charging process, the pressure in the pipe falls below the pressure of the cylinder, and the flow into the cylinder actually stops and reverses for a short time. This causes a lower charging of the cylinder. Later in the process, the pressure is high at the intake port, but by that time, the intake port is closing, and it is not so effective and the charging over all is poorer. In cycle 1, the cylinder mass ratio at the closing of the intake port was 0.882, whereas at the end of cycle 2, the cylinder mass ratio was only 0.811 at port closing, a reduction of 8%. The end of the two cycles is quite similar, as it is primarily produced from the pulse generated with the initial opening and charging of the intake port, and subsequent cycles will be very similar to cycle 2.

A summary of this effect is shown in Figure 7.33, which shows the effect of pipe length on the cylinder mass ratio for the conditions of the previous examples. Two curves are shown. The first is that where the gas in the intake pipe is stationary. Here, it can be seen that after a pipe length of about 800 mm, there is no effect of pipe length on the cylinder charging. The explanation is simply that the expansion wave created when the intake port opens does not have enough time to reach the open end, be reflected, and arrive back at the intake port before it closes. A simple calculation confirms this. The port is open for 180 crank angle degrees, so the wave must propagate from the intake port to the open end and back again in less than the opening time.

For an order of magnitude calculation, it is assumed that the wave travels at the speed of sound. As pointed out in Equation (7.50), the waves travel at the speed of sound plus the local gas velocity. For the outward wave, the gas velocity is 0, but not for the inward. However, since the gas velocity is still relatively small compared to sonic velocity, it is assumed that the propagating velocity is close to sonic velocity. Then the length for which the wave travels 2 lengths of the intake pipe and arrives at the intake

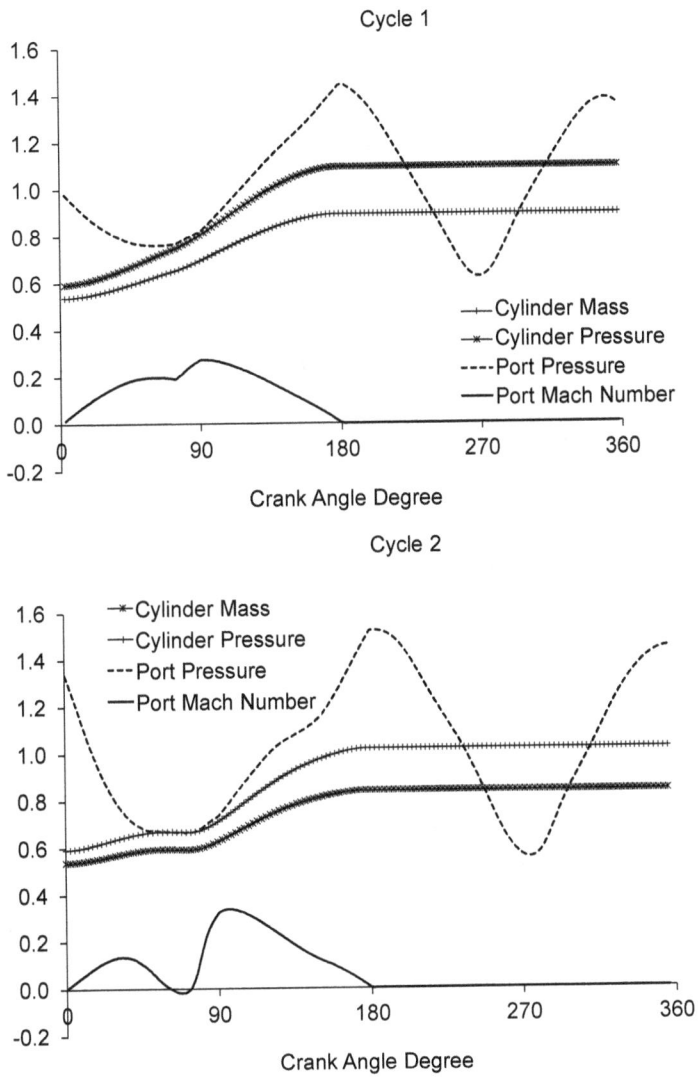

Figure 7.32: Cylinder and port pressure in atm, port Mach number and relative cylinder mass for the conditions of the previous example with a pipe length of 400 mm. Cycle 1 is the cycle with stationary air, and Cycle 2 is the subsequent cycle, where the wave motion from the previous cycle is used as the starting condition.

port just as it closes in 180 crank angle degrees is:

$$2L = \Delta t \cdot u_g \tag{7.55}$$

$$L = \frac{\Delta t \cdot u_g}{2} = \frac{\Delta \theta \cdot u_g}{2 \cdot 6 \cdot N} \tag{7.56}$$

$$\approx \frac{180 \cdot 340}{2 \cdot 6 \cdot 6000} = 850mm \tag{7.57}$$

This is in good agreement with the detailed calculations. When the pipe is longer than this, the wave generated by the cylinder opening can no longer affect the filling process. However, the wave still continues to propagate in the closed pipe, and as was shown above, the wave motion can affect the subsequent cycle. Thus, Figure 7.33 shows that there is a pronounced effect of the intake pipe length on the charging process for the cylinder for other cycles than the very first. Pipe lengths of about 300 mm and 600 mm give significantly better charging than other pipe lengths.

Similar results can be shown for the exhaust process, as the blowdown process creates a strong wave in the exhaust port that resonates and affects the air exchange in a similar manner as the results above. This is of greatest interest in the two-stroke engine, since both ports are open during the scavenging process, and the exhaust system can be de-

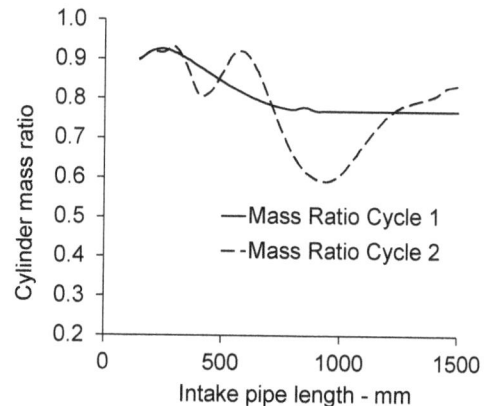

Figure 7.33: The effect of intake pipe length on the simulated example for an equivalent two stroke engine speed of 6000 rpm.

signed as to provide low pressure pulses during the major portion of the scavenging process to remove as much exhaust gas as possible. Then a high pressure pulse in the exhaust is desired late in the scavenging process to prevent intake air from being short circuited into the exhaust pipe. For four-stroke engines, the effect is not as pronounced. The initial portions of the blowdown process are not affected by exhaust pressure as long as the pressure in the cylinder is more than about twice the exhaust port pressure and the flow choked. Although the remainder of the exhaust process is quite dependent on the gas dynamics in the exhaust system, the exhaust stroke effectively removes the greater part of the exhaust gases from the cylinder by displacement, until there is on the order of 5 % residual gas left in the cylinder at the end of exhaust stroke and start of the intake stroke. Even a doubling for example, of the residual mass from 4 to 8 % would only give about a 4 % change in the air charge at full load. For this reason, manifold dynamics have the greatest effect on intake systems for four-stroke engines, and it is here most effort has been concentrated in past developments. While the benefits obtained by optimizing an exhaust system are helpful, the intake system is the most effective in terms of producing more power.

7.7 Effect of Manifold Tuning on Torque and Power

Figure 7.34 shows a comparison of the calculated and measured filling efficiencies of the GM Quad 4 engine, as calculated with the program described in erference [86]. It can be seen that there is good agreement in both the magnitude of the filling efficiency and the fact that the filling efficiency peaks at a speed of about 4000 rpm.

These results, along with those given in the paper for the exhaust process give confidence that the program is capable of correctly predicting phenomena in intake and exhaust systems. It should be pointed out that the simulation is only constructed to calculate the dynamics in single runners between pressure reservoirs and cylinders with varying volumes and valve areas, as determined by camshaft characteristics.

With this background, a series of computer simulations has been made for the basic the geometry of a 4 cylinder engine consisting of a bore of 91.9

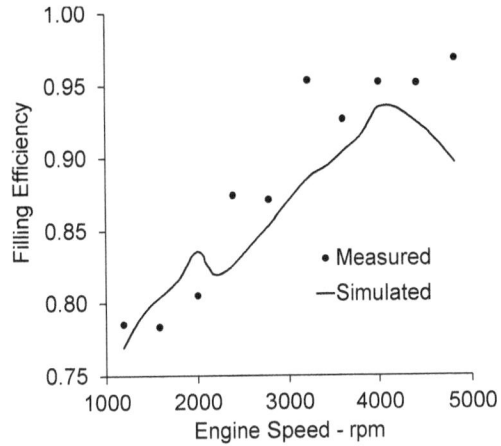

Figure 7.34: A comparison between experimental filling efficiencies and those calculated by an intake manifold simulation program for a 4 cyl. engine.

mm, a stroke of 85.1 mm, a connecting rod length of 135.9 mm and a compression ratio of 9.5:1. The engine has 2 intake valves with a diameter of 35.5 mm and 2 exhaust valves with a diameter of 30.0 mm. Some summaries of the effects of various parameters on engine performance are given in the following. The manifold used for these calculations was a simple straight pipe with constant diameter of 41.3 mm for the intake and 73.0 mm for the exhaust. The large diameter of the exhaust was used to simulate the junction of the runners into one large tube. As will be shown in the following, the engine performance is weakly dependent on the exhaust system for the dimensions used. The weak dependence is mainly due to the fact that the exhaust system primarily influences the residual fraction and the overlap process, both of which normally do not have major effects on engine performance. A smaller exhaust manifold diameter would increase the effects of the exhaust system.

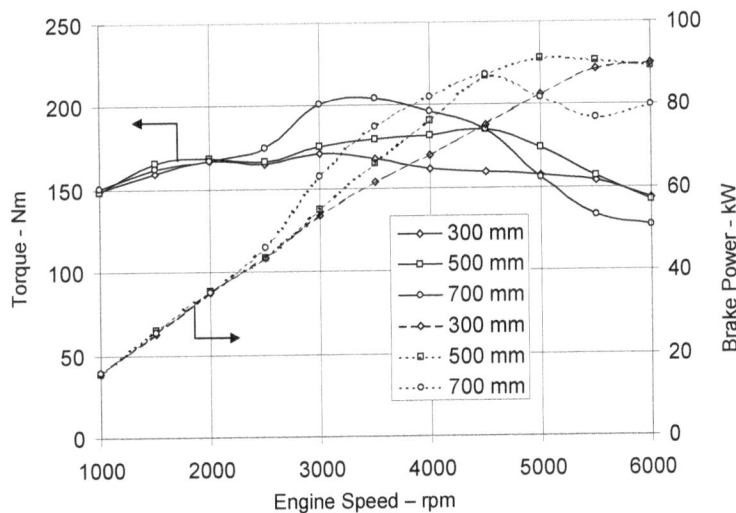

Figure 7.35: The effect of intake manifold length on engine torque and power at WOT. (Torque - solid lines, Power - dotted).

7.7.1 Intake Manifold Length

As is generally known to be the case, the length of the intake manifold has an important impact on engine performance. This is shown in Figure 7.35, for the simple manifold and the above engine parameters. The results show that a runner length of about 700 mm between the intake valve and the plenum gives a much higher torque than the other lengths in the speed range between 2500 rpm and 4500 rpm. At speeds above 4500 rpm the runner length of 500 mm gives the highest torque and power. At speeds below 2500 rpm, there is not a great difference in the performance of any of the manifolds. At the lower speeds, gas velocities are small and the pressures available from the kinetic energy of the gas is also small. At the highest speeds, all of the manifold lengths begin to show lower torque and volumetric efficiencies, as the effects of friction and shorter valve opening times reduce volumetric efficiency.

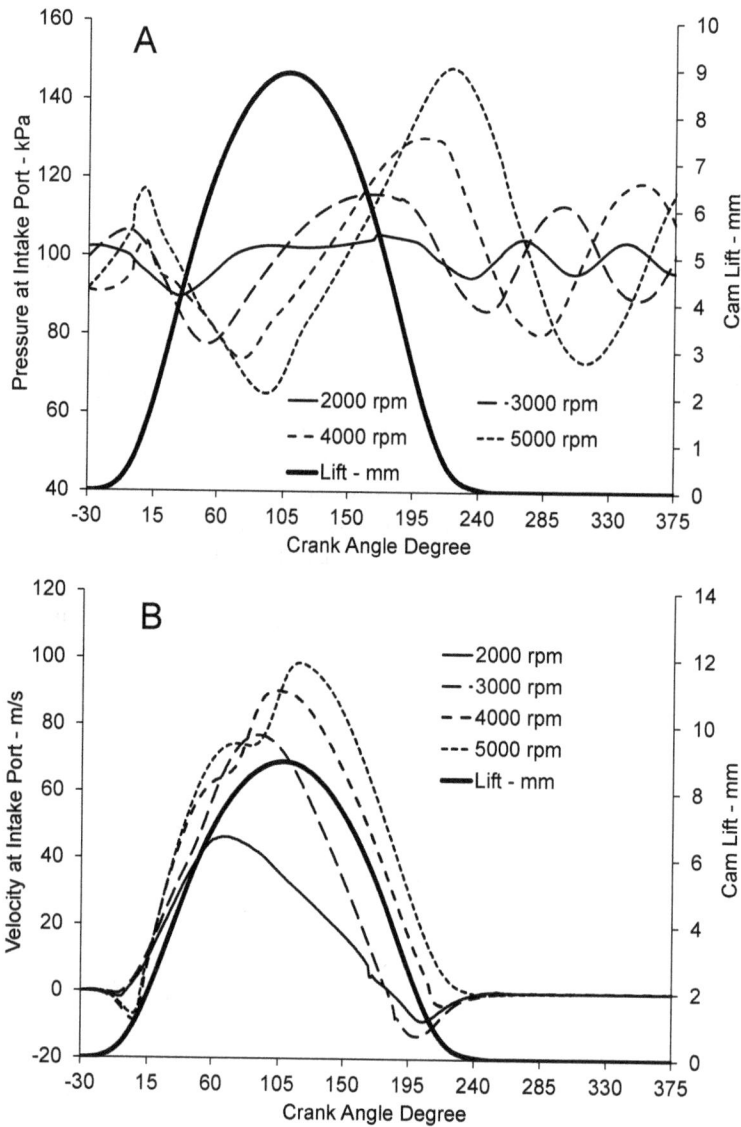

Figure 7.36: Pressure time (A) and velocity time (B) histories as a function of engine speed near the intake valve for a 505 mm intake runner.

An explanation of the effects of speed on the volumetric efficiency and hence the torque can be seen from the pressure time history of the manifold pressure at the intake port. This is shown in Figure 7.36A for the manifold length of 700 mm. Shown are the pressure and velocity time histories for 4 engine speeds along with the intake valve lift profile. Of major importance is the pressure in the intake port near the time that the valve closes. At 2000 rpm, the pressure is low near the time of intake closing, as a low-pressure wave is in the vicinity of the port. At 3000 rpm, the pressure at intake closing is higher, since the phasing has changed at the increased speed, and there is now a high-pressure wave near the port. The highest torque and volumetric efficiency occurs between 3500 and 4500 rpm.

Figure 7.36A shows that for these conditions, there is a high-pressure wave at the port, and that at 4000 rpm and 5000 rpm, the pressure at intake closing is nearly the same. The volumetric efficiency begins to fall around 5000 rpm because of flow losses, and the torque falls even more because of higher friction losses at the high engine speed. The differences in phasing of the pressure waves at the different speeds can be clearly observed.

Figure 7.36B shows the velocities at the same location. As pointed out earlier, the pressure history alone does not always give a complete indication of the flow situation. However, there is a difference in the results of Figures 7.36A and 7.36B as compared to the results of Blair shown earlier. The difference is that in Figure 7.36 it is the engine that determines the flow, which is driven by the movement of the piston, rather than being completely determined by the wave dynamics in the intake pipe. This is apparent in Figure 7.36B, where the velocity curve has an overall shape similar to that of the instantaneous piston speed.

The waves do have an effect, however, as can be seen by the decreases in the velocity at the higher engine speeds near the middle of the intake process. It is around this time that a low-pressure wave arrives near the port, and the low pressure tends to reduce the flow rate into the cylinder. At the end of the intake process (after BDC), the piston is moving upwards, which will tend to give a negative flow rate. Because of the high-pressure wave being reflected by the closing valve, the airflow remains positive for nearly the entire intake process at the higher speeds, resulting in a larger volumetric efficiency and giving more engine torque. Note that even though the pressure near the port is higher for the engine speed of 5000 rpm, the maximum pressure occurs too late to improve the volumetric efficiency, and in fact, the volumetric efficiency starts to fall between 4000 and 5000 rpm, due to the slightly later arrival of the pressure wave, and the reduced time to fill the cylinder.

Note also that the velocities are larger as the speed goes up. This is a natural consequence of the positive displacement of the engine. The piston displaces a given amount of air per intake stroke, and at the higher speeds this occurs in a shorter time, and results in higher intake system velocities. As shown earlier, there is a direct connection between the velocity and the associated pressure change (Bernoulli's equation or the First Law of Thermodynamics), and as the speed goes up, the magnitude of the pressure pulsations increases. This is the reason that in spite of the phasing differences of the waves that occur at the lower speeds, Figure 7.35 does not show much difference between the manifold lengths, as the pressure waves are not large enough to make a significant difference in the air charging.

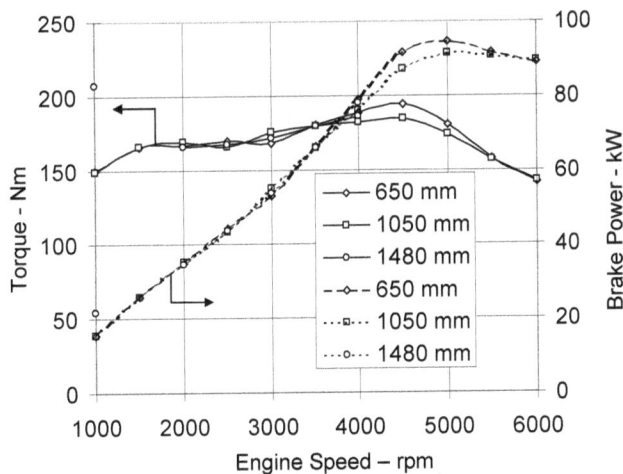

Figure 7.37: The effect of exhaust manifold length on torque and power at WOT. (Torque - solid lines, Power - dotted).

7.7.2 Exhaust Manifold Length

The effect on exhaust manifold length on full torque and power is shown in Figure 7.37. The effect is very small, as mentioned previously. Since the intake and exhaust processes are nearly completely separated in the positive displacement 4-stroke engine, the primary way the exhaust influences the intake is through the residual fraction. It is more important at higher engine speeds, since there is higher pressure at the end of the exhaust stroke, (see Figure 7.13), and a higher residual fraction. For 4-stroke engines, the intake remains the most important portion of the air exchange process. For 2-stroke engines, however, the entire charging and scavenging process is controlled by the gas dynamic process and there is a direct connection between the intake and exhaust ports since they are open simultaneously for a substantial portion of the scavenging process. Blair's books give an excellent description of these processes and their effects [87], [14], [15].

7.7.3 Intake Manifold Diameter

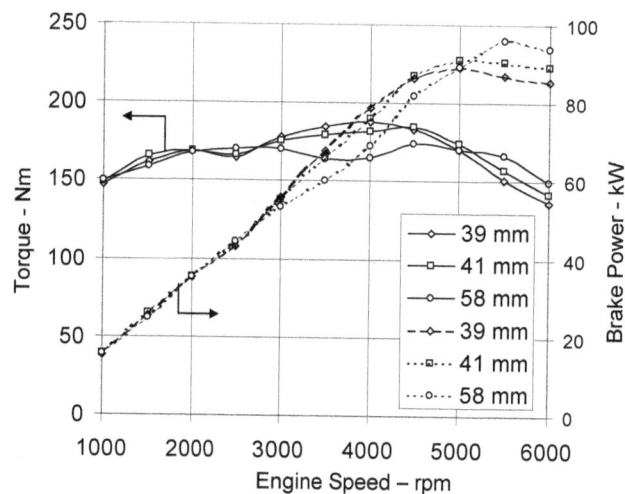

Figure 7.38: The effect of intake manifold diameter on torque and power at WOT. (Torque - solid lines, Power - dotted).

The effect of intake manifold diameter on full load torque and power is shown in Figure 7.38. The manifold length is the intermediate length (500 mm) as shown in the previous figures. It can be seen that the effects of manifold dynamics are larger when the manifold diameter is reduced. The smaller diameter gives higher velocities, which in turn give stronger pulses. The volumetric flow rate to the engine is determined by the instantaneous piston speed, and conservation of mass results in higher gas velocities in the smaller manifolds, when the volumetric flow rate is maintained. The gas velocities will also have some effect on the phasing of the pressure pulsations, as can be seen from Figure 7.38. As was shown earlier, the wave propagation velocity is the sum of the sonic velocity and the local gas velocity in the wave. Thus, for smaller diameter pipes, the waves will propagate faster. It should be noted that in Figure 7.38, the largest diameter gives twice the area as the intermediate diameter. The velocities in the pipe are therefore half those of the reference manifold, and the volumetric efficiency and torque curves approach those that would be expected without any form of tuning of the intake manifold. Although the use of a smaller intake runner diameter will increase the magnitude of the pressure oscillations, the higher velocities and larger surface to volume ratios will have a negative effect on the fluid friction, and will limit the benefits of charge air tuning in a practical application. The higher torque of the largest diameter intake manifold at higher speeds is an indication of this.

7.7.4 Extended Cam Duration

Figure 7.36 showed a dynamic interaction between the closing of the intake valve and the waves in the intake manifold system. It is, therefore, natural to expect that the cam profile will have an effect on the charging process. For the earlier discussion without regard to pressure pulsations in the manifolds, it is expected that earlier opening and later closing of the valves will result in better performance at higher speeds and worse performance at lower speeds.

To illustrate the magnitude of these effects along with the pressure pulsations, a modified camshaft profile was used. It was made by simply extending duration of maximum valve lift by 10° of crank shaft rotation. The profile was expanded symmetrically around the middle of the profile. A comparison of the standard camshaft and the extended duration camshaft is shown in Figure 7.39. The extended camshaft was used in connection with the reference intake manifold length of 500 mm. The previous results indicated that

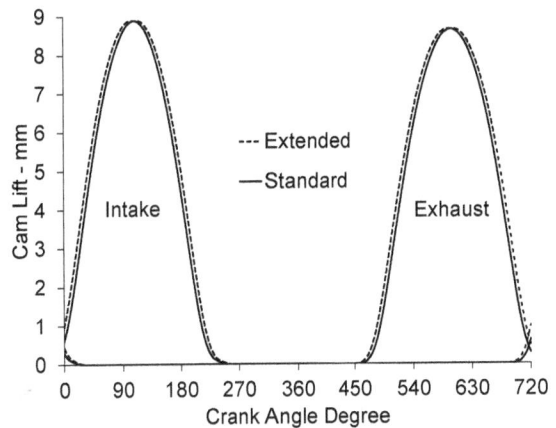

Figure 7.39: Lift profile for the extended duration camshaft as compared to the original camshaft for the simulated engine.

the timing of the high pressure pulse at the end of the intake process had a large effect on the volumetric efficiency and hence the torque of the engine. It was seen that as the speed increased from 4000 to 5000 rpm, the pressure pulse began to arrive too late to achieve maximum effect on the cylinder pressure at intake closing. Therefore, it would be expected that a delayed closing of the intake valve would prove beneficial at higher engine speeds.

This benefit is indeed shown in Figure 7.40, and at engine speeds above 4500 rpm, the extended duration camshaft gives a higher torque. However, the extended camshaft results in poorer performance at the lower speeds. Between 2500 and 3500 rpm, the differences appear to be due to phasing differences. That is, the whole curve appears shifted towards a higher speed in relation to the basic camshaft. In addition, at the very lowest speeds, the extended camshaft gives significantly lower torque and power. This is due to the upward piston movement near intake valve closure pushing the cylinder gasses back into the intake manifold. At the low speeds, the pressure pulsations are small, and do not play a dominant role in the charging process. Thus, with fixed cam timing, a low speed sacrifice is paid for the high speed improvement achieved with the longer duration.

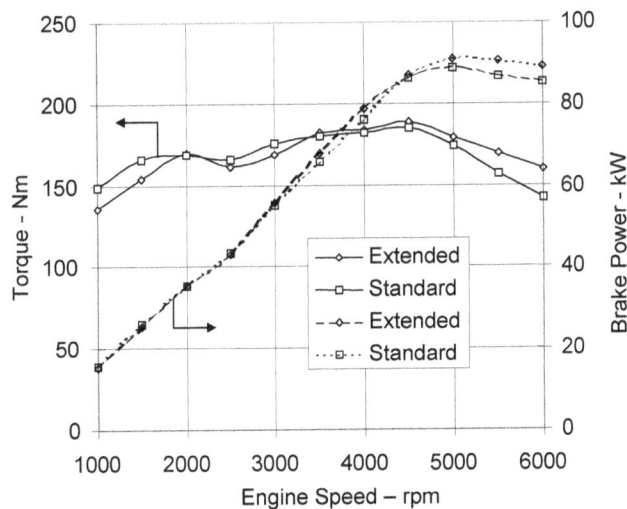

Figure 7.40: The effect of extended cam duration on torque and power at WOT. (Torque - solid lines, Power - dotted).

7.7.5 Practical Systems

Due to the resonance in the intake runner between the plenum and the cylinder, it might be expected that the optimum torque would be obtained with an infinitely variable runner length, that could be adjusted for each engine speed. However, the results of the previous section show that the resonance occurs over a fairly broad range of speeds. For example, Figure 7.35 shown that the torque curves for manifold lengths of 700 and 500 mm cross at about 4500 rpm. A good compromise between the two lengths could be obtained by using a runner length of 700 mm at engine speeds below 4500 rpm, and a length of 500 mm above 4500 rpm. While this does not give the absolute largest amount of torque, it does give a much broader range of high torque than using only one length. This idea has been used on some production automobiles.

The first system shown has been used by Opel on a 6 cylinder in-line SI engine, shown in Figure 7.41. The system consists of 6 runners of equal length extending into a chamber connecting all 6 runners. There are 2 runners leading out of this chamber to the throttle plate area, which acts as a second reservoir. In the middle of the chamber there is a simple valve which divides the chamber in half when closed. On the bottom of Figure 7.41, the valve is open, and the chamber acts as a plenum, and the system resonates with a short length. As the previous calculations have shown, the shorter system is best for the higher speeds. On the top of Figure 7.41, the valve is shown closed. In this case the effective volume of the plenum is much smaller and the resonating length of the runners is that between the valves and the throttle body.

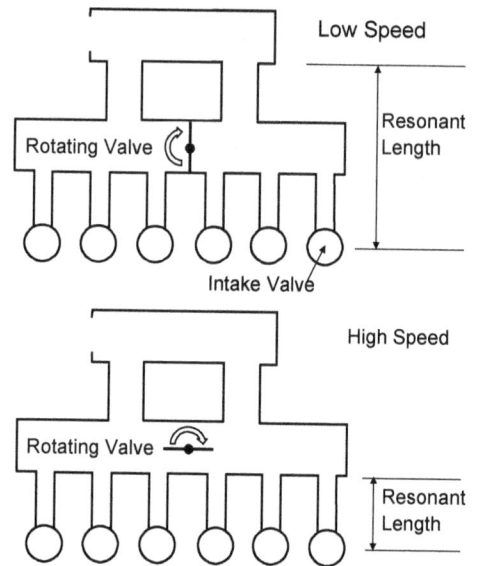

Figure 7.41: Principle of the dual length system on a 6-cylinder inline engine of Opel.

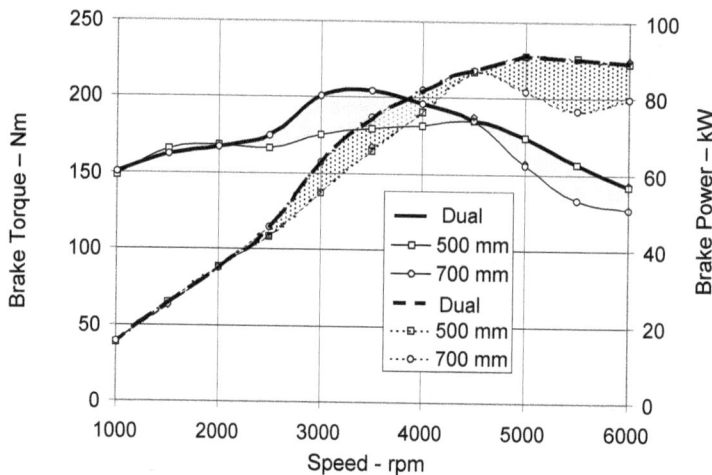

Figure 7.42: Torque and power curves for an engine with two effective intake manifold lengths. The heavy lines show the resulting torque and power curves, and the shaded areas indicate the gains at lower and higher speeds from using the longer and shorter lengths.

The resulting torque curves for the engine of the previous section are shown in Figure 7.42. With the valve open and the resonance occurring between the chamber and the ports, (short resonance length) the highest torque is obtained at speeds between 4500 and 6000 rpm. With the valve closed, the torque is higher at basically all speeds below 4500 rpm. By simply switching between the two lengths, a favorable torque curve is obtained.

The other system shown has been used by Audi. In this case, the engine is a V6 configuration. The space requirements for the manifold in this type of engine do not allow the use of a long system such as used in the Opel engine. Instead, curved runners are used, such that they lie between the banks of the V6. A schematic of this is shown in Figure 7.43. The manifold is nearly round, which enables runners

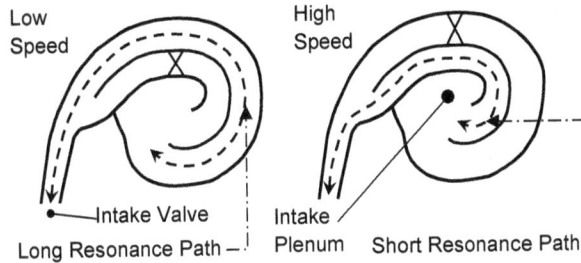

Figure 7.43: Schematic cross section intake manifold system used on the Audi V6 engine.

of substantial length to be used, while maintaining a compact engine configuration. The curvature of the runners has little effect on the resonance, and the one-dimensional concepts presented earlier in this chapter can be satisfactorily applied.

The air comes into the manifold in the plenum in the middle, in an axial flow from one end. With the valve open in the short runner, the air can take a shorter path length between the effective plenum and the intake port. This is the high speed condition, it gives the highest torque at higher speeds. When the shorter path is closed, and the air must take the longer pathway before reaching the intake valve. This configuration gives a longer effective runner length and provides higher torque at the lower speed, in the same fashion as the Opel engine in the previous example.

These examples indicate that it is possible to achieve substantial benefits from intake manifold tuning, and that variable length systems can be constructed by simple means. Due to the broadness of the torque peaks achieved at a given runner length, it is possible to achieve an elevated, but flat torque curve with only two different runner lengths. For the engines shown, the benefits were found at speeds over 2500 rpm. Manifold tuning effects are limited at lower engine speeds, due to lower gas velocities and reduced pressure wave magnitudes.

7.8 Problems

Problem 7.1

An engine has the following data: 4-stroke, speed = 1500 rpm, bore = 150 mm, stroke = 200 mm, number of cylinders = 12, volumetric efficiency = 0.85 and indicated efficiency = 0.30.

The engine is operated with either gasoline or generator gas. At the measurement location for the intake temperature and pressure, all the fuel has evaporated when operated on gasoline. Operating conditions for gasoline and gas operation are:

	Gasoline	Gas
Fuel/air ratio	0.08	0.84 kg/kg
Lower heating value	43,000	5,500 kJ/kg
Mole weight	113	24,7
Intake temp.	380	380 K
Press.	0.95	0.95 bar
Humidity	0.02	0.02 kg/kg

Find the indicated power when operated on both gasoline and generator gas. With gasoline operation, the mechanical efficiency = 85%, and the friction power is assumed to be the on gas as well as gasoline. Find the brake power and brake specific fuel consumption in both cases. (Gasoline : IP 385 kW; BP 327 kW; $bsfc$ = 328 g/kWh. generator-gas: IP 270 kW; BP 212 kW; $bsfc$ = 2780 g/kWh)

Problem 7.2

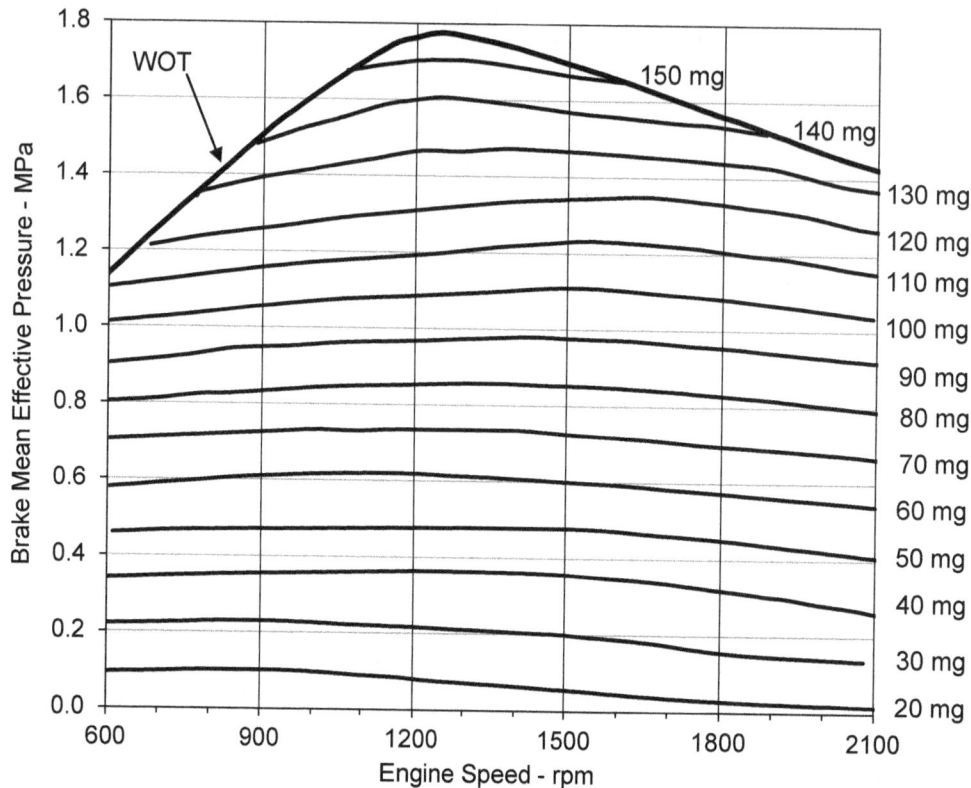

Figure 7.44: Engine fuel delivered per injection for problem 7.2.

There are natural gas powered busses that operate with rebuilt diesel engines, but with spark ignition and a homogeneous mixture of natural gas and air in the intake. The purpose of this problem is to estimate the consequences of these alterations. For that basic version, the engine in Figure 7.44 can be chosen. The engine has the following specifications: 6 cylinder, direct injection, Bore = 121 mm, Stroke: 140 mm Compression ratio: 18:1 Max power = 235 kW at 2100 omdr/min, with a brake specific fuel consumption of 208 g/kWh. The maximum torque = 1350 Nm at 1200 rpm with a brake specific fuel consumption of 195 g/kWh.

To operate on natural gas without knocking, the compression ratio must be reduced to 12.5:1 and the turbocharger removed. For emissions purposes, the engine is to be operated with a lean mixture, at a fuel air equivalence ratio of 0.8.

Natural gas has a representative composition of $C_1H_{3.6}$, with a lower heating value of 47,000 kJ/kg. A volumetric efficiency of 0.91 can be assumed for natural gas version of the engine for all engine speeds. Assume that the intake temperature is 35°C, the intake pressure 94 kPa, and the absolute humidity 0.015 kg water/kg dry air.

The new compression ratio is achieved by removing material from the top of the piston. How many cubic cm must be removed from each piston to achieve the new compression ratio? If the indicated efficiency on natural gas is 0.40, calculate the brake power on natural gas at full load and 1500 rpm. After the changes in the construction, it can be assumed that the friction mean effective pressure = 210 kPa. (45 cm^3, $BP = 68.6$ kW)

Problem 7.3

A 4-stroke spark ignition engine with carburetor operates with a fuel air equivalence ratio of = 1.15. The intake pressure is 94 kPa and the volumetric efficiency 0.91. The molecular weight of gasoline is 114 kg/kmol and the stoichiometric mixture ratio 0.067. Half of the gasoline is evaporated in the intake pipe, where the temperature is measured at 335K. The air is dry. The engine is altered to operate with gasoline injection directly into the cylinder, which eliminates the need to heat the intake pipe, and therefore the intake temperature falls to $T_{in} = 305$K. Find the ratio between the engine's air capacity before and after the changes, as the effect of gasoline evaporation after the measuring point can be neglected. (1.058)

Problem 7.4

A 4 cylinder, 4-stroke spark ignition engine is to operate on propane, C_3H_8. Propane has a vapor pressure of 5.5 bar at 20°C, and therefore is completely evaporated in the intake manifold. The engine is to operate with an excess air ratio of 1.25. The compression ratio is 10.5:1. The indicated efficiency is 82% of the value from an ideal, adiabatic Otto fuel air process (equilibrium cycle). The engine has a mechanical efficiency of 0.80 and a volumetric efficiency of 0.88 for an intake pressure of 97 kPa and intake temperature of 30°C. The exhaust pressure is 110 kPa. Propane has a lower heating value of 45.9 MJ/kg. The engine has a bore of 95 mm and stroke of 100 mm. 1. Find the brake power for full load at 3000 rpm.

2. Find the torque.

3. Find the brake specific CO_2 emission in g/kWh.

4. Find the indicated mean effective pressure.

($BP = 50.8$ kW, $T = 161.7$ Nm, $bsfc = 244$ g/kWh, $bsCO_2 = 730.6$ g/kWh, $imep = 895.6$ kPa)

Chapter 8

Two Stroke Engines

8.1 Background

Two-stroke engines are used in an effort to obtain higher specific power, that is, more kW per liter of displacement. The idea is to perform a work producing process each revolution of the engine instead of every other revolution as in the 4-stroke cycle. For the same mean effective pressure, a 2-stroke engine would produce twice the power of a 4-stroke engine:

$$BP = \frac{bmep \cdot V_d \cdot N}{60} \tag{8.1}$$

The indicated mean effective pressure was earlier defined as the work per unit displacement per cycle for the portion of the engine cycle from BDC at the start of compression to BDC at the end of expansion. Ideally, one could expect the same mean effective pressure from a 2-stroke or 4-stroke engine. Some equations will be presented in this chapter to show that. There are some factors that often result in a lower mean effective pressure for 2-stroke engines. The first, which applies to all 2-stroke engines, is that the air exchange process must occur within a single revolution. This means in practice that one must "give up" a portion of the displacement volume to the air exchange process. During the air exchange, the cylinder is open to the intake, exhaust or both, which means that compression and expansion cannot occur and little work can be produced during this scavenging period around bottom dead center.

The second is that the scavenging process limits the amount of fresh air entering the engine, and hereby the amount of fuel that can be burned. The air exchange process does not occur through positive displacement as in the 4-stroke engine, but instead through flow related processes. The air exchange in a 2-stroke engine is, therefore, much more sensitive to the design of the flow system than for a 4-stroke engine, and in general more difficult to conduct effectively, particularly in small engines. For large 2-stroke diesel engines, exhaust valves are used instead of ports, and proper matching of turbochargers allows an efficient air exchange process.

The third factor, which is most applicable to carbureted gasoline engines, is that 2-stroke engines use the crankcase to supply air for scavenging. The bottom side of the piston is used as a pump for the air, and this limits the amount of air per cycle that can be pumped through the engine to about the displacement volume. With this low flow and an often not too effective scavenging process, there is often a large amount of residual gas in the cylinder at the end of scavenging, which also reduces the amount of air available in the cylinder for combustion

The basic processes in a 2-stroke engine were shown in Figure 1.2. While the ports are closed, the engine operating principles are the same as those for four-stroke engines previously discussed. The scavenging process will determine the values of temperatures, pressures and residual gas fractions at the start of the actual compression. The engine shown is an SI engine, though 2-stroke diesel engines are also common. The difference, in principle, is that in a diesel engine, only air is inducted, and a fuel injector replaces the spark plug.

The air exchange process in a 2-stroke engine is different from that of the 4-stroke engine. In the latter, positive displacement of the air and exhaust gasses is achieved through piston movement on the intake and exhaust strokes. In the former, the air exchange is dictated by flows into the cylinder, inside the cylinder and out through the exhaust port. The piston is nearly stationary at the bottom of the cylinder when the fresh charge is blown through the engine to achieve scavenging. In the scavenging process it is necessary to generate pressurized intake air to be available when the intake port opens. Some common types of scavenging are shown in Figure 8.1. The main goal of the scavenging process is

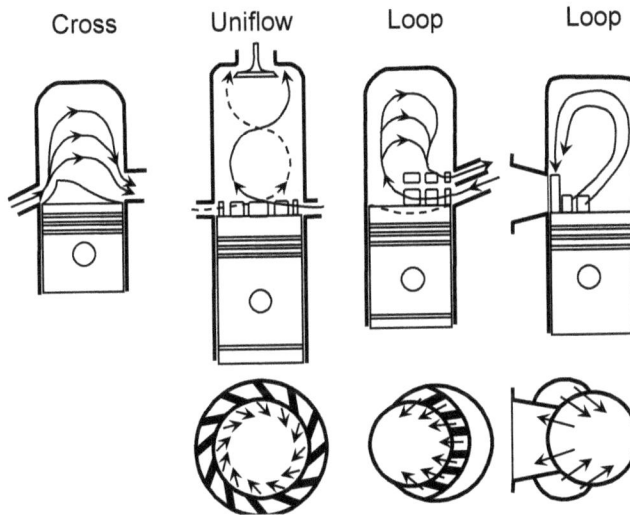

Figure 8.1: Commonly used scavenging systems in 2-stroke engines.

to remove the combustion products and recharge the cylinder with fresh air or a fresh air/fuel mixture. One of the simplest types of scavenging is the cross scavenging concept, seen on the left of Figure 8.1. In this concept, there is an intake port on one side of the cylinder and an exhaust port on the other side. There is a deflector on the piston, such that the intake flow is directed towards the top of the cylinder and away from the exhaust port. This is intended to prevent the intake charge from going directly out of the exhaust port (*short circuiting*) and instead to displace the burned charge. This type of scavenging leads to a very simple construction of the ports, and is used on very low cost engines.

A variation on the cross scavenging idea is that of the loop scavenging, also shown in Figure 8.1. As with the cross scavenging idea, both intake and exhaust flow occur through ports in the cylinder walls. In the loop scavenging concept, the incoming charge is directed towards the opposite cylinder wall from the intake ports, and its inertia used to keep it away from the exhaust ports, which are typically located on the same side of the cylinder as the intake ports. When the exhaust ports open prior to intake opening, the exhaust gasses flow out of the cylinder in the opposite direction to the incoming charge. Later, the incoming charge flows towards the opposite wall, makes a loop and then flows towards the exhaust port. A central idea is to keep the intake charge away from the exhaust port for as long a time as possible. This improves displacement of the combustion products and reduces the short circuiting (the flow of fresh charge directly out the exhaust ports).

In both of these system, the fluid flow situation is critical to good scavenging. There is little piston motion to move the charge. In the case of gasoline engines with these types of scavenging, the energy for the intake flow is obtained from compression of the gasses in the crankcase by the piston. The intake air flows into the crankcase before entering the cylinder through the transfer port. The inlet of the crankcase is typically sealed with a reed valve such that when the piston moves toward top dead center, the pressure in the crankcase drops and the valve opens, allowing air to flow in. When the piston reverses its direction, the reed valve is pressed against its seat by the increasing crankcase pressure, and the pressure in the crankcase increases until the intake port is opened by the piston, and air flows into the cylinder. Another system is a port through the skirt of the piston that opens to allow filling of the crankcase. Figure 8.2 shows a typical pressure time diagrams for the cylinder and crankcase of a 50 cc 2-stroke engine operating as a compression ignition engine. It can be seen that as the cylinder pressure increases due to compression, the crankcase on the opposite side of the piston has a decreasing pressure.

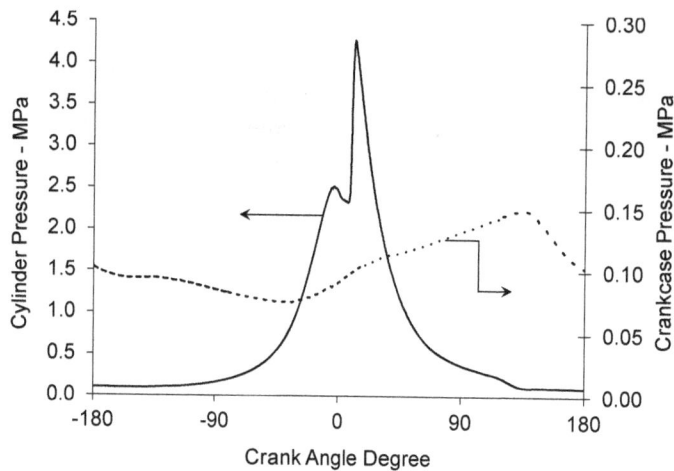

Figure 8.2: Pressure - crank angle diagrams for the cylinder (solid line) and crankcase (dashed line) of a single cylinder 2-stroke engine using the crankcase and a ported piston to provide scavenging air.

Figure 8.3 shows $p-V$ diagrams for the cylinder and crank case of the engine in Figure 8.3. This engine has a ported piston that opens the crankcase to intake air to allow crankcase filling. The piston port is open near TDC, and air fills the crankcase. As the piston moves downward (cylinder expansion) the crankcase air is compressed until the scavenging port in the cylinder wall opens and lets the compressed crankcase air charge enter the cylinder. The crankcase pressure falls as the piston moves upward to compress the gasses in the cylinder. Near TDC, the piston port opens air flows into the low pressure of the crankcase, and the cycle is repeated. The pressure at the end of the crankcase filling process is very dependent on the geometry of the intake system, and pressure wave dynamics are important here.

The work performed on the crankcase gasses corresponds to an mean effective pressure of 43.3 kPa. This is a pumping loss and can be considered part of the engine friction. In a four stroke engine, the crankcase pressure is maintained close to atmospheric, and the pumping work in the crankcase is usually neglected. A potential disadvantage of these scavenging systems is the use of ports in the cylinder walls to determine timing. The first requirement of these systems is that the exhaust port must open before the intake port, with sufficient time between the openings to allow the cylinder pressure to decrease to the level of the intake pressure. But this means that on the upward stroke of the piston, the exhaust port closes after the intake port, and so there is a period of time where the fresh charge can "leak" into the exhaust system, decreasing charging of the engine and giving less power. Exhaust pipe tuning can help this if there is a high pressure pulse at the exhaust port at the time both ports are open. There is also the possibility of an auxiliary valve in the exhaust port to close off the cylinder, but this has been difficult to accomplish in a simple yet reliable fashion.

The other common scavenging type is uniflow scavenging, which is especially used on larger diesel engines. Most of these systems use an exhaust valve in the cylinder head combined with intake ports at the bottom of the cylinder. This increases the complexity and weight of the engine due to the valve train components. However, it has the major advantage that the exhaust opening and closing are not determined by port position alone. That is, the exhaust timing with a valve can be altered relative to the intake ports such that the exhaust valve can be closed before the intake ports close. This allows higher cylinder pressure at the start of compression and more power, especially in the case where the intake pressure is above the exhaust pressure, as with engines using blowers, and with many turbocharged engines.

With the 2-stroke engines, it is possible to supply extra air in order to achieve better scavenging. This is practical with an engine that injects fuel into the cylinder directly, after the scavenging process is completed. The two kinds of engines that operate like this are the diesel engine and the direct injection gasoline engine. In these engines, only air is blown through the engine during scavenging, so there is no major loss in fuel, nor is there any increase in air pollution due to the increased scavenge air flow when the flow is increased. The blower does, of course, require power to operate. As long as the pressure differences are not too large, the blower can be made to operate with acceptable efficiency, as discussed

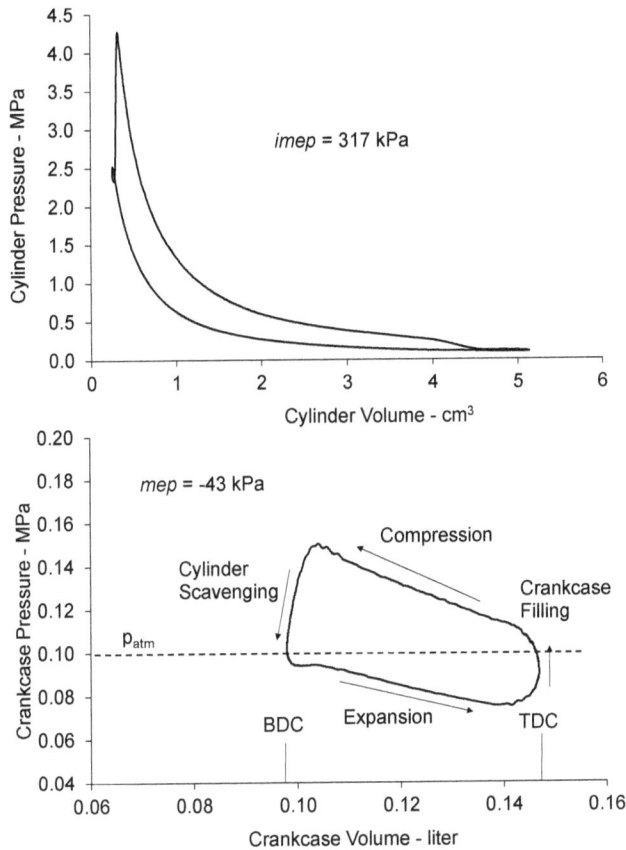

Figure 8.3: Pressure volume diagrams for the cylinder (upper) and crankcase (lower) of a single cylinder 2-stroke engine using the crankcase and a ported piston to provide scavenging air.

in Section 11.2. The blowers used normally have too low efficiency for higher levels of pressure charging.

With a carbureted engine, however, the use of extra scavenging air would result in a large increase in fuel consumption since more fuel would simply be blown through the engine along with the air not trapped in the cylinder. Even with the delivery of the ideal air charge, this is a major source of exhaust emissions and a reason for high fuel consumption of carbureted 2-stroke engines.

Some examples of 2-stroke engines are shown in Figure 8.4. The diesel engine uses a blower to improve air scavenging, otherwise there is too much residual gas at the end of scavenging and the power limited. This increases the size of the engine, and costs power to operate, which increases fuel consumption. As will be seen later, the basic efficiency of a 2-stroke diesel should be of a comparable level to that of a four-stroke. Extra arrangements for improved air flow are a disadvantage for 2-stroke diesel engines especially if they are turbocharged, since they increase size, weight and cost. Medium sized diesel 2-stroke diesel engines are basically no longer produced.

One application where the 2-stroke diesel engine is dominant is that of larger marine diesel engines. An example of such an engine is shown in Figure 8.5. The largest current marine engines operate at about 80-90 rpm, have a bores of about 1 meter and strokes of over 3 meter and operate with a mean effective pressures of about 19 - 20 bar. In a 14 cylinder version, the largest engine produces 87 MW at a bsfc between 162 and 170 g/kWh. This corresponds to a brake thermal efficiency of 50 - 53%. The bmep is similar to that of heavy duty truck engines, and combined with the 2-stroke cycle, produces a large amount of power. The engines are turbocharged, and proper design/selection of the turbocharger is an important of the scavenging process, and the pressure difference between the compressor and turbine provides the scavenging flow, instead of the blower shown in the above example. Uniflow scavenging shown in Figure 8.1 is used in this type of engine.

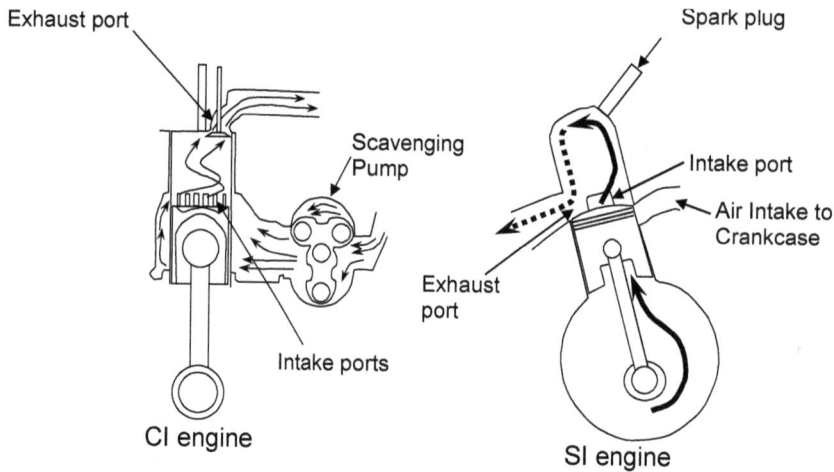

Figure 8.4: Examples of 2-stroke engines. An older DI diesel engine with a blower to aid scavenging (left) and a small SI engine for a scooter or power tool application.

8.2 Scavenging Parameters

With this background, it is of interest to look at the parameters that are used to describe the scavenging/air exchange process in 2-stroke engines. The definitions used are based on the book by Blair [87]:

1. *The Delivery Ratio*, which denoted by R_d, and defined as the ratio of the flow rate of air measured at the intake of the engine to the ideal flow of air which would fill the displacement volume of the engine at atmospheric density at the given speed.

$$R_d = \frac{\dot{m}_a}{\dot{m}_{a,id}} = \frac{\dot{m}_a}{\frac{\rho_{atm} V_d N}{60}} \tag{8.2}$$

2. *The Scavenging Ratio*, denoted by R_s, and defined as the ratio of the flow rate of air measured at the intake of the engine to the ideal flow of air which would fill the entire cylinder volume of the engine at the density of the scavenging air at the given speed.

$$R_s = \frac{\dot{m}_a}{\dot{m}_{a,id}} = \frac{\dot{m}_a}{\frac{\rho_s V_{cyl} N}{60}} \tag{8.3}$$

The scavenging density, ρ_s can be calculated from the temperature and pressure of the scavenging air at the intake to the engine. The total cylinder volume is used because in principle the entire volume could be filled with fresh charge. This is opposed to the 4-stroke engine, where it is the displacement of the piston that determines the ideal amount of air entering the engine. The cylinder volume is usually calculated from the displacement volume and the compression ratio.

$$V_{cyl} = V_d \left(\frac{\varepsilon}{\varepsilon - 1} \right) \tag{8.4}$$

Thus the scavenging ratio can be calculated as:

$$R_s = \frac{\dot{m}_a}{\dot{m}_{a,id}} = \frac{\dot{m}_a}{\frac{\rho_s V_d \varepsilon N}{60 (\varepsilon - 1)}} \tag{8.5}$$

An alternative definition of the scavenging density was proposed by Taylor and Taylor [6]. This is based on the idea of scavenging with an engine having both intake and exhaust ports. The opening and closing of the ports is defined by the piston position, and since the exhaust port must open before the

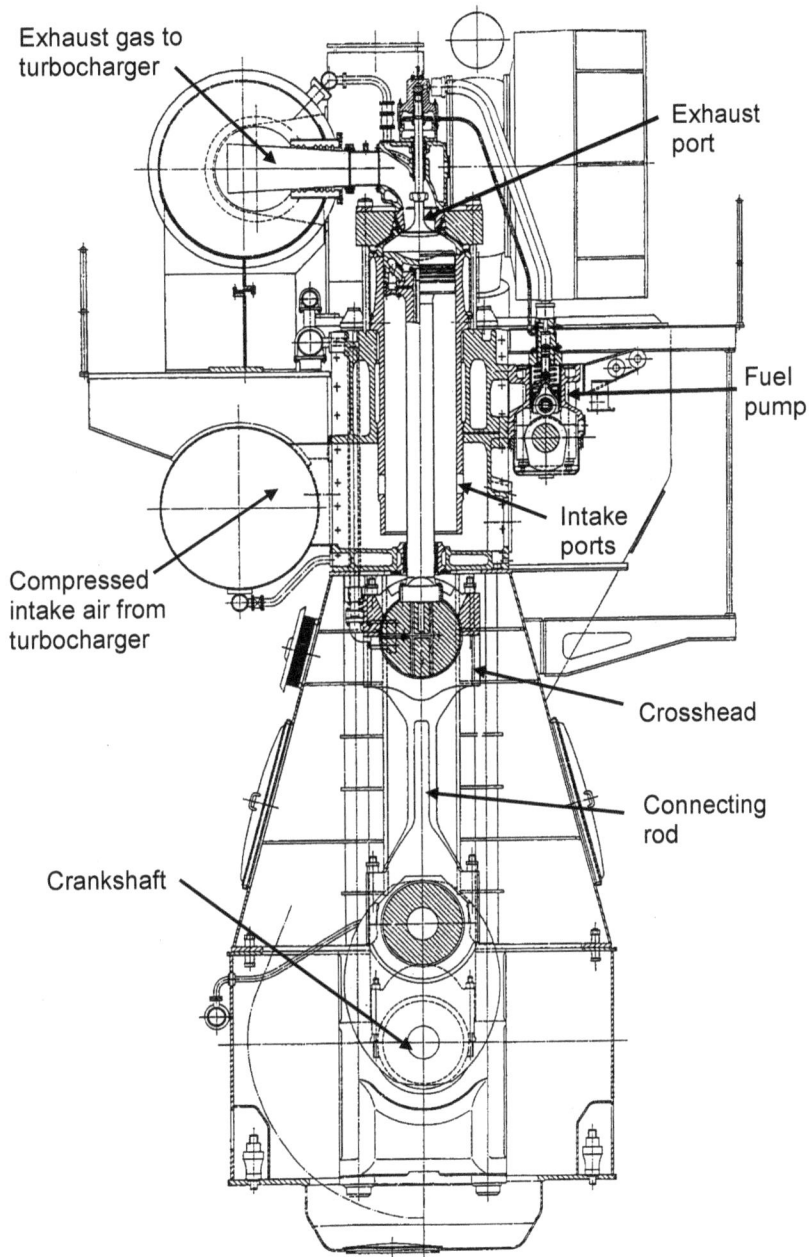

Figure 8.5: A large 2-stroke Marine diesel engine, the engine shown has a bore of 60 cm and a stroke of 2.29 m. Courtesy MAN Diesel, used by permission.

intake to let the high pressure gasses out, the exhaust port also closes after the intake port does. In this case, it is the exhaust pressure that theoretically determines the density of the cylinder gasses at the end of the air exchange process. However, the temperature of the air in the cylinder is determined by the temperature of the intake gasses. With this reasoning, a scavenging density can be calculated using the intake temperature and the exhaust pressure:

$$\rho_{sc,Tay} = \frac{p_{exh}}{RT_{in}} \tag{8.6}$$

Taylor and Taylor also discuss a correction of the intake airflow for water vapor and evaporated fuel. This correction is the same as used for four-stroke engines, and is discussed in Section 7.2. In the following this correction is not considered.

3. The *Trapping Efficiency*, η_{tr} is defined as the ratio of the mass of fresh charge trapped in the cylinder at the completion of the scavenging process to the mass of air supplied to the intake of the engine.

$$\eta_{tr} = \frac{\dot{m}_a'}{\dot{m}_a} \tag{8.7}$$

where: $\dot{m}_a'$ represents that portion of the air flow measured at the inlet of the engine that remains trapped inside the cylinder at the completion of the scavenging process.

Due to mixing of the fresh charge and residual gasses, the residual charge is not displaced completely as the fresh charge enters, and some fresh charge leaves the engine without being trapped in the engine. Even with a scavenging ratio greater than 1, the trapping efficiency of the engine will normally be less than one.

4. The *Charging Efficiency*, η_{ch}, is defined as the ratio of the fresh charge trapped in the cylinder to the ideal capacity of the engine. Then:

$$\eta_{ch} = \frac{\dot{m}_a'}{\rho_s V_d \cdot \frac{\varepsilon}{\varepsilon - 1} \frac{N}{60}} \tag{8.8}$$

Note that there is a difference of opinion as to which reference volume should be used when calculating the charging efficiency. Some sources, for example the SAE Handbook, use the displacement volume and others use the total cylinder volume, including the compression volume, as shown here.

From the above definitions, it can be seen that:

$$\frac{Trapped}{Ideal} = \frac{Supplied}{Ideal} \cdot \frac{Trapped}{Supplied} \quad \Rightarrow \quad \eta_{ch} = R_s \cdot \eta_{tr} \tag{8.9}$$

The charging efficiency is the closest counterpart to the volumetric efficiency used in the case of four-stroke engines. Normally in a four-stroke engine, it is assumed that all the intake air is trapped in the cylinder, though this is not strictly true in some cases, particularly in a turbocharged engine with valve overlap.

5. The *Scavenging Efficiency*, η_s, is defined as the ratio of the fresh air charge trapped in the cylinder to the total cylinder charge at the end of scavenging. The charge consists of the trapped air mass, m_a' and the residual gas mass, m_r Then:

$$\eta_s = \frac{m_a'}{m_a' + m_r} \tag{8.10}$$

The scavenging efficiency indicates the effectiveness of the scavenging process in terms of removing residual gasses. In an actual engine, it is difficult to determine, since one has to know the masses at the time of exhaust port closing. Using the definition of the residual fraction, f, from four-stroke engines, the scavenging efficiency can be seen to be $\eta_s = 1 - f$. When calculating the amount of air available for combustion in an engine with lean combustion, one needs to take into account that there is air in the residual gas. Methods from Section 3.1.2 can be used for this purpose. This is no different than common usage in four stroke engines. Blair [14] uses an additional parameter called *purity* to determine the oxygen concentration in the cylinder charge. Here, it is assumed that the combustion products result from stoichiometric combustion, and that anything in excess is air. This "air" is then added to the trapped air from the scavenging air, and the ratio of that to the total trapped cylinder mass is called the purity. The two methods are equivalent.

For maximum performance, the goal is to achieve a high value of the charging efficiency. This can be done either by increasing the amount of air sent through the engine (increasing the scavenging ratio) by increasing the pumping capacity of the system, or by increasing the efficiency with which the air is retained in the cylinder (trapping efficiency). Mounting a blower in the intake system of the engine is one example of how the air flow to the engine can be increased. The disadvantages of this are the cost, increased power consumption, and in the case of carbureted engines, an increased amount of unburned fuel blown directly into the exhaust. This gives high fuel consumption and hydrocarbon emissions. Blowers have been, therefore, only used in diesel engines, since the scavenging gas is only air, and scavenging air that is not trapped in the cylinder does not represent a degradation in either efficiency or emissions. The pressure difference for these blowers is usually modest, giving small power requirements for the blowers.

The blower power can be obtained from the same relationship as a compressor, Equation (11.6):

$$P_{bl} = \frac{\dot{m}_a \cdot c_p \cdot T_{in}}{\eta_{bl}} \left(\left[\frac{p_{in}}{p_{ex}} \right]^{\frac{\gamma-1}{\gamma}} - 1 \right) \tag{8.11}$$

where the blower efficiency is defined as the ratio of the isentropic compression work to the actual compression work.

The other way to obtain a higher charging efficiency is to improve the trapping efficiency. This is the method needed for carbureted engines, since large delivery ratios give efficiency and emissions problems. This is achieved by careful design of the intake and exhaust ports that determines the flow of the scavenging gasses in the cylinder. Trapping efficiency can also be improved by the careful design of the exhaust system to effectively utilize the pressure pulsations that naturally arise there. The book by Blair [87], has an extended description of the scavenging process for 2 stroke engines.

Clearly, it is preferred to obtain a good charging efficiency by improving the trapping efficiency. There are, however, limits to this, a main reason being a large amount of mixing due to the occurrence of turbulent jets in the intake process. In fact, one ideal scavenging model is based on the concept of perfect, instantaneous mixing of the fresh charge and residual gasses in the cylinder, and is the basis for empirical scavenging models, which will be shown in Section 8.4.

8.3 Power and Efficiency

In determining the power of the engine, it is only the fuel trapped in the cylinder during the scavenging process that can be used to produce power. In a similar fashion to the four-stroke engine, one can write[1]:

$$IP = \dot{m}'_a \cdot FA' \cdot H_u \cdot \eta'_i \tag{8.12}$$

where: $FA' = \frac{\dot{m}'_f}{\dot{m}'_a}$ is the fuel air ratio in the cylinder based on the amounts of fuel and air trapped in the cylinder when the scavenging process is over. The notation refers to quantities trapped in the cylinder at the time of last port closing on the compression stroke. Also:

$$\eta'_i = \frac{IP}{\dot{m}'_f \cdot H_u} \tag{8.13}$$

The scavenging process has a large effect on engine power and, for carbureted engines, the efficiency. This is schematically illustrated in Figure 8.6. This figure shows the fate of the fuel in the case of an engine with a carburetor and the case of in-cylinder injection (diesel or GDI) engine. In the carbureted engine, a portion of the fuel goes directly through the engine together with the non-trapped air. In the direct injection engine, all the fuel is injected into the cylinder when it is closed to the surroundings. Thus for injected engines, the trapped fuel flow is equal to the fuel flow measured at the intake to the engine.

---

[1]Strictly speaking, in a 4-stroke engine, one should also refer to the trapped air charge and not the air supplied, but normally these two are nearly identical.

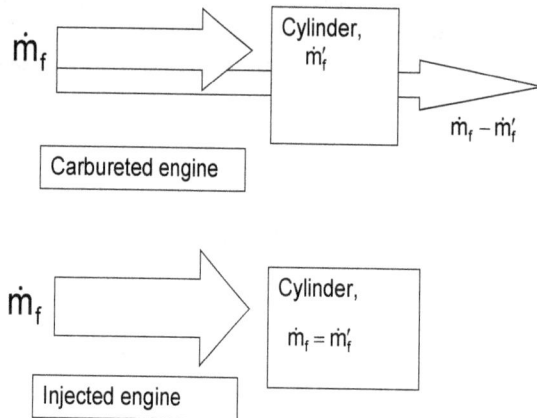

Figure 8.6: Flow path of the fuel for carbureted and injected 2-stroke engines.

For an injected engine, the trapped fuel is equal to the supplied fuel, so: $\eta_i = \eta_i'$ and since $\dot{m}_a' = \dot{m}_a \cdot \eta_{tr}$:

$$FA' = \frac{FA}{\eta_{tr}} \tag{8.14}$$

Then for an injected engine:

$$FA' \cdot \eta_i' = \frac{FA \cdot \eta_i}{\eta_{tr}} \tag{8.15}$$

For a carbureted engine, $FA = FA'$, and $\eta_i = \eta_i' \cdot \eta_{tr}$, which gives the same result as Equation (8.15). Thus Equation (8.12) is valid for both injected engines and those with carburetors, and can also be written:

$$IP = \dot{m}_a' \cdot FA' \cdot H_u \cdot \eta_i' \tag{8.16}$$

$$= \dot{m}_a' \cdot FA \cdot H_u \cdot \frac{\eta_i}{\eta_{tr}} \tag{8.17}$$

Using the definitions of the scavenging parameters, an equation for the power can be written in different ways:

$$IP = \eta_{ch} \cdot \rho_{sc} \cdot V_d \left(\frac{\varepsilon}{\varepsilon - 1} \right) \cdot \frac{N}{60} \cdot H_u \cdot \frac{FA \cdot \eta_i}{\eta_{tr}} \tag{8.18}$$

Since $\eta_{ch} = R_s \cdot \eta_{tr}$, Equation (8.18) can also be written:

$$IP = R_s \cdot \rho_{sc} \cdot V_d \left(\frac{\varepsilon}{\varepsilon - 1} \right) \cdot \frac{N}{60} \cdot H_u \cdot FA \cdot \eta_i \tag{8.19}$$

Using the relationship between power and mean effective pressure:

$$imep = \eta_{ch} \cdot \rho_s \left(\frac{\varepsilon}{\varepsilon - 1} \right) \cdot H_u \cdot FA' \cdot \eta_i' \tag{8.20}$$

or:

$$imep = R_s \cdot \rho_s \left(\frac{\varepsilon}{\varepsilon - 1} \right) \cdot H_u \cdot FA \cdot \eta_i \tag{8.21}$$

Equation (8.20) is directly comparable with the analogous equation for the four-stroke engine (Equation (8.22)), since all quantities there are considered to be based on the air trapped in the cylinder (the entire airflow is assumed trapped in four-stroke engines):

$$imep = \eta_v \cdot \rho_{in} \cdot H_u \cdot FA \cdot \eta_i \quad 4 - Stroke \tag{8.22}$$

In terms of engine power, Equations (8.16) - (8.21) are valid for both carburetor and cylinder injection engines and show that there is no difference in power between a carbureted engine and an injected engine due to the fact that there is fuel lost with the scavenging air in a carbureted engine. The reason is that it is only the trapped air that determines the amount of fuel that can be burned and therefore the power. The fuel lost with ineffective scavenging will only affect the efficiency of the engine. This is shown in the following.

The definition of the indicated specific fuel consumption is:

$$isfc \equiv \frac{\dot{m}_f}{IP} = \frac{\dot{m}_a \cdot FA}{IP} \tag{8.23}$$

Using the definitions of the scavenging parameters, the expression for the isfc can be written:

$$isfc = \frac{R_s \cdot \rho_{sc} \cdot V_d \left(\frac{\varepsilon}{\varepsilon-1}\right) \cdot \frac{N}{60} \cdot FA}{\eta_{ch} \cdot \rho_{sc} \cdot V_d \left(\frac{\varepsilon}{\varepsilon-1}\right) \cdot \frac{N}{60} \cdot H_u \cdot FA' \cdot \eta_i'} \tag{8.24}$$

For specific fuel consumption in g/kWh and heating value in kJ/kg, Equation (8.24) can be written:

$$isfc = \frac{3.6 \cdot 10^6}{\eta_{tr} \cdot \frac{FA'}{FA} \cdot H_u \cdot \eta_i'} \tag{8.25}$$

For an injected engine, Equation (8.14) applies, and so:

$$isfc = \frac{3.6 \cdot 10^6}{H_u \cdot \eta_i'} \tag{8.26}$$

For a carbureted engine, $FA = FA'$ and so:

$$isfc = \frac{3.6 \cdot 10^6}{H_u \cdot \eta_{tr} \cdot \eta_i'} \tag{8.27}$$

Thus, as expected, the carbureted engine is penalized for "throwing away" a substantial amount of fuel during the scavenging process.

The indicated efficiency used in the above equations is that based on the amount of fuel trapped in the cylinder. Thus, methods like those used for four-stroke engines can be applied (Air cycles, simple simulations and such) can be used to estimate power and efficiencies for the trapped fuel. In the 2-stroke engine, one should distinguish between the geometrical compression ratio defined by the compression volume and displacement volume, and the "effective" expansion ratio. The latter is primarily defined by the time when the exhaust port opens and the scavenging process begins. When the pressure falls to exhaust pressure, work production for all intents and purposes ceases.

8.4   Scavenging Models

As discussed, a major difference between 2-stroke and four-stroke engines is the scavenging process, which determines the state of the cylinder gasses at the start of the effective compression: temperature, pressure and residual fraction. Since the 2-stroke engine is not a positive displacement device other methods must be used to model the air exchange process than with four-stroke engines. A first approximation is the complete mixing process. The assumptions in this case are that an incremental amount of mass entering the cylinder is instantaneously mixed with the contents of the cylinder and that the cylinder remains at constant volume. If the total mass in the cylinder during the scavenging process is called m_c and the mass of fresh mass in the cylinder m_f, the fraction of fresh mass in the cylinder at any time is $x \equiv \frac{m_f}{m_c}$. If a small amount of fresh mass, δm enters the cylinder, a corresponding amount of mass must leave if steady state is assumed. The leaving mass will carry an amount of fresh mass with it that is equal to $\delta m \cdot x$. Then the rate of change of the fresh mass in the cylinder is:

$$\delta m_f = \delta m \cdot 1 - \delta m \cdot x \tag{8.28}$$

Table 8.1: Empirical models for Blair's scavenging model for 2-stroke gasoline engines.

Type	M	C
Ported uniflow, very good	-1.7827	0.2094
3 Port loop, good	-1.6709	0.1899
5 Port loop, quite good	-1.6993	0.3053
5 Port loop, bad scavenging	-1.3516	0.1435
Cross Scavenging, Normal	-1.0104	-0.1170
Racing Cross with radial ports	-1.6325	0.1397

Dividing by the total mass to obtain the fresh mass fraction,

$$dx = (1 - x) \frac{dm}{m_c} \tag{8.29}$$

Then integrating:

$$x = 1 - e^{-\left(\frac{m}{m_c}\right)} \tag{8.30}$$

where m is the total amount of air that has entered the cylinder. Since x is the amount of fresh charge relative to the total charge (ideal mass), and the volume has been assumed to be constant, the cylinder mass corresponds to the ideal mass. This means that x can be set equal to the charging efficiency, and $\frac{m}{m_c}$ can be set equal to the scavenging ratio. Then using Equation (8.9).

$$\eta_{ch,id} = 1 - e^{-R_s} \tag{8.31}$$

$$\eta_{tr,id} = \frac{1}{R_s}\left(1 - e^{-R_s}\right) \tag{8.32}$$

These equations give proper trends for the changes in the charging and trapping efficiencies as a function of delivery ratio, and could be used as a very first approximation.

However, there is a difference in the scavenging relationships for different types of scavenging systems. Empirical relationships have been developed, and Equations (8.31) and (8.32) form the basis for models of different scavenging systems. Blair [87] has measured the scavenging efficiency for a number of 2-stroke engines with different porting systems.

The basic form of Blair's model is:

$$\eta_s = 1 - exp\left(M \cdot R_s + C\right) \tag{8.33}$$

where M and C are empirical constants, whose values are given in Table 8.1 for scavenging of several types of 2-stroke engines. In a newer edition of his book [14], Blair has an improved model which uses three empirical parameters instead of 2. The trends and magnitudes are quite similar to this model, so the older version is presented here for simplicity.

The scavenging characteristics predicted by the values in Table 8.1 are plotted in Figure 8.7 and compared to the ideal mixing case. The model is for the scavenging efficiency, which is based on the amount of trapped mass in the cylinder. Blair also points out that the methods used to obtain the scavenging efficiency, result in a value based on volume and not mass. Thus, though results are given in a simple form, they are not straightforward to use in estimating scavenging parameters for an actual engine, since the trapped mass at the end of the scavenging process is not known. This is because the piston moves up towards the cylinder head through the scavenging process, and some of the mass may be expelled from the cylinder by this process. Also, pressure oscillations in the intake and exhaust systems will affect the cylinder pressure at the closing of the exhaust port, affecting the trapped mass. The relation to the charging efficiency, then, can be difficult to estimate. The scavenging efficiency is useful for simulation work, since it gives the residual amount in the cylinder, which is needed to determine the heat release, as in Equation (2.109) for example. The residual fraction is equal to $1 - x$.

Figure 8.8 shows the ratio of volume in the cylinder to the maximum cylinder volume as a function of degrees after bottom dead center. This shows how much of the volume has been "lost" through

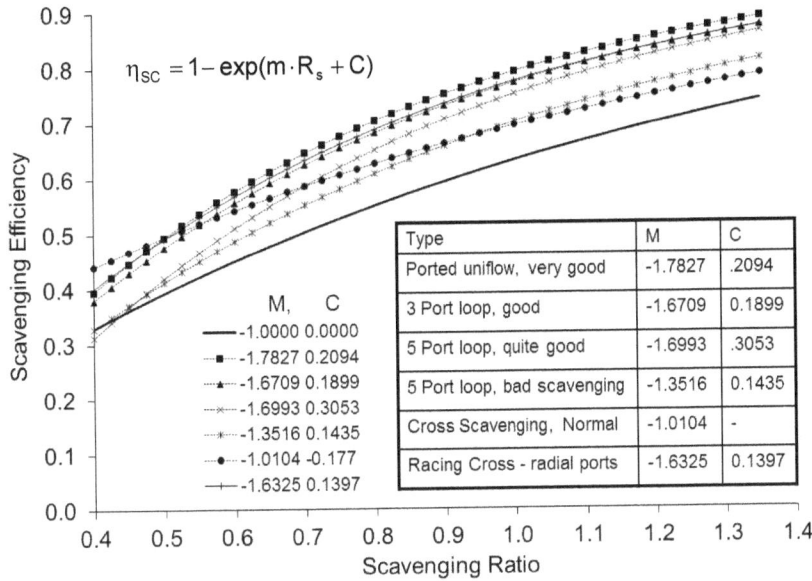

$$\eta_{sc} = 1 - \exp(m \cdot R_s + C)$$

M, C
——— -1.0000 0.0000
---■--- -1.7827 0.2094
---▲--- -1.6709 0.1899
---✱--- -1.6993 0.3053
---✻--- -1.3516 0.1435
---●--- -1.0104 -0.177
---+--- -1.6325 0.1397

Type	M	C
Ported uniflow, very good	-1.7827	.2094
3 Port loop, good	-1.6709	0.1899
5 Port loop, quite good	-1.6993	.3053
5 Port loop, bad scavenging	-1.3516	0.1435
Cross Scavenging, Normal	-1.0104	-
Racing Cross - radial ports	-1.6325	0.1397

Figure 8.7: Scavenging efficiency as a function of scavenging ratio as predicted by the model of Blair and compared to the ideal mixing model ($M = 1, C = 0$).

displacement by delaying the closing of the exhaust port. If the exhaust port closes at 70° ABDC, the cylinder volume is about 72 ⇔ 75% of the total cylinder volume at BDC for the compression ratios shown. With a constant pressure in the cylinder, this would correspond to a loss of about $20 - 25\%$ of the air in the cylinder. Effectively, the ports will probably act as closed slightly earlier, as the area become small before closing, and restricts the flow. Assume a scavenging ratio of about 0.9. Blair's results show that the scavenging efficiency is on the order of 0.7 for engines with typical scavenging. This means that the total cylinder mass is about 73% of the mass to fill the total cylinder volume, and 70% of the cylinder mass is fresh air, giving a charging efficiency of about 0.51. In a properly scavenged engine, this estimated value is expected to be low. The perfect mixing scavenging model gives a scavenging efficiency of 0.59 for this scavenging ratio, which in the ideal case, is equal to the charging efficiency. So as a first rough estimate in the lack of better information, one could use the perfect mixing scavenging model to estimate the charging efficiency using the scavenging ratio. More accurate methods requiring computer simulations are presented in Blair's books.

According to correlations, the scavenging for all the different types and qualities for loop scavenged engines are better than those of the ideal mixing case, but follow a similar trend. The best scavenging results are for the uniflow system, which is better than the very best loop scavenging results. The cross scavenged engine has a good scavenging efficiency for low scavenging ratios, but later becomes inferior to the other scavenging types. For small, cheap engines where scavenging flow is not optimized and scavenging ratios expected to be low, this can be a satisfactory choice. A reason why it is poorer at higher scavenging ratios is that it is more susceptible to short circuiting flow than the loop scavenged idea. The reason that the ideal model gives approximately correct values for the actual engine is that real scavenging processes are better than perfect mixing, but that there is a loss of air due to the late closing of the exhaust port that reduces the charging efficiency.

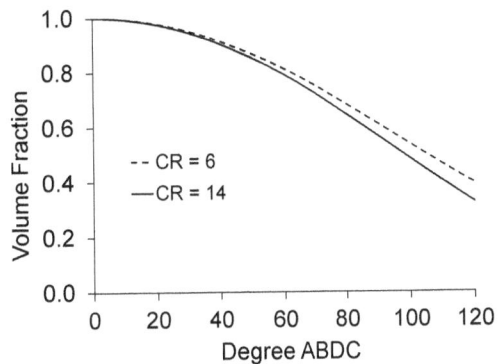

Figure 8.8: The ratio of the cylinder volume at a given crank angle to the total cylinder volume at BDC for two compression ratios. The crank angle denotes the time at which the exhaust port closes.

The effectiveness of the scavenging process is dependent on the engine design, especially that of the ports and associated flow system. Figure 8.9 shows representative measured trapping efficiencies for a 2-stroke spark ignition engine. Note that the trapping efficiency is dependent on the engine speed, this is due to resonance phenomena in the flow in the intake and exhaust systems.

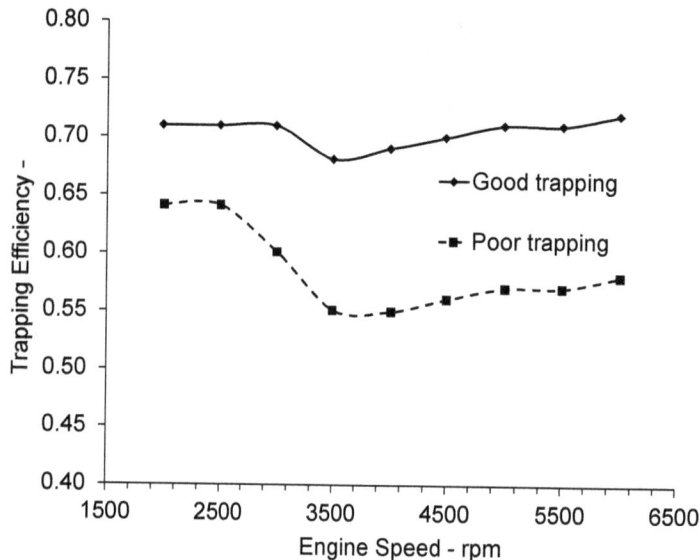

Figure 8.9: Typical trapping efficiencies for a 2-stroke SI engine efficiency as a function of engine speed for good trapping and poor trapping [88].

Example

As a first estimate of the performance of an SI 2-stroke engine consider the following example. Consider an engine equipped with a carburetor having "quite good 5 port loop scavenging" characteristics as shown in Figure 8.7. Assume that the engine is operating at a stoichiometric fuel air ratio of 0.068 on gasoline with a lower heating value of 44000 kJ/kg, with a geometric compression ratio of about 9, and an effective expansion ratio of about 6. Using the fuel air cycle model at this fuel air ratio, and assuming that the real cycle indicated efficiency is about 80% of this value, the indicated thermal efficiency based on the trapped fuel, $\eta_i' \approx 0.3$. Assume also that the scavenging ratio is about 0.8, since crankcase scavenging is used.

For the scavenging efficiency, Equation (8.33) gives:

$$\eta_s = 1 - exp\left(M \cdot R_s + C\right) = 1 - exp\left(-1.6993 \cdot 0.8 + 0.3053\right) = 0.652$$

If the cylinder volume at effective exhaust port closing is about 0.75 of the total cylinder volume, then an estimate of the charging efficiency would be $\eta_{ch} \approx 0.652 \cdot 0.75 = 0.489$. The perfect mixing equation with a scavenging ratio of 0.8 gives a value for the charging efficiency of 0.551. So a reasonable estimate for the charging efficiency is about 0.520. This is consistent with an average trapping efficiency of 0.65, indicated in Figure 8.9 where from Equation (8.9) $\eta_{ch} = \eta_{tr} \cdot R_s = 0.80 \cdot 0.65 = 0.52$

Then using Equation (8.20):

$$imep = \eta_{ch} \cdot \rho_s \left(\frac{\varepsilon}{\varepsilon - 1}\right) \cdot H_u \cdot F' \cdot \eta_i' = 0.520 \cdot 1.18 \left(\frac{9}{8}\right) \cdot 44000 \cdot 0.068 \cdot 0.3 = 620kPa.$$

Two stroke friction is typically lower than that of 4-stroke engines, and varies with speed. Assume that it is on the order of 100 kPa. Then the expected *bmep* of the engine is on the order of 520 - kPa.

The mechanical efficiency is:

$$\eta_m = \frac{620 - 100}{620} = 0.839$$

And the $bsfc$:

$$bsfc = \frac{3.6 \cdot 10^6}{H_u \cdot \eta_{tr} \cdot \eta_i' \cdot \eta_m} = \frac{3.6 \cdot 10^6}{H_u \cdot \frac{\eta_{ch}}{R_S} \cdot \eta_i' \cdot \eta_m} = \frac{3.6 \cdot 10^6}{44000 \cdot \frac{0.520}{0.8} \cdot 0.3 \cdot 0.839} = 500 g/kWh$$

The 2-stroke engines listed in Table 1.3 are high performance versions, and have $bmep$ on the order of 850 - 900 kPa. On the other hand, some 2-stroke engines for chain saws have maximum $bmep$ of the order of 300 - 500 kPa, so as an average value, the above estimate is reasonable.

A summary of the status and problems with 2-stroke SI engines in shown in Figure 8.10. First, the

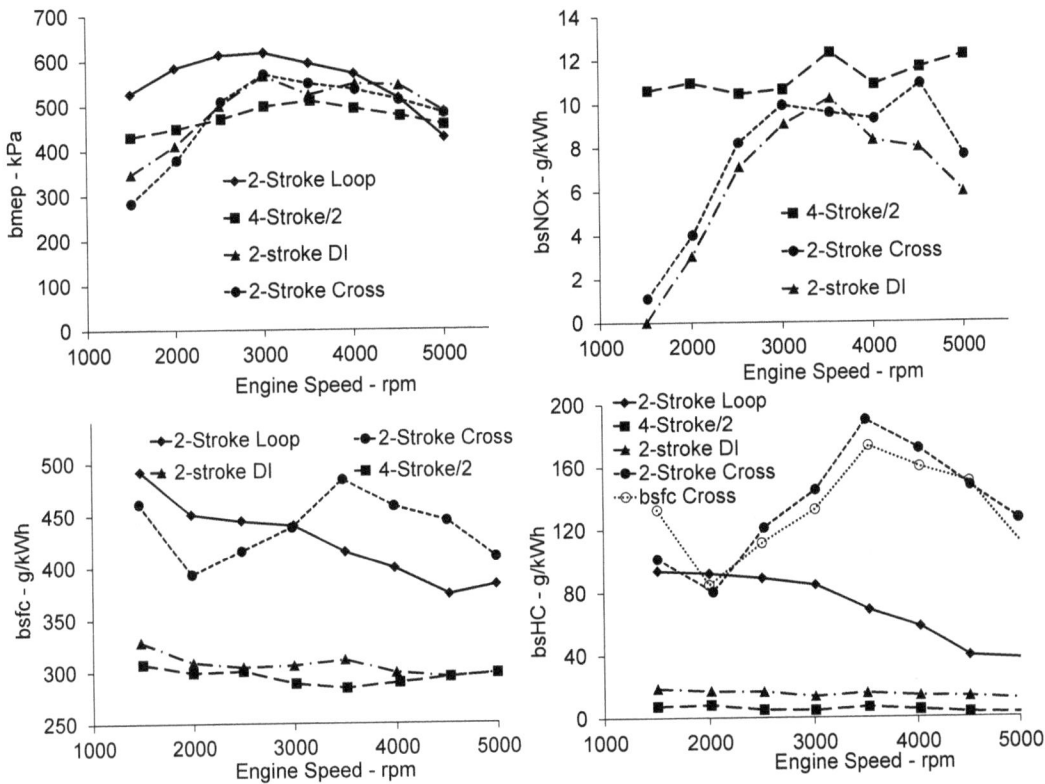

Figure 8.10: Full load performance, fuel consumption and emissions for different versions of 2-stroke engines compared to a 4-stroke SI engine with electronic fuel injection. [88]

results show that with good scavenging, the estimate of the $bmep$ from the previous example is reasonable. For most conditions, the $bmep$ is high enough to provide more power for a given displacement than the 4 stroke engine (note that the $bmep$ for the 4-stroke engine is divided by 2 to enable the comparison). This translates to a smaller engine and lower weight, which in a vehicle application, helps fuel economy. However, the $bsfc$ curves show a large difference in fuel economy for the different engine types. The engines with premixed fuel and air have a fuel consumption that is 30 to 50 % higher than the direct injected 2 stroke engine and the 4 stroke engine. This is due to the fuel lost in the scavenging process, as will be shown in the following. If a 2-stroke engine is to compete with a 4-stroke engine on the basis of fuel consumption, direct injection is a must. In the case of direct injection, the fuel consumption is close to that of the 4-stroke engine, though a little higher.

The scavenging process plays a large role in determining the emissions of a 2-stroke engine. It is for reasons of emission control that 2-stroke engines are not found in passenger cars and similar sized vehicles. The results for the NO_x emissions show quite low values at low power levels for the carbureted and direct injected 2-stroke engines. These are associated with low $bmep$ values, where the scavenging ratio is reduced by throttling. This means lower charging and scavenging efficiencies which have been shown above to give increased residual gas fraction. This functions in a similar manner to EGR, giving lower combustion pressures and temperatures, which also keeps NO_x emissions low. At the higher speeds for these engines, the NO_x emissions of the 2 and 4-stroke engines are of the same magnitude.

A more dramatic illustration of the problems with scavenging is seen from the $bsHC$ emissions, which are extremely high for the premixed engines. That this is due to scavenging can be seen from the results of the cross-scavenged engine. At 3500 rpm, the $bsfc$ of the premixed engine is about 480 g/kWh, compared to about 310 for the direct injection engine, a difference of 170 g/kWh. At the same condition, the $bsHC$ of the premixed engine is about 185 g/kWh while that of the direct injection version is about 15. The latter emissions are those formed in the cylinder, since no fuel escapes with direct in-cylinder injection. The difference of 170 g/kWh is a direct indication of the fuel lost in scavenging and is basically equal to the increase in fuel consumption. An independent measurement of the trapping efficiency of the 2-stroke engine gives a value of 0.57 at this condition. Assuming that the fuel not appearing as hydrocarbon in the exhaust was trapped, an estimate of the trapping efficiency can be obtained:

$$\eta_{tr} \approx \frac{total\ consumption - exhaust}{total\ consumption} = \frac{480 - 185}{480} = 0.61$$

There should be a correction made for the amount of hydrocarbons formed the combustion process, but this is not normally known for an engine. This simple method gives a good approximation to the trapping efficiency. The use of exhaust gas composition can be used in a more rigorous fashion to determine trapping efficiencies.[89]

Unfortunately, the 2-stroke SI engine cannot benefit from the benefits of the three way catalyst as can be done with the 4-stroke engine. If operated at stoichiometric conditions, a large amount of fuel would be found in the exhaust along with the air necessary to burn in due to the scavenging process. An active catalyst would ignite this fuel, leading to rapid overheating and destruction of the catalyst.

In the case of the direct injected engine, there is still a significant amount of air which goes through the engine untrapped. If combustion occurs under stoichiometric conditions, the exhaust will still be lean (oxygen present) due to the extra scavenging air. This means that NO_x reduction by a 3-way catalyst would not be possible. If the exhaust was to be stoichiometric, the combustion would have to be very rich to compensate, and there would still be a lot of energy released on the catalyst as CO and H_2 burn, and fuel consumption would be about the same as the premixed engine. The catalyst would still overheat when the CO and H_2 burn. Emissions control is a major problem facing the 2-stroke SI engine. If not solved, these engine can disappear as standards become tighter. Some modern small 2-stroke gasoline engines for scooters have been equipped with electronic fuel injection, with significantly reduced HC emissions as a result.

An interesting new application of a small 2-stroke engine utilizes a mo-ped engine, modified to operate as a direct injection diesel engine [90]. Compression ignition operation is made possible through the use of a higher compression ratio, and the use of DME as a fuel. DME has the fuel vaporization characteristics necessary to prevent the accumulation of liquid fuels on the cylinder walls and pistons of such a small engine. Smoke free operation on DME is also achieved in this size of engine, whereas other more conventional diesel fuels have extremely high smoke emissions, and are not practical. An engine map and maximum power are shown in Figure 8.11. Power is limited by the scavenging process, and matching of the injector (a modified GDI injector) with the combustion chamber air motion. Emissions levels comparable to European standards for much larger stationary engines were achieved. Maximum brake efficiency of 30%, corresponding to a $bsfc$

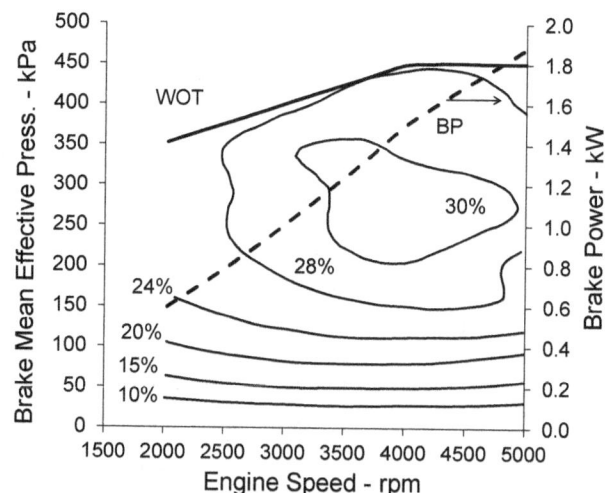

Figure 8.11: Performance results for a 50 cm^3, 2-stroke diesel engine operating on DME.

of 282 g/kWhr on a typical diesel fuel, was achieved over a wide operation range. The purpose of the engine was initially to provide a small, lightweight high efficiency engine for use a mileage marathon vehicle, but the engine could be considered for other applications for small weight, high efficiency.

8.5 Summary

The 2-stroke engine has the potential to produce large amounts of power per unit engine mass or volume. It is this characteristic that has made it attractive for applications where weight is very important, and fuel consumption less important. These areas include hand-held tools and small vehicles such as 2 or 3 wheelers. A premixed 2-stroke engine has much too high fuel consumption to be competitive on all but the smallest vehicles and applications. The major obstacle facing 2-stroke SI engines at the present is the emissions challenge.

The largest, most efficient engines made today are large 2-stroke diesel engines used in ships, with efficiencies over 50%. They too are being subjected to increasingly strict emissions requirements, but due to size and steady state operation, there are more emission control options available.

8.6 Problems

Problem 8.1

A single cylinder 2-stroke spark ignition engine with carburetor has a cylinder bore = 85 mm and stroke = 110 mm and is operated at 2500 rpm, where the following measurements were taken:

$IP = 17$ kW, $FA = 0.08$, Intake temperature = 345K, trapped indicated efficiency, $\eta_i' = 0{,}29$, compression ratio = 8 and intake pressure = 1.00 bar. The lower heating value = 43000 kJ/kg. Find the changing efficiency. ($\eta_{ch} = 0.568$.)

Problem. 8.2

A 6-cylinder 2-stroke diesel engine with cylinder diameter = 150 mm, stroke = 180 mm and compression ratio = 15 gives the following results during testing: Brake power = 440 kW at N = 2000 rpm for an exhaust pressure = 1.01 bar and intake temp. = 350K. The delivery ratio = 1.40 and the fuel consumption = 0,027 kg/sec at an excess air ratio = 1.50. The fuel heating value = 44 MJ/kg and the stoichiometric mixture ratio = 0,067.

1. Calculate the total air consumption neglecting humidity. 2. Calculate the charging efficiency. 3. Calculate the brake thermal total efficiency. ($\dot{m} = 0.959$ kg/s; $\eta_{ch} = 0.88$; η_e 0.,37)

Problem 8.3

A 4-cylinder 2-stroke diesel engine has a cylinder diameter of 125 mm, a stroke of 150 mm, and a compression ratio of 21. At the intake port, the average temperature is 71C and the pressure is 1.05 bar. The engine operates at 2000 rpm and the measured air mass flow is 0.335 kg/s. The brake specific fuel consumption is 188 g/kWh, and brake power is 121 kW. A sample of the cylinder gasses taken after combustion, but before the opening of the exhaust port indicates a fuel/air ratio of 0.041. Find the charging efficiency, delivery ratio and trapping, and compare with the values from a perfect mixing scavenging process. ($\eta_{ch} = 0.563$; $\eta_{tr} = 0.460$; $R_d = 1.22$; Ideal: $\eta_{ch} = 0.706$; $\eta_{tr} = 0.577$; $R_s = 1.22$)

Problem 8.4

A 1-cylinder 2-stroke spark ignition engine with spark ignition has the following specifications. Bore: 85 mm, Stroke: 70 mm, speed: 3000 rpm, compression ratio (geometric): 7,9, Compression ratio (effective value after exhaust port closing): 6,7, Fuel/air ratio: 0,0661, $bmep$ 591kPa, $bsfc$ 395 g/kWh, brake specific hydrocarbon (HC) emission 120 g/kWh, Delivery ratio = 0,72, Temperature at the intake port = 40C, intake pressure = 110 kPa, and fmep = 80 kPa, including pumping work in the crankcase.

It can be assumed that all the hydrocarbons in the exhaust gasses occur solely due to scavenging of fresh air and fuel during the air exchange process. All other hydrocarbons are assumed to be burned in the cylinder. The fuel and air are completely mixed before they enter the cylinder. Under these conditions, calculate trapping efficiency of the air exchange process. Calculate the engine's actual charging efficiency and compare it with that for an ideal, perfect mixing process. What is the indicated efficiency based on the amount for fuel that is retained in the cylinder during scavenging. Compare this value with that for an ideal Otto fuel-air process with a compression ratio corresponding to the engine's effective compression ratio based on exhaust port closing. ($\eta_{tr} = 0{,}696$, $\eta_s = 0{,}501$, $\eta_{s,id} = 0.513$, $\eta_i' = 0.328$, $\eta_i'/\eta_{th}' = 0.826$)

Chapter 9

Friction

9.1 Introduction

An important factor in the determination of the brake efficiency of an internal combustion engine is the loss of energy that is involved in facilitating the movement of the mechanical parts of the engine. In addition, there are losses involved with the movement of the air in and out of the engine. The energy to overcome these losses must come from the work done on the piston top, *i. e.* the indicated work. One can then call any energy that is "lost" between the indicated work and the brake output as friction. This may be more general than the more common idea of friction between two surfaces and such, but is an advantageous definition for engines. Given this idea, the basis relationship for performance is obtained, Equation (1.5), which in the sense just discussed, might just as well serve as a definition of engine friction.

9.1.1 Sources of Friction

There are many sources of friction in engines. Some of them are listed here:

- Piston-Crank Components

  - Piston and piston rings rubbing on the cylinder wall
  - Journal bearings on crankshaft and connecting rods
  - Piston pin friction

- Valve train friction

  - Camshaft bearings
  - Valve friction
  - Gear or belt losses to drive the camshaft

- Fluid Flow Losses

  - Pressure losses in the intake and exhaust pipes
  - Pressure losses across the valves
  - Pumping losses in the crankcase
  - Part load throttling in SI engines

- Auxilliary/Accessory items

  - Electrical generator
  - Oil pump
  - Fuel injection pump
  - Water pump/Cooling fan

These classifications are, somewhat arbitrary, and are only intended to give an indication of the many different types of losses that are associated with engine operation.

9.1.2 Significance of Friction

Friction always plays a role in determining the efficiency and output of an engine, but in some cases, it has a greater significance. Some of these tendencies are discussed in the discussion of engine maps and characteristics in Chapter 12. When an engine is operating at full load, friction is least significant. As an illustration of some orders of magnitude, consider a large, turobcharged CI engine with a *bmep* of 1800 kPa, an *fmep* of 200 kPa, and an indicated thermal efficiency of 0.46. Assume that a 5% reduction in friction is achievable. How much does this affect brake thermal efficiency at full load and at a condition where the engine requires a bmep of 400 kPa? Results are shown in Table 9.1.

Table 9.1: Example to illustrate the importance of friction for a turbocharged CI engine at full load and part load.

	Full load		Part load	
	Base	Reduced	Base	Reduced
imep - kPa	2000	2000	600	590
fmep - kPa	200	190	200	190
bmep - kPa	1800	1810	400	400
η_m	0.90	0.905	0.667	0.678
η_e	0.414	0.416	0.307	0.319
Relative change in η_e	-	+0.56%	-	+1.7%

Assume the friction is not load dependent. For the full load, the indicated mean effective pressure is held constant, as this is determined by combustion conditions and the maximum amount of fuel injected. The lower friction results in a slightly larger maximum power, and small increase in brake thermal efficiency. This may not be that important in an application where full load is only occasionally used, but in a racing engine with close competition, this could make a big difference in the results.

At part load, it is the *bmep* that is held constant, since it is determined by load requirements. The lower friction requires a correspondingly lower *imep* to provide the same output. Here, since the friction is a much larger portion of the output, the improvement in fuel economy achievable by lowering friction is substantially larger than at full load. It can be seen, though, that to achieve a major reduction in fuel economy through friction reductions, a very large reduction in the friction is required. Many radically different construction principles have been proposed to this end. However, the large reduction in friction required to make a significant change in efficiency, and the cheap reliable construction of the slider crank mechanism have made it difficult for new mechanical engine design concepts to compete with the conventional slider-crank mechanism. The Wankel engine is one that has had some limited success.

9.2 Mechanical Friction

Moving parts in engines are always lubricated, so the classic dry friction characterized with a friction coefficient is not relevant. The lubricating situations encountered are:

- Hydrodynamic friction. This is encountered in journal bearings and between pistons and the cylinder walls, where two surfaces with a lubricant between move relative to each other, and a fluid film is built up between them that keeps the surfaces separated. Hydrodynamic lubrication can often be analyzed by solving the Reynolds' equation for the situation at hand.

Figure 9.1 schematically shows the situations between two lubricated surfaces. From the definition of viscosity, the force, F, required to move the plates is the frictional force for shear stress τ and viscosity μ:

$$F = \tau \cdot A = A \cdot \mu \frac{du}{dy} = A \cdot \mu \frac{U}{\delta y} \tag{9.1}$$

It is the hydrodynamic film which determines the friction, and the viscosity of the lubricant plays an important role in determining the friction. Of the types of mechanical friction in engines, this is the dominating type, though the surfaces may be round instead of flat. Equation (9.1) establishes

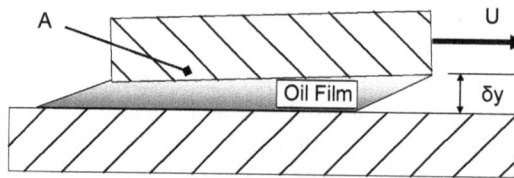

Figure 9.1: Sketch of the situation for two lubricated surfaces with an oil film in between with a relative velocity, U, and a surface area A, and a film thickness, δy.

an important relationship for engine friction, namely that the friction force (therefore work per revolution) is proportional to the velocity. The velocities of the moving parts in an engine are proportional to the engine speed, and thus due to hydrodynamic friction, the major trend of the friction mean effective pressure in engines is a linear increase with speed. The total friction is a combination of many surfaces with different geometries, where the basic dependence on speed can be traced to Equation (9.1). As will be seen later, there is also a smaller higher order dependence on engine speed due to fluid mechanical losses, but the linear (viscosity) term is the most important in determining speed dependence.

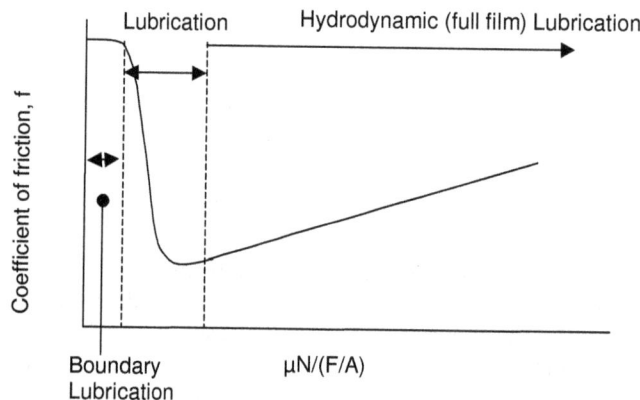

Figure 9.2: The Stribeck diagram, which indicates the different friction regimes. μ, is the viscosity, N the engine speed, F the applied force on the bearing over surface area, A.

The nature of hydrodynamic lubrication is shown in the classic Stribeck diagram, Figure 9.2, which shows the different regimes for the coefficient of friction (friction force divided by normal force) as a function of the loading parameter, $\dfrac{\mu \cdot N \cdot A}{F}$ for a journal bearing, or film lubricated sliding plates. The linear dependence of the friction coefficient with increasing speed can be seen in the hydrodynamic lubrication region. When the speed or viscosity becomes too low, or the force becomes too high to maintain a full film of lubricant between the surfaces, there is a transition to the boundary lubrication region, and the friction between the surfaces increases, as does the wear.

- Boundary friction. When the relative velocity between two lubricated surfaces is insufficient to maintain a lubricating oil film, the two surfaces can contact each other at high points on the surfaces. This gives something approaching dry friction and a dramatic increase in friction and wear. This is most likely to occur in a reciprocating mechanism, where the velocity periodically becomes zero. Boundary friction does not lend itself to mathematical analysis.

An indication of the process that occurs as lubrication goes over to the boundary region can be obtained by looking at some results of molecular dynamics calculations [91]. The technique involves calculating the intermolecular forces between a large number of molecules that are located between two surfaces that are forced together. As the surfaces are squeezed together the lubricant arranges itself into layers between the surfaces. Eventually, there will be only one layer of molecules between the surfaces (here only a small area is considered). The resistance to squeezing the molecules out from between the surfaces increases as the molecular size increases. The larger molecules also cover a larger portion of the surface when the last layer is squeezed out. This situation is shown in Figure 9.3, where the arrangement of the molecules and atoms can be seen for the last molecular layer to be squeezed out from between the surfaces for three different hydrocarbons. Also shown is the results of wear with a machine that measures wear between a ball reciprocating on a plate in the presence of a lubricant. The figure shows that the larger molecules cover more of the surface and give lower wear.

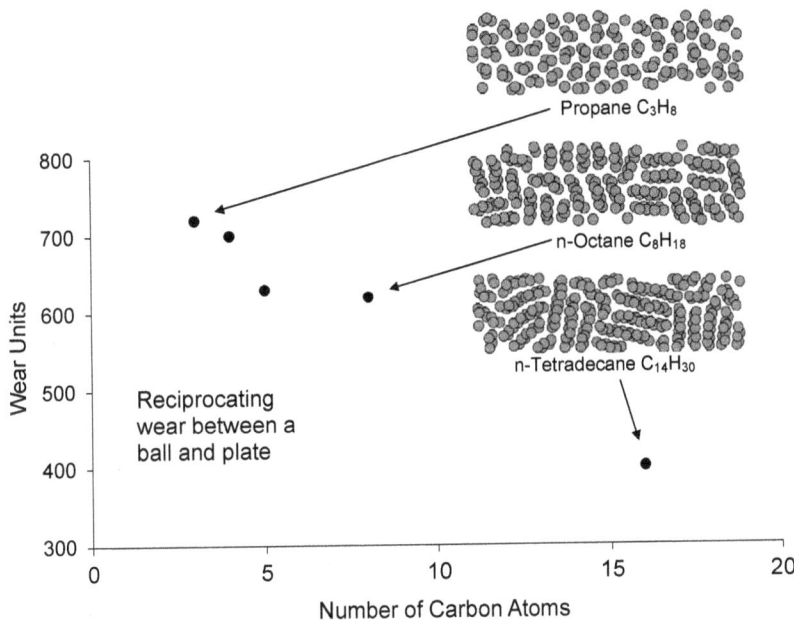

Figure 9.3: Surface wear between a reciprocating ball and plate for different paraffinic hydrocarbons, and the calculated number of carbon atoms on the surface for the last layer of lubricant squeezed out between two surfaces.

The wear is important in a reciprocating system, since there is always a situation where the velocity is zero, and the Stribeck parameter also becomes zero, and the mixed or boundary lubrication situation arises. This is the situation when the piston stops at TDC, for example, but the procedures and tests used to obtain these results were actually developed to investigate and evaluate wear in diesel fuel injection systems with reciprocating pumps. Lubrication of these pumps has proven to be increasingly a problem with modern low sulfur fuels required for low emissions and exhaust gas after treatment. The wear in such situations is not necessarily connected to the viscosity of the lubricant, and the characteristics of a lubricant regarding wear and friction in the reciprocating system are called *lubricity*. This property is commonly improved in diesel fuels by the addition of a substance containing long chain molecules that adhere to the pump surfaces and remains between the two surfaces, even when the velocity approaches zero.

- Rolling friction. This type of friction is encountered where one part rolls over another, for example in ball or needle bearings. This is often more important in transmission systems, where such bearings are used for shafts, though they can be found in the engine mechanisms in some two stroke or special engines. One reason that these low friction bearings are not used on engine shafts is that of problems with assembly. The race of the bearing must be very smooth, and is built as a one-piece ring that cannot be separated for assembly as the classic journal bearing can be for crankshafts and connecting rods. Effective friction coefficients of under 0.05 are observed with these bearings.

9.3 Methods for Determining Friction

Some of these have been discussed other places in the text, and are mentioned here in order to summarize and compare them.

9.3.1 Indicator Diagrams

The first method, and the only theoretically accurate method, is that of measuring the *indicated power* of the engine with a pressure transducer and then subtracting the brake power from the indicated power calculated by an integration of the $p-V$ diagram. First the indicated work is obtained from the integration of the $p-V$ diagram using Equation (1.4) and this is converted to mean effective pressure with Equation (1.8). Then the *imep* can be converted to power with Equation (1.9). The brake power is found from a dynamometer test where the torque is measured, using Equation (1.3). The friction power can then be determined from Equation (1.5).

Advantages - First of all, the method is theoretically correct. The other advantage, is that it is possible to distinguish the friction associated with the 4-stroke air exchange from the other sources of friction, since this is readily available from the $p-V$ diagram.

Disadvantages - Instrumentation. An accurate pressure transducer must be installed in a suitable location in the combustion chamber so that pressure measurements are not disturbed by non-combustion related pressure oscillations. The pressure transducer must not be sensitive to coolant or combustion chamber temperature. For multi-cylinder engines, it is common to only measure the pressure in one combustion chamber, but to obtain the best accuracy, the pressure in all cylinders should be measured, as there may be cylinder to cylinder variations in fuel and air quantities, timings and other variables that could affect performance.

The other disadvantage is the need for a very precise indication of piston position. This normally requires a crank angle position sensor mounted at a known crank angle position. This is very important, as an error of only 1 degree can give an error on the order of 2-4 % in the indicated work. The effect of correct location of the piston position is shown in Table 9.2, for the $p-V$ data shown in Figure 3.37. The timing was changed $\pm 1°$ from the correct position. The error in imep ranges from 2 to 4 %, depending on the spark timing.

Table 9.2: Effect of crank angle position on the *imep* from the integration of the $p-V$ diagram for a CFR engine. (See Figure 3.37)

Timing	32° BTC	20° BTC	2° BTC
correct placement	769 kPa	801 kPa	727 kPa
+ 1° error	803 kPa	827 kPa	742 kPa
	(+4.4 %)	(+3.2 %)	(+2.1 %)
- 1° error	735 kPa	776 kPa	712kPa
	(-4.4 %)	(-3.1 %)	(-2.1 %)

In terms of friction measurement, this error is amplified when the brake power is subtracted, and depending on the mechanical efficiency, can be more than 3 times the error in the indicated work. For example, assume that at a spark timing of 20° BTC the dynamometer measurement of the toruqe results in a *bmep* of 551 kPa. For the correct crank angle indicator location, this gives an *fmep* of 250 kPa. If there is an error of +1° in the location of the crank angle position sensor, the friction mean effective pressure found would be $827 - 551 = 277kPa$, corresponding to an error of 10.4%, or about 3 times the error in the *imep*. In spite of these precautions, this method is the only way to get the correct friction.

9.3.2 Motoring Method

The next method is the *motoring method*. Here, the engine is operated and measurements of the brake performance taken. Immediately following this test, the ignition and/or fuel are shut off, and an electrical motor is used to turn the engine over at the same speed and throttle position as the previous powered test. The power consumption required to turn over the engine in this procedure, the so-called *motoring power* is said to be equal to the friction power.

The main advantages to this method are that it does not require any modification to the engine to install a pressure transducer[1]. As such, unmodified production engines, for example, can be tested for friction using the motoring method. The results are normally immediately available and do not require anything but the most elementary data processing. The motoring method was used for many years before small, rapid and accurate pressure sensors were developed, and is still in common use.

There are two major disadvantages to the method. The first is the requirement for a powerful electric motor attached to the engine/dynamometer system. The most common technique here is an electric motor generator set, in which the change between operation as a motor or generator is quickly facilitated. As mentioned in Section 6.6.2 this technique is used for transient testing of emissions for heavy duty vehicles. The rapid load response of such emissions testing units is normally not required for ordinary performance testing. Units have been produced, where an electric motor is coupled to the same shaft as an eddy current dynamometer, but the motor generator set is a preferred method, as a motor combined with an eddy current dynamometer would cost a similar amount to a motor-generator set. Depending on the type of engine, the electrical power needed is on the order of 20% of the maximum engine power.

The other disadvantage of the motoring method is accuracy. When the engine is motored, conditions are not the same as in the fired engine, and therefore, the friction will be different. There are two main areas of concern. The first is the lack of combustion. This affects the cylinder pressure which ultimately affects bearing loads (see Chapter 13). Lower cylinder pressures will give less friction loss in the bearings. Additionally, the cylinder pressure pushes the compression rings on the piston out against the cylinder wall to promote sealing. This also affects ring friction, as will be discussed later. So with motoring operation, bearing friction is lower than during real operation due to lower cylinder pressures without combustion.

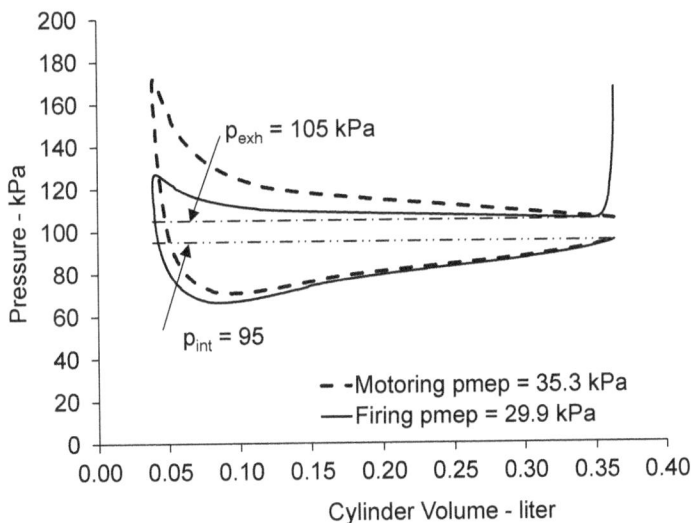

Figure 9.4: Simulated p-V diagrams of the air exchange process for one cylinder of a 1.3 liter SI engine operating at 30000 rpm, compression ratio 9.3, motoring and firing at stoichiometric FA. The valve timings are intake opening at 20° BTDC, intake closing at 80° ABDC, exhaust opening at 90° BBDC, and exhaust closing at 20° ATDC.

There is another factor which compensates for this error for 4-stroke engines, and that is air exchange. In Chapter 7, the importance of the blowdown process in assisting the air exchange process was mentioned. It was found that a substantial amount of the cylinder charge leaves the cylinder around BDC due to high cylinder pressure, and that this reduces cylinder pressure on the exhaust stroke because there is less mass to push out of the cylinder. In the motoring test, the pressure when the exhaust valve opens is

---

[1]Pressure transducers mounted in spark plugs are available for spark ignition engines, but much care must be exercised to prevent pressure oscillations in the connector between the cylinder and the transducer that could disturb the pressure measurement.

close to atmospheric, and in any event, much lower than when there is combustion. Then there is much more air to be pumped out of the engine on the exhaust stroke, resulting in higher pressure there, and a larger amount of pumping work.

Figure 9.4 shows a computer simulation of the pumping loops of an SI engine operating at 3000 rpm and motored at the same speed and throttle position. Under normal operation, the pressure shortly after BDC drops rapidly in the blowdown process, but the high cylinder pressure causes in increase in the pumping work at the start of the exhaust stroke. As shown in Figure 7.14, a significant portion of the cylinder mass is expelled during the blowdown process. However, later on the exhaust stroke, the pressure remains near the exhaust manifold pressure. The situation with the motored engine is reversed. The pressure at BDC is close to the exhaust pressure, but there is no blowdown, and all of the gas remains trapped in the cylinder. At the start of the exhaust stroke, the pumping loss is low. Later on near the end, however, the pumping work is high, since there is a lot of mass to be pressed through a small valve area. When the total process is considered, the pumping loss for the motored engine is about 20% higher than that of the fired engine.

These two effects compensate for each other somewhat, but there is no guarantee that the effects cancel completely, and this is actually quite unlikely. However, the motoring results are close enough to the actual friction to be useful, and certainly can indicate trends in friction. Figure 9.5 shows the motoring test results for a 1.8 liter spark ignition engine over the entire operating range. The friction increases with engine speed. At low speeds, the friction increases with decreasing load, a reflection of increased pumping work with part load throttling. At higher speeds and loads, the dependence on load is lower. Here, the greater mass flow through the intake system components results in higher gas velocities in the intake manifold and pipes increase, which in turn increase the fluid friction losses there and compensate for lower pumping losses at high loads.

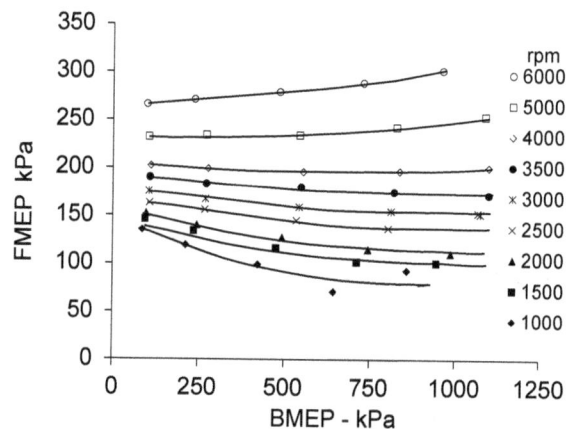

Figure 9.5: The motoring mean effective pressure for 1.8 liter SI engine as a function of speed and load.

9.3.3 Willans Line

A third method that can be used is the Willans line method. This method was described in Section 1.4, and will only be briefly mentioned here. In addition to being used to estimate engine friction, the Willans line is a convenient way to demonstrate fundamental factors concerning engine efficiency. Since it is based on the concept of a constant indicated thermal efficiency, and a friction which is not dependent on load, it can only be used in a limited number of cases. It is, therefore, most useful for a diesel engine, where the indicated efficiency tends towards a constant value at light loads, and the friction is not greatly dependent on load. In diesel engines, the thermal efficiency typically decreases somewhat at full load, due to longer combustion duration. Then the Willans line should be used for the lowest loads, in the region where it is straight. Figure 9.6 shows this. With modern computer controlled diesel engines, which can have special control strategies for emissions control, care must be exercised, that the thermal efficiency is constant in a region where the extrapolation is performed. Some engines also use throttling to regulate EGR at light loads, and this has an impact on friction. This is particularly used on light duty vehicle engines, where a light duty operation is very important for emissions testing and regulation. For engine development purposes, the Willans line method should probably be regarded as the last choice because of accuracy concerns. For diesel engines where there is no other option, it can be used with the above cautions. It is possible to obtain friction estimates with the Willans line method using data from an engine map using the $bsfc$ and BP to obtain the mass flow of the fuel, as illustrated in Figure 1.15. Figure 9.6 shows Willans line friction data obtained from a 2 liter, turbocharged direct injection diesel engine.

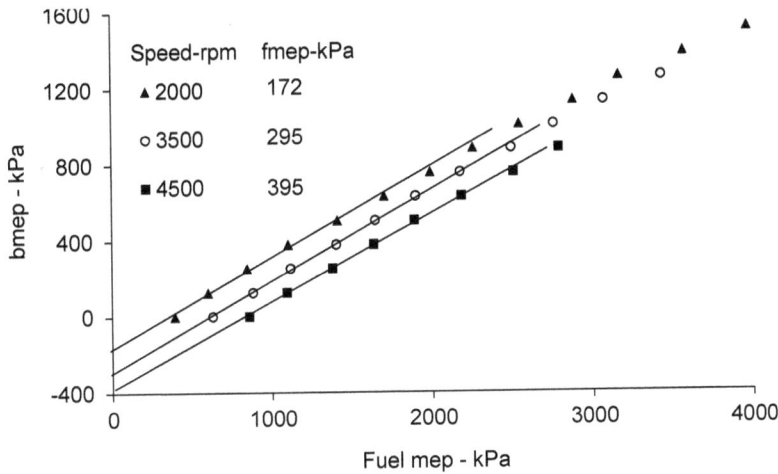

Figure 9.6: The Willans line analysis for a 2 liter, 4-cylinder turbocharged direct injection diesel engine.

9.4 Journal Bearings

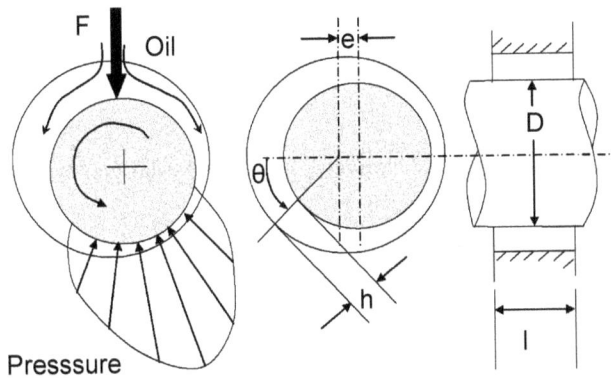

Figure 9.7: Left: A journal bearing showing the oil inlet flow and the pressure distribution that supports the load, F, middle: the eccentric shaft, right: a cross section of the bearing.

Figure 9.7 shows a cylindrical journal bearing, as is used for the connecting rod, the main bearings and camshafts. The position of the drawing on the left is at a different rotational angle than the two on the right. The journal is normally made of steel, and the bearing consists of a replaceable insert of a special alloy. Lubricant is pumped between the two, and as the journal rotates relative to the bearing, viscous forces build up a pressure distribution, that supports the weight of the load. The following is a simplified calculation of the losses in a journal bearing under steady state conditions. More accurate and detailed methods can be found in tribology texts.

If one considers a small element of area on the shaft, dA, shown in Figure 9.7 there is a friction force dF_f:

$$dF_f = \mu \cdot \frac{U}{h} dA = \mu \cdot \frac{U}{h(\theta)} \cdot l \cdot \frac{D}{2} \cdot d\theta \tag{9.2}$$

where μ is the dynamic viscosity and U is the surface velocity of the rotating shaft. This assumes that there is a linear velocity profile between the shaft and the bearing, and that there are no effects in the axial direction. There is a difference between the diameter of the journal and the inside diameter of the bearing, d_r, which can also be referred to in terms of a radial clearance, $C_r = d_r/2$. The journal does not lie in the center of the bearing. Figure 9.7 shows this as the distance, e. The eccentricity can also be referred to as a relative eccentricity, $\varepsilon \equiv e/C_r$. The distance between the bearing and the journal varies around the journal according to the relationship:

$$h = C_r \cdot (1 + \varepsilon \cdot \cos\theta) \tag{9.3}$$

The total friction force can be found by integrating the differential force around the periphery of the journal:

$$F_f = \int_0^{2\pi} \mu \frac{U \cdot l \cdot D}{2 \cdot C_r \cdot (1 + \varepsilon \cdot \cos\theta)} \cdot d\theta = \mu \frac{U \cdot l \cdot D}{C_r} \cdot \frac{\pi}{\sqrt{1 - \varepsilon^2}} \tag{9.4}$$

The power consumption in the journal is given by the product of the force and the velocity:

$$P_f = F_f \cdot U = \mu \frac{U^2 \cdot l \cdot D}{C_r} \cdot \frac{\pi}{\sqrt{1 - \varepsilon^2}} \tag{9.5}$$

This method underestimates the actual losses in the bearing, particularly for large values of ε. For typical engine bearings, the geometry is such that this error is acceptable. Equations (9.4) and (9.5) indicate that the losses are directly proportional to the viscosity of the lubricant, as well as the instantaneous relative eccentricity, ε. The viscosity of the oil is quite dependent on the temperature and type of oil used, and the viscosity can vary by a factor of ≈ 3 for a reasonable range of engine oil operating temperatures. Therefore, it is important to have an accurate estimate of the temperature. The relative eccentricity ε, is also dependent on the viscosity. It can be shown that the relative eccentricity is a function of the bearing load, F.

$$F = \mu \cdot \frac{U \cdot l^3}{4C_r^2} \cdot \frac{\varepsilon}{(1 - \varepsilon)^2} \cdot \sqrt{\pi^2 (1 - \varepsilon^2) + 16\varepsilon^2} \tag{9.6}$$

If the load and the viscosity of the bearing are known, the relative eccentricity can be found from Equation (9.6).

The dynamic viscosity, μ, has SI units of $\frac{N \cdot s}{m^2} = \frac{kg}{m \cdot s}$. In the literature, the units of Poise or centiPoise are often found. 1 Poise is equal to 1 $\frac{g}{cm \cdot s}$. To obtain a dynamic viscosity in $\frac{kg}{m \cdot s}$, a value in centiPoise must be multiplied by a factor of 0.001.

The other form of viscosity often encountered is the kinematic viscosity, ν, which is equal to $\frac{\mu}{\rho}$, where ρ is the density of the fluid. A commonly encountered unit of kinematic viscosity is the Stoke, which is equal to 1 $\frac{cm^2}{s}$. To obtain ν in the SI units of $\frac{m^2}{s}$, a value in centiStokes must be multiplied by a factor of 10^{-6}.

Figure 9.8 shows a plot of Equation (9.6). When the load is high, the eccentricity approaches 1, meaning that the bearing comes very close to the journal, and that the oil film is very thin. This could lead to surface contact and extraordinary wear, and the bearing parameters such as length, width and clearance must be chosen along with the proper viscosity to insure operation without excess wear.

The simplified calculation outlined here is conducted in the following manner. First, the effective lubricant viscosity must be obtained. Assume that the inlet temperature of the oil to the bearing is known, as well as the inlet pressure/lubricant flow. Due to the friction work in the bearing, the temperature of the oil will increase with passage through the bearing. Some of this energy is conducted away through the metal parts, while the rest causes an increase in the oil temperature. The viscosity is

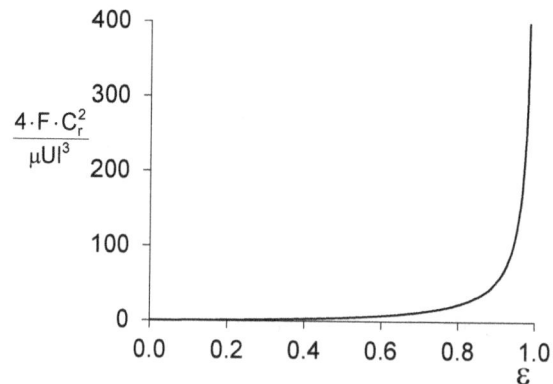

Figure 9.8: Load parameter as a function of the relative eccentricity in a journal bearing from Equation (9.6).

determined by the average temperature of the oil and the temperature dependence of the viscosity. The temperature dependence of the viscosity is typically of the form for standard oils:

$$\mu = K \cdot \exp\left(\frac{A}{B+T}\right) \tag{9.7}$$

where K, A and B are empirical constants.

Then the relative eccentricity must be determined. The calculation can be based on an average bearing load throughout the entire cycle to obtain an estimate of the effective eccentricity.

The journal actually moves inside the bearing throughout the engine cycle, and the eccentricity will not be constant. Using techniques such as computer simulation, discussed in Section 2.5, and bearing loads, discussed in Section 13.1, the forces can be determined throughout the cycle, and then the relative eccentricity can be determined throughout the cycle using models of hydrodynamic lubrication films [92]. Those results can be used to determine an average relative eccentricity throughout the cycle, and Equation (9.5) can then be used to calculate the bearing power loss.

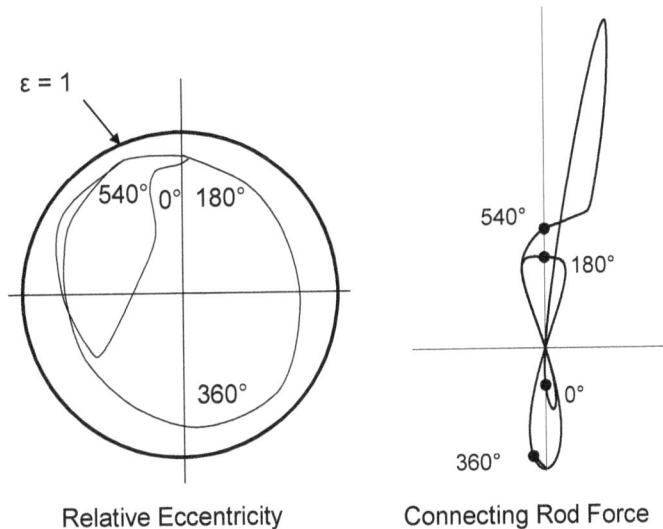

Relative Eccentricity   Connecting Rod Force

Figure 9.9: The load on the connecting rod bearing and the schematic representation of the relative eccentricity of the bearing for a spark ignition engine, adapted from [92]. 0° corresponds to TDC and the start of the expansion stroke/start of combustion.

Figure 9.9 shows the tendency of the results of such calculations for the large end of a connecting rod as a function of engine crank angle position for a spark ignition engine [92]. The eccentricity is nearly constant, and the rod moves relative to the journal during the portion of the cycle when the load is predominantly inertial, but when the pressure load becomes dominant, there is a significant change in the location and eccentricity. At higher speeds, the inertial loads become more significant relative to the pressure loads, this can be seen using the methods of Chapter 13.

9.5 Plane Bearings

Another kind of lubrication situation arises in engines, as two plane surfaces move relative to one another. Examples of this are the compression rings of the piston sliding along the cylinder wall, and a portion of the piston skirt moving along the cylinder wall, bearing the load from the side thrust of the connecting rod, as discussed in Section 13.1. In large engines, lubrication of the crosshead is also of this type.

Figure 9.10 shows a plane bearing, where there is a bearing load, F acting on the sliding block, just behind its center. The motion causes a wedge of oil to be built up between the sliding block and the surface. For an infinitely long bearing, a simplified hydrodynamic calculation can be used to obtain the following relation between the bearing load and the thickness of the oil film, and to obtain the friction

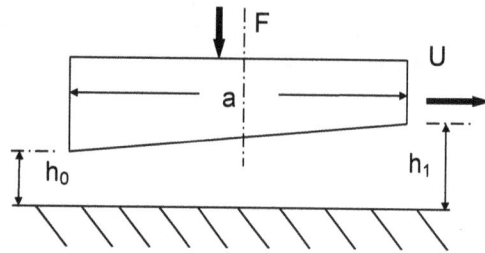

Figure 9.10: A plane bearing with a wedge shaped oil film.

coefficient.

$$F_l = \frac{6 \cdot \mu \cdot U \cdot a^2}{h_o^2} \left[\ln(1 + m) - \frac{2m}{m + 2} \right] \tag{9.8}$$

where: F_l is the bearing force per unit length perpendicular to the direction of motion, and $m = (h_1 - h_o)/h_o$. The length of the bearing, a, determines the value of F_l, and the width is needed to find the total friction force, assuming that the bearing is wide enough that Equation (9.8) is applicable. For a piston, for example, it is usually acceptable to assume that a certain fraction of the circumference of the piston is in contact with the cylinder wall, typically on the order of 20-25%.

The friction coefficient is given by:

$$f = K_1 \sqrt{\frac{\mu \cdot U}{F_l}} \tag{9.9}$$

K_1 can be shown to be a weak function of m, as indicated by the following table:

m	K_1
0.6	2.14
2	1.81 (minimum value)
10	2.09

The friction force per unit length, F_f, is then:

$$F_f = f \cdot F_l \tag{9.10}$$

And the power required to overcome the friction is equal to the product of the friction force and the velocity. Average values are used for estimates.

Special tests have shown that an oscillating piston behaves like the plane bearing shown in Figure 9.10. The piston is located in the cylinder in a position such that the greatest distance between the piston and the cylinder is found at the front edge of the piston. An oil film is found on the pressure side of the piston, but this is not the case on the other side. Since there is an oil film, then hydrodynamic lubrication will take place, as the low amount of wear on the piston typically indicates. Piston friction forces can then be described by Equations (9.8) and (9.9). In practical cases, a modified form of Equation (9.9) may be used, for example by adding an empirical constant.

The bearing force on the piston arises from the reaction of the connecting rod. This is due to the cylinder pressure and inertial forces acting on the piston, and the angle of the connecting rod. At TDC and BDC, the connecting rod is in line with the cylinder axis, and there is no side force. This is fortunate, since there is no motion to sustain the oil film. The compression piston rings are forced out towards the cylinder wall by the pressure behind them. This can be a problem especially near TDC during combustion, since the gas forces are large, and the motion small. Further exacerbating this problem is the higher temperatures near the top of the cylinder liners, which reduces the viscosity of the oil. Boundary lubrication can occur here, and the top of the cylinder is normally the location of the largest amount of wear in the cylinder bore.

The oil control rings have only their internal tension acting to force them towards the cylinder wall, so the forces there are typically lower. They are used to control the amount of oil reaching the compression rings, and can also be considered to be hydrodynamically lubricated.

Example

Consider a plane bearing surface that has a width of 3 cm and a length of cm 10 perpendicular to the direction of motion, and that Equation (9.8) is applicable. A load of 75 kN is applied to the bearing, which moves at a velocity of 15 m/s. Assume that m = 2, and that an SAE30 oil is used to lubricate the bearing, which is maintained at a temperature of 100°C. The oil has a specific gravity of 0.84. Determine the clearance height, h_o.

From Table 9.3, the kinematic viscosity is 16.33 cSt, which is equal to $1.633 \cdot 10^{-5} \text{m}^2/\text{s}$. The dynamic viscosity is then $\mu = \rho \cdot \nu = 1.633 \cdot 10^{-5} \cdot 840 \text{kg/m}^3 = 0.01372$ kg/m-s. The force per length is $75000/0.1 = 750000$ N/m. Solving Equation (9.8), $h_o = 1.2086 \cdot 10^{-5}$m. From the definition of m, $h_1 = 3.6260 \cdot 10^{-5}$m.

The friction force can also be found from Equations (9.9) and (9.10).

$$ f = K_1 \sqrt{\frac{\mu \cdot U}{F_l}} = 1.81 \sqrt{\frac{1.633 \cdot 10^{-5} \cdot 15}{750000}} = 0.000948 $$

and

$$ F_{fric} = f \cdot F_l \cdot L = 0.000948 \cdot 750000 \cdot 0.1 = 71.1 N $$

9.6 Oil Viscosity

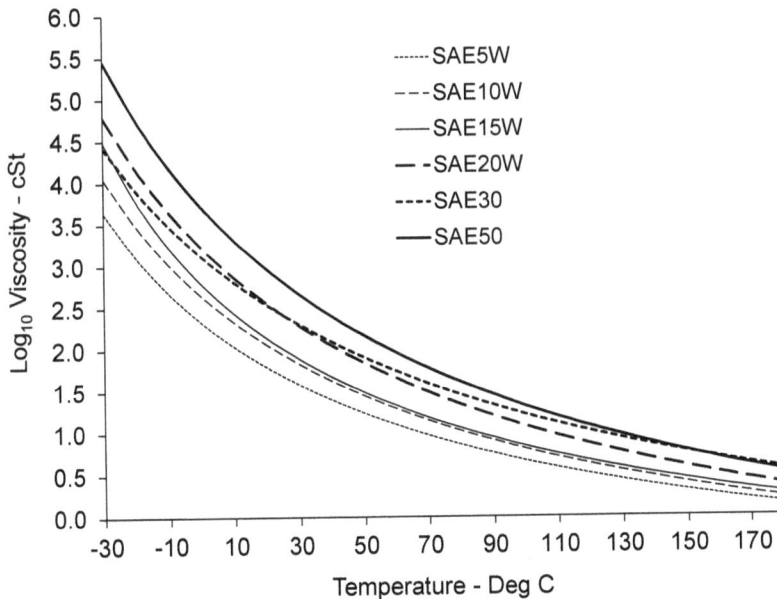

Figure 9.11: The kinematic viscosity of several SAE grades of lubricating oil as a function of temperature. The viscosity is in centiStokes, where $1 \text{cSt} = 1 \text{mm}^2/\text{s}$.

The viscosity of the lubricating oil is a very important variable in the above calculations. There are different grades of oil for different service conditions. Low viscosity is needed for cold weather starting, while a "high" viscosity at engine operating temperatures is needed to provide adequate lubrication. The Society of Automotive Engineers International has established standard grades of lubricating oil. Engines are developed using oils of these specifications, and the operator should use the same grade to obtain satisfactory operation. The dependence of viscosity on temperature is shown in Figure 9.11 for different SAE grades of oil. There are also so-called multi-grade oils, for example SAE10W-30, which has the viscosity of SAE10W at low temperatures, and that of SAE30 at high temperatures. This is accomplished by adding certain polymers to the oil that affect its viscosity dependence on temperature. Lubricating oil was originally obtained from petroleum and still is to a large extent today. However, various forms of synthetic, chemically pure lubricants have appeared on the market in recent years.

Table 9.3: Kinematic viscosity in centistokes of different grades of lubricating oil as a function of temperature.

Temperature - °C	SAE5W	SAE10W	SAE15W	SAE20W	SAE30	SAE50
-30	4319	10985	29722	60219	25991	278503
0	214.4	445.2	622.6	1697	1319	5198
25	48.29	87.29	102.5	261.7	261.3	646.4
50	17.28	27.96	30.79	69.30	80.57	146.8
100	4.595	6.323	6.848	11.88	16.33	20.53
125	2.925	3.789	4.149	6.419	9.273	10.34
150	2.030	2.500	2.779	3.883	5.819	5.901
175	1.502	1.772	2.002	2.556	3.938	3.701

Some viscosity values from Figure 9.11 for certain temperatures are listed in Table 9.3

In addition to viscosity improvers, lubricating oil contains additives for various functions such as detergents, corrosion and oxidation inhibitors, pour point suppressants and dispersants to keep particulate matter in suspension. The ability of the oil to neutralize acid or bases is also important in engine service.

9.7 Friction Correlations

In the design of an engine, it is important to be able to determine the friction of all the components of the engine, in order to establish where to concentrate improvement efforts. For more general use, such as determining general performance, estimating changes in performance and fuel economy, and simulations, it is sufficient to be able to determine the friction of the engine. Since the design of modern engines is generally the same for a given type of engine, for example turbocharged DI diesel engines, or naturally aspirated SI engines, correlations are available which give reasonable estimates.

The book by Heywood [7] presents correlations for SI engines and CI engines. The first of these to be shown is the correlation for SI engines. It is based on motored data for wide open throttle operation. The correlation for $fmep$ as a function of speed in kPa is:

$$fmep = 97 + 15 \cdot \left(\frac{N}{1000}\right) + 5 \cdot \left(\frac{N}{1000}\right)^2 \tag{9.11}$$

This correlation includes the mechanical friction, and the pumping losses associated with wide open throttle operation, but not the part load contribution to pumping loss. An SI engine correlation including pumping from motoring data is given in Equation (5.23).

The next correlation is for CI engines. There are two types of CI engines involved, that give different friction. The dependence on speed is roughly the same for both types, and the friction mean effective pressure differs by only a constant. The correlation for the motoring friction of direct injection CI engines in kPa is:

$$fmep = 75 + 48 \cdot \left(\frac{N}{1000}\right) + 0.4 \cdot S_p^2 \tag{9.12}$$

for DI (direct injection) engines and:

$$fmep = 144 + 48 \cdot \left(\frac{N}{1000}\right) + 0.4 \cdot S_p^2 \tag{9.13}$$

for IDI (indirect injection) engines. Piston speed, S_p, is in m/s.

A comparison of the predictions of Equation (9.11) with experimental measurements with the motoring method for SI engines is shown in Figure 9.12. There are two engines shown. One is a relatively modern (about the year 2000) 4 cylinder, 1.8 liter engine with 4 valves per cylinder, and the second is an older design (about 1980) 4 cylinder, 1.1 liter engine with 2 valves per cylinder.

First of all, it can be seen that the *fmep* for both engines is of a similar level. Second, the Heywood correlation gives a good estimate of the engine friction, and its dependence on speed. The friction of the engine with 2 valves per cylinder is slightly higher than that with 4 valves per cylinder, an indication that the pumping losses may be lower with more valves, though the difference is smaller at high speed, which is not expected. In both cases, it appears as if the speed dependence with the Heywood correlation predicts a stronger increase with speed at the very highest speeds than the data indicate. For general design purposes, the correlation can be used, though for specific engines it would be desirable to adjust the parameters slightly if good data is available.

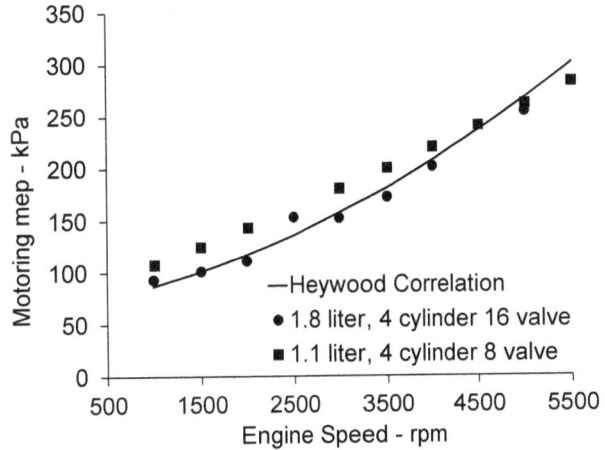

Figure 9.12: A comparison of motored friction data with the predictions of the Heywood correlation for SI engines at wide open throttle for two different engines.

The correlation for DI diesel engines is illustrated in Figure 9.13A, where it is compared to data for several direct injection diesel engines. The first is a 2.15 liter, 4-cylinder turbocharged engine, the friction mean effective pressure determined from an engine map and Willans line analysis [2]. The second is a 2 liter 4-cylinder turobcarged engine using engine data and detailed Willans line analysis. A third engine is a 1 liter, 2-cylinder engine of older design, with bore and stroke of 85 mm. The fourth engine is a 12-liter, 6-cylinder turbocharged engine, typical of truck or bus applications. The results show that the Heywood correlation can be used to realistically estimate friction for a variety of sizes and technologies.

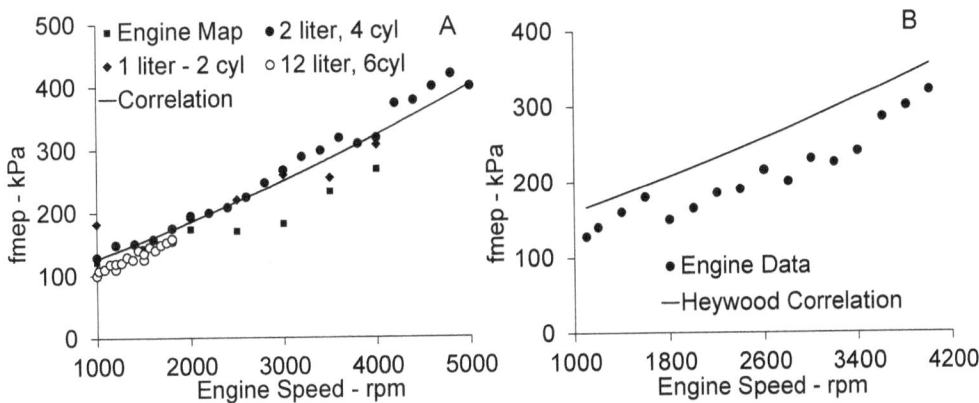

Figure 9.13: The friction mean effective pressure for several direct injection diesel engines compared to the predictions of the Heywood DI correlation - A and The friction mean effective pressure from a Willans line analysis compared to the predictions of the Heywood correlation for a 1.6 liter, 4-cylinder turbocharged indirect injection diesel engine - B.

Figure 9.13B shows the comparison between the predictions of Equation (9.13) and measurements made with a Willans line analysis of a 1.6 liter, turbocharged IDI engine. The trend with respect to speed is well predicted, but the correlation values are too high by about 40-50 kPa, and the measurements are actually closer to the DI correlation. Heywood's correlation was based on engines from the 1960's so it may be that they had a higher friction.

For any of the engines, the friction predictions in terms of trends with speed are satisfactory. It may be necessary to adjust for the absolute level by changing the constant term in the correlation equations.

9.8 Role of Engine Components in Friction

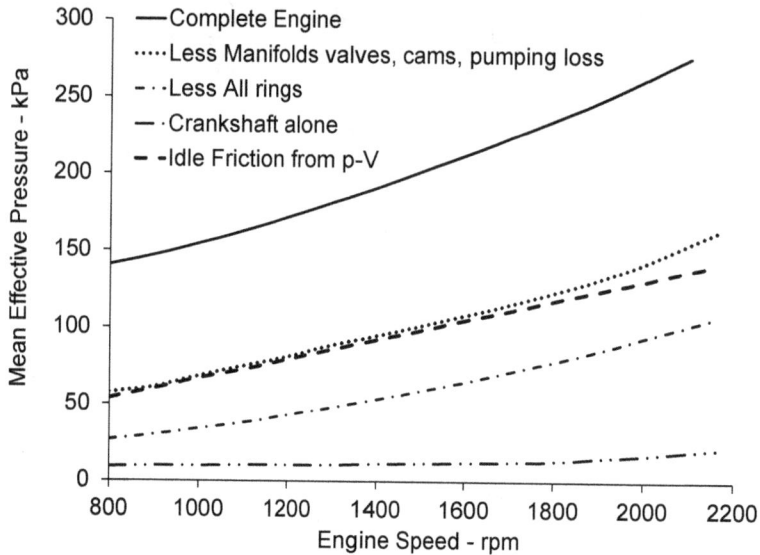

Figure 9.14: Engine friction for a 6-cylinder diesel engine measured with the motoring method as engine components are progressively removed. Adapted from [36].

In a classic study of friction, Brown [36] made some very accurate measurements of friction determined from pressure volume diagrams, and compared the results to motoring tests of a complete engine and an engine where various components were removed one-by-one. The results for a 6 cylinder diesel engine are shown in Figure 9.14. The dashed line for idle friction indicates actual mechanical friction determined from the p-V diagram, where the pumping losses are measured directly. It agrees quite closely to the motoring mep for the engine with the manifolds, valves and camshaft removed, which should reduce the pumping losses to a negligible value.

Further removal of components shows that the oil and water pumps do not give major losses. The piston rings give a fairly constant contribution to the $fmep$ of about 50 kPa, though the pressure forces, which normally add to the spring tension of the rings, are not present here. A big difference is noted when the pistons and rods are removed, and only the crankshaft present, indicating that the piston is a major contributor to the mechanical engine friction.

Vickery [92] performed a similar study of the effect of different components on the motoring friction in a spark ignition engine. By removing various groups of components, he obtained the results shown in Figure 9.15. The fan, water pump and electrical equipment contributed with about 20 % of the total friction. In more modern engines, fans are electrically driven and controlled by thermostats. Their power consumption is lower than those of engines at the time of Vickery's study. The tests were conducted at wide open throttle, but the losses associated with the air exchange process are still significant, as was the case with the diesel engine. At part load, pumping losses are high with the SI engine, and at idle conditions an additional contribution on the order of 80 kPa to the $fmep$ can be expected from this source [7].

The above results are shown on the basis of the distribution of the losses in Figure 9.16. Since the original data were obtained in somewhat different configurations, the results were replotted to estimate the three main systems, the pumping components; the piston rings and connecting rods; and the crankshaft, camshaft and oil pump combination. The effects of auxiliary equipment such as fuel pumps, generators, *etc.* were not included.

Though not precise, both studies give approximately the same distribution as a function of engine speed. The shaft related friction is about 15 to 20% of the total, the piston and rods contribute with about 20 to 30% of the total. These relative contributions increase somewhat with speed. The pumping

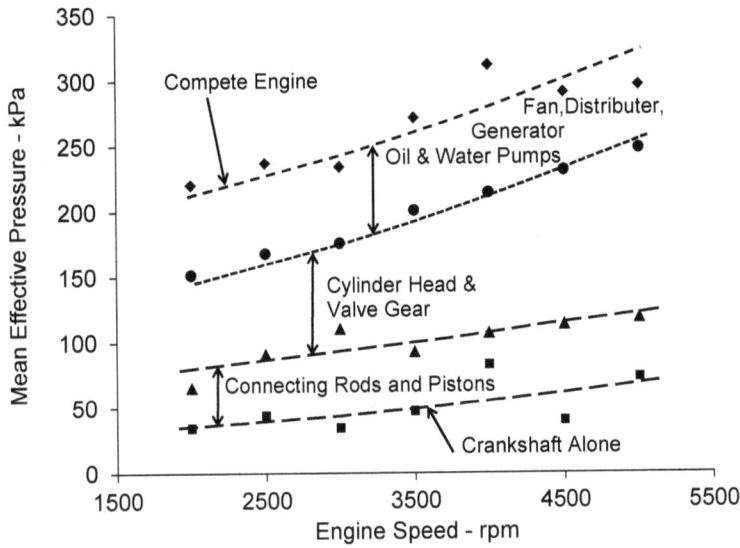

Figure 9.15: Engine friction for a 4-cylinder spark ignition engine measured with a motoring method as engine components are progressively removed. Adapted from [92].

portion is on the order of 60 to 50% of the total, its relative magnitude decreasing slightly with speed. These results give an estimate of the relative friction contributions. There are additional effects, such as the part load pumping in SI engines, and the increase of bearing loads with cylinder pressure that will affect the results in an individual engine. Depending on a variety of parameters, an increase in the fmep of a highly turbocharged engine due to increased cylinder pressure is estimated to be on the order of 10%.

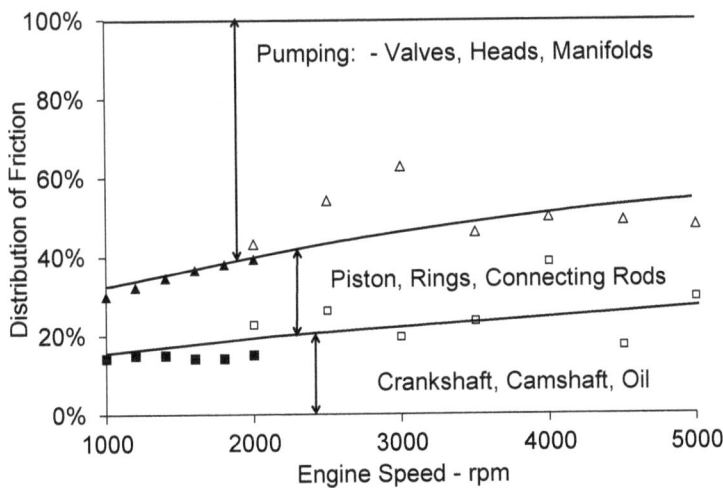

Figure 9.16: The relative contribution of various components to motoring friction as a function of engine speed. Open symbols from the SI engine data of [92] and closed symbols from the CI engine data of [36].

9.9   Problems

Problem 9.1

Determine the friction mean effective pressure of engine in Problem 1.1 at 1500 rpm. Assume that the indicated efficiency is constant for brake mean effective pressure less than 0.4 MPa. Calculate the indicated efficiency at 1500 rpm and a brake mean effective pressure of 1 MPa. Assume that the friction is not dependent on load for a given engine speed, and that the lower heating value is 42500 kJ/kg. ($fmep \approx 200$ kPa. $\eta_i \approx 0.51$)

Problem 9.2

Consider an 8-cyl. 4-stroke spark ignition engine with bore and stroke = 100 amd 90 mm at 4000 RPM. The engine is lubricated with an SAE 30 oil with an bearing entrance temperature of 110°C. The oil viscosity is a function of temperature:

$$\mu = 0.03868 \cdot \exp \frac{1090}{(T_c + 95)}$$

where $T - c$ is in °C, and μ in cP Data for the engine's bearings are:

Bearing data:	Number	Width	Diameter	Total clearance
Main bearing	5	38 mm	70 mm	0.075 mm
Connecting rod bearing	8	25 mm	64 mm	0.070 mm

1. Calculate temperature increase for lubricating oil when passing through the bearings, where it can be assumed that half of the frictional heating is conducted to engine block, and that the average value for the relative eccentricity is 0.8. The specific heat of the oil is 2.30 kJ/kg-K, the density is 825 kg/m3 (c_p and ρ independent of temperature), and the mass flow of lubricating oil to the bearings is:

Main bearings	0,02 kg/sec per bearing
Connecting rod bearings	0,008 kg/sec per bearing

(The oil to the connecting rods has passed through the main bearings!)

2. Calculate the engine's power and mean effective pressure loss due to bearing friction, estimate the power of the oil pump, and convert the latter to a "mean effective pressure". It can be assumed that the pumps delivery capacity is 30% larger than the oil flow to the main bearings. Oil is delivered to the bearings at a pressure of 4 bar.

3. Calculate the engine's power and mean effective pressure losses from piston friction (excl. ring friction), if it is assumed that the oil film consists of 25% of the piston perimeter for the full length of the piston skirt (75 mm). The piston's slanted position is given by $m = (h_1 - h_o)/h_o = 2.0$ and the minimum film thickness is 10% of the piston clearance of 0.2 mm. The mean temperature of the oil film can be assumed to be 90°C and the temperature increase due to friction can be neglected.

($\Delta T_{total} = 13.5°C; P_{bearing} = 5021W; mep_{bearing} = 0.266bar; P_{pump} = 126W, p_{pump} = 0.0068bar; P_{piston} = 6616W; mep_{piston} = 0.351$ bar.)

Problem 9.3

A 4-cylinder 4-stroke combustion engine has a bore and stroke of 80 and 78 mm. The engine operates at 4200 rpm with an indicated mean effective pressure = 9.6 bar.

The following information can be used to calculate the mechanical loss: Temperature in the main bearing ll0°C(Ave. temperature) Temperature of the cylinder walls 95°C (Ave. temperature). The oil viscosity can be calculated as in Problem 9.2. The oil density is 825 kg/m³, the specific heat = 2.3 kJ/kg-K and the flow is 0.015 kg/s to each main bearing. The force between the piston and cylinder per unit length of the perimeter can be calculated by:

$$F_l = K_a \cdot \mu \cdot S_p \frac{a^2}{h_o^2}$$

where: a is the piston height of 62 mm h_o is the smallest oil film thickness between the piston and cylinder and is $= 0.015$ mm. K_a is a factor determined by the angle of piston in the cylinder and is $= 0.6$. S_p is the mean piston speed in m/s. The friction coefficient can be calculated as $f = 5.3 \cdot 10^{-3} \cdot \sqrt{\mu}$. It can be assumed that the oil filled area covers 25% of the piston perimeter. The friction power in a main bearing can found from:

$$P_{fric} = F \cdot U = \mu \cdot U^2 \cdot L \cdot D_L \frac{K_b}{C_r}$$

Where: F is the friction force on shaft surface. U is the tangential speed of the shaft surface. L is the length of the bearing $= 33$ mm. D_L is the diameter of the bearing $= 56$ mm. C_r is the radial bearing clearance $= 0.03$ mm. K_b is a constant $= 5.0$. There is a main bearing between each cylinder, 5 bearings in all.

It can be assumed that the losses in the main bearing and the between the cylinders and pistons comprise 40% of the total loss, (air exchange losses included).

Calculate: 1. The total friction loss. 2. The mechanical efficiency. 3. The temperature increase in the oil in a main bearing when half of the friction power goes to heating. (Loss $= 10.05$ kW; $\eta_m = 0.809$; $\Delta T = 5.38$ °C.)

Chapter 10

Heat Transfer

10.1 Introduction

Heat transfer is an important aspect of engine operation and design. It affects the thermal efficiency of the engine through heat losses during especially the combustion and expansion processes. It affects the volumetric efficiency of the engine, as cold incoming gasses are heated by warm engine surfaces, reducing the charged mass. It affects the mechanical strength of parts such as pistons and valves, where their strength is lessened by high temperatures. It is related to friction and lubrication, as friction energy must be removed from the engine through heat transfer, and the heat transfer processes determine lubricating oil temperature, a key factor in determining the durability of the engine. Heat transfer in the exhaust system determines the energy available to operate turbochargers, and the temperature of exhaust after-treatment devices, crucial to meeting modern emission control standards. The over all cooling of the engine is important as the cooling system, either water or air, determines requirements for air flow to the engine, important in vehicle design. The heat transfer away from engine surfaces to either the cooling water or air must be adequate to keep the temperature of all engine parts in a satisfactory operating range. The performance of heat exchangers is important for overall engine cooling and for the cooling of engine intake air after compression in pressure charged engines.

10.2 Heat Transfer Modes

Three types of heat transfer are generally considered in engineering systems, all three play significant roles in engine design and operation.

- *Conduction.* This is heat transfer through materials due to a temperature gradient, and is governed by Fourier's Law:

$$\dot{Q}_x = -k \cdot A_x \frac{\partial T}{\partial x} \tag{10.1}$$

  where: $\dot{Q}_x$ is the heat flux in the x direction in W, k is the thermal conductivity of the material in W/m-K, and T the temperature of the material. Conduction of energy in the piston is a very important topic in determining the temperatures of the material. Pistons are usually made of aluminum, which loses strength rapidly with higher temperatures. Conduction through cylinder liners is important in connection with determining surface temperatures and the temperature of the oil film lubricating the piston and rings.

- *Convection.* This is heat transfer involving the motion of fluids. This type of heat transfer is normally described by what is called Newton's law of cooling:

$$\dot{Q} = h \cdot A \left(T - T_\infty \right) \tag{10.2}$$

where: $\dot{Q}$ is the heat flux from a surface in W, h is the heat transfer coefficient in W/m²-K, A is the area of the surface, T the temperature of the surface, and T_∞ the temperature of the fluid surrounding the surface.

Most convection in engines occurs through forced convection, where there is an externally imposed flow field determining the flow over the heat transfer surface. The heat transfer coefficient is usually related to flow and fluid properties through correlations of dimensionless parameters:

- Nusselt number : $Nu_x = \frac{h \cdot x}{k}$, where k is the thermal conductivity of the fluid, and x a characteristic distance.

- Reynolds number: $Re_x = \frac{\rho v x}{\mu}$ where ρ is the fluid density, x a characteristic distance, v the velocity of the flow, and μ the dynamic viscosity of the fluid.

- Prandtl number : $Pr = \frac{\mu \cdot c_p}{k}$, where k, μ and c_p are properties of the fluid. Properties effects are given in the Prandtl number.

The convection correlations are typically of the form:

$$Nu_x = f\left(Re_x, Pr\right) \tag{10.3}$$

typically,

$$Nu_x = C \cdot Re_x^a, Pr^b \tag{10.4}$$

where a, b and C are empirically determined constants.

- *Radiation.* This is heat transfer between substances through the transfer of energy in the form of electromagnetic radiation. Radiation plays an important role especially in the heat transfer from diesel engines during combustion. There is also some radiation heat transfer from exhaust systems and blocks to the surroundings, though this is not usually important in regard to performance. An expression for the radiation heat transfer between two surfaces that only "see" each other is:

$$\dot{Q}_{1 \to 2} = \frac{\sigma T_1^4 - \sigma T_2^4}{\frac{1 - \epsilon_1}{\epsilon_1 A_1} + \frac{1}{A_1 F_{12}} + \frac{1 - \epsilon_2}{\epsilon_2 A_2}} \tag{10.5}$$

where: σ is the Stefan-Boltzmann constant, $56.7 \cdot 10^{-12} kW/m^2 - K^4$, ϵ_i is the emissivity of body i, A_i is the radiating surface of body i, and F_{12} is the shape factor, that is the fraction of energy radiated from surface 1 that reaches surface 2. In an engine environment, the emissivity of most surfaces is close to one, with the exception of polished parts. In the combustion chamber, it has been shown that the diesel flame has an emissivity quite close to one, and the piston and cylinder heads also have a high emissivity, due to the layer of soot and deposits on the surfaces. In spark ignition, there is little soot in the flames and their emissivity is normally very much lower than in diesel engines, and radiation heat transfer does not play much of a role during combustion in SI engines.

10.3   Steady State Heat Transfer

10.3.1   Heat Transfer Resistances

As pointed out in Section 1.1, heat transfer is not required for internal combustion engine operation. But it cannot be avoided in practice. When gas temperatures occur in excess of 2000K, mechanical parts can easily become too hot and fail unless properly cooled. Heat transfer in engines typically involves gas cooling through solid walls to an cooling fluid, either air or a mixture of water and anti-freeze fluid, such as ethylene glycol. A classical situation is shown in Figure 10.1, where three heat transfer resistances are shown, where $Q = R \cdot \Delta T$. The first is the convection resistance on the gas side, R_g, where:

$$R_g = \frac{1}{h_g A_{wg}} \tag{10.6}$$

and h_g is the gas side heat transfer coefficient, and A_{wg} is the surface area on the gas side. The heat transfer coefficient is the effective value, and may contain a radiation component, especially in diesel combustion. The next resistance is that of the wall, R_w, where:

$$R_w = \frac{l}{k_w A_w} \tag{10.7}$$

and l is the thickness of the wall, k_w the thermal conductivity of the wall, and A_w the cross section of the wall. It is assumed here that the wall is thin and the heat flow is one dimensional. If not, 2- or 3-dimensional conduction heat transfer analysis is needed. This is usually the case where the "wall" is the piston. The final resistance is that on the coolant side, R_c, where:

$$R_c = \frac{1}{h_c A_{wc}} \tag{10.8}$$

and h_c is the heat transfer coefficient, and A_{wc} the surface area on the coolant side.

This is a convective resistance, determined by the flow of either the liquid or air coolant. In the case of the former, this resistance is usually much lower than that on the gas side, and the two areas are close to equal. However, in the case of air cooling, the coolant side heat transfer coefficient will be of a similar magnitude to the gas side, and a larger area is needed to keep the wall temperature down. This can be seen by the use of fins on the outside of most air-cooled engines. An air-cooled engine has the advantage that no separate fluid is needed, eliminating problems with leaks and potential freezing, and there is no need for an additional heat exchanger. On the other hand, on a large number of air cooled engines, an air fan is required to provide adequate air circulation, and this typically costs more energy than pumping a liquid. In terms of engine construction, the liquid-cooled engine typically is cast with internal passages to provide for the flow of coolant. Internal passages are not needed for air cooling, though fins are normally needed as mentioned above. Because of weight, price and size considerations, the smallest engines are quite often air cooled, though air cooling has been used on engines in the 100 kW range, and formerly on large aircraft piston engines.

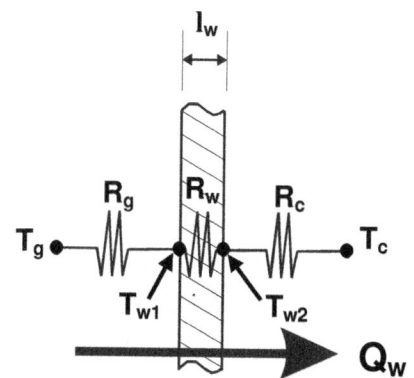

Figure 10.1: A schematic diagram of the heat transfer through an arbitrary wall in an engine.

It is of interest to look a little closer at these resistances. To do that, some information is needed about some of the heat transfer characteristics. A review article in the literature gives an excellent review of all aspects of engine heat transfer as of the late 1980's[93]. In this article, measurements are cited indicating that average in-cylinder gas film heat transfer coefficients can be in the vicinity of 500 W/(m^{2}K) and average heat flux in the cylinder liner over the cycle on the order of 250 kW/m^2 for engines with bore/stroke sizes on the order of 100 mm.

Example

Assume that the cylinder liner is 5 mm thick and made of steel with a thermal conductivity of about 45 W/(m-K). In a water cooled engine it is quite difficult to know the flow in the complicated geometry of the cooling system, so a rough estimate is made using water in turbulent flow over a flat plate with a characteristic distance of 10 cm (the approximate size of the stroke or the bore) and a velocity of 10 m/s. This gives a heat transfer coefficient on the order of $5 \cdot 10^4$W/(m^{2}K). Then assuming water cooling and that the areas are equal on both sides of the wall and calculating on the basis of 1 m^2:

$$R_g = \frac{1}{h_g A_{wg}} \approx \frac{1}{500 \cdot 1} = 2 \cdot 10^{-3} \frac{K}{W}$$

$$R_w = \frac{l}{k_w A_w} \approx \frac{0.005}{45 \cdot 1} = 1.1 \cdot 10^{-4} \frac{K}{W}$$

$$R_c = \frac{1}{h_c A_{wc}} \approx \frac{1}{5 \cdot 10^4 \cdot 1} = 2 \cdot 10^{-5} \frac{K}{W}$$

It can be seen then, that there is about an order of magnitude difference between each resistance in the following decreasing order: cylinder gasses, the wall, and the coolant film. For a typical cyclic average heat flux on the order of 250 W/m^2 according to Borman and Nishiwaki [93], and assuming a coolant temperature of 100$°C$, this gives an average gas temperature of about 630$°C$, an inner cylinder liner temperature of about 130$°C$, and a liner coolant side surface temperature of about 105$°C$. These numbers are only estimates, and will change for different operating conditions and locations. The important point is that the main resistance to heat transfer is the gas film on the inside of the combustion chamber, with the wall having a heat transfer resistance on the order of a tenth that of the cylinder gas, and the coolant film having a resistance an order of magnitude lower than that for a liquid cooled engine. For an air-cooled engine, the heat transfer coefficients for the coolant will be much lower than for liquid, and so the ratios of resistances will to a large extent be determined by the area of the external cooling surface. That is, the number of fins, their size, and their arrangement. It is very likely that the air-cooled engine will have higher surface temperatures than a liquid-cooled engine, though the low environmental temperature for air cooling compared to a typical liquid coolant temperature of about 100$°C$ helps some.

One of the first needs for heat transfer information from an engine is that of determining coolant needs. In a vehicle application, this is important in connection with the placement and size of the heat exchanger used to cool the engine. Both air and liquid cooling are used on vehicle engines, with the tendency for air cooling on smaller engines and liquid cooling on the larger engines. The selection of heat exchangers is an important topic with respect to the overall engine cooling and in the case of supercharged or in the case of turbocharged engines, the selection of the after cooler. Considerations such as heat transfer performance, pressure loss, size and weight as a function of cost are very important. Methods for analysis of heat exchangers are well covered in relevant heat transfer texts, and will not be covered here.

10.3.2 Measurement of Steady State Heat Transfer

Steady state heat transfer can be measured by conducting an energy balance on the engine, or by measuring coolant flow and temperatures in an engine. In case of direct measurement, it is necessary to thermally insulate the engine itself from the surroundings, so that all of the energy is removed in the coolant, and not by radiation and conduction to the surroundings. In this case, the following energy balance can be used as a check on accuracy, since it should be satisfied in the ideal case.

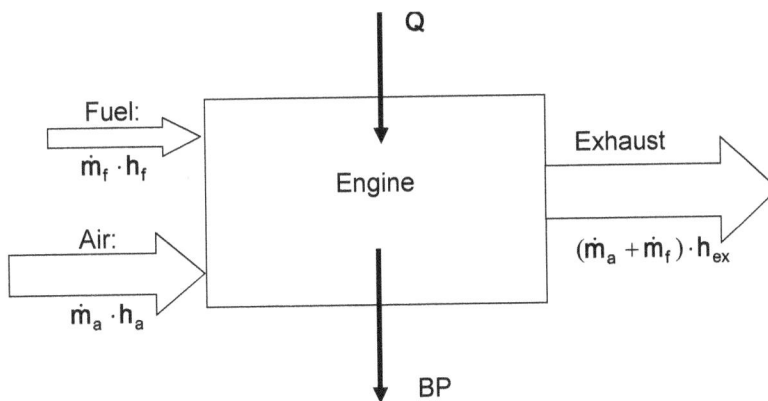

Figure 10.2: A control volume for and engine showing energy flows.

Based on the sketch in Figure 10.2, the energy balance for an engine can be written:

$$\dot{Q} - BP = \dot{m}_{ex}h_{exh} - \dot{m}_a h_a - \dot{m}_f h_f \tag{10.9}$$

where the enthalpies include heats of formation, as in Equation (3.55). The enthalpy of the exhaust gas components includes a thermal (sensible) contribution and a chemical contribution, and for very accurate measurements, an exhaust gas analysis is needed. The need is greater in SI engines operating with rich mixtures, as the unburned hydrocarbons, CO and H$_2$ can contain a significant fraction of the input chemical energy. For engines operating lean, this contribution is small. A reasonable approximation to

Equation (10.9) is can be made using the lower heating value at atmospheric temperature, and using atmospheric temperature as a reference for sensible enthalpy:

$$\dot{Q}_{in} + \dot{Q}_{loss} = \dot{m}_f H_u + Q_{loss} = BP + \dot{m}_e c \int_{T_{in}}^{T_{ex}} \bar{c}_{p,ex} dT + \sum \dot{m}_{i,ex} H_{u,i} \qquad (10.10)$$

where: $\dot{m}_{i,ex}$ is the mass flow rate of an exhaust component, i, and $H_{u,i}$ it's heating value, $\bar{c}_{p,ex}$ is the average specific heat of the exhaust gasses. The individual mass flows of the exhaust components can be calculated from measured exhaust composition. Most emissions measuring instruments measure in volume percent, so the mass flow rate of specie i can be calculated as:

$$\dot{m}_{i,ex} = \dot{m}_{ex} x_i \frac{M_i}{M_{ex}} \qquad (10.11)$$

where x_i is the mole fraction of specie i on a wet volume basis, the M is molecular weight.

$\dot{Q}_{loss}$ is net result of the heat transfer to the coolant and losses from non-insulated components to the surroundings. It should have a negative value in accordance with the standard sign convention where heat transfer to a system is assumed positive. It includes the heat transfer to the coolant from the gasses to the engine parts as well as the friction produced by the engine, so if the actual heat transfer is desired, one must correct for the friction, either from measurements or an estimate from the correlations in Chapter 9.

When determining heat losses through an energy balance then, it is necessary to know the fuel and air flow rates, the heating value of the fuel, the exhaust temperature and the exhaust composition. Measurement of exhaust temperature is simple in practice, but the interpretation is difficult. The amount of heat loss determined by the energy balance is also affected by the location of the exhaust temperature measurement, and heat losses in the system up to the exhaust temperature measurement location are included in $\dot{Q}_{loss}$. For the heat transfer within the cylinder, corrections are needed for the losses between the exhaust valve and the temperature measurement, discussed later in this chapter.

The other problem with the exhaust temperature measurement is that the closer to the cylinder the measurement occurs, the greater the temporal variation in both the flow rate and temperature. Figure 7.14 shows typical variations in the flow out of the exhaust port in the absence of pressure oscillations. At the same time, the exhaust temperature varies due to the expansion in the cylinder under the blowdown process. Thus, to correctly determine the correct energy in the exhaust gasses, one needs to integrate the product of the mass flow and the enthalpy to determine the value of the enthalpy flow. This requires instantaneous values of the exhaust temperatures. At the present state of the art, devices such as thermocouples or thermisters of adequate strength are too slow to record these temperature changes, and difficult radiation methods must be used to measure the radiation from the CO_2 or water in the exhaust gas. Normally the errors here are accepted.

10.3.3 Steady State Heat Transfer Correlation

A classic correlation for the total, steady state heat transfer from engines was presented by Taylor and Taylor [6]. The basic form and behavior of this correlation is still correct, and is useful for looking at trends, and estimating effects of variables on heat loss. It is the only method cited for steady-state heat transfer in a very recent review article [94]. The heat transfer is described by Newton's law of cooling, where the characteristic area is selected as the total piston area of all cylinders, and the temperature difference is that between the engine coolant at T_c and a cycle average gas temperature, T_g.

$$\dot{Q} = h \cdot A_p (T_g - T_c) \qquad (10.12)$$

The heat transfer coefficient is written in the form of the Nusselt number based on the cylinder bore, B,

$$Nu_B = \frac{h \cdot B}{k_g} \qquad (10.13)$$

and a correlation is obtained in terms of a Reynolds number:

$$Nu_B = aRe^m \qquad (10.14)$$

where the Reynolds number is calculated as:

$$Re_B = \frac{\dot{m}_g \cdot B}{\mu_g \cdot A_p} \tag{10.15}$$

and a and m are empirical constants, $\dot{m}_g$ is the total gas flow through the engine combustion chamber, and the subscript g refers to the properties of the combustion gasses.

For a range of older engines, Taylor and Taylor have suggested the values of $a = 10.4$ and $m = 0.75$. For specific engines, a is suggested to vary from about 7.5 to 15, with 10.4 being an average value.

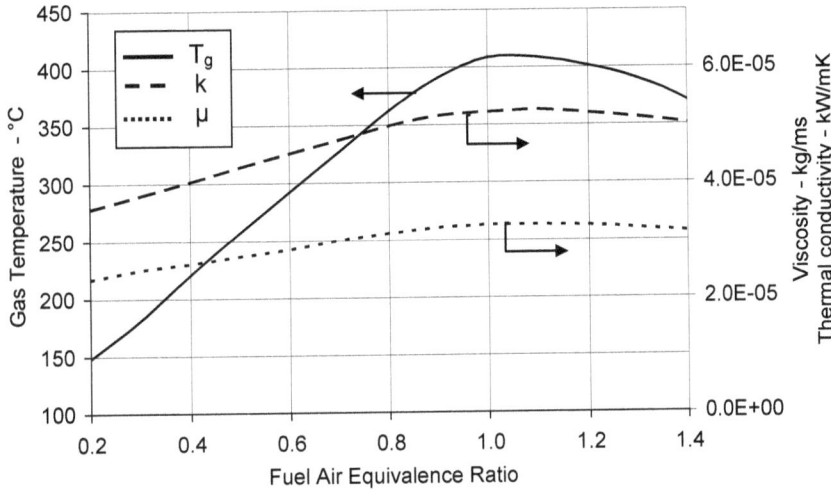

Figure 10.3: The average gas temperature, viscosity and thermal conductivity as a function of fuel air equivalence ratio, to be used in the correlation of Equation (10.14). The gas temperature is based on an intake temperature of 300K, for other intake temperatures, Taylor and Taylor recommend adding $0.35 \cdot (T_{in} - 300)$ to the value from the above curve.

To use the correlations, the gas properties must be known. In the development of the correlation, it was found that μ_g k_g and T_g all can be written as functions of fuel air equivalence ratio. Those functions, converted to SI units, are shown in Figure 10.3. T_g is also a weak function of the exhaust pressure, as shown in Taylor and Taylor.

Some care should be used in interpreting the values of the heat transfer coefficient and gas temperatures. The heat transfer coefficient is based on a reference area, A_p, that is readily available for calculation purposes, that is the piston area. It has some relation to, but is not really, the true heat transfer area. Additionally, the heat transfer area varies with time, as shown in Equation (2.110), so the results from the Taylor and Taylor correlation are averaged through spatial and temporal variations. As shown in Figure 10.1, the heat transfer includes the effect of three heat transfer resistances involving: the gas film and area, the wall thickness, conductivity and area, and the coolant film and area. The gas temperature is also an average value, found from experimental heat transfer measurements, which gives an measured time averaged heat transfer.

10.3.4 Effect of Heat Transfer on Efficiency

An important aspect for the heat loss is its effect on the thermal efficiency. In Section 2.4.2 it was shown that the heat loss has an effect on the temperatures and pressures throughout the cycle and, therefore, on the thermal efficiency. The nature of this trend can be shown from Taylor and Taylor's steady state heat transfer correlation. In the efficiency calculation, the important factor is the heat loss through heat transfer in relation to the total energy input to the engine.

For a 4-stroke engine, the input energy to the engine is given by the product of the fuel flow and the heating value:

$$\dot{Q}_{in} = \dot{m}_f \cdot H_u = \eta_v \frac{\rho_{in} V_d N}{120} FA \cdot H_u \tag{10.16}$$

Using this, and equations (10.12) through (10.15) with $a = 10.4$, the fraction of input energy lost to the coolant, f_{HT}, can be written:

$$f_{HT} = 10.4 \frac{k_g A_p^{0.25}}{B^{0.25} \mu_g^{0.75}} (T_g - T_c) \left(\frac{120}{\eta_v \rho_{in} V_d N} \right)^{0.25} \frac{(1 + FA)^{0.75}}{FA \cdot H_u} \qquad (10.17)$$

Consider the case of an SI engine operating at wide open throttle with variable speed. If the volumetric efficiency doesn't change much, then Equation (10.17) can be approximated by:

$$f_{HT} \approx \mathcal{K} N^{-0.25} \qquad (10.18)$$

where $\mathcal{K}$ represents the product of a number of terms which are, or are close to being, constant.

The relationship has some important implications with respect to performance and economy. According Equation (10.18), the fraction of energy lost to the cylinder walls decreases as speed increases. It basically does so because the exponent in Equation (10.14) is less than 1. Though the total heat transfer increases with engine speed, the fraction of energy lost through heat transfer decreases and the indicated thermal efficiency increases as the engine speed increases. Things such as combustion changes or engine adjustments can have additional effects here, but this general trend exists. This has an effect on engine maps, as discussed in Section 12.3.2. Figure 10.4 shows the indicated efficiency of an SI engine operating at full load with constant, stoichiometric fuel air ratio and MBT timing. The tendency is as indicated by Equation (10.18), though the results fit better with an exponent in Equation (10.14) on the order of 0.80. The same trend can

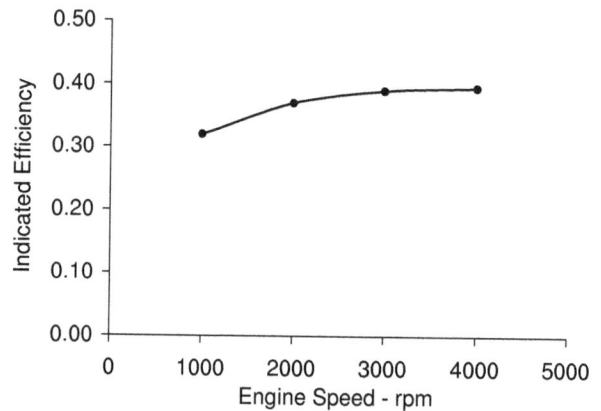

Figure 10.4: The indicated efficiency of a 1.3 liter, 4-stroke SI engine with a compression ratio of 8.8:1 running at wide open throttle with a stoichiometric mixture.

be seen in Figure 7.5, though there are slight variations in the fuel air ratio there.

Equation (10.17) can be used to estimate the general trends in heat losses with respect to different engine variables. Note that a reduced heat loss is expected to give an improved indicated efficiency. It is not possible, however, to quantitatively determine how much this improvement will be, since the correlation used here is for the average heat transfer through the entire engine cycle, and the distribution of the time dependent heat loss throughout the entire cycle is needed to determine the actual efficiency improvement. Cycle simulations, as introduced in Section 2.5 can be used to give better estimates.

Some general trends can be shown for different operating variables and engine types. Figure 10.5 shows the estimated fraction of input energy lost to heat transfer for a spark ignition engine running at full throttle with varying speed, and with varying load at constant speed. As indicated above, the relative heat loss decreases with increasing speed, and at the higher speeds, there is not much effect, similar to the effect on indicated efficiency shown in Figure 10.4. This gives confidence that the trends are reasonable. Absolute values will depend on the coefficient, a, and exponent m, the former varying by about ±50%. Figure 10.5 also shows that the relative amount of heat loss decreases with increasing load for an SI engine, in addition to pumping losses and lower compression ratio, this is also a reason why part load operation with SI engines gives lower efficiency than CI engines.

Diesel engine operation can also be investigated with the Taylor and Taylor correlation. It is difficult to estimate a full load test with a diesel engine, since the full load mixture ratio is adjusted with respect to smoke and torque requirements. These limits are quite engine specific, and so the full load tendency will not be shown here. The relative heat loss for part load operation is shown as a function of fuel air ratio for two different 4-stroke CI engines in Figure 10.6. The first is a 4 cylinder, 1.8 liter naturally aspirated engine, while the second is a 4 cylinder, 1.9 liter turbocharged and aftercooled engine. In Figure 10.6, actual engine data for volumetric efficiency, intake conditions and fuel air ratios were used in Equation (10.17). The trends are generally similar, with no significant difference between the two types of engine.

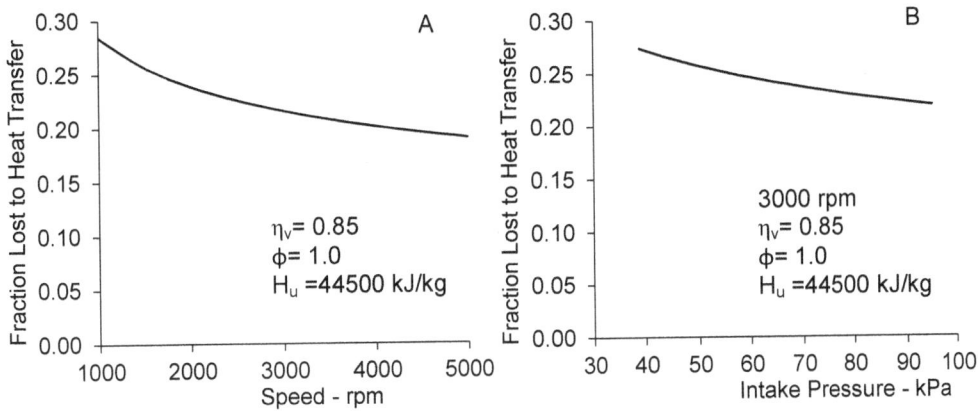

Figure 10.5: The estimated fraction of input energy lost to heat transfer as a function of speed (A) and load (B) for a 1.5 liter (80mm bore, 75mm stroke, 4 cyl) 4-stroke SI engine, operating with a stoichiometric mixture using Equation (10.17).

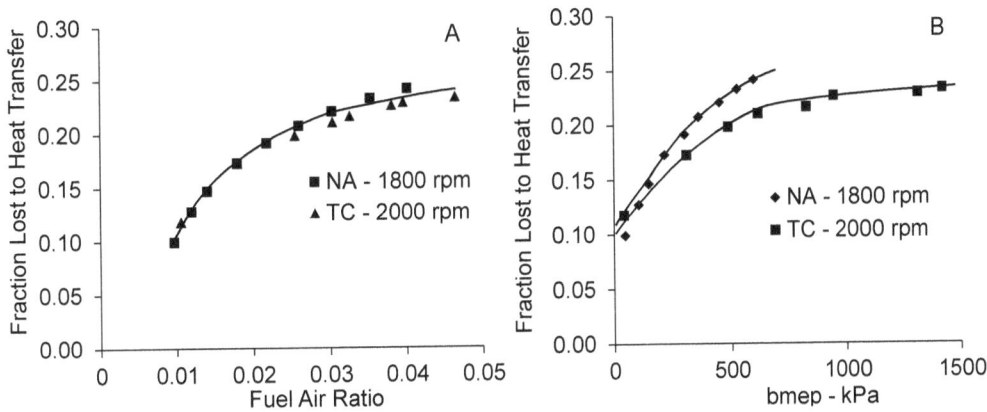

Figure 10.6: The estimated fraction of input energy lost to heat transfer as a function of fuel air ratio and bmep for a 1.8 liter naturally aspirated direct injection diesel engine (A) and a 1.9 liter turbocharged after-cooled direct injection diesel engine (B), using Equation (10.17).

This is because the heat transfer in the Taylor and Taylor correlation is strongly related to the average gas temperature, which in turn is a function of fuel air ratio for lean mixtures. Of note is the result that the relative heat losses are very low at part load operation. This is due to the lean mixtures which give low average gas temperatures, in contrast to the SI engine, where the stoichiometric mixture gives a high gas temperature, regardless of the load. This is an additional advantage to diesel engine operation at part load compared to spark ignition engines.

But with regard to load, the fuel air ratio is only part of the story with a turbocharged engine. Because of the extra air, a turbocharged engine will produce more power at a given fuel air ratio, and this shows that a turbocharged engine will have an even lower percentage heat loss at a given bmep than a naturally aspirated engine, an additional help for better efficiency. This is clearly indicated in the right hand side of Figure 10.6. Aftercooling in the case of turbocharged engines is also beneficial, as it reduces the average cycle temperature.

The general relationships can be used to investigate the relative effects of some engine design variables on efficiency. Since the gas heat transfer described by Equations (10.12) to (10.15) is dependent on the surface area, it is of interest to see what the effect of the overall cylinder geometry is on the relative heat loss. In particular, the heat transfer is related to the surface area, and the total fuel mass/energy input is related to the cylinder volume, so the effect of surface to volume ratio is of interest. Two variables which will have an effect on this are the absolute engine size, and the bore to stroke ratio.

Figure 10.7 shows the effects of these two variables on the fraction of energy lost to heat transfer for an SI engine operating at full throttle with a stoichiometric fuel air ratio. The results show that a large engine size reduces the percentage heat loss, thus giving an improvement in efficiency. Also, a larger bore to stroke ratio reduces the relative heat loss thereby improving efficiency. It is well known in practice that larger engines have higher efficiency.

But there are other issues to be dealt with in engine design than just the heat loss. For example, a large bore size in a spark ignition engine will give a longer flame travel, and thereby increase the tendency to knock. The possible need for knock control measures such as lower compression ratio or retarded timing could reduce or eliminate benefits of lower heat losses. Mechanical issues are also involved. For example, a larger bore would give larger loads on the crankcase components due to the larger piston area, as indicated by Equations (13.17) and (13.18). A thicker piston crown could also be needed. The optimal engine design is a compromise between many factors; size, weight, price, fuel economy, durability just to name some. But the trends indicated here are helpful in determining the effects of at least some of the variables.

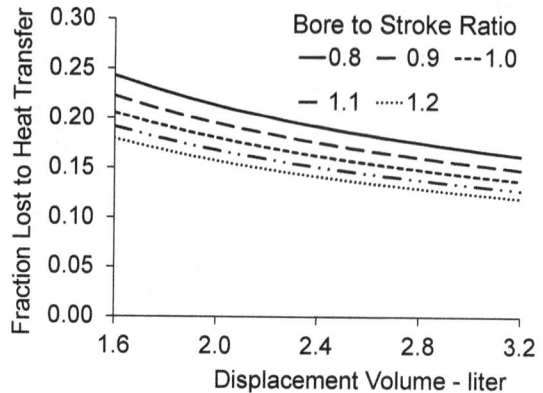

Figure 10.7: The estimated fraction of input energy lost to heat transfer for a spark ignition engine as a function displacement volume for varying bore to stroke ratio, for operation at 3000 rpm, stoichimetric mixture and full load.

10.4 Unsteady Heat Transfer

10.4.1 Heat Transfer Coefficient Correlations

In the time dependent simulations of engines, the instantaneous heat loss to the cylinder walls is an important part of the in-cylinder processes. There are several correlations that have been used estimate this heat loss, and they are discussed in the following.

Nusselt Heat Transfer Correlation

The very first heat transfer correlation for engines is that developed by Nusselt in 1923. It has been used for transient heat transfer modelling, though it was originally not intended for this purpose. The heat transfer is separated into two modes, convection and radiation, and an effective heat transfer coefficient is calculated for both parts. The convective portion has the form:

$$h_c = 5.41 \cdot 10^{-3} \left(1 + 1.24 S_p\right) \left(p^2 T\right)^{\frac{1}{3}} \tag{10.19}$$

The radiation portion is given by:

$$h_r = \frac{4.21 \cdot 10^{-12}}{\left(\frac{1}{\epsilon_g} + \frac{1}{\epsilon_w - 1}\right)} \frac{\left(T^4 - T_w^4\right)}{\left(T - T_w\right)} \tag{10.20}$$

In these expressions, the heat transfer coefficients, h_c and h_r are in units of MW/m²-K, the pressure, p in MPa, the temperature in K, the mean piston speed, S_p in m/s. The emissivities of the gas and walls, ϵ_g and ϵ_w are dimensionless. Borman and Nishiwaki [93] give some alternative values for the constants in Equation (10.19). As with the rest of these correlations, constants can be adujsted to fit experimental data if it is present.

Eichelberg Heat Transfer Correlation

Another old correlations is the Eichelberg equation from 1939 [16]. Eichelberg actually measured the transient heat transfer throughout the cycle. It is converted to the basis of heat transfer per crank angle degree, and is given in the following. It is very simple to use, and as was seen in Section 2.5, it gives results similar to newer correlations.

$$h = \frac{4.11 \cdot 10^{-5}}{N} \left(S_p\right)^{\frac{1}{3}} \left(p \cdot T\right)^{\frac{1}{2}} \tag{10.21}$$

where: h is in kJ/m²-CA°-K, p is the cylinder pressure in kPa, S_p is the mean piston speed in m/s, and T is the mass average cylinder temperature in K.

Annand Heat Transfer Correlation

In an effort to estimate the effects of radiation heat transfer separate from convection, Annand has developed an additional heat transfer correlation [18]. It has the following form, essentially the same as the Nusselt correlation:

$$\dot{Q}_{HT} = A_{HT} \left[a \frac{k_g}{B} Re^b \left(T - T_w\right) + c\sigma \left(T^4 - T_w^4\right) \right] \tag{10.22}$$

where: T_w is a representative cylinder wall temperature, and the Reynolds number, $Re = \frac{S_p B}{\nu_g}$, and a c, and c are empirical constants. Finol and Robinson [94] mention values of a between 0.25 and 0.8, $b = 0.7$ and $c = 0.57$ for diesel engines and $c = 0.075$ for spark ignition engines, the latter reflecting the lower radiation from SI engine combustion. For a 2-stroke diesel engines, suggested values are $a = 0.76, b = 0.64$ and $c = 0.54$

For both the Nusselt and Annand correlations, the use of bulk average temperature for the radiation contribution is inaccurate, based on recent knowledge of diesel engine combustion. A better way to estimate the radiation contribution from diesel diffusion flames would be to use the local combustion temperature of a constant pressure flame at the combustion conditions, since it is the hot particles in the flame that cause most of the radiation, though very recent work indicates that gas radiation is more important than previously thought [95]. That means that there is a much larger contribution from radiation already from the very first part of the combustion than would be predicted by Equation (10.22). The problem is to determine an equivalent radiating area, definitely not an easy task. Later on in the combustion, the bulk average temperature and the flame temperature approach each other, but especially at low loads there is a substantial difference. There is no known correlation that attempts to do this somewhat formidable task, so at the present the Nusselt/Annand form is the best available.

Woschni Heat Transfer Correlation

While the Eichelberg correlation above is simple to implement, it is considered by many to be out of date, and the results may not be accurate. In modern engines, the Woschni correlation is most commonly used [17]. It is often corrected somewhat to match experimental measurements for specific engines. It is generally regarded to be better than the Eichelberg, though the simulation calculations shown in Section 2.7.6 indicate that for the CFR engine, at least, there is not much difference. This was also found for the simulation of heat loss in a small 2-stroke diesel engine.

The Woschni correlation is based on a simple Nusselt number - Reynolds number correlation, which uses the cylinder bore as the characteristic distance, and a velocity that is dependent on the mean piston speed and the pressure rise caused by combustion. This divides the heat transfer process throughout the engine cycle into different stages, and includes the influence of combustion on the heat transfer. The correlation has the following form:

$$h = 3.26 \cdot B^{-0.2} \cdot p^{0.8} \cdot T^{-0.55} \cdot w^{0.8} \tag{10.23}$$

Where: h is the heat transfer coefficient in W/m²-K, B is the cylinder bore in m, p is the instantaneous pressure in kPa, T is the instantaneous temperature in K, and w is the characteristic velocity in m/s.

The velocity is expressed in the following relationship:

$$w = \left[K_1 \cdot S_p + K_2 \frac{V_d T_r}{V_r p_r} (p - p_{motoring}) \right] \tag{10.24}$$

For the gas exchange process, Woschni recommends that $K_1 = 6.18$ and $K_2 = 0$. After intake closing and prior to combustion on the compression stroke, $K_1 = 2.28$ and $K_2 = 0$, For the period after the start of combustion and before the opening of the exhaust valve, $K_1 = 2.28$ and $K_2 = 3.24 \cdot 10^{-3}$.

The properties with the subscript, r, in Equation (10.24) can be chosen at any convenient crank angle position after the closing of the intake valve and the start of combustion. Note that the relationship of these three terms in this equation makes them proportional to the inverse of the trapped cylinder mass.

The primary difficulty with implementing the Woschni equation is that the motoring pressure is included. If used in connection with a heat release analysis, one would have to run the engine at the same conditions without combustion, which is not practical unless a motoring dynamometer is available. Otherwise a simulation is needed. Since the correlation is empirical, and the accuracy not guaranteed, it is suggested that the motoring pressure can be estimated. One simple method is to use the ideal gas law from the condition just before combustion and conduct an isentropic compression calculation.

$$p_{motoring} \approx p_{ref} \left(\frac{V_{tot}}{V} \right)^{\gamma} \tag{10.25}$$

where: V_{tot} is the total cylinder volume at the time when the reference cylinder pressure p_{ref}, is known, V is the actual cylinder volume at a given crank angle degree, γ the specific heat ratio, and p_{ref} the measured cylinder pressure at the start of compression.

A value of the specific heat ratio should be used so that the motoring pressure is always lower than the actual pressure at the start of combustion. Though near the start of combustion the difference is small, and the error probably negligible.

10.4.2 Heat Transfer Measurements

To determine the effect of heat transfer on cycle efficiency, it is necessary to know how the heat transfer varies throughout the engine cycle. In the simple simulation of Section 2.5, it was assumed that the wall was at one uniform temperature throughout. This is not really the case, and the pistons and exhaust valve typically are hotter than other parts such as the intake valve and cylinder liner, particularly at the bottom of the latter. The temperatures of the surfaces of the combustion chamber also vary throughout the cycle.

Figure 10.8 from Lefeuvre, et. al. [97], [96] shows the variation of wall temperature throughout the engine cycle for a turbocharged 4-stroke diesel engine at various locations throughout the cylinder. The temperatures were measured with what is called surface thermocouples. These consist of a threaded plug of material with heat transfer properties close to those of the cylinder wall. Most are used with steel liners and heads. A hole is drilled through the plug and fitted with an insulating cylinder, and a wire of a material that makes a thermocouple junction with the plug material is passed through the inside of the insulation. Finally, a very thin layer of the material of the inside wire is vacuum deposited over the end of the plug and wire. The thermocouple junction is formed in the annular ring of the plug. With a thickness of only a few μm, the temperatures measured are very close to the surface temperature and respond very rapidly.

The highest temperatures inside the combustion chamber were found in the cylinder head, where the gasses are exposed to the high temperature combustion and strong convection and radiation heat transfer (piston temperatures were probably higher, but were not measured). The wall thickness of the cylinder head between the combustion chamber gasses and the coolant is often thicker than the cylinder liner thickness due to construction techniques, materials and strength consideration. This contributes to higher surface temperatures due to increased thermal resistance of the wall compared to a cylinder liner, for example. The cylinder liner temperatures decrease as one moves away from the TDC location, as these locations are only briefly exposed to combustion products, and they are exposed to low gas temperatures at the end of expansion and during the start of the compression stroke.

The temperature at the top of the liner, location 3 in Figure 10.8, is important in relation to engine wear and lubrication. At TDC, the piston comes to a halt and the conditions are poor for hydrodynamic

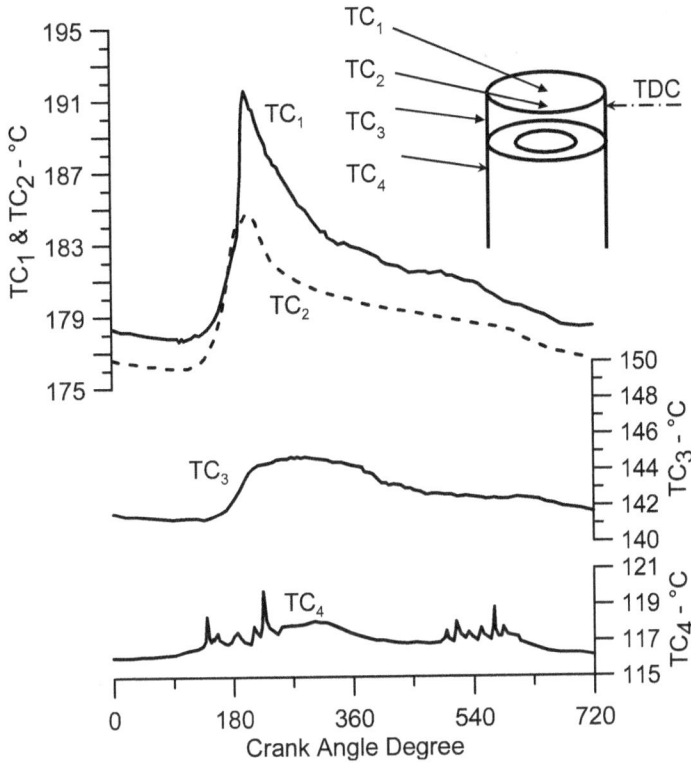

Figure 10.8: The variation of combustion chamber wall temperature throughout the cycle at various locations for a turbocharged 4-stroke engine with bore and stroke of 114.3 mm at a speed of 2000 rpm. Adapted from [96].

lubrication. The wear situation is aggravated by high temperature, as this reduces oil viscosity. If the temperature becomes too high, long term oxidation, deposits *etc.* could become a problem. The temperature at the bottom of the liner is close to the coolant temperature of $88 \pm 5°$ C.

The trace at location 4 is interesting, in that the time of passage of the piston rings can be observed by the spikes in the trace. The spikes indicate an increase in heat transfer due to the close contact between the rings and the cylinder wall. Piston rings then, actually play a role in piston cooling. After the rings pass upwards, the cylinder wall is exposed to the piston surface and oil film. Depending on piston skirt length, the cylinder wall can also be exposed to the crankcase gasses. This period is not relevant for the simulation of the condition of the combustion chamber gasses, but illustrates the complexity of the over all heat transfer process for the engine.

From the wall surface temperature-time histories, it is possible to calculate the instantaneous heat flux to the surface of the combustion chamber. The method is based on the one-dimensional transient heat conduction (Laplace) equation:

$$\frac{\partial T}{\partial t} = \frac{k}{\rho c_p} \frac{\partial^2 T}{\partial x^2} \qquad (10.26)$$

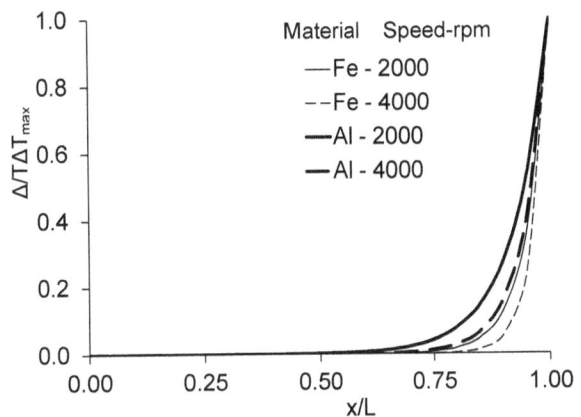

Figure 10.9: The amplitude of oscillation of the temperature throughout a solid wall with a constant temperature at $\frac{x}{L} = 0$ and a sinusoidal variation at $\frac{x}{L} = 1$ for a 5 mm iron wall and a 7 mm aluminium wall at equivalent engine speeds of 2000 and 4000 rpm.

Equation 10.26 can be solved by different methods, for example an analytic solution where the measured temperature is converted to a Fourier series, or through numerical methods. These are described in the paper by Borman and Nishiwaki [93]. The instantaneous heat flux is determined from the temperature gradient with respect to distance at the wall of the combustion chamber, as in Equation (10.1).

While there are temperature swings of on the order of 10°C at the surface, these temperature variations are quickly damped out as one looks into the wall toward the coolant, and the surface temperature at the coolant side is essentially constant. This characteristic is shown for a simplified case where there is a sinusoidal temperature variation with time on one surface of a thin wall, while the other end is held at a constant temperature. There is an analytical solution to this problem [98] that can be readily calculated for engine conditions. Figure 10.9 shows that for representative materials and conditions, the temperature oscillations only occur close to the combustion chamber surface, and the coolant side temperature at $x = 0$ remains constant. For thicker walls, the relative penetration would be even smaller. These results emphasize the importance of making surface thermocouples quite thin.

The temperature profiles from Figure 10.8 were analyzed with Equation (10.1) to give the heat transfer rates. These are shown in Figure 10.10. As expected, the heat transfer rates correlate with the time-

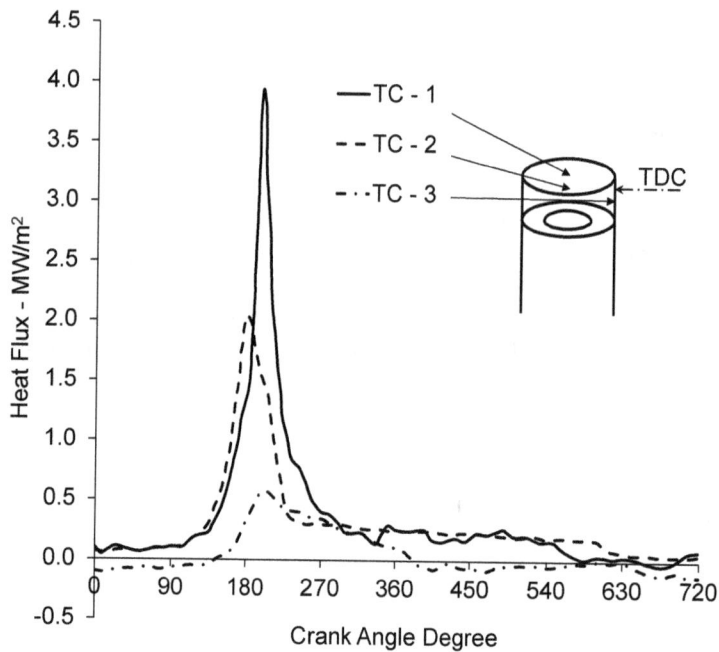

Figure 10.10: The variation of combustion chamber surface heat transfer throughout the cycle at various locations for the data of Figure 10.8.

averaged level of the wall temperature, which indicates the steady state, or average value of the heat transfer. The magnitude of the surface temperature swings indicate the variation in the heat transfer throughout the cycle. The highest variations in the heat flux are located in cylinder head during combustion. Since the piston passes the top wall thermocouples after about 35°, they are only exposed to the high temperature combustion gases after the majority of the heat release has occurred.

10.4.3 Radiation in Diesel Engines

As expected from the nature of the diesel flame, radiation plays a significant role in the heat transfer, especially in the period around combustion. In Section 3.6 the diffusion flame in the diesel engine was shown to contain a large number of soot particles, which burn in a thin zone between the spray and the surrounding air. In classic diffusion flames, the combustion occurs at a mixture ratio close to stoichiometric, and with a temperature close to the adiabatic flame temperature. The burning particles emit a large amount of radiation, which is apparent from combustion pictures.

Flynn, et. al. [99], [100] studied radiation from diesel engine combustion for the same diesel engine as

LeFeuvre, *et. al.* [97], [96] by use of radiation measurements at several different wave lengths. It has been shown that the radiation has a distribution with respect to wave length in the infra-red region similar to that of a classic black body, with an emissivity that approaches 1 during the period of maximum heat release, and then tapers off throughout the combustion period. The results from this study at a condition very close to those for Figures 10.8 and 10.10 are shown in Figure 10.11.

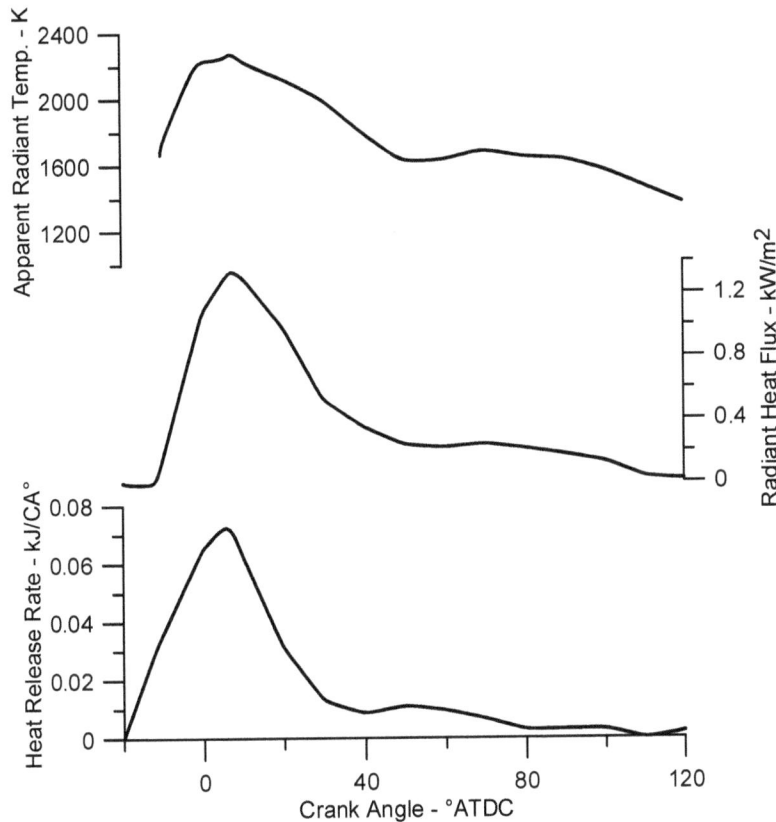

Figure 10.11: The intensity of radiation heat transfer, apparent radiation temperature and rate of heat release in a turbocharged direct injection diesel at a fuel air equivalence ratio of 0.439, 2000 rpm, data from [99].

One of the first pieces of information available is the magnitude of the heat transfer from radiation compared to the total heat transfer. The radiation was measured in a line across the cylinder, parallel to the surface of the cylinder head. So the most relevant results from Figure 10.10 are those of thermocouples 1 and 2, with an average maximum heat flux of about 3 MW/m^2, as compared to the radiation component at about 1.4 MW/m^2. Then for this engine, about 40 percent of the heat transfer during combustion is due to radiation, a factor that should not be ignored. Other studies indicate a similar magnitude.

A second point of interest is related to the modelling of the radiation component. Both the Nusselt and Annand correlations include a radiation contribution, which is calculated from the mass average temperature ($T_{ma} = pV/(mR)$). But the effective radiation temperature very quickly comes up to a value of around 2000K, in the vicinity of the flame temperature for a stoichiometric mixture, as was discussed in Section 3.4. The mass average temperature will be much below that, especially at the start of combustion, when the pressure rise due to combustion is small. Even at the end of combustion, for a fuel air equivalence ratio of 0.44 as shown here, the mass average temperature would be only near a stoichiometric flame temperature for the combustion of a lean homogeneous mixture. Thus, the predictions using the currently available correlations for radiation should not be able to accurately reproduce the real radiation heat transfer behavior.

In order to see what that means radiation component means with respect to combustion, Figure 10.12 shows the net accumulated heat release from a direct injection diesel engine as a function of the accumulated amount of energy injected into the engine, where the energy injection rate and combustion conditions were quite similar, and the combustion efficiency was taken into consideration [46]. Two fuels are compared: diesel fuel which has a radiant flame, and DME, which does not form carbon particles during combustion, and therefore has a non-luminous flame with low radiation. At the end of the curve, the net heat release from the diesel fuel is about 5% lower than that from the DME. This difference can be attributed to the difference in the radiation heat transfer between the two fuels, since the flow situation was similar and convection heat transfer should be about the same. This means that there is more energy available for generation of work in the case of the low radiating flame, with the potential of a higher thermal efficiency.

Based on NO_x emissions, Egnell also estimated the effective mixture ratio of combustion for diesel fuel and DME and found it to be slightly rich, in agreement with the model of the diffusion flame burning near a stoichiometric mixture presented in Section 3.6. So the radiation heat transfer observed agrees well with the latest observations of diesel combustion. The particles have been found to burn in a diffusion flame at a nearly stoichiometric mixture, close to the adiabatic flame temperature. These hot particles give off radiation that approaches black body conditions. The radiation starts early in the combustion process as soon as the diffusion flame is established. The local flame temperature is high, even though a mass average cylinder temperature would be too low to predict much radiation with older correlations. The

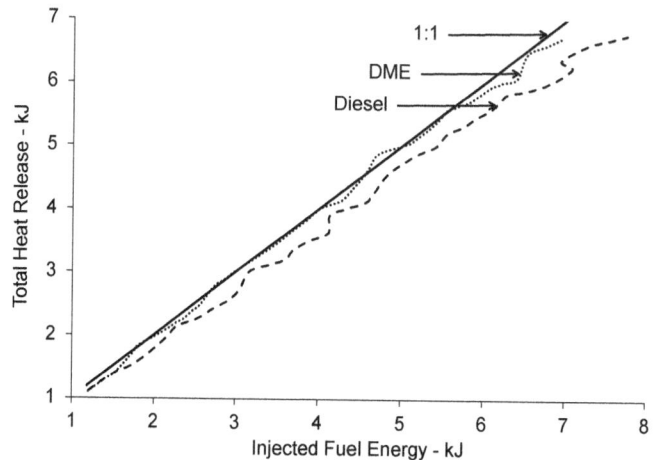

Figure 10.12: The net accumulated heat release in a DI diesel engine as a function of the amount of energy injected for DME and diesel fuel. Net heat release is the combined effect of combustion and heat loss, adapted from [46].

radiation with diesel fuel apparently causes a loss of approximately 5-10% of the fuel energy during combustion. This was seen from a comparison of net accumulated heat release in an engine operating with radiating and non-radiating flames. Very recent data indicates that gas radiation plays a more significant role than previously expected, particularly at the longer infra-red wave lengths [95].

10.5 Heat Transfer in Exhaust Systems

In modern engines, exhaust after-treatment systems such as catalysts, and particle filters are common. Their operation is dependent on the exhaust temperature. In addition, heat losses in the exhaust system affect turbine inlet temperatures, an important variable in turbocharger operation. Therefore, it is of interest to be able to calculate heat losses in exhaust systems when designing and evaluating such systems. Exhaust flows near the engine ports are transient in nature, as shown in Figure 7.14. There is a also a large amount of turbulence generated by the flow out of the exhaust valve, and this has an effect on the heat transfer. Additionally, exhaust ports are not straight, and curvature has an effect on flow patterns and hence the convective heat transfer.

Given this complexity, exhaust port heat transfer is normally determined from empirical correlations of the Nusselt number as a function of the Reynolds number:

$$Nu_D = C \cdot Re_D^m \qquad (10.27)$$

where: D is the exhaust port or pipe diameter, and C and m are empirical constants. The Reynolds number is usually calculated on the basis of the mass flow rate, using the continuity equation:

$$Re_D = \frac{4\dot{m}_{ex}}{\pi \mu D} \qquad (10.28)$$

Flow in the port section can be modelled with a correlation throughout the cycle with correlations for opened and closed valves [101], and these results were used in a computer simulation to give equivalent values of $C \approx 0.1$ and $m \approx 0.8$ averaged over and engine cycle. In a straight section of exhaust pipe downstream from the port, values of $C = 0.0774$ and $m = 0.769$ were found [86]. It was also shown that this correlation gives a good estimate of the heat transfer used in a transient situation, where the flow varies throughout the engine cycle [86].

A recent study of a straight portion of the exhaust pipe showed that the distance from the inlet of the section has an influence on the heat transfer [102]. The correlation for the Nusselt number is:

$$Nu_D \left(1 + \frac{0.075 Re_D^{0.25}}{x/D} \right) 0.76 Re_D^{0.44} \tag{10.29}$$

where: x is the distance from the inlet of the exhaust port section. These results are compared to those of Meisner [86] in Figure 10.13. The results are in general agreement for low Reynolds numbers, but

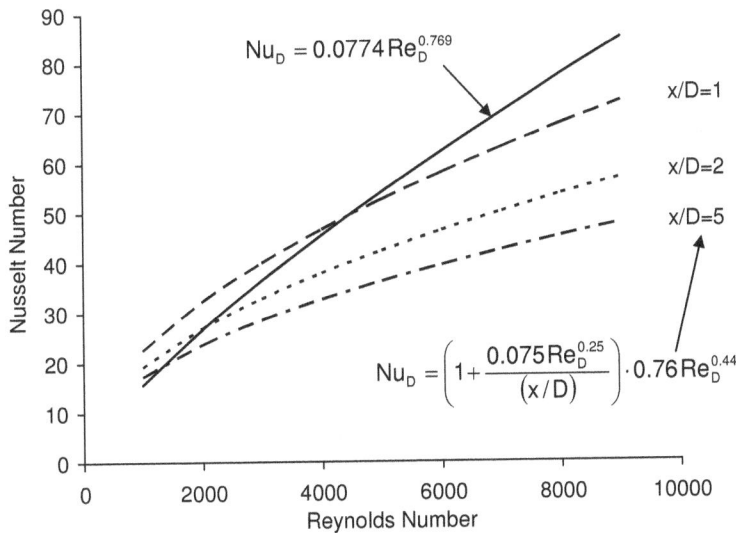

Figure 10.13: A comparison between two different correlations for heat transfer in the straight portion of an exhaust pipe.

there are significant differences in the values of the two. For higher Reynolds numbers, the correlation from Equation (10.29) is preferred, since the other studies had lower Reynolds numbers. In any event, a significant degree of uncertainty will be encountered in making estimates of exhaust gas heat transfer.

Example:

The following example is intended to show the effects of heat transfer in exhaust pipes. Consider a 4-stroke, SI engine, with $B = S = 85$ mm, operating with a filling efficiency of 0.90, with standard air at 101 kPa, 20°C at a speed of 2000 rpm. The fuel air ratio is stoichiometric, $FA = 0.068$. Consider also a section of the exhaust pipe with a diameter of 35 mm, where the flow from only one cylinder passes, and the cylinder walls are at 20°C. To establish the temperature as a function of pipe length, a control volume with length Δx is considered, as shown in Figure 10.14. The flow is assumed to be steady, and enters the pipe section at a temperature of $T_1 = 700$°C. Kinetic energy is assumed to be negligible. In the diagram, P refers to the wetted perimeter of the pipe, and A to the cross-section area, and Q is shown as positive into the control volume in accordance with the standard sign convention. An energy balance on the system give the following:

$$\dot{m}_x h_{x+\Delta x} + \hat{h} \cdot P \cdot \Delta x \left(T_\infty - T \right) = \dot{m}_{x+\Delta x} h_{x+\Delta x} \tag{10.30}$$

where: $\hat{h}$ is the heat transfer coefficient.

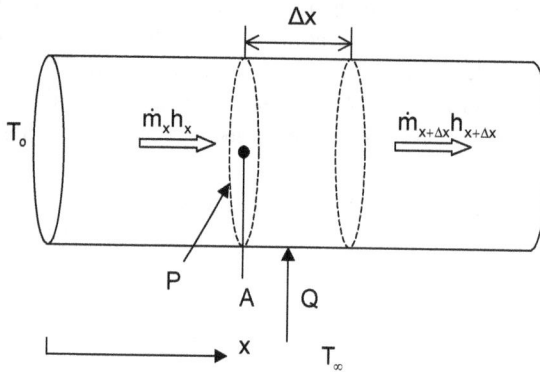

Figure 10.14: A control volume for calculating the heat loss in an exhaust pipe.

Assuming the specific heat is constant and there are no chemical reactions, Equation (10.30) can be rearranged:

$$\frac{h_{x+\Delta x} - h_{x+\Delta x}}{\Delta x} = \frac{\hat{h} \cdot P (T_\infty - T)}{\dot{m}} = c_p \frac{T_{x+\Delta x} - T_{x+\Delta x}}{\Delta x} \tag{10.31}$$

Then, letting $\Delta x \to 0$, one obtains:

$$\frac{dT}{dx} = \frac{\hat{h} \cdot P (T_\infty - T)}{\dot{m} c_p} \tag{10.32}$$

With a known inlet temperature, this equation can be solved to determine the temperature as a function of distance from the entrance to the exhaust pipe. If the correlation of Equation (10.27) is used, the heat transfer coefficient is constant and Equation (10.32) and be solved directly to give:

$$T - T_\infty = (T_0 - T_\infty) \exp \left[-\frac{\hat{h} P x}{\dot{m} c_p} \right] \tag{10.33}$$

In the case where Equation (10.29) is used to calculate the heat transfer coefficient, Equation (10.32) must be integrated as a function of the length of the pipe.

To obtain an estimate of the temperature change in the exhaust gas Equation (10.33) is used.

The mass flow is calculated as:

$$\begin{aligned}
\dot{m}_{tot} &= \dot{m}_a + \dot{m} = \dot{m}_a \cdot (1 + FA) \\
&= (1 + FA) \cdot \eta_f \rho_o V_d \frac{N}{120} \\
&= (1 + 0.068) \cdot 0.9 \cdot \frac{101}{0.287 \cdot 293} \cdot \frac{\pi}{4} \cdot 0.085^2 \cdot 0.085 \frac{2000}{120} \\
&= 9.28 \cdot 10^{-3} kg/s
\end{aligned}$$

The exhaust properties are estimated to be those for air, since about 75% of the exhaust gas is nitrogen, and the correlations are not precise. For a temperature of 700°C, the viscosity can be found in reference books to be about $41.1 \cdot 10^{-6}$ kg/m-s, the specific heat equal to 1136 J/kg-K and the thermal conductivity equal to 0.0658W/m-K.

The Reynolds number is:

$$Re_D = \frac{4 \dot{m}_{ex}}{\pi \mu D} = \frac{4 \cdot 9.28 \cdot 10^{-3}}{\pi \cdot 41.1 \cdot 10^{-6} \cdot 0.035} = 8207$$

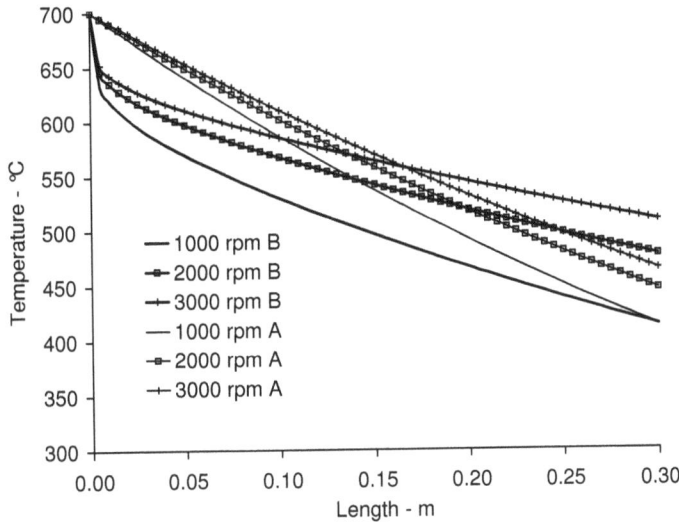

Figure 10.15: Comparison of exhaust temperatures for the example using the two heat transfer correlations: Case A; Equation (10.27) with A= 0.0774 and b= 0.769, and Case B; Equation (10.29) for 3 engine speeds.

Using the simple correlation from Figure 10.13, the heat transfer coefficient is:

$$\hat{h} = Nu\frac{k}{D} = 0.0774 \cdot Re_D^{0.769}\frac{k}{D}$$

$$= 0.0774 \cdot 8207^{0.769}\frac{0.0658}{0.035}$$

$$= 149 W/m - K$$

Assume a length of pipe $= 1$ cm $= 0.01$ m. Then:

$$T = (T_0 - T_\infty)\exp\left[-\frac{\hat{h}Px}{\dot{m}c_p}\right] + T_\infty$$

$$= (700 - 20)\exp\left[-\frac{149 \cdot \pi \cdot 0.035 \cdot 0.01}{9.28 \cdot 10^{-3} \cdot 1136}\right] + 20$$

$$= 687.3°C$$

Then in just a length of 1 cm, the temperature falls by about 17 °C.

To compare the two correlations, Equation (10.29) was used in combination with Equation (10.32), and the temperature found by numerical integration with a commercial software package. The results are shown in Figure 10.15 for 3 engine speeds with the same filling efficiency and inlet temperature to the exhaust pipe. The correlation with the entry length factor gives a very rapid initial decrease in temperature, due to the factor x/D in the correlation for small values of x. Farther downstream, the two correlations give similar temperatures. As was the case for overall heat transfer, the relative heat transfer decreases with engine speed and the temperatures are, therefore, higher at the highest speeds. Which correlation gives the best results is uncertain, since in actual engine systems, it can be difficult to define the location for entry length effects, and in all correlations of this type, there is uncertainty and variation from engine to engine.

Chapter 11

Pressure Charging

11.1 Introduction

The basic idea of pressure charging is to increase the amount of air flowing through a given engine. This allows more fuel to be burned and more power produced for any given engine size. The basic power equations for any type of engine show that by increasing the intake air pressure/density, the power of an engine may be increased. One of the most common ways of accomplishing this is through the use of a compression device at the intake of the engine.

There are different types of compression devices used to accomplish this as well as methods of driving them. The first device used historically was a mechanically driven compressor, where the power was provided by a mechanical connection between the compressor and the outlet shaft of the engine. Such a technique is commonly called *supercharging*. Different types of compressors have been used in this connection. One of the disadvantages of this method is that power is taken from the engine output shaft to power the compressor.

When developments in the area of turbomachinery had advanced sufficiently, it became practical to use a new method to provide power to drive the compressor. In this case, a turbine is placed in the engine exhaust system, and the hot exhaust gasses provide power to a turbine, which is connected directly to the compressor, without any mechanical connection to the engine output shaft. This technique is referred to as *turbocharging*, and is very prevalent in modern diesel engines. The exhaust gas energy is used to replace the mechanical energy required when a compressor is mechanically driven. This gives more power and a better efficiency since no engine shaft power is used for the compression. In addition, it allows use of centrifugal compressors, which are small, but can deliver large amounts of air at high pressures.

11.2 Roots Blower

There are two basic kinds of compressors used for supplying higher pressure air to engines: *positive displacement* and *turbocompressors*. In the former, a device ingests a fixed quantity of air per revolution and sends it on, in this case to the engine. A 4-stroke engine is also a positive displacement device, and the compressor is sized to deliver more air than the engine can "accept". The actual compression can take place in the compressor (for example in the Lysholm compressor) or at the outlet of the compressor, as occurs in the Roots blower.

There are some advantages to a positive displacement device, in that they tend to match the general flow characteristics of the positive displacement engine over a wide operating range. In principle, an ideal Roots blower would give the same increase in torque over the entire operating range of an engine, if it were not for the fact that the volumetric efficiencies of these devices are differing functions of speed and pressure ratio and the compression efficiency of the compressor changes with operating conditions.

A Roots blower is shown schematically in Figure 11.1. The concept is simple, the rotors have a special geometry, either a cycloid curve or an involute, gear-like profile, such that there is always very close "contact" between the rotors. Actually, there is a small clearance and a little leakage, but the close

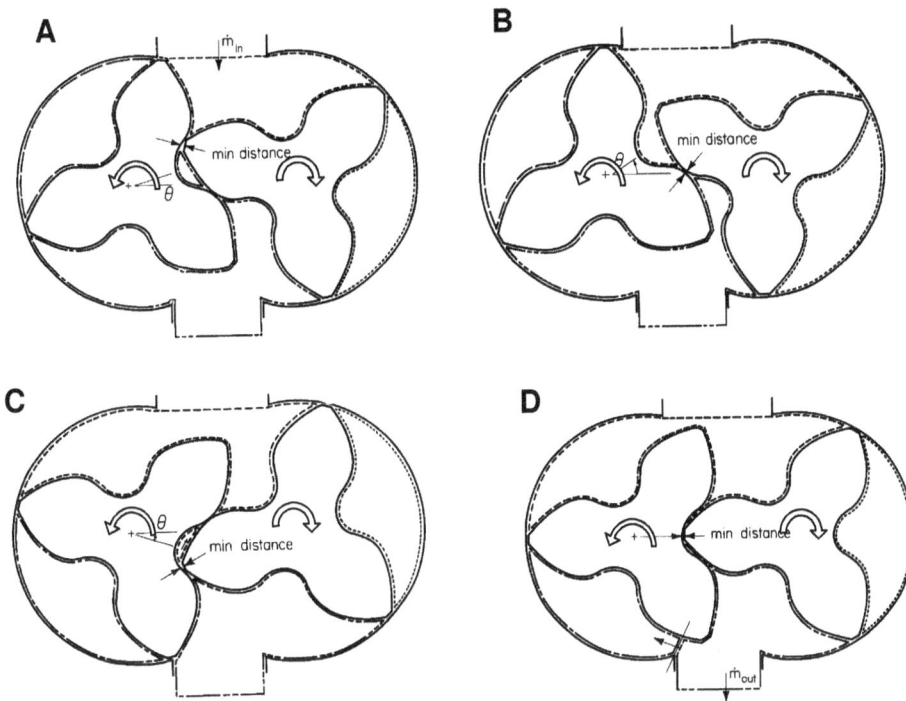

Figure 11.1: An end-on view of a Roots blower. In picture A, the left hand rotor is just closing and trapping a volume of intake air, at B the air volume is trapped, and is moved to the point of opening shown in C. At D the outlet port is open.

tolerances function like a seal. There are also very small clearances between the rotor tips and the rotor ends. The rotors are connected with gears, and rotate in opposite directions.

On the inlet side, the volume between the rotor and the housing is filled with atmospheric air, and as the trailing rotor tip passes the inlet port, a volume of air is trapped. Later in the cycle, the leading rotor tip passes the outlet port, and the air is then "pushed" into the outlet, or in this case, the intake of the engine or an air cooler before the intake manifold. There is no compression until the outlet port of the blower opens, and the trapped air is compressed by the high pressure air in the engine intake system. The compression occurs because the engine has a smaller flow capacity than the blower. The compression occurs irreversibly with mixing at constant volume and is not very efficient. The efficiency relative to an isentropic process gets worse as the pressure rise increases. Thus, the name blower, as the device was initially intended to move air with little compression. However, as long as compression ratios are under about 1.8:1 the efficiency is acceptable.

A performance, or compressor, map of a Roots blower is shown in Figure 11.2. A compressor map is a plot with the pressure ratio of the compression as the ordinate, the flow rate as the abscissa, with lines of constant compressor speed and compressor efficiency. These are the fundamental characteristics of the compressor, and can be used to match the compressor with an engine and determine how much work is required to accomplish the compression.

Note that the for the Roots blower, the compressor efficiency is fairly low, and decreases as the pressure ratio goes above about 1.5 This is a reasonable pressure level for spark ignition engines, as pressure ratios above this begin to give problems with engine knock. Note also that the lines of constant speed are almost vertical. If there was no leakage they would be vertical, as it would be expected to deliver a given volume of air per revolution if the intake and outlet were totally separated. But as the outlet pressure increases, there is more and more leakage back towards the intake and the volumetric efficiency drops.

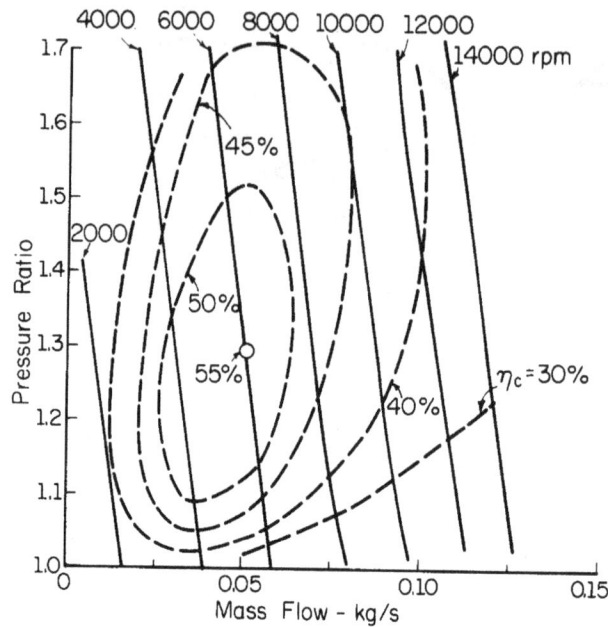

Figure 11.2: A compressor map for a Roots blower.

11.3 Centrifugal Compressor

The principle behind a centrifugal compressor is quite simple. The velocity of an air stream is increased by adding kinetic energy. This is done with a rotor, where the air enters, is guided along the rotor by vanes, and leaves the rotor at a high velocity due to centrifugal force. This high velocity air is decelerated in a well designed diffuser and, according to Bernoulli's principle, the reduction in velocity gives an increase in pressure.

Figure 11.3 shows a typical rotor for a centrifugal compressor. The air comes in axially at the center, and the blades change the direction of the flow such that it flows radially outward at the outer diameter. The gas velocity is related to the rotor velocity. After leaving the rotor, the air passes through a region of increasing area called a diffuser, so that the flow is decelerated and the pressure increases. The diffuser usually has a scroll-like shape and its outlet is connected to a heat exchanger for cooling the exit air or to the engine intake manifold if air cooling is not used.

Figure 11.4 shows a compressor map for a centrifugal compressor that is used on a 2 liter diesel engine. The first obvious point

Figure 11.3: The rotor for a centrifugal compressor.

is that it has a very different shape than that of the positive displacement compressor. The lines of constant speed are curved, and as the pressure ratio increases at constant compressor speed, the flow decreases. Eventually a maximum pressure ratio is reached. If the flow is reduced below this point, a condition called surge or stall arises. Basically, because the velocity is not high enough to be reduced to a stagnation pressure above that in the receiver. This condition is called *stalling* or surging and is to be avoided, since operation to the left of the stall limit can result in damage to the compressor, as the flow can change direction and flow into the compressor exit instead of out. At high flow rates for a given speed, the curves become vertical, and the flow approaches a choked condition.

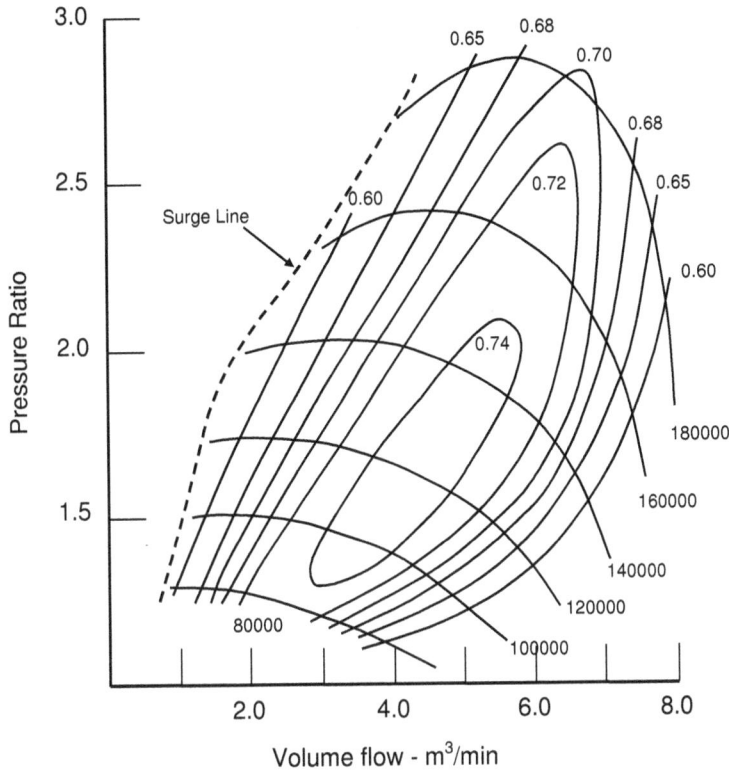

Figure 11.4: An example of a map for a centrifugal compressor. Speeds in rpm.

The next point is that the pressure ratios are much higher than those shown for the positive displacement blower in Figure 11.2. This is due to the principle used of converting velocity to pressure. Note also that the maximum pressure ratio increases roughly quadratically with the rotational speed. This characteristic does not really match that of a positive displacement piston engine, so some attention must be paid to this.

Another point is that the efficiency of the compressor is higher than that of positive displacement devices, also at high pressure ratios. This is because the process of decelerating high velocity air to a low velocity and high pressure is close to isentropic. A large part of the losses are due to boundary layer friction, and to flow losses when the flow direction is not always optimal with respect to the blade geometry. Not shown in the figures is the fact that the centrifugal compressor is also much smaller and lighter than a positive displacement device.

Finally, since the centrifugal compressor and radial turbine are both devices based on the transfer of kinetic energy to and from gasses, they have fundamentally similar characteristics, and go well with each other when coupled together on a turbocharger.

Given all these factors, where are the different types of devices used? Positive displacement blowers have been used for increasing the scavenging ratio in two stroke diesel engines, which is not really pressure charging. They are used in high performance spark ignition engines because the flow characteristics match well and they often deliver higher pressure at low engine speed than centrifugal compressors. This gives an advantages with low speed torque. As mentioned above, the pressure ratios are not so large on spark ignition engines to avoid knock, so the efficiency penalty is acceptable.

Centrifugal compressors are dominant in turbocharged engines. There are a few engines operating with mechanically driven centrifugal compressors, but one problem is the great difference between the compressor and engine speed, especially for smaller engines. An important early application was their use as superchargers on combat aircraft in World War II. The use of turbocharging is extensive in diesel engines since the high cylinder temperatures and pressure which accompany intake air compression are conducive to good combustion. There are fewer turbocharged spark ignition engines, mainly due to knock considerations. It is also more difficult to obtain high pressure ratios at very low engine speed, and many turbocharged spark ignition engines have suffered from low torque at low engine speed.

In the following, the theoretical basis of compressors and turbines will be presented, and the two devices coupled as a turbocharger.

11.4 Compressor Calculation

Figure 11.5 shows a schematic diagram of a compressor, where work is supplied, and the air is compressed from a pressure, p_1, and temperature, T_1, to a higher temperature, T_2 and pressure p_2. The diagrams refers to conditions where the velocity is low and kinetic energy negligible, that is a distance away from the rotor.

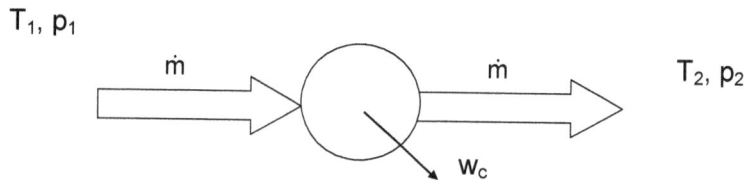

Figure 11.5: A schematic diagram of the flow and states of a compressor.

In an ideal compressor, the model for the compression process is an adabatic compression[1]. The theoretical, or isentropic, work for compression from state 1 to 2 per unit mass can be written:

$$-w_{c,s} = h_{2,s} - h_1 \tag{11.1}$$

where $w_{c,s}$ is the compression work performed per unit mass in the isentropic compression process, and the subscript s refers to conditions after isentropic compression.

For an ideal gas with constant specific heat, this can be written as:

$$-w_{c,s} = c_p \left(T_{2,s} - T_1\right) = c_p \cdot T_1 \left[\left(\frac{p_2}{p_1}\right)^{\frac{\gamma-1}{\gamma}} - 1 \right] \equiv c_p \cdot T_1 \cdot Y_c \tag{11.2}$$

Since the power is the work performed per unit time:

$$P_{c,i} = \dot{m}_c \cdot c_p \cdot T_1 \cdot Y_c \tag{11.3}$$

where $P_{c,i}$ is the power requirement for an ideal compressor with a flow of ideal gas $= \dot{m}_c$.

The efficiency for a real compressor is calculated as the ratio of the isentropic power to the real power required for compression from the same stagnation pressure. The power that is actually used to power the compressor is:

$$P_c = \dot{m}_c \left(h_2 - h_1\right) = \dot{m}_c \cdot c_p \left(T_2 - T_1\right) \tag{11.4}$$

where T_2 denotes the actual temperature at the completion of the real, non-isentropic, compression process.

The compressor efficiency is then:

$$\eta_c = \frac{P_{c,i}}{P_c} = \frac{\dot{m}_c \cdot c_p \cdot T_1 \cdot Y_c}{\dot{m}_c \cdot c_p \cdot (T_2 - T_1)} = \frac{T_1 \cdot Y_c}{T_2 - T_1} \tag{11.5}$$

The actual power requirement is obtained from this as:

$$P_c = \frac{\dot{m}_c \cdot c_p \cdot T_1 \cdot Y_c}{\eta_c} \tag{11.6}$$

---

[1]This is considered to be the ideal for a rapid compression, though actually the lowest compression work is obtained with isothermal compression, which is not practical in engine systems

The compression temperature is:

$$\eta_c \left(T_2 - T_1 \right) = T_1 \cdot Y_c \rightarrow T_2 = T_1 \left(1 + \frac{Y_c}{\eta_c} \right) \tag{11.7}$$

The compressor efficiency is a complicated function of the flow and pressure ratio, and is most often presented in the form of a compressor map as shown in Figure 11.4. In a later section, the mathematical description of a compressor map will be presented.

Equation (11.7) is a very important relationship, since it can be used to calculate the input temperature to the engine itself or to an intercooler. The inlet temperature plays an important role in determining engine power, since it affects intake density and volumetric efficiency, as seen in Chapter 7.

11.5   Constant Pressure Turbine

By carefully designing the lengths and diameters of the exhaust port runners, it is possible to obtain more turbine power by utilizing the kinetic energy of the exhaust gasses. This method was widely used in early turbocharged engines with low compressor and turbine efficiencies, but is not used very much in modern engines. The most common procedure is to collect the exhaust gasses in a receiver, that is a chamber with a relatively large volume, where the kinetic energy of the exhaust gasses is converted to a high, nearly constant pressure. The exhaust gasses are then expanded across the turbine of the turbocharger down to approximately atmospheric pressure. Work is obtained in this process, and is commonly used to power the air compressor in a turbocharger. In many cases, the power available is larger than that needed to power the compressor, and the turbine power is reduced by leading some of the exhaust gasses through a turbine by-pass (waste gate), or less frequently, expanding the exhaust gasses through two turbines, the second of which is used to produce additional power from the engine, improving its efficiency. The latter method, called turbo-compounding, can improve engine efficiency by over 10 % in practice, but is complicated, and is not used in on-road engines. Turbo-compounding is used on large marine diesel engines, where an electrical generator is attached to the turbine shaft, and the power produced is electrical instead of mechanical.

Figure 11.6 shows a schematic diagram of a turbine, where work is obtained as the gas is expanded from a pressure, p_5, and temperature, T_5, to a lower temperature, T_6 and pressure p_6. State 5 represents the exhaust gas leaving the engine manifold. The points on the diagram refer to conditions where the velocity is low and kinetic energy negligible. The work per unit mass that can theoretically be obtained

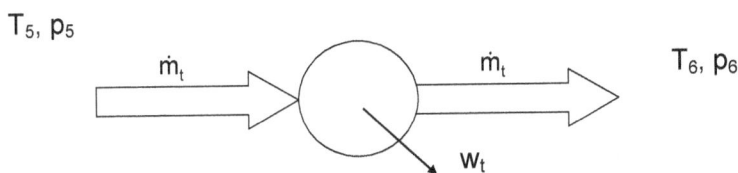

T₅, p₅ — $\dot{m}_t$ — $\dot{m}_t$ — T₆, p₆ — w_t

Figure 11.6: A schematic diagram of the flow and states of a turbine.

by isentropic expansion from the turbine inlet pressure p_5 to the outlet pressure p_6 is:

$$w_{t,s} = h_5 - h_{6s} \tag{11.8}$$

where h_{6s} is the enthalpy per unit mass after isentropic expansion from the static pressure p_5 to the exhaust pressure p_6.

If the exhaust gas is considered to be an ideal gas with constant specific heat, the ideal work per kg is:

$$w_{t,s} = c_p' \cdot \left(T_5 - T_{6s} \right) = c_p' \cdot T_5 \left(1 - \frac{T_{6s}}{T_5} \right) = c_p' \cdot T_5 \left(1 - \left(\frac{p_6}{p_5} \right)^{\frac{\gamma'-1}{\gamma'}} \right) \equiv c_p' \cdot T_5 \cdot Y_t \tag{11.9}$$

Where c_p' and γ' refer to exhaust gas conditions.

Since power is work per second, the turbine power is:

$$P_{t,s} = \dot{m}_t \cdot c_p' \cdot T_5 \cdot Y_t \tag{11.10}$$

The efficiency for a real turbine is calculated as the ratio of the actual power to the isentropic power obtained by expansion from the same stagnation pressure. The power that is actually obtained from the turbine is then (assuming an ideal gas):

$$P_t = \dot{m}_t \left(h_5 - h_6\right) = \dot{m} \cdot c_p' \left(T_5 - T_6\right) \tag{11.11}$$

where T_6 denotes the actual temperature at the completion of the real, non-isentropic, expansion process.

The turbine efficiency is then:

$$\eta_t = \frac{P_t}{P_{t,s}} = \frac{\dot{m}_t \cdot c_p' \cdot (T_5 - T_6)}{\dot{m}_t \cdot c_p' \cdot T_5 \cdot Y_t} = \frac{T_5 - T_6}{T_5 \cdot Y_t} \tag{11.12}$$

The actual power requirement is obtained from this is:

$$P_t = \eta_t \cdot \dot{m}_t \cdot c_p' \cdot T_5 \cdot Y_t \tag{11.13}$$

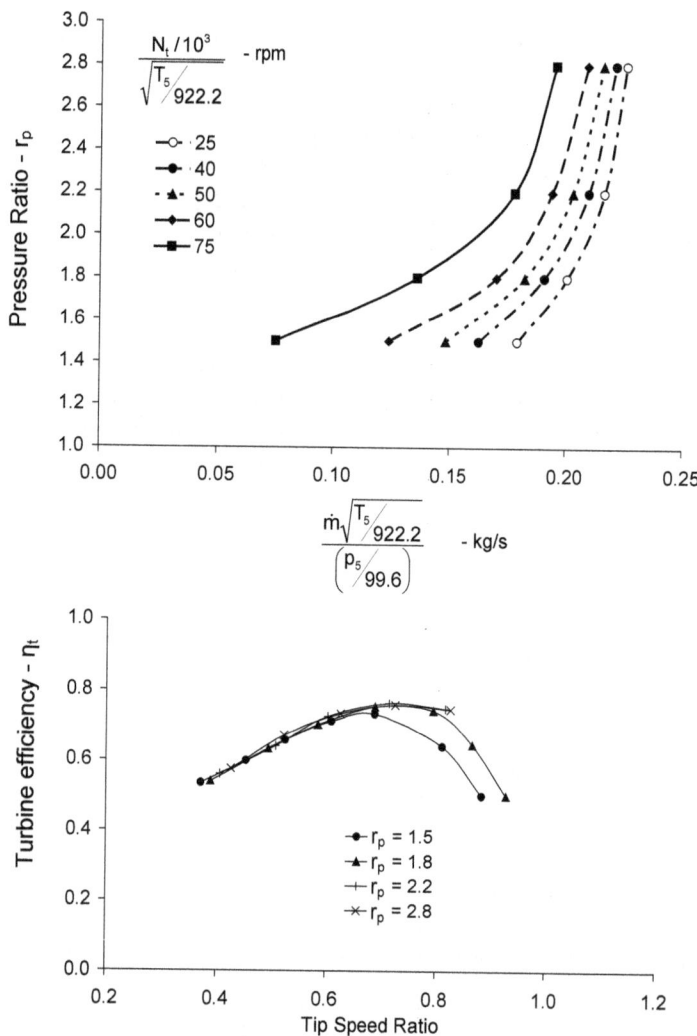

Figure 11.7: The flow rate and efficiency of a radial turbine as a function of pressure ratio and speed for a turbine with a diameter of 12.7 cm (T in K).

Figure 11.7 shows the characteristics of a radial turbine of a size suitable for a heavy duty engine for a highway truck. The mass flow and speed are corrected to standard temperatures and pressures. For use with a given condition, the pressures and temperatures actually encountered can be used to convert to the actual conditions. The mass flow is a function of both the pressure ratio and the rotor speed. Though not shown here, the mass flow is obviously also a function of the nozzle area. The nozzle area is the reduced flow area at the inlet to the turbine wheel, where the exhaust gas is expanded to a high velocity as it impinges on the turbine wheel.

It is very important that the proper turbine area is selected when matching a turbocharger to an engine. This is done by selecting the flow area in the nozzle ring. Many modern turbochargers are equipped with exchangeable nozzle rings for easy changes. If the turbine area is too large, the pressure drop across the turbine will be too small, with a smaller power according to Equation (11.13). For a given mass flow rate, the turbine power increases as the nozzle area decreases, due to higher pressure drop. This has the disadvantage of increasing the pressure at the exhaust valve, which has a negative effect on power and efficiency if allowed to become too large. In practice, a rule of thumb says that the optimum nozzle area is in the vicinity of 5 to 10 % of the piston area for diesel engines.

The efficiency results for the radial turbine are shown using a special variable for the turbine speed. This variable is the tip speed ratio, which is the ratio of the tip of the rotor in the turbine to the isentropic expansion velocity from the stagnation pressure p_5 to the atmosphere pressure to the velocity. From the 1st law of thermodynamics, the expansion velocity can be calculated as:

$$v_s = \sqrt{2c_p T_5 \left[1 - \left(\frac{p_o}{p_5} \right)^{\frac{\gamma-1}{\gamma}} \right]} \qquad (11.14)$$

The tip speed is simply calculated as:

$$U_t = \frac{\pi D_r \cdot N_r}{60} \qquad (11.15)$$

where the subscript, r, refers to the rotor. The tip speed ratio, r_{tip} is defined as the rotor tip speed divided by the expansion velocity. When plotted in terms of the tip speed ratio, the efficiency in a centrifugal turbine is nearly independent of the pressure ratio. This is shown in Figure 11.7.

11.6 Turbochargers

It is common to couple an exhaust turbine together with a centrifugal compressor to form an aggregate without mechanical connection to the engine crankshaft. The turbine powers the compressor to increase the airflow to the engine, enabling an increase in engine power. It is found that at full loads, when the engine has its largest need for air, the greatest amount of energy is available to power the turbine. A sketch of a turbocharger system on an engine is shown in Figure 11.8. The various components will be explained in the following. Intercoolers and waste gates are not found on all applications, but are very common.

In a stationary condition, when the compressor and turbine are connected, the turbine power output (minus a very small mechanical loss) must equal the compressor power requirement. Then for a constant pressure turbine:

$$P_c = P_t \cdot \eta_m = \frac{\dot{m}_c \cdot c_p \cdot T_1 \cdot Y_c}{\eta_c} = \dot{m}_t \cdot c_p' \cdot T_5 \cdot Y_t \cdot \eta_t \cdot \eta_m \qquad (11.16)$$

where η_m is turbocharger mechanical efficiency, normally $\eta_m \approx 1$.

In the case of a diesel (or other cylinder-injected) engine, the turbine flow is greater than the compressor flow by the amount of the fuel injected into the cylinder. In addition, many turbochargers are regulated with a waste gate, which under certain conditions bypasses a portion of the flow around the turbine nozzle to reduce power output and limit the intake pressure provided by the compressor. That is $\dot{m}_t = \dot{m}_a + \dot{m}_f - \dot{m}_{waste\ gate}$

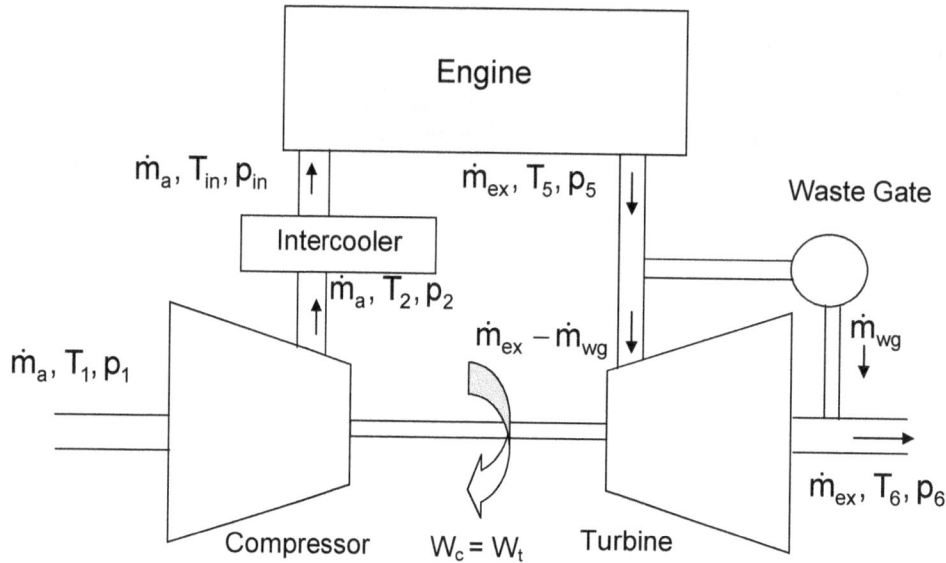

Figure 11.8: A sketch of the components to be found in turbocharging systems.

Equation (11.16) can be rearranged to:

$$\eta_H = \eta_t \cdot \eta_c \cdot \eta_m \cdot \frac{\dot{m}_t}{\dot{m}_c} = \frac{c_p \cdot T_1 \cdot Y_c}{c_p' \cdot T_5 \cdot Y_t} \tag{11.17}$$

where η_H is the ratio between the turbocharger's theoretical consumption and the theoretical output per unit mass. It is called the group efficiency. By multiplying Equation (11.17) by $\frac{T_5}{T_1}$, a parameter is obtained which is only a function of the compressor and turbine pressure ratios, $\frac{p_2}{p_1}$ and $\frac{p_6}{p_5}$, under which the turbocharger operates:

$$\eta_H \cdot \frac{T_5}{T_1} = \frac{c_p}{c_p'} \frac{\left[\left(\frac{p_2}{p_1}\right)^{\frac{\gamma-1}{\gamma}} - 1\right]}{\left[1 - \left(\frac{p_6}{p_5}\right)^{\frac{\gamma'-1}{\gamma'}}\right]} \tag{11.18}$$

This relationship is shown in Figure 11.9 for $\gamma = 1.40$, $\gamma' = 1.34$, $c_p = 1.00$ and $c_p' = 1.13$ kJ/kg-K.

Instead of using Figure 11.9, Equation (11.18) can be rewritten to solve for the pressure ratio over the compressor as a function of the other relevant parameters.

$$\frac{p_2}{p_1} = \left\{1 + \eta_t \cdot \eta_c \cdot \eta_m \cdot \frac{T_5 \cdot \dot{m}_t \cdot c_p'}{T_1 \cdot \dot{m}_c \cdot c_p}\left[1 - \left(\frac{p_6}{p_5}\right)^{\frac{\gamma'-1}{\gamma'}}\right]\right\}^{\frac{\gamma}{\gamma-1}} \tag{11.19}$$

In practice, the pressure ratio across the turbine is adjusted by selecting the proper size of the turbine nozzle ring.

11.7 Waste Gate

Due to the characteristics of the centrifugal compressor and the turbine, it is common for the intake pressure to become too high at higher engine speeds and loads when performance is satisfactory at lower speed. Therefore, a means of limiting the output pressure of the turbocharger is often required. The principle of the method can be seen from Equation (11.19), where there is a dependence of compressor

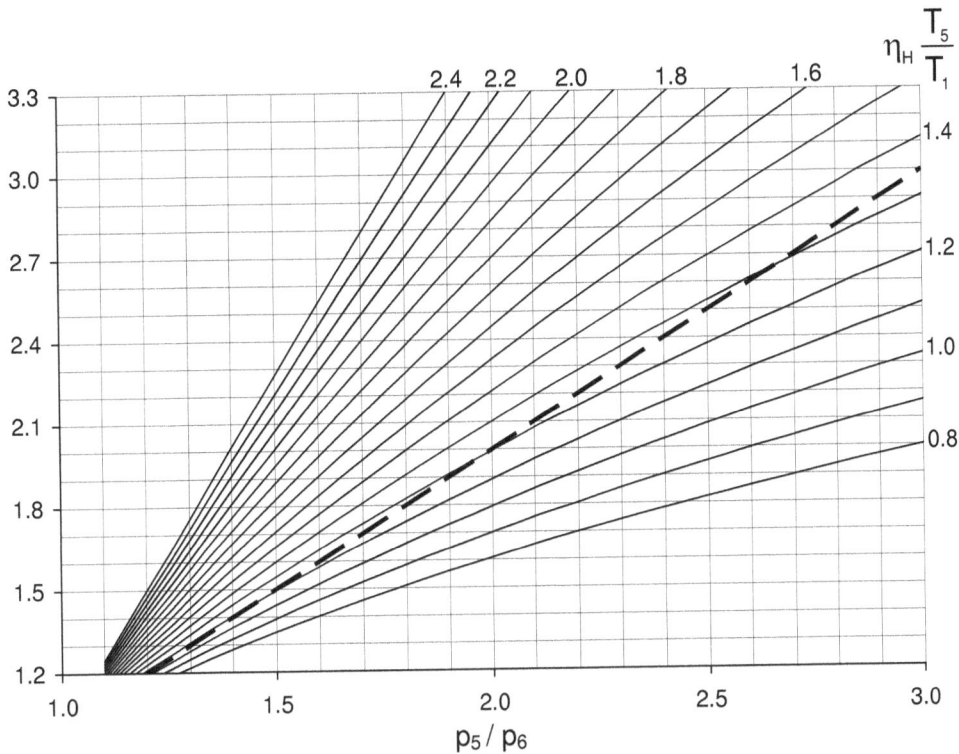

Figure 11.9: Compressor pressure ratio as a function of turbine pressure ratio for a constant pressure turbine. The dashed line represents equal pressure. Above that the intake pressure is greater than the exhaust pressure (positive scavenging).

pressure ratio on the mass flow through the turbine. If other parameters are held constant and the turbine mass flow reduced, the pressure ratio for the compressor will also be reduced.

This idea is used to control compressor outlet pressure. In the turbine, there is a passageway, parallel to the expansion over the turbine wheel, where exhaust gasses can be bypassed around the turbine wheel. This path is controlled by a valve, which is equipped with a means for varying its area. This arrangement is called a waste gate, as the valve function like a gate, and the flow bypassed through the valve is "wasted", in that it does not produce power. A waste gate is illustrated in Figure 11.10.

Waste gates are commonly controlled by a simple mechanical arrangement, consisting of a spring loaded diaphragm connected to the waste gate valve by a rod, or simple mechanical link. The diaphragm is exposed to the outlet pressure of the compressor, such that the pressure force opposes the spring. When the spring preload is overcome, the diaphragm moves, raising the valve from its seat, allowing a bypass flow around the turbine. The pressure at which the waste gate opens is determined by the preload on the spring, and the increase in pressure as the valve is opened further is controlled by the spring constant. Properly designed, the waste gate holds the intake pressure nearly constant once the opening conditions for the waste gate are reached.

The use of a waste gate permits a turbocharger to be chosen that provides adequate air for good torque at low speeds without damaging the engine at high speeds. It is especially common on smaller engines

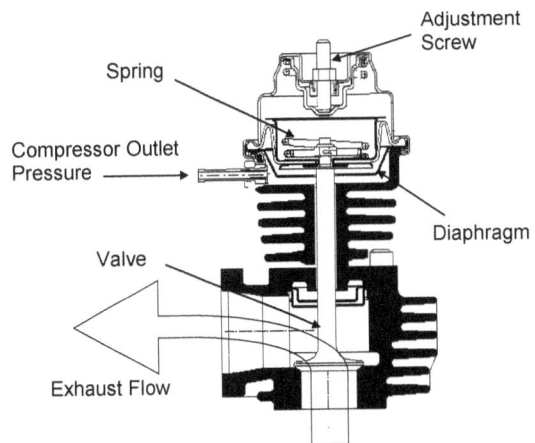

Figure 11.10: The waste gate, used for limiting the intake pressure of a turbocharged engine.

operating over a wide speed range. For larger engines that often operate over a smaller speed range, waste gates are often not used.

11.8 Compressor Condition for a 4-Stroke Engine

In the above, it is assumed that the compressor and turbine efficiencies are known. Usually this is not the case in advance, as the efficiencies depend on the flow rate and speed, that is the operating state, in a non-linear manner. This section shows the iterative process for determining the operating condition of a mechanically driven compressor on a engine. The procedure for a turbocharger is more complicated, since iteration of the turbine is involved, and engine conditions depend on both compressor and turbine. As an example of the iterative process, the following shows how to determine the operating state for a compressor that is mechanically connected to a 4-stroke engine. This means that the compressor speed is known. In a turbocharger, the speed is determined as part of the iteration process. For one method of iteration, see for example the four quadrant diagram method of Winkler [103].

It is assumed that a compressor map is available, and that the compressor operates with a constant known speed. For a mechanically driven compressor, the compressor speed is typically a constant multiple of the engine speed.

The task at hand is to determine the compressor operating point that corresponds to the operating point of the engine. The calculation here will assume the engine is operating at wide-open throttle. The process is started by assuming a pressure ratio across the compressor. When the pressure ratio and speed are known, the compressor efficiency and mass flow can be read off the compressor map. From the pressure ratio and efficiency, the outlet temperature of the compressor can be found through Equation (11.7). The stationary condition is found when the air flow to the engine, calculated from the outlet conditions of the compressor, corresponds to the compressor flow rate read off of the compressor map.

The engine airflow is found by the following method:

1. Engine intake pressure:

$$p_{in} = p_o \left(\frac{p_2}{p_1} \right)_{map} \tag{11.20}$$

2. The engine intake temperature is found from Equation (11.7), with the possibility of a correction in the case that an aftercooler is used to reduce the intake temperature.

3. Engine airflow (for a 4-stroke engine):

$$\dot{m}_a = \eta_v \frac{p_{in}}{RT_2} \cdot \frac{V_d \cdot N}{120} \tag{11.21}$$

Remember that the volumetric efficiency can be a function of the intake air temperature and make necessary adjustments (see Chapter 7).

4. Compare the airflow read off the compressor map with that calculated from Equation (11.21) If these agree, the correct solution has been found. If not, another pressure ratio is assumed and the process repeated until the two flows agree.

When the operating point has been established, one can then use the power equation for the engine to calculate the output and efficiency. In the case of a mechanically driven compressor, the compressor power requirement can be calculated from Equation (11.21). It must be subtracted from the output power of the engine to obtain the correct output. The use of a compressor also has a small effect on the pumping work of the engine, as the intake pressure is higher than atmospheric, and the exhaust pressure basically atmospheric. So some of the work put into the compressor is put back into the engine on the intake stroke. This amount of work needs to be calculated from the $p - V$ diagram for the cylinder pressure. A first order estimate would be a mean effective pressure equal to the difference between the intake manifold pressure and the exhaust pressure.

For a complete turbocharger, assuming that one has experimental data for the turbine equivalent to the compressor map, one has to assume a compressor condition, including the operating speed and then find the operating point as explained above. This gives the engine mass flow. For a diesel engine there is a connection between the temperature increase between the intake and exhaust ports and the fuel air ratio. Typically, the temperature increase across the engine from intake port to exhaust port is a nearly linear function of the fuel air ratio. See Figure 6.19 for example. This can be used to find the exhaust temperature, which in turn is equal to the turbine inlet temperature. From the mass flow and speed, the turbine efficiency and pressure ratio can be found in a similar fashion to the compressor calculation, assuming the turbine characteristics are at hand. The turbine power is obtained from Equation (11.13). If the turbine power matches the compressor power requirement, the condition is satisfied. Otherwise, a new compressor (turbocharger) speed is needed and the iteration must be repeated.

If the turbine and compressor efficiencies are known in advance, that method is more simple. One can use the four quadrant diagram method described in Winkler [103]. This would also give a good starting point for the overall iteration described above, provided the efficiencies are reasonable. Detailed matching of engine and turbocharger can be done in the case where mathematical models for the turbocharger components are available. This is shown in detail in Section 11.11

11.9   Example of a Turbocharged Engine

In order to demonstrate the changes that occur when an engine is turbocharged the following example is presented.

Consider a 4-stroke, directed injection diesel engine, with a displacement of 5 liters, and operating at a speed of 2500 rpm. Assume that the combustion process is limited to a maximum fuel air ratio of 0.04, and that the engine has an indicated efficiency of 0.45. The engine friction is 215 kPa when the engine is not turbocharged. The volumetric efficiency is 0.85. Assume standard diesel fuel with a heating value of 42000 kJ/kg, and that atmospheric conditions are 20°C, 1 atm.

In the naturally aspirated form, the engine has the following performance:

Using Equation (1.31),

$$
\begin{aligned}
BP &= \eta_v \rho_o \frac{V_d N}{60x} FA \cdot H_u \cdot \eta_i - FP \\
&= 0.85 \cdot 1.2 \cdot \frac{0.005 \cdot 2500}{120} 0.04 \cdot 42000 \cdot 0.45 - \frac{215 \cdot 0.005 \cdot 2500}{120} \\
&= 80.3 - 22.4 = 57.9 \ kW
\end{aligned}
$$

and

$$
\begin{aligned}
bmep &= \frac{120 \cdot BP}{V_d \cdot N} = 556 kPa \\
\eta_{m,eng} &= \frac{BP}{IP} = \frac{57.9}{80.3} = 0.721 \\
bsfc &= \frac{3.6 \cdot 10^6}{\eta_i \cdot \eta_m \cdot H_u} = \frac{3.6 \cdot 10^6}{0.45 \cdot 0.721 \cdot 42000} = 264 g/kWh
\end{aligned}
$$

Now assume that a turbocharger is used, and that the compressor efficiency = 0.71, and the turbine efficiency = 0.70. The nozzle ring on the turbine has been selected such that the pressure ratio over the turbine = 2.0. Assume that because of increased cylinder pressure, the friction increases by 6%. The engine exhaust temperature is 450°C. Then new intake conditions must first be established.

For the compressor, the compressor outlet pressure is found by Equation (11.19), using atmospheric conditions of 1bar and 20°C:

$$
\begin{aligned}
\frac{p_2}{p_1} &= \left\{ 1 + \eta_t \cdot \eta_c \cdot \eta_{m,tc} \cdot \frac{T_5 \cdot \dot{m}_t \cdot c'_p}{T_1 \cdot \dot{m}_c \cdot c_p} \left[1 - \left(\frac{p_6}{p_5} \right)^{\frac{\gamma'-1}{\gamma'}} \right] \right\}^{\frac{\gamma}{\gamma-1}} \\
&= \left\{ 1 + 0.70 \cdot 0.71 \cdot 1.0 \cdot \frac{(450 + 273) \cdot 1.04 \cdot 1.13}{293 \cdot 1 \cdot 1.001} \left[1 - \left(\frac{1}{2} \right)^{\frac{1.34-1}{1.34}} \right] \right\}^{\frac{1.4}{1.4-1}} \\
&= 2.08
\end{aligned}
$$

A pressure on the order of 2 atmospheres is common in modern engines. In order to find the engine intake conditions, the compressor outlet temperature must be found using Equation (11.7):

$$
\begin{aligned}
T_2 &= T_1 \left(1 + \frac{Y_c}{\eta_c}\right) \\
&= 293 \left(1 + \frac{2.08^{\left(\frac{1.4-1}{1.4}\right)} - 1}{.71}\right) \\
&= 389K
\end{aligned}
$$

One could correct the volumetric efficiency as a function of the intake temperature, but that will not be done in this case. It is also assumed that the indicated thermal efficiency is unchanged, and that the same fuel air ratio is used. The new intake density is:

$$
\rho_{in} = \frac{p_{in}}{RT_{in}} = \frac{208}{0.287 \cdot 389} = 1.86 kg/m^3
$$

and

$$
BP = 0.85 \cdot 1.86 \cdot \frac{0.005 \cdot 2500}{120} 0.04 \cdot 42000 \cdot 0.45 - \frac{1.06 \cdot 215 \cdot 0.005 \cdot 2500}{120}
$$
$$
= 124.5 - 23.7 = 101 \; kW
$$

and

$$
bmep = \frac{120 \cdot 101}{.005 \cdot 2500} = 967 kPa
$$
$$
\eta_{m,eng} = \frac{101}{124.5} = 0.811
$$
$$
bsfc = \frac{3.6 \cdot 10^6}{0.45 \cdot 0.811 \cdot 42000} = 235 g/kWh
$$

As a result of the turbocharging, the power has increased 74 %, the mechanical efficiency has increased 11.3 %, and the bsfc decreased 11%. The reason for the decrease in the bsfc is a better mechanical efficiency. The indicated efficiency is constant and the friction nearly constant. Therefore, almost all the extra work performed is available as useful output.

On the negative side, it must be remembered that the cylinder pressure on the compression stroke has increased by a factor of two. Therefore, the engine must be more sturdily constructed, and the combustion process may have to be altered to avoid excessive cylinder pressures, which could lower the indicated efficiency.

11.10   Charge Air Cooling

As seen in the above example, the compression results in a high engine intake temperature, which has a negative effect on increasing the intake air density. An obvious solution to this is to cool the air between the outlet of the compressor and the intake of the engine. Such a process is called *intercooling, after cooling* or *charge air cooling*, and the device called an intercooler, after cooler or charge air cooler. In transport engines, atmospheric air is a common cooling medium, although engine coolant can be used. A common factor used to describe the the cooling process is the effectiveness, denoted α and defined as:

$$
\alpha \equiv \frac{actual\ temperature\ drop}{max\ possible\ temperature\ drop} = \frac{T_{comp,out} - T_{engine,in}}{T_{comp,out} - T_{coolant}} \tag{11.22}
$$

The effectiveness of the after-cooler is the result of a compromise between performance, cost, weight/size and pressure drop. Typical values are on the order on 0.6 to 0.7 for air-to-air cooling. To see the effect of charge air cooling, one can apply an air-to-air heat exchanger to the previous example. Assume that the coolant is atmospheric air and that the effectiveness is 0.65. Pressure drops in the cooler are neglected.

The new engine intake temperature now is:

$$T_{in} = T_{comp,out} - \alpha \left(T_{comp,out} - T_{coolant} \right)$$
$$= 389 - 0.65 \left(389 - 293 \right)$$
$$= 327K$$

Using this temperature in the above calculations, it is found that the power increases and the fuel consumption is improved slightly.

$$\rho_{in} = \frac{p_{in}}{RT_{in}} = \frac{208}{0.287 \cdot 327} = 2.216 kg/m^3$$

and

$$BP = 0.85 \cdot 2.216 \cdot \frac{0.005 \cdot 2500}{120} 0.04 \cdot 42000 \cdot 0.45 - 23.7$$
$$= 148.5 - 23.7 = 125 \ kW$$

and

$$bmep = \frac{120 \cdot 125}{.005 \cdot 2500} = 1200 kPa$$
$$\eta_{m,eng} = \frac{125}{148.5} = 0.842$$
$$bsfc = \frac{3.6 \cdot 10^6}{0.45 \cdot 0.842 \cdot 42000} = 226 g/kWh$$

The results are summarized in Table (11.1). By the use of turbocharging and charge air cooling with the same displacement volume, maximum power was increased by 116% while fuel consumption decreased by 14%. As pointed out above, some changes are needed in the engines to accommodate the increased pressures, but the benefits are well worth the effort. In the above, the fuel air ratio has been held constant, and therefore there is an implicit increase in the fuel injection per cycle. Pump adjustments will be required when adding charge air cooling to an existing engine. Watson and Janota [104] have shown that with aftercooling, if the fuel ratio is held constant, the heat transfer per cycle remains constant when using charge air cooling. This is consistent with the heat transfer model of Taylor and Taylor discussed in Chapter 10. Therefore, there is no substantial increase in the thermal loading of the engine with charge air cooling. There has been a continuing trend towards turbocharged and now aftercooled diesel engines, and these technologies are being adapted to smaller and smaller engines. A key element in the smaller diesel engines is electronically controlled, common rail fuel injection, as well as improved combustion chamber design.

Table 11.1: Calculated results for estimation of the effects of turbocharging and charge air cooling on a 5 liter DI diesel engine operating at 2500 rpm for a constant $FA = 0.04$.

Parameter	Naturally Aspirated	Turbocharged	Turbocharged/air cooled
IP - kW	80.3	124.7	148.5
FP - kW	22.4	23.7	23.7
BP - kW	57.9	101	125
$imep$ - kPa	771	1195	1430
$bmep$ - kPa	556	967	1200
$bsfc$ -g/kWh	264	235	226
$\eta_{m,eng}$	0.721	0.811	0.842
air flow - g/s	125	194	231
fuel flow - g/s	5	7.76	9.24

11.11   Turbochager Modeling

When analyzing operation of a turbocharged engine, it is advantageous to be able to model the operation of the turbocharger in a relatively simple mathematical way. That is to say, it is useful to have mathematical equations that describe the interdependencies of the flow, pressure ratio, speed and efficiency of both the turbine and the compressor. This enables the use of modern analytical tools, that allow the simultaneous solution of nonlinear algebraic equations. The turbocharger models can be coupled with relatively simple engine models and used for control purposes, as well as selection and matching of engine and turbocharger components, that is , mean value engine models. The following presents such mathematical models. Only radial compressors are modelled here, as they are the type used in turbocharging for engines up to the largest sizes. Further, the model is developed for turbines without variable nozzle angle, though it could be a starting point for the development of a model for a variable nozzle angle.

11.11.1 Compressor Model

The following model was developed for use in engine control systems, but can be used for additional purposes. A more complete description and background can be found in Reference [64]. The model is based on two dimensionless parameters. The first is the head coefficient, Ψ:

$$\Psi = c_p \cdot T_0 \cdot \frac{(p_2/p_0)^{\frac{\gamma-1}{\gamma}} - 1}{0.5 \cdot U_c^2} \tag{11.23}$$

where U_c is the tip speed of the compressor wheel:

$$U_c = D_c \cdot \pi \cdot N_c/60 \tag{11.24}$$

where N_c is the rotational speed of the compressor in rpm, and D_c the compressor diameter.

The second parameter is the dimensionless flow coefficient, Φ:

$$\phi = \frac{\dot{m}_c}{\rho_0 \frac{\pi}{4} D_c^2 \cdot U_c} \tag{11.25}$$

where ρ_0 is the compressor inlet density.

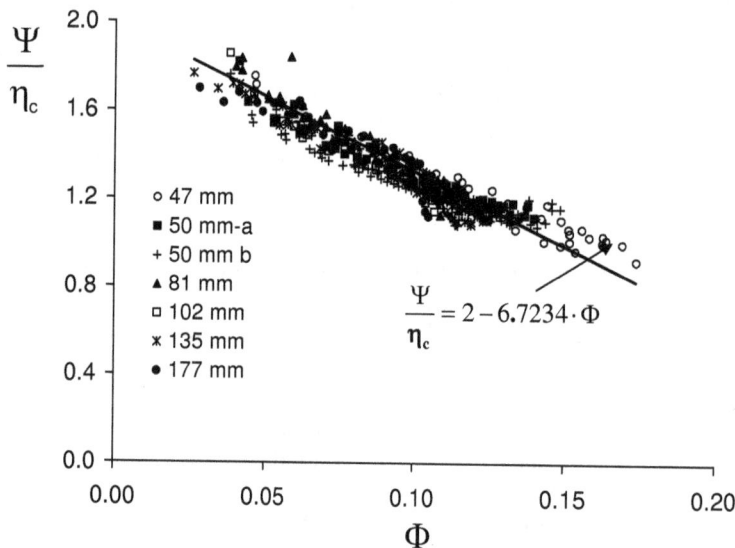

Figure 11.11: Dimensionless pressure ratio divided by compressor efficiency as a function of the dimensionless flow for a series of commercial turbocharger radial compressors with different rotor diameters.

It can be shown that for compressors, a simple relationship exists between the head coefficient, the efficiency and the flow coefficient. Figure 11.11 shows the relationship between the term obtained by dividing the compressor head parameter by the compressor efficiency and the dimensionless flow coefficient for a range of sizes of commercial compressors used for turbocharging engines. These compressors all have a similar blade angle. For different blade angles, the curve might be slightly different, but the nearly linear relationship will still be observed. For the geometrically similar compressors, the result is linear, and one curve can be used for all the compressors. A linear fit gives the equation shown in the figure.

A second useful relationship is that between the head coefficient and the flow coefficient. This function is shown in Figure 11.12 for the same compressors as were shown in Figure 11.11. There is more variation in this relationship for the different sizes of compressors. The head coefficient is higher for the larger compressor diameters, indicative of a higher efficiency with the larger diameters. But for a given size, the pressure ratio as a function of flow can be modelled fairly simply. For most compressors, an empirical model of the form:

$$\Psi = \Psi_o \cdot \left(1 - \exp\left(\frac{\Phi - \Phi_o}{\tau_\Psi} \right) \right) \tag{11.26}$$

can be used, where Ψ_o, Φ_o, and τ_Ψ are empirical parameters. In some cases at high speed (to the right in Figure 11.12), the curves diverge at different speeds, and some of these parameters may be fit as a function of compressor speed in order to give a more accurate fit. However, the compressors used for engines are capable of producing pressure ratios in excess of 3:1, whereas many engines are limited to a boost ratio on the order of 2:1 in order to maintain mechanical reliability. Thus, these higher speeds are not usually reached in practice. In fact, in many cases it is of acceptable accuracy to assume that the value of Ψ remains constant at its maximum value. If this is possible, modelling is much simpler.

The two correlations given are for results over the entire compressor map, and the entire map has thus been reduced to two equations. If one knows the compressor speed and pressure, then the flow and compressor efficiency can be obtained from Equations 11.23 and 11.25. The constants for the empirical equations given in the model can readily be obtained using conventional compressor maps, generally available for turbocharger manufacturers.

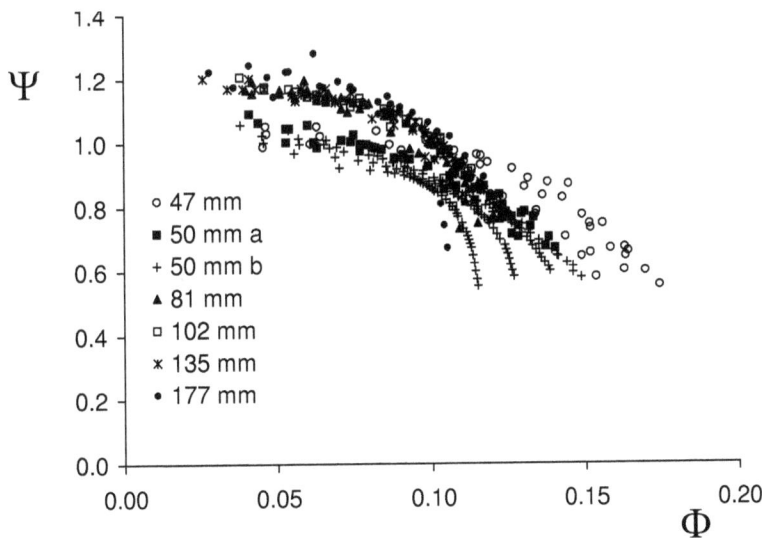

Figure 11.12: Dimensionless pressure ratio as a function of the dimensionless flow and compressor diameter for a series of commercial turbocharger radial compressors with different rotor diameters.

11.11.2 Turbine Model

In this section, an semi-empirical model will be presented for radial flow (centrifugal) turbines. This type of turbine is most common for engines used for on-road vehicles. The model will also be for a fixed nozzle geometry. Some newer turbines on turbochargers have a variable nozzle angle, which primarily affects the efficiency. The ideas presented here could be used as a starting point for a model of this type of device. Large 2-stroke marine diesels use axial turbines, but they are not discussed here. References [104], and [105] give examples of the flow and efficiency characteristics of axial turbines.

Figure 11.7 shows typical characteristics for a radial turbine. This turbine will be used to demonstrate the model. Methods will be similar for other radial turbines. The efficiency is basically a function of the tip speed ratio alone, r_{tip}, and the figure suggests a simple correlation of efficiency as a function of the tip speed ratio. A third order polynomial gives a reasonable fit to the curve, though other curves may be used.

$$\eta_t = 0.8725 r_{tip} + 2.2578 r_{tip}^2 - 2.7968 r_{tip}^3 \tag{11.27}$$

Figure 11.13 shows an example of a third order polynomial fit to the data.

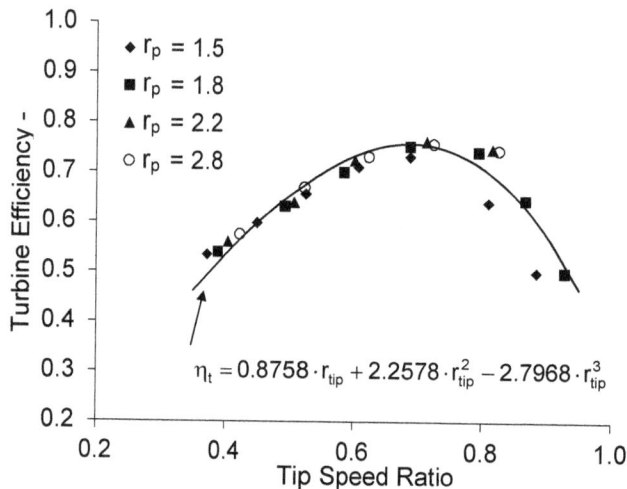

Figure 11.13: An example of a curve fit for the turbine efficiency of the the turbine shown in Figure 11.7.

As seen in Figure 11.7, the turbine flow is a function of the turbine speed and pressure ratio. A characteristic of flow in a radial turbine that can be seen is that at high pressure ratios, the flow chokes. That is, at a given turbine speed the flow approaches a maximum value, where the flow is no longer dependent on the pressure ratio. A second feature is that for no flow, there is still a pressure ratio across the turbine. This is because at no flow, the turbine wheel acts as a compressor, and increases pressure. A turbine flow model has been developed making use of these ideas. First, the choked mass flow rate and the pressure ratio at no flow were determined from the turbine data. Their variations are shown in Figure 11.14 as a function of reduced turbine speed, $N_t\sqrt{T_5/922}$, where T_5 is the turbine inlet temperature in K. The mass flows are corrected for compressible flow. That is,

$$\dot{m}_{cor} = \dot{m}_t \cdot \frac{\sqrt{T_5/922}}{p_5/100} \tag{11.28}$$

for pressure in kPa. The empirical equations for the variations of these parameters are:

$$\dot{m}_{choke,corr} = 0.2273 - 4.796 \cdot 10^{-6} \cdot \left(\frac{N_t}{\sqrt{T_5/922}}\right)^2 \tag{11.29}$$

$$r_{p,0} = 1 + 5. \cdot 10^{-5} \left(\frac{N_t}{\sqrt{T_5/922}}\right)^2 \tag{11.30}$$

The final element of the flow model is the variation of the flow with respect to the pressure ratio. The ratio of the corrected turbine flow rate to the critical corrected mass flow rate is shown as a function of the difference between the turbine pressure ratio and the no-flow pressure ratio in Figure 11.15. There is a small dependence of this ratio on the corrected turbine speed. Physically, the flow depends on the pressure ratios across both the nozzle ring of the turbine, and the pressure drop across the rotor. An empirical equation was developed that gives a good representation of the flow, while being simple to use.

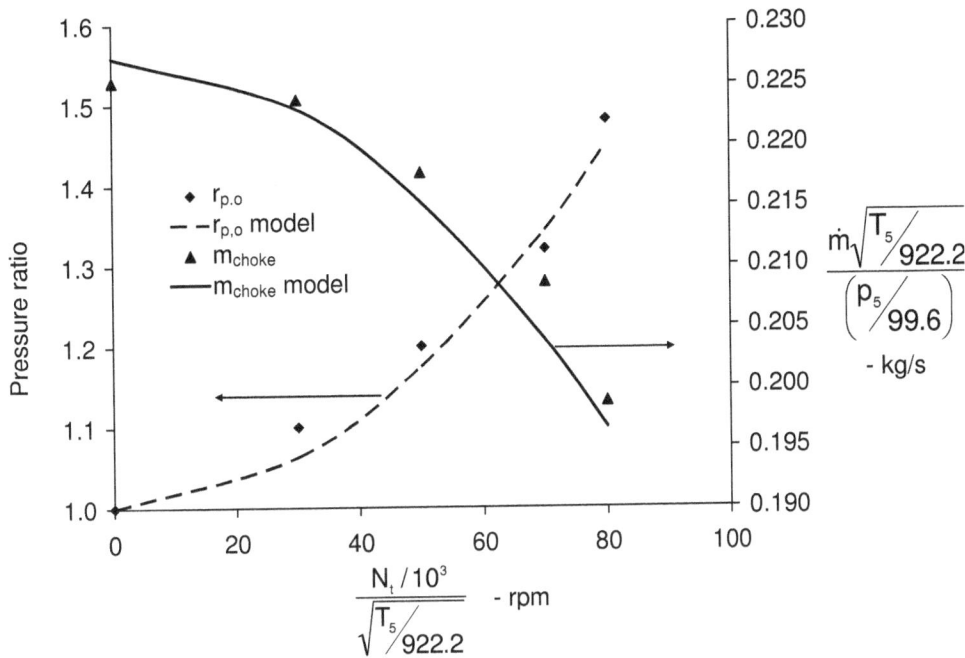

Figure 11.14: Variation of the choked mass flow and no-flow pressure ratio as a function of reduced turbine speed for the turbine shown in Figure 11.7.

The function chosen was the Weibe function, as was used in modelling the heat release function. In this case, the function has the form:

$$\frac{\dot{m}_{cor}}{\dot{m}_{choke}} = 1 - \exp\left(-\alpha \cdot (r_p - r_{p,0})^{[b_w+1]}\right) \tag{11.31}$$

In Figure 11.15 the model curves are also shown for the same corrected turbine speeds. The value of the constant b_w is -0.15, and to make a minor correction for the speed dependence, the parameter, α was made a function of the turbine speed:

$$\alpha = 3.3197 - 0.0138 \cdot \frac{N_t}{\sqrt{T_5/922}} \tag{11.32}$$

It can be seen that the empirical function gives a good representation of the flow rate ratio as a function of pressure drop and turbine speed. The combined flow model is able to reproduce the flow as a function of speed and pressure ratio to within a few percent over the operating range shown in Figure 11.7.

Since the model is empirical, the constants will most likely have to be adjusted for different sized turbines. In particular, the critical flow rate will be strongly related to the turbine flow area. Theoretically, the flow should be linearly proportional to that area. In practice, it is difficult to obtain detailed turbine information from manufacturers, but for modelling purposes, the flow in Equation (11.29) should be proportional to the turbine flow area. This could be used, for example, to find relative changes needed in flow area when matching the turbocharger to the engine.

Equations (11.28) through (11.32) then give a mathematical description for the turbine in Figure 11.7. Since the equations are non-linear, they need to be solved with an equation solver, several of which are available in modern computer software packages. In addition, the procedure shown in developing the model gives a methodology by which models of radial turbines in general can be developed. The functions are generally simple, since the selection of important variables has a basis in physical processes. Since the functions are empirical, however, other functions could be used should they be more advantageous. It is recommended that such functions be well behaved, and that high order polynomials should be avoided. High accuracy fitting of data may not be warranted, since turbine data is generally difficult to obtain, and often contains significant variability. The most important point, as was discussed in engine simulation, it that the trends with respect to relevant variables are correct.

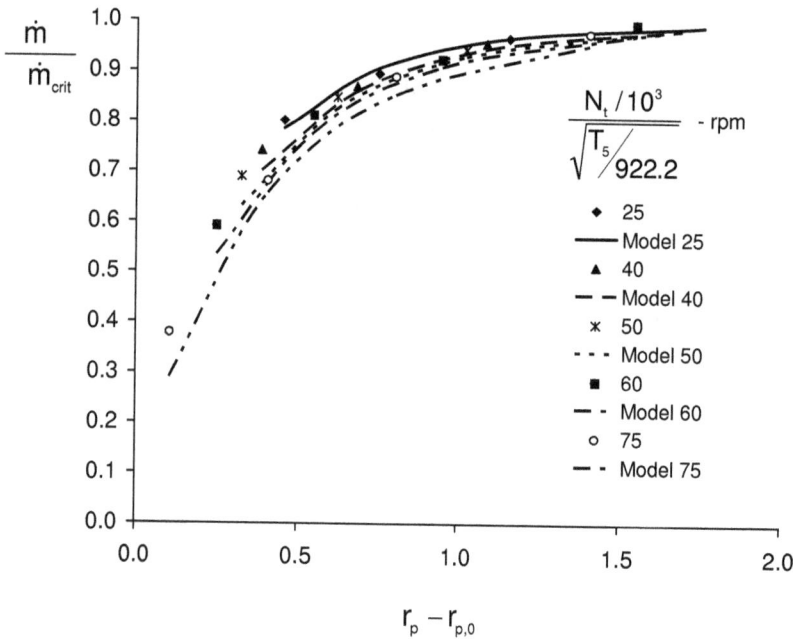

Figure 11.15: Ratio of turbine flow to choked flow as a function of pressure ratio for the turbine shown in Figure 11.7.

11.11.3 Waste Gate Model

The classical waste gate is a mechanical device, shown in Figure 11.10. As such, it is relatively simple to model. With modern electronic controls, it is possible to control the waste gate by other means, but they will not be covered here. A schematic diagram of a waste gate is shown in Figure 11.16. The intake

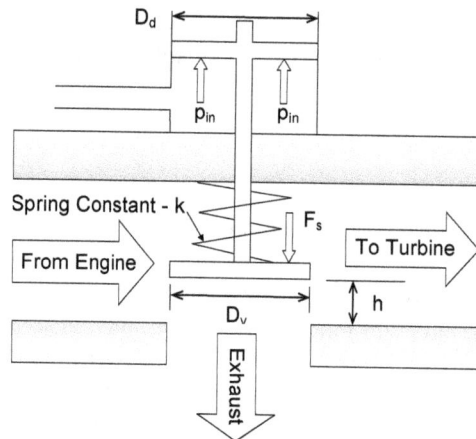

Figure 11.16: A schematic diagram of the function of a waste gate.

pressure, p_{in} acts on the diaphragm with diameter, D_d giving a force equal to

$$F_d = p_{in}\frac{\pi}{4}D_d^2 \tag{11.33}$$

The flow area is equal to the product of the lift of the valve and the area of the head of the valve:

$$A_f = \pi D_v h \tag{11.34}$$

The flow through the waste gate can be modelled using the compressible flow relations, Equations (5.3) or (5.5) with appropriate flow coefficient. The pressure difference is that across the turbine. Since the intake manifold pressure is typically limited to about 2 atmospheres, and the exhaust pressure is usually close to the intake pressure, choked conditions can readily occur for the flow over the waste gate, and the equation for choked flow, Equation (5.5) must be used.

The waste gate starts to open when the intake pressure exerts a force on the diaphragm that is equal to the pre-load force on the spring, F_0. Denoting this pressure p_{set}, the pre-load force can be determined:

$$F_0 = (p_{set} - p_0)\frac{\pi}{4}D_d^2 = (p_{set} - p_0)A_d \tag{11.35}$$

The opening height of the waste gate is then determined by a force balance on the shaft of the waste gate:

$$h = 0 \; ; \qquad p_{in} \le p_{set} \tag{11.36}$$

$$h = \frac{(p_{in} - p_0) \cdot A_d - F_0}{k} \; ; \qquad p_{in} > p_{set} \tag{11.37}$$

In the case of electronic control of the waste gate, a different function for the height of the waste gate as a function of the intake pressure would be needed, depending on the type of pressure sensor and valve actuator used. The principle is the same however. The value of the spring constant can be changed to determine the rise in the intake pressure when it exceeds the opening pressure. The higher the spring constant, that is the stiffer the spring, the greater the rise in the intake pressure as the load is increased.

When the flow through the waste gate has been determined, it must be subtracted from the flow through the exhaust port in order to determine the flow through the turbine:

$$\dot{m}_t = \dot{m}_{ex} - \dot{m}_{wg} \tag{11.38}$$

11.12   Modeling a Turbocharged Engine

With the above mathematical models for the compressor and the turbine, it is possible to estimate the fuel consumption and performance of an engine-turbocharger system. The model presented here is a steady state model. For transient operation, differential equations are needed for the slowly varying parameters such as engine speed, turbocharger speed, and intake manifold pressure. A mean value model for a turbocharged diesel engine could be based on these results, but is out of the scope of this text. Principles similar to those presented for the SI engine MVEM can be used (See Section 5.6.3), and a transient turbocharger model is needed, which relates the difference in compressor and turbine torques to the angular acceleration of the turbocharger, using its moment of inertia.

The equations in the previous section that describe the operation of the turbocharger components are non-linear, and so it is necessary to use an non-linear algebraic equation solver. One thing that should be noted when using non-linear equations solvers in the connection is that if initial values to start the iteration process are too far from the final values, the equation solving routine may be unstable or find roots that do not have physical meaning. Therefore, it is important give good initial estimates of all the values in the equations, and to place limits on physical variables, in order to keep the equation solving routines stable. For example, one should limit the compressor and turbine efficiencies to a maximum value of 1.0 or possibly a little less, since they physically will not exceed this value. Similarly, lower limits should be placed on variables. For example, T_2, the temperature after compression should not be lower that the atmospheric temperature, so that could be used as a lower limit.

The engine to be simulated will be a direct injection diesel engine, with charge air cooling, and a waste gate. There are various levels of simulation that can be used to simulate a turbocharged engine. In this case, the following general procedure will be used. It will be assumed that the indicated thermal efficiency is a known function of engine variables. Secondly, it will be assumed that the engine friction follows the Heywood correlation, Equation (9.12), and is independent of the cylinder pressure. Engine load variations are simulated through the use of a variable fuel air ratio up to a maximum value for all engine speeds. As in a real engine, the choice of the maximum fuelling rate is a design variable, and is used to achieve the desired full load torque curve of the engine. Pressure losses in the charge air cooler are assumed to be negligible, and it is assumed to operate with a constant effectiveness.

Diesel Exhaust Gas Temperature Model

One of the most important variables determining the operating condition of a turbocharger is the exhaust gas temperature at the inlet to the turbine. Winkler has presented a simple model for exhaust gas temperature of a diesel engine [103]. It is based on a energy balance between the intake and exhaust ports of the engine:

$$(T_5 - T_{in})\dot{m}_{ex}c_{p,ex} = (1 - \eta_i - \eta_k)\dot{m}_f H_u \tag{11.39}$$

where: the subscript ex refers to exhaust conditions, and the variables are numbered as in Figure 11.8. The variable η_k is the fraction of the input energy that is lost to the coolant during the in-cylinder processes. The equation means that of the fuel input energy, some is used for indicated work, indicated by η_i. Another portion is lost to the coolant, indicated by η_k. The remainder is left in the exhaust gasses at the temperature T_5. Using fundamental definitions and relationships, Equation (11.39) can be written[2]:

$$\frac{T_5}{T_{in}} = 1 + \frac{1 - \eta_i - \eta_k}{\eta_i} \cdot \frac{A_p \cdot imep \cdot S_p/4}{T_{in} \cdot \dot{m}_{ex} \cdot c_{p,ex}} \tag{11.40}$$

For modelling purposes, it is necessary to know how the variable η_k behaves as a function of other variables. An analysis was made from engine data for a turbocharged 2 liter DI diesel engine. In this case, all the variables in Equation (11.40) except η_k were known from measurements, and so it was possible to solve for η_k. It was found that the coolant fraction was mostly dependent on fuel air ratio and engine speed. Therefore, a simple empirical correlation was developed to calculate η_k:

$$\eta_k = (0.23 + 3.6 \cdot FA) \cdot (1000/N)^{0.65} \tag{11.41}$$

The results are shown in Figure 11.17, where in the graph on the left, the coolant fraction has been multiplied by $1000/N$ raised to the power 0.65 and plotted as a function of fuel air ratio for varying load and engine speed. At low values of the fuel air ratio, a large spread is observed, but at these conditions, the temperature differences are small and difficult to measure, and so large relative errors are found. In

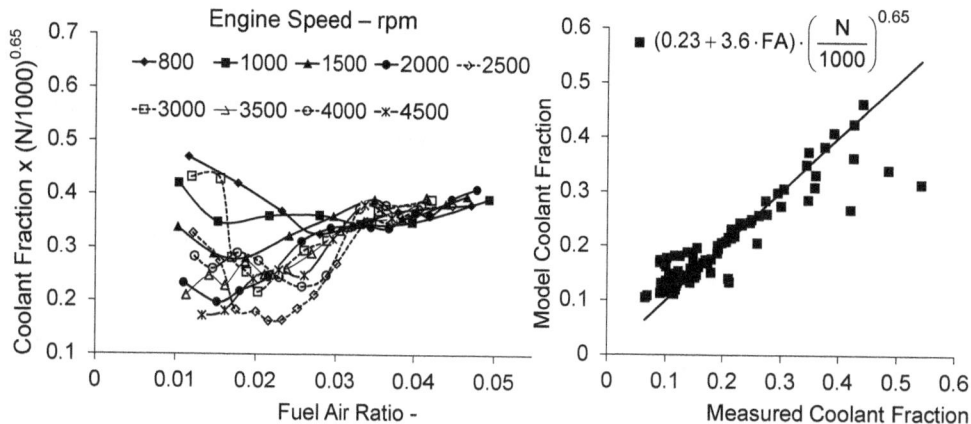

Figure 11.17: Development of a correlation for the fraction of energy lost to the coolant of a 2 liter, 4-cylinder turbocharged DI diesel engine.

the curve on the right, the coolant fraction predicted by Equation (11.41) is plotted as a function of the measured value. The agreement is acceptable within the accuracy of the data, and indicates that the correlation should give acceptable trends and values.

With development of the exhaust gas temperature model, adequate sub-models have been supplied to estimate the performance of an engine. In order to simulate an actual engine, the indicated thermal efficiency has been assumed to be a function of the fuel air ratio, based on some measurements on a 2 liter

---

[2]The derivation of this equation is a good exercise for the student.

DI diesel engine. It was found that the indicated thermal efficiency decreased at higher fuel air ratios due to the longer combustion duration. Due to combustion modification for emission control, this engine also had lower indicated thermal efficiency at lower loads (lower FA). Therefore, an empirical function was developed which shows this trend:

$$\eta_i = 0.45 + 3 \cdot FA - 75 \cdot FA^2 \qquad (11.42)$$

The function is shown in Figure 11.18. The model here is rather arbitrary, and one could include effects of other variables such as engine speed, and if data were available, the effects of timing, EGR, *etc.* Knowledge of this behavior must be obtained from an experimental engine data base, or from reliable engine cycle simulation programs. These sources can provide information about how the indicated efficiency is a function of engine variables, and these dependencies can be implemented into the model as illustrated here with the indicated thermal efficiency. The main emphasis in this model is to show the interaction between engine operation and the behavior of several subsystems, such as compressor, turbine and waste gate.

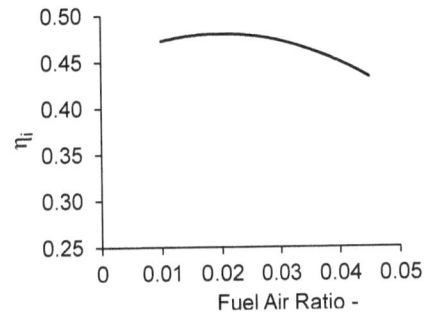

Figure 11.18: The simulated indicated thermal efficiency as a function of fuel air ratio for a turbocharged diesel engine.

In summary, the complete turbocharger model consists of the compressor model in Section 11.11.1, the turbine model in Section 11.11.2 and the waste gate model in Section 11.11.3 combined with the equation resulting from setting the compressor power equal to the turbine power, Equation (11.19). To couple the turbocharger with the engine and determine performance, equations are required for:

- Friction - Equation (9.12)
- *imep* - Equation (1.33)
- Engine flow - Equation (7.10),
- Intake temperature - Equations (11.7) and (11.22)

When the engine operating flows are determined, the power and fuel consumption can be determined by standard engine relationships in Chapter 1.

Example

As an example to show the results of the model, calculations were made on a 4-stroke, 6 cylinder, direct injection diesel engine. The engine has a bore of 131 mm and a stroke of 125 mm, for a displacement volume of 10.1 liters. It was assumed that the engine has a volumetric efficiency of 0.90, independent on the intake temperature. Charge air cooling is assumed, with no pressure loss in the cooler, which has an effectiveness of 0.7 at all conditions. The indicated thermal efficiency is given by Equation (11.42), and the coolant loss fraction by Equation 11.41. The turbine has a diameter of 116.8 mm, and has the characteristics shown in Figure 11.7. The efficiency is modelled by Equation (11.27). All parameter variations for the turbine model are given by the equations in Section 11.11.2. The compressor diameter is 114.3 mm, and compressor can be characterized by the equations:

$$\psi = (2 - 2.14 \cdot \phi) \cdot \eta_c \qquad (11.43)$$

$$\psi = 1.38 \cdot \left(1 - \exp\left(\frac{\phi - .15}{.02} \right) \right) \qquad (11.44)$$

The waste gate valve has a diameter of 25mm, and the membrane on which the intake pressure acts has a diameter of 50 mm. The spring in the waste gate has a spring constant with a value of 30000 N/m, and the preload on the membrane is such that the forces from the intake pressure and the preload on the spring balance when the intake pressure is equal to 190 kPa.

The equations were solved for varying fuel air ratio at different engine speeds using the EES equation solving program. Figure 11.19 shows the simulated performance of the engine for part load operation at a speed of 2000 rpm. First note in the upper curves that at a bmep of about 1100 kPa, there is change in the slopes of the intake and exhaust pressure curves. This begins at an intake pressure of 190 kPa, determined by the pre-load on the waste gate spring. These pressures continue to rise as the *bmep* increases, but as a slower rate, as the waste gate opens more and more to reduce turbine flow. The compressor outlet temperature has a similar characteristic. The effect of charge air cooling can be seen in that the intake temperature is lower than T_2, and is fairly close to the atmospheric temperature of 293K.

In the middle set of curves, the opening of the waste gate can be seen by the increasing difference between the exhaust port flow and the flow through the turbine, starting from the same load. There is still an increase in the intake pressure after the waste gate opens, but the rate of increase with load has been reduced significantly. If a lower final pressure is desired, then the characteristics of the waste gate must be changed, that is, a different spring must be installed. The influence of waste gate opening can also be seen on the compressor speed (equal to the turbine speed), and its rate of increase with load is also lower after the waste gate opens.

Given that the indicated efficiency is nearly constant, then for increased power after waste gate opening, the fuel flow rate must continue to rise in a continuous fashion as the *bmep* increases. However, since the waste gate opening limits the amount of air through the turbocharger, the FA curve as a function of *bmep* will turn up as the power is increased above the point where the waste gate opens. This can also be seen in slope of the fuel air ratio curve in the middle set of curves. That the fuel air ratio curve as a function of *bmep* is not constant at lower loads is a reflection that the indicated thermal efficiency was assumed to be

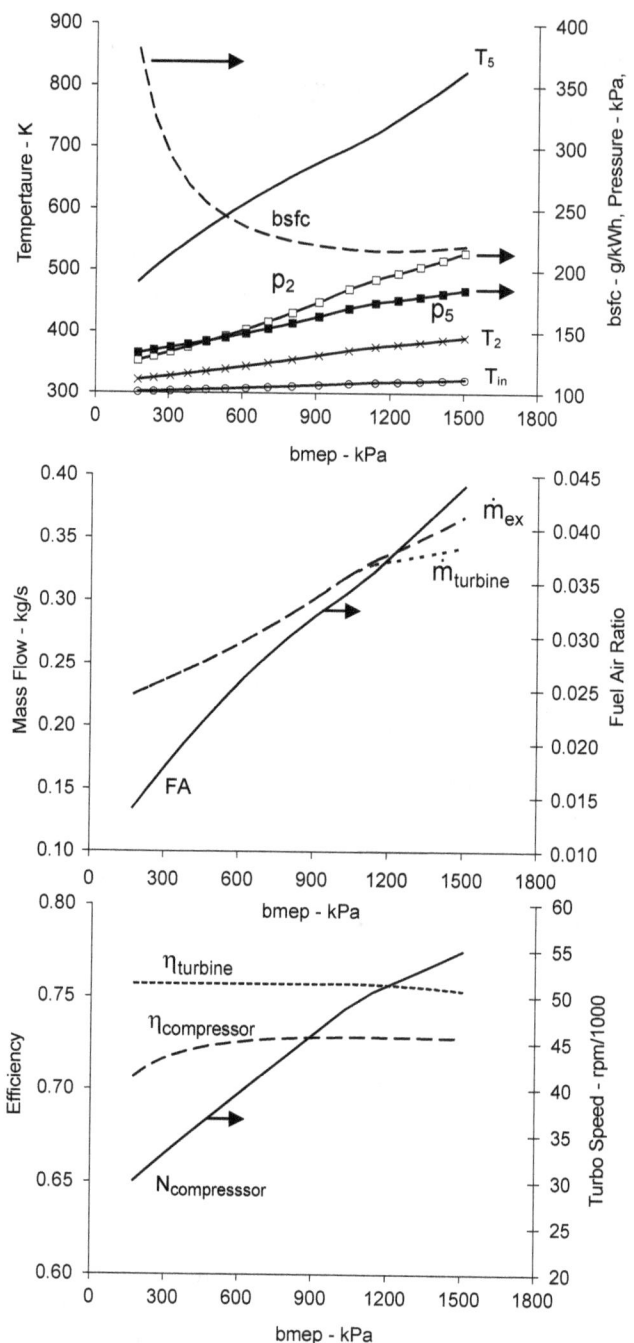

Figure 11.19: Simulated part load characteristics of a 10 liter, 4-stroke turbocharged diesel engine with charge air cooling at an engine speed of 2000 rpm.

a function of the fuel air ratio, with a maximum efficiency value at $FA \approx 0.025$, which occurs at a *bmep* of about 600 kPa.

A consequence of this is that there is an increase in the slope of the curve of the exhaust temperature, T_5, as a function of *bmep* when the waste gate opens. There is a steadily increasing need for more fuel as the load keeps increasing, but because of the waste gate opening, there is not so much air to absorb this energy, and the temperature increases even more with load.

The *bsfc* exhibits the classic behavior, in that at *bmep* = 0, the specific fuel consumption is infinite. The reduction in *bsfc* with increasing load is due to increasing mechanical efficiency, since the friction is constant. At the highest loads, the *bsfc* starts to increase again, which is due to the behavior of the indicated thermal efficiency. The latter was assumed to decrease at the highest fuel air ratios, as explained above.

The curves in Figure 11.19 were obtained using the model, and the trends seen are observed in actual turbocharged 4-stroke engines. Typically, one can obtain information on engine performance and fuel consumption parameters fairly readily, but information on detailed pressure, temperature measurements and turbocharger operating parameters is only rarely available, and requires a comprehensive measuring program. Such a simulation can be very instructive in obtaining information about how an engine performs, and understanding the interactions of many parameters involved in engine performance.

11.13   Summary

Pressure charging provides the means to increase the output of an engine by supplying more air than the engine would consume for atmospheric intake conditions. When the engine delivers a significant increase in power for the same mechanism, friction only increases a limited amount, and an increase in efficiency is obtained as well, through a better mechanical efficiency. Changes in indicated efficiency are normally small, and in the case of spark ignition engines, the thermal efficiency may decrease due to measures needed to prevent knock when the cylinder temperatures and pressures are increased.

Because diesel combustion is improved with higher temperature and pressure, turbocharging and charge air cooling has become dominant in diesel engines of all sizes. Because of knock limitations, the use of pressure charging in spark ignition engines has been limited. Additional information on turbocharging of engines can be found in other books, for example, Zinner [106] and Watson and Janota [104].

11.14   Problems

Problem 11.1

During testing of a 6-cylinder 4-stroke diesel engine with bore = 240 mm and stroke = 310 mm at standard conditions of p = 1.0 bar and T = 25°C, the following results were measured at a speed of 750 rpm: Brake torque = 4057 Nm; Friction torque = 794 Nm; Brake specific fuel consumption = 218 g/kWh; Brake specific air consumption 5371 g/kWh; Lower heating value 42700 kJ/kg. Calculate the indicated efficiency, the brake mean effective pressure and the volumetric efficiency.

The engine is then equipped with a turbocharger and charge air cooler with the following conditions: Turbine efficiency = 0.75; Compressor efficiency = 0.80; Charge air cooler effectiveness = 0.63; Coolant temperature = 20°C; Turbine inlet pressure = 1.65 bar: Turbine inlet temperature 500°C.

Calculate the turbocharged engine's brake mean effective pressure and brake specific fuel consumption with the same fuel/air ratio, when it is assumed that engine friction increases by 5% due to the change in cylinder pressure. It can be assumed that $\gamma = 1.34$ and $c_p = 1.129$ kJ/kg-K for exhaust gasses. ($\eta_i = 0.462$; bmep = 606 kPa; η_v =0.774; bmep = 1282 kPa; bsfc = 200 g/kWh.)

Problem 11.2

For a 6-cylinder 4-stroke diesel engine with bore = 100 mm, stroke = 120 mm and compression ratio = 16, the brake torque = 292 Nm, and the friction torque = 73,5 Nm at a speed of 3000 rpm. The air consumption is 0.13 kg/s, the fuel/air equivalence ratio = 0.67, the lower heating value 43000 kJ/kg and the stoichiometric fuel/air ratio = 0.066. The intake conditions are 20°C and 1.01 bar.

With the same fuel/air ratio, the engine is to be operated at an altitude of 2500 m, where the pressure is 0.76 bar and temperature = 0°C. Calculate the engine's brake power at altitude relative to the power when tested at atmospheric conditions.

The engine is then equipped with a turbocharger. The compression process in the compressor can assumed to occur according to a polytropic process with polytropic exponent = 1.6. Calculate the required pressure ratio over the turbine when the engine gives the same power at 2500 m altitude as when tested at atmospheric conditions. The effect of the changes in intake conditions on volumetric efficiency can be neglected.

Calculate the necessary turbine inlet pressure, when the exhaust gas temperature is 550°C and the turbocharger group efficiency = 55%. For air, $c_p = 1.004$ kJ/kg-K and $\gamma = 1,4$; for exhaust gas $c_p = 1.129$ kJ/kg-K and $\gamma' = 1.34$. ($BP_2/BP_1 = 66.4/91.7$; $p_6/p_5 = 0.820$; $p_5 = 0.927$ bar)

Problem 11.3

The blower in Figure 11.2 is installed on a 2-liter 4-stroke SI engine engine and operates at 2 times the engine speed. At the basis condition (without blower) at 3000 rpm, the engine operates with a stoichiometric mixture of octane. The engine has a volumetric efficiency of 0.75, an indicated efficiency of 0.37, and a friction mean effective pressure of 150 kPa. Intake pressure and temperature are 100 kPa and 20°C.

Calculate the brake power after installation of the blower at 3000 rpm under the assumption that the engine's friction mean effective pressure and indicated efficiency are unchanged. ($BP = 48.2$ kW)

For engine operation at full load, calculate the engine's intake pressure as a function of engine speed from 1000 to 6000 rpm.

Problem 11.4

An 8-cylinder, 4-stroke direct injection diesel engine has a cylinder bore of 102 mm, stroke of 100 mm and compression ratio of 18:1. The engine is naturally aspirated and has a maximum torque of 410 Nm at 2150 rpm. It is assumed that the engine friction is given by Heywood's correlation, Equation (9.12). Assuming that the engine's indicated efficiency = 0.48 and the volumetric efficiency = 0.90, determine the fuel air ratio and brake specific fuel consumption at maximum load. The lower heating value is 42.5 MJ/kg, and assume atmospheric temperature and pressure of 20°C and 101 kPa.

The engine is to be turbocharged such that at full load at 2150 rpm, the torque is increased to 650 Nm. In combination with turbocharging, a charge air cooler is to be used. The exit temperature of the charge air cooler is regulated to be constant at 45°C. The turbocharger compressor has an efficiency of 0.73 and the turbine has an efficiency of 0.76. Assume that at the friction increases by 5% when the turbocharger is used and that the indicated efficiency is unchanged. Determine the required pressure ratio over the compressor in order to be able to achieve a torque of 650 Nm for a fuel air ratio of 0.04. Determine the pressure ratio over the turbine if the exhaust temperature = 440 °C. Use the same exhaust gas properties as in Figure 11.9. ($FA = 0.0448$, $bsfc = 221$ g/kWh, $rp_c = 1.723$, $rp_t = 1.589$)

Chapter 12

Engine Characteristics

12.1 Full Load Curves

When using engines in different applications, it is important to couple the engine and application together in the most effective way. Of primary importance is the ability of the engine to perform the work required in a given task. Therefore, one first looks at the full load curves of the engine, that is, the power, torque and specific fuel consumption as a function of engine speed for the fully loaded condition. Here it can be seen if the engine is able to satisfy power and speed requirements.

Figure 12.1 shows the full load performance curves of a 1.8 liter spark ignition engine. As discussed in Section 7.4, the effects of changing volumetric efficiency, friction and percentage heat loss play important roles in shaping these curves. The brake specific fuel consumption of SI engines at wide open throttle is often higher than the lowest values, as the full load fuel air mixture can be richer than stoichiometric in order to achieve maximum power.

By turbocharging the spark ignition engine, the air flow through the engine and therefore the power, can be increased. However, due to the increased tendency for the engine to knock, the pressure of the combustion end gasses must be limited, and the compression ratio usually is decreased for turbocharged versions of SI engines. Figure 12.2 shows the full load power and torque curves for a naturally aspirated SI engine, and a turbocharged version of the same engine. There is no significant increase in torque at the lowest speeds, due to the characteristics of the compressor, which does not give a large pressure increase at low operating speeds. At higher speeds, the

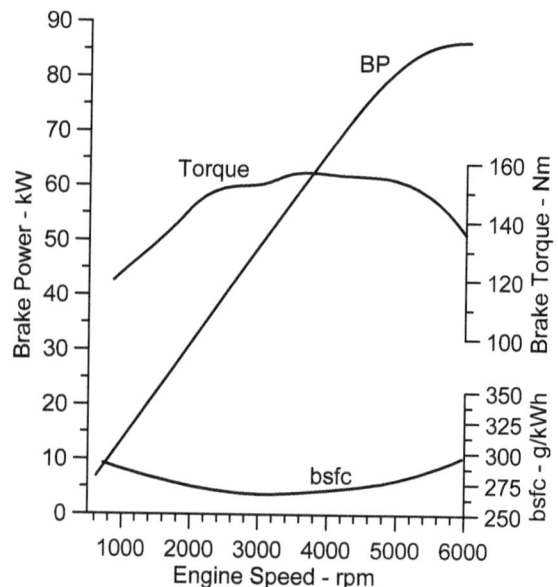

Figure 12.1: Torque, power and brake specific fuel consumption for 1.8 liter spark ignition engine.

difference between the torque of the two versions of the engine is roughly constant. This is the region where an unregulated turbocharger would supply too large an amount of air, so a waste gate is utilized to normally used to maintain an approximately constant maximum intake pressure. From Equation (1.32), a constant increase in intake pressure pressure would give a nearly constant increase in bmep.

When turbocharging an SI engine, the efficiency is affected in two directions. First, it is typical to lower the compression ratio by for example, 9.1 to 7.5:1, to reduce knock. This will reduce the indicated thermal efficiency by a factor of about 0.93 according to the trends predicted by the Fuel-Air cycle calculation in Section 2.4. In addition, it might be necessary to retard the spark timing somewhat to help avoid

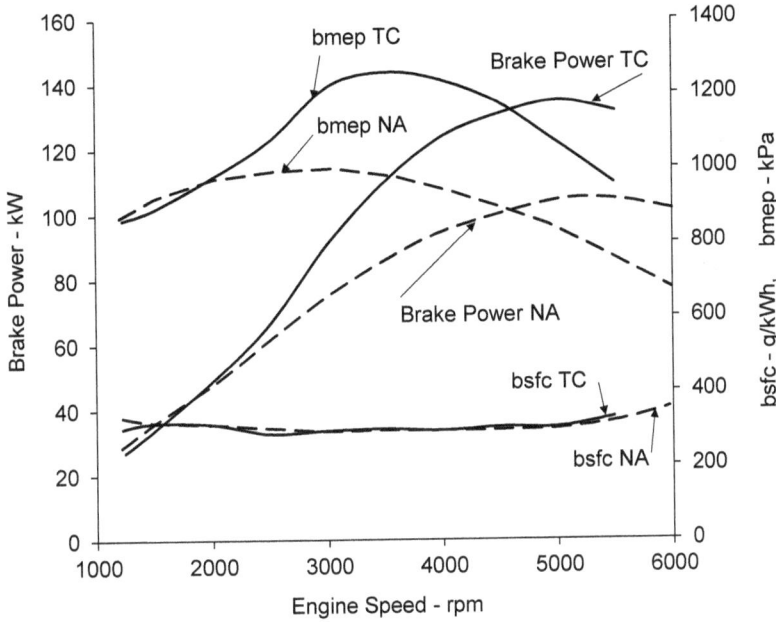

Figure 12.2: Full load performance of a 3 liter V6 engine in turbocharged and naturally aspirated versions. Adapted from [107].

knock. On the other hand, the fact that the engine size remains the same while the power output is increased will mean that the friction power is about the same while the brake power is higher, increasing the mechanical efficiency, mitigating at least some of the effects of the lower indicated efficiency. In the above example, these effects nearly cancel each other through the entire speed range at full load.

Nearly all modern diesel engines are equipped with turbochargers, which affect the full load characteristics of these engines. Figure 12.3 shows the full load performance of a 4.25 liter turbocharged, DI diesel engine, suitable for light utility vehicle use. As in the case of the SI engine, turbocharger effects are different at various speeds, where low speeds give low boost, and not as much fuel can be added without deteriorating performance and causing excessive particulate emissions. At higher speeds there is excess air, the waste gate limits intake pressure and the torque is primarily determined by the amount of fuel injected. This engine is typical with a region of essentially constant torque, followed by falling torque at higher speeds, where one begins to approach constant power conditions. Difficulties in achieving good combustion at higher speeds with DI engines, may also play a role in determining the torque characteristics at higher speeds. The difference in the torque between maximum torque and the torque at maximum power is called torque rise. It is important in many applications, in that when operating above the maximum torque speed, if the load is increased, tending to make engine speed decrease, the torque increases to counteract that, and prevents the engine from stalling. This

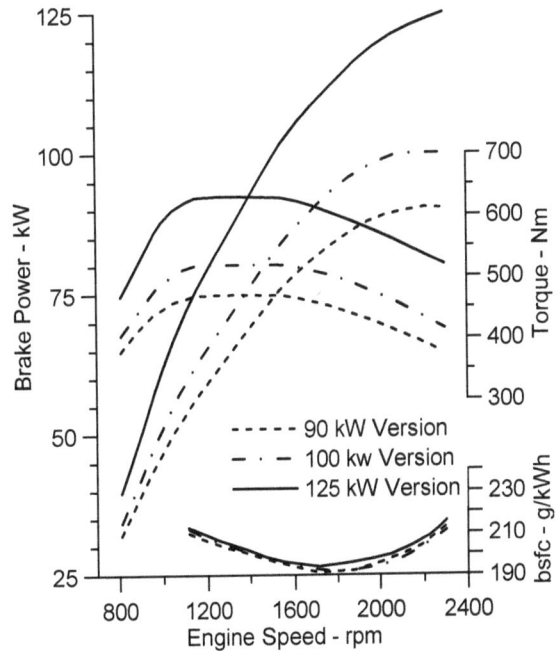

Figure 12.3: Torque, power and brake specific fuel consumption for 3 versions of the 4.25 liter direct injection diesel engine. Adapted from [108].

is especially important in off-road applications.

It is common for engine manufacturers to produce different versions of engines for a basis displacement, often by varying the turbocharger used. Full load curves readily show which version of an engine is appropriate for a given task. Three versions of the engine in Figure 12.3 are available in different maximum power ratings. For all versions of the engine the best fuel consumption is slightly above 190 g/kWh at an engine speed of about 1700 rpm.

Full load characteristics for a large turbocharged DI diesel engine are shown in Figure 12.4. This engine could be used in a heavy duty vehicle or bus. The torque curve of this engine exhibits the torque rise shown in the previous example. By adjusting the fuel injection amount at full load as a function of speed, the maximum torque curves can be tailored to a specific application with turbocharged diesel engines, since there is sufficient air available to ensure good combustion. Eventually, mixing problems and high fuel consumption limit maximum speeds in diesel engines to lower values than SI engines. This engine has good fuel

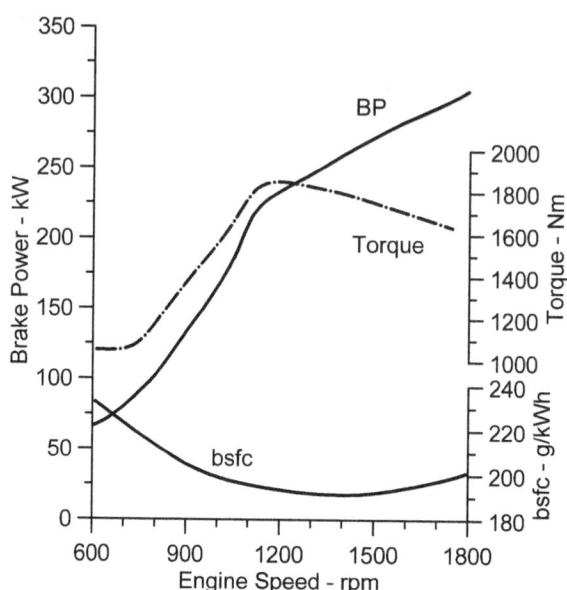

Figure 12.4: Full load characteristics for a 6 cylinder, 12 liter, turbocharged, DI diesel engine.

consumption, corresponding to an maximum brake thermal efficiency of about 0.45 at full load. Large CI engines general run over a smaller speed range than small ones, and large heavy duty vehicles typically have more gears than light utility vehicles or passenger vehicles.

12.2   Part Load Curves

Not so commonly seen in the literature, but also important and often measured, are part load curves, that is engine performance when the speed is held constant and the load varied, typically from no load to maximum. In the following, some examples of part load curves will be show, and some parameters in addition to fuel consumption will be shown, as they help to clarify engine operation.

The usual independent parameter in part load curves is the torque or *bmep*. In application terms, the most important parameter of interest is the fuel consumption, or *bsfc*, since there is sufficient power available, given that the engine has been chosen from its maximum load characteristics. Figure 12.5 shows typical values of *bsfc* as a function of *bmep* for A: 3 versions of SI engines and a modern, DI diesel engine, and B. 2 turbocharged and 2 naturally aspirated DI and IDI diesel engines, all for a mean piston speed of about 7.5 m/s. The general shape of such curves is the same for all engines, and if the results were extended to *bmep* = 0, the *bsfc* would be infinite. In curve set A, the homogeneous charge 1.1 liter engine has the highest fuel consumption, due to a slightly lower compression ratio and smaller size than the homogeneous charge 1.8 liter engine, which is more of a high performance type. This is also evidenced by the larger maximum bmep for the 1.8 liter engine. The maximum *bmep* values of 950 to 1050 kPa are typical for a naturally aspirated SI engines with homogeneous charge. The *bsfc* for both of these engines shows a distinct increase just before maximum load, as the fuel air ratio is changed from the stoichiometric value for operating a three-way catalyst, to a richer value for maximum power. According to the tendencies shown in Figure 2.13, the richer than stoichiometric fuel air ratio gives lower thermal efficiency/higher fuel consumption. The 1.3 liter GDI engine has better fuel consumption than the two homogeneous charge SI engines below a *bmep* of about 600 kPa, above which the fuel consumption increases drastically. The better fuel consumption at low loads is primarily due to lack of pumping losses, and possible a higher compression ratio, made possible by the GDI combustion process. The upturn is due to difficulties in preparing a combustible mixture from the spray where it is extremely

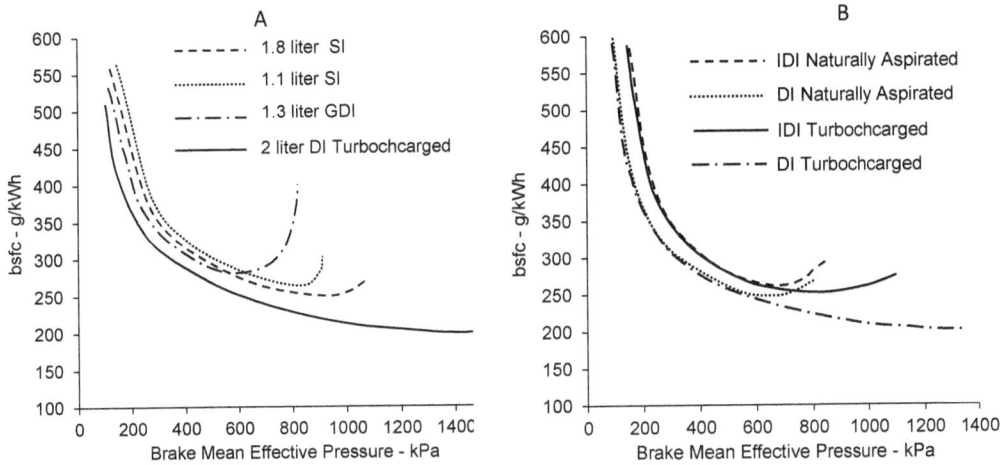

Figure 12.5: Part load fuel consumptions for different engines at a piston speed of about 7.5 m/s. A: 3 SI engines and an turbocharged DI diesel engine, B: Turbocharged and naturally aspirated Indirect and Direct Injected diesel engines.

difficult to mix all the fuel with air in the time available. This problem also limits the maximum *bmep* of this engine to a lower value than the homogeneous charge engines. In practice, the injection process can be changed to give homogeneous operation, and the curve at higher loads would resemble those of the 2 SI engines. The DI diesel engine has better fuel consumption than the other engines due to lack of pumping, a higher compression ratio, and larger maximum *bmep* because of the extra air available with the turbocharger. Figure 12.2 shows a higher maximum *bmep* for the turbocharged SI than the non-turbocharged SI, but still lower than the DI diesel, since knock in the SI limits the amount of boost than can be used.

Figure 12.5B compares fuel consumption for some diesel engines. The first comparison is that between the InDirect Injection and the Direct Injection types. The IDI has higher fuel consumption throughout the entire operating range. This is due to the pumping loss between the pre-chamber and main chamber, and the higher heat loss due caused by the high velocity combustion jet leaving the pre-chamber and impinging on the piston. This higher fuel consumption is a primary reason that the IDI engine has largerly disappeared since about the year 2000.

The other comparison is in Figure 12.5B is between turbocharged and naturally aspirated engines. For each type of combustion system, the fuel consumption at lighter loads is basically unaffected by the presence of the turbocharger. The effect of the turbocharger can be seen in that the maximum *bmep* for the engines increases substantially when turbocharged. There is an accompanying improvement in fuel consumption at these higher loads, and that is basically due to the improved mechanical efficiency. The friction of the engine is only slightly affected by the increased cylinder pressures, and so the brake engine efficiency increases at loads greater than the maximum of the naturally aspirated engine. Additionally, the the range of *bmep*'s where

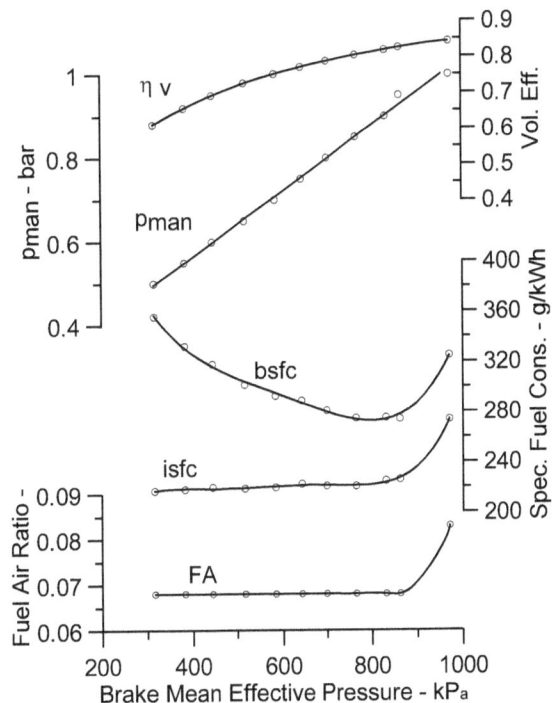

Figure 12.6: Part load engine operating characteristics for a 1.1 liter, SI engine at 3000 rpm.

fuel consumption is close to the minimum value is much wider with the turbocharged engines, giving a better range of good fuel consumption, a useful characteristic for vehicular applications.

Part load curves are also of interest in illustrating the operating principles of the engines. Figure 12.6 shows the operating parameters of a conventional 1.1 liter SI engine. The curve at the bottom shows that the engine operates at a stoichiometric mixture, except at the maximum power condition. Given the constant fuel air ratio and speed, as expected the indicated efficiency ($isfc$) only changes a little until the mixture is made rich at full load. The bsfc falls as explained previously, mechanical friction being nearly constant and pumping work falling as load increases, giving an increase in mechanical efficiency with load. That the pumping portion of the friction decreases with load is shown by the intake manifold pressure, which is proportional to the $bmep$, as indicated by the relationships from Chapter 1. The exhaust pressure, not shown, remains nearly constant for all loads. The volumetric efficiency increases somewhat with $bmep$, as discussed in Section 7.1. Thus the basic concept of homogeneous charge SI engines is seen, power is varied by changing the amount of air flow to the engine, keeping the ratio of fuel and air constant, and requiring a spark ignition system to ignite the homogeneous charge.

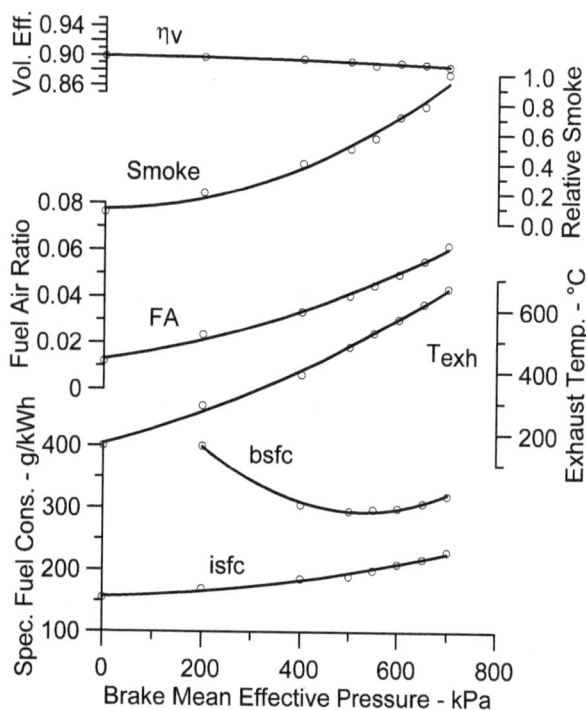

Figure 12.7: Part load engine operating characteristics for a 1.17 liter naturally aspirated Direct Injection, single cylinder oil test diesel engine at 2000 rpm.

Figure 12.7 shows part load characteristics for a naturally aspirated direct injection single-cylinder, oil testing diesel engine. The engine shows the most fundamental operating characteristics of a diesel engine, and is the diesel engine equivalent to the engine in Figure 12.6. Without a turbocharger, the air flow to the engine remains nearly constant, as shown by the volumetric efficiency, which changes by less than 2 percent over the entire load range. As the load is increased, the increasing fuel results in an increasing fuel air ratio, clearly shown in Figure 12.7. Also shown is the indicated specific fuel consumption, which is lower at the lightest loads, and increases as the load increases. This is an indication of a longer combustion period as more fuel is injected, resulting in a poor indicated efficiency, as the combustion extends further into the expansion stroke with higher loads. See, for example Figure 3.47. With only small changes in the indicated efficiency and volumetric efficiency, The equations of Section 1.2.5 say that for a constant indicated efficiency and $fmep$ the fuel air ratio should be proportional to the $bmep$. In this engine, the indicated efficiency drops with load ($isfc$ increases), and so the fuel air ratio, plotted as a function of $bmep$, should have an increasing slope, which is the case in Figure 12.7. With a constant inducted air mass, the increasing fuel gives rise to increasing temperatures, and as a result, the trend of the exhaust gas temperature closely follows that of the fuel air ratio. The $bsfc$ shows the expected trends, with the effect of increasing mechanical efficiency dominating all other factors.

The final parameter shown in Figure 12.7 is a relative smoke value, related to particulate emissions. Since the engine is of an older design, the absolute values are totally unacceptable for a 21st century engine, but the trend shows an additional reason for the disappearance of the naturally aspirated DI diesel engine. As the load increases, smoke increases directly with the load, and is highest at the full load.

The fuel air ratio here can be seen to be close to the stoichiometric value of about 0.068, and there is neither adequate time nor air to complete combustion without unacceptable particulate emissions. One of the reasons the prechamber (IDI) engines were dominant in small vehicle use until about the year 2000, was their better smoke characteristic, and the ability to operate at higher speeds than a DI engine without excessive smoke, at the sacrifice of a somewhat lower efficiency. With these older engines, it was the allowable amount of smoke that was the limiting factor in the maximum *bmep* of the engine. Figure 12.17 shows some smoke limits for a similar design, the numbers refer to the darkness on a piece of filter paper, through which a given volume of exhaust gas have been drawn. The computer controlled, modern turbocharged DI engine has changed this as it delivers an excess of air, and it is not necessary to operate at fuel air ratios so close to stoichiometric. In addition, modern turbocharged DI diesel engines have very high injection pressure, computer controlled fuel injection, often with more than one injection per cycle (see Figure 3.50), and improved combustion chamber design, achieved with the help of CFD design methods. Though the particulate emissions are much lower with modern DI engines, their tendency is still a steady increase with load.

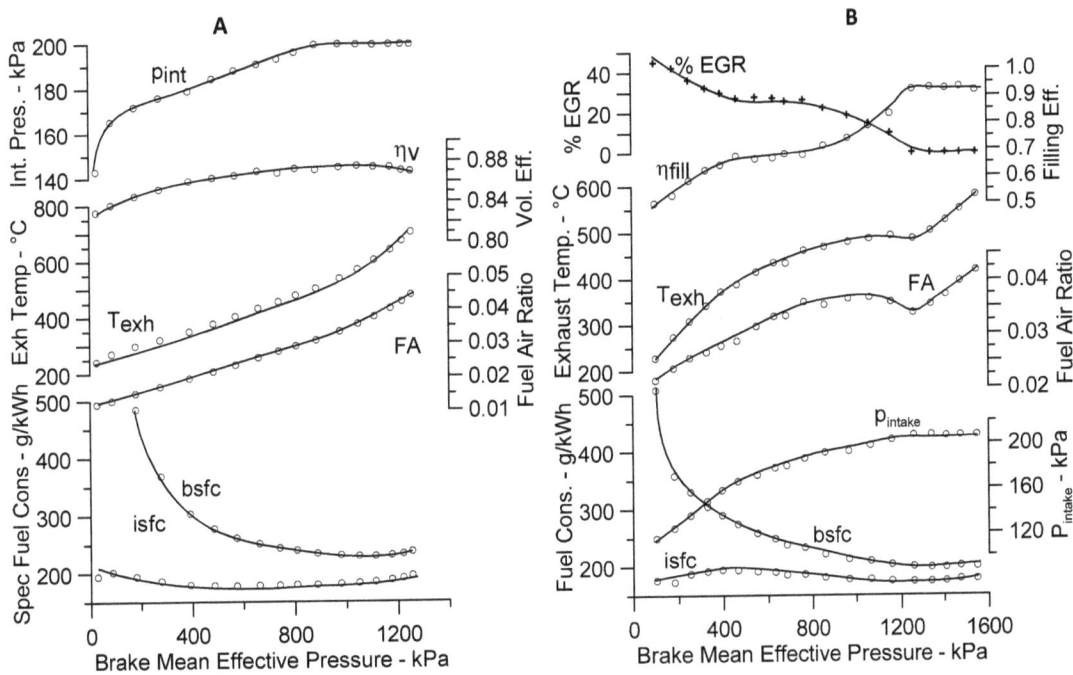

Figure 12.8: Part load engine operating characteristics for a 2 liter turbocharged Direct Injection, 4 cylinder diesel engine operating at 4000 - A and 2400 rpm - B.

Figure 12.8 shows part load curves for a modern, turbocharged direct injection diesel engine. Two versions are shown: A: operation at 4000 rpm, without EGR implementeted to reduce the NO_x emission. The other version, B, shows the engine with EGR. Version A is closer in operation to that of Figure 12.7. The main difference in operation is the effect of the turbocharger, as shown by the intake pressure. At light loads, the exhaust temperature is low, and as indicated by Equation (11.19), the output pressure of the compressor is low. As the load increases, the intake pressure increases until the maximum value allowed by the waste gate is achieved. The fuel air ratio does not increase as rapidly with load as was seen in Figure 12.7, since the air charge per intake stroke now increases as the load increases. The actual amount of fuel injected per cycle increases nearly in proportion to the *bmep*, but the increase in fuel air ratio with *bmep* is slower than for the naturally aspirated engine up to the point where the maximum intake pressure is reached (the waste gate opens). After this, air flow remains essentially constant, and the slope of the fuel air ratio curve increases to a value similar to that of a naturally aspirated engine. The exhaust temperature temperature follows the tendency of the fuel air ratio.

For control of NO_x emissions, EGR is commonly used in modern diesel engines. The effect of this on part load characteristics is shown in Figure 12.8B. The amount of EGR is shown as a percentage of the total mass flow through the engine. It is applied most heavily at the lightest load, and gradually reduced as load increases, until there is no more EGR at the highest loads. Since SCR methods are often used to reduce NO_x, and they are most efficient at higher exhaust temperatures, the EGR can raise the exhaust temperature and increase the effectiveness of the SCR system.

The influence of the EGR can be seen on the filling efficiency, here defined as the actual flow of air divided by the ideal flow of air for the given intake pressure and temperature. Since the EGR fills a portion of the intake charge, there is less air flow through the engine with EGR, as shown by the low filling efficiency with EGR, where the filling efficiency is the air flow divided by the air flow if the engine was filled with air at intake conditions. Not shown, the volumetric efficiency defined on the basis of total flow - air plus EGR - remains at a nearly constant value of that without the EGR. But since essentially the same amount of fuel is needed to provide a given power, and there is less air flow, so the fuel air ratios are higher than those of the engine without EGR for a given *bmep*. When the EGR rate goes to zero, the fuel air ratio actually drops, then proceeds to change at a rate similar to the engine without EGR. The exhaust temperature closely follows the trend of the fuel air ratio, as was the case in the engines without EGR. The high exhaust temperature affects the operation of the turbocharger, and the intake pressure increases faster with load at lower bmep than it does for the case with no EGR. Since EGR dilutes the charge and reduces the oxygen concentration for combustion, there is a negative effect on the fuel consumption, best seen with the *isfc*, which is high at the loading where EGR is applied.

12.3   Engine Maps

To predict vehicle performance and fuel consumption under all conditions, and to match an engine to an application in which the load and speed vary, a more complete description of the engine is required. This is given in the so called *engine map*. This is a combination of full load and part load curves and shows curves of constant brake specific fuel consumption plotted on a set of axes, where some form of engine load, normally torque or *bmep* is used for the ordinate, and the engine speed or mean piston speed, is used for the abscissa. An example of an engine map is shown in Figure 12.9 for a newer turbocharged, 2.15 liter 4-stroke diesel engine with a variable geometry turbocharger. The engine map is a plot of the engine's brake specific fuel consumption as a function of the engine's torque (brake mean effective pressure in this case) and the engine speed. Since the brake mean effective pressure is proportional to the torque, either of the parameters can be used. The mean effective pressure is more of interest in terms of comparing engine performance as a function of size, while the torque is most useful in a specific application. Sometimes the engine speed is expressed in terms of the average piston speed, useful in comparing engines of different sizes.

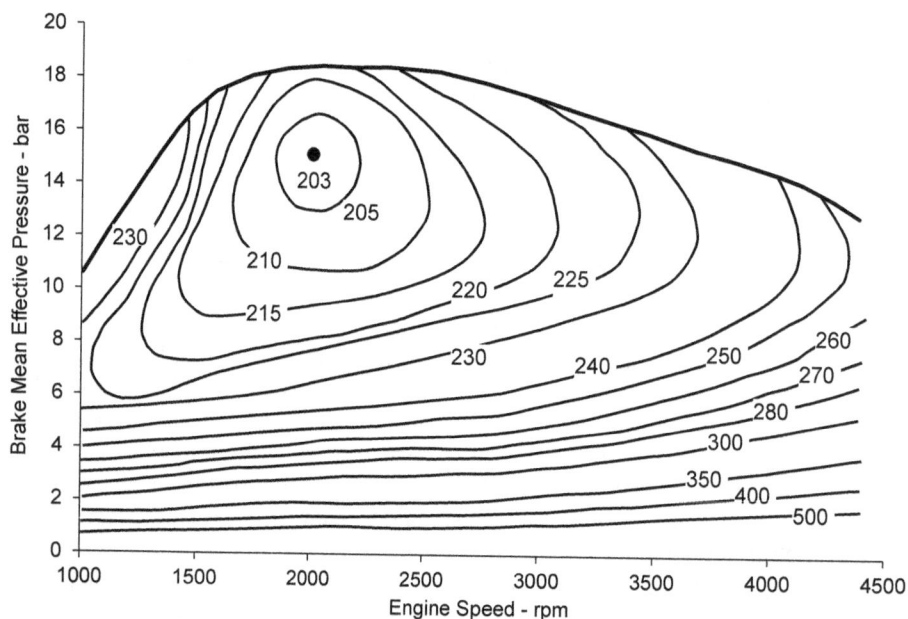

Figure 12.9: Engine map for the 2.15 liter, turbocharged diesel engine. Adapted from [2].

12.3.1 Simple Engine Model

While an engine map may look complicated, it is based on only a few parameters, and is no more than a plot of the brake thermal efficiency over the engine operating range, since:

$$bsfc = \frac{3.6 \cdot 10^6}{H_u \cdot \eta_e} = \frac{3.6 \cdot 10^6}{H_u \cdot \eta_i \cdot \eta_m} \tag{12.1}$$

The fuel consumption is more important to a user than efficiency, since it is directly related to fuel costs. Since the indicated thermal efficiency appears directly, factors such as compression ratio, injection/ignition timing, heat losses and fuel air ratio for SI engines will be important for the shape of the engine map. The mechanical efficiency will be determined by engine friction. To show the basic nature of an engine map, a very simple map will be presented. The engine chosen is a naturally aspirated diesel engine, with a constant indicated efficiency and a fmep that is only a function of engine speed, described by Heywood's correlation. The equations that are needed to establish an engine map in this case are:

Basic engine characteristics: The following two equations determine essentially the entire shape of the map:

$$\eta_i = 0.46 \tag{12.2}$$

$$fmep = 75 + 0.048 \cdot N + 0.04 \cdot \left(\frac{S \cdot N}{30}\right)^2 \ kPa \tag{12.3}$$

Maximum Torque: The load is limited by assuming that the fuel air ratio may not exceed a limit value of 0.045 in order to maintain combustion quality. This defines the maximum torque (bmep) curve. It is assumed that the volumetric efficiency is constant.

$$m_a = \eta_v \cdot V_d \cdot \rho_{in} \tag{12.4}$$

$$FA = \frac{m_f}{m_a} \tag{12.5}$$

$$\eta_v = 0.85 \tag{12.6}$$

Load Variation: The load is changed by varying the volume of fuel injected, as is done in practice. The maximum value of the fuel injected is determined by the limiting fuel air ratio. The speed range was specified for typical values of an engine of this size (reasonable piston speed):

$$m_f = V_f \cdot \rho_f \tag{12.7}$$

$$V_{f,min} \leq V_f \leq 3.26 \cdot 10^{-8} \ m^3 \tag{12.8}$$

$$1000 \leq N \leq 3000 \tag{12.9}$$

Engine Output: Given the indicated efficiency and friction assumed above, the basic engine performance is determined from the power equation, where the *imep* is from Equation (1.37):

$$imep = \frac{m_f \cdot H_u \cdot \eta_i}{V_d} \tag{12.10}$$

$$bmep = imep - fmep \tag{12.11}$$

$$\eta_m = \frac{bmep}{imep} \tag{12.12}$$

$$bsfc = \frac{3600000}{H_u \cdot \eta_i \cdot \eta_m} \tag{12.13}$$

Basic parameters: Some basic parameters are needed:

$$\begin{aligned}
B &= 0.090 \ m \\
S &= 0.095 \ m \\
V_d &= \frac{\pi}{4} \cdot B^2 \cdot S \ m^3 \\
\rho_f &= 850 \ kg/m^3 \\
\rho_{in} &= 1.2 \ kg/m^3
\end{aligned} \tag{12.14}$$

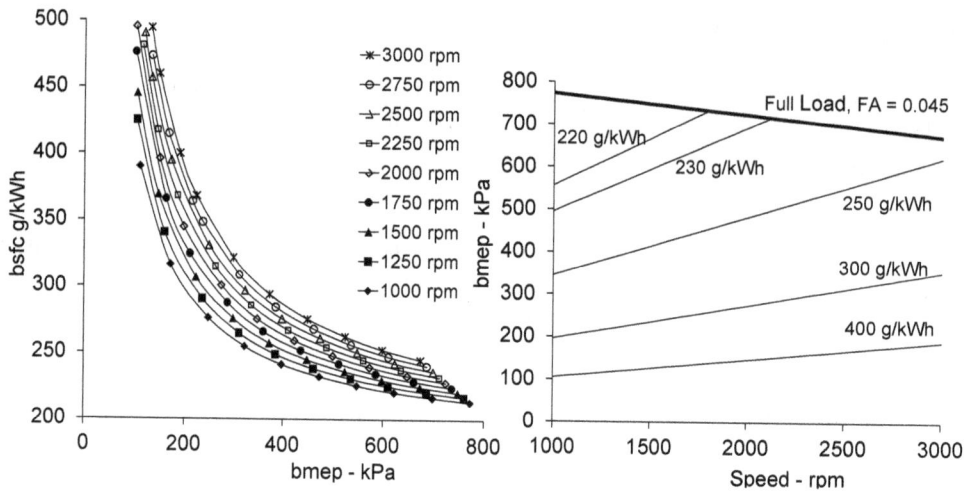

Figure 12.10: Fuel consumption curves (upper) and the resulting engine map (lower) for the simple engine model.

The $bsfc-bmep$ curves and the resulting engine map for the above example are shown in Figure 12.10. The friction is only a function of speed, and so as a result, the mechanical efficiency decreases as the load decreases at a given speed, and the $bsfc$ approaches infinity at low loads. This is more apparent from the part load fuel consumption curves, but is also seen on the engine map by the larger numerical differences between the roughly equidistant $bsfc$ curves as the $bmep$ approaches zero. This is a feature seen on every piston engine map.

The second feature to be noted in Figure 12.10 is that on the engine map, the lines of constant $bsfc$ have a positive slope as the speed increases. Said another way, at a constant $bmep$, the $bsfc$ increases as the speed increases. This again can only be caused by the mechanical efficiency in the present example, and is a reflection of the fact that the engine $fmep$ increases as the speed increases. This feature is also a characteristic of piston engine maps.

12.3.2 Factors Modifying the Simple Model

These features are the basis for all engine maps. There are many factors that will cause the map for any given engine to have a different appearance, but the features above will always be there and form the basis. The variations will be due to many factors, most of which affect the indicated efficiency as the operating point changes. Some factors are:

Heat losses

In all operating engines, there are heat losses to the combustion chamber walls. These are a function of the speed, load and mixture ratio, and they will have an effect on the indicated efficiency. Typically, the fraction of heat lost to the walls decreases as the engine speed and load increase. Heat loss effects are often responsible for the slope of the constant $bsfc$ curves on the map being negative at low speeds, as friction is low, and a significant increase in indicated efficiency is often observed, due to lower relative heat loss with increasing speed. At higher speeds, friction becomes more important, and the slope of the curve becomes positive as shown above. This normally results in the best efficiency being found at a speed between the lowest speed and that of maximum torque.

Combustion Timing

For reasons of knock or emissions control, the spark timing in SI engines may not be the MBT value. Similarly, in CI engines, injection timing may be adjusted to lower exhaust emissions or maintain maximum cylinder pressure limits. With computer control of engines, there is a great deal of flexibility here, and this can have an effect on the local nature of the engine map. These changes affect the indicated thermal efficiency.

Mixture Ratio

Especially at full load, SI engines are often set to run rich, resulting in higher power, but also higher fuel consumption. Very lean mixtures can affect the thermal efficiency in a positive way, if combustion rates can be maintained. The fuel air ratio in CI engines is a function of load, and is not an independently controllable variable as is the case with the SI engine.

Pumping Losses

In homogeneous charge spark ignition engines, the portion of friction loss due to pumping increases as the load decreases, this increases fuel consumption at part load. This can not readily be seen from the shape of the engine map, but affects the values. A diesel engine is normally not throttled, so pumping losses are small, even with turbocharged engines, where intake and exhaust pressure follow each other through the load range.

Combustion Duration

Diesel combustion at high loads often takes longer than at lower loads in terms of crank angle degrees, and often is reflected in increased fuel consumption of diesel engines at full loads. The use of turbocharging helps this, as will be seen in maps to follow.

The above are some of the more important factors affecting the shape and values on the fuel consumption on the engine map. Other factors can be encountered.

Some of these influences can be seen from an engine map made from the example engine in Chapter 11. This engine has some elements in common with the engine model used to generate Figure 12.10. These are a constant friction mean effective pressure and a limiting fuel air ratio. Two major differences are that the indicated thermal efficiency of the engine in Chapter 11 was assumed to be a function of the fuel air ratio, and that the air flow process was modified by the turbocharger, waste gate and charge air cooler.

The map for the example engine of Chapter 11 is shown in Figure 12.11 for speeds between 1000 and 3000 rpm. The first difference to be seen is the maximum *bmep* curve. The maximum *bmep* is larger than for the naturally aspirated engine of the simple model, and increases with speed as the turbocharger output increases. At a speed of about 1700 rpm, the *bmep − rpm* curve flattens out some as the waste gate opens. In accordance with the higher intake pressure of the turbocharged engine, the maximum *bmep* is much higher at the same fuel air ratio because more fuel is added per cycle. In an actual engine, the fuel injected per stroke would decrease at higher speeds, to give the torque rise discussed earlier.

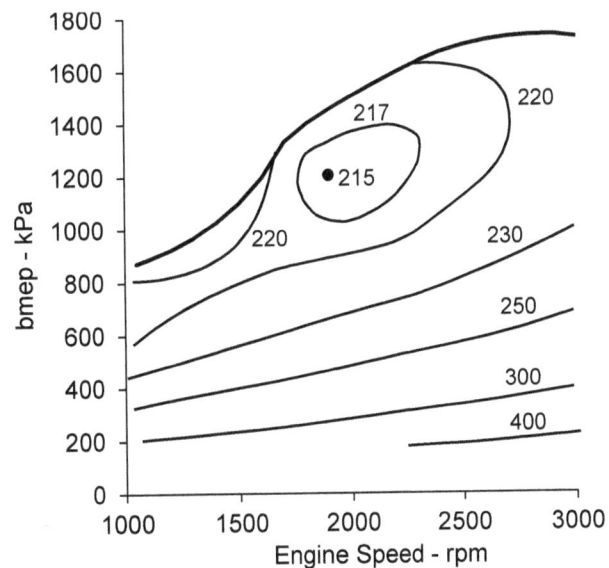

Figure 12.11: The engine map for the simulated 10 liter DI diesel engine with turbocharger and waste gate (*bsfc* in g/kWh).

Regarding fuel consumption, at light loads, the *bsfc* curves are essentially linear, similar to the naturally aspirated simple engine. At higher loads (higher FA), the *bsfc* reaches a minimum before full load is achieved, and then increases again as the load approaches its maximum value. At high loads, the mechanical efficiency is nearly constant, and the indicated thermal efficiency plays an important role in determining the shape of the $bsfc$ contours on the map. Note that above about $bmep = 600\text{kPa}$ there is not much change in the value of the $bsfc$, the values of the lines being close to each other. At the lower speeds, the same behavior is observed, but cannot be seen on the map because of the choices of $bsfc$ levels shown. At 1200 rpm, for example, a minimum $bsfc$ of about 219 g/kWh is found for $bmep$ between 700 and 750 kPa, after which it increases to 224 g/kWh at the maximum $bmep$ of 924 kPa.

Other factors, not included in the model illustrated would be the effect of engine speed on the percentage heat lost in-cylinder (η_i) and the reduction of η_i with speed due to mixing limitations in the combustion process. These effects will make actual CI engine maps differ from this example. See Figure 12.9 for example.

In practice, an engine map is normally determined by testing the engine in a series part load tests for the entire speed range of the engine. The map is then constructed by identifying the operating speeds and loads at which specific fuel consumption values are found, and then plotting them on the torque-speed axes.

The engine map represents the operation of the engine over its complete operating range in terms of speed and load, and the fuel consumption of the engine can be determined for any operating point. Since the power of the engine is a function of the speed and the torque (mean effective pressure) the power can be determined at any operating point, as shown in the following equation.

$$BP = \frac{2\pi TN}{60000} = \frac{bmep \cdot V_d \cdot N}{60000x} \tag{12.15}$$

where: BP = brake power, T = engine torque in N-m, N = engine speed in rpm, $bmep$ = brake mean effective pressure in kPa, V_d = engine displacement volume in liters, $x = 1$ for a two stroke engine and $x = 2$ for a 4 stroke engine.

The fuel consumption is then simply calculated using from the $bsfc$:

$$\dot{m}_f = BP \cdot bsfc \tag{12.16}$$

Therefore, according to Equation (12.15) lines of constant power can be represented on the engine map by hyperbolic curves. This reminds us that a given engine power may be obtained by different combinations of speeds and loads, within the confines of the engine map. At part load, there is a large degree of freedom, and it is the job of the applications engineer to choose the engine/vehicle system which offers the optimum performance of the vehicle in terms of fuel economy and vehicle performance.

The full load curve and a typical set of experimental part load test curves are shown in Figure 12.12 for a conventional 1.1 liter SI engine. This engine operates at stoichiometric conditions except at full load, where the fuel air ratio is increased to give maximum power. The test data were taken for given manifold pressures at the various speeds. Specified values of torque are also commonly used in engine testing.

The engine map resulting from these data is also shown in the lower half of Figure 12.12. The fuel consumption lines on the map are nearly equally spaced geometrically, but the part load curves emphasize that the difference in values between these lines increases as the load decreases. A curve of constant power of 15 kW has been drawn on the engine map in Figure 12.12.

There are several important features with engine maps. The first is the full load curve. For any given engine, there is a full load condition at a given speed above which the engine can not be operated. This is the same as the torque curve in the full load performance curves shown above, even though it may be given in terms of bmep. It is an important feature of any engine, and is often tailored to meet vehicle specifications. The influence of the torque curve on driving characteristics of vehicles will be described in more detail in Chapter 15.

For spark ignition engines, which normally are not turbocharged, the breathing characteristics discussed in Chapter 7 play a dominant role in determining the torque curve. As shown there, the curve is typically rounded in nature, with a maximum torque in the vicinity of half the speed at which maximum power is achieved. This is predominantly determined by the valve timing of the engine. In modern gasoline engines, more extended use is being made of tuned manifolds, and the torque curves often have peaks where gas dynamic resonance increases or decreases the volumetric efficiency relative to that of a non-tuned engine. In addition, turbocharged or supercharged engines may have flatter torque curves, or

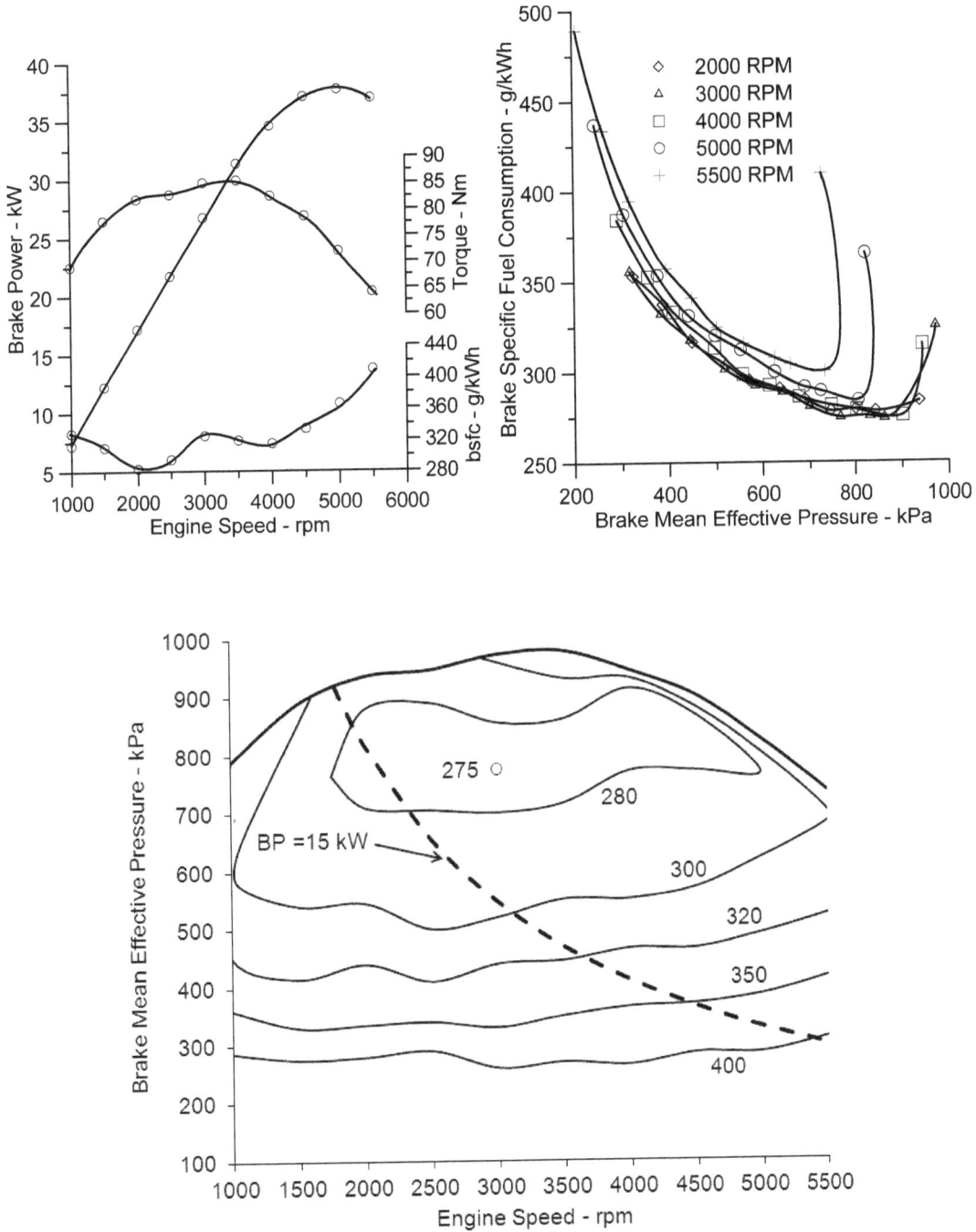

Figure 12.12: The full load curve and a series of part load fuel consumption (g/kWh) test results for a 1.1 liter spark ignition engine (upper), and the resulting engine map (lower). A line of constant power of 15 kW is shown in the engine map.

sharply rising torque curves at low speeds with the curves flattening out at high speeds, due to the limiting of the maximum cylinder pressure. The latter is common for turbocharged engines, where the turbocharger is not effective at low speeds but the turbocharger output must be restricted with a waste gate or another device at high speeds to prevent overloading.

The general shape of the fuel consumption curves is similar to that of the simple model. The biggest difference is at the low speeds, where heat loss effects play a larger role in actual engines. With any engine, the brake specific fuel consumption rises sharply as the load is decreased for all speeds. The changes in indicated efficiency are small with respect to the relative changes in mechanical efficiency in the low load area.

With this in mind, it can be seen that it is always a good idea in terms of efficiency, to operate an engine at a high degree of loading. However, another condition that is usually observed on engine maps is that the fuel consumption increases as operation comes close to the full load condition. This arises from different causes in SI and diesel engines. In spark ignition engines, it is due to a richer mixture for the full power condition, for slightly greater power than a stoichiometric mixture. The decreased oxygen also protects the engine components from oxidation at the high exhaust temperatures found with full load. This strategy is normally applied only for loads greater than approximately 85 % of full load, and results in an increase in fuel consumption. The increase in $bsfc$ is typically on the order of 10-15% of the minimum value at a given speed. This can best be seen at the highest load in the part load curves in the upper portion of Figure 12.12.

The fuel consumption also tends to increase near full load with diesel engines as well. However, in this case, it is usually a deterioration of the combustion process which is the cause. Typically, the fuel injection period and subsequent second phase of the combustion process become longer as the load is increased in the direct injection diesel engine. The energy released from the last portion of the fuel to burn then cannot be expanded through the entire expansion stroke of the engine, causing a decrease in efficiency/increase in $bsfc$. This last portion of the fuel must be burned in a condition with lower oxygen concentration and in falling temperatures and pressures. As a result, the particulate emissions/smoke produced also increase. Historically, the torque curves of diesel engines have been determined at least to some degree, by a smoke limit, that is, the maximum power is restricted by limiting the amount of fuel injected to an amount which does not produce more than a given amount of soot. Turbocharging now supplies extra air and full load performance for modern turbocharged engines is improved relative to non-turbocharged engines, though the increase in $bsfc$ near full load is still commonly seen, as will be shown in subsequent engine maps. The use of common rail injection with variable injection pressure also helps to improve full load performance.

In either engine type - SI or CI, the best efficiency at a given speed is normally obtained by operating the engine at a high load, but not completely at full load.

12.4 Spark Ignition Engines

Figure 12.12 shows the engine map for a 1.1 liter spark ignition engine, and some of the part load fuel consumption curves from which it is obtained. Using reasonable values for engine parameters, one can obtain a representative value expected for the maximum $bmep$ of a 4 stroke spark ignition engine without turbocharging. If we assume standard atmospheric conditions, a volumetric efficiency $\eta_v = 0.9$, as the engine has a tuned manifold, an indicated thermal efficiency $\eta_i = 0.36$ (0.8 times the ideal fuel air cycle efficiency at compression ratio of 9.5:1), a friction mean effective pressure = 180 kPa (from Heywood's correlation at 3000 rpm), a stoichiometric fuel air ratio FA_s =0.068 and a heating value $H_u = 44500$ kJ/kg, we obtain for the $bmep$:

$$bmep = \eta_v \cdot \rho_{in} \cdot FA \cdot H_u \cdot \eta_i - fmep = 0.9 \cdot 1.2 \cdot 0.068 \cdot 44500 \cdot 0.36 - 180 = 996 kPa$$

This is in reasonably good agreement with the values shown in Figure 12.12.

Figure 12.13 shows an engine map for an SI engine with and without turbocharging. The fuel consumptions are similar throughout the engine operating range. The minimum fuel consumption of the naturally aspirated engine is slightly lower than that of the turbocharged engine, indicating that the effects of reducing the compression ratio and retarding spark timing to avoid knock are slightly more

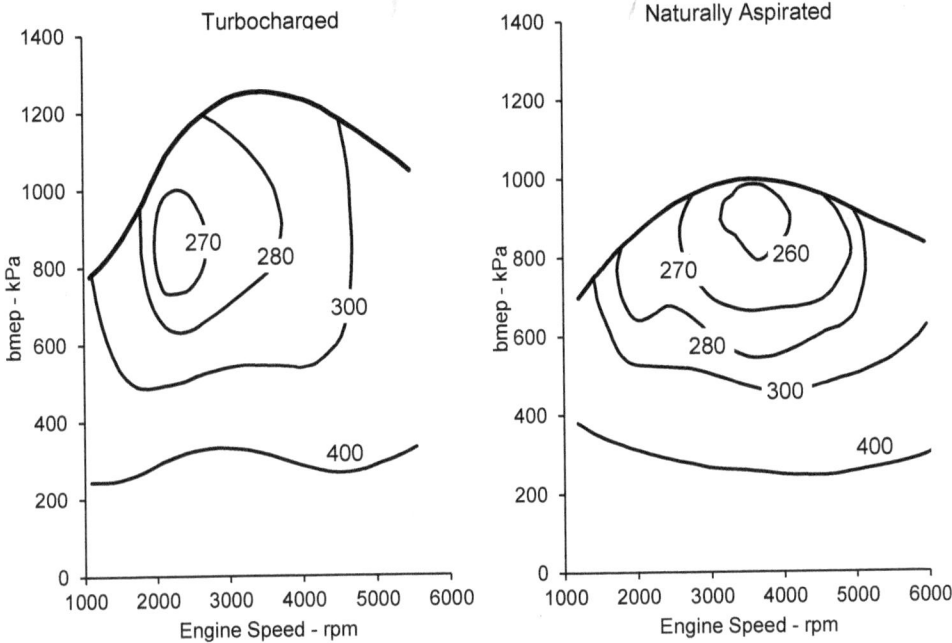

Figure 12.13: Engine maps of a spark ignition engine in turbocharged and naturally aspirated forms. The results are scaled to the torque and speed specifications of the engine in Figure 12.2, using normalized characteristics of [104]. Fuel consumption in g/kWh.

important than better mechanical efficiency. The main benefit of turbocharging the SI engine is then an increased power density, and there is little, if any, advantage with regard to fuel consumption.

An alternative form of pressure charging used with spark ignition engines is that of a positive displacement compressor or blower driven by the engine crankshaft, or supercharger, as shown in Section 11.2. This has two advantages compared to turbocharging. The first is there is no delay associated with the compressor operation, as it is directly coupled to the engine crankshaft. The second is that since it is a positive displacement device, as is a piston engine, the air flow characteristics are similar. A blower will increase intake pressure if its volumetric flow is larger than that of the engine. The pressure increase can be regulated by both the size of the compressor and its speed relative to the engine. This can be seen be equating the air flow through the compressor to that of the engine:

$$\eta_{v,c}\rho_o\frac{V_cN_c}{60} = \eta_{v,e}\rho_{int}\frac{V_dN_e}{120} \tag{12.17}$$

Solving for the intake manifold pressure:

$$p_{in} = p_o\frac{\eta_{v,c}}{\eta_{v,e}}\frac{T_{atm}}{T_{int}}\frac{V_c}{V_e}\frac{N_c}{N_e} \tag{12.18}$$

The volumetric efficiencies change slightly with operating conditions. The engine intake temperature is more important, as it increases with increasing compression. Therefore, an intercooler is usually used to reduce the increase in intake temperature with supercharging. Equation (12.18) indicates that for a given speed ratio between the blower and the engine, the intake pressure should be roughly constant at all engine speeds. This means that there will be a high intake pressure at low speeds, something not seen with the turbocharged engine. This can be an advantage regarding low speed torque, but extra care must be taken to avoid knocking, especially at low speeds.

Another consideration is that at part load, the compressor is not needed to supply the extra air to the engine. It is very difficult to mechanically couple and de-couple the compressor at operating speeds, so in practice a air by-pass system for the compressor air is used. This does not avoid the main drawback of the compressor, and that is the mechanical power used to drive the device. The Roots blower type of compressor is not very efficient at high pressure ratios, but for the SI engine, this is not a critical problem, as the amount of pressure increase is limited by knock, thus maximum intake pressures (and torques) are similar in turbocharged and supercharged SI engines.

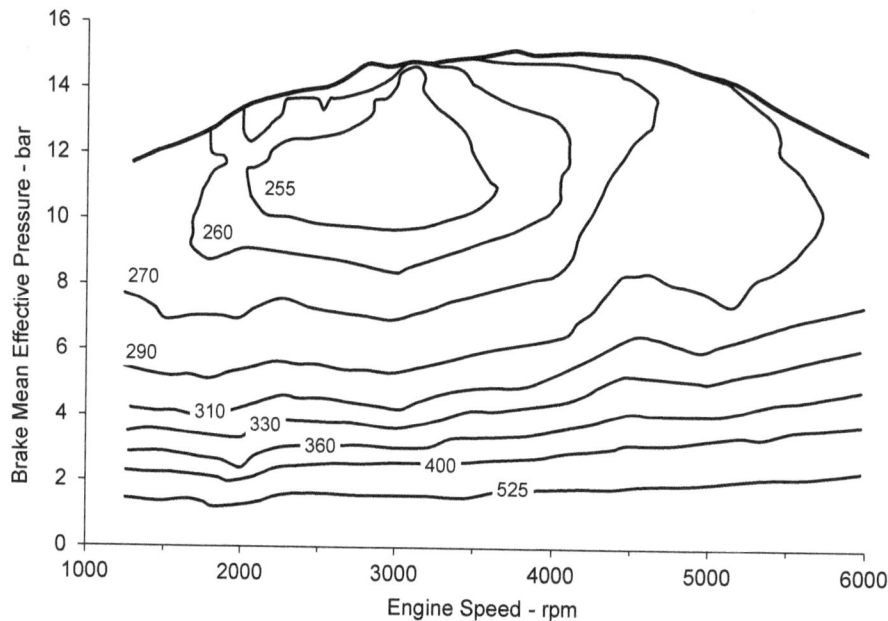

Figure 12.14: Engine map for 2 liter, 4 cylinder SI engine with a Roots type compressor. Adapted from [109]. Fuel consumption in g/kWh.

Comparing to Figure 12.13, it can be seen that the maximum *bmep* at speeds below about 3000 rpm for the engine with the positive displacement compressor (supercharged) is higher than that of the turbocharged engine. This is a substantial advantage with the use of the compressor, as its flow characteristics match that of a piston much better than the turbo charger. The fuel consumptions for the supercharged engine are shown to be lower than those of the turbocharged version, but the supercharged engine is also much newer. For similar versions, the supercharged engine does not exhibit a significant fuel consumption advantage, and is most likely slightly worse.

12.5 Diesel Engines

12.5.1 Indirect Injection Diesel Engine

For many years, it was only possible to operate small, high speed diesel engines required for light duty vehicle operation by using a divided chamber diesel engine. Either a symmetrical prechamber, or a swirl generating chamber with tangential entering air is used for the injection of fuel and the initiation of combustion. This initial combustion increases the prechamber pressure and forces the burning gasses and fuel into the main combustion chamber on the top of the piston where this burning jet creates a lot of mixing, and combustion can be completed with the remaining air in the main chamber. This process requires a later timing than in the DI engine, in order to avoid reverse flow of combustion gasses back into the prechamber on compression. This has a negative influence on the indicated efficiency (see the discussion in Section 2.1). While this type of combustion system gave low smoke operation at higher speeds, some losses are associated with increased heat transfer because of the flows through the connecting passage between the two chambers, and the impact of the burning jet on the piston. Thus, even though a compression ratio of over 20 is needed for these engines, they have a thermal efficiency lower than direct injection (DI) diesel engines operating with a single combustion chamber with fuel directly injected into the single combustion chamber.

The prechamber/swirl chamber engines, commonly called indirect injection (IDI) diesels do have an advantage over spark ignition engines because of the higher compression ratio and an additional advantage at part load due to the absence of intake throttling, an advantage shown by all diesel engines. Their efficiencies lie between those of SI engines and direct injection diesel engines. In recent years, developments in computer control and injection systems have all but replaced IDI engines with DI engines for road transport.

Naturally Aspirated Engine

An engine map for a naturally aspirated IDI diesel engine is shown in Figure 12.15. First of all, it can be seen that the maximum *bmep* of the naturally aspirated or non-turbocharged (NA) diesel engine is lower than the spark ignition engine of Figure, 12.12which is of a similar displacement. Typically for the prechamber engine, the fuel air ratio is limited by smoke to a value of about 0.055. In addition, the thermal efficiency will be increased compared to the SI engine, due mainly to the higher compression ratio, but the compression ratio improvement is reduced somewhat by the later, less efficient combustion and the higher heat loss. Experience shows that an indicated thermal efficiency of around 0.40 - 0.43 is not unreasonable for an IDI engine. The volumetric efficiency will be of a similar magnitude to that of the SI engines with similar speeds. The *fmep* of this engine type is higher than other diesel engine types due to higher compression ratio and pumping of the air in the prechamber. Heywood's correlation, Equation (9.13), predicts an *fmep* of about 315 kPa at 3000

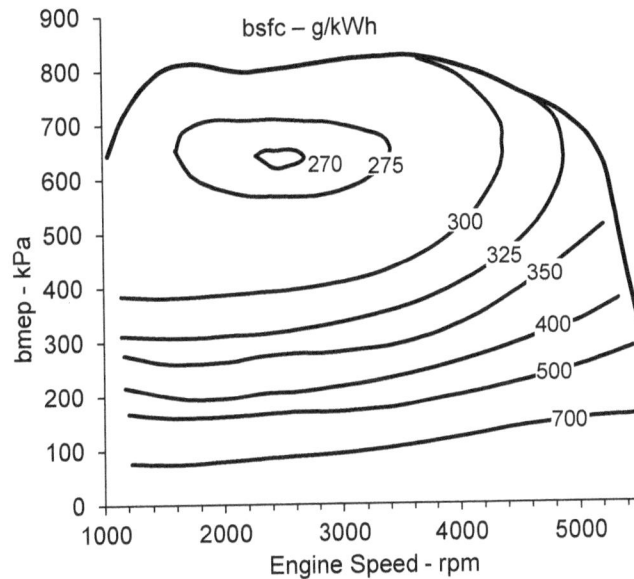

Figure 12.15: Engine map for a 4-cylinder naturally aspirated IDI diesel engine, with a displacement of 1.47 liters. Fuel consumption in g/kWh, adapted from [11]

rpm. This is higher than that for the DI engine because of the extra work pumping the air in and out of the prechamber. Using these values, and estimate of the maximum *bmep* for the naturally aspirated IDI engine gives:

$$bmep = \eta_v \cdot \rho_{in} \cdot FA \cdot H_u \cdot \eta_i - fmep = 0.90 \cdot 1.2 \cdot 0.055 \cdot 42500 \cdot 0.43 - 315 = 770 kPa$$

This value is reasonably close to that of Figure 12.16. An estimate of the *bsfc* gives:

$$bsfc = \frac{3.6 \cdot 10^6}{H_u \cdot \eta_i \cdot \eta_m} bsfc = \frac{3.6 \cdot 10^6}{42500 \cdot 0.43 \cdot \frac{770}{770 + 315}} = 277 g/kWh$$

which is also close to the value in Figure 12.16. At part load, the efficiency difference between this engine and the SI engine should be greater, but it is difficult to make accurate comparisons, since the *bsfc* - *bmep* curves become quite steep at part load.

Turbocharged Engine

As expected, Figure 12.16 shows the turbocharged IDI engine to give a higher *bmep* than the naturally aspirated engine, due to the extra air which makes the combustion of more fuel possible. At low speeds, the increase in *bmep* is modest, due to the characteristics of the centrifugal compressor used in the turbocharger. In this device, which converts kinetic energy into static pressure, the pressure is proportional to the velocity (speed) squared, and thus the pressures achieved by turbocharging are much higher at higher speeds. This same characteristic was seen in the case the turbocharged SI engine above. The turbocharged engine in Figure 12.16, is equipped with a waste gate to maintain a maximum allowable

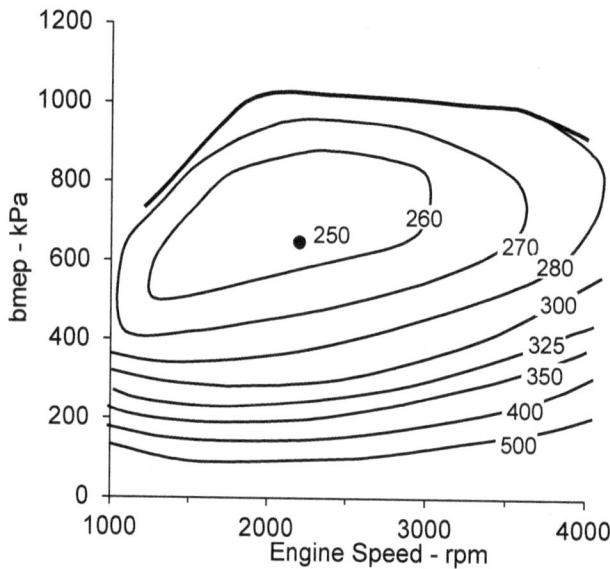

Figure 12.16: Engine map for a 4-cylinder turbocharged indirect injection industrial diesel engine, with a displacement of 1.6 liters. Fuel consumption in g/kWh.

intake pressure for the purpose of controlling maximum cylinder pressures. The torque falls off at higher speeds due to the increase in the frictional power as the engine speed increases, and possibly due to a lower fuel air ratio, as combustion becomes more difficult at higher engine speeds. The minimum $bsfc$ of the turbocharged engine is lower, since indicated thermal efficiency and the $fmep$ are about the same, but the brake power larger due to the extra air and fuel. This produces more power for the same mechanism, giving a better mechanical efficiency of the engine. This particular engine was designed for industrial application, and has a lower maximum $bmep$ than more modern turbocharged DI engines for vehicles, and does not show the torque rise more commonly seen with vehicular engines.

12.5.2 Direct Injection Diesel Engine

Naturally Aspirated Engine

The most efficient engines in operation today are direct injection diesel engines. They have the advantage of a high compression ratio, typically 15-18, and a single compact combustion chamber with low heat losses. In a non-turbocharged DI engine, the first portion of the combustion (about 30% of the fuel) occurs close to constant volume, giving an improvement over the prechamber engine, where the combustion is generally later and slower. This gives a higher indicated thermal efficiency for the DI engines. On the other hand, compared to IDI engines, it is more difficult to limit the smoke at higher loads and speeds, and the maximum fuel air ratios are somewhat lower.

One can estimate that the indicated thermal efficiency of the DI engine will be about 1.25 times that of the stoichiometric SI engine with a compression ratio of 9:1 based on equilibrium cycle analysis (higher compression ratio, leaner mixture). This gives an indicated thermal efficiency of about 0.45, which is typical. According to Heywood [7], the DI engine has a lower friction than the IDI engine, by about 70 kPa at the same speed. Because of the nature of the DI combustion process, it is necessary to reduce the maximum fuel air ratio to about 0.045 at full load to limit smoke. The volumetric efficiency of the DI engine would be slightly lower than the IDI engine, since the swirl needed for good combustion is generated by causing the intake air to rotate. Using these assumptions, the estimated maximum $bmep$ of the naturally aspirated DI engine at 2000 rpm becomes:

$$bmep = \eta_v \cdot \rho_{in} \cdot FA \cdot H_u \cdot \eta_i - fmep = 0.85 \cdot 1.2 \cdot 0.042 \cdot 42500 \cdot 0.45 - 170 = 738 kPa$$

$$bsfc = \frac{3.6 \cdot 10^6}{H_u \cdot \eta_i \cdot \eta_m} bsfc = \frac{3.6 \cdot 10^6}{42500 \cdot 0.45 \cdot \frac{738}{738 + 170}} = 231 g/kWh$$

Both of these values agree fairly well with the values shown in Figure 12.17 For a naturally aspirated diesel engine. The *bmep* is about the same as a naturally aspirated IDI engine. The effects of a lower maximum fuel air ratio are compensated for by the higher thermal efficiency and lower friction. The former is due to a more advantageous timing of combustion and lower heat losses. The *bsfc* improvement is due to the better indicated thermal efficiency and slightly lower friction. Note that the maximum piston speed is about 11 m/s for the DI engine, lower than the maximum value of about 14 m/s for the IDI engine. Thus, compared to the naturally aspirated IDI engine, the naturally aspirated DI engine will have a similar torque, but a lower power, since the maximum speed will be less. This was another reason for using IDI engines for road vehicles before the advent of the small, turbocharged DI engines. The engine shown in Figure12.17 is an older model, used for off-road applications, and smoke limits are shown. Acceptable smoke limits were used to define maximum torques of these engines.

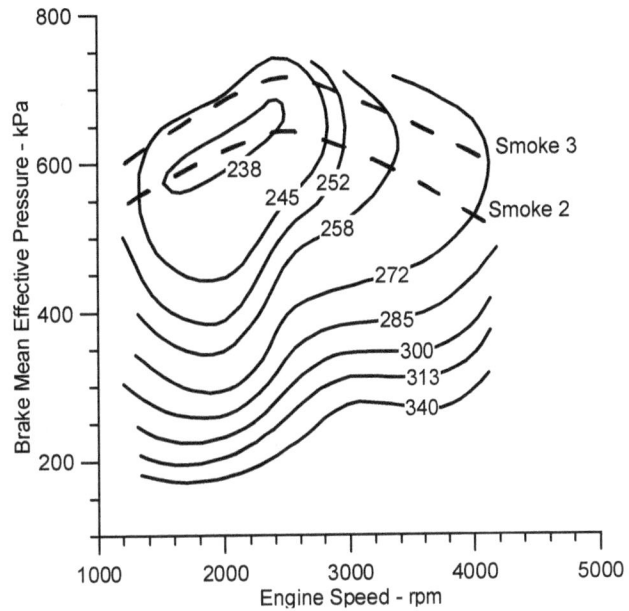

Figure 12.17: Engine map for a 2-cylinder, 1 liter naturally aspirated direct injection diesel engine, bore = stroke = 85 mm. Fuel consumption in g/kWh.

Turbocharged Engine

One way to maintain the high efficiency of the DI engine and to compensate for the lower power is to make use of turbocharging, and a large percentage of the DI diesel engines produced today use turbocharging, often combined with charge cooling in order to increase intake air density, reduce the thermal loading on the engine, and improve exhaust emissions.

The characteristics for a large, turbocharged and intercooled, direct injection diesel engine is shown in Figure 12.18. For this engine, the maximum *bmep* is about 1600 kPa, as compared to the naturally aspirated value of about 740. This will give a significant increase in the mechanical efficiency. If we assume that a naturally aspirated engine has an *fmep* of 170 kPa, and a *bmep* of 740 kPa, this corresponds to a mechanical efficiency of 0.813. Assuming the *bmep* increases to 1600 kPa through turbocharging, and the friction increases with 10% due to higher cylinder pressures, the new mechanical efficiency is:

$$\eta_m = \frac{bmep}{imep} = \frac{bmep}{bmep + fmep} = \frac{1600}{1600 + 187} = 0.895$$

Assuming no change in the indicated thermal efficiency, compared to the naturally aspirated engine, the minimum *bsfc* would be improved:

$$bsfc_{min} \approx 220(0.81/0.895) = 200 g/kWh$$

This is in good agreement with the values shown in Figure 12.18, and shows the advantages of turbocharging on both power and efficiency (as well as the dangers of assuming a constant mechanical efficiency when correcting for changes in engine conditions!!).

It is for the reason of this very low fuel consumption that DI engines have become the completely dominant form of power plant for applications where fuel costs are a large part of vehicle operation. In recent years, technology has improved considerably, and the benefits of direct injection operation are now being realized in the small high speed engines needed for passenger vehicle operation, giving an even greater reduction in fuel consumption than the IDI engine.

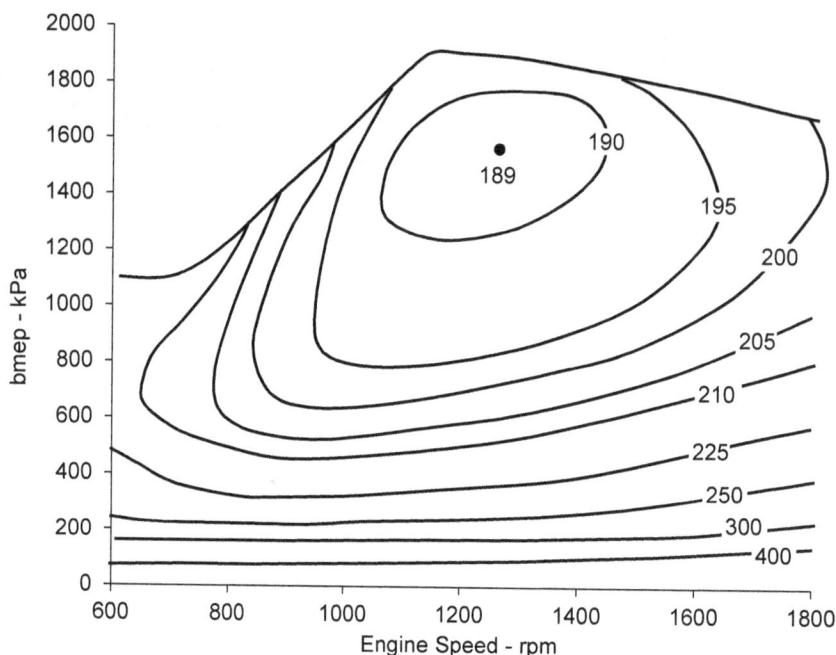

Figure 12.18: Engine map for a turbocharged DI diesel engine for heavy duty application with a displacement of 12 liters. Fuel consumption in g/kWh.

A second example of an engine map for a turbocharged, intercooled, direct injection diesel engine is shown in Figure 12.19. This is the engine for which the part load curves were shown in Figure 12.8. Of special interest is the area at intermediate speeds between 2000 and 3000 rpm and part load conditions for which EGR has been applied to reduce NO_x emissions in order to help meet exhaust emission standards. When compared to some of the previous maps, it can see that in this region, fuel consumption is increased, and the constant specific fuel consumption lines are not as simple. The extent to which legislative emissions requirements affect fuel consumption depends on the specific engine and/or vehicle application, and the choice of emission control technology, as well as the fundamental design of the engine and control system. Since engines historically were developed with economy as the main focus, emission control most often results in an increase in fuel consumption.

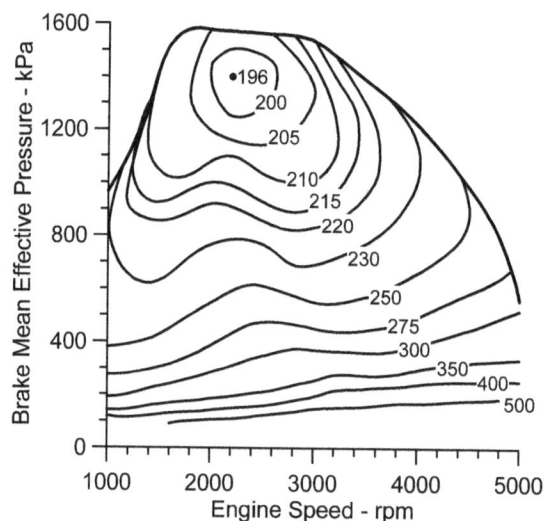

Figure 12.19: Engine map for a 2 liter, 4 cylinder liter turbocharged DI diesel engine for light duty application. Fuel consumption in g/kWh.

12.6 Two Stroke SI Engines

In applications where a cheap, light engine is needed, and fuel price is not too important, two stroke engines are used widely. In terms of vehicle applications, this has been in the area of two wheel vehicles, although this application has come under intense pressure, since this kind of engine normally has an extremely high emission of unburned fuel. Figure 12.20 shows the engine maps for one of these engines,

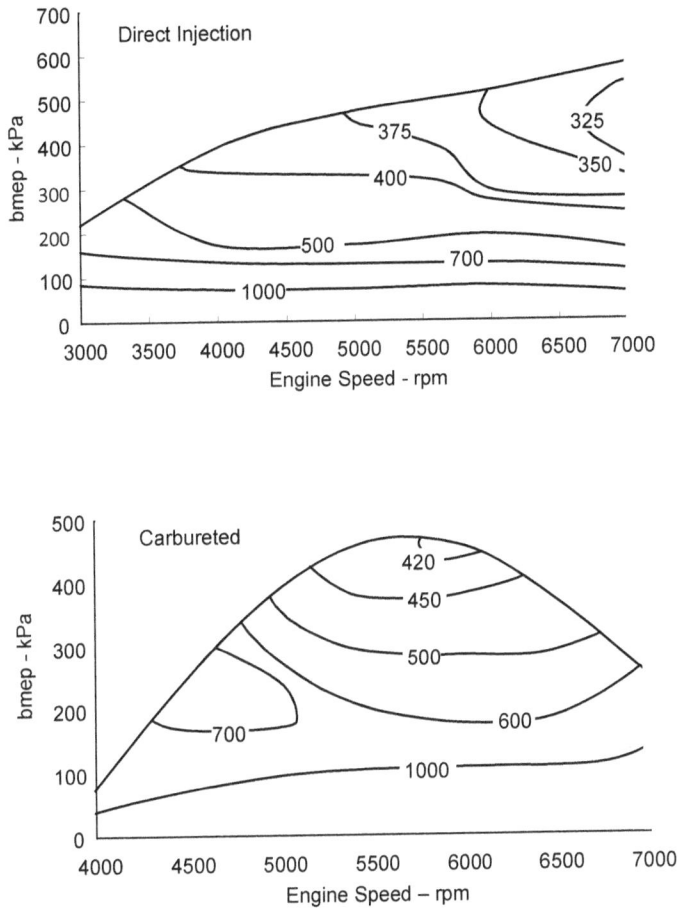

Figure 12.20: Engine maps for two versions of a cross-scavenged two cylinder two stroke spark ignition engine with displacement of 50 cc, and an effective compression ratio of 6.2:1. The version with a carburetor is the lower map, and the version with in-cylinder fuel injection the upper. Fuel consumption in g/kWh.

and a potential solution to the emissions problem. The engine with the carburetor, the lower set of curves, is typical of engines in application for mopeds or scooter, where emissions standards are lenient or not yet in force. It can be seen that at lower speed the *bmep* is very low for this engine. This is due to the low efficiency of the air exchange operation. The old exhaust gasses are pushed out by (and mixed with!) the incoming charge of fuel and air and as a result, up to 35% or so of the gasses in the cylinder at the time of combustion consists of residual gasses which do not contribute to the combustion process. A rough estimate of 0.65 times the maximum *bmep* of 900 kPa for a 4 stroke engine, gives a *bmep* of 585 kPa, which would be further lowered by the typically low effective compression ratio of these engines. This is close to the value shown in Figure 12.20.

In addition, on the order of 1/3 of the gasses flowing through the carburetor can escape directly to the exhaust pipe during the scavenging process, and never be trapped in the combustion chamber. This is the primary cause of the high HC emissions of the carbureted two-stroke engine, and a big contributor to the very high *bsfc* of this engine, with a minimum value of 375 g/kWh shown in Figure 12.20. A solution to this is the use of a special fuel injection system which injects the fuel directly into the cylinder, where it is mixed with the air, and not lost to the exhaust. This is the configuration shown in the "Direct Injection" engine. In this small engine, the direct injection provides more power at a lower fuel consumption.

For a larger engine, a similar comparison was shown in Figure 8.10. The fuel injection engine ("2 Stroke DI") does not show more power, since the amount of air available for burning fuel is not increased by the fuel injection system. On the other hand, the only fuel supplied is that needed to combine with the available air trapped in the combustion chamber. Thus the minimum bsfc of the engine 2 stroke

DI engine is 280 g/kWh, close to that of conventional 4 stroke spark ignition engines. The difference in the fuel consumption between the carburetted ("2 Stroke loop") and injected versions of the engine is essentially the HC emission of the engine, in this case on the order of 150k/kWh, greatly in excess of the 4 stroke spark ignition engine, totally unacceptable in modern engines.

The engine map for a small 2-stroke diesel engine operating on dimethyl ether was shown in Figure 8.11.

12.7   Problems

Problem 12.1

Determine the amount of fuel injected per cycle in mm³ for the engine in Figure 12.8 as a function of $bmep$ for both cases. Assume a fuel density of 0.84 g/cm³. A: $V_f \cong 31.2mm^3$, B: $V_f \cong 29.3mm^3$

Problem 12.2

Convert the indicated specific fuel consumption in 12.7 to indicated thermal efficiency. Determine the friction mean effective pressure and the brake power of the engine in Figure 12.7 for maximum load. ($\eta_i = 0.544 \rightarrow 0.367, fmep = 274, BP = 13.7$ kW).

Problem 12.3

Compare the exhaust temperature as a function of fuel air ratio for the engines of Figures 12.7 and 12.8 at 2400 rpm to that of Figure 6.19.

Problem 12.4

Use the model of Problem 5.8 and construct an engine map for that engine from 1000 rpm to 6000 rpm.

Chapter 13

Forces and Balancing

13.1 Forces in a Single Cylinder Piston Engine

For the determination of the inertia forces in piston engines, simplifying assumptions are often used, such that irregularities in the motion of the piston are ignored. The motion of the connecting rod can be seen as a combination of the linear motion of the center of gravity and of the rotational movement on an axis through the center of gravity. The moments associated with this assumption are usually neglected.

To simplify the calculation of the inertial effects, the crank mechanism is simplified to an idealized system which dynamically corresponds to the real system. The effective mass of the crank arm referred to the radius of the crankshaft, r, is:

$$m_c = m_{cp} + 2m_a \cdot \frac{r_1}{r} \qquad (13.1)$$

where: m_{cp} is the mass of the crank pin, m_a is the mass of the crank arm and r_1 is the distance from the main bearing axis to the center of gravity of the arm.

The mass of the connecting rod, m_{cr} is considered to be referred to the ends of the connecting rod, such that the center of gravity is maintained at the same location. Then:

$$m_{cro} = m_{cr} \cdot \frac{s_r}{L} \qquad (13.2)$$

$$m_{crr} = m_{cr} \cdot \frac{s_o}{L} \qquad (13.3)$$

where: m_{cro} is the portion of the mass referred to the small end of the connecting rod (oscillating), m_{crr} is the portion of the mass referred to the large end of the connecting rod (rotating), s_o = the distance from the center of gravity to the small end of the connecting rod and s_r the distance from the center of gravity to the large end of the connecting rod.

The center of gravity can be found by calculation, for example in a CAD program, or if drawings are not available, by a weighing procedure. The rod is weighed in two positions, the first measurement being the weight of the entire connecting rod, F_t and the second measurement one in which one end of the rod is supported on the scale, and the other end of the rod is supported on a fixed mount at the same horizontal height as the scale platform. If the weight on the scale in the latter case is F_2, and the total length of the connecting rod, L, then a balance of moments around the center of gravity of the connecting rod gives:

$$s = L \cdot \frac{F_t - F_2}{F_t} \qquad (13.4)$$

where s is the distance from the center of gravity to the end of the connecting rod weighed in the horizontal position.

Then the total equivalent rotating mass for connecting rod and piston in one cylinder is:

$$m_r = m_c + m_{crr} \qquad (13.5)$$

and the oscillating mass is:

$$m_o = m_p + m_{cro}$$

(13.6)

where: m_p is the mass of the piston and pin. If there is a crosshead, its mass is added to the oscillating mass.

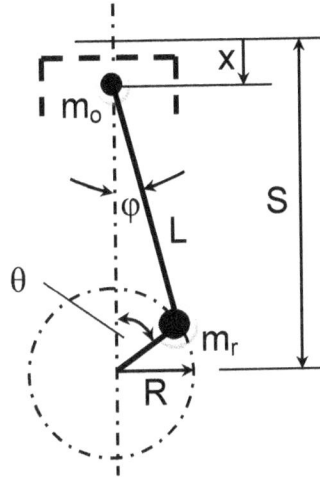

Figure 13.1: A schematic diagram of the slider crank mechanism for a piston engine.

The system for one piston can be shown simply in Figure 13.1. If the crankshaft is rotating at a constant rotational speed, ω rad/s, the force on the rotating masses is:

$$F_r = m_r \cdot r\omega^2$$

(13.7)

The direction of the forces are in the outward direction from the the center of the crankshaft towards the position of the crank pin. The rotating forces can be readily cancelled by counterweights.

The oscillating forces are different. While the crankshaft rotates at a constant speed, and its position is given by angular position, the reciprocating motion is not as simple, due to the link posed by the connecting rod. The standard crank angle, θ, is usually defined as zero when the piston is at top dead center, that is, when the cylinder volume is at a minimum. The angle ϕ is defined as the angle between the center line of the cylinder and the connecting rod, as shown in Figure 13.1. Denoting the distance of the piston from TDC as x, the following expressions can be derived from the geometry in Figure 13.1.

$$x = L + R - [R \cdot \cos(\theta) - L \cdot \cos(\phi)]$$

(13.8)

$$R \cdot \sin(\theta) = L \cdot \sin(\phi)$$

(13.9)

$$x = R(1 - \cos(\theta)) + L\left(1 - \sqrt{1 - \sin^2(\phi)}\right)$$

(13.10)

$$= R(1 - \cos(\theta)) + L\left(1 - \sqrt{1 - \frac{R^2}{L^2}\sin^2(\theta)}\right)$$

(13.11)

The binomial series can be used to expand the square root term:

$$(x + y)^n = x^n + nx^{n-1}y + \frac{n(n-1)}{2!}x^{n-2}y^2 + \frac{n(n-1)(n-2)}{3!}x^{n-3}x^3 + \ldots$$

(13.12)

It can be shown that the volume is closely approximated by using just the first two terms of the binomial series. While higher order terms can be included at the price of a complex calculation, the higher order terms decrease in importance quite rapidly, and it is common only to be concerned with the second term, that is the second order effects. Then:

$$x = R(1 - \cos\theta) + \frac{R^2}{2L}\sin^2\theta$$

(13.13)

This equation leads to the volume expression used in the simulation section, Equation (2.106). The instantaneous piston velocity is obtained differentiating Equation (13.13):

$$v = \frac{dx}{dt} = \frac{dx}{d\theta}\frac{d\theta}{dt} = \left[R\sin\theta + \frac{R^2}{2L}\sin 2\theta\right]\omega \tag{13.14}$$

and the acceleration is obtained by differentiating a second time:

$$a = \frac{dv}{dt} = \left[R\cos(\theta) + \frac{R^2}{L}\cos 2(\theta)\right]\omega^2 \tag{13.15}$$

The force on the oscillating parts is along the axis of the cylinder and of the magnitude:

$$F_o = m_o \cdot a = m_o\omega^2\left[R\cos(\theta) + \frac{R^2}{L}\cos 2(\theta)\right] \tag{13.16}$$

Equation (13.16) shows that the forces have components of different frequencies. Of course, there are higher order frequencies, but these have been neglected here. A more detailed analysis will show that the third term in the series has a magnitude of about 1/50 of the second order term, so for most purposes, the inclusion of the second order is enough. Equation (13.16) also shows that the second order forces oscillate at twice the frequency of the main forces. The direction of the forces is along the axis of the cylinder, and changes with the motion of the piston. For example, the force on the oscillating components is directed towards the crankshaft when the piston is at top dead center. Physically, this means that the piston motion is opposed as it approaches TDC, and the force is acting to pull the piston away from the TDC position. This means that there is a tensile force in the connecting rod. The gas pressure opposes this stress, and thus in a 4-stroke engine, the tensile force on the connecting rod is highest during the air exchange process, where there is no pressure difference over the top of the piston to help decelerate it.

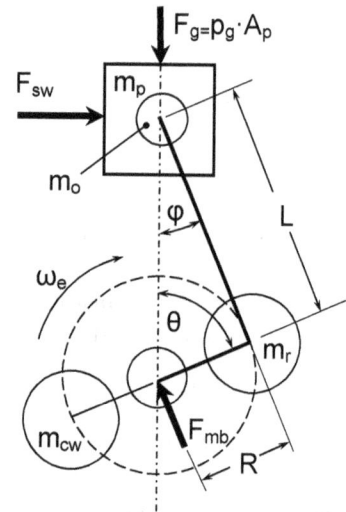

Figure 13.2: The forces between the engine block and the piston-crankshaft mechanism in a single-cylinder engine.

The use of the above to determine forces involved in the piston-crankshaft mechanism will be shown in the following. Consider a single cylinder engine as shown in Figure 13.2, where the crankshaft arm length, R, $= S/2$, and the connecting rod length is L. The crankshaft is at an arbitrary angle, θ, and the pressure in the cylinder is p_g. The crankshaft is equipped with a counterweight, which exactly balances the rotating mass of the connecting rod, and it is assumed that the arms to the rotating mass and the counterweights are dynamically equivalent and therefore in balance. The engine is assumed to be rotating in a clockwise direction, with a speed $\omega_e = \frac{2\pi N}{60}$ rad/s.

The forces on the total system of piston, connecting rod and crankshaft are shown in Figure 13.3. Here, a section of the connecting rod is shown, and the internal forces indicated. This is done so as to determine the force in the connecting rod. Then taking the upper half of the system shown in Figure 13.3 and drawing a free body diagram, the forces in the connecting rod can be determined. The force polygon for this system is shown in the upper right of Figure 13.3. From this, the force in the connecting rod can be calculated to be:

$$F_{cr} = \frac{p_g \cdot A_p - (m_o + m_p) \cdot a_p}{\cos\phi} \tag{13.17}$$

The sidewall force on the piston is:

$$F_{sw} = F_{cr} \cdot \sin\phi \tag{13.18}$$

Note that the direction of the connecting rod force is in the direction of the axis of the connecting rod, given by the angle ϕ. This force is now known, and can be transferred to the free body system of the

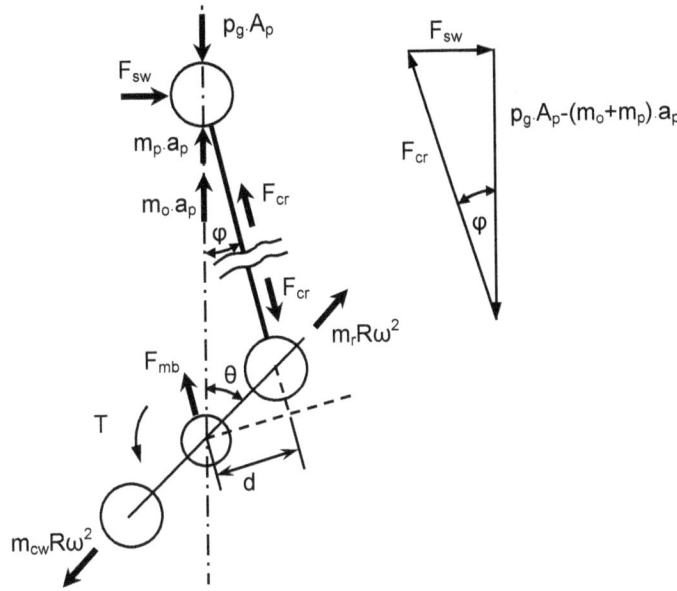

Figure 13.3: Free body diagrams for the piston-pin-reciprocating mass of the connecting rod system, and for the crankshaft-rotating mass of the connecting rod. Also shown in the force polygram for the piston.

lower half of Figure 13.3. It is seen that the total reaction force in the main bearings is equal to the force from the connecting rod. But since the forces are not co-linear, there is a moment, which taken around the center of the crankshaft, is equal to:

$$T_R = F_{cr} \cdot d = F_{cr} \cdot R \cos\left(90° - \theta - \phi\right) \tag{13.19}$$

In steady operation, this torque is balanced by the resistance torque of the load, and is equal in magnitude to the output torque of the engine. It is then straight forward to calculate the torque of the engine throughout the operating cycle, if the cylinder pressure is known as a function of the crank angle. One of the useful purposes of the cycle simulations similar to those of Section 2.5 is to provide this information, though a more complete simulation with intake and exhaust pressures due to flow through the valve would be preferred. Experimental data obviously can be used. A simple expression for the torque variation throughout a cycle can be written:

$$T(\theta) = \frac{p_g \cdot A_p - (m_o + m_p) \cdot a_p}{\cos\phi} \cdot R \cdot \cos\left(90° - \theta - \phi\right) \tag{13.20}$$

where the piston acceleration, a_p is given by Equation (13.15), and the angle ϕ can be obtained from:

$$\phi = \sin^{-1}\left(\frac{R}{L}\sin\theta\right) \tag{13.21}$$

Figure 13.4 shows the cylinder pressure and instantaneous torque on the crank shaft of a CFR engine calculated from experimental data and Equation (13.20). The average torque is 37.3 N-m. Since the integration was taken over 720°, this value includes the torque for the pumping loop. If the pressure is set equal to zero in Equation (13.20), the average torque will be zero.

The main bearing load determined above is of interest in relation to the durability and friction in the main bearings. The piston side thrust from Equation (13.18) is of interest in relation to the friction between the piston skirt and the cylinder wall.

Other forces of interest can also be obtained from these results. From a free body diagram for the upper part of the connecting rod, the force on the piston pin can be determined. This and the corresponding force polygon are shown in Figure 13.5. The two components of the piston pin reaction force are calculated

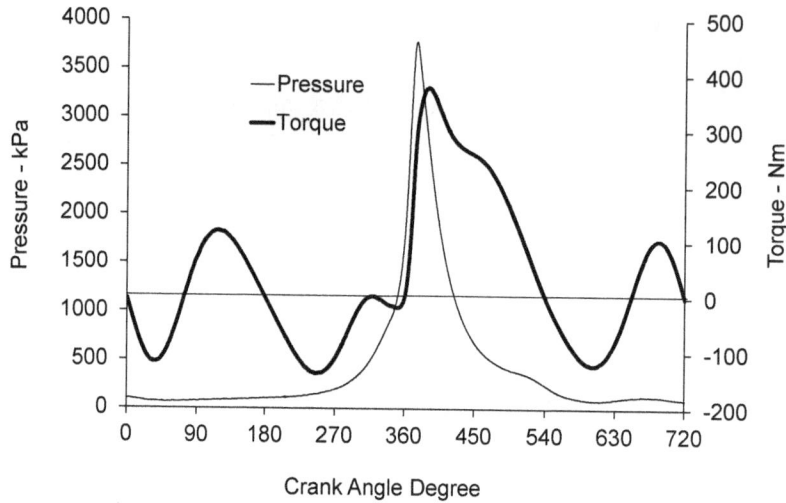

Figure 13.4: The cylinder pressure and instantaneous torque on the crankshaft for a CFR engine with full load, operating at 1800 RPM, with a spark timing of 20° BTDC.

to determine the magnitude and direction of the force on the piston pin:

$$F_{pp,y} = F_{cr} \cos\phi + m_o \cdot a_p \tag{13.22}$$

$$F_{pp,x} = F_{cr} \sin\phi \tag{13.23}$$

$$F_{pp} = \sqrt{F_{pp,y}^2 + F_{pp,x}^2} \tag{13.24}$$

$$\beta = \tan^{-1} \frac{F_{pp,y}}{F_{pp,x}} \tag{13.25}$$

The remaining force of interest is the force on the crank arm, where the connecting rod is attached. This is obtained from a free body diagram and force polygram for the large end of the connecting rod. The inertial force here is that due to the rotating portion of the connecting rod. Its acceleration is determined by the rotational speed of the engine, and is of constant magnitude assuming the engine speed is constant. Then:

$$F_{cp,y} = F_{cr} \cos\phi - m_r R\omega^2 \cos\theta \tag{13.26}$$

$$F_{cp,x} = -F_{cr} \sin\phi - m_r R\omega^2 \sin\theta \tag{13.27}$$

$$F_{cp} = \sqrt{F_{cp,y}^2 + F_{cp,x}^2} \tag{13.28}$$

$$\gamma = \tan^{-1} \frac{F_{cp,y}}{F_{cp,x}} \tag{13.29}$$

Example

Consider an engine with the following specifications: Bore = 80 mm, stroke = 75 mm, connecting rod length = 160 mm, connecting rod mass 1.4 kg, piston mass including pin = 1 kg. The center of gravity of the connecting rod is 40 mm from the large end. The engine is operating at 3000 rpm. Find the forces on the main bearings, the piston side thrust, the torque, and the forces on the piston pin and the crank throw arm at a crank angle position of 60° ATDC for a cylinder pressure of 1500 kPa.

The oscillating and reciprocating masses of the connecting rod can be found using Equations (13.2) (13.3) :

$$m_{cro} = m_{cr} \cdot \frac{s_r}{L} = 1.4 \cdot \frac{40}{160} = 0.35 kg$$

$$m_{crr} = m_{cr} \cdot \frac{s_o}{L} = 1.4 \cdot \frac{120}{160} = 1.05 kg$$

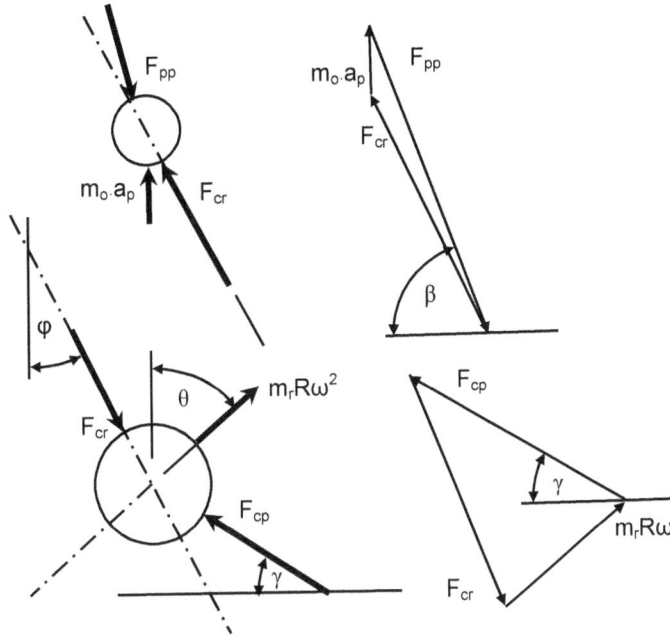

Figure 13.5: Free body diagrams for both ends of the connecting rod.

Then engine speed in radians per second is:

$$\omega = \frac{2\pi N}{60} = \frac{2\pi 3000}{60} = 314.1 s^{-1}$$

The piston acceleration is calculated with Equation (13.15):

$$a = \left[R\cos\theta + \frac{R^2}{L}\cos 2\theta \right]\omega^2$$

$$= \left[.0375\cos(60) + \frac{0.0375^2}{.160}\cos(120) \right] 314.1^2 = 1417 m/s^2$$

The pressure force on the top of the piston is:

$$F_g = A_p \cdot p_g = \frac{\pi}{4}B^2 \cdot p_g = \frac{\pi}{4} \cdot 0.08^2 \cdot 1.5 \cdot 10^6 = 7540 N$$

From the force polygon for the reciprocating mass at the piston, the connecting rod force and the piston side thrust can be determined using Equations (13.17) and (13.18):

$$F_{cr} = \frac{p_g \cdot A_p - (m_o + m_p) \cdot a_p}{\cos\phi} = \frac{7540 - (0.35 + 1.0) \cdot 1417}{\cos 11.7} = 5746 N$$

where:

$$\phi = \sin^{-1}\left(\frac{R}{L}\sin\theta \right) = \sin^{-1}\left(\frac{0.0375}{0.160}\sin 60 \right) = 11.7°$$

$$F_{sw} = F_{cr} \cdot \sin\phi = 5746 \cdot \sin 11.7 = 1165 N$$

Then the main bearing reaction force is equal to the force in the connecting rod, that is $F_{mb} = 5746 N$.

The torque at this crank angle position can be found using Equation (13.19):

$$T_R = F_{cr} \cdot d = F_{cr} \cdot R\cos(90 - \theta - \phi) = 5746 \cdot 0.0375\cos(90 - 60 - 11.7) = 204.6 Nm$$

The force on the piston pin is found from the force polygon in Figure 13.5:

$$F_{pp,y} = F_{cr}\cos\phi + m_o \cdot a_p$$
$$= 5746\cos 11.7 + 0.35 \cdot 1417 = 6122N$$
$$F_{pp,x} = F_{cr}\sin\phi$$
$$F_{pp,x} = 5746\sin 11.7 = 1165N$$
$$F_{pp} = \sqrt{F_{pp,y}^2 + F_{pp,x}^2}$$
$$F_{pp} = \sqrt{6122^2 + 1165^2} = 6232N$$
$$\beta = \tan^{-1}\frac{F_{pp,y}}{F_{pp,x}}$$
$$\beta = \tan^{-1}\frac{1165}{6122} = 10.8°$$

Similarly, the force on the crank arm pin can be found:

$$F_{cp,y} = F_{cr}\cos\phi - m_r R\omega^2 \cos\theta$$
$$F_{cp,y} = 5746\cos 11.7 - 1.05 \cdot 0.0375 \cdot 314.1^2 \cos 60 = 3684N$$
$$F_{cp,x} = -F_{cr}\sin\phi - m_r R\omega^2 \sin\theta$$
$$F_{cp,x} = -5746\sin 11.7 - 1.05 \cdot 0.0375 \cdot 314.1^2 \sin 60 = 4531N$$
$$F_{cp} = \sqrt{F_{cp,y}^2 + F_{cp,x}^2}$$
$$F_{cp} = \sqrt{3648^2 + 4531^2} = 5817N$$
$$\gamma = \tan^{-1}\frac{F_{cp,y}}{F_{cp,x}}$$
$$\gamma = \tan^{-1}\frac{3684}{4531} = 39.1°$$

13.2   Inertial Forces for Multi-cylinder Piston Engines

There have been many geometrical configurations of the cylinders of piston engines through the years. One can speak of two general limiting types, an in-line engine and a "star" formed engine, and variations in between these. In the in-line engine, all the cylinder axes are parallel and in the same plane parallel to the crankshaft, the crank throws are distributed in a circle when viewed from the end of the crankshaft. The crank throws form what can be called a crankshaft polygon. In a regular star engine (for example an older aircraft piston engine), all the cylinders lie in the same plane perpendicular to the crankshaft. This type of engine was dominant on larger aircraft up to the time of the turbojet engine. Since then, star engines have basically not been produced. The V-engine is an intermediate form between the two extremes.

In an in-line engine, the resultant inertia force is found by vector addition of all the individual forces. This is most conveniently seen by looking at the crankshaft from one end. Since there is a length to the crankshaft in an in-line or V engine, moments arise along the length of the crankshaft. These are best seen from a side view of the crankshaft and will be discussed later.

13.2.1   Rotating Forces

The resultant rotating force for the entire engine is the vector sum of all the force vectors for the rotating masses in all cylinders.

$$\vec{F}_r = \sum_{i=1}^{n_{cyl}} \vec{F}_{i,r} \tag{13.30}$$

Since the motion is strictly rotational, there is no higher order component of the rotating forces. In crankshaft configurations where there is symmetry, the rotational forces cancel each other. Counter weights can be also used to balance the rotating forces, this is helpful in reducing the bending moments within the crankshaft.

13.2.2 Oscillating Forces

It was found in the previous section that the forces originating from the oscillating masses in a single cylinder can be expressed by Equation (13.16). For a multi-cylinder engine, the resultant force of the oscillating components can be calculated in the following way. Take as an example, a 4-cylinder engine that has the cylinder arrangement shown in Figure 13.6.

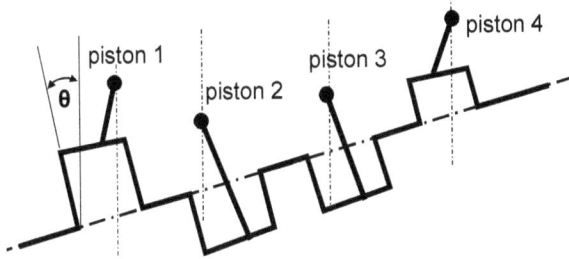

Figure 13.6: The arrangement of the crank throws in a 4 cylinder, in-line engine.

The inertial forces for all the cylinders combined can be calculated as follows. The direction is in along the vertical axis of Figure 13.6.

$$
F = (m_{o,1} + m_{o,4})\omega^2 \left[R\cos\theta + \frac{R^2}{L}\cos 2\theta \right]
$$

$$
+(m_{o,3} + m_{o,3})\omega^2 \left[R\cos(\theta + 180) + \frac{R^2}{L}\cos(2\theta + 2\cdot 180) \right] \tag{13.31}
$$

Using $\cos(\theta + 180) = -\cos\theta$ and $\cos(2\theta + 360) = \cos 2\theta$, this can be rewritten:

$$
\frac{F}{m_{o,tot}R\omega^2} = \left[\cos\theta - \cos\theta + 2\frac{R}{L}\cos 2\theta \right] = \left[\frac{2R}{L}\cos 2\theta \right] \tag{13.32}
$$

Now the terms in θ in Equation (13.32) cancel. These are the first order terms, and this means that the first order oscillating forces are balanced in this arrangement. However, the terms in 2θ do not cancel. This means that the second order oscillating forces do not balance in this cylinder arrangement, giving rise to engine vibrations with a driving force of twice the engine frequency. The normalized forces from the above are shown in Figure 13.7. The first order forces are shown for the cylinder pairs. Since cylinder 1 and 4 are at the same angular position, the forces are in phase. Cylinders 2 and 3 also have the forces in phase, but they are shifted by 180 degrees from the other cylinders. Therefore, the sum of the two components is zero.

For the second order forces, the frequency is doubled, and all the forces are in phase. The magnitude of the force will depend on the mass, as can be seen from the dimensionless force used in Figure 13.7. It is possible to cancel the second order forces with an device to generate a force of the same magnitude and frequency but opposite direction as the second order forces. This device, called the Lancaster balancer, uses two shafts geared together to rotate in opposite directions, at twice the engine speed. Each shaft has a weight which gives a rotational force, $mR\omega^2$, equal to half of the unbalanced second order force. The concept is shown in Figure 13.8. Since the shafts rotate in opposite directions, the horizontal forces will always cancel each other, but the vertical forces will be added. In modern automobile engines, this idea is not used, as it adds to the weight, size, complexity and cost of an engine. Instead, the engine mounting system is designed to reduce the transmission of the 2nd order vibrations to the rest of the vehicle. A Lancaster balancer could be used in a larger stationary 4-cylinder in-line engine, where the forces may be larger, and the complexity not such a burden.

For a 3 cylinder in-line 4-stroke engine, the crank throws are normally placed at an angle of 120° from each other. A crankshaft layout of this type is shown in Figure 13.9. In a 6 cylinder, in-line 4-stroke

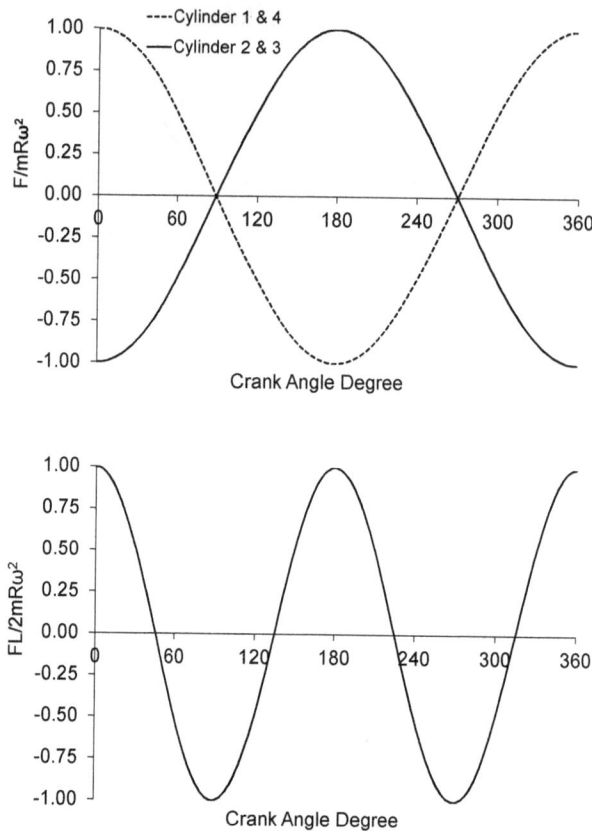

Figure 13.7: The first and second order forces for the 4 cylinder engine with the crankshaft of Figure 13.6. The sum of the first order forces is zero. The second order force curve is valid for the individual forces or the total force.

engine, the cylinders are also spaced at a separation of $120°$. A typical axial arrangement corresponds to that with two elements as shown in Figure 13.9 joined end-to-end, symmetrical between pistons 3 and 4. The following analysis is then valid for the inertial forces of either the 3- or 6-cylinder 4-stroke engine, though not moments. Denoting the position of piston 1 as θ, and assuming θ positive in the clockwise direction, the resultant inertial force is:

$$F = m_{o,1}\omega^2 \left[R\cos\theta + \frac{R^2}{L}\cos 2\theta \right]$$

$$+ m_{o,2}\omega^2 \left[R\cos(\theta + 120) + \frac{R^2}{L}\cos(2\theta + 2\cdot 120°) \right]$$

$$+ m_{o,3}\omega^2 \left[R\cos(\theta + 240°) + \frac{R^2}{L}\cos(2\theta + 2\cdot 240°) \right] \tag{13.33}$$

Assuming equal masses for each cylinder, this can be rewritten:

$$F = m_{o,1}R\omega^2 \left[\cos\theta + \cos(\theta + 120) + \cos(\theta + 240)\right]$$

$$+ \frac{m_{o,1}R^2\omega^2}{L}\left[\cos 2\theta + \cos(2\theta + 240) + \cos(2\theta + 480)\right] \tag{13.34}$$

$$= m_{o,1}R\omega^2 \left[\cos\theta + \cos(\theta + 120) + \cos(\theta + 240)\right]$$

$$+ \frac{m_{o,1}R^2\omega^2}{L}\left[\cos 2\theta + \cos(2\theta + 120) + \cos(2\theta + 240)\right] \tag{13.35}$$

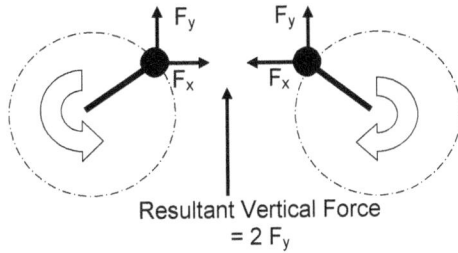

Figure 13.8: The concept of the Lancaster balancer. Two shafts with unbalanced weights rotate in opposite directions to give a force in opposition to the unbalanced second order inertial forces in a four-cylinder in line engine.

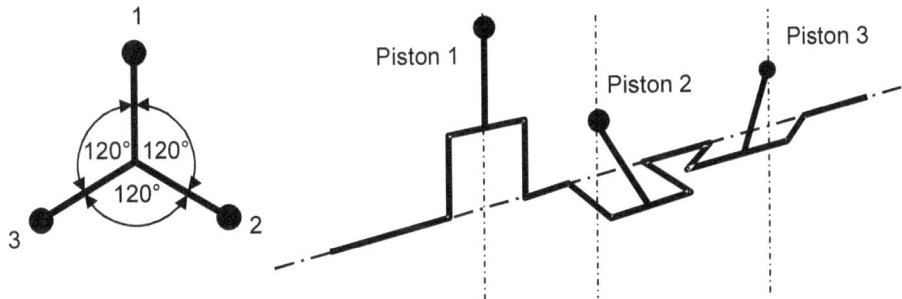

Figure 13.9: The arrangement of the crank throws in a 3 - cylinder, in-line engine.

These expressions can be evaluated by taking a reference coordinate at the arbitrary angle θ. This means that the cosine of the reference position is always one. Now the other angles will always be either at an angle of $\pm120°$ from this reference, and $\cos(120) = \cos(-240) = 0.5$, which means that the sum of the terms inside the square brackets in Equation (13.35) is equal to zero for both the first and second order terms. Then, the 3 cylinder, inline engine has no unbalanced first or second order inertial forces. The same applies to an in-line 6-cylinder 4-stroke engine, with cylinder spacing of 120°.

These forces can also be seen in a graphical context. The inertial forces for the oscillating masses always occur along the cylinder axes, but the magnitude at any time can be obtained as the projection on the cylinder axis of a imaginary rotating vector with the maximum magnitude and the appropriate frequency. It is simplest to examine the 1st and 2nd order forces separately.

For the 1st order forces, the imaginary vector has the same direction as the crank throws, or in other words, the direction is shown by the polygon diagram of the crankshaft, which could also be called the 1st order polygram. It is seen that the vector diagram is formed in a similar fashion as the diagram of the rotating forces, assuming that the vectors for the individual cylinders are taken in the same order. For the 2nd order oscillating forces the direction of the imaginary vectors is not directly determined by the crankshaft polygon. The forces are proportional to $\cos(2\theta)$. The direction of the forces is therefore given by a polygram, whose angles are twice as large as those of the crankshaft polygram, what could be called a 2nd order polygram. Figure 13.10 shows the crankshaft polygram and vector diagram for 1st order forces in a 3-cylinder piston engine in two cases: one with equal oscillating masses and the other where the oscillating mass of one of the cylinders is larger that the other two. The engine, therefore, has a resultant 1st order force, R_1 that is not zero. In the shown position, the instantaneous value is $R_1 \cos\gamma$ when the crankshaft has rotated clockwise through the angle γ, the inertial force (that is, the component in the direction of piston motion) is at its maximum value. The resultant of the rotating forces is equal to zero, since it is assumed that the only difference in the masses is in the mass of the piston or piston pin on cylinder 3.

Figure 13.11 shows a crankshaft polygram ("star") diagram, 2nd order polygram diagram and vector diagrams for a 6-cylinder 2-stroke engine with regular ignition order. All the resultant forces are zero, which can also be seen from the crankshaft polygram and the 2nd order polygram. The forces balance each other. Note that the angles here are 60° here instead of 120° for the 4-stroke engine, because all

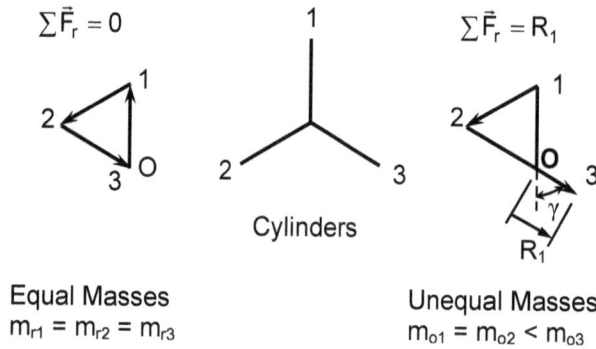

Figure 13.10: Cylinder polygrams for first order forces in a 3-cylinder in-line engine.

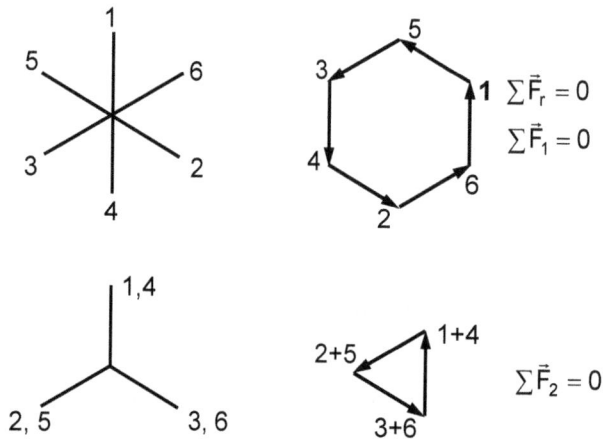

Figure 13.11: Cylinder polygrams for first and second order forces in a 6-cylinder, 2-stroke in-line engine.

the cylinders must have ignition (piston at TDC) once each revolution in a 2-stroke engine.

13.2.3 Moments Due to Oscillating forces

Since the inertial forces are distributed along the length of the crankshaft, moments arise. The magnitude of the moments depends on the magnitude and direction of the forces, the distance between cylinders and the number of cylinders. The distinction is made between rotating moments, moments from 1st, 2nd and higher order forces, and partially between outer and inner moments.

An inner moment is understood to be the sum of the moments from the individual inertial forces on one side of a given slice through the engine. The outer moment is understood to be the sum of the moments on both sides of the slice. In the cases where the sum of the forces is zero, the outer moment does not change with a parallel displacement of the moment axis. The moment is thus said to be "free".

The main bearings absorb the moments, with the forces being transferred to the engine supports. The forces on the main bearing along with the reactions from the base determine the bending moments on the engine support. The outer moments affect, and are absorbed by, the base. If the outer moments are zero, the reaction forces are also zero. The bending moments on the mount and the inner moments are identical in this case. Inner moments are important, as they are related to the bending stresses to be found in the different parts of the crankshaft.

As with the determination of the forces, the moments can be calculated mathematically and shown in vector diagrams at a given position to determine the sum of the moments. As an example, the three cylinder engine in Figure 13.9 can be used. The situation for the calculation of moments is shown in the corresponding figure, where the oscillating forces and the distance between the cylinders are shown in

Figure 13.12. Only the first and second order terms will be considered. It is assumed that the rotating

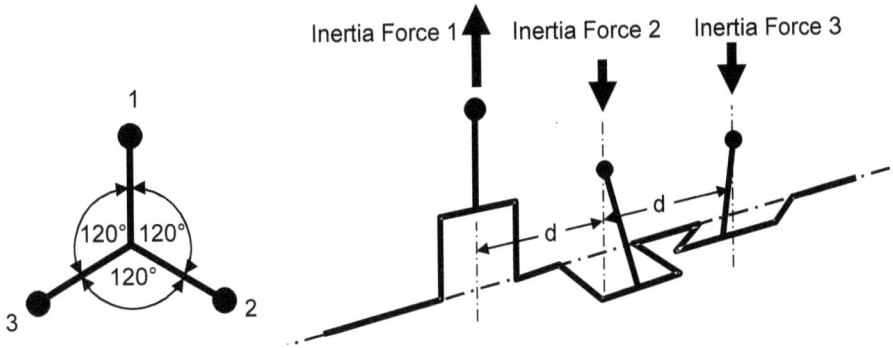

Figure 13.12: The inertial forces and distance between the crank throws of a 3 - cylinder, in-line engine.

forces are balanced by counterweights on the crankshaft, so only oscillating forces are considered here.

In this example, moments are taken at the center of connecting rod on cylinder one. Then there are only two moments to consider, from cylinders 2 and 3, and the masses are considered equal. For the first order terms where θ is the angular position of cylinder 1:

$$\sum M_I = d \cdot m_{o,1} R \omega^2 \left[\cos(\theta + 120) + 2d \cos(\theta + 240) \right] \tag{13.36}$$

And for the second order terms:

$$\sum M_{II} = d \cdot m_{o,1} \omega^2 \frac{R^2}{L} \left[\cos(2\theta + 2 \cdot 120) + 2d \cos(2\theta + 2 \cdot 240) \right] \tag{13.37}$$

The first and second order moments calculated from Equations (13.36) and (13.37) are shown in Figure 13.13 for one crankshaft revolution, where they have been normalized by the terms $d \cdot m_{o,1} R \omega^2$ and $d \cdot m_{o,1} \omega^2 R^2 / L$ for 1st and 2nd order moments respectively.

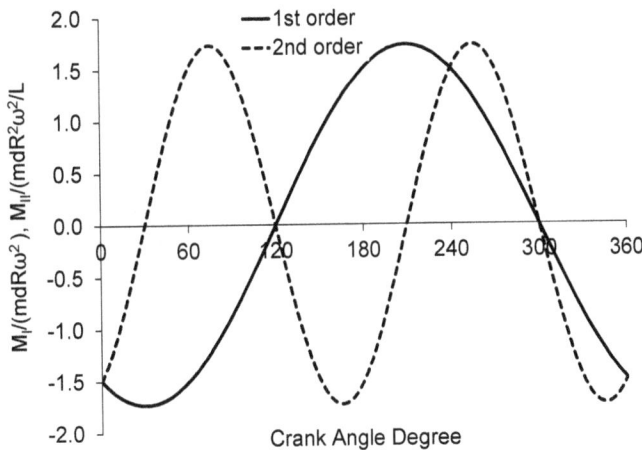

Figure 13.13: The normalized first and second order moments for the 3 cylinder engine with the crankshaft of Figure 13.9.

The calculation can also be shown graphically in a vector diagram, as in Figure 13.14, where the 1st order moments are shown on the left and the 2nd order moments on the right. The condition is shown at an arbitrary angle, θ, for the position of cylinder 1. Since the forces are cosine functions, the magnitude is determined from the component along the cylinder axis of a rotating vector with the magnitude as given in Equation (13.35). Moments can be depicted in the same way, where the magnitude of the moment is equal to the product of the magnitude of the force and the distance from the moment axis. Then

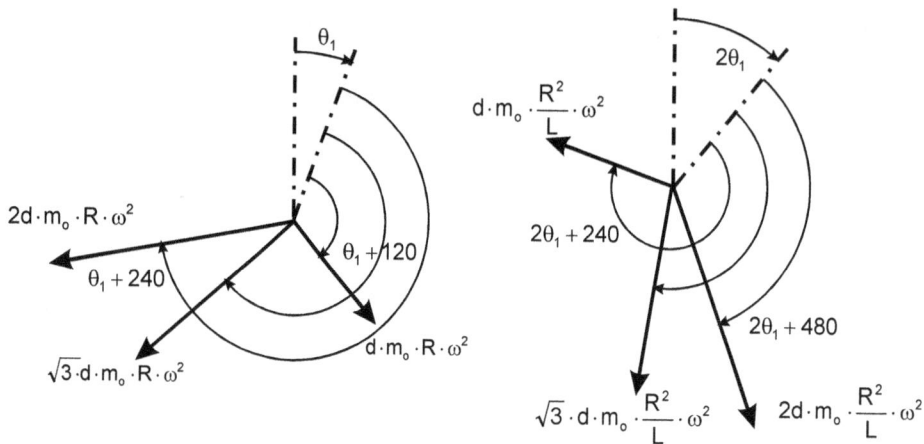

Figure 13.14: Vector representation of the first and second order moments for the 3 cylinder engine with the crankshaft of Figure 13.12.

by taking the projection of the resultant of the moments on the cylinder axis, the total moment can be determined. Taking the component of the resultant moment vector, one obtains:

$$M_I = \sqrt{3} \cdot d \cdot m_{o,1} R \omega^2 \cos(\theta + 150) \tag{13.38}$$

And for the second order terms:

$$M_{II} = \sqrt{3} \cdot d \cdot m_{o,1} \omega^2 \frac{R^2}{L} \cos(2\theta - 150) \tag{13.39}$$

A plot of these equation, normalized by the maximum value of the moment is the same as that shown in Figure 13.13.

The magnitude of the rotating moment is again determined by the component of the resultant vector. Since the forces lie in the plane of the cylinder, the moment does as well.

Figure 13.15 shows the moment diagram for a 5-cylinder in-line piston engine. It is assumed that there are two counterweights of the same mass located at cylinder 1 and cylinder 5. Note that the moments for cylinders 2 and 4 are of the same magnitude, but of different directions, since the forces are the same, but they act in different directions due the angles of the crank throws. In the cylinder polygon, cylinder 1 is in a vertical direction. The moment is drawn horizontally in the moment diagram, and the moments are all rotated 90 ° relative to the cylinder polygon. This is arbitrary, other sources show the moments in the same direction as the cylinder polygon. As long as all moments have the same relative orientation to the cylinder direction, the determination of the resulting moment will be unaffected.

General Analysis

Forces and moments can be analyzed in a general fashion for a multi-cylinder engine [84]. Again, only primary and secondary forces and moments are considered, as in Equation (13.16). An analysis can be performed on an in-line engine, which can be extended to V-engines by considering them to consist of 2 in-line engines with a constant angular displacement with respect to crank angle degree. An in-line engine has a number of cylinders, each of which has an angular displacement, ϕ_i from the first cylinder and the distance from the axis of any cylinder to the axis of the first cylinder is a_i. Since the cylinders lie in the same plane, the forces are in the same direction, that is in the plane of the cylinders. It is assumed that the masses of all the reciprocating parts are the same for each cylinder.

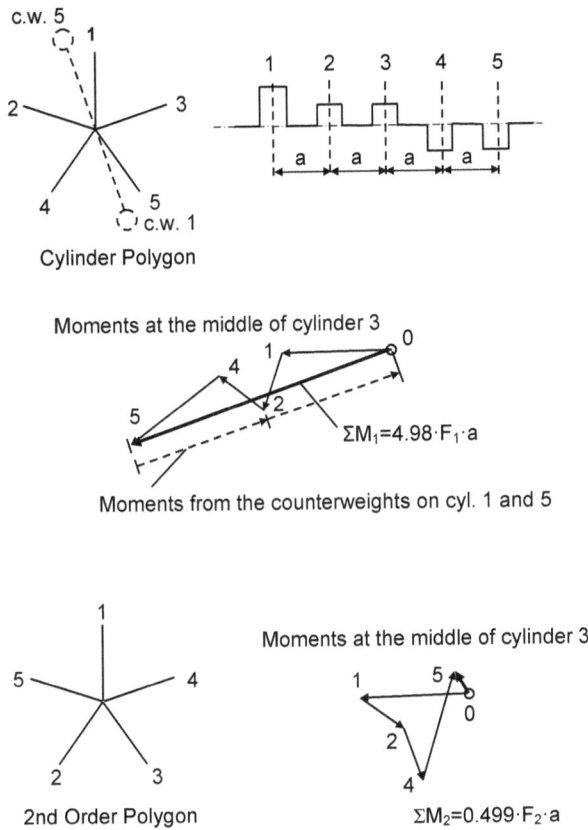

Figure 13.15: Moment diagrams for a 5-cylinder in-line engine.

Then for equal oscillating mass in all cylinders, the sum of the primary forces can be written:

$$
\begin{aligned}
\sum F_p &= \sum m_o R\omega^2 cos\theta_i \\
&= m_o R\omega^2 \sum \cos\left(\theta_1 + \phi_i\right) \\
&= m_o R\omega^2 \sum \left[\cos\theta_1 \cos\phi_i - \sin\theta_1 \sin\phi_i\right] \\
&= m_o R\omega^2 \left[\cos\theta_1 \sum \cos\phi_i - \sin\theta_1 \sum \sin\phi_i\right]
\end{aligned}
\tag{13.40}
$$

And similarly for the secondary forces, one obtains:

$$
\sum F_s = m_o \frac{R^2}{L}\omega^2 \left[\cos 2\theta_1 \sum \cos 2\phi_i - \sin 2\theta_1 \sum \sin 2\phi_i\right]
\tag{13.41}
$$

The total unbalance or shaking force, is the sum of the two components:

$$
F = F_p + F_s
\tag{13.42}
$$

Since the values of ϕ_i are determined by the geometry of the engine, the unbalance can be determined from them and a first order unbalance with have the frequency of the engine rotational speed, and a second order unbalance will have twice that frequency.

In order for the first order forces to be in balance:

$$
\sum \cos\phi_i = 0
\tag{13.43}
$$

$$
\sum \sin\phi_i = 0
\tag{13.44}
$$

Similarly, in order for the second order forces to be in balance:

$$\sum \cos 2\phi_i = 0 \tag{13.45}$$

$$\sum \sin 2\phi_i = 0 \tag{13.46}$$

The line of action of any resultant shaking force is not known from the above analysis, but can be determined by an analysis of the moments of the engine. Using cylinder 1 as a reference, and knowing the distances of each cylinder from cylinder 1, the resulting moments can be determined using the same method as above. Then the primary, or first order, moment is:

$$\sum M_p = \sum F_{p,i} a_i = \sum m_o R \omega^2 a_i \cos \theta_i$$
$$= m_o R \omega^2 \left[\cos \theta_1 \sum a_i \cos \phi_i - \sin \theta_1 \sum a_i \sin \phi_i \right] \tag{13.47}$$

And similarly for the secondary moment, one obtains:

$$\sum M_s = m_o \frac{R^2}{L} \omega^2 \left[\cos 2\theta_1 \sum a_i \cos 2\phi_i - \sin 2\theta_1 \sum a_i \sin 2\phi_i \right] \tag{13.48}$$

The total moment is the sum of the first and second order components:

$$M = M_p + M_s \tag{13.49}$$

If there is a shaking force from Equation (13.42), it acts at the distance a_R from the reference plane of the first cylinder:

$$a_R = \frac{F}{M} \tag{13.50}$$

It is possible for the shaking force to be zero with a non-zero moment. This moment will tend to make the engine move in an "end over end" fashion.

The location of the resultant line of action of the unbalanced force can be constant or a function of the engine crank angle position. Unless the line of action in this case passes through the center of gravity of the engine, a similar "end over end" shaking moment will exist.

Equations, (13.40), (13.41), (13.64) and (13.48) all can be represented by the component of a vector rotating with either engine speed for first order force or moment, or else with twice the engine speed for second order force or moment. These equations all have the general form:

$$P = A \cos \alpha - B \sin \alpha \tag{13.51}$$

Then in order to use the trigonometric relationship,

$$\cos(\alpha + \beta) = \cos \alpha \cos \beta - \sin \alpha \sin \beta \tag{13.52}$$

one can write:

$$\frac{P}{Q} = \frac{A}{Q} \cos \beta - \frac{B}{Q} \sin \beta \tag{13.53}$$

$$\frac{A}{Q} = \cos \alpha \tag{13.54}$$

$$\frac{B}{Q} = \sin \alpha \tag{13.55}$$

where:

$$\tan \alpha = \frac{\sin \alpha}{\cos \alpha} = \frac{B}{A} \tag{13.56}$$

To determine the magnitude of Q,

$$\sin^2 \alpha + \cos^2 \alpha = 1 \tag{13.57}$$

which gives:

$$\frac{A^2}{Q^2} + \frac{B^2}{Q^2} = 1$$
$$Q = \sqrt{A^2 + B^2} \tag{13.58}$$

Then an equivalent version of Equation (13.51) can be written:

$$P = \sqrt{A^2 + B^2} \cos(\alpha + \beta) \tag{13.59}$$

where α is determined by Equation (13.56)

Then the first and second order forces and moments for an in-line engine can be written as follows:

For the primary force:

$$F_p = m_o R \omega^2 \sqrt{\left(\sum \cos \phi_i\right)^2 + \left(\sum \sin \phi_i\right)^2} \cos(\theta_1 + \alpha_1) \tag{13.60}$$

where: $\alpha_1 = \arctan \dfrac{\sum \sin \phi_i}{\sum \cos \phi_i}$

For the secondary force:

$$F_s = \frac{m_o R^2 \omega^2}{L} \sqrt{\left(\sum \cos 2\phi_i\right)^2 + \left(\sum \sin 2\phi_i\right)^2} \cos(2\theta_1 + \alpha_2) \tag{13.61}$$

where: $\alpha_2 = \arctan \dfrac{\sum \sin 2\phi_i}{\sum \cos 2\phi_i}$

For the primary moment:

$$M_p = m_o R \omega^2 \sqrt{\left(\sum a_i \cos \phi_i\right)^2 + \left(\sum a_i \sin \phi_i\right)^2} \cos(\theta_1 + \alpha_3) \tag{13.62}$$

where: $\alpha_3 = \arctan \dfrac{\sum a_i \sin \phi_i}{\sum a_i \cos \phi_i}$

For the secondary moment:

$$M_s = \frac{m_o R^2 \omega^2}{L} \sqrt{\left(\sum a_i \cos 2\phi_i\right)^2 + \left(\sum a_i \sin 2\phi_i\right)^2} \cos(2\theta_1 + \alpha_4) \tag{13.63}$$

where: $\alpha_4 = \arctan \dfrac{\sum a_i \sin 2\phi_i}{\sum a_i \cos 2\phi_i}$

The direction of the forces is along the cylinder axis.

As an example, using the engine of 13.9 one has:

$\phi_1 = 0, \phi_2 = 120, \phi_3 = 240, a_2 = d, a_3 = 2d$.

Then:

$$\sum \cos \phi_i = \cos 0 + \cos 120 + \cos 240 = 1 + (-0.5) + (-0.5) = 0$$
$$\sum \sin \phi_i = \sin 0 + \sin 120 + \sin 240 = 0 + \left(\frac{\sqrt{3}}{2}\right) + \left(-\frac{\sqrt{3}}{2}\right) = 0$$

And the primary forces are 0. For the secondary forces:

$$\sum \cos 2\phi_i = \cos 0 + \cos 240 + \cos 480 = 1 + (-0.5) + (-0.5) = 0$$
$$\sum \sin 2\phi_i = \sin 0 + \sin 240 + \sin 480 = 0 + \left(-\frac{\sqrt{3}}{2}\right) + \left(\frac{\sqrt{3}}{2}\right) = 0$$

So the secondary forces are also balanced.

For the first order moment:

$$\sum a_i \cos \phi_i = 0 \cos 0 + d \cos 120 + 2d \cos 240$$
$$= 0 \cdot 1 + d \cdot (-0.5) + 2d \cdot (-0.5) = -1.5d$$
$$\sum a_i \sin \phi_i = 0 \sin 0 + d \sin 120 + 2d \sin 240$$
$$= 0 + d \left(\frac{\sqrt{3}}{2} \right) + 2d \left(-\frac{\sqrt{3}}{2} \right) = - \left(\frac{d\sqrt{3}}{2} \right)$$

Then using Equation (13.59):

$$\sum M_p = m_o R \omega^2 \left[-1.5d \cdot \cos \theta_1 + d \frac{\sqrt{3}}{2} \sin \theta_1 \right]$$
$$= \sqrt{3} \cdot d \cdot m_o R \omega^2 \cos(\theta_1 + 210°)$$

where:

$$\alpha_3 = \arctan \frac{\sum a_i \sin \phi_i}{\sum a_i \cos \phi_i} = \arctan \frac{\frac{-d\sqrt{3}}{1.5}}{-1.5d} = 210°$$

And for the second order moment:

$$\sum a_i \cos \phi_i = 0 \cos 0 + d \cos 240 + 2d \cos 480$$
$$= 0 \cdot 1 + d \cdot (-0.5) + 2d \cdot (-0.5) = -1.5d$$
$$\sum a_i \sin \phi_i = 0 \sin 0 + d \sin 240 + 2d \sin 480$$
$$= 0 \cdot d + d \left(-\frac{\sqrt{3}}{2} \right) + 2d \left(+\frac{\sqrt{3}}{2} \right) = \left(\frac{d\sqrt{3}}{2} \right)$$

$$\sum M_s = \frac{m_o R^2 \omega^2}{L} \left[-1.5d \cos 2\theta_1 - d \frac{\sqrt{3}}{2} \sin 2\theta_1 \right]$$
$$= \sqrt{3} \cdot d \cdot \frac{m_o R^2 \omega^2}{L} \cos(2\theta_1 + 30°)$$

where:

$$\alpha_4 = \arctan \frac{\sum a_i \sin \phi_i}{\sum a_i \cos \phi_i} = \arctan \frac{\frac{d\sqrt{3}}{1.5}}{-1.5d} = 210° = -150°$$

These results are the same as those shown in Figure 13.14.

13.2.4   Inertial Forces and Moments for V-Engines

First Order Forces

Since any unbalanced force or moment for an inline engine can be represented by a component of a rotating vector, the situation for a V engine can be analyzed in a straightforward manner. The method shown in the following is based on a presentation from Küttner [110].

The cylinder arrangement for a V-engine is shown in Figure
13.16. The planes of the cylinders are assumed to be symmetrical with respect to a vertical axis. The position of the
crankshaft is denoted by θ_c, with the zero position being mid-
way between the two cylinder banks, with the angle between
the banks denoted as γ. Each cylinder bank has forces and
moments that can be determined by the methods for in-line
engines shown previously. The forces in each cylinder bank act
in the plane of that bank, and the magnitudes of the forces
and the moments are given by trigonometric functions of an
angle, where the angle is the angle of the crankshaft relative
to a cylinder, typically the first cylinder, with respect to its
top dead center position. Then for the left cylinder bank, that
angle, θ_L can be written in terms of the crank shaft angle and
the angle between the two cylinder banks:

Figure 13.16: Geometry and definitions
for a V-engine.

$$\theta_L = \frac{\gamma}{2} + \theta_c \tag{13.64}$$

Similarly for the right bank:

$$\theta_R = \frac{\gamma}{2} - \theta_c \tag{13.65}$$

where θ_c is taken to be positive in the clockwise direction.

Consider the first order forces. As shown in Equation (13.60), the first order unbalanced forces for the
left and right cylinder banks can be written as:

$$F_{L,I} = P_{L,I} \cos\left(\theta_L + \alpha_{L,1}\right) = P_{L,I} \cos\left(\theta_c + \frac{\gamma}{2} + \alpha_{L,1}\right) \tag{13.66}$$

$$F_{R,I} = P_{R,I} \cos\left(\theta_R + \alpha_{R,1}\right) = P_{R,I} \cos\left(\theta_c - \frac{\gamma}{2} + \alpha_{R,1}\right) \tag{13.67}$$

Where $P_{L,I}$ and $P_{R,I}$ are constants determined by the dimensions of the oscillating components and the
configurations of the cylinders. If there is the same number of cylinders with the same configuration in
each bank, these values are normally equal. This will be assumed in the following and denoted P_I. It is
also assumed that $\alpha_{L,1} = \alpha_{R,1} = \alpha_1$. Although it is not necessary, the results are less complicated and
better illustrate the principles. A new variable can be defined to combine the crankshaft position and
the angle α_1: $\theta_{F,I} = \theta_c + \alpha_1$.

The final forces can be determined by combining the horizontal and vertical components of the forces
from each cylinder bank. For the horizontal forces, they act in opposite directions, and the positive
directions are assumed to be to the right and upwards in Figure 13.16. Then the horizontal component
of the combined first order forces, $F_{h,I}$ is:

$$
\begin{aligned}
F_{h,I} &= \left[P_I \cos\left(\theta_{F,I} - \frac{\gamma}{2}\right) - P_I \cos\left(\theta_{F,I} + \frac{\gamma}{2}\right)\right] \sin\left(\frac{\gamma}{2}\right) \\
&= P_I \sin\left(\frac{\gamma}{2}\right) \left[\cos\left(\theta_{F,I} - \frac{\gamma}{2}\right) - \cos\left(\theta_{F,I} + \frac{\gamma}{2}\right)\right] \\
&= P_I \sin\left(\frac{\gamma}{2}\right) \left[-2 \sin\theta_{F,I} \sin\left(\frac{-\gamma}{2}\right)\right] \\
&= 2 P_I \sin^2\left(\frac{\gamma}{2}\right) \sin\theta_{F,I}
\end{aligned}
\tag{13.68}
$$

Similarly for the vertical forces:

$$
\begin{aligned}
F_{v,I} &= \left[P_I \cos\left(\theta_{F,I} - \frac{\gamma}{2}\right) + P_I \cos\left(\theta_{F,I} + \frac{\gamma}{2}\right)\right] \cos\left(\frac{\gamma}{2}\right) \\
&= P_I \cos\left(\frac{\gamma}{2}\right) \left[\cos\left(\theta_{F,I} - \frac{\gamma}{2}\right) + \cos\left(\theta_{F,I} + \frac{\gamma}{2}\right)\right] \\
&= P_I \cos\left(\frac{\gamma}{2}\right) \left[2 \cos\theta_{F,I} \cos\left(\frac{-\gamma}{2}\right)\right] \\
&= 2 P_I \cos^2\left(\frac{\gamma}{2}\right) \cos\theta_{F,I}
\end{aligned}
\tag{13.69}
$$

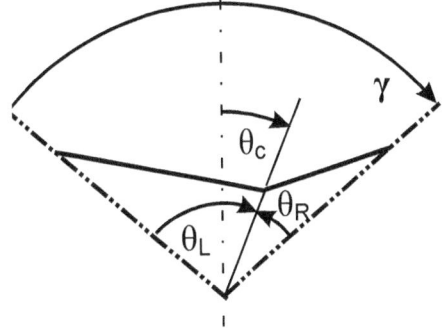

From the components, the magnitude of the resulting force can be determined:

$$
\begin{aligned}
F_I &= \sqrt{F_{h,I}^2 + F_{v_I}^2} \\
&= 2P_I \sqrt{\left[\sin^2\left(\frac{\gamma}{2}\right)\sin\theta_{F,I}\right]^2 + \left(\cos^2\left(\frac{\gamma}{2}\right)\cos\theta_{F,I}\right)} \\
&= 2P_I \sqrt{\sin^4\left(\frac{\gamma}{2}\right) + \cos^2\theta_{F,I}\left[\cos^4\left(\frac{\gamma}{2}\right) - \sin^4\left(\frac{\gamma}{2}\right)\right]} \\
&= 2P_I \sqrt{\sin^4\left(\frac{\gamma}{2}\right) + \cos^2\theta_{F,I}\cos\gamma} \\
&= 2P_I \sqrt{\sin^4\left(\frac{\gamma}{2}\right) + \cos^2\left(\theta_c + \alpha_1\right)\cos\gamma}
\end{aligned}
\tag{13.70}
$$

The angle at which the force acts relative the to center line between the cylinders, β_1, is determined by the relative magnitudes of the two components, and is given relative to the middle of the V in a clockwise direction:

$$
\begin{aligned}
\tan\beta_1 &= \frac{F_{h,I}}{F_{v,I}} \\
&= \frac{\sin^2\left(\frac{\gamma}{2}\right)}{\cos^2\left(\frac{\gamma}{2}\right)} \frac{\sin\theta_{F,I}}{\cos\theta_{F,I}} \\
&= \tan^2\left(\frac{\gamma}{2}\right)\tan\theta_{F,I} \\
&= \tan^2\left(\frac{\gamma}{2}\right)\tan\left(\theta_c + \alpha_1\right)
\end{aligned}
\tag{13.71}
$$

Second Order Forces

The second order forces can be established in a similar manner. From Equation (13.61), the second order unbalanced forces for the left and right cylinder banks can be written as:

$$
F_{L,II} = P_{L,II}\cos 2\theta_L = P_{L,II}\cos\left(2\theta_c + \alpha_{L,2} + \gamma\right)
\tag{13.72}
$$
$$
F_{R,II} = P_{R,II}\cos 2\theta_R = P_{R,II}\cos\left(2\theta_c + \alpha_{R,2} - \gamma\right)
\tag{13.73}
$$

where $P_{L,II}$ and $P_{R,II}$ are constants determined by the dimensions of the oscillating components and the configurations of the cylinders. A similar new variable can be defined to combine the crankshaft position and the angle α_2: $2\theta_{F,II} = 2\theta_c + \alpha_2$.

Then assuming the left and right banks have the same geometry, the horizontal component of the combined second order forces, $F_{h,II}$ is:

$$
\begin{aligned}
F_{h,II} &= \left[P_{II}\cos\left(2\theta_{F,II} - \gamma\right) - P_{II}\cos\left(2\theta_{F,II} + \gamma\right)\right]\sin\left(\frac{\gamma}{2}\right) \\
&= 2P_{II}\sin\left(\frac{\gamma}{2}\right)\sin\gamma\sin 2\theta_{F,II}
\end{aligned}
\tag{13.74}
$$

Similarly for the vertical forces:

$$
\begin{aligned}
F_{v,II} &= \left[P_{II}\cos\left(2\theta_{F,II} - \gamma\right) + P_{II}\cos\left(2\theta_{F,II} + \gamma\right)\right]\cos\left(\frac{\gamma}{2}\right) \\
&= 2P_{II}\cos\left(\frac{\gamma}{2}\right)\cos\gamma\cos 2\theta_{F,II}
\end{aligned}
\tag{13.75}
$$

The above can be rearranged to give:

$$
\begin{aligned}
F_{II} &= \sqrt{2}P_{II}\sqrt{\cos^2 2\theta_{F,II}\left(\cos 2\gamma + \cos\gamma\right) + \sin^2\gamma\left(1 - \cos\gamma\right)} \\
&= \sqrt{2}P_{II}\sqrt{\cos^2\left(2\theta_c + \alpha_2\right)\left(\cos 2\gamma + \cos\gamma\right) + \sin^2\gamma\left(1 - \cos\gamma\right)}
\end{aligned}
\tag{13.76}
$$

And the angle at which the force acts relative the to center line between the cylinders β_2 is:

$$\tan \beta_2 = \tan 2\theta_{F,II} \tan \gamma \tan \frac{\gamma}{2} \tag{13.77}$$

$$= \tan (2\theta_c + \alpha_2) \tan \gamma \tan \frac{\gamma}{2}$$

First Order Moments

A similar analysis can be applied to moments using Equation (13.62). For the first order moments:

$$M_{L,I} = Q_{L,I} \cos (\theta_L + \alpha_{L,3}) = Q_{L,I} \cos (\theta_c + \alpha_{L,3} + \gamma) \tag{13.78}$$

$$M_{R,I} = Q_{R,I} \cos (\theta_R + \alpha_{R,3}) = Q_{R,I} \cos (\theta_c + \alpha_{R,3} - \gamma) \tag{13.79}$$

where $Q_{L,I}$ and $Q_{R,I}$ are constants determined by the dimensions of the oscillating components and the configurations of the cylinders. Taking the components in the vertical and horizontal directions:

$$M_{h,I} = (M_{R,I} - M_{L,I}) \sin \frac{\gamma}{2}$$

$$= \left[Q_{R,I} \cos \left(\theta_c + \alpha_{L,3} - \frac{\gamma}{2} \right) - Q_{L,I} \cos \left(\theta_c + \alpha_{L,3} + \frac{\gamma}{2} \right) \right] \sin \frac{\gamma}{2} \tag{13.80}$$

$$M_{v,I} = (M_{R,I} + M_{L,I}) \cos \frac{\gamma}{2}$$

$$= \left[Q_{R,I} \cos \left(\theta_c + \alpha_{R,3} - \frac{\gamma}{2} \right) + Q_{L,I} \cos \left(\theta_c + \alpha_{L,I} + \frac{\gamma}{2} \right) \right] \cos \frac{\gamma}{2} \tag{13.81}$$

And the final moment and direction are given from the components,

$$M_I = \sqrt{M_{h,I}^2 + M_{v,I}^2} \tag{13.82}$$

$$\tan \beta_3 = \frac{M_{h,I}}{M_{v,I}} \tag{13.83}$$

Where β_3 is the angle relative the to center line between the cylinders at which the moment acts.

In determining the moments, a common reference must be taken for the distance along the crankshaft, for example, the middle of the first cylinder on the left cylinder bank. If the moments are different for each cylinder bank, because of axial displacement of the cylinders in the case of non-free moments, then Equations (13.82) and (13.83) can be solved directly. In some cases, the location will not matter, that is if the moment is "free" as described in section 13.2.3. If that is the case, then for the same geometry in each bank, $Q_{L,I} = Q_{R,I}$, and the analysis follows as in the previous case. In addition, for free moments, $\alpha_{L,3} = \alpha_{R,3} = \alpha_3$ and a new variable can be defined to combine the crankshaft position and the angle α_3: $\theta_{M,I} = \theta_c + \alpha_3$. The following is then valid for free moments with cylinder banks of the same geometry.

$$M_{h,I} = 2Q_I \sin^2 \frac{\gamma}{2} \sin (\theta_{M,I}) \tag{13.84}$$

$$M_{v,I} = 2Q_I \cos^2 \frac{\gamma}{2} \cos (\theta_{M,I}) \tag{13.85}$$

which gives:

$$M_I = 2Q_I \sqrt{\sin^4 \left(\frac{\gamma}{2} \right) + \cos (\theta_{M,I}) \cos \gamma}$$

$$= 2Q_I \sqrt{\sin^4 \left(\frac{\gamma}{2} \right) + \cos (\theta_c + \alpha_3) \cos \gamma} \tag{13.86}$$

and

$$\tan \beta_3 = \tan^2 \frac{\gamma}{2} \tan (\theta_{M,I}) = \tan^2 \frac{\gamma}{2} \tan (\theta_c + \alpha_3) \tag{13.87}$$

Second Order Moments

The situation for second order moments is similar to that of first order:

$$M_{L,II} = Q_{L,II} \cos(2\theta_L + \alpha_{L,4}) = Q_{L,II} \cos(2\theta_c + \alpha_{L,4} + \gamma) \tag{13.88}$$

$$M_{R,II} = Q_{R,II} \cos(2\theta_R + \alpha_{R,4}) = Q_{R,II} \cos(2\theta_c + \alpha_{R,4} - \gamma) \tag{13.89}$$

where $Q_{L,II}$ and $Q_{R,II}$ are constants determined by the dimensions of the oscillating components and the configurations of the cylinders, and $\alpha_{L,4}$ and $\alpha_{L,4}$ are the angles from Equation (13.63).

Taking the components in the vertical and horizontal directions:

$$M_{h,II} = (M_{R,II} - M_{L,II}) \sin\frac{\gamma}{2}$$

$$= [Q_{R,II} \cos(2\theta_c + \alpha_{R,4} - \gamma) - Q_{L,II} \cos(2\theta_c + \alpha_{L,4} + \gamma)] \sin\frac{\gamma}{2} \tag{13.90}$$

$$M_{v,II} = (M_{R,I} + M_{L,I}) \cos\frac{\gamma}{2}$$

$$= [Q_{R,II} \cos(2\theta_c + \alpha_{R,4} - \gamma) + Q_{L,II} \cos(2\theta_c + \alpha_{L,4} + \gamma)] \cos\frac{\gamma}{2} \tag{13.91}$$

And the final moment and direction are given from the components,

$$M_{II} = \sqrt{M_{h,II}^2 + M_{v,II}^2} \tag{13.92}$$

$$\tan\beta_4 = \frac{M_{h,II}}{M_{v,II}} \tag{13.93}$$

Where β_4 is the angle relative the to center line between the cylinders at which the moment acts.

Again assuming that the second order moments are free, and that the cylinder arrangement is the same on both banks and setting $2\theta_{M,II} = 2\theta_c + \alpha_4$, it can be shown that:

$$M_{h,II} = \sqrt{2}Q_{II} \sin\frac{\gamma}{2} \sin\gamma \sin(2\theta_{M,II}) \tag{13.94}$$

$$M_{v,II} = \sqrt{2}Q_{II} \cos\frac{\gamma}{2} \cos\gamma \cos(2\theta_{M,II}) \tag{13.95}$$

Which gives:

$$M_{II} = \sqrt{2}Q_{II}\sqrt{(\cos 2\gamma + \cos\gamma)\cos^2(2\theta_{M,II}) + \sin^2\gamma(1 - \cos\gamma)}$$

$$= \sqrt{2}Q_{II}\sqrt{(\cos 2\gamma + \cos\gamma)\cos^2(2\theta_c + \alpha_4) + \sin^2\gamma(1 - \cos\gamma)} \tag{13.96}$$

and

$$\tan\beta_4 = \tan\frac{\gamma}{2}\tan\gamma\tan(2\theta_{M,II}) = \tan\frac{\gamma}{2}\tan\gamma\tan(2\theta_c + \alpha_4) \tag{13.97}$$

Using the above results then, the first and second order forces and moments for a V-engine can be determined. Other geometries, for example, a "W"- engine with 3 rows of cylinders can be examined using the same principles. In modern engines, in-line and V-engines are dominant.

Examples

To illustrate the above, two examples are shown. The first will be a two cylinder engine, where the forces are unbalanced. Here the first and second order forces will be analyzed. In the single cylinder bank, $\alpha_1 = 0°$.

Consider a two cylinder V engine with a angle between the cylinders of 80°. The forces are calculated with Equation (13.16), from which it can be seen that $P_I = m_o R\omega^2$ and $P_{II} = m_o \frac{R^2}{L}\omega^2$. Assume that the crank angle position is at 20 degrees. Then:

$$\frac{F_{L,I}}{P_{L,I}} = \cos\left(\theta_c + \frac{\gamma}{2}\right) = \cos\left(20° + \frac{80°}{2}\right) = 0.5$$

$$\frac{F_{R,I}}{P_{R,I}} = \cos\left(\theta_c - \frac{\gamma}{2}\right) = \cos\left(20° - \frac{80°}{2}\right) = 0.940$$

$$\frac{F_{h,L,I}}{P_I} = \frac{F_{L,I}}{P_I}\sin\frac{\gamma}{2} = 0.500\sin 40° = 0.321$$

$$\frac{F_{v,L,I}}{P_I} = \frac{F_{L,I}}{P_I}\cos\frac{\gamma}{2} = 0.500\cos 40° = 0.383$$

$$\frac{F_{h,R,I}}{P_I} = \frac{F_{R,I}}{P_I}\sin\frac{\gamma}{2} = 0.940\sin 40° = 0.602$$

$$\frac{F_{v,R,I}}{P_I} = \frac{F_{R,I}}{P_I}\cos\frac{\gamma}{2} = 0.940\cos 40° = 0.720$$

Then:

$$\frac{F_{h,I}}{P_I} = \frac{F_{h,R,I}}{P_I} - \frac{F_{h,L,I}}{P_I} = 0.602 - .321 = 0.281$$

$$\frac{F_{v,I}}{P_I} = \frac{F_{v,R,I}}{P_I} + \frac{F_{v,L,I}}{P_I} = 0.383 + .720 = 1.103$$

The result can be found:

$$\frac{F_I}{P_I} = \sqrt{\frac{F_{h,I}}{P_I}^2 + \frac{F_{v,I}}{P_I}^2} = \sqrt{0.281^2 + 1.103^2} = 1.138$$

$$\tan\beta_1 = \tan\frac{F_{h,I}}{F_{v,I}} = \frac{0.281}{1.103} \Rightarrow \beta_1 = 14.3°$$

Alternatively, Equations (13.70) and (13.71) can be used:

$$\frac{F_I}{P_I} = 2\sqrt{\sin^4\left(\frac{\gamma}{2}\right) + \cos^2\theta_c\cos\gamma} = 2\sqrt{\sin^4 40° + \cos^2 20°\cos 80°} = 1.138$$

$$\tan\beta_1 = \tan^2\left(\frac{\gamma}{2}\right)\tan\theta_c = \tan^2 40°\tan 20° \Rightarrow \beta_1 = 14.3°$$

The solution is shown graphically in Figure 13.17. Note that for a 90° V, the force would not be a function of the crank angle, since $\cos 90° = 0$ when calculating using Equation (13.70).

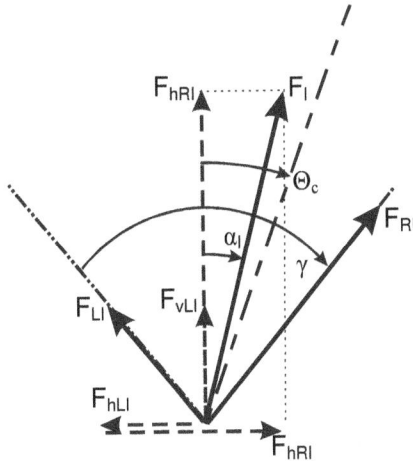

Figure 13.17: Graphical solution for the first order forces for a V-2 engine.

The calculation for the second order forces is similar. Since the forces are dependent on twice the angular displacements, $2\theta_L = 2\left(\theta_c + \frac{\gamma}{2}\right) = 2\theta_c + \gamma$ and $2\theta_R = 2\left(\theta_c - \frac{\gamma}{2}\right) = 2\theta_c - \gamma$. Then using the same engine as for the first order forces, but with a crank angle of 15°:

$$\frac{F_{L,II}}{P_{L,II}} = \cos(2\theta_c + \gamma) = \cos(2\cdot 15° + 80°) = -0.342$$

$$\frac{F_{R,II}}{P_{R,II}} = \cos(2\theta_c - \gamma) = \cos(2\cdot 15° - 80°) = 0.643$$

The components in the horizontal and vertical directions are:

$$\frac{F_{h,L,II}}{P_{II}} = \frac{F_{L,II}}{P_{II}} \sin\frac{\gamma}{2} = -0.342 \sin 40° = -0.220$$

$$\frac{F_{v,L,II}}{P_{II}} = \frac{F_{L,II}}{P_{II}} \cos\frac{\gamma}{2} = -0.342 \cos 40° = -0.262$$

$$\frac{F_{h,R,II}}{P_{II}} = \frac{F_{R,II}}{P_{II}} \sin\frac{\gamma}{2} = 0.643 \sin 40° = 0.413$$

$$\frac{F_{v,R,II}}{P_{II}} = \frac{F_{R,II}}{P_I I} \cos\frac{\gamma}{2} = 0.643 \cos 40° = 0.492$$

Then:

$$\frac{F_{h,II}}{P_{II}} = \frac{F_{h,R,II}}{P_{II}} - \frac{F_{h,L,II}}{P_{II}} = 0.413 - (-0.220) = 0.633$$

$$\frac{F_{v,II}}{P_{II}} = \frac{F_{v,R,II}}{P_{II}} + \frac{F_{h,L,II}}{P_{II}} = 0.492 + (-.262) = 0.230$$

The result can be found:

$$\frac{F_{II}}{P_{II}} = \sqrt{\frac{F_{h,II}}{P_{II}}^2 + \frac{F_{v,II}}{P_{II}}^2} = \sqrt{0.633^2 + 0.230^2} = 0.674$$

$$\tan\beta_2 = \tan\frac{F_{h,I}}{F_{v,I}} = \frac{0.633}{0.230} \Rightarrow \beta_2 = 70°$$

Alternatively, Equations (13.76) and (13.77) can be used:

$$\frac{F_{II}}{P_{II}} = \sqrt{2}\sqrt{\cos^2 2\theta_c \left(\cos 2\gamma + \cos\gamma\right) + \sin^2\gamma\left(1 - \cos\gamma\right)}$$

$$= \sqrt{2}\sqrt{\cos^2\left(2\cdot 15°\right)\left(\cos\left(2\cdot 80°\right) + \cos 80°\right) + \sin^2 80°\left(1 - \cos 80°\right)}$$

$$= 0.674$$

$$\tan\beta_2 = \tan 2\theta_c \tan\gamma \tan\frac{\gamma}{2} = \tan 2\cdot 15° \tan 80° \tan\frac{80°}{2} \Rightarrow \beta_2 = 70.0°$$

The solution is shown graphically in Figure 13.18.

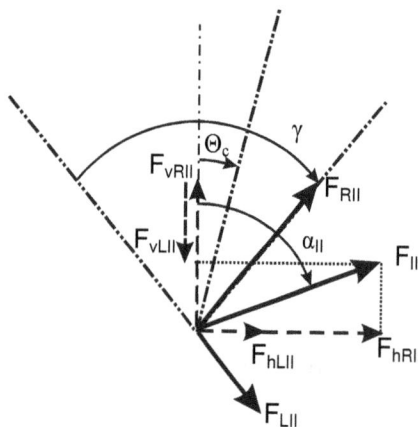

Figure 13.18: Graphical solution for the second order forces for a V-2 engine.

A second example is used for the moments, the engine with the crankshaft of Figure 13.12 will used. The results can be calculated with vector diagrams as shown above, or through the use of Equations (13.86) and (13.87) for the first order moments and Equations (13.96) and (13.97) for the second order moments. The moments for a single row of cylinders are given by Equations (13.38) and (13.39). In the latter, the phase angle is $-150°$ which is equivalent to $+210°$. From these equations, it can be seen that $Q_I = \sqrt{3} \cdot d \cdot m_{o,1} R\omega^2$ and $Q_{II} = \sqrt{3} \cdot d \cdot m_{o,1}\omega^2 \frac{R^2}{L}$. It can be shown that the both the first and second moments for this engine are free moments, that is, it does not matter at which location the moments are taken. Then it does not matter if one cylinder bank is displaced with respect to the other along the end of the crankshaft.

Consider the case of the first order moments when $\gamma = 90°$.

$$M_I = 2Q_I\sqrt{\sin^4\left(\frac{\gamma}{2}\right) + \cos(\theta_c + \alpha_3)\cos\gamma}$$

$$= 2Q_I\sqrt{\sin^4(45°) + \cos(\theta_c + 150°)\cos 90°}$$

$$= 2Q_I\sqrt{\left(\frac{1}{\sqrt{2}}\right)^4 + \cos(\theta_c + 150°)\cdot 0} = Q_I = \sqrt{3}\cdot d\cdot m_{o,1}R\omega^2$$

$$\tan\beta_3 = \tan^2\frac{\gamma}{2}\tan(\theta_c + \alpha_3)$$

$$= \tan^2 45°\tan(\theta_c + \alpha_3) = \tan(\theta_c + \alpha_3)$$

This means that the magnitude of the first order moment of the $90°$ V-6 engine is constant and equal to the magnitude of one of the cylinder banks, and it rotates at engine speed with the same angular displacement from the center of the cylinders as the individual cylinders have from their axis.

For an engine with a $60°$ V, a similar calculation will show that the moment is a function of the crank angle, with a maximum value of $M_I = 1.5 \cdot d \cdot m_{o,1}R\omega^2$, which occurs when $\cos(\theta_c + \alpha_3) = 1$.

For the second order moment with a $90°$ V-6 engine:

$$M_{II} = \sqrt{2}Q_{II}\sqrt{(\cos 2\gamma + \cos\gamma)\cos^2(2\theta_c + \alpha_4) + \sin^2\gamma(1 - \cos\gamma)}$$

$$= \sqrt{2}Q_{II}\sqrt{(\cos 2\cdot 90° + \cos 90°)\cos^2(2\theta_c + \alpha_4) + \sin^2 90°(1 - \cos 90°)}$$

$$= \sqrt{2}Q_{II}\sqrt{(-1 + 0)\cos^2(2\theta_c + \alpha_4) + 1\cdot(1 - 0)}$$

$$= \sqrt{6}d\cdot m_{o,1}\omega^2\frac{R^2}{L}\sin(2\theta_c + \alpha_4)$$

$$\tan\beta_4 = \tan\frac{\gamma}{2}\tan\gamma\tan(2\theta_c + \alpha_4)$$

$$= \tan 45°\tan 90°\tan(2\theta_c + \alpha_4) \Rightarrow \beta_4 = -90°$$

The second order moment then lies in the horizontal plane ($90°$ from the midline of the engine). The actual direction can be determined by looking at the signs of the horizontal and vertical components of the vectors.

13.3 Problems

Problem 13.1

A 4-stroke engine has the following dimensions: Cylinder bore: 80 mm; Stroke: 75 mm; Connecting rod length: 160 mm; Connecting rod center of gravity: 40 mm from the large end; Piston mass: 700 g; Piston pin mass: 150 g; Connecting rod mass: 975 g. At $50°$ after Top Dead Center the cylinder pressure is 1200 kPa The engine operates with 5000 rpm

Calculate: a. The instantaneous torque b. The side thrust from the piston on the cylinder wall. c. The force on the piston pin

Problem 13.2

The engine in Problem 13.1 is a 5-cylinder in-line engine, which is shown in the drawing. The distance between cylinders is 100 mm Calculate the 1st and 2nd order unbalanced inertial forces and moments. Assume that the rotating forces are balanced.

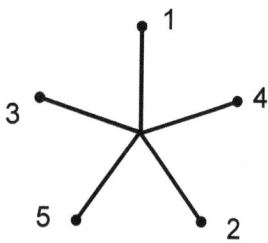

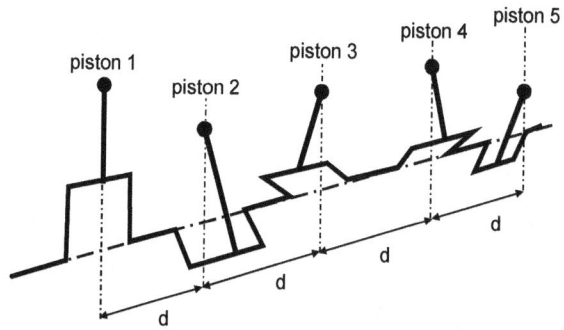

Chapter 14

Driving Resistances

14.1 Introduction

When a vehicle moves, different forms of resistance occur which have to be overcome by the power of the engine. The sum of these resistances determines the forces required to move the vehicle. From the interaction between the driving resistance and the engine power, a vehicle's propulsion power, fuel consumption, maximum speed, ability to climb, and acceleration characteristics can be determined.

In order to determine the necessary power requirements to satisfy performance demands, it is necessary to clarify and determine the total driving resistance. It can be divided up into the following 5 parts:

1. Transmission resistance W_{tr}
2. Total rolling resistance W_r
3. Aerodynamic resistance W_a
4. Gradient resistance W_g
5. Acceleration resistance W_{ac}

This total driving resistance must be overcome by the power source, normally a petrol, diesel, or gas powered engine.

Of these resistances, W_{tr} , W_r, and W_a are significantly affected by the design and speed of the vehicle, while W_g and W_{ac} are determined as well by the type of driving being performed, that is acceleration and/or operation on a grade. For different transport conditions the resistances can have differing importance, but in principle, it is not possible to avoid the resistances when a vehicle's performance is to be considered from an energy point of view.

The above 5 resistances are of differing character, but in most cases it is appropriate in calculations to refer all the considered driving resistances to the contact surface between the tires and the road. The same is the case for the driving force of the power source.

Denoting the equivalent driving force delivered at the engine but referred to the driving wheel as D_T, the following equilibrium condition is valid during driving:

$$D_T = W_{tr} + W_r + W_a + W_g + W_{ac} \tag{14.1}$$

The actual force left to move the vehicle at the wheels when the transmission losses are subtracted, D_t is:

$$D_t = D_T - W_{tr} = W_r + W_a + W_g + W_{ac} \tag{14.2}$$

In some cases it is appropriate to use the corresponding powers, where the power is equal to the force times the speed:

$$BP \cdot \eta_{tr} = P_e = D_t \cdot v = (W_r + W_a + W_g + W_{ac}) \cdot v \tag{14.3}$$

where η_{tr} is the efficiency of the system's transmission, v the vehicle velocity in m/s, and BP is the power of the engine at the flywheel.

14.2 Transmission Resistance

As long as the vehicle is self-propelled, which is assumed in the following, a form for resistance occurs which is directly dependent on the vehicle construction. That is the total friction resistance in the transmission mechanism that is necessary to transmit power or force from the powertrain to the road surface.

The transmission resistance includes all the restrictive forces that occur on the way from the power source's clutch flange to the driving wheels, and an oversight of the involved factors can be obtained by considering a truck transmission for example, as shown in Figure 14.1.

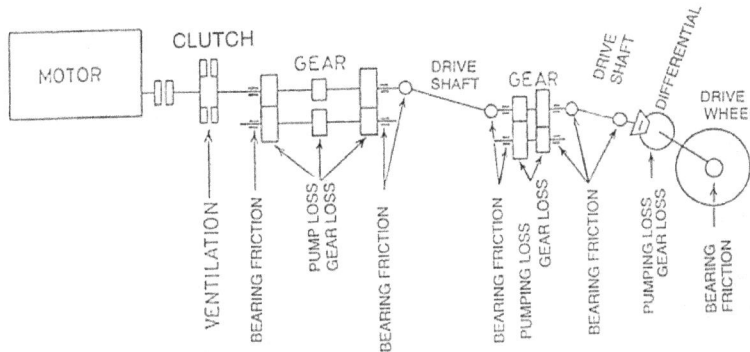

Figure 14.1: The various components in the transmission system. This is for a larger vehicle, for a passenger car, there would only be one gear set.

It is expected that the losses shown will vary with the speed and torque (the transmitted power) as well as the surrounding temperature. Further, it is clear that the different elements exhibit differing loss characteristics. For example, the torque loss due to ventilation of the clutch and oil pumping in the gear are proportional to the speed squared, while they are independent of the transmitted power. Bearing friction is slightly dependent on the speed and strongly dependent on the transmitted torque. The gearing loss is dependent on the torque as well as the speed

In modern mechanical (standard) automobile transmissions, all of the above losses are of such a modest size that for the most part, and for the purpose of vehicle dynamic considerations, it is completely satisfactory to include them as an aggregated loss, for example as previously mentioned in the form of an efficiency. One normally assumes that a conventional transmission such as that shown, can be expressed as:

$$W_{tr} = D_T(1 - \eta_{tr}) \tag{14.4}$$

with the efficiency:

$$0.85 < \eta_{tr} < 0.95 \tag{14.5}$$

which in itself may be dependent on operating conditions and the quality and viscosity of the lubricant.

As representative values of transmission efficiencies for the different components can one assume:

Clutch (ventilation loss)	0.990-0.995
Gear Box (pumping and gear friction)	0.990-0.980
Drive shaft(pumping loss)	0.990
Differential (pumping and gear friction)	0.990-0.980
Wheel friction (friction loss)	0.995

It is a characteristic of the conventional mechanical vehicle transmissions that the average efficiency approaches 95%, which indicates a very good economy. This condition is undoubtedly the most important reason that other forms of transmission with apparent advantages have a difficult time being accepted. In the case of automatic transmissions, efficiencies can be lower, due to losses in the torque converter. This loss can be dependent on operating conditions, as it is now common to lock the torque converter under stable operating conditions in order to reduce the losses there and improve vehicle fuel economy.

Figure 14.2 shows a representative values of the torque conversion ratio and the efficiency of a hydraulic torque converter. The efficiency and torque ratio are functions of the operating conditions, different from that of manual transmissions, where the torque ratio is fixed for a given gear ratio, and the efficiency near 100%. In all the following, a manual transmission will be assumed. The methods for determining the various performance and fuel economy parameters will be the same for automatic transmissions as for manual transmissions, but the detailed knowledge of the torque conversion ratio and efficiency would be required at any given operating condition to obtain specific values. Manual transmissions are still very common in European automobiles, the automatic transmission is dominant in North America.

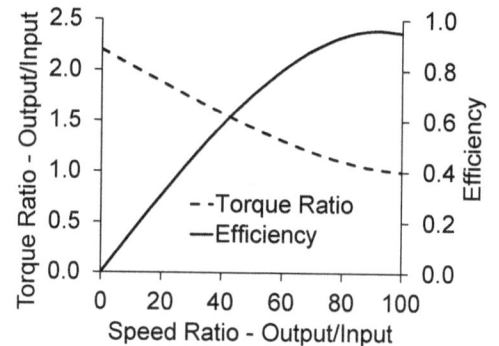

Figure 14.2: Typical torque conversion ratio and efficiency characteristic for a hydraulic torque converter.

14.3 Rolling Resistance

The rolling resistance is that involved with the rotation of the wheels of the vehicles. This can include bearing friction from the wheel bearing, but for road vehicles, by far the main portion of rolling resistance is due to the resistance encountered in the deformation of the tires. The tire is deformed as a result of different influences, and the *Rolling Resistance* is normally defined as the energy loss in the tire on account of the deformation and the resulting hysteresis. By hysteresis is understood the amount of energy that the tire will absorb when it is compressed and expanded in the course of a revolution. In addition, the surface friction in the contact surface (5-10%) and the aerodynamic friction (1-3 %) play small roles. But several studies indicate that about 95% of the rolling resistance is due to hysteresis losses in the tire materials and structure

The tire supports the load on the vehicle, through a pressure distribution on the tire, as illustrated in Figure 14.3. A tire has a complicated structure, with reinforcement in the form of belts and threads, acting in different directions. As might be expected then, there is a connection between the deformations in all three co-ordinate directions. For example, studies have shown that a force in the horizontal direction will introduce a vertical deformation.

Depending on the character of the wheels and the road, and the driving conditions as well, a larger or smaller contact surface is formed, which is deformed in accordance with the road surface at driving speed. If the wheel and road were perfectly elastic, the total normal force (wheel pressure) would be symmetrically distributed around a vertical plane that contains the wheel axle, as shown in Figure 14.3.

This is actually not the case in practice. Depending on conditions, the distribution will not be symmetrical, which is also shown in Figure 14.3 case b. This will give rise to a negative torque - rolling resistance, since the result force from the pressure distribution does not act through the center of the axle during operation. It is obvious that the smaller the deformation and the closer the material is to the ideal elastic conditions, the lower the rolling resistance.

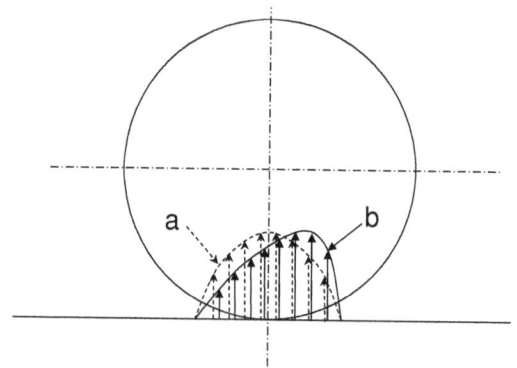

Figure 14.3: The pressure distribution exerted on a vehicle tire by the road.

The load gives rise to deformation in the tire wall. One can consider several possible conditions, which are illustrated in Figure 14.4.

a) The imposed deformation disappears completely when the unloading occurs. The process is completely elastic. In other words, completely reversible, and the working loss is equal to zero for a revolution of a wheel. This is shown in Figure 14.4a.

b) The plastic deformation caused by the loading remains unchanged after the load is removed. This

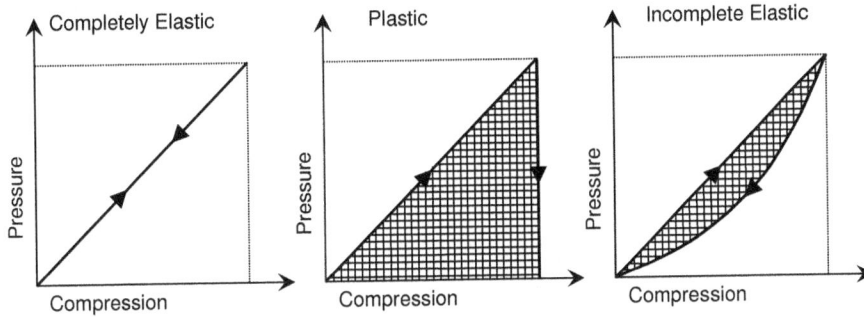

Figure 14.4: Different conditions for the compression/deformation of a tire as a result of pressure forces from the road.

is shown in Figure 14.4b.

c) In reality, a tire is not completely elastic or non-elastic and in practice a loading will therefore show a deformation that lies between the conditions shown in Figures a) and b). For a loading and subsequent unloading there must be an energy loss- a hysteresis loss, which is shown by the area between the loading and unloading curves in Figure 14.4c.

14.3.1 Deformation of Automobile Wheels

The automobile wheel (air-rubber ring) is in principle a reinforced rubber torus that is suspended and held to the wheel rims (and therefore to the wheel disc) with an internal inflation pressure. The inflation pressure is typically on the order of 1-10 bar, depending on the wheel size and use. The most noticeable deformation of the wheel is the radial or vertical compression due to the wheel pressure. This deformation is a necessary prerequisite for driving on non-level roads, since it prevents the vertical acceleration of the solid center of the wheel from attaining unacceptably high values. The acceleration at the center of the wheel is transmitted to the suspension system. The tire surrounds irregularities smaller than the instantaneous contact surface and acts as a pressure spring for larger irregularities which correspond to waves in the roadway. This reduces the load on other parts of the suspension system and the rest of the vehicle, and results in a smoother ride for the passenger.

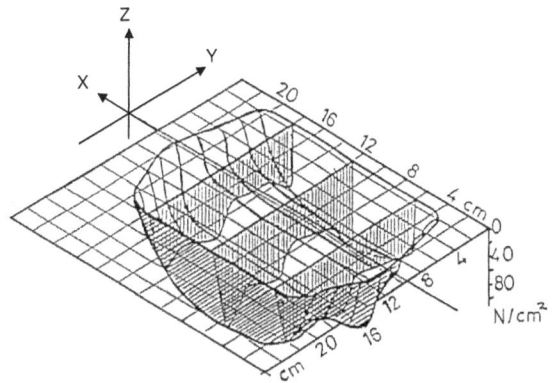

Figure 14.5: An example of the pressure distribution on the contact surface of a loaded tire. The tire pressure is 6.5 bar, and the vehicle is moving at a steady speed of 18.3 km/h with a load of 33700 N.

The static vertical deformation is naturally dependent on the inflation pressure, since the tire doesn't have much stiffness of its own. The tire will flatten out until the contact surface area times the inflation pressure is equal to the wheel pressure (normal force on the wheel divided by the final contact area). This is not exactly the case, since the tire has a certain stiffness of its own, dependent on the shape and construction of the tire, which gives a resistance in addition to the pressure. The actual tire's deformation under given conditions can only be found by experiment.On account of the tire's own stiffness and the variation of this, (depending on the nature of the reinforcement) the distribution of the surface pressure over the contact face of the tire is not uniform. The variation for the actual wheel depends on the operating condition (acceleration, braking, driving in a curve). Figure 14.5 shows an example of a pressure distribution on a loaded tire.

In an operating vehicle, the deformation process is dynamic, and there can be a speed dependence involved in hysteresis. Static results can be used to show the general tendencies. The vertical deformation and the horizontal deformations for static loadings of two types of tire are shown in Figure 14.6. The vertical deformation is shown on the x-axis, and the horizontal deformation is shown as one of the curve parameters, along with the inflation pressure.

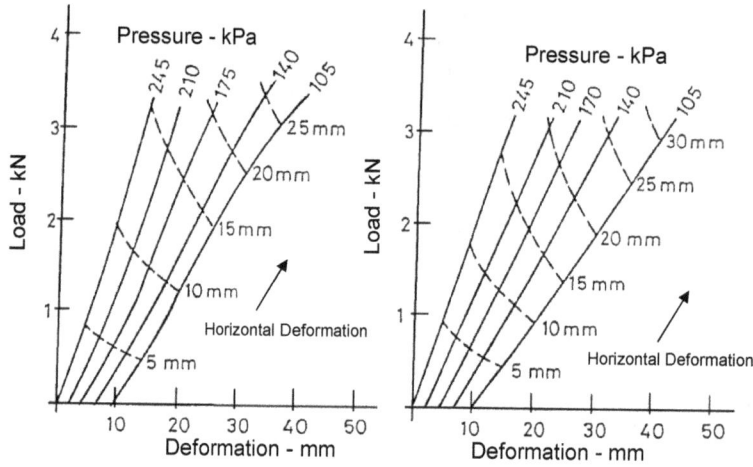

Figure 14.6: Tire load as a function of vertical deformation, with horizontal deformation perpendicular to the wheel and tire pressure as parameters for two types of tires. The figure on the left is for a 13 inch (33cm) cross ply tire, the figure on the right for a 13" radial tire.

The vehicle's or wheel's rolling resistance is, therefore, dependent on the elasticity, plasticity, and friction characteristics of the parts, and it is understandable that with incomplete elasticity the larger the flexibility of the parts, the greater the rolling resistance. The form and the area of the "hysteresis wing" will depend on the material and the deformation history and, therefore, the characteristics of the wheel that are dependent on the hysteresis vary with the deformation frequency, that is with the wheel's rotational speed. But in all cases, the hysteresis wing's area is a measure of the work that is converted to heat in the tire and a measure of the actual rolling resistance.

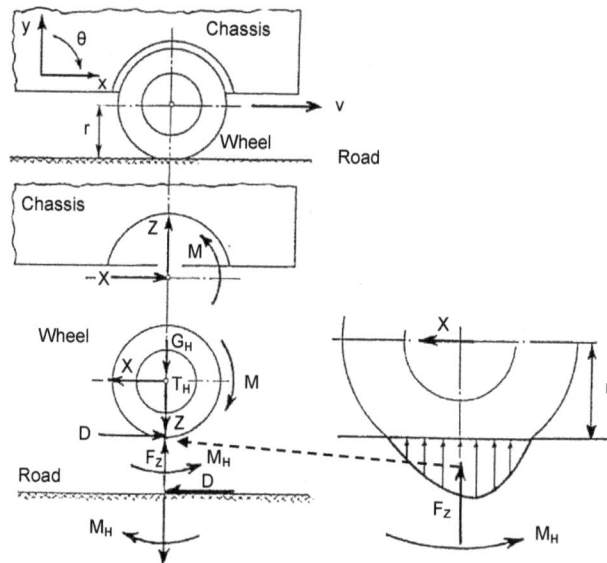

Figure 14.7: Forces, moments and pressure distribution involved in the movement of tires on vehicles.

In order to describe a wheel's motion, one must write the equations of motion for the planar system shown in Figure 14.7. The center of gravity T in the axis of rotation moves in the direction of the x-axis and in addition turns through angle θ. Considering the wheel and the chassis separately, the reaction forces X and Z the horizontal and vertical force between the wheel and the chassis, G_H, the weight of the tire and wheel F_Z, and D, the tractive force between the wheel and the roadway is obtained. In addition there are the moments M, originating primarily from the engine, $D \cdot r$, due to the tractive force at the wheel radius, r. An additional moment, M_H, is found, since the reaction force from the pressure distribution is not symmetrical with respect to the wheel axis. This eccentricity causes a moment between the tire's contact area and the road.

For a wheel mass m and the accompanying weight $m \cdot g$ and moment of inertia I, one has the following equations of motion for the horizontal, vertical and rotational accelerations respectively:

$$m \cdot \ddot{x} = m \cdot a = D - X \tag{14.6}$$
$$m \cdot \ddot{z} = F_Z - (Z - G_H) \tag{14.7}$$
$$I \cdot \ddot{\theta} = M - D \cdot R - M_H \tag{14.8}$$

Considering the case where the wheel is rolling without any form for acceleration, the translational and rotating acceleration are zero, that is: $\ddot{x} = 0$ and $\ddot{\theta} = 0$ and the equations give

$$D = X \tag{14.9}$$
$$M = D \cdot r + M_H \tag{14.10}$$

If the driving moment, M, is assumed to be nil and if there is no braking moment or bearing friction, there is nonetheless - as experience also shows with a push cart - a horizontal force, $-X$, on the wheel. It must be applied to overcome the resistance, $-D$. This resistance is called the rolling resistance and is normally denoted as W_r. That is to say:

$$-D = W_r \tag{14.11}$$

14.3.2   Rolling Resistance Coefficient

Through experiments it has been found that the rolling resistance especially varies with the wheel load, F_Z, and can often be assumed to be a linear function such that:

$$W_r = f_r \cdot F_Z \tag{14.12}$$

where f_r is a dimensionless proportionality constant, called the rolling resistance coefficient.

Since the situation is being considered without driving or braking moments, one obtains for $M = 0$:

$$-D \cdot r = M_H = W_r \cdot r \tag{14.13}$$

or:

$$W_r = \frac{M_H}{r} \tag{14.14}$$

In connection with rolling resistance, one often works with the power that is absorbed by the tire element during rolling, that is:

$$P_{loss} = W_r \cdot v = f_r \cdot F_Z \cdot v \tag{14.15}$$

Figure 14.8 shows rolling resistance as a function of tire inflation pressure and loading. This curve verifies as well, that one in practice can usually approximate rolling resistance as a linear function of the loading, though at high speeds this is not so accurate.

The power required to overcome rolling resistance for single wheels with two different kinds of tires is shown in Figure 14.9

In the above, rolling resistance for a single wheel is treated, but to determine the necessary driving force or power, it is necessary to show the total rolling resistance for the entire vehicle.

From the expression: $W_r = f_r \cdot H$ one can use the following for the entire vehicle:

$$W_r = f_r \cdot G_S = f_r \cdot m_o \cdot g \tag{14.16}$$

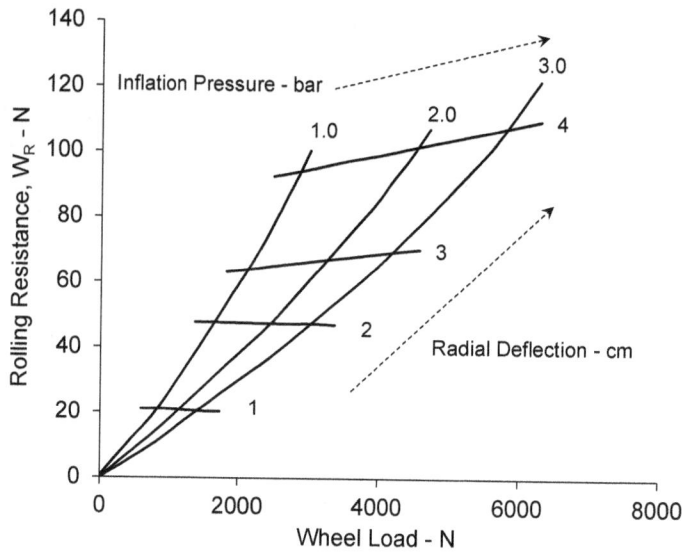

Figure 14.8: Rolling resistance of a tire as a function of tire inflation pressure and wheel loading for a type 65-13 tire at 30 km/h.

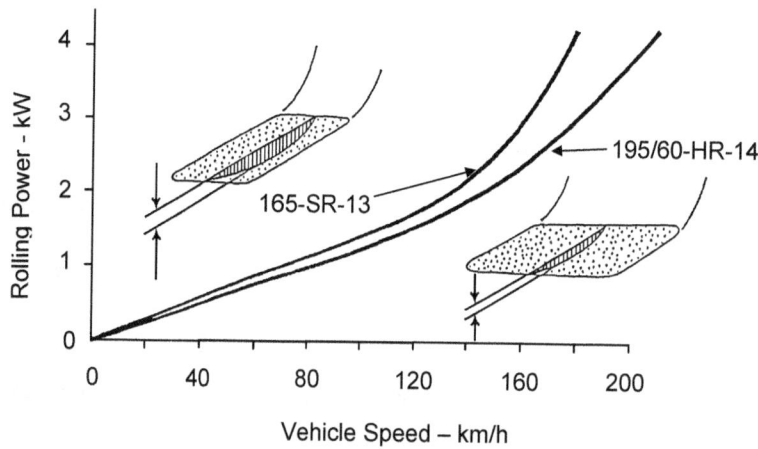

Figure 14.9: Power consumed by rolling resistance of a wheel as a function of vehicle speed for two tire types.

where: G_S is the vehicle's total weight, m_o is the vehicle's total mass, and H is the force for one wheel.

But Equation (14.16) is only valid if one can assume the same rolling resistance coefficient for all the wheels. Expressed in power, one has:

$$P_r = f_r \cdot G_S \cdot v = f_r \cdot m_o \cdot g \cdot v \qquad (14.17)$$

where: v is the speed in m/s.

In principle, one can speak of different road surfaces for the wheels for each axle, different types of tire, tread pattern, profile depth and loading, and in this case it can be relevant to use an expressions such as:

$$W_{r,tot} = f_{r,F} \cdot H_F + f_{r,B} \cdot H_B \qquad (14.18)$$

where: $W_{r,tot}$ is the total rolling resistance, $f_{r,F}$ is the rolling resistance for the front wheels, $f_{r,B}$ is the rolling resistance for the back wheels, The most general expression allows for different loads; $H_{F,i}$ and different rolling resistance coefficient $f_{r,i}$ at each wheel:

$$W_{r,tot} = \sum_{i=1}^{N_{wheels}} f_{r,i} \cdot H_{F,i} \tag{14.19}$$

The given rolling resistance coefficient, f_r, is dimensionless and is neither directly measurable nor constant. The determination must therefore be made by measuring of the actual rolling resistance for the actual wheel with varying operating conditions. Such measurements have shown that the rolling resistance coefficient depends on a large number of parameters, partly for wheels in general and partly for the actual wheel. The difficulty with the determination of the different factors arises because the rolling resistance in practice is not constant, and is physically dependent on a series of factors:

Reinforcing pattern and material; Surface material; Tire diameter, shape, width and inflation pressure; wheel pressure; vehicle speed; tire temperature, transfer of power/braking forces.

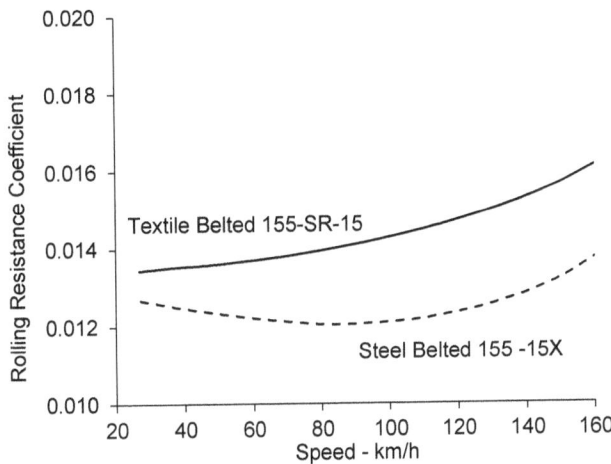

Figure 14.10: Rolling resistance coefficient versus speed for 2 different types of tires.

Figure 14.10 shows the rolling resistance coefficient for 2 different types of tire, determined as average values for a large number of measurements. The rolling resistance coefficients shown here have been found by testing passenger car tires. For trucks, the following resistance coefficient is smaller as a rule - if one considers the same type of tire - on account of their significantly higher working pressure.

Gillespie gives typical values for rolling resistance coefficients for different vehicles and driving surfaces [111]. They are shown in Table 14.1

Table 14.1: Typical values for rolling resistance coefficients from Gillespie [111]

Vehicle type	Concrete	Surface Medium Hard	Sand
Passenger Car	0.015	0.08	0.30
Heavy Truck	0.012	0.06	0.25
Tractor	0.02	0.04	0.2

Gillespie also gives an expression for the rolling resistance coefficient as a function of speed:

$$f_r = f_o + 3.24 \cdot f_s \left(\frac{V}{160} \right)^{2.5} \tag{14.20}$$

Where V is the speed in kph, and f_0 and f_s are coefficients given in Figure 14.11 as a function of tire inflation pressure. The data from Gillespie have also been fit to the following expressions:

$$f_s = 0.02 - 0.02 \left(1 - \exp\left(-0.00934 \left(p - 68.93 \right) \right) \right) \tag{14.21}$$

$$f_o = 0.02 - 0.0127 \left(1 - \exp\left(-0.010155\left(p - 68.93\right)\right)\right) \tag{14.22}$$

Equations (14.21) and (14.22) are valid for $68.9 \le p \le 350 kPa$, and p is the gage tire pressure in kPa.

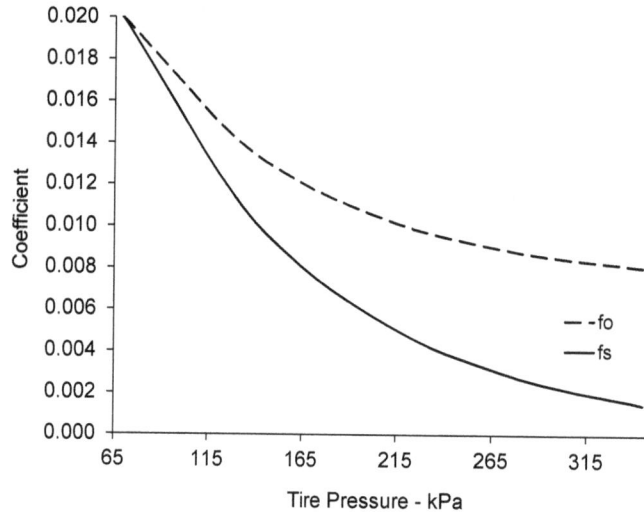

Figure 14.11: Coefficients for Equation (14.20). [111].

Gillespie also cites estimates of rolling resistance coefficients of heavy duty vehicles. These results are:

$$f_r = (0.0041 + 0.00000256 \cdot V) \cdot C_h \tag{14.23}$$
$$f_r = (0.0066 + 0.00000288 \cdot V) \cdot C_h \tag{14.24}$$

For the speed, V, in kph.

Equation (14.23) is valid for radial tires and Equation (14.24) is valid for bias-ply tires. The coefficient C_h is a function of the road surface:

$C_h =$ 1.0 for smooth concrete
$C_h =$ 1.2 for worn concrete, brick, cold asphalt
$C_h =$ 1.5 for hot asphalt

14.4   Air Resistance

The motion of a vehicle affects the air surrounding it. Some parts of the air mass surrounding a vehicle will be set in motion by the motion of the vehicle, and to do this requires a certain amount of energy, which the moving body must deliver. This is the basic reason for air resistance. The importance of aerodynamics in the design of conventional road vehicles increased sharply after the oil crises of the 1970's. Vehicular resistance to air motion has been reduced significantly since then, particularly for passenger cars. The basis for aerodynamic considerations with a road vehicle is also the theory that is used for aircraft construction but one must be aware that treatment of vehicle chassis with regard to aerodynamics would hardly be possible without an extended use of experiments, and often experiments that are difficult to conduct.

The reasons that experimental results are so necessary for road vehicles are:

1. The flow around the body of the vehicle is three-dimensional, while the flow round wings is mainly two dimensional.

2. For practical reasons, the chassis of a road vehicle cannot be as effective as the thin aircraft body.

3. In contrast to an aircraft, the road vehicle must follow a stationary surface: the road

4. The aircraft can be divided into a body and wings.

Even though great progress has been made in the area of computational fluid dynamics, for most practical purposes it must be recognized that the exact analysis of the complex air flow around a vehicle is still not realizable. However, recent improvements in computational fluid dynamics have made it an increasingly useful tool for determining and improving the air resistance of vehicles.

When a body moves through air, a resistance to its movement will be created, since at each instant the vehicle displaces a given amount of air. The path that the air particles describe is normally characterized with streamlines, denoted by lines in the plane of the paper. Figure 14.12 shows a theoretical, inviscid and therefore resistance-free flow, where regardless of the shape of the body, the streamlines join again behind the body, and thereafter move undisturbed (ideal flow potential flow). This type of ideal flow

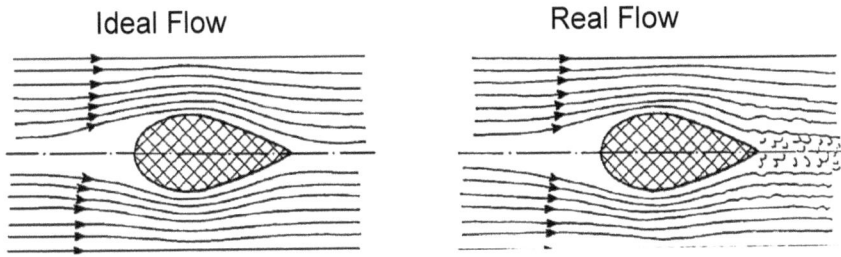

Figure 14.12: A schematic view of ideal, streamline flow around an obstacle and real flow with separation.

cannot be realized in practice, since air is viscous. This causes the well-ordered flow around even a well-rounded body to be disturbed. When the air is suddenly is forced to the side, up and/or down, it is impossible to avoid swirling flow, even for a streamlined vehicle.

The air has mass, which gives rise to inertial forces when the flow speed and/or direction is changed. From Bernoulli's equation, one finds that when a fluid in motion is stopped, the static pressure rises. In this situation one speaks of a dynamic air pressure:

$$p = \frac{1}{2}\rho v^2 \qquad\qquad (14.25)$$

This is an expression for the kinetic energy and its conversion into potential energy.

The dynamic pressure can originate when the moving air is stopped by sort kind of obstacle and the speed is reduced to zero. The relationship can be expressed by converting the kinetic energy to air pressure. Since the velocity of importance in Equation (14.25) is the relative velocity between a surface and the air, if a surface area is moved through the air with the same relative velocity, the pressure at the surface would be the same as given by Equation (14.25). If the entire surface is exposed to air brought to a standstill on the surface (stagnation) then the force would be given by:

$$F = P \cdot A = A \cdot \frac{1}{2}\rho v^2 \qquad\qquad (14.26)$$

where v is the relative velocity between the stagnant air and the moving vehicle and A is the area exposed to the pressure. In the case of a moving body - for example a vehicle - the dynamic pressure is created,

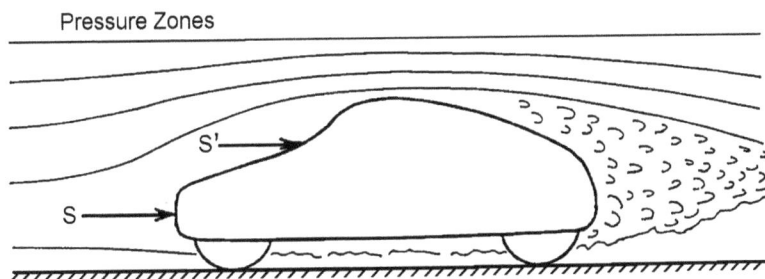

Figure 14.13: Flow around a vehicle, showing pressure zones and wake formation.

as shown in Figure 14.13 in the pressure zone S, where the flow stagnates. Another pressure zone S' can be characterized by the front windscreen, but the pressure there is not the stagnation pressure given by Equation (14.25), since the flow does not stop completely. Other pressure high areas that can be formed are caused by things such as lights, rear view mirrors, roof carriers.

Table 14.2 gives the importance of the relative speed on the pressures developed. The table gives the dynamic pressure for varying speed, altitude, h, and therefore the air density, ρ (kg/m) at various speeds.

Table 14.2: Stagnation pressure as a function of speed and altitude. Pressure in Pa.

Speed v -m/s	v - km/h	h=0m $\rho = 1.27$	h=1000m $\rho = 1.15$	h=2000m $\rho = 1.03$	h=3000m $\rho = 0.93$
0	0	0	0	0	0
10	36	63.7	57.4	51.6	46.4
20	72	255	230	206	189
30	108	573	515	464	417
40	144	1018	917	825	741
50	180	1592	1433	1288	1158
60	216	2285	2062	1855	1670

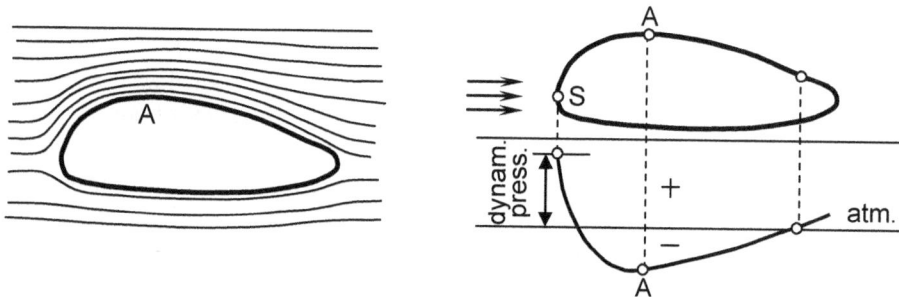

Figure 14.14: Pressure distribution arising from flow around a streamlined object.

When the air particles meet an obstruction - for example a streamlined object - they are deflected in a direction tangent to the surface and move away from it under the influence of centrifugal force. Therefore, a low pressure area will be created between the air particles and the surface. Typically, a pressure distribution will be created, as shown in Figure 14.14. In zone S (stagnation zone like that shown in Figure 14.13), the speed is zero and the pressure is therefore the greatest. In zone A, the speed is the greatest and the pressure correspondingly lowest, symbolized by the smallest distance between the stream lines. After the largest cross sectional area of the object - corresponding to an increase of the flow cross sectional area - the speed falls and the pressure increases. Here, ideal flow is shown, where the streamlines rejoin and there is no wake.

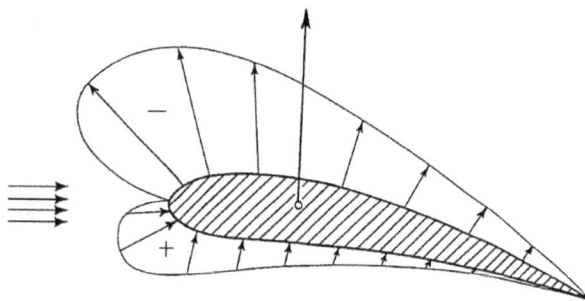

Figure 14.15: Pressure distribution arising from flow around an airplane wing, showing low pressure above and high pressure below.

If one now considers the shape of the chassis as similar to that of the wing of an aircraft, the static pressure is of decisive importance. For the flow conditions for a slightly rounded wing, the flow is divided at the pressure zone's point of attack as shown in Figure 14.15.

On the upper side of the wing, the flow is accelerated, which means low pressure. On the lower side of the wing, the flow is decelerated, which means high pressure. Experiments have shown that the low pressure is the lowest where the profile is the widest, that is where the velocity is the greatest. Then the pressure in the low pressure area increases (becomes closer to atmospheric) moving in a direction towards the tail. The pressure on the wing's under side has a similar development, but is significantly smaller, and there is, as is well known, a net upwards force. The pressure difference in Figure 14.15 is greatest near the front of the wing and becomes smaller towards the back.

Figure 14.16 shows the scaled pressure distribution for a actual vehicle. Of note are the low pressure regions in front of the front wheels, the high pressure region on windshield and nearby rising surfaces, and the large region of low pressure at the top of the vehicle. There is a large upwards force, since the low pressures act on horizontal surfaces. This force is opposed to the weight of the vehicle, and lessens the force on the wheels that gives traction to the vehicle. It can in general be said that for small vehicles the upwards force is small, and can easily be negative. On high performance vehicles, the use of "spoilers", or inverted wings, are often to create a downwards force to improve traction.

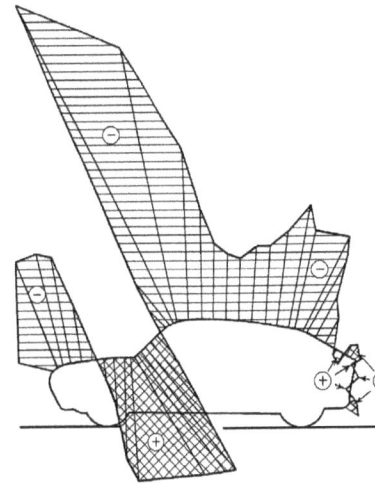

Figure 14.16: Pressure distribution arising from flow around a vehicle.

14.4.1 Pressure or Shape Resistance

As shown earlier, at the top of a wing or vehicle, there is a region where the velocity is high and the pressure low. Following this, there is a region in the streamline flow region where the velocity starts to decrease and the pressure starts to increase as one moves towards the back of the vehicle. This is a result of the shape of the object, giving an external pressure field as shown for example in Figure 14.16.

To show the consequences of this, a simplified version of the flow over an airfoil shape is shown in Figure 14.17. The pressure and pressure gradient are shown schematically at the top of the figure. The positive pressure gradient (pressure increasing as distance along the surface increases) acts on the boundary layer. In boundary layer theory, it is normally assumed that the pressure in the boundary layer is uniform at any location. This means that there are no pressure gradients in the direction perpendicular to the surface. Thus, the free stream pressure is imposed upon the whole boundary layer. When the pressure increases in the direction along the surface, this opposes the motion of the boundary layer, causing a decrease in velocity. Without friction, the decrease in velocity would match that imposed by the pressure rise, according to Bernoulli principles. In the boundary layer, however, the friction has caused a loss of kinetic energy, and the fluid near the surface has little kinetic energy to oppose the pressure increase. Then at some point after the pressure gradient becomes positive, the flow near the surface on the boundary layer will be reversed, causing the boundary layer to become detached, or separate from the surface. This is shown for the boundary layer profile at the right of Figure 14.17. The resulting flow behind the vehicle becomes complex, and a significant amount of energy is lost in the wake produced. This results in a low pressure behind the vehicle, and given the high pressure on the front of the vehicle, cause a drag force due to the pressure difference.

By this process, the "external flow", which actually moves as a frictionless flow, is prevented from "closing itself" behind the body in question, and the boundary layer causes an energy exchange and causes the formation of varying sizes of vortices in the wake. This so-called shape resistance , which comprises about 70-80 % of the total air resistance, is mainly created by the dynamic pressure at the front part of the vehicle and the low pressure that the vortices create, which in return depends on the boundary layer. The pressure in the boundary layer will increase with distance. But if the body is reduced in size sufficiently slowly, the pressure in the external flow, in spite of the presence of the boundary layer, will increase so slowly that the boundary layer mainly can follow the shape of the object - then the vortex

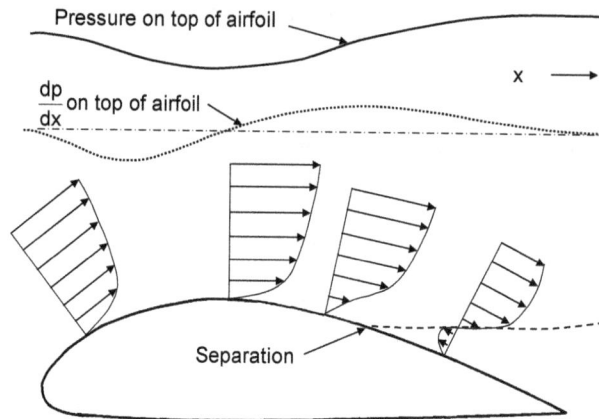

Figure 14.17: Schematic illustration of the separation of the boundary around a shaped object.

field or wake can be relatively small and the air resistance modest For a more sudden area reduction, the pressure in the external flow will increase so strongly that the boundary layer's reverse flow will form a powerful vortex.

14.4.2 Surface or Friction Resistance

This portion of the air resistance is created by the viscous forces on the surface of the vehicle, caused by the motion the air that passes over the surface of the vehicle. An air boundary layer on the order of 10 mm is created during driving. The boundary layer, which increases in the direction of the flow, creates shear stresses, which in turn cause a surface resistance. This is the largest at the front end of the object

For laminar flow in the boundary layer, a tangential force per unit area, F_l, is created, which can be expressed as:

$$F_l = \mu \frac{v}{a} \tag{14.27}$$

where μ is the viscosity (for air $\mu = 18 \cdot 10^{-6} Ns/m^2$), v is the flow speed and a is the boundary layer thickness.

The discussion of surface friction over flat plates can be found in many books on fluid mechanics, and it will only be recalled here that the shear forces, F_D, are correlated to the flow through the use of a drag coefficient, C_D, defined as:

$$C_D \equiv \frac{F_D}{\frac{\rho}{2} \cdot v_\infty^2 \cdot A_D} \tag{14.28}$$

Where A_D represents the surface area over which the flow at the velocity v_∞ is passing.

The form of this resistance is similar to that of the total air resistance, Equation (14.32) that will be discussed later. For the present, only the form of the expression is required. Using Equations (14.32) and (14.28), the ratio of the frictional drag force to the total aerodynamic resistance can be obtained:

$$\frac{F_D}{W_a} = \frac{C_D \cdot A_D}{C_W \cdot A_f} \tag{14.29}$$

where C_W is the air resistance coefficient, and A_f is the frontal area of the vehicle.

Consider a passenger car. It can be shown from fluid mechanics, that for flow over flat plates at speeds and dimensions corresponding to a this type of vehicle, the value of C_D is in the order of 0.003, and the total surface area about 25 m². As will be shown later, the frontal area is on the order of 2.5 m², and the resistance coefficient on the order of 0.35. Then putting these values into Equation (14.29):

$$\frac{F_D}{W_a} = \frac{0.003 \cdot 25}{0.35 \cdot 2.5} \approx 0.1$$

it is found that the frictional resistance is about 10-15 % of the total aerodynamic resistance of the vehicle. The boundary layer though, has a great significance for the flow on the body of the vehicle, since it has an influence on the external flow, that is the flow that determines the pressure, or shape resistance.

14.4.3 Induction Resistance

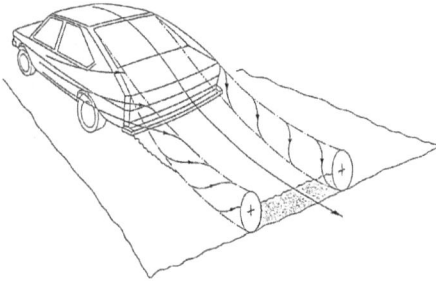

Figure 14.18: Vortices generated by the difference between the low pressure at the top of the vehicle and the high pressure at the bottom of the vehicle

The induced resistance is a result of the pressure distribution that is formed around the vehicle during driving. Normally the pressure under the vehicle will be higher than the pressure over it, since air velocities are low and there are surface friction forces. The pressure will try to equalize itself by the air flow outside of the body from the underneath the vehicle to the low pressure area at the top.

With this, two vortices are formed (induced), as shown in Figure 14.18, whose energy is taken from the vehicle and thus manifests itself as a resistance. The larger the pressure difference between the under side and the over side, the larger the induced resistance will be. It typically comprises only about 0-10 % of the total air resistance.

Inner Resistance or through Flow

The inner resistance is that portion of the air resistance that is characterized by the necessary air for cooling of engine parts and to the ventilation system. Depending on how appropriately the intake and exhaust openings are built and placed as well as the varying part of opening and closing of the fan system, the air through-flow gives a loss. The flow for cooling has the potential for significant losses, since there is a complicated flow path between the entrance to the radiator and the exit of the engine compart-

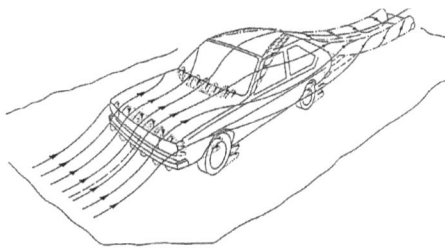

Figure 14.19: The flow due to ventilation and cooling needs of a vehicle.

ment, normally out the bottom. There has a been a trend towards smaller openings for cooling air, so that only the required amount of air is admitted and the losses kept to a minimum. The flow for cooling and ventilation is shown in Figure 14.19. Today, a complicated ventilation system is also needed because of increasing demand for air conditioning in the vehicle. The inner air resistance comprises about 10-20 % of the total air resistance of a passenger vehicle.

14.4.4 Air Resistance Equation

The above mentioned air resistance contributions cannot be separately determined in practice, and, therefore, it is not possible through calculations or separate experiments to determine how large a change in the total air resistance any change in details can achieve. Normally it can be said that the observed energy loss will to a large extent increase with increasing speed, but how much fuel consumption will increase is especially dependent on the shape of the chassis and cross section area.

On the assumption that pressure forces are the largest part of the aerodynamic resistance, one can use Equation (14.25) as the basis for the calculation of the observed air resistance, W_a, which can be expressed as being proportional to:

$$W_a \propto A \cdot \frac{1}{2}\rho v_r^2 \qquad\qquad (14.30)$$

Table 14.3: Typical frontal areas of different types of vehicles

Truck	$A_f = 4.0 - 9.0 m^2$
Busses	$A_f = 4.5 - 7.0 m^2$
Passenger cars over 1.5 ton	$A_f = 2.5 - 4.0 m^2$
Passenger cars under 1.5 ton	$A_f = 1.5 - 3.0 m^2$
Motorcycles	$A_f = 0.7 - 1.0 m^2$

where ρ is the air density, $v_r = v \pm v_w$ is the relative speed between the air and the vehicle, v is the vehicle speed (m/s) and v_w the component of the air speed (wind) in the direction against the vehicle's direction of motion (m/s). It should be noted that although the effect of relative wind motion is included in the velocity, the wind direction, especially side wind, can have a significant on flow around a vehicle and, therefore, the air resistance. The area, A used in Equation (14.30) is taken as the frontal area of the vehicle.

The air resistance increases with the square of the relative speed v. Now the dynamic pressure per unit area and its corresponding force if the entire frontal area is exposed to that pressure is:

$$W_a' = \frac{1}{2}\rho\, v_r^2 \cdot A_f \tag{14.31}$$

where A_f is the vehicle's frontal area, that is the cross section of the vehicle perpendicular to the direction of forward motion of the vehicle.

In the case where only a vehicle's general dimensions (height, width, length etc.) are known, A_f can be calculated from $A_f \approx 0.9 \cdot s \cdot h$, where s is the track (width between the tires) and h is the height of the vehicle, or from $A_f \approx (0.9 \cdot h)(0.9 \cdot b)$, where b is the width of the body of the vehicle.

Though the basis for the calculation is the stagnation pressure, the previous sections have shown that this pressure does not act on the entire frontal area, and there are other resistances as well. Therefore, to take into consideration the effect of the vehicle's shape and the nature of the surface, it is necessary to correct with a shape factor, that is air resistance coefficient, C_w, often called a drag coefficient. The commonly used expression for the determination the the vehicle's air resistance is then:

$$W_a = \frac{1}{2}\rho\, v_r^2 \cdot A_f \cdot C_w \tag{14.32}$$

where $\frac{1}{2}\rho\, v_r^2$ is the stagnation pressure dependent portion, and $A_f \cdot C_w$ is the vehicle dependent portion. C_w is often regarded constant, and independent of the vehicle speed. In the case of side wind, though, C_w is known be dependent on wind direction.

Due to the 2nd order dependency, the air resistance is the dominant factor at high speeds. In the case of high speed vehicles, the air resistance coefficient is, therefore, of very great significance.

14.4.5 Air Resistance Coefficient

The air resistance of vehicles is a very complicated matter and one can find entire books on the subject, for example Hucho [112]. In the following, some typical values of for the aerodynamic resistance parameters for a variety of road vehicles are given. There are many references available for more specific information concerning the detailed discussion of the influence of various parameters on the air resistance of vehicles. The values presented here are to give an idea of the current status of the the aerodynamic resistance of vehicles, and to give representative values for calculations. Table 14.4 shows typical values for frontal air resistance coefficient, frontal areas, and their product as a function of vehicle size for recently produced passenger cars. The resistance coefficients are generally smaller for the large size vehicles, but the areas are larger, so the product of the two does not increase so much with vehicle size.

Table 14.5 lists the range of values for the air resistance coefficient for a variety of road vehicles. The larger vehicles have larger coefficients, since they typically have a large flat frontal surface, where pressures close to the stagnation pressure act on the surfaces. Design details play an important role in determining the air resistance, small changes as rounding vehicle edges, or changing windshield angles can have a significant effect on the air resistance of a vehicle. Significant improvements have been made

Table 14.4: Typical values of air resistance coefficients and frontal areas for modern passenger cars as a function of vehicles weight.

Mass - kg	$A_f - m^2$	C_w	$C_w A_f - m^2$
830	1.76	0.36	0.634
850	1.87	0.33	0.617
1050	1.98	0.29	0.574
1300	2.09	0.29	0.606
1550	2.15	0.33	0.710

in vehicle aerodynamics since the oil crises of the 1970's. In 1980, air resistance coefficients typically ranged from 0.39 to 0.50 and by the mid-1990's, this range was lowered to about 0.29 to 0.36. Since air resistance is the most significant factor in determining vehicle power requirements at speeds over about 60 km/h, this has resulted in significant fuel consumption reductions at highway speeds. Other types of vehicles have also had significant improvements made in their air resistance characteristics. In general, the larger and older the vehicle, the higher the air resistance coefficient.

Table 14.5: Range of air resistance coefficients for different types of road vehicles.

Type of Vehicle	Range of C_w
Passenger Car	0.27 - 0.45
Delivery Vans	0.33 - 0.51
Buses	0.41 - 0.65
Trucks	0.48 - 0.75
Trucks with trailers	0.55 - 0.95

Motorcycles are much smaller than passenger cars, but because of the design and the lack of an enclosure for the rider, have much higher air resistance coefficients than cars. Typical values of air resistance coefficients for motorcycles with sitting riders are in the range of 0.48 to 0.55, and the range of $C_w A_f$ values is on the order of 0.38 to 0.45. The values depend on the position and size of the rider.

14.5 Driving Condition

The three types of driving resistance, transmission, rolling and aerodynamic, that have been discussed up to now correspond to the necessary driving force for moving the vehicle with a given speed on a flat horizontal road. This is a standard driving condition, that is characterized by:

- Driving on a horizontal road surface

- Driving with no wind

- Driving with no acceleration.

The driving force is then defined by

$$D_T \cdot \eta_{tr} = D_t = W_r + W_a \tag{14.33}$$

$$D_T = \frac{W_r + W_a}{\eta_{tr}} \tag{14.34}$$

At most driving conditions, the power of the engine is greater than that needed to operate the vehicle. This extra power or torque, gives an indication of the ability of the vehicle to accelerate or ascend a grade. Therefore, on the basis of this, a graphical picture is often used to show the excess power determined by:

$$BP \cdot \eta_{tr} = P_e = D_t \cdot v = (W_r + W_a) \cdot v \tag{14.35}$$

where: BP is the engine power and P_e is the effective power at the wheel.

An example of the excess power available for a passenger car is shown in Figure 14.20. In this case, the engine of Figure 14.22 was applied to a vehicle with a mass of 1300 kg, $C_w A_f = 0.71 m^2$, $f_r = 0.013$, a tire diameter of 0.254 m, a transmission efficiency of 0.90, and an overall gear ratio, $\theta_{tot} = 2.94$. Assuming the engine speed of 6061 rpm does not exceed structural limits, the maximum vehicle speed where the total resistance is equal to the power at the wheels is 197 km/h. At speeds below this, there is excess power available at the wheels which could be applied to acceleration or to climbing a grade. Note that at low speeds the aerodynamic and rolling power requirements are on the same order of magnitude, but that at high speeds the aerodynamic power requirement is dominant.

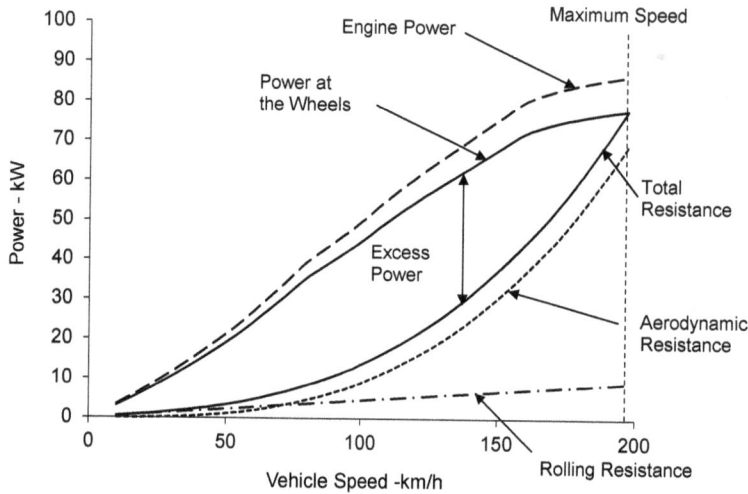

Figure 14.20: The power available to drive a vehicle, the power required to drive the vehicle on a flat road with no wind, and the excess power available as function of vehicle speed.

14.6 Operating Resistance

The standard driving condition is an ideal condition, and depends on the vehicle resistances that are to significant degree also dependent on the nature of the vehicle, size, tires, *etc.* In practice, the vehicle is found in a variety of operating conditions, where the road gradient and the acceleration play important roles. That is, they are important with respect to determining the driving resistance, which depends on the actual vehicle operation and, therefore, the surroundings in which the vehicle is found: the vehicle's driving pattern

14.6.1 Gradient Resistance

Gradient resistance is in principle very simple to treat, since at each point of the path of the vehicle, it is given by the component of the force at the center of gravity in the direction of motion of the vehicle:

$$W_g = m_o \cdot g \cdot sin(\alpha) \qquad (14.36)$$

where α is the gradient, or angle of inclination.

The situation is shown in Figure 14.21. Gradient resistances effects on the dynamics of a vehicle are modest in a low country like Denmark, but even in a relatively flat country there is a not insignificant influence on fuel consumption, since the potential energy stored in the vehicle during the climbing of a hill can only be recovered to a

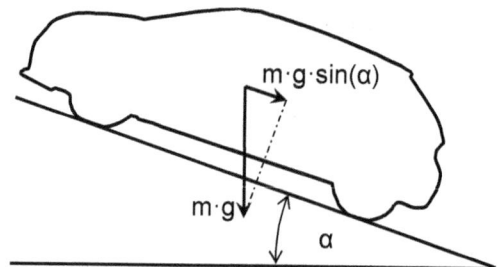

Figure 14.21: The forces involved when a vehicle climbs a grade.

modest degree. It is, therefore, common practice to establish allowable gradients for public highways, depending on the capacity of the road and the speed potential. The higher the speed, the lower the maximum allowed gradients. Motorways are build for the highest speeds, and consequently have the lowest gradients. Maximum gradients of 4 to 6% are typical for motorways. In extreme mountainous terrain, this can only be accomplished through the construction of tunnels. For common highways, maximum gradients are higher, typically 8 to 10%. In common mountainous roads, gradients range from 8 to 15%, and in some cases up to 20% as seen on the Stalheimkleiva road in Norway. There are a few streets in the world, where the gradient is as high as 30% but these are few and the stretches are quite short. The grades given are not based on the sine of the angle of inclination, but rather the tangent. As Table (14.6) shows, the error resulting from assuming that the sine is equal to the tangent is less than 2% for gradients under 20%, and the following approximate expression is used:

$$sin(\alpha) \approx tan(\alpha) = \frac{p}{100} \tag{14.37}$$

or:

$$W_g = m_o \cdot g \cdot tan(\alpha) = \frac{m_o \cdot g \cdot gradient\ in\%}{100} \tag{14.38}$$

If the gradient is, for example, given to be 8%, the road rises 8 meters for a horizontal line of 100 meter.

Table 14.6: Percentage error between sin and tan functions

%	tan α	α - deg	sin α	Error in %
5	0.05	2.82	0.0494	1.2
10	0.10	5.67	0.0987	1.3
15	0.15	8.50	0.1478	1.5
20	0.20	11.33	0.1965	1.75
25	0.25	14.00	0.2419	3.4
30	0.30	16.67	0.2868	4.6
	0.35	19.33	0.3311	5.7

14.6.2 Acceleration Resistance

When a body with mass, m_o, is accelerated, the required force can be written from Newton's second law as:

$$F_f = -m_o \frac{dv}{dt} = -m_0 a \tag{14.39}$$

since the force is in the opposite direction as the acceleration. It is most common to write for the vehicle:

$$W_{ac} = m_o a \tag{14.40}$$

Now the accelerating system has contributions from several different components. In addition to the integrated translational movement of the vehicle, a series of rotational forces are involved, related to elements such as the flywheel, gears, drive shaft, wheels, *etc.* So it is necessary to include these contributions as a supplement to the vehicle's rest mass. The rotational contribution to acceleration is in addition to the conventional rest mass of a system without rotating parts.

To find the contribution of the rotating parts to acceleration resistance, consider an arbitrary transmission part with the polar moment of inertia, I_x, located in the transmission at a place where the speed ratio in relation to the power axle is $\theta_{xd} = \omega_x/\omega_d$, where ω_x is the part's rotational speed in radians/sec, and ω_d that of the drive axle. The respective rotating parts supplementary mass can be found through the use of the angular momentum equation:

$$M = I \frac{d\omega}{dt} \tag{14.41}$$

which gives:

$$M_x = W_{ax} \cdot r \frac{1}{\theta_{xd}} = I_x \frac{d\omega_x}{dt} \tag{14.42}$$

$$W_{ax} = I_x \frac{\theta_{xd}}{r} \frac{d\omega_x}{dt} \tag{14.43}$$

where W_{ax} is the wheel force on the road needed to rotationally accelerate the object x in the drive train, and r is the tire radius.

From:

$$\omega_x = \theta_{xd} \cdot \omega_d = \theta_{xd} \frac{v}{r} \tag{14.44}$$

One finds:

$$\frac{d\omega_x}{dt} = \frac{\theta_{xd}}{r} \frac{dv}{dt} = \frac{\theta_{xd}}{r} \cdot a \tag{14.45}$$

and:

$$W_{ax} = I_x \left[\frac{\theta_{xd}}{r} \right]^2 \frac{dv}{dt} = I_x \cdot a \left[\frac{\theta_{xd}}{r} \right]^2 \tag{14.46}$$

Defining an equivalent mass $m_x = W_{xd}/a$, one obtains:

$$m_x = I_x \left[\frac{\theta_{xd}}{r} \right]^2 \tag{14.47}$$

and considering all rotating masses, with associated moment of inertia and speed ratio, the total acceleration resistance for an element that rotates becomes:

$$W_{ac} = \left[m_o + \sum I_x \cdot \left(\frac{\theta_{xd}}{r} \right)^2 \right] \cdot a \tag{14.48}$$

The first term in this equation corresponds to the acceleration of the component if it does not rotate, and the second term corresponds to the rotational acceleration of the element if it does not accelerate linearly. Often, the moments of inertia for the various parts are not known. Gillespie has given an empirical equation for estimating the contribution of rotating components to the total acceleration resistance of a vehicle [111]:

$$\frac{m_o + m_{rot}}{m_o} = 1 + 0.04 + 0.0025 \theta_{tot} \tag{14.49}$$

where θ_{tot} is the over all gear ratio, m_o is the rest mass of the vehicle, and m_{rot} is the equivalent inertial mass of the rotating components.

14.7 Calculation of a Vehicle's Fuel Consumption

A calculation of a vehicle's fuel consumption, even for driving on a flat road with constant speed, is complicated by the fact that the efficiencies are not constant, but depend on factors such as load, the instantaneous speed, and corresponding engine speed, which in turn are dependent on the chosen gear.

Considering given gear selection and a given speed for a given vehicle, the power and fuel consumption can be calculated when the engine's power and efficiency curves are determined. This information is obtained in the engine map, discussed in Chapter 12.

At a given engine speed, determined from the vehicle speed and gear used, the corresponding given maximum torque does not need to be used in the situation under consideration. The operating point of the engine will be at an arbitrary location between the abscissa and the maximum torque curve on the engine map. As mentioned earlier, the engine power in kW is given by the product of the torque and the corresponding speed, that is:

$$BP = T \cdot \omega_e \rightarrow P_e = \frac{2\pi T N}{60000} \tag{14.50}$$

where ω_e is the engine's angular speed (rad/s or s^{-1}), T the torque in Nm and N is the engine speed in revolutions per minute.

If the power is maintained constant, the product of T and N is also constant, which corresponds to a hyperbola on the engine map. The efficiency through which the engine is able to produce the desired force, varies with engine load, shown by the lines on the engine map, that is, iso-consumption curves connecting points with the same specific fuel consumption.

Considering driving resistance in the standard operating condition, no gradient and no acceleration the transmission, rolling and aerodynamic resistances must be overcome:

$$W - W_{tr} = W_r + W_a \tag{14.51}$$

One has for constant speed to a good approximation:

$$W - W_{tr} = f_r \cdot m_o g + \frac{1}{2}\rho v^2 \cdot A_f \cdot C_w \tag{14.52}$$

Corresponding to a required engine power, BP:

$$BP \cdot \eta_{tr} = D_t \cdot v = \frac{T_w}{r} \cdot v \tag{14.53}$$

where T_w is the torque at the wheel. For constant speed:

$$BP = \left(f_r \cdot m_o g + \frac{1}{2}\rho v^2 \cdot A_f \cdot C_w \right) \frac{v}{\eta_{tr}} \tag{14.54}$$

The above is for driving on a flat horizontal road. For constant speed, it corresponds to the necessary engine torque of:

$$T = \frac{\left(f_r \cdot m_o g + \frac{1}{2}\rho v^2 \cdot A_f \cdot C_W \right) \cdot r}{\eta_{tr}} \cdot \frac{1}{\theta_{tot}} \tag{14.55}$$

where the total gear ratio is θ_{tot}.

To refer the driving resistance to the engine, it is necessary to take into consideration the speed and the corresponding engine rotational speed, which is determined by:

$$v = r \cdot \omega = \frac{2\pi N}{60} \cdot \frac{r}{\theta_{tot}} \tag{14.56}$$

$$= \frac{Speed\ in\ km/hour}{3.6} \tag{14.57}$$

By locating the engine condition in torque and speed required to overcome the driving resistance at the chosen gear ratio on the engine map, one is able to read the necessary power and corresponding brake specific fuel consumption for the given speed and engine speed which it requires.

During acceleration:

$$BP = \left(f_r \cdot m_o g + \frac{1}{2}\rho v^2 \cdot A_f \cdot C_w + m_{eq} \cdot a \right) \frac{v}{\eta_{tr}} \tag{14.58}$$

Where the mass, m_{eq} is the equivalent mass including the rotational contribution calculated in Equation (14.48), and a is the acceleration (negative when decelerating):

$$m_{eq} = m_o + \sum I_x \cdot \left(\frac{\theta_{xd}}{r} \right)^2 \tag{14.59}$$

For brake power in kW and the specific fuel consumption, $bsfc$, in gram/kWh, the hourly consumption, B_t is:

$$B_t = \frac{bsfc}{1000} \cdot BP \quad kg/h \tag{14.60}$$

or, if the density of the fuel is ρ_f kg/liter, the volumetric fuel consumption, B_s is:

$$B_s = \frac{bsfc}{1000} \cdot \frac{BP}{\rho_f} \quad liter/h \tag{14.61}$$

With test driving, fuel consumption is often given in terms of liter per 100 km or km per liter corresponding to the conversion conditions:

$$FC_1 = \frac{kph \cdot \rho_f \cdot 1000}{BP \cdot bsfc} \quad km/l \tag{14.62}$$

or,

$$FC_2 = \frac{bsfc}{10 \cdot kph} \cdot \frac{BP}{\rho_f} \quad l/100km \tag{14.63}$$

where kph is the vehicle speed in terms of km/h, and ρ_f the density of the fuel in kg/liter.

Example

Consider a vehicle with a mass of 1100 kg, rolling resistance coefficient, $f_r = 0.013$, $C_w = 0.35$ and a frontal area A_f of $2.15m^2$ and a transmission efficiency of 0.90 using the engine with the engine map shown in Figure 14.22.

At a vehicle speed of 100 km/h this gives a vehicle power requirement of 13.6 kW, and an engine power requirement of 15.1 kW, which is about 18% of the engine's maximum power. This power requirement is drawn in the engine map in Figure 14.22 as a hyperbola to show all possible engine operating points that can supply this power.

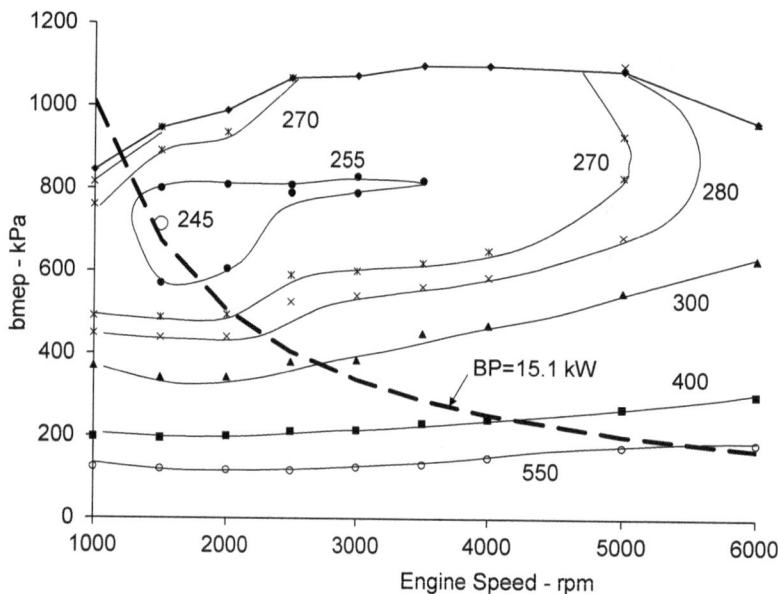

Figure 14.22: Engine map for the example calculation.

The diagram shows that it is most efficient to obtain the power at a very low engine rotational speed. The faster the engine operates to produce the necessary power, the more fuel it uses per unit of power produced. As was shown earlier, this is a reflection of high friction at high speed. In terms of economy then, it is important to drive in a high gear (low ratio). For a gear ratio that gives this power at an engine speed of 1400 rpm, the engine's fuel consumption is very close to its minimum value of about 245 gram fuel per kWh. At 6000 rpm is it about 560 gram fuel pr. kWh. There is a factor of more than two difference between the fuel consumption obtained by achieving the required 15.1 kW at the optimum efficiency condition and the worst.

If the engine produces the required 15.1 kW at 1400 rpm, one obtains fuel consumption or mileage of about 20 km per liter fuel. For a speed of 4000 rpm, the fuel consumption is about 12.5 km per liter, and if the power is produced at the highest speed, the consumption increases to about 8.7 km per. liter.

At the very best condition, the engine in Figure 14.22 uses 245 gram fuel per kWh of mechanical energy produced. This corresponds to an efficiency, of about 33 percent. These numbers are obtained with constant speed, But this particular operating point is not normally achieved under normal driving conditions. It only occurs under conditions of high acceleration or steep gradient at low speed. The more often the engine operates in this point, the better from an economy point of view. In practice, the smaller the engine, the more frequently it operates at higher load conditions. Full load is often not to be preferred for economy reasons, as spark ignition engines often use richer mixtures. Diesel engines sometime have a lower efficiency at maximum load, due to slower and less efficient combustion. Modern turbocharged diesel engines have better full load efficiency than their non-turbocharged predecessors. The impact of gearing choice on fuel consumption and performance is discussed extensively in Chapter 15.

14.8 Problems

Problem 14.1

The power requirements for a medium size passenger car to overcome rolling resistance, air resistance and losses in the transmission system on a flat road are given below:

constant speed (km-t)	60	80	100
required engine power (kW)	6.7	12.4	20.9

The engine is a 4-stroke spark ignition engine with total displacement volume 1618 cm^3 (max power 51.4 kW at 5300 RPM, which gives a top speed of 142 km/h). The gasoline used has density of 740 kg/m^3 and lower heating value of 43500 kJ/kg. It is assumed that the engine's indicated efficiency is constant at 36% irrespective of speed and load, and that the total loss to friction and pumping work gives a constant mean effective pressure loss of 2.3 bar. 1) Calculate fuel consumption in km/liter at the constant speeds of 60, 80 and 100 km/t. The gearing is such that 1000 RPM for the engine corresponds to the speed of 28.4 km/t.

2) An overdrive gear is installed that reduces the engine's engine speed by 20% at a given vehicle speed (that is, 800 RPM engine speed corresponds to 28.4 km/t). The transmission losses are unchanged. Calculate the fuel consumption at 80 km/t in overdrive gear in km/liter.

3) In a more powerful version of the vehicle, a V-8 motor corresponding to 2 of the original engine is used (that is, the total displacement volume is 3236 cm^3, maximum power 103 kW). The gearing is also changed such that the maximum power is developed at the vehicle's maximum speed (183 km/t, at 5000 RPM). Assume that the power is linearly proportional to engine speed up to maximum power. Calculate the fuel consumption at 80 km/t, assuming that efficiency and loss correspond to the 4-cylinder engine. (1: 14,57 km/l; 12,3 km/l; 10,12 km/l. 2: 13.42 km/l 3: 9.998 km/l)

Chapter 15

Gearing and Performance

There are several aspects of vehicle operation that are affected by the selection of the gear ratio. The first of these is the *acceleration characteristics* of the vehicle. In order to perform acceptably, a road vehicle must have an acceleration that is sufficient to keep the vehicle from becoming an obstacle in traffic.

A second is the *maximum speed* of the vehicle. Given standards on highways, the vehicle must be capable of operating at an adequate maximum speed, again to perform acceptably with the rest of traffic. It is also important that the vehicle maintain an acceptable speed when ascending grades.

For overall performance, it is advantageous to be able to apply the *available power* of the engine over the entire speed range. The *tractive force* applied to the wheels and the transition between gears is strongly affected by the torque characteristics of the engine and the number of gears.

Gear ratio also has a major effect on *fuel consumption*. For normal driving conditions, it is desired to use as little fuel as possible, both from environmental and economics points of view.

The effects of gear ratios on these performance criteria, and methods for selecting gear ratios will be discussed in the following sections. The examples shown will be for standard transmissions, with manual gear shift and clutch, but the same principles apply to automatic transmissions. For these, the efficiencies and torque transfer ratios are more complicated, but the vehicle requirements for performance and efficiency are the same.

15.1 Powertrain Considerations With Engine Powered Vehicles

Historically, most road vehicles have been powered by combustion engines. There are many types and sizes of vehicles with differing purposes, yet on the highways each has certain requirements in terms of speed, acceleration and climbing ability. The driving characteristics of the vehicles are determined by the powertrain. That is, the entire system from engine to wheel (clutch, transmission, differential and tires), as was shown in Figure 14.1. Implicit in all the development is the interest in performing the necessary driving functions while using the least amount of fuel, and emitting the smallest amount of air pollutants. The final design is always a compromise between all these criteria.

In the future, there will be other types of powertrains in the highway transportation system. Since about the year 2000 it has been possible to purchase hybrid automobiles, which combine combustion engines, electrical motors and batteries for storage of electrical energy, and electric vehicles have been in production since about 2010. The design and optimization of the powertrain for such a hybrid vehicle is substantially more complicated that that for a combustion engine alone, since there are more degrees of freedom for the designer. For non-road vehicles, other systems have been used. As an example, diesel electric locomotives have used a diesel engine to provide power to electric traction motors on the wheels of the locomotive. Electrical vehicles have existed since the beginning of self-propelled vehicles.

In the following, only the combination of combustion engines and "Conventional" transmissions will be considered. What is to be described is the combination of the torque and fuel consumption characteristics of the propulsion unit (here the combustion engine), the characteristics of the rest of the powertrain (wheels, gears, *etc.*) to determine the performance and fuel consumption of the vehicles. The goal is a basic understanding of how to combine these. With this background, the reader should have a good start towards understanding the more complicated powertrains systems that may appear in the future.

There are some fundamental disadvantages of the internal combustion engine when used in a vehicle power train:

1. The maximum power is roughly proportional to the working frequency, that is the maximum torque is nearly constant over the operating speed range.

2. The engine does does not produce power at a working frequency that is less than about:

$$\omega_{min} \simeq 0.2 \cdot \omega_{max} \tag{15.1}$$

The lower speed limit of the rotational speed necessitates that the connection between the driving machine and the driven when must be broken when the vehicle is to stand still and must be re-established when the vehicle is to move again. This is done with a clutch in a standard transmission or through the slippage in the torque converter of an automatic transmission when driving and the gear is not in neutral.

The driving machine must be then supplemented with a clutch, that during the start of the vehicle from a complete stop and also within certain limits, functions as a continuously variable gear (through slippage in the clutch or torque converter) for a short period of time. The clutch is in itself not sufficient for gearing variations, since a vehicle should be able to move itself satisfactorily under widely variable driving conditions, as well as exhibiting a reasonably good economy. This is not realizable in a vehicle with a single gear, since the driving units relative torque does not have a fairly high value, such a series electric motor does.

The gasoline and diesel engines must therefore in addition to a clutch be equipped with a gear(s) with a variable speed ratio, within certain limits such that the engine torque can be amplified as required. Unfortunately, the torque requirement cannot be decided uniquely on the basis of a given driving requirement, since it depends on a series of arbitrarily varying conditions, that in reality only can be described through statistical methods. The most important variables are gradients, that is, the geographical conditions and the acceleration requirements caused by traffic or the traffic pattern.

15.2 Tractive Force Diagram

One of the most important factors for the choice of the gear ratio in an automobile transmission is the torque characteristic of the engine in question, and the actual vehicle's driving resistances. The torque of the engine is transmitted through the drive train to the driving wheel of the vehicle. The force exerted by the wheel on the rolling surface is responsible for the motion of the vehicle, and is called the tractive force. The tractive force diagram can be considered to be a graphical presentation of the previously discussed driving and operation resistance equilibrium:

$$D_T = W_t + W_r + W_a + W_g + W_{ac} \tag{15.2}$$

The transmission resistance, W_t, rolling resistance, W_r and air resistance, W_a must all be overcome for any driving situation. For all intents and purposes they are dependent on the vehicle and the driving speed. After these resistances have been overcome, any power that may be left over can be used for acceleration or grade climbing if needed. It is therefore useful to separate the resistances such that:

$$D_T - W_t - W_r - W_a = D_f = W_g + W_{ac} \tag{15.3}$$

If the value of the engine torque at full load is used to calculate the force on the wheel, the left-hand side of the Equation (15.3) gives an expression for the portion of the tractive force that is available for acceleration and climbing, that is, "The free tractive force":

$$D_f = W_g + W_{ac} \tag{15.4}$$

Under situations where less than full power is required, the output of the engine is reduced, so that the free tractive force is zero. The diagrams in the following will show the traction force resulting from engine operation at full load. This establishes limits of operation, other conditions requiring less power being generally available, though with some restrictions to speed in connection with changing gears, as will be shown in the following as well.

Assuming that the transmission resistance can be calculated with a simple constant transmission efficiency and using the driving resistances from Equations (14.16), and (14.32) one has:

$$\frac{T_e \cdot \theta_i}{r_w} \cdot \eta_{tr} - m_o \cdot g \cdot f_r - \frac{\rho \cdot v^2 \cdot A_f}{2} \cdot C_w = D_f \tag{15.5}$$

where: T_e is the engine torque θ_i is the instantaneous total gear ratio in the transmission and η_{tr} is the transmission efficiency.

When comparing different vehicles, it is convenient to have the equations in dimensionless form, where the tractive force is divided by the gravitational force (weight) of the vehicle, $m_o \cdot g$:

$$\frac{D_T}{m_o \cdot g} - \frac{W_t}{m_o \cdot g} - \frac{W_r}{m_o \cdot g} - \frac{W_a}{m_o \cdot g} = \frac{D_f}{m_o \cdot g} = \frac{W_g}{m_o \cdot g} + \frac{W_{ac}}{m_o \cdot g} \tag{15.6}$$

or:

$$f_T - w_t - w_r - w_a = f_f = w_g + w_{ac} \tag{15.7}$$

The term f_f in equation (15.7) is called the *unit free tractive force*. For practical purposes, it is convenient to have the *free tractive force* as a function of driving speed in km/h or as a function of the relative driving speed, V_x/V_{max}. The basis for the unit tractive force diagram is naturally the engine's torque characteristic. This speed torque curve can be converted to a (V, D_t) curve or (V, D_f) curve using the the transmission's total gear ratio as a parameter.

For evaluating the choice of gear ratios one uses the so-called tractive force diagram, which gives a total picture of the available tractive force as a function of the speed for the respective gears, as shown in 15.1a and b. Figure 15.1a shows the force in Newtons, and is well suited to the evaluation of a definite case, but for comparison of different vehicles it can be advantageous to use the dimensionless form as in Figure 15.1b. At this point, it can be seen that the engine full load torque curve plays an important role

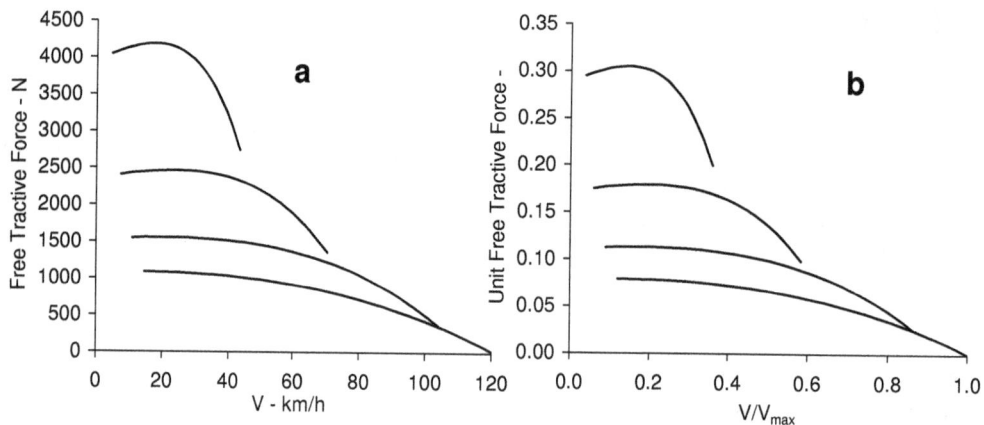

Figure 15.1: Typical free tractive force-vehicle speed diagrams, a) absolute values of force in Newtons and speed in km/h, and b normalized with vehicle weight (Unit tractive force) and maximum vehicle speed.

in performance. The positive term on the left hand side of Equation (15.5) is proportional to the engine torque. Especially at low speed, the aerodynamic resistance is low, and the engine torque is multiplied by a large number due to the gear ratio in the first gear. As the vehicle speed and engine speed are related by a constant factor for a given gear ratio, the free tractive force curve in first gear is nearly proportional to the engine torque.

The importance of the gear ratio in converting the torque and speed can also clearly be seen in Equation (15.5) and Figure 15.1. As the gear is changed from the lowest to highest, the numerical value of the gear ratio becomes smaller. This means that at a given vehicle speed, the engine runs slower. Thus, the tractive force curves extend to higher vehicle speeds before the maximum allowable engine speed is reached when changing to higher gears. Additionally, since the numerical value of the gear ratio becomes smaller when going to higher gears, there is less amplification of the engine torque when it is applied to the wheels, and the maximum value of the tractive force curves decreases in the higher gears. At maximum speed, the free tractive force may become zero, if the engine torque is just adequate to move the vehicle before the maximum engine speed is reached. Maximum vehicle speed can also be limited by maximum engine speed, and in such a case the free tractive force would not be zero at the maximum vehicle speed.

The total gear ratio can normally be defined as:

$$\theta = \frac{input\ part\ displacement}{output\ part\ displacement} \tag{15.8}$$

and for rotating parts for example:

$$\theta = \frac{\omega_{in}}{\omega_{out}} = \frac{N_{in}}{N_{out}} \tag{15.9}$$

Or taking into account the conservation of energy:

$$\theta = \frac{T_{out}}{T_{in}} \tag{15.10}$$

The relationship between vehicle and engine speeds is:

$$v = \frac{2\pi}{60} \cdot N_w \cdot r_w \tag{15.11}$$

$$= \frac{2\pi}{60} \cdot N_e \cdot r_w \frac{1}{\theta_{tot}} \quad m/s \tag{15.12}$$

$$= \frac{2\pi}{60} \cdot N_e \cdot r_w \cdot 3.6 \frac{1}{\theta_{tot}} \quad km/h \tag{15.13}$$

In addition, the normalized tractive force applied to the wheels to provide motion is:

$$f_t = T_e \cdot \frac{\theta_{tot}}{r_w} \frac{\eta_{tr}}{m_o g} \tag{15.14}$$

That is:

$$f_f = f_t - w_r - w_a$$
$$= T_e \cdot \frac{\theta_{tot}}{r_w} \frac{\eta_{tr}}{m_o g} - f_r - \frac{1}{2}\rho v^2 \cdot A_f \cdot C_w \frac{1}{m_o g} \tag{15.15}$$

or for BP in kW:

$$f_t = \frac{D_t}{m_o g} = \frac{1000 \cdot BP \cdot \eta_{tr}}{v \cdot m_o g} \tag{15.16}$$

$$f_f = \frac{1000 \cdot BP}{v} \frac{\eta_{tr}}{m_o g} - f_r - \frac{1}{2}\rho v^2 \cdot A_f \cdot C_w \frac{1}{m_o g} \tag{15.17}$$

Consider an engine with the characteristics in the Table 15.1 used with a vehicle having the specifications in the same table. The full load performance data are shown in Figure 15.2, and the fuel consumption of the engine is shown in the engine map of Figure 15.3, which gives data for the entire engine operating range, not just full load.

The characteristic torque curve form, with the maximum near the middle of the speed range, as shown separately in Figure 15.2, is an expression of technical compromises and choices made when designing the engine. As seen in Chapter 7, camshaft and intake manifold design affect the torque curve in spark ignition engines, and in some cases, the use of supercharging or turbocharging is an additional factor.

Table 15.1: Engine torque and power, and vehicle data for sample calculations. Vehicle data: mass, $m_o = 1400$ kg, tire radius, $r_w = 0.29$, length, $L = 2,45m$, air resistance coefficient, $C_w = 0.45$, Frontal Area, $A_f = 3.0m^2$, $\eta_{tr} = 0.9$.

Speed - rpm	Torque - Nm	Power - kW
500	114	6.0
1000	116	12.1
1500	117	18.4
2000	118	24.7
2330	118.2	29.7
2500	118.1	30.9
3000	116	36.4
3500	111.2	40.6
4000	102.3	42.9
4160	98.6	43.0
4500	90.6	42.4
4800	82.1	41.2

The torque curve shown has a greater drop in torque at high speeds than is typical for modern spark ignition engines. In terms of drivability, it will be seen that this type of torque curve has some advantages, especially with few gears in a transmission. As an approximation, it can be stated that the curve becomes flatter and the maximum torque moves toward higher engine speed with increasing mechanical efficiency. With this arises a certain conflict, since the matching of the vehicle's need in practice becomes more difficult when the maximum torque remains flat with increasing engine speed.

For diesel engines, the basic shape of the torque curve is accomplished through the regulation of the maximum fuel injection amount. This is under the condition of adequate air in the cylinder, and turbocharging plays an important role in the determination of the level of the torque curve in diesel engines. The decease in torque at the highest speeds for the engine in Figure 15.2 resembles that often seen in heavy duty diesel engines. Torque curves for a variety of engine types are shown in Chapter 12.

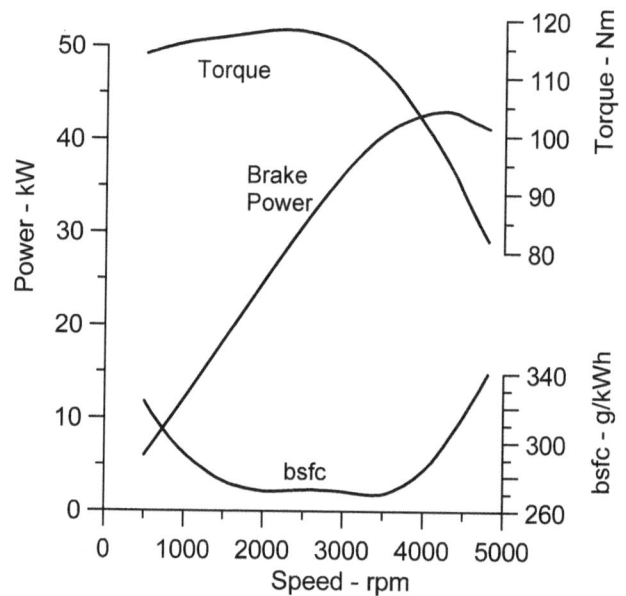

Figure 15.2: Full load performance of example engine.

For the engine and vehicle in Table 15.1 it is possible to calculate the tractive force for the vehicle as a function of vehicle speed. One possibility is to consider that the engine operates at the maximum power, and to find the tractive force available at different engine speeds. To do this, it is necessary to make some assumptions about the gear ratio.

The maximum engine power of 43.0 kW is obtained at a speed of 4166 revolutions per minute. In the following, it will be assumed that an infinitely variable gear be used, such that the engine always operates at this condition of full power. For the vehicle to operate at 10 kph at this engine speed, the total gear ratio must be 48.1:1. The torque applied to the driving wheel is then 48.1 times the engine torque. Similarly, at a speed of 140 kph, the gear ratio must be 3.44:1, which results in a much lower torque on the wheel.

The results are shown in Figure 15.4, where the total tractive force applied to the wheel is shown in Figure 15.4a and the unit tractive force applied to the wheel is shown in Figure 15.4b. Note that this is the force applied to the wheel, no consideration has been taken of driving resistances here yet. This is

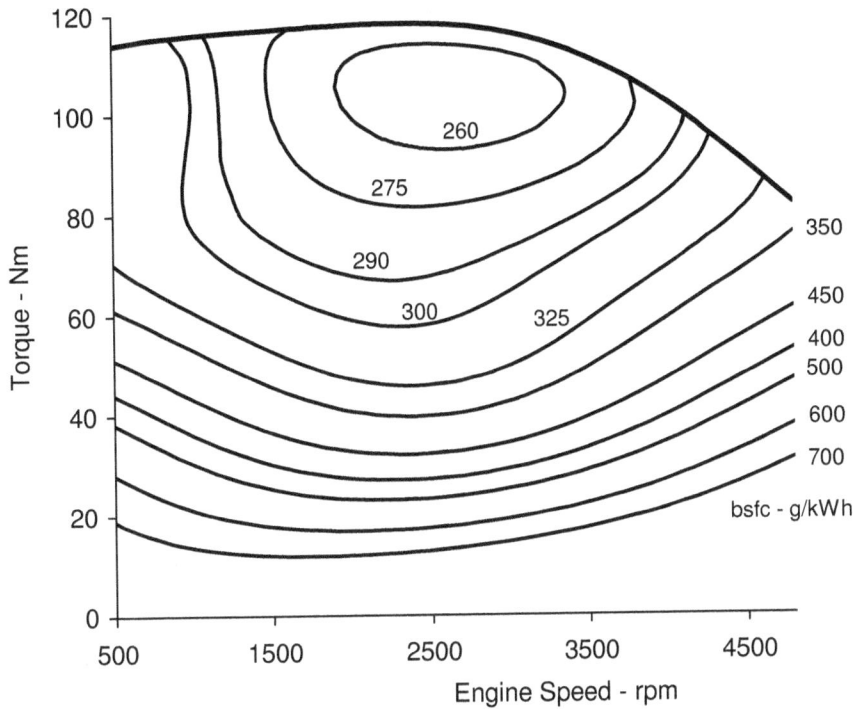

Figure 15.3: Engine fuel consumption map for the example engine.

the force available to overcome the driving resistances.

In practice, it is not possible to operate a vehicle with a continuously variable gear ratio over this range, although the continuously variable transmission (CVT) has been under development for a number of years. In current vehicles, one selects or designs a transmission to suit the driving needs and economic limitations of the vehicle at hand. The number of gears in the transmission ranges widely, from 3 gears in simple passenger cars to over 15 gears in large, heavy duty vehicles. Recent developments in passenger car transmissions has led to larger number of gears, in order to provide better fuel economy while maintaining good performance.

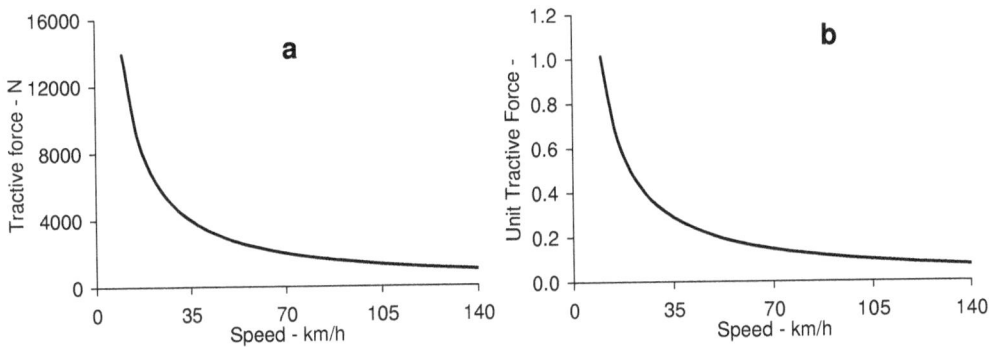

Figure 15.4: Tractive force available at the wheel for the engine and vehicle of Table 15.1 operating at the full power condition of 43.0 kW at 4166 rpm, with an infinitely variable gear ratio.

15.3   The Acceleration Properties of a Vehicle

This section will discuss the influence of gearing on starting acceleration of a vehicle. If the driving resistances are less than the engine's effective tractive force as applied to the wheels, the vehicle can accelerate. The amount of acceleration possible is important for determining the performance of the vehicle. Thus, it is necessary to illustrate the connection between characteristics of the engine, the properties of the vehicle and performance. In this section, the determination of the acceleration performance will be shown. Since the torque on the wheels is highest, and the driving resistance lowest in the lowest gear at low vehicle speeds, only low speed acceleration will be considered here. The principles can be applied at any speed, though.

When the vehicle is in operation, torque is transferred from the engine to the wheels through the drive components. When the vehicle accelerates, a portion of the torque is used for acceleration of the rotating parts of the vehicle and engine, and the remainder used for a linear acceleration. That is, the acceleration of rotating components requires not only that they be set in to translational motion, but that they are rotated as well. That this is the case is easily seen by observing the effort required to accelerate a rotating bicycle wheel when it is not in contact with the ground. Even though the vehicle (bicycle) does not move, energy is required to accelerate the wheel.

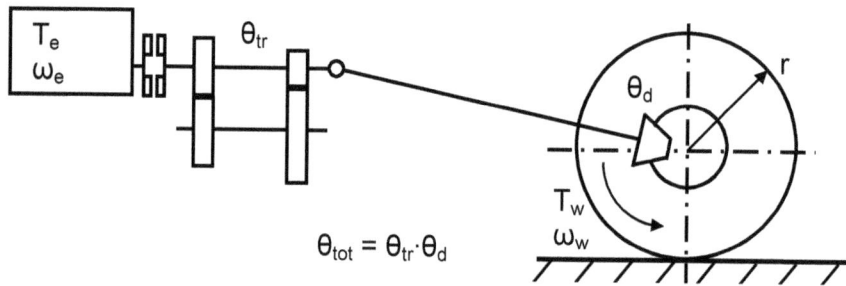

Figure 15.5: Schematic view of the engine transmission system showing gear ratios of the components of the powertrain.

Figure 15.5 shows the components of the drive train for a simple vehicle with a mechanical transmission and a single differential. In the following, it will be assumed that it is placed in a vehicle with a rest mass, m_o, which includes the mass of the rotating parts but without any contribution from their rotation. In a powertrain, the gear ratio, θ normally refers to the ratio of the rotational speed of a component on the normal power receiving end. For example, in a transmission it is equal to the ratio of the input shaft speed over the output speed. It is common to refer to the overall gear ratio, θ_t which is equal to the rotational speed of the engine divided by the rotational speed of the driving wheel.

Since gears must have the same pitch (number of teeth/diameter) to operate properly, the diameter of a gear is proportional to the number of teeth it has. For two gears that mesh and turn, the velocity at the point of contact is the same for each gear. Then since $V = R \cdot \omega$, then $r_1 \cdot \omega_1 = r_2 \cdot \omega_2$, and using the same pitch, $n_{t,1} \cdot \omega_1 = n_{t,2} \cdot \omega_2$, where n_t represents the number of teeth on the gear. Also, by equating the work performed by the first gear to that received by the second, the relationship between the torques on the two gears can be obtained.

$$F \cdot \Delta x = \frac{M_1}{r_1} \cdot \Delta x \tag{15.18}$$

$$= \frac{M_2}{r_2} \cdot \Delta x \tag{15.19}$$

The relationships are given in Equations (15.20) and (15.21), where α_{12} corresponds to number of teeth on the first gear divided by the number of teeth on the second gear.

$$\alpha_{12} = \frac{\omega_2}{\omega_1} \tag{15.20}$$

$$= \frac{M_1}{M_2} \tag{15.21}$$

If the gear ratio is such that the output speed is reduced, that is lower than the input speed, the output moment is then greater than the input moment, and *vice versa*.

When dealing with the rotating parts, it is convenient to refer them to a single rotating component. For example, in Figure 15.5, there are 3 possible rotational speeds: the wheel speed, the drive shaft speed and the engine speed. For the vehicle calculations, the rotational inertia effects will be referred to the rotational speed of the wheel. This is because the wheel rotational speed is proportional to the speed of the vehicle (assuming no slippage).

Consider that the wheel rotates with a speed ω_w and that the wheel has a radius r_w. This is of course related to the vehicle speed, as $v = r_w\omega_w$. From basic mechanics, the relation between the moment/torque applied to the wheel and the acceleration is:

$$T = I \cdot \alpha = I \cdot \frac{d\omega}{dt} \tag{15.22}$$

where α is the angular acceleration and I the moment of inertia.

Assume that some component, x, in the drive train rotates with a rotational speed, ω_x, such that

$$\theta_{xw} = \frac{\omega_x}{\omega_w} \tag{15.23}$$

Starting from the engine and assuming that it is accelerating, the torque at the clutch is obtained by subtracting the torque required for the acceleration of the engine:

$$T_c = T_e - I_e \cdot \alpha_e \tag{15.24}$$

The next position in the drive train is at the outlet of the transmission, or the drive shaft. Here, the engine torque is multiplied by the gear ratio of the transmission. Assuming that all the inertia is on the input side, that is, rotating at engine speed, and compensating for transmission loss:

$$T_d = (T_c - I_t \cdot \alpha_e) \cdot \theta_{tr} \cdot \eta_{tr} \tag{15.25}$$

Finally, the torque is amplified and the speed reduced at the differential, giving the torque at the axle. The torque on the axle is then use to accommodate the rotational acceleration of the wheel, and the remainder is left for the translational acceleration of the vehicle.

$$T_a = (T_d - I_d \cdot \alpha_d) \cdot \theta_d = r_w F_w + I_w \alpha_w \tag{15.26}$$

where F_w is the force of the wheel acting at the road surface.

The above can be solved for the tractive force at the wheel, in which it is assumed that all the losses are applied to the engine torque through a transmission efficiency, η_{tr}. In addition, for the rotation of the wheel,

$$v = r_w\omega_w \tag{15.27}$$

Which gives:

$$a = r_w\alpha_w \tag{15.28}$$

Then the force on the wheel resulting when energy has been taken to accommodate the acceleration is:

$$F_w = \frac{T_e\theta_{tr}\eta_{tr}}{r_w} - \left[(I_e + I_t) \cdot \eta_{tr} \cdot \theta_{tot}^2 - I_d \cdot \theta_d^2 - I_w\right] \frac{a}{r_w^2} \tag{15.29}$$

Setting the tractive force equal to the sum of the rolling, aerodynamic and gradient resistances and the product of the equivalent accelerating mass and its acceleration:

$$\frac{T_e\theta_{tr}\eta_{tr}}{r_w} = m_o + \left[(I_e + I_t) \cdot \eta_{tr} \cdot \theta_{tot}^2 - I_d \cdot \theta_d^2 - I_w\right] \cdot a + W_r + W_a + W_g \tag{15.30}$$

where T_e is the engine torque, which is considered to be constant, and the I's, are the moments of inertia of the engine, transmission, differential and wheels. Then in addition to the rest mass, m_o, there is a term that represents the rotational acceleration of the individual components. This means for example, that not only must the engine be laterally translated, its components must also be set in rotation as the vehicle starts to accelerate. The rotational contribution is proportional to the square of the gear ratio of the individual components. In the drive train, the engine usually has the largest moment of inertia, and in low gears the overall gear ratio has a large numerical value. So engine inertia is a very important parameter in determining vehicle acceleration at low speeds.

There are two main rotating masses that are significant: a) The equivalent rotating mass of the engine and the other input component's rotating masses that rotate with the engine speed: b) The wheels and the other output component's rotating masses that rotate with the wheel's rotational speed: ω_w. This is dominated by the inertia of the wheels.

In Equation (15.30) the engine torque is multiplied by the overall gearing ratio, and the rotational inertia deduction increases by the square of the gear ratio. There must, therefore, exist an optimal gearing ratio with consideration to the start acceleration in the 1st gear, which gives maximum acceleration, even though the driving force with constant speed still will increase with increasing gear ratio.

At any given speed, the acceleration can be calculated by rewriting Equation (15.30):

$$a = \frac{\frac{T_e \theta_{tr} \eta_{tr}}{r_w} - W_r - W_a - W_g}{\left[m_o + I_e \cdot \left(\frac{\theta_{tot}}{r_w} \right)^2 \cdot \eta_{tr} + \frac{I_w}{r_w^2} \right]} \tag{15.31}$$

In the case of acceleration from a stop, the gradient and aerodynamic resistances are not included, as the speed is low and the road assumed level. The transmission inertia is included with the engine, and differential inertia neglected.

Consider a vehicle with the following data:

$T_e = 100$ Nm, $m_o = 1200$ kg, $r_w = 0.3$ m, $\theta_{tot} = 14.6 : 1$, $\eta_{tr} = 0.90$, $f_r = 0.017$, $I_w = 2.4 \, kg - m^2$ per wheel, $I_e = 0.40 \, kg - m^2$

Assuming the overall gear ratio, θ_{tot}, to be variable, the acceleration is found to be:

$$a = \frac{100 \frac{\theta_{tot} \cdot 0.9}{0.3} - 1200 \cdot 9.81 \cdot 0.017}{1200 + 0.4 \cdot \left(\frac{\theta_{tot}}{0.3} \right)^2 \cdot 0.9 + \frac{4 \cdot 2.4}{0.3^2}} \tag{15.32}$$

$$= \frac{75 \cdot \theta_{tot} - 50.03}{326.7 + \theta_{tot}^2} \tag{15.33}$$

Figure 15.6 shows the shape of the curve of acceleration as a function of overall gear ratio, and indicates that a maximum acceleration on the order of $2.0 m/s^2$ is obtained for an overall gear ratio of about 18:1. Mathematically, the exact overall gearing ratio that gives the greatest acceleration can be found by differentiating the acceleration function, $a = f(\theta_{tot})$, Equation (15.31). This gives:

$$\frac{da}{d\theta} = \frac{\left(m_o + \frac{I_w}{r_w^2} + I_i \cdot \left(\frac{\theta_i}{r_w} \right)^2 \cdot \eta_{tr} \right) T_e \cdot \left(\frac{\eta_{tr}}{r_w} \right)^2 - \left(\frac{T_e \theta_i}{r_w} - m_o \cdot g \cdot f_R \right) 2 I_i \cdot \left(\frac{\theta_i}{r_w} \right)^2 \cdot \theta}{\left(m_o + \frac{I_w}{r_w^2} + I_i \cdot \left(\frac{\theta}{r_w} \right)^2 \cdot \eta_{tr} \right)^2} \tag{15.34}$$

The maximum point is found from setting $\frac{da}{d\theta} = 0$, the numerator of Equation (15.34) equal to zero. The solution is:

$$\theta_{tot,amax} = \frac{m_o \cdot g \cdot f_r \cdot r_w}{T_e \cdot \eta_{tr}} + \sqrt{\left(\frac{m_o \cdot g \cdot f_r \cdot r_w}{T_e \cdot \eta_{tr}} \right)^2 + \frac{m_o r_w^2 + I_w}{I_i \eta_{tr}}} \tag{15.35}$$

which has only one value, since the ratio cannot be negative:

Equation (15.35) is often simplified by defining two constants:

$$m_m = \frac{m_o \cdot g \cdot f_r \cdot r_w}{T_e \cdot \eta_{tr}}$$

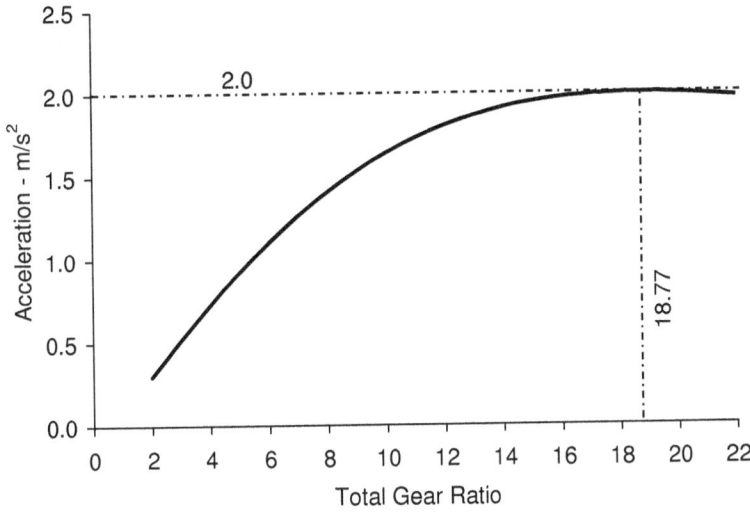

Figure 15.6: The acceleration of the vehicle in the example at low speed as a function of the overall gear ratio.

and

$$i_r = \frac{m_o r_w^2 + I_w}{I_i \eta_{tr}}$$

such that

$$\theta_{tot}^2 - 2m_m\theta - i_r = 0 \tag{15.36}$$
$$\theta_{tot} = m_m \pm \sqrt{m_m^2 + i_r} \tag{15.37}$$

The term m_m is the ratio of the force required to overcome the rolling resistance to the force applied to the wheels from the torque of the engine (T_e/r_w). The term i_r involves the inertia of the engine and components and the mass of the vehicle.

For the example under consideration:

$$i_r = \frac{1200 \cdot 0.3^2 + 2.4 \cdot 4}{0.4 \cdot 0.9} = 326.7$$
$$m_m = \frac{1200 \cdot 9.81 \cdot 0.017 \cdot 0.3}{100 \cdot 0.9} = 0.667 \tag{15.38}$$

and

$$\theta_{tot} = 0.667 + \sqrt{0.667^2 + 326.7} = 18.77 : 1$$

corresponding to 1.3 times the chosen θ_1. Since the curve of Figure 15.6 is relatively flat near the peak, the final choice is not that critical. The acceleration for the ratio of 14.7 is $1.936 m/s^2$, while that for 18.7 is $2.000 m/s^2$, a difference of only 3.3% Thus there is no great advantage to selecting the 18.7:1, and indeed an extra gear might be needed for this, or larger steps required between gears if extra gears are not used. The former increases vehicle expense without adding acceleration performance, and the latter sacrifices drivability without substantially increasing performance.

It is also apparent that the term m_m is not very important, at least for this case, where it is much smaller than the term i_i. Assuming that the term m_m is negligible, one obtains an overall gear ratio of 18.1. To a first approximation, the overall gear ratio can be estimated as:

$$\theta_{tot} \approx \sqrt{i_r} \tag{15.39}$$

The price for amplification of the engine torque is the increase in the equivalent inertia of the engine, which increases with the square of the gear ratio.

A corresponding graphical determination of 1st gear's maximum ratio for achieving maximum acceleration is shown in Figure 15.7 for different values of m_m and i_r. The relative importance of the rolling resistance and the inertia terms can be seen. In a case like that considered, where the chosen total gear

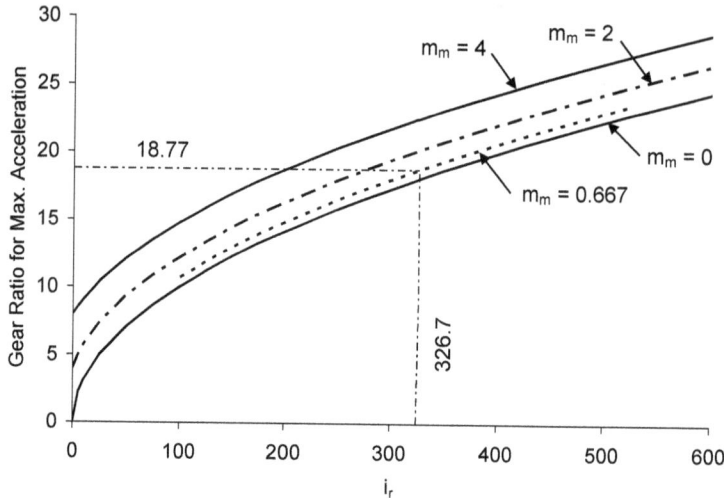

Figure 15.7: The gear ratio required for maximum acceleration of a vehicle as a function of the parameters m_m and i_r from Equation (15.36).

ratio is relatively large, and where the estimated rotating mass is also large, it may be necessary to make a correction, in an effort to achieve maximum acceleration. In other cases, the chosen total gearing ratio can be so small and the rotating masses so small, that it is irrelevant to try to achieve maximum start acceleration.

If 1st gear is chosen on the basis of maximum start acceleration, it will additionally be of interest to examine if the ratio chosen lies in an area, in which the use of a reasonable tire coefficient of friction value indicates that wheel spin is avoided.

15.4 Gear Ratio Selection

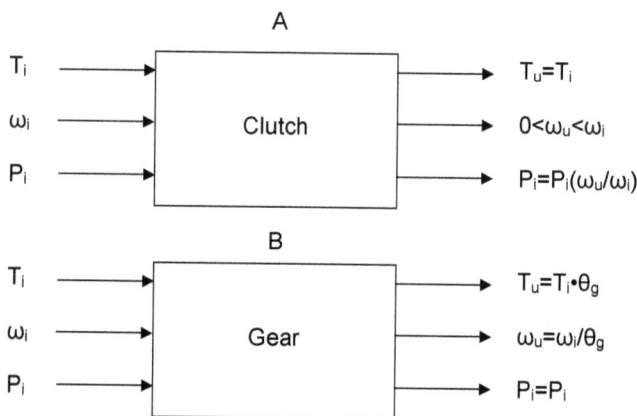

Figure 15.8: Schematic diagram of the operation of clutches and gears for a vehicle.

In order to take advantage of the combustion engine's economic advantages, it is necessary to supplement it with a clutch-gear unit that can compensate for the weakness mentioned in Section 15.1. This transmission system gives the conversion ratio for the clutch shown in Figure 15.8A, which can be regarded as an angular speed transformer.

The transmission system determines the conversion ratio for the gearing, shown in Figure 15.8B, and can be regarded as a torque converter. In principle, there is nothing the matter with combining them in one unit, but until now, the concepts of clutch and gear have been used separately.

It is easy to imagine that a continuously variable gear with a sufficiently large variation range would be the ideal gear type, and that such a gear would also be able to overtake the role of the clutch, if the variation range is so large that $\theta_{max} \Rightarrow \infty$. Such an arrangement is technically possible, but to date has not been feasible on an economic basis.

If one, for example, wishes to evaluate a power train in connection with a given vehicle, one can initially consider the use of an ideal step-less gear, that is, a gear with a continuous, variable speed ratio, and a sufficiently large variation range. In this special case, the driving range is naturally only limited by the constant power curve that corresponds to the prime mover's maximum brake output. The maximum tractive force can be determined on the basis of this brake power, since:

$$D_t \cdot v = P_{max} \tag{15.40}$$

That is, a hyperbola in the tractive force diagram

Figure 15.4a shows the tractive force (available) from $BP_{max} = 43.0$ kW. Dividing both sides of the last mentioned equation with the vehicle weight, $m_o g$ one obtains the picture of the unit tractive force, as shown in Figure 15.4b. Note that the forces in Figure 15.4a and b do not go to zero, as would be the case for the free tractive force. Here, the rolling and aerodynamic losses have not been subtracted.

If the engine is kept at the speed that gives the maximum power, (4166 rpm) all driving conditions that lie on the limiting hyperbola are reachable by selecting the ideal gear ratio. The total operation area in the power-velocity diagram represents the area under the power curve that can, through the use of the ideal gear, be covered by regulation of the torque, that is by regulating the output of the engine with the throttle. One can also let the engine operate with constant speed, that is to be regulated along a vertical line in the power diagram, as shown in Figure 15.9a for example, with $N_{P,max} = 4166$ rpm.

The ideal gear also makes it possible, for driving conditions that demand lower power, to regulate the engine to the speed that gives the lowest fuel consumption at the power level in question, that is the largest efficiency.

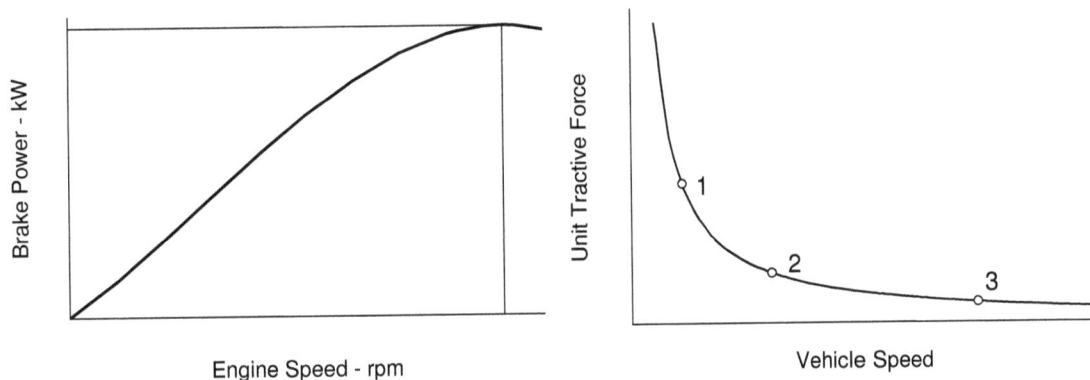

Figure 15.9: Schematic engine power curve and unit tractive force curve, showing the three possible operating points at full engine power.

Now the use of the maximum power over the whole speed range can only be realized with an ideal, stepless gearing system. With a fixed gear ratio, the maximum engine power is obtained at an engine/vehicle speed determined by this ratio. Therefore, for a fixed gear, the maximum power point has one intersection on the unit tractive curve for each gear ratio.

If one then uses, for example, a 3-gear transmission, one can utilize the maximum power, BP_{max}, for only three points on the driving condition diagram. These points are shown by the numbers 1, 2 and 3 in Figure 15.9. The maximum power curve is the hyperbola. As the gear ratio is reduced (higher gear) the vehicle moves faster at a given engine speed, and the torque amplification is reduced. That is, the tractive force is lower in the higher gears.

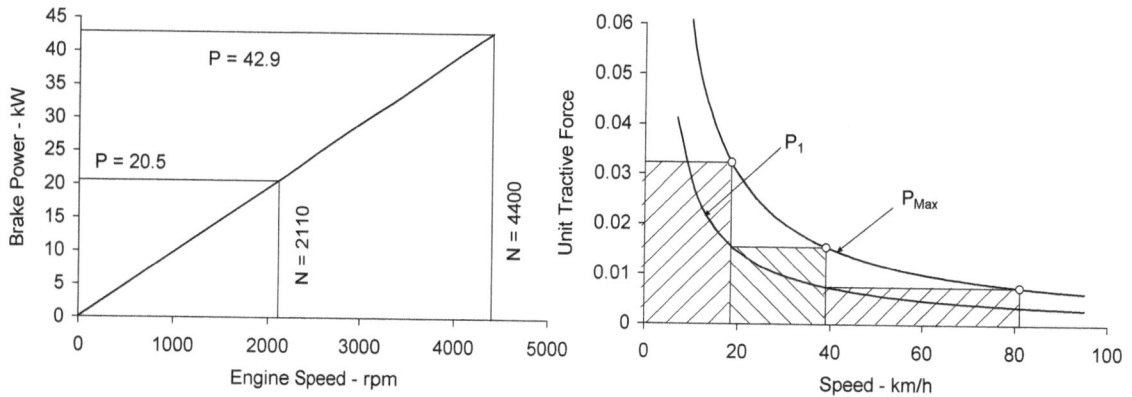

Figure 15.10: Schematic engine power curve and unit tractive force curve, for constant engine torque, showing possible driving conditions as determined by the torque.

The rest of the driving area must be covered by an engine speed regulator. Not all operating points below the hyperbola in the above figure are accessible with a 3-gear transmission. The limitation of the area, for example between 1st and 2nd gear, cannot be the BP_{max} hyperbola, but a curve that lies lower, and whose path is naturally determined by the torque curve of the engine. This will be illustrated in a simplified form by assuming that the BP_{max} curve is proportional to the speed, that is, by assuming constant torque. The power curve then becomes linear with speed, and one obtains the unit tractive force shown in Figure 15.10.

For this assumption with a given gear ratio, the engine can handle all the driving speeds and loads in the cross hatched driving area of Figure 15.10, limited by the points that correspond to the number of gear steps, where the engine's maximum power is utilized. Only a power represented by the hyperbola P_1 can be utilized in the entire driving area. The engine's maximum power can only be used in the three circled points on the P_{max} hyperbola.

If the vehicle is operated at full load up the the maximum power curve in first gear, and then changed to second gear, the following happens: First, the engine speed drops because of the gear change. Slip in the clutches and torque converter reduce the harshness of this change, as well as falling engine speed during the gear change.

The other thing that happens is a decrease in the tractive force applied to the wheels. Even though the engine is still producing maximum torque, the smaller value of the gear ratio gives a lower magnification of engine torque, thus a lower tractive force. In order to have smoother more flexible operation, it is desirable to have the difference between P_1 and P_{max} in the preceding figure as small as possible.

A new concept can be introduced called "The drivability acceptability", which is determined from:

$$\eta_K = \frac{P_l}{P_{max}} \tag{15.41}$$

In order to bring η_K up to 1 with the use of a gearbox with a reasonably small number of gear steps, there is only the possibility of improving the engine such that over an adequately large speed range it has constant power. This means that the engine's characteristic drawn in a speed (v) - tractive force (f) diagram has a hyperbolic nature. This translates to what is called a rising torque curve. That is, at full load, when the engine speed decreases, the torque increases. This torque rise is especially the case in heavy-duty turbocharged diesel engines. The more closely the engine power approaches the shape of the hyperbola on the $f - v$ diagram, the fewer steps are necessary in the gear box to achieve smooth full load operation.

An ideal situation is shown in Figure 15.11, where it is assumed that the engine torque curve is such that the engine power is constant between the speeds denoted by points "a" and "b" in the figure. When building or choosing the gear box, the torque curve's fall has a large significance since, according to the two following figures, one obtains the best approximation to the maximum power hyperbola, one naturally when the fall of the torque curve has a hyperbolic nature. If one looks at things from the point of view of the engine manufacturer, an effort is exerted to prevent the characteristic in the torque speed curve from

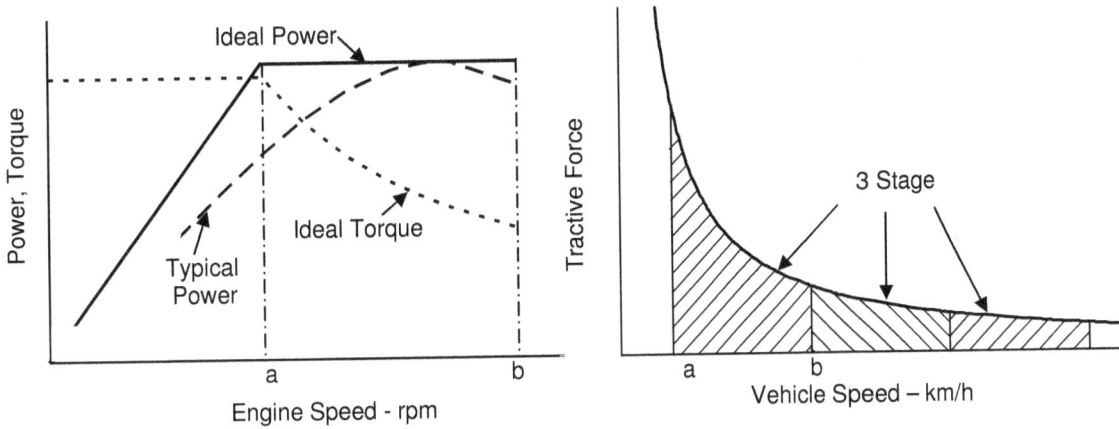

Figure 15.11: Schematic engine power curve and unit tractive force curve, in the ideal case where the engine power is constant at higher speeds.

approaching a hyperbola, since this tendency just means that the engine's mean effective pressure falls with increasing speed. This gives low efficiency due to the high relative amount of friction, especially in the case of a naturally aspirated engine. The situation is not quite so bad for a turbocharged engine, since the mechanical efficiency is higher, due to a high indicated mean effective pressure, and a friction mean effective pressure on a level comparable to that of a naturally aspirated engine. Therefore, with large turbocharged engines, efficiency loss associated with a decreasing torque characteristic is smaller than for non-turbocharged engines.

This conflict must, of course, lead to compromise solutions, when it is a question of the combination of an engine and a gearbox for typical road vehicles, it is natural to attempt to make the vehicle smooth and easy to operate. The situation is different for racing vehicles, where the maximum possible power must be obtained from the engine, and maximum wheel torque desired up to the gear change. Such vehicles are, less smooth than a passenger car, and are typically equipped with a large number of gear steps.

If one considers a typical gasoline engine and wants the greatest possible η_K value, it can be seen from Figure 15.12 that η_K does not become 1.00 through the use of a small number of gear steps in the entire speed range, since the power hyperbola and the curve for the engine's mean effective pressure curve in different directions, and the 2 curves usually intersect at one only point, that is for a given speed in each gear step.

Considering the $BP_{N,max}$ hyperbola (the power for full load operation at maximum engine speed) through point A on the full load brake mean effective pressure curve ($bmep_{max}$) in Figure 15.12, point A is determined by the power at the engine's maximum speed ($BP_{N,max}$). The curve for this power will go through another point on the full load $bmep$ curve, namely point B. By operating only in this area for each single gear step, one obtains a relatively good approximation to the hyperbola for maximum power, for the shape of this torque curve. One can call the ratio between N_{max} and N_B the ideal interval in the gear ratio. If the gear ratios are chosen such that the speed would not drop below N_B during a change, the engine operates in a condition near to constant power, giving smooth gear changes with few gears.

Considering instead $BP_{T,max}$, that is, the power curve through the point of maximum torque, one finds the intersection C, which characterizes the absolute maximum interval. Dropping to a speed below N_C on changing gears would result in a lower torque, giving poorer acceleration, and accentuating the gap between the tractive force curves for the gear changes, seen in Figure 15.10 for example. In order to obtain good performance, then the gear ratios should be within the limits:

$$\frac{N_{max}}{N_B} < \alpha < \frac{N_{max}}{N_C} \tag{15.42}$$

where α is the interval.

If this demand is satisfied for the entire operating range, it normally means that the number of steps in the gear box will be larger than that which can be allowed for the purpose of good economy. In typical

467

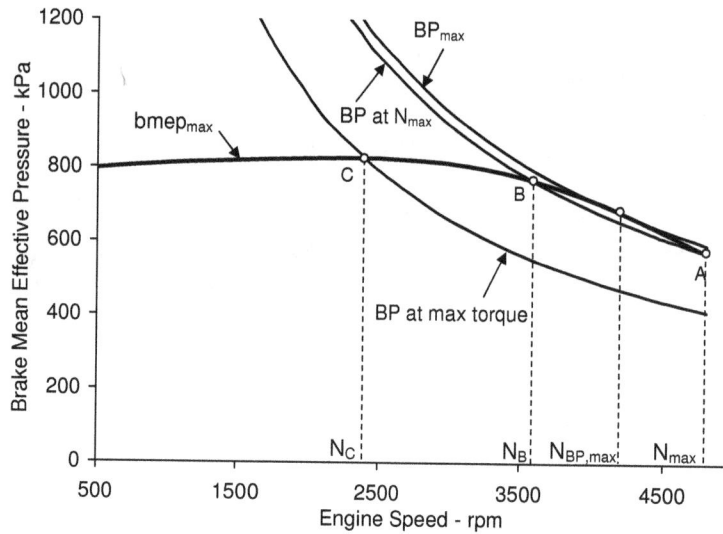

Figure 15.12: Different power conditions for consideration of gear ratio interval selection.

passenger cars, and smaller utility vehicles normally only 3, 4 or 5 steps are used. The disadvantages that accompany a choice of α outside of the limit are very significant in the higher gears, and it is common to let α increase with the lower gears, if the choice of 3-4 gears does not satisfy the given demand.

For the practical planning of the gear ratios in a gear box, one uses the previously mentioned unit tractive force, f_f. Since it is reference point C that represents the greatest tractive force, here it is the tractive force and not the power one works with. Therefore, it is reasonable to choose just that reference point C as the origin of the determination of the gear ratios.

Acceleration and climbing ability are strongly influenced by the maximum torque of the engine. Gear selection should take this into consideration for determining performance.

For the example in Table 15.1, one finds for the maximum engine torque condition that $T_{e,max} \Rightarrow BP = 29700$ W. These values can be used for selecting gear ratios. Using Equation (15.16) at the maximum torque condition for v in m/s, one obtains:

$$f_f = \frac{29700}{v} \frac{0.9}{1400 \cdot 9.81} - 0.017 - \frac{1}{2} 1.23 \cdot 3.0 \cdot 0.45 \frac{v^2}{1400 \cdot 9.81} \tag{15.43}$$

$$= \frac{1.946}{v} - 0.017 - 6.045 \cdot 10^{-5} \cdot v^2 \tag{15.44}$$

This gives the results shown in Figure 15.13. The desired free tractive force for acceleration can be chosen in the lowest gear. In the highest gear, the acceleration and climbing ability at maximum torque, $T_{e,max}$, for a relevant operating speed can also be decided. The size of the free tractive forces must of course be determined according to the use of the vehicle, but for typical passenger cars and trucks, they are typically set to 30-35 % and 5-7 %. respectively. These limits decide the maximum and minimum gear ratios desired for operation. The lines parallel to the abscissa corresponding to these percents give as the intersections with the unit tractive force curve the speeds V_1 and V_D. For the example given, $V_1 = 22$ km/h and $V_D = 75$ km/h and the corresponding overall gear ratios are 11.9:1 and 3.5:1. Since the gear ratio for a transmission in the highest gear is on the order of 1, the latter ratio is obtained through the selection of the gear ratio in the differential.

The ratio between the gearing ratio in 1st gear and the highest gear is:

$$\frac{\theta_1}{\theta_H} = \frac{v_H}{v_1} = \frac{v_D}{v_1} \tag{15.45}$$

and if $\theta_H = 1$,

$$\theta_1 = \frac{v_H}{v_1} = \frac{v_D}{v_1} \tag{15.46}$$

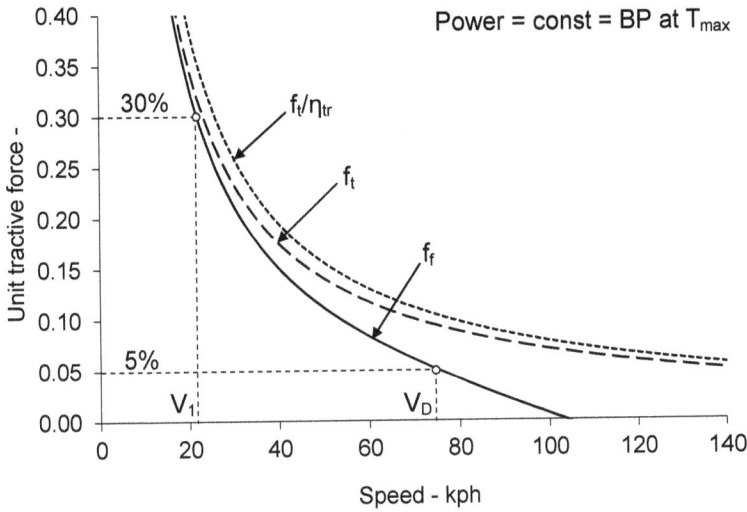

Figure 15.13: Unit tractive force as a function of vehicle speed for the example.

From the section on gradient resistance, the determined v_1 gives the direct unit inclination, since:

$$\frac{W_g}{m_o g} = \sin\alpha \cong \tan\alpha \tag{15.47}$$

which means that the determined climbing ability is related to the unit tractive force for the fully loaded vehicle. It can naturally be discussed where it is justified in all cases to use this relatively high unit tractive force, since passenger cars in particular are seldom fully loaded.

15.5 Maximum Speed

The choice of the unit tractive force in the highest gear, in reality does not depend much on the desired ability to climb, but to a larger extent on the characteristics of the engine around the maximum power condition, as well as the smoothness of driving one wishes and the desired top speed. With reference

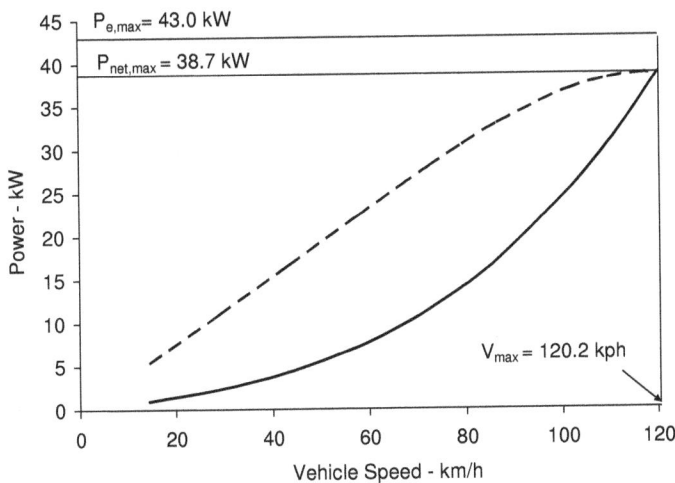

Figure 15.14: Maximum vehicle speed is determined by the intersection of the vehicle loading curve and the power curve of the engine at the specified gear ratio.

to Figure 15.14, the upper limit for a vehicle's total gearing ratio can be determined graphically on the basis of achieving the highest possible top speed, apparent as the intersection between a horizontal line corresponding to $BP_{max} \cdot \eta_{tr}$ and the driving resistance curve for $P_{vehicle} = (W_R + W_A)v$. For the example under consideration we have from the previous table on a flat road with no wind:

$$BP_{max} = 43000W \Rightarrow BP_{max} \cdot \eta_{tr} = 43000 * 0,9 = 38700W \tag{15.48}$$

$$P_{vehicle} = P_r + P_a = \left(1400 \cdot 9.81 \cdot 0.017 + 1/2 \cdot 1.23 \cdot 3.0 \cdot 0.45v^2\right) \cdot v \tag{15.49}$$

$$P_{vehicle} = (233.5 + 0.830 \cdot v^2) \cdot v \quad (W) \tag{15.50}$$

and for $\theta_{tot} = 3.79 : 1(\theta_{4th\ gear} = 1 : 1)$ The curves meet at a speed of:

$$v = \frac{2\pi}{60} \cdot 0.290 \cdot \frac{1}{3.79} \cdot 4166 = 33.4m/s = 33.4 \cdot 3.6 = 120.2km/h \tag{15.51}$$

This assumes that a gearing ratio has been chosen that causes the intersection to correspond to $BP_{max} \cdot \eta_{tr}$. This can be determined from:

$$T_{e,max} \cdot \eta_{tr} \cdot \frac{\theta_{tot}}{r_w} = m_o g \cdot f_R + \frac{1}{2}\rho \cdot A_f \cdot C_w \cdot \left(\frac{v}{3.6}\right)^2 \tag{15.52}$$

$$= m_o g \cdot f_R + \frac{1}{2}\rho \cdot A_f \cdot C_w \cdot \left(\frac{\pi}{30}\right)^2 \cdot N^2 \cdot \left(\frac{r_w}{\theta_{tot}}\right)^2 \tag{15.53}$$

$$T_{e,max} \cdot \eta_{tr} = \frac{m_o g \cdot f_R}{\theta_{tot}/r_w} + \frac{1}{2}\rho \cdot A_f \cdot C_w \cdot \left(\frac{\pi}{30}\right)^2 \cdot N^2 \cdot \left(\frac{1}{\theta_{tot}/r_w}\right)^3 \tag{15.54}$$

that is, a third order equation with respect to θ_{tot}/r_w. For the parameters shown and a wheel radius of 0.29 m, the solution to Equation (15.54) is $\theta_{tot} = 3.79$.

Top speed is, of course, primarily dependent on the engine's maximum power, but in addition, it is somewhat dependent on the gearing condition in the highest gear, and the θ-value (gear ratio) can influence:

- The vehicle's smoothness in the highest gear

- Maximum speed

- Noise level for motor way driving for example

- Fuel consumption for driving on for example the motorway.

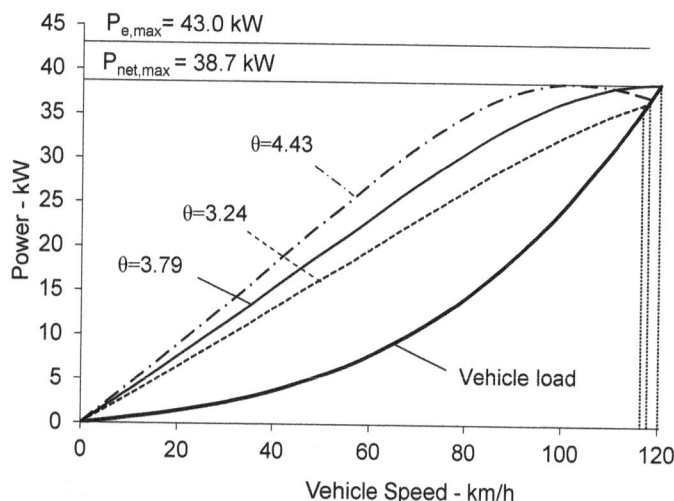

Figure 15.15: Maximum speed of the example vehicle for three different gear ratios, $\theta_1 = 3.79$, maximum engine power achieved, $\theta_2 = 4.43$, maximum engine speed achieved, and $\theta_3 = 3.24$, an economy gearing.

Figure 15.15 shows the driving resistance power requirement and the engine power delivered for three different gearing ratios: $\theta_1 = 3.79$, $\theta_2 = 4.43$ and $\theta_3 = 3.24$. Using the gearing ratio θ_2, the vehicle's maximum speed will be 118.3 km/h, achieved at the engine's maximum speed. It is normally the case, though, that the maximum speed can not be used with continuous operation, and the maximum speed attained here is, therefore, not that which one calls the motorway speed or maximum practical speed. On the other hand, this gearing ratio will provide a relatively large power surplus for acceleration in an important working area, and therefore a smooth driving characteristic. This must normally be paid for with a greater noise level and higher fuel consumption.

The gearing ratio θ_1 where the engine delivers its maximum available power at the maximum vehicle speed, gives the highest maximum speed of $V_{max} = 120.2$ km/h, which is not significantly higher than that for θ_2. The maximum speed here corresponds to the speed of the peak of the engine power curve, but not to the maximum speed of the engine, which with this gearing gives a vehicle speed 132 km/h. This speed could in principle be achieved by going down hill. The slightly higher maximum speed is paid for with a somewhat smaller power surplus in the working area of 60-110 km/h.

If one considers the normal medium vehicle speed range, a lower engine speed gives a better fuel consumption and less noise, and this can be achieved to an even greater extent with an even smaller gearing ratio, for example θ_3. Here, the engine's power curve intersects the resistance power requirement with a lower power, and there will be a lower top speed, but the speed difference is very small, and the maximum speed obtained with this gear is 117.7 km/h. The tendency in later years has been more in the direction of a "modest" overdrive. For example, in a 5-gear transmission, θ_3 corresponds to the 5th gear, which gives best economy. Maximum speed is often encountered in 4th gear in a 5-gear transmission (θ_1 or θ_2). The reasons for this can be seen in Figure 15.15.

The dependence of the maximum vehicle speed on the overall gear ratio is shown in Figure 15.16. Also shown are the corresponding engine speed and engine power at the maximum speed condition as the gear ratio is varied.

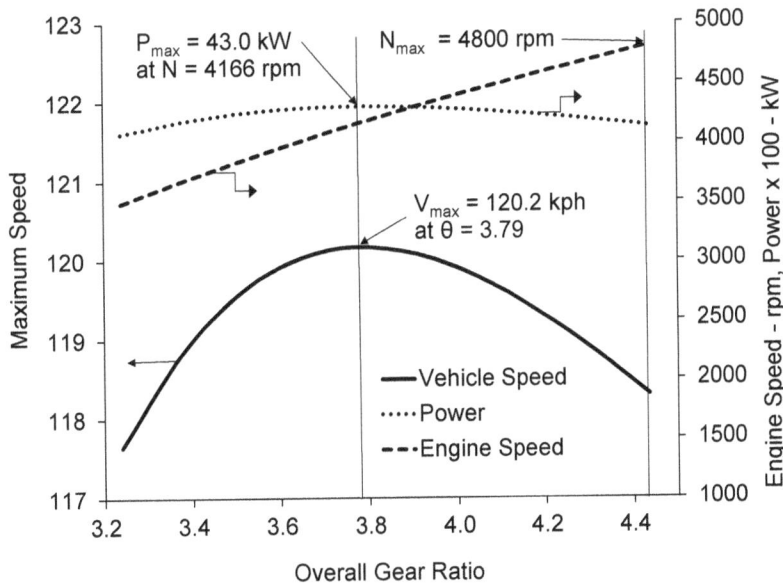

Figure 15.16: Maximum vehicle speed and the corresponding engine speed and power for the example vehicle as a function of overall gear ratio.

15.6 Gear Ratio and Economy

In the previous section various criteria for choosing the gearing ratio have been discussed:

1. Maximum speed
2. Maximum tractive force

3. Use of the available power in the entire speed range

4. Maximum acceleration

But in addition, the choice of the gearing ratio will have an influence on the fuel consumption or fuel economy. Even though the fuel consumption often must be said to have a relatively small significance for the lifetime economy of an entire vehicle, it is one of the areas of great interest to the vehicle users.

The calculation of vehicle fuel consumption was described in Section 14.7. A vehicle's fuel consumption, or more precisely the specific fuel consumption, that is, the consumption per unit of distance traveled, is for the first part dependent on the work consumed for the road length in question, that is the unit road length:

$$E_l = \int_0^l \sum W \cdot ds \tag{15.55}$$

where W is the total driving resistance.

Since it is known that the driving resistance is a function of the vehicle, weather conditions and driver, the work required per unit road length will be a variable quantity. When it is additionally considered that the engine, which must provide the propulsive work, has an efficiency that is very dependent on the load, it is clear that the specific fuel consumption of a vehicle is not a constant, but to a large degree dependent on the use of the vehicle, driving style, and service, and also the operational possibilities, such as the gear choice.

Figure 15.2 shows a typical set of performance curves for a spark ignition engine. The reasons for the shape of the curves have been discussed in Chapter 12. A typical full load curve consists of engine torque, power and brake specific fuel consumption as a function of engine speed, for the throttle in the wide open position. As seen in Figure 15.2, an engine's full load fuel consumption curve normally has a minimum value at a speed that lies near the speed where the engine delivers the greatest torque/mean effective pressure. For example, in this case there is a value of 286 g/kWh at full load.

But as is shown in Chapter 12, the specific fuel consumption varies markedly over the engine operating range, and is significantly higher at part load than full load due to throttling losses in SI engines and lower mechanical efficiency. For normal steady state operation, a vehicle engine operates at engine power output much lower than its maximum capability.

If one wishes to analyze fuel consumption on the basis of a chosen gearing ratio, it is necessary to clarify the fuel consumption with the possible part load conditions for the respective speeds. This information is available on the engine map. Figure 15.17 shows an engine map for an example engine with curves drawn for

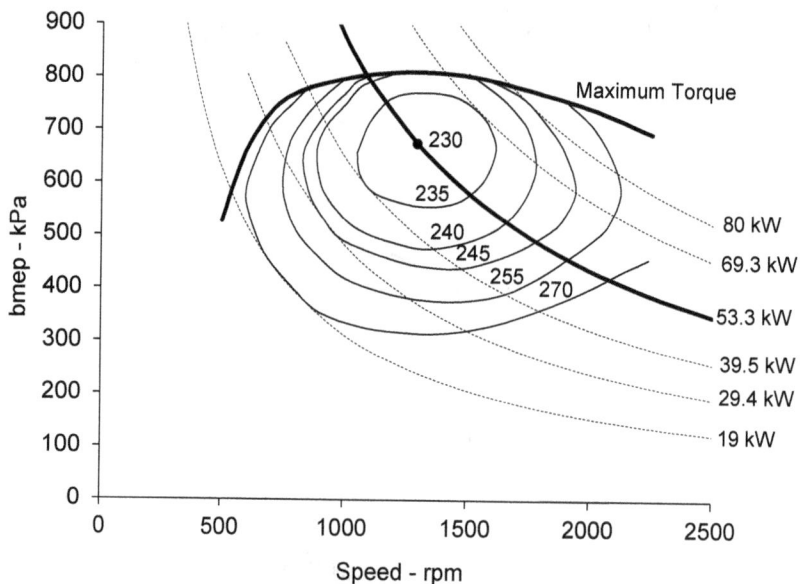

Figure 15.17: The engine map for a 7.3 liter CI engine, showing a line of constant power of 53.3 kW through the point of minimum brake specific fuel consumption.

different values of constant specific fuel consumption and brake power. Values are typical of a naturally aspirated indirect injection diesel engine. The lines of constant power are hyperbolas on the *bmep*-speed curve, since brake power is proportional to the product of the speed and the *bmep*. It can be seen from this diagram that the absolutely lowest specific fuel consumption of 230 g/kWh occurs at the following condition:

$$N = 1290 \ rpm; \ bmep = 637 kPa \ BP = 53.3 kW$$

The minimum fuel consumption is achieved at a power lower than the maximum power of 98 kW at the maximum engine speed of 2300 rpm. Since the vehicle requires varying power as a function of vehicle speed, an alternate way to look at ideal operation is to find the most efficient engine operating point for each engine power.

Looking at the engine map then, the lowest fuel consumption at a given power is found at the point where that power hyperbola is tangent to a line of constant fuel consumption. For a given power, any movement away from this point will result in a higher fuel consumption. A line can be drawn through these tangent points, and this line shows how the engine would have to be regulated in order to achieve the lowest possible consumption in the entire operation area. An example is given in Figure 15.18 by the curve A-A. It cannot, therefore, be correct to use a continuously variable gear ratio to maintain a constant engine speed. In order to achieve the lowest consumption in the entire operating range, the engine and gear ratio must both be regulated as functions of vehicle speed.

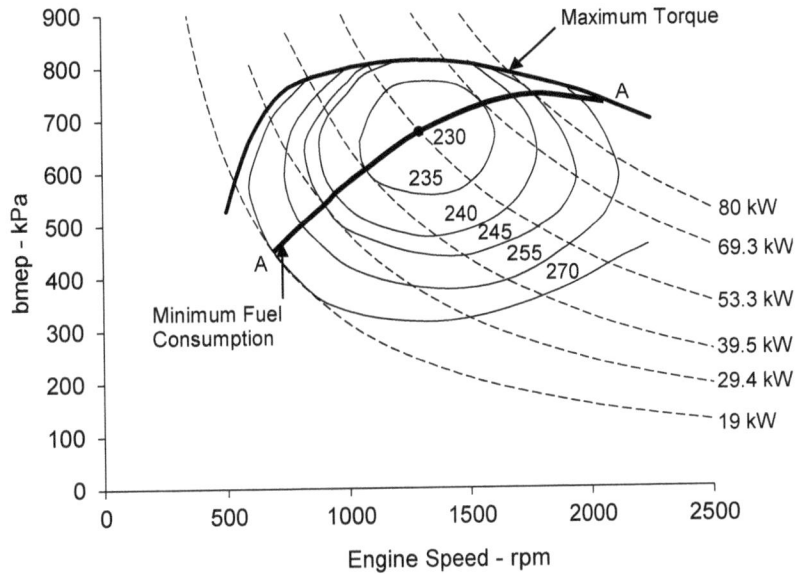

Figure 15.18: Minimum brake specific fuel consumption for any power for the engine of Figure 15.17.

For operation at a given power, there is a connection between the engine performance and the vehicle performance, depending on the gear ratio chosen.

For the engine:

$$BP = \frac{2\pi T_e N_e}{60000} \Rightarrow T_e = \frac{60000 BP}{2\pi N_e} \tag{15.56}$$

The vehicle and engine speeds are related through the gear ratio chosen by:

$$v = \frac{2\pi r_w N_w}{60} = \frac{2\pi r_w N_e}{60 \cdot \theta} \tag{15.57}$$

The unit tractive force can be written as a function of the vehicle speed, and either the engine torque in Equation (15.15), or as a function of engine power in Equation (15.17). Note that the gear ratio does not appear in Equation (15.17). It varies according to:

$$\theta = \frac{2\pi r_w N_e}{60 \cdot v} \tag{15.58}$$

Figure 15.19A shows the unit free traction force for lines of constant power, using the powers from minimum fuel consumption curve of Figure 15.18. The calculation was performed for a vehicle with the following characteristics: mass = 3200 kg, rolling resistance coefficient = 0.015, tire radius = 0.35 m, $A_f \cdot C_w = 4.5 m^2$. Figure 15.19B shows the total gear ratios associated with the tractive forces. At the lowest speed, very high gear ratios result.

If one always wanted the engine to operate at maximum efficiency, then it would be necessary to operate at the engine condition for maximum efficiency ($bmep = 637$ kPa at 1290 rpm). Though not practical, this could be accomplished by the use of an ideal, continuously variable transmission available. What this would entail for a given vehicle can be seen in Figure 15.19, which shows the unit free tractive

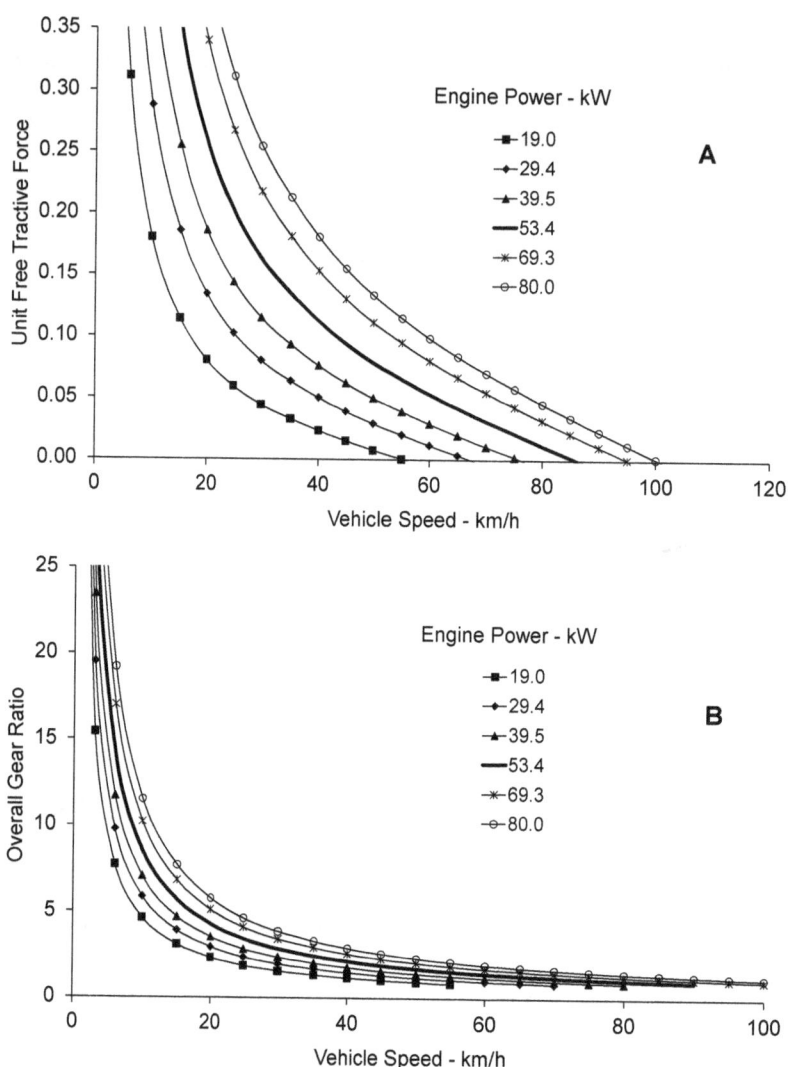

Figure 15.19: A - Free unit tractive force for lines of constant power for the engine used in Figure 15.17. B - Gear ratios required for operation at the conditions in A.

force as a function of speed for constant powers. Stable operation on a level load would only be achieved where the unit free tractive force is equal to 0. In this figure this corresponds to a speed of about 85 km/h for the power of 53.4 kW. Since the tractive force is greater than zero for speeds below this, the vehicle would either have to be accelerating or else driving up a slope if engine operation was to be at the point of maximum efficiency. It would also be possible to operate going down a slope at a speed greater than 85 km/h.

It is clearly not reasonable to consider such operation, but the point is, that at any vehicle speed, there is a power required. For steady state operation, it is of interest to find the gear ratio that allows the vehicle to operate with the lowest fuel consumption. In other words, for any given engine power, what combination of speed and load gives the lowest fuel consumption for the vehicle? The ratio of engine speed to vehicle speed would then give the resulting gear ratio, and if it was possible to continuously adjust the gear ratio, minimum fuel consumption would be obtained for any given speed.

Though the continuously variable transmission, CVT, has not proven itself yet in durable economic operation, so it will only be used as a model here. It can be used as a reference to see how much of a fuel consumption penalty is paid when using conventional transmissions with fixed gear ratios.

If one instead considers a conventional gearbox with 3-4 gears, it is not possible to regulate the engine along a single curve that corresponds to the lowest fuel consumption for all powers. The significance of not being able to operate at precisely the minimum fuel consumption can be found by determining the operating range permissible for given percentage loss in fuel consumption.

Consider the condition for the engine under consideration, where the minimum fuel consumption is 230g/kWh. A 6 percent loss in fuel consumption corresponds to a specific fuel consumption of 243.8 g/kWh. Then by following the constant power curve from the point of minimum fuel consumption to its intersections with the curve of $bsfc = constant = 243.8$, the points of 6% increased fuel consumption can be found. By doing this for the range of engine speed, the variation limits for a loss of 6% can be drawn.

Curves for acceptable margins for deviation from the ideal curve, of 2%, 4% and 6%, are shown on the engine map in Figure 15.20. That is, one obtains areas inside of which one should work in order to limit the fuel consumption penalty. For this engine map, there is a reasonably large region where the fuel consumption loss is less than 6%. Unfortunately, this area does not coincide very well with the typical road load curve, which starts at low $bmep$ for low speed, and the follows a quadratic curve in the speed.

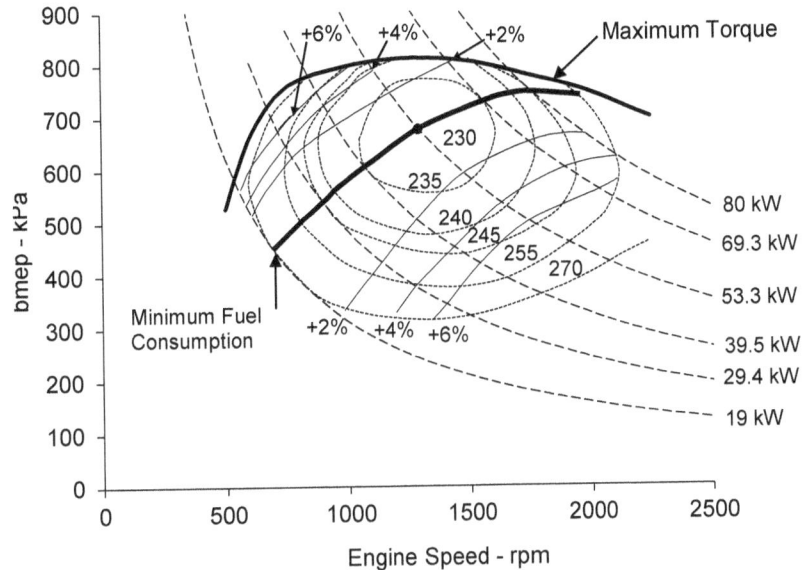

The operating points from the engine map can then be transferred to a tractive force diagram. This has been done for the engine of Figure 15.17,

Figure 15.20: Operating range for different acceptable penalties in specific fuel consumption relative to the minimum value at an given engine power for the engine of Figure 15.17.

with the following vehicle characteristics: mass = 3200 kg, rolling resistance coefficient = 0.015, tire radius = 0.35 m, $A_f \cdot C_w = 4.5m^2$, and a differential gear ratio of 2.90:1. Consider, for example, a gearbox with 3 gears: 2.62:1. 1.6:1, and 1.00. Then for a given gear ratio, the vehicle speed is proportional to the engine speed, as given by Equation (15.11), and the engine torque at any speed is found on the engine map from the intersection of the engine speed and the line through the points of minimum fuel consumption, or lines at a given deviation from the minimum fuel consumption. Given the vehicle characteristics, Equations (15.15) or (15.17) can be used to calculate the free tractive force for the conditions giving the lowest fuel consumption (100%) as well as the limiting curves for ±2% on the unit tractive force diagram, as shown in Figure 15.21. Note that the lines in Figure 15.21 do not correspond to full load conditions, but to operation at minimum fuel consumption conditions, or given deviations from it. It is not possible to operate the engine at maximum efficiency and maximum power at the same time, as can be seen in the figure since the lines at minimum efficiency do not extend all the way to the maximum power line of 98 kW.

In this way, one can obtain an impression of how close the gear intervals should be to each other in order to cover the most common operating area with 102% fuel consumption. Otherwise, by maintaining the number of gears and gearing interval one can read how much the fuel consumption increase would be. In addition, the figure shows which areas are the most undesirable. There is only shown one example here, and one naturally has to be aware of the fact that the appearance of the consumption characteristics to a large extent can vary from one engine to another and thereby from one type of car to another. It is the methodology that is most important here.

Transferring the curves of constant consumption on the engine map to the driving condition diagram, one obtains a set of curves of different levels for each of the gear intervals used, and with that a certain impression of the consumption with different driving conditions. Such a diagram does not, though, give

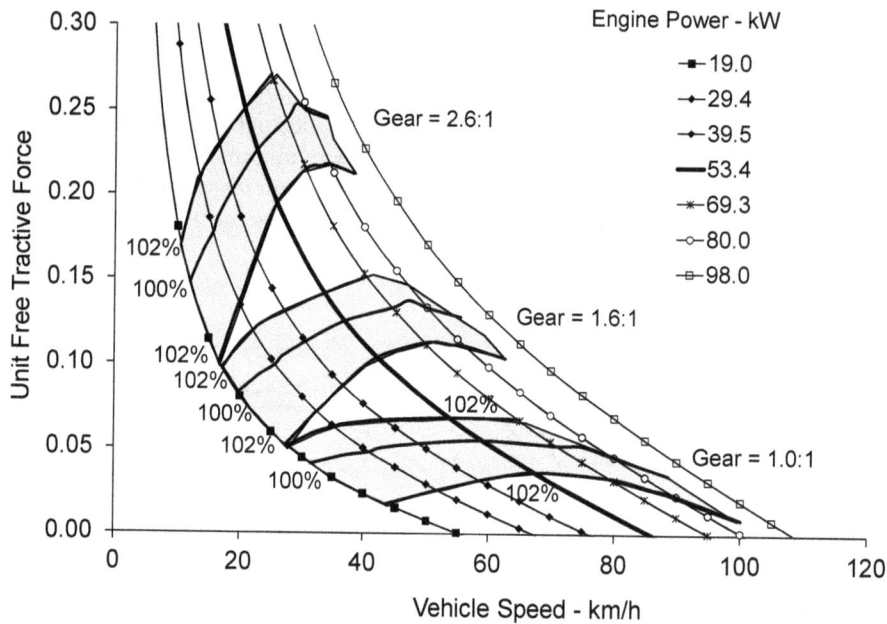

Figure 15.21: Free unit tractive force as a function of vehicle speed for a vehicle operating at the power giving minimum fuel consumption for that engine speed. Also shown are the limits for a 2 percent increase from the minimum fuel consumption, for the engine of Figure 15.17.

direct information on the vehicle's specific consumption, which is the quantity in which the user of the vehicle is most interested

In order to obtain that information, the consumption curves must be transformed in such a manner that the final consumption curves in the driving condition diagram give the consumption in, for example, liter per kilometer or liter per 100 km. The fuel consumption can naturally also be given by the specific stretch of road, for example in kilometer per liter fuel. Figure 15.22 shows an engine's constant fuel consumption curves transferred to 4th gear in a driving condition diagram, and the last figure shows the specific consumption of the length of road in km/litre for a steady speed in 4th gear.

This is done by assuming a speed and bmep/torque, and again using Equation (15.15) or (15.17) to determine the unit free tractive force. Figure 15.22A shows the engine specific fuel consumption ($bsfc$), while Figure 15.22B shows the vehicle fuel consumption in km/l, calculated as per Section 14.7. The condition corresponding to steady driving on a level road is given by the horizontal axis, that is the condition where the unit free tractive force is equal to zero. The conditions above the curve are allowable driving conditions, where the vehicle can either accelerate, climb a gradient, or a combination of both. For reference, the unit tractive force is shown for a lower gear, $\theta_{tot} = 6.30 : 1$

Plotted in this way, one can see whether vehicle load or engine efficiency contributes most to fuel consumption. For example, note that for large values of the unit free tractive force, the engine efficiency is high, (low $bsfc$) but the vehicle fuel economy is low (few km/l). This is because the power requirement due to acceleration or gradient, overcomes the increase in engine efficiency with the higher load.

In order to retain legibility, the fuel consumption curves are drawn in the example under consideration for 4th gear alone, but the other gears will naturally give a corresponding picture. If the complete diagram is drawn, it is possible to determine which gear is the most efficient for a given driving condition. It can readily be seen that when there are only relatively few gears available, it will be advantageous at any given time to use the highest of the possible gears. For a given speed, the fuel consumption deteriorates when the load increases even though the engine efficiency increases. The highest vehicle efficiency is obtained for the a low engine efficiency in this case. For a given tractive force, fuel consumption increases with speed due to the larger power requirements, primarily aerodynamic, while the engine efficiency does not change that much for most conditions. The techniques used in the chapter enable the reader to evaluate strategies for improving this situation.

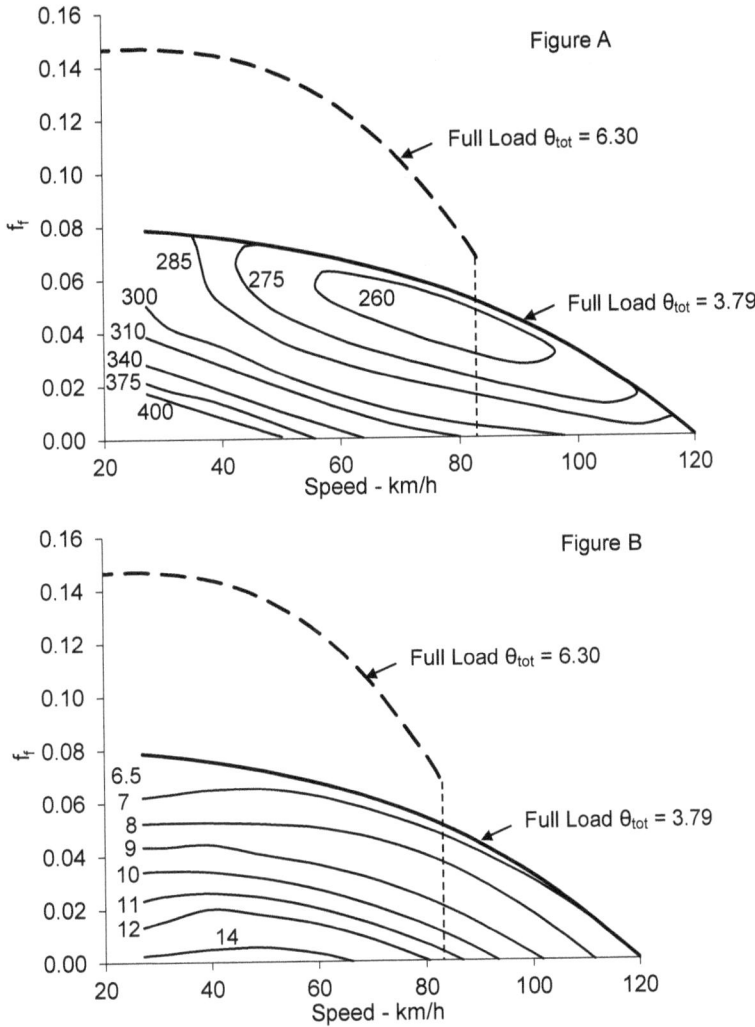

Figure 15.22: Engine specific fuel consumption (Figure A, in g/kWh) and vehicle fuel consumption (Figure B, in km/l), for a vehicle in 4th gear($\theta_{tot} = 3.79$). Engine of Figure 15.2 and vehicle of Table 15.1.

Figure 15.3 shows the engine fuel consumption map for the engine used in the example calculation. Using this map, the fuel consumption rate for the engine over its entire operating range can be calculated. For any given power with ρ_f in g/liter:

$$\dot{Q}_f = \frac{BP \cdot bsfc}{\rho_f} (liter/hour) \tag{15.59}$$

In Figure 15.23 the vehicle power requirement for no wind on a flat road, P_W, is shown as a function of engine speed for the two gearing ratios. For the gear ratio that gives maximum engine power, it can be seen that the engine operates at a higher speed. For a given engine power, higher speed means higher friction, lower torque, and therefore lower efficiency. Changing to the "Overdrive" gear, the engine speed is reduced at any given vehicle speed, giving higher load and better efficiency.

In this example at 80 km/h, a "normally-geared" vehicle will have a fuel consumption of 7.2 liter per hour and with overdrive 6.6 liter per hour, corresponding to:

Normal-gear: $(7.2/80) \cdot 100 = 9.0 \ liter/100 \ km$

"Overdrive": $(6.6/80) \cdot 100 = 8.25 \ liter/100 \ km$, a reduction of 8.3%.

For a speed of 115 km/h, it is similarly found that:

Normal-gear: $(16.3/115) \cdot 100 = 14.2 \ liter/100 \ km$

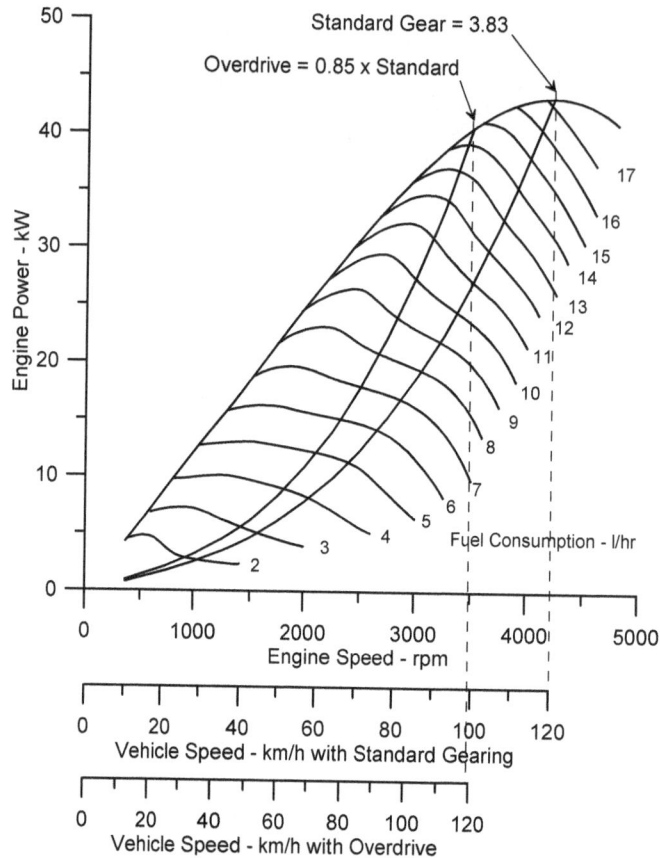

Figure 15.23: Fuel consumption for the example vehicle for two different gears, standard and "overdrive". The power shown is the engine power.

"Overdrive": $(14.8/115) \cdot 100 = 12.9 \ liter/100 \ km$, a reduction of 9.2%.

This reduction in fuel consumption gives in turn a reduction in surplus power to use for climbing and acceleration. Ignoring possible transmission efficiency changes, which an extra set of gears could cause, one has:

Surplus power at 80 km/h:

Normal-gear: 32.3- 15.0 = 17.3 kW, "Overdrive": 26.8 - 15.0 = 11.8 kW

Surplus power at 115 km/h:

Normal-gear: 38.6 - 36.4 = 2.2 kW, "Overdrive": 0 kW

15.7 Number of Gears and Gear Interval

As shown earlier, the total interval in the gearbox is:

$$\theta_{tot} = \theta_1 \frac{v_H}{v_1} \tag{15.60}$$

The next choice must be the selection of the number of gear intervals. It is clear that for most vehicles this choice must be a compromise, since different and partially conflicting conditions are important, the most important being:

1. *Cost and Weight* indicate naturally, that one use as few gears steps as possible.

2. *Driveability* demands many gear steps on the other hand.

3. *Operation* - in as much as it is a manual transmission - indicates few gear steps.

4. The *Engine Characteristics* of a modern engine indicate many gear steps, as a consequence of the nearly flat torque curve.

No matter how many gear steps are used, as a first approximation, it would seem desirable to distribute the conversion ratios in accordance with a quotient series. This is because such a distribution gives the possibility of using the same conversion interval between gears and thereby the same power interval no matter what the speed interval of the vehicle (See Figure 15.12). A classic solution for the quotient series, can be expressed:

$$\frac{\theta_1}{\theta_2} = \frac{\theta_2}{\theta_3} = \frac{\theta_3}{\theta_4} = \ldots\ldots\ldots \tag{15.61}$$

Table 15.2 gives the resulting intervals for a transmission where ratio is 3.22:1 in first gear for a three

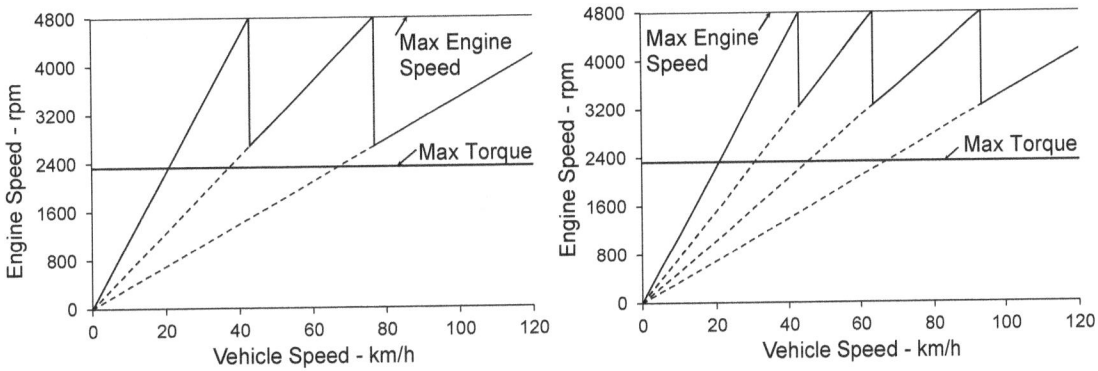

Figure 15.24: Engine speed as a function of vehicle speed for 3 and 4 gear quotient series of Table 15.2, $\theta_d = 3.79 : 1$.

and four speed transmission. Figure 15.24 shows the relationship between engine and vehicle speed using the quotient series for these gear ratios. As a construction guideline, it can be assumed, as mentioned

Table 15.2: Examples of quotient series for gear ratios for a transmission where ratio is 3.22:1 in first gear.

3-Gear Transmission	4-Gear Transmission
$\theta_1 = 3.222 : 1$	$\theta_1 = 3.222 : 1$
$\theta_2 = 1.80 : 1$	$\theta_2 = 2.18 : 1$
$\theta_3 = 1 : 1$	$\theta_3 = 1.48 : 1$
—	$\theta_4 = 1 : 1$

earlier, that the ratio, α, between consecutive gear ratios (in this case a constant) cannot be larger than that determined from $N_{max} = N_A$ to $N_{e,max} = N_c$, and is usually larger than the ratio of the maximum power speed $N_{e,max}$ to the maximum engine speed, N_{max}. This gives the two limits (See Figure 15.12):

$$I: \quad \alpha_i = \frac{N_A}{N_B} \tag{15.62}$$

$$II: \quad \alpha_{max} = \frac{N_A}{N_C} \tag{15.63}$$

I: Consider now a gearbox with z gears and using a quotient series for the gear ratios. It is found that for the final overall gear ratio as the upper limit:

$$\alpha_i^{z-1} = \theta_1 = \frac{v_H}{v_1} \rightarrow z_1 = 1 + \frac{\ln(\theta_1)}{\ln(\alpha_1)} \tag{15.64}$$

where z is the number corresponding to the ideal interval and $\alpha_i = (N_A/N_B)$.

The lower limit is found with z_{min} and $\alpha_{max} = (N_A/N_C)$.

II:

$$\alpha_{max}^{z-1} = \theta_1 = \frac{v_H}{v_1} \rightarrow z_{min} = 1 + \frac{\ln(\theta_1)}{\ln(\alpha_{max})} \tag{15.65}$$

For the Engine of Figure 15.2, the maximum torque speed, N_C, is 2330 rpm, the speed where the maximum speed curve cuts the *bmep* speed curve, N_B, is 3590 rpm and the maximum speed, N_A, is 4800 rpm, that is the gear ratio change factor should be between the limits:

$$\frac{4800}{3590} < \alpha < \frac{4800}{2330}$$

and so $\alpha_{min} = \frac{4800}{3590} = 1.34$ and $\alpha_{max} = \frac{4800}{2330} = 2.06$

If in accordance with Figure 15.13, 30% and 5% gradients in 1st and the highest gears are chosen, the total conversion ratio is:

$$\theta_1 = \frac{v_H}{v_1} = \frac{74}{23} = 3.217 : 1$$

and the number of gears is:

$$z_{max} = 1 + \frac{\ln(3.217)}{\ln(1.337)} = 1 + \frac{1.169}{0.290} = 5.023$$

$$z_{min} = 1 + \frac{\ln(3.217)}{\ln(2.060)} = 1 + \frac{1.169}{0.7227} = 2.618$$

And there is then the possibility of using 3, 4, or 5 gears. The conversion ratios based on a quotient series in the 3 cases are:

3. gears:

$$\alpha_3 = \sqrt{3.217} = 1.794$$
$$\theta_1 = 3.217, \ \theta_2 = 1.794, \ \theta_3 = 1.0$$

4. gears:

$$\alpha_3 = \sqrt[3]{3.22} = 1.476$$
$$\theta_1 = 3.217, \ \theta_2 = 1.476 \cdot 1.476 = 2.179,$$
$$\theta_3 = 1.476, \ \theta_4 = 1.0$$

5. gears:

$$\alpha_5 = \sqrt[4]{3.217} = 1.339$$
$$\theta_1 = 3.217, \ \theta_2 = 1.389 \cdot 1.389 \cdot 1.389 = 2.404,$$
$$\theta_3 = 1.339 \cdot 1.339 = 1.794, \ \theta_4 = 1.339, \ \theta_5 = 1.0$$

The corresponding curves for the unit free tractive force using the example engine are given in Figures 15.25A and B, and Figure 15.26A. In this example the vehicle mass, $m_o = 1400kg$, the rolling resistance coefficient $f_r = 0.017$ the air resistance coefficient, $c_w = 0.45$, the frontal area, $A_f = 3m^2$, the transmission efficiency, $\eta_{tr} = 0.9$, the differential gear ratio, $\theta_d = 3.79$, and the wheel diameter, $r_w = 0.29m$.

For the gear steps where the rotating engine and flywheel mass can be neglected on account of low engine speeds, the acceleration ability is readily obtained:

$$f_f \cdot m_o g = m_o \cdot a \Rightarrow \ a = f_f \cdot g \tag{15.66}$$

Since the overall gear ratio in first gear is fixed at 3.22:1, all the first gear curves are the same. Especially at the lower speeds where the driving resistance is a weak function of speed, the relation relation between the engine torque and the unit tractive force is quite apparent. The figures also give the speed areas where the demand for working in the power area of N_B/N_{max} is not satisfied. This is seen by the magnitude in the drop of the unit free tractive force between two gears from the maximum vehicle speed at a given gear down to the next gear at the same vehicle speed (vertical dotted lines). These conditions are naturally the most unsatisfactory with the use of the 3 speed transmission, where is the largest unsuitable speed area with highest/next gear: θ_2/θ_3. There is a larger drop in the tractive force with gear change for fewer gears, which results in a more dramatic decrease in acceleration upon gear changes for fewer gears.

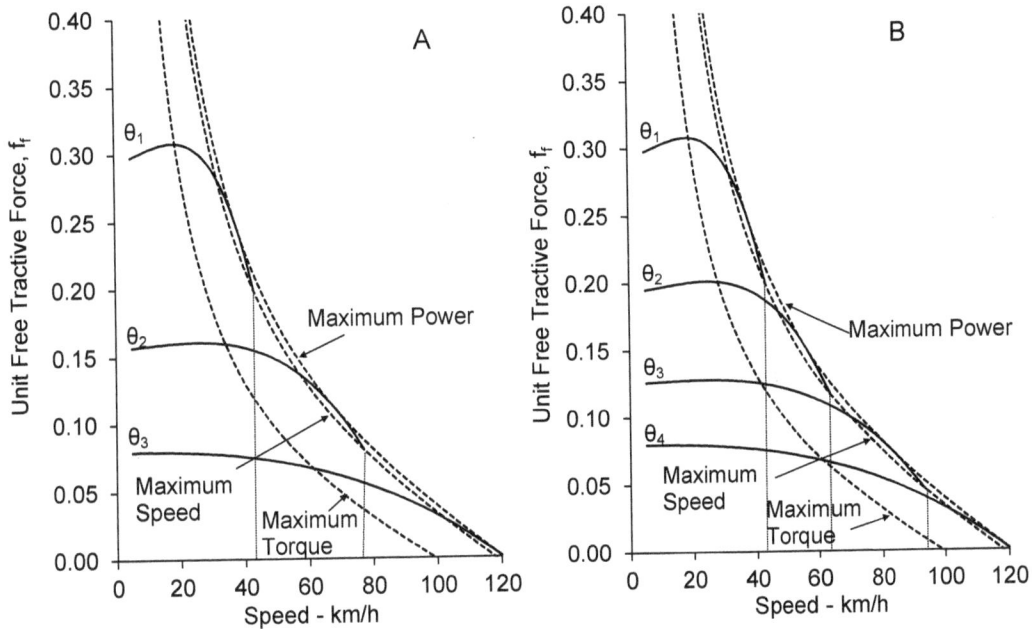

Figure 15.25: Unit free tractive force as a function of speed for vehicle of Table 15.1 and engine of Figure 15.2 with A: 3-gear and B: 4 gear transmission based on the quotient series.

For the case of the particular engine torque curve, where the torque falls off substantially at the higher speeds, and the 5 gear transmission, (Figure 15.26A), there are very few conditions where the maximum power cannot be applied to the vehicle. A similar condition is obtained with larger engines, which typically also have a torque curve that falls off at high speeds. (See Chapter 12). It is possible to achieve better conditions in the lower speed area by choosing the interval between θ_2 and θ_3 smaller than the calculated value of 1.794, but this will result in a greater step between θ_1 and θ_2.

The gear ratios presented by the method above can give a good starting point for gear ratio selection. In practice, the gear ratios should be determined for each engine/vehicle combination and its intended use. Considerations of fuel economy and drivability need to be taken into account and the gear ratios determined in the final analysis by on the road testing.

In order to show typical values of the gear ratios used in actual on road vehicles, a survey was made of the gear ratios for the manual 5 speed transmissions of a large number of light duty/passenger vehicles for the model year 1996. The 5 speed manual transmission is a popular choice for transmissions for the European market, where automatic transmissions are still not as dominant in passenger cars as on the North American market.

The vehicles range in curb weight from 630 to 1890 kg, and maximum torques from 53 to 300 Nm. The average torque to curb weight ratio is 0.135 Nm/kg, with a range from 0.204 to 0.083. Spark and compression ignition engines are both included, with values for naturally aspirated and turbocharged engines represented. The 5- speed transmission is the most commonly offered transmission in this range of vehicles. The 5th gear is typically chosen for good cruising fuel economy. The average gear ratios, maximum and minimum values for these vehicles are shown in Table 15.3 as well as the overall gear ratios from engine to wheel, including the differential.

The average gear and differential ratios from Table 15.3 have been used in connection with the engine of Figure 14.22 to show the unit free tractive force as a function of speed for a typical modern SI engine

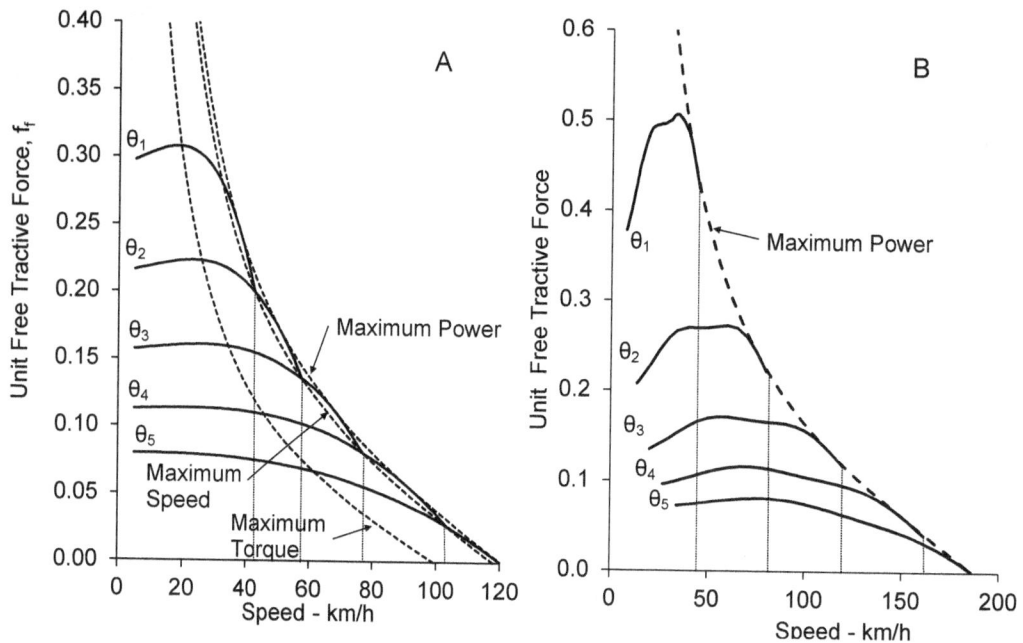

Figure 15.26: Unit free tractive force as a function of speed for vehicle of Table 15.1 and engine of Figure 15.2 with a 5-gear transmission based on the quotient series and B: the engine of Figure 14.22 using average values of gear ratios for 1996 model passenger cars.

Table 15.3: Average, maximum and minimum values of the transmission gear and differential gears ratios for a selection of 80 passenger vehicles with 5-speed manual transmissions from the model year 1996.

Gear	Average	Maximum	Minimum	Average θ_{tot}
1	3.54	4.08	3.09	14.0
2	1.99	2.29	1.74	7.87
3	1.33	1.54	1.12	5.26
4	0.986	1.16	0.830	3.89
5	0.780	0.970	0.640	3.08
Differential	3.96	5.08	3.07	-

powered passenger car. The 1.8 liter, 4-stroke engine produces a brake power of 86.4kW at 6000 rpm, with a maximum torque of 155.4 Nm at 5000 rpm. The maximum torque curve vs speed of this engine is flatter than that of the example engine shown previously, particularly at the highest speeds. For the example shown here, the vehicle parameters are a weight of 1430 kg (simulated vehicle, 2 passengers and some fuel), frontal area of $2.01 m^2$, air resistance coefficient of 0.32, and rolling resistance according to Equation (14.20). The tire radius is 0.28 m, and the transmission efficiency assumed to be 0.91.

For the gear ratio in first gear, there is a large free unit tractive force, and in accordance with Equation (15.66), an acceleration of about one half "g" can be expected, or the ability to climb a grade approaching 50 percent. When shifting from first to second gear, there is a greater drop in the unit tractive force, than for the quotient series (see Figure 15.26A). First gear is then used primarily for a good acceleration, and there is a large drop in the unit tractive force when shifting to second gear. In the higher gears, there is a smaller drop when shifting, and due to the drop off in torque at the very highest engine speeds, operation at maximum engine power is possible at nearly all speeds above about 100 km/h.

There is also a smaller difference in the unit tractive forces between 4th and 5th gear for the average gear ratio values, compared to the quotient series. Particularly 5th gear is chosen for good fuel consumption when cruising, so 4th gear is most likely chosen based on performance or drivability criteria. Though not the case here, in some vehicles, the maximum vehicle speed is achieved in fourth gear rather than fifth. Were the quotient series to be used along with the same overall gear ratios in first and fifth gears, the free unit tractive force curves would be more evenly spaced in the vertical direction.

15.8 Problems

Problem 15.1

In Automobile Revue, 1998 the following data for a Ferrari F355 can be found: Engine: 4-stroke, V-8, cylinder diameter = 85 mm, Stroke = 77 mm, compression ratio = 11.1:1. The engine gives a maximum power of 280 kW at 8250 rpm, and has a maximum torque of 363 Nm at 6000 rpm and operates on gasoline.

The transmission has 6 gears with the following ratios: I - 3,07; II - 2,16; III - 1,61; IV - 1,27; V - 1,03; VI - 0,84. The differential has a ratio of 4,3.

The car has a length of 425 cm, a width of 190 cm and height of 117cm with an empty weight of 1450 kg. Additionally, the maximum speed is 295 km/h, and that vehicle speed at an engine speed of 1000 rpm is 35 km/h in VI gear. The acceleration is given as 0 - 97 km/h in 4.6 s.

It is estimated that the car's rolling resistance coefficient is 0.011 at low speed and 0,015 at maximum sped. In addition the drag coefficient is estimated to be 0.33, and the transmission efficiency = 90%. a. Find the brake mean effective pressure at when the engine operates with maximum torque. b. What is wheel radius? c. Calculate car's acceleration in Gear-I for the conditions in question a. and a masse of 1550 kg. Assume that in Gear-I, the equivalent acceleration mass of the rotating parts amounts to 50% of vehicles unloaded masse. d. If the vehicle has a constant acceleration corresponding to that in Question c., how long will it take for the vehicle to accelerate up to 97 km/h from 0 km/h? e. Compare the engine's maximum power at the wheels with vehicle power required for operation at maximum speed. (a: $bmep = 1305$ kPa, b: $r = 0.335$m, c: $a = 5.43$ m/s^2, d: $t = 4.65$ s, e: $P_{hjul} = 213$ kW, $BP = 252$kW)

Problem 15.2

A passenger car has a 4-stroke, 4-cylinder spark ignition engine that has a bore and stroke of 78,5 and 82 mm respectively. The compression ratio is 9,6:1. The engine map is found in Figure 15.27:

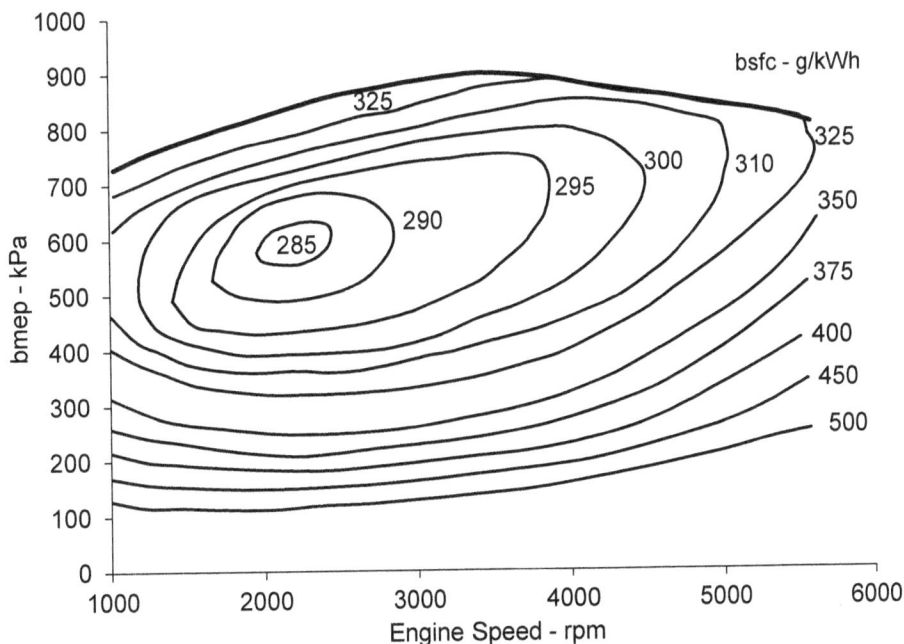

Figure 15.27: Engine map for Problem 15.2.

In the version of the vehicle, the frontal area = 1.75 m^2, air resistance coefficient = 0.33, and the rolling resistance coefficient = 0.012. The gear ratios are:

The transmission efficiency = 0,91. The empty weight = 920 kg, and it can be assumed to have a load of 150 kg. The wheel diameter = 54 cm. The rotating inertia can be estimated by Equation (14.49).

Gear	Ratio
1	3.42
2	1.95
3	1.36
4	1.05
5	0.85
Differential	3.56

For a constant speed of 90 km/h on a flat road with no wind, Calculate the fuel economy in liter/100km in 5-gear. For these conditions, what is the maximum grade the car can drive up and still maintain the same speed and gear?

The engine has a 3-way catalyst and operates with a stoichiometric mixture ratio. If the volumetric efficiency is 0.88, calculate the engine's intake pressure for operate at 90 km/h on a flat road. At 90 km/h on a flat road in 5th gear, the driver shifts down into 4th gear to accelerate with wide-open throttle. If the acceleration from that condition is assumed to be constant, how long will it take the car to accelerate up to a speed of 110 km/h? Is the vehicle capable of operating in 5th gear with maximum engine speed with constant speed on a flat road with no wind? (4,85 1/100km, 4,03°, 36,1 kPa, 6,27s, the engine lacks 3,7 kW if $bmep = 810$ kPa at 5600 rpm).

Problem 15.3

The engine in Figure 7.44 is used in a truck with the following specifications: Chassis weight (empty): 7500 kg, Maximum load: 16000 kg, frontal area = 9.66 m^2 The gear ratios are: Gear 1 = 10.18; Gear 2 = 7.16; Gear 3 = 5.04; Gear 4 = 3.75; Gear 5 = 2.72; Gear 6 = 1.92; Gear 7 = 1.35; Gear 8 = 1.00. Differential ratio = 3.78, Tire diameter =1.,01 m, Air flow coefficient, $C_w = 0.8$ based on the and the rolling resistance coefficient = 0.012. The transmission efficiency = 0.92. The inertia of the rotating components in first gear corresponds to 7.5 % of the weight of the unloaded vehicle. Assume a fuel specific gravity of 0.835.

1. Find the maximum speed of the vehicle if it is determined by the maximum speed of the engine. 2. For a fully loaded vehicle on a flat road with no wind, is it possible to operate at the speed from question a? 3. On a flat road with no wind, find the trucks fuel consumption in km/liter with a maximum load and a speed of 80 km/h in 8th gear. 4. On a flat road, find the maximum acceleration of the fully loaded vehicle in 1st gear with a speed of 5 km/h. (106 km/h, yes- excess power of 40.3 kW, ≈2.78 km/l, ≈3.35 m/s^2).

Chapter 16

Gasoline Direct Injection (GDI) Engines - Case Study

The GDI engine has appeared in production recently due to a combination of factors. The basic idea is operating with a unthrottled, stratified, lean mixture, with spark ignition. This concept shows improvements in engine fuel economy compared to conventional SI engines, that can be estimated using basic engine principles. This chapter presents an introductory feasibility analysis of the concept. It also helps to show the connection between engine and vehicle.

The development of this type of engine has been underway for a long time. However, it is only recently that improvements in technology have made this a practical application that could be put into production. The technologies involved are:

- Improved experimental methods to study flow and combustion in engines

- Improved calculational methods, primarily CFD

- Improved electronic fuel injection combined with computer control technology

The story of this engine is presented as a form of case study, to indicate the connection between basic engine principles to investigate the feasibility of a technology.

16.1 Introduction and Historical Background

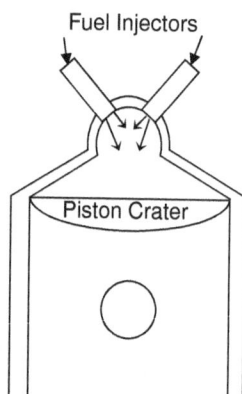

Figure 16.1: An early version of a Direct Injected Gasoline Engine.

Gasoline Direct Injection (GDI) is a modern name for an old concept. Even before the Second World War, engineers were working of the concept of an engine that would operate on gasoline but with some of the advantages of diesel engines. Figure 16.1 shows the basic concept of such an engine, built by Ricardo.

The engine injected the fuel directly into the combustion chamber, and used a form of prechamber to keep the fuel concentrated near the spark plug during combustion. In this way, the mixture is stratified. That is, it is divided into zones of different mixture strength. Ideally, the mixture is approximately stoichiometric near the spark plug, and consists of air alone farther away. With this ideal arrangement, the overall mixture ratio of the engine is lean, which gives a better indicated thermal efficiency because of more advantageous thermodynamic properties of the cylinder gasses (recall the ideal air cycles and the effect of specific heat ratio).

In addition, if the mixture can be separated into two zones with different stoichiometry, all the burnable mixture could be concentrated into the one zone and all the air left over in the other. The amount of fuel could then be changed by changing the size of the fuel-containing zone alone. This would eliminate the need to throttle the air charge, since the direct relationship between the fuel and the air as used in a homogeneous charge engine would no longer be required. The elimination of throttling would improve engine efficiency, especially at low loads, where pumping work is a significant fraction of the engine work.

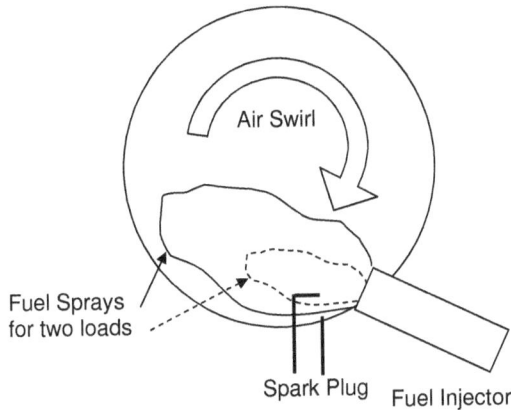

While this sounds simple, in practice it has been very difficult to accomplish. In the 1950's, the TCP engine developed by Texaco, tried to accomplish this by injecting fuel into the cylinder air, which was set into a rotary motion, shown in Figure 16.2.

The concept was basically to increase the size of the fuel air zone by lengthening the injection duration. While the engine ran, it was not possible to get adequate combustion over a wide range of operating conditions. At high loads, the mixing takes too long and the engine has a tendency to smoke. At light loads, the mixing tends to be too fast, and the hydrocarbon emissions and fuel consumption of the engine increase because of a large amount of fuel being overly mixed, and too lean to burn well.

Figure 16.2: The Texaco Combustion Process stratified charge engine.

In the 1960's, Ford Motor Co. developed the engine called PROCO "PROgrammed COmbustion". The engine was intended to accomplish the same goals as the TCP engine, but used a bowl in the piston and a centrally located fuel injector in an effort to contain the mixture and achieve proper stratification. This engine was the subject of significant research and development, but never was put into production. The concept of the PROCO engine is shown in Figure 16.3.

Many variations of the gasoline direct injection engine have been proposed through the years. All of those developed before the 1980's were unable to adapt the injection process to the wide range of change engine conditions. Since the 1980's some developments have occurred which dramatically changed the outlook for the GDI engine.

The first of these was the wide spread use of computer technology to control engines. The introduction of the three-way catalyst in the mid 1980's required the use of computers to control the ignition and fuel injection processes for the engine. This led to more advanced and flexible engine control technology and improved fuel injection. The successful operation of a GDI engine requires a great flexibility in engine control.

Key developments also occurred in the area of fluid dynamics. One of these was the improvement in Computational Fluid Dynamics. New multi-dimensional computer codes were developed to more accurately predict the in-cylinder behavior of gas flow and fuel sprays. Through the use of these programs, a better

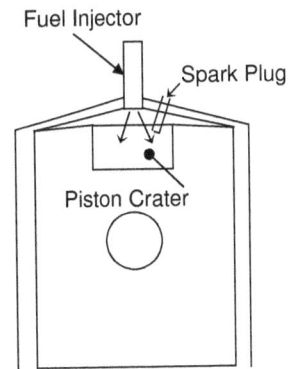

Figure 16.3: A schematic diagram of the the PROCO engine concept.

understanding of the flow and mixing processes inside the cylinder was obtained, and the CFD codes made it possible to more rapidly evaluate the various concepts for controlling fuel and air mixing. In addition, as a way to evaluate computer methods, new methods were developed for visualizing the flow inside of operating engines, predominantly based on laser technology. These methods make it possible to understand the flow inside of engines and the effect of changes.

It is still not possible to run the GDI engine as a truly stratified charge engine over its entire operational range, and at many conditions, particularly high load, the GDI engine must be operated essentially the same as a conventional homogeneous charge engine. Modern fuel injector and computer technology makes this possible.

16.2 Technical Background

Since homogeneous charge spark ignition (SI) engines have been operating successfully for many years, they question arises as to why one would want to develop a gasoline direct injection engine. To answer that, one can first look at the applications and reasons for the popularity of the standard SI engine. The primary use of SI engines today is in the area of passenger cars. These engines are popular because of their high power to weight ratio and historically lower cost than diesel engines. Until the recent development of small turbocharged direct injection diesel engines, SI engines have had a higher power to weight ratio because they can operate at stoichiometric fuel air mixture ratios. This enables them to completely utilize all the air in the cylinder. Since the air is less dense, it is the limiting factor in determining the power output/mean effective pressure of the engine.

In addition, spark ignition engines typically operate over a much higher speed range than diesel engines. The higher operating speed also contributes to the high power output of SI engines, in that at a given torque, a higher engine speed gives a higher power. The use of spark ignition gives a more precise control of the start of ignition than the compression ignition process in the diesel, giving better control over the timing of the combustion than with compression ignition.

For passenger cars, the engines do not operate at maximum load very much of the time, but at lighter loads. Thus, passenger car engines are not built to operate for long periods of time at the maximum power, which helps to reduce their weight and cost. This is important in the passenger car application, where initial vehicle cost plays a dominating economic role.

While spark-ignited homogeneous combustion is advantageous in the above respects, it has a disadvantage, in that the mixture at the spark plug must always be at or near the stoichiometric fuel air ratio. Therefore, to regulate the output of the engine, (that is, the fuel input) it is necessary to regulate the air input simultaneously with the fuel. At light loads in conventional SI engines, the density of the incoming charge in a homogeneous charge engine is reduced to reduce the amount of fuel that is drawn into the cylinder. The throttling used to reduce intake pressure causes pumping losses (see Section 2.2), which reduce engine efficiency, and are especially important at light loads, where the energy required to overcome the friction of the engine is of the same magnitude as the output of the engine. In addition, the lower the output of the conventional SI engine, the more the intake air must be throttled, and the greater the efficiency loss.

The diesel engine has an advantage at light loads, since its heterogeneous combustion process allows regulation of the fuel amount independently of the amount of air in the cylinder. On the down side of this is the combustion process, which is limited by mixing rates in the engines. Thus, the diesel engine is limited in its maximum output by both speed and load restrictions. At high speed, the fuel and air cannot be mixed fast enough, and at high load, it is not realistic to try to get the fuel to mix with all of the air in the cylinder. The engine begins to smoke badly long before that. Modern diesel engines compensate for this through the use of turbocharging, which is much better suited to diesel engines than to SI engine. GDI engines operating at full load with non-homogenous operation have similar problems to those of diesels: it is difficult to obtain good mixing and as a result there are problems with the formation of soot and particulate emissions.

In an effort to combine the good part load fuel consumption of an unthrottled engine with good specific power output of a homogeneous charge engine, attention in recent years has again been focussed on what has been called the direct injection stratified charge engine (DISC). This is a bit of a misnomer, since the diesel engine also fulfills this description. Therefore, the latest popular term, Gasoline Direct Injection (GDI) is more appropriate.

The GDI engine has the potential to fulfill the objective of good part load efficiency by operating with non-homogeneous (stratified) combustion at light loads, and then switching over to a homogeneous combustion at heavier loads. This type of operation was not realistic at the time of development of the TCP and PROCO engines, due to the lack of appropriate computer based engine technology. Today, it is possible and the first GDI powered vehicles were put into production in the late 1990's and utilized this technique.

Table 16.1: Vehicle specifications for the example calculation

Engine:	1.8 liter, 4 cylinder, 4-stroke
Volumetric efficiency:	0.80
Compression Ratio Throttled:	9.5:1
Compression Ratio GDI:	11:1
Vehicle:	Mass = 1200 kg
	Frontal area 1.9 = m^2
	Drag coefficient = 0.36
	Rolling resistance coefficient = 0.013
	Differential gear ratio = 3.72:1
	Tire diameter = 0.56m
Fuel:	Heating value = 44.343 MJ/kg
	Density = 701 g/liter
Transmission gears:	3.5, 1.95, 1.27. 0.97, 0.77

16.3 GDI Example

Given the above, it appears most reasonable to look at the GDI engine at light loads, since that where this type of engine is most likely to work satisfactorily, and where the biggest gains in efficiency due to pumping loss reduction are possible. As an example, consider a vehicle with the specifications shown in Table 16.1. A higher compression ratio is used for the GDI version, since the fuel injection just before ignition reduces the tendency of the engine to knock due to the short time that the end gasses exist.

For the vehicle traveling at a speed of 80 km/h, the total resistance is the sum of the aerodynamic resistance and the rolling resistance:

$$F_{tot} = F_r + F_A = f_R \cdot m \cdot + \frac{1}{2} \cdot \rho \cdot C_D \cdot A_F \cdot v^2$$

$$= 0.013 \cdot 1200 \cdot 9.8 + \frac{1}{2} \cdot 1.2 \cdot 0.36 \cdot 1.9 \cdot \left(\frac{80}{3.6}\right)^2 = 356N$$

The power required to move the vehicle is

$$P = F_{tot} \cdot v = 356 \cdot \frac{80}{3.6} = 7900W = 7.9kW$$

Given a transmission efficiency of 90 %, the power required from the engine is then 8.78 kW.

To estimate engine performance, the following procedure is used, utilizing equations from previous chapters. To calculate the wheel and engine speeds from the vehicle speed, v, in m/s:

$$N_w = \frac{60v}{2\pi r_w}$$
$$N_e = N_w \cdot \theta_{tot}$$

The brake mean effective pressure is obtained from the engine power and speed:

$$bmep = \frac{120 \cdot BP}{V_d \cdot N_e}$$

The friction and pumping works are obtained from correlations of Chapter 9 for full load, and the pumping work estimated from data shown in the book by Heywood [7]. The indicated mean effective pressure can be calculated:

$$imep = bmep + fmep_{wot} + fmep_{pumping}$$

For the throttled engine, the intake pressure determines the $imep$, provided the indicated thermal efficiency and intake air temperature are known. It this example the indicated thermal efficiencies from the

Table 16.2: Comparison of conventional spark ignition (SI) engine and GDI engine powered vehicle with specifications of Table 16.1 in two gears at a constant speed of 80 km/hr. Fuel properties iso-Octane

Engine type	SI	GDI	SI	GDI
Gear	4	4	5	5
Engine speed - rpm	2735	2735	2171	2171
bmep - kPa	214	2144	270	270
WOT friction - kPa	175	175	153	153
Estimated pumping mep - kPa	66	17	62	17
imep - kPa	455	406	485	440
Indicated efficiency (η_i from FA cycle * 0.8)	0.37	0.44	0.37	0.44
Intake pressure - kPa	45.1	90	48.1	90
Fuel/Air ratio	0.0662	0.0252	0.0662	0.0272
Fuel/Air equivalence ratio	1.00	0.380	1.00	0.411
Mechanical efficiency	0.47	0.527	0.556	0.613
Brake thermal efficiency	0.174	0.232	0.206	0.270
bsfc - g/kWh	467	350	394	300
Air flow - kg/s	0.0170	0.0340	0.0144	0.0270
Fuel flow - g/s	1.14	0.854	0.962	0.734
Fuel consumption - km/liter	13.7	18.2	16.2	21.2
Fuel consumption liter/100km	7.31	5.48	6.18	4.71
Percent improvement in liter/100km	-	24.9	-	23.7

Fuel Air Equilibrium are found for relevant compression and equivalence ratios and multiplied by 0.8. The intake temperature is assumed to be 30°C:

$$p_{in} = \frac{imep \cdot R \cdot T_{in}}{\eta_v \cdot FA \cdot H_u \cdot \eta_i}$$

For the unthrottled, GDI engine, it is the fuel air ratio that determines the imep, since the air flow is determined by the speed with intake pressure near atmospheric.

$$FA = \frac{imep \cdot R \cdot T_{in}}{\eta_v \cdot p_{in} \cdot H_u \cdot \eta_i}$$

The fuel flow for either engine is then:

$$\dot{m}_f = FA \frac{\eta_v \cdot p_{in} \cdot V_d \cdot N}{120 \cdot R \cdot T_{in}}$$

The vehicle mileage in km/liter is:

$$km/l = \frac{v \cdot \rho_f}{1000\dot{m}_f}$$

for v in m/s, ρ_f in kg/liter, and $\dot{m}_f$ in kg/s. The engine brake thermal efficiency is:

$$\eta_e = \eta_i \cdot \eta_m = \eta_i \frac{bmep}{imep}$$

and the brake specific fuel consumption:

$$bsfc = \frac{3.6 \cdot 10^6}{H_u \cdot \eta_e}$$

Table 16.2 shows an estimate of the engine operating conditions and the resulting fuel consumption for the vehicle at this speed. The calculations above show that there is a potential for improvement on the order of 25 percent for operation at light loads. On the other hand, this advantage disappears in the case where the engine must be operated at stoichiometric conditions, since the engine operating conditions

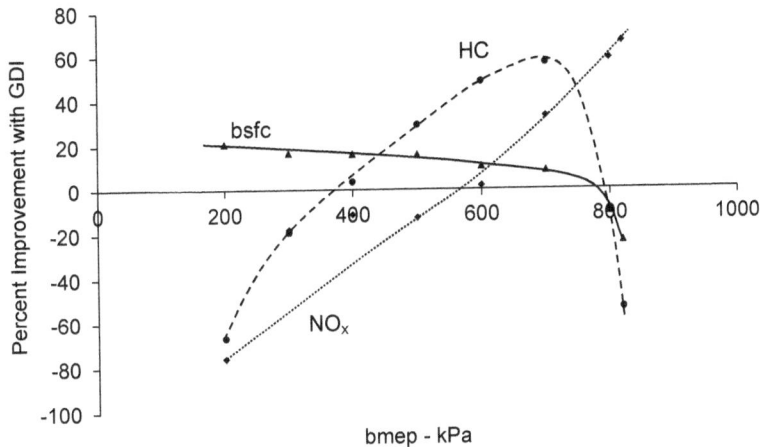

Figure 16.4: Relative changes in *bsfc*, hydrocarbon and oxides of nitrogen emissions for a GDI engine at 1500 rpm. Calculated from the data of reference [113]

are basically the same, that is, unthrottled operation. The improvement obtained from operation with a GDI engine will then be quite dependent on the loading cycle, and the larger the percentage of operation at light load, the larger the expected improvement.

The improvement comes from the following factors:

1. The elimination of the throttling process and reduction in pumping work

2. A higher indicated thermal efficiency due to the leaner overall mixture

3. A higher compression ratio with the GDI engine

The advantage of engine operation at a lower speed (higher gear) for the same vehicle speed is also seen. This holds true for both homogeneous engines and those with stratified charge. When the engine runs slower, it has a lower friction and higher brake mean effective pressure, giving a better mechanical efficiency.

These calculations give results a bit higher than, but of the same magnitude as, some experimental results for a GDI engine operating at 1500 rpm with the compression ratios of Table 16.1 as shown in Figure 16.4. The advantages and disadvantages can be seen here. First, there is a substantial efficiency benefit at light load, though exhaust emissions suffer here. The NOx emission was lowered by about 75% through the use of EGR at part load, but the use of a 3-way catalyst is not applicable due to the excess air ratio being greater than 1.0. Reference [113] also indicated that the GDI engine has significant particulate emissions at higher loads.

16.4 Practical Aspects

Since there appear to be fuel consumption benefits to be attained with a GDI engine, how can this be done in practice? The first GDI design to be put into production is that from Mitsubishi, which is shown in Figure 16.5.

It is based on what is called the tumble based design. Rather than rotating around the axis of the cylinder, the air is set in motion in a direction perpendicular to that axis. In the PROCO and TCCS systems, air swirl was predominately used to control the processes. In the new process, the piston is used to help control the motion of the spray. In fact, the spray is reflected off of the piston top and directed towards the spark plug. An additional benefit is that the high temperature of the piston assists in the evaporation of the fuel. In addition to the piston crater, Figure 16.5 shows that there is a "squish" area to the left of the piston crater. This area supplies additional air motion near the end of the compression process, which helps in directing the air toward the spark plug and controlling the combustion process.

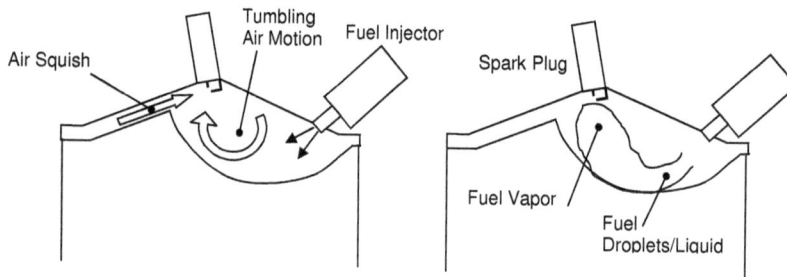

Figure 16.5: A scehmatic diagram of the GDI combustion system put into production by Mitsubishi.

There is a combination of elements required to the GDI engine to operate successfully.

1. The proper, tumbling, air motion in the cylinder shown as air swirl in Figure 16.5. It is accomplished through the geometry of the intake port, where the air comes in predominantly along one side of the cylinder, reflects off the piston top and comes up on the other side. This motion, much like the tumbling motion in a washing machine, is preserved through the compression stroke, and as can be seen from Figure 16.5, is in the proper direction to send the fuel spray towards the spark plug for ignition.

Without tumble, the fuel does not reach the spark plug in time to have a good ignition, and fuel economy suffers from combustion too late in the expansion process. If the tumble motion is too large, one runs the risk of the fuel being "blown past" the spark plug, with poor ignition as a result. The optimum solution requires the adjustment of several parameters, tumble ratio (ratio of tumble rpm to engine rpm), spark timing, injection timing, injector location, piston geometry, squish speed, injection pressure and geometry and other factors. The effective design and operation of a GDI engine is not a simple matter.

2. The second requirement is the piston shape. The bowl in the piston serves to direct the fuel spray towards the spark plug in combination with the tumbling motion described in (1). A suitable piston temperature must also be achieved to provide adequate evaporation of the fuel without damage to the metal from thermal loading, or excessive cracking of the fuel on the surface.

3. The fuel injection must be able to evaporate the fuel quite rapidly. This is accomplished by the use of a swirl injector. As opposed to the lower pressure injectors that are used for intake port injection and throttle body injectors (a few bar), The high pressure swirl nozzle used for the GDI engine uses a fuel pressure of around 75 bar. In addition to the higher injection pressure, the fuel enters the nozzle tip with a high tangential velocity. These effects help to break up the spray, and as a result, a finely atomized fuel spray with a wide spray angle is produced, as needed for good combustion with the GDI engine.

The electronic fuel injector can also be used to provide homogeneous operation by injecting the fuel near the end of the intake stroke. Due to the air motion in the cylinder, the fuel can be nearly completely evaporated and mixed by the time of combustion. The high injection pressure helps this process.

16.5 Strategy for Load Variation

Since the performance and the emissions of the GDI engine deteriorate significantly at higher loads and speeds, it is necessary to use a combination of stratified charge and homogeneous charge operation. The GDI engine is operated in the stratified charge mode at part load and lower speed. When the load and/or speed increase such the satisfactory performance or emissions are not attainable, the operation is changed to stoichiometric mixture, homogeneous charge operation. Here a catalyst can be used in the 3-way mode. Part load emissions are problematic, especially those of NOx, and EGR is a commonly used option. CO and HC emissions can be reduced by using the catalyst to oxidize them, for unthrottled operation there is excess air to aid the oxidation, but exhaust temperatures are lower than with stoichiometric operation.

The changes in operating conditions requires an operating strategy. That is, it is necessary to decide what conditions are suitable for stratified charge operation, and what conditions are best suited to homogeneous charge operation. Figure 16.4 showed that the greatest benefits are gained at part load operation, and the improvement decreases at the load increases. At wide-open throttle, the GDI engine operates as a homogeneous combustion engine, and there is in principle no difference between the two engines at full load. As with diesel engines, the combustion of a spray becomes much more difficult at higher engine speeds as well. The details of the strategy will be determined by the characteristics of the given engine, but the strategies presented to date all resemble those shown in Figure 16.6.

Three regions are shown, the unthrottled mode at lower loads and speed, a stoichiometric region, where standard SI combustion is attained, and 3-way catalyst operation possible, and a rich region for maximum power at full load. One of the implications of the mixed mode operation is that there needs to be a throttle in the engine, even though the engine operates in an "unthrottled" mode at light loads. For engine regulation at medium to heavy loads, the airflow must be adjusted with load variations, since the mixture is stoichiometric. The throttle must be electronically controlled simultaneously with the injection rate, since it is important to control engine operation between for changes between the homogeneous and non-homogeneous modes of operation. Making this mode change without excess emissions or undesirable engine surging is a significant challenge to the engine control system.

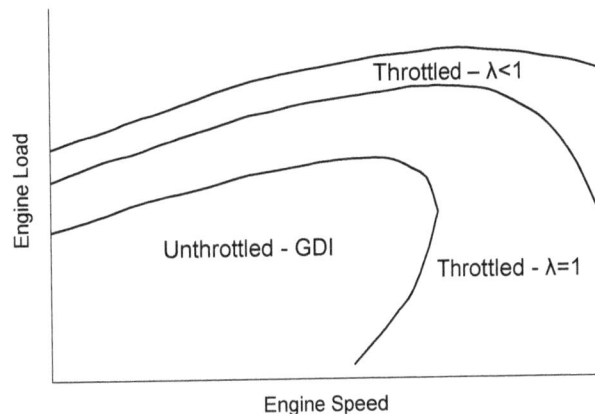

Figure 16.6: The strategy for varying the fuel air ratio over the operating range of a GDI engine.

There are problems with NO_x emissions in the unthrottled region, since there is a limit to the amount of EGR that can be used, and the three-way catalyst will not work for the operating conditions where the fuel air ratio is not stoichiometric. Some of the techniques proposed are the so-called lean NO_x catalyst, which reduces NO_x emissions even in lean exhaust gasses, and some form of NO_x trap, which stores the NO_x until it can be removed from the exhaust gasses under more advantageous conditions. Both of these technologies are new and require the modern low sulphur fuels.

16.6 Conclusion

The GDI engine is presented here, as a case study in engine development. It demonstrates the use of basic engine principles to improve fuel economy, while still satisfying increasingly stringent emissions control regulations. The implementation of modern technologies and development tools (CFD calculation, electronic injection, electronic engine control and catalyst technology, for example) are applied to an old concept, to make a viable product.

The GDI engine presents the possibility to improve the fuel consumption of passenger cars engines, or other gasoline powered engines. This is especially true for engines that operate at part load of a large portion of the time. At full load, GDI engines operate like conventional spark ignition engines, although the in-cylinder injection process improves the knock resistance of the engine and allows operation at higher compression ratios. Since the fuel economy improvement is dependent on engine operating condition, the fuel economy improvement encountered in a vehicle in practice will be strongly dependent on the driving cycle. The more operation at light loads, the more improvement when going to GDI. The experimental results achieved indicate that the maximum improvement would be less than 20 %.

A future challenge for this type of engine will be meeting more stringent emissions standards, since the operation under higher efficiency, lean conditions precludes the use of the three way catalyst. Other options for emissions control, especially for NO_x, are not as effective. There may also be a problem with the formation of particulate emissions with this type of engine.

Chapter 17

Appendices

17.1   Appendix 1 - Thermodynamic properties

The following tables present thermodynamic properties for combustion products and some hydrocarbons. All substances are treated as ideal gasses. The properties in the tables have been calculated using the correlations developed by NASA as described in the publication: CAP: A Computer Code for Generating Tabular Thermodynamic Functions from NASA Lewis Coefficients , by Michael J. Zehe, Sanford Gordon, and Bonnie J. McBride, NASA/TP2001-210959/REV1, currently available at the website; https://www.grc.nasa.gov/WWW/CEAWeb/TP-2001-210959-REV1.pdf

Correlation coefficients and the facility to build tables of thermodynamic properties for many chemical species have been made available by NASA, the current website being: https://www.grc.nasa.gov/WWW/CEAWeb/ceaThermoBuild.htm

The enthalpy in the tables is based on the standard reference condition of the chemical elements in their natural state at 1 bar pressure, 25°C.

$$h(T) = h_f^\circ(298.15) + \int_{298.15}^{T} c_p dT \tag{17.1}$$

where $h_f^\circ(298.15)$ is the standard enthalpy of formation, and is found in the tables as the value of the enthalpy at the temperature of 298.15K.

The entropy at other pressures can be found by:

$$s(T,p) = s^\circ(T) + R \ln \frac{p}{p_o} \tag{17.2}$$

where $s^\circ(T)$ is the table value, and p_o is 1 bar.

When calculating isentropic processes for mixtures of gasses, it is the total entropy of the mixture which remains constant, not necessarily the entropy of the individual species.

Note that the values in J/mol are the same as in kJ/kmol, 298K in the tables is actually 298.15K.

Table 17.1: Thermodynamic properties for methane, CH_4

T	c_p	c_v	h	u	$h_T - h_{298}$	$u_T - u_{298}$	s°
K	J/mol-K	J/mol-K	J/mol	J/mol	J/mol	J/mol	J/mol-K
298	35.69	27.38	-74597	-77076	0	0	186.4
300	35.76	27.44	-74531	-77025	66	51	186.6
400	40.62	32.30	-70727	-74053	3869	3022	197.5
500	46.58	38.27	-66371	-70529	8225	6547	207.2
600	52.69	44.37	-61407	-66395	13190	10680	216.2
700	58.54	50.23	-55843	-61662	18754	15413	224.8
800	64.01	55.70	-49712	-56363	24885	20713	233.0
900	69.06	60.75	-43054	-50537	31542	26538	240.8
1000	73.67	65.36	-35914	-44228	38683	32848	248.3
1100	77.85	69.53	-28335	-37480	46262	39595	255.5
1200	81.62	73.31	-20358	-30335	54239	46741	262.5
1300	85.03	76.71	-12022	-22831	62574	54245	269.1
1400	88.09	79.78	-3364	-15003	71233	62072	275.6
1500	90.86	82.55	5587	-6885	80183	70191	281.7

Table 17.2: Thermodynamic properties for ethylene, C_2H_4

T	c_p	c_v	h	u	$h_T - h_{298}$	$u_T - u_{298}$	s°
K	J/mol-K	J/mol-K	J/mol	J/mol	J/mol	J/mol	J/mol-K
298	42.89	34.57	52498	50019	0	0	219.3
300	43.06	34.75	52577	50083	80	64	219.6
400	53.03	44.72	57380	54054	4882	4035	233.3
500	62.46	54.15	63164	59007	10666	8988	246.2
600	70.67	62.36	69831	64843	17333	14824	258.3
700	77.72	69.40	77259	71439	24762	21421	269.8
800	83.84	75.53	85344	78693	32846	28674	280.5
900	89.23	80.92	94003	86521	41506	36502	290.7
1000	93.92	85.61	103167	94853	50670	44834	300.4
1100	98.04	89.73	112770	103624	60272	53606	309.5
1200	101.66	93.34	122759	112782	70261	62763	318.2
1300	104.82	96.51	133087	122278	80589	72259	326.5
1400	107.59	99.27	143710	132070	91212	82052	334.4
1500	110.01	101.70	154593	142122	102095	92103	341.9

Table 17.3: Thermodynamic properties for ethane, C_2H_6

T	c_p	c_v	h	u	$h_T - h_{298}$	$u_T - u_{298}$	s°
K	J/mol-K	J/mol-K	J/mol	J/mol	J/moh$_T - h_{298}$l	J/mol	J/mol-K
298	52.50	44.18	-83848	-86327	0	0	229.2
300	52.72	44.41	-83750	-86245	97	82	229.5
400	65.45	57.14	-77847	-81173	6001	5154	246.4
500	77.91	69.60	-70671	-74828	13177	11499	262.4
600	89.17	80.85	-62305	-67294	21542	19033	277.6
700	99.12	90.80	-52881	-58701	30967	27626	292.1
800	107.91	99.59	-42521	-49172	41327	37155	305.9
900	115.70	107.39	-31332	-38815	52516	47512	319.1
1000	122.53	114.22	-19412	-27726	64436	58600	331.7
1100	128.53	120.22	-6852	-15998	76996	70329	343.6
1200	133.79	125.48	6270	-3707	90118	82620	355.0
1300	138.38	130.07	19884	9075	103732	95402	365.9
1400	142.39	134.08	33927	22287	117775	108614	376.3
1500	145.89	137.58	48345	35874	132193	122201	386.3

Table 17.4: Thermodynamic properties for propane, C_3H_8

T	c_p	c_v	h	u	$h_T - h_{298}$	$u_T - u_{298}$	$s°$
K	J/mol-K	J/mol-K	J/mol	J/mol	J/mol	J/mol	J/mol-K
298	73.59	65.27	-104675	-107154	0	0	270.3
300	73.95	65.64	-104539	-107033	136	121	270.8
400	93.98	85.67	-96140	-99465	8536	7689	294.8
500	112.56	104.25	-85793	-89951	18882	17204	317.8
600	128.73	120.42	-73708	-78697	30967	28457	339.8
700	142.65	134.34	-60122	-65942	44553	41212	360.7
800	154.75	146.44	-45239	-51890	59436	55264	380.6
900	165.38	157.06	-29221	-36704	75454	70450	399.4
1000	174.61	166.29	-12210	-20524	92466	86630	417.3
1100	182.68	174.36	5663	-3482	110339	103672	434.4
1200	189.72	181.41	24291	14314	128967	121469	450.6
1300	195.86	187.55	43578	32769	148253	139924	466.0
1400	201.21	192.90	63438	51798	168113	158952	480.7
1500	205.88	197.56	83797	71326	188473	178480	494.7

Table 17.5: Thermodynamic properties for benzene, C_6H_6

T	c_p	c_v	h	u	$h_T - h_{298}$	$u_T - u_{298}$	$s°$
K	J/mol-K	J/mol-K	J/mol	J/mol	J/mol	J/mol	J/mol-K
298	81.93	73.62	82876	80397	0	0	269.1
300	82.51	74.20	83028	80534	152	137	269.7
400	112.78	104.47	92821	89495	9945	9098	297.6
500	138.56	130.25	105431	101274	22555	20877	325.6
600	159.38	151.06	120367	115378	37490	34981	352.8
700	176.12	167.81	137171	131351	54295	50954	378.7
800	189.86	181.55	155491	148840	72615	68443	403.1
900	201.39	193.07	175070	167587	92194	87190	426.2
1000	211.02	202.71	195706	187392	112830	106995	447.9
1100	219.24	210.92	217230	208085	134354	127687	468.4
1200	226.26	217.95	239514	229537	156638	149140	487.8
1300	232.29	223.98	262450	251641	179573	171244	506.2
1400	237.48	229.17	285945	274305	203069	193908	523.6
1500	241.98	233.66	309923	297452	227047	217055	540.1

Table 17.6: Thermodynamic properties for cyclo-hexane, C_6H_{10}

T	c_p	c_v	h	u	$h_T - h_{298}$	$u_T - u_{298}$	$s°$
K	J/mol-K	J/mol-K	J/mol	J/mol	J/mol	J/mol	J/mol-K
298	105.34	97.02	-123295	-125773	0	0	297.4
300	106.11	97.79	-123099	-125593	196	180	298.0
400	148.63	140.32	-110365	-113691	12929	12083	334.4
500	188.66	180.34	-93462	-97620	29832	28154	371.9
600	223.39	215.08	-72813	-77801	50482	47972	409.5
700	252.62	244.30	-48969	-54789	74326	70985	446.2
800	277.02	268.71	-22450	-29102	100844	96672	481.5
900	297.43	289.11	6303	-1180	129597	124593	515.4
1000	314.41	306.09	36921	28607	160216	154381	547.6
1100	328.67	320.35	69096	59951	192391	185724	578.3
1200	340.64	332.33	102579	92602	225873	218375	607.4
1300	350.78	342.47	137164	126355	260458	252129	635.1
1400	359.41	351.10	172685	161045	295979	286818	661.4
1500	366.82	358.50	209005	196534	332300	322308	686.5

Table 17.7: Thermodynamic properties for toluene, n-C_7H_8

T	c_p	c_v	h	u	$h_T - h_{298}$	$u_T - u_{298}$	s°
K	J/mol-K	J/mol-K	J/mol	J/mol	J/mol	J/mol	J/mol-K
298	103.28	94.96	50168	47689	0	0	320.2
300	103.95	95.64	50359	47865	192	176	320.8
400	139.39	131.07	62552	59226	12384	11537	355.6
500	170.26	161.95	78080	73923	27912	26234	390.2
600	195.74	187.43	96423	91435	46255	43746	423.5
700	216.60	208.29	117075	111255	66907	63566	455.3
800	233.93	225.61	139626	132975	89459	85286	485.4
900	248.56	240.25	163771	156288	113603	108599	513.8
1000	260.90	252.59	189263	180949	139095	133260	540.7
1100	271.44	263.13	215894	206748	165726	159060	566.1
1200	280.48	272.17	243502	233525	193334	185836	590.1
1300	288.25	279.93	271948	261140	221780	213451	612.8
1400	294.94	286.63	301116	289476	250948	241787	634.5
1500	300.74	292.43	330907	318436	280739	270747	655.0

Table 17.8: Thermodynamic properties for n-heptane, n-C_7H_{16}

T	c_p	c_v	h	u	$h_T - h_{298}$	$u_T - u_{298}$	s°
K	J/mol-K	J/mol-K	J/mol	J/mol	J/mol	J/mol	J/mol-K
298	165.17	156.86	-187772	-190251	0	0	428.1
300	165.98	157.67	-187465	-189960	306	306	429.1
400	210.70	202.38	-168629	-171955	19142	19142	483.0
500	252.07	243.76	-145445	-149602	42327	42327	534.5
600	287.46	279.14	-118418	-123406	69354	69354	583.7
700	317.16	308.84	-88144	-93964	99628	99628	630.3
800	342.24	333.92	-55140	-61791	132632	132632	674.4
900	363.61	355.29	-19819	-27302	167952	167952	715.9
1000	381.56	373.25	17468	9154	205240	205240	755.2
1100	396.91	388.60	56406	47261	244178	244178	792.3
1200	410.65	402.34	96797	86820	284569	284569	827.4
1300	422.98	414.67	138490	127681	326261	326261	860.8
1400	434.09	425.78	181353	169713	369124	369124	892.6
1500	444.15	435.84	225273	212802	413044	413044	922.9

Table 17.9: Thermodynamic properties for n-octane, n-C_8H_{18}

T	c_p	c_v	h	u	$h_T - h_{298}$	$u_T - u_{298}$	s°
K	J/mol-K	J/mol-K	J/mol	J/mol	J/mol	J/mol	J/mol-K
298	187.77	179.46	-208741	-211220	0	0	467.3
300	188.70	180.38	-208392	-210887	348	333	468.5
400	239.63	231.32	-186973	-190299	21768	20921	529.8
500	286.68	278.37	-160604	-164761	48137	46458	588.4
600	326.77	318.46	-129872	-134860	78869	76359	644.3
700	360.17	351.86	-95474	-101294	113267	109926	697.3
800	388.11	379.79	-58019	-64671	150721	146549	747.3
900	411.67	403.36	-17997	-25480	190744	185740	794.4
1000	431.36	423.04	24187	15873	232928	227092	838.8
1100	448.57	440.25	68202	59057	276943	270276	880.7
1200	463.65	455.34	113830	103853	322571	315072	920.4
1300	476.89	468.57	160871	150063	369612	361282	958.1
1400	488.51	480.19	209153	197514	417894	408733	993.8
1500	498.72	490.40	258526	246054	467266	457274	1027.9

Table 17.10: Thermodynamic properties for CO

T	c_p	c_v	h	u	$h_T - h_{298}$	$u_T - u_{298}$	s°
K	J/mol-K	J/mol-K	J/mol	J/mol	J/mol	J/mol	J/mol-K
298	29.14	20.83	-110530	-113009	0	0	197.7
300	29.14	20.83	-110476	-112971	54	39	197.8
400	29.34	21.03	-107554	-110880	2976	2129	206.2
500	29.79	21.48	-104600	-108757	5931	4252	212.8
600	30.44	22.12	-101589	-106578	8941	6431	218.3
700	31.17	22.86	-98509	-104329	12021	8680	223.1
800	31.90	23.58	-95356	-102007	15175	11002	227.3
900	32.57	24.26	-92132	-99614	18399	13395	231.1
1000	33.18	24.86	-88844	-97158	21686	15851	234.5
1100	33.71	25.39	-85499	-94645	25031	18365	237.7
1200	34.17	25.85	-82105	-92082	28426	20928	240.7
1300	34.57	26.25	-78667	-89476	31863	23533	243.4
1400	34.91	26.60	-75193	-86833	35337	26176	246.0
1500	35.21	26.90	-71686	-84158	38844	28852	248.4
1600	35.47	27.16	-68152	-81454	42378	31555	250.7
1700	35.70	27.39	-64593	-78727	45937	34282	252.9
1800	35.90	27.59	-61012	-75978	49518	37031	254.9
1900	36.08	27.77	-57413	-73210	53117	39799	256.8
2000	36.24	27.93	-53797	-70425	56734	42584	258.7
2100	36.38	28.07	-50165	-67625	60365	45384	260.5
2200	36.51	28.20	-46521	-64812	64010	48197	262.2
2300	36.62	28.31	-42864	-61986	67666	51023	263.8
2400	36.73	28.42	-39196	-59150	71334	53859	265.4
2500	36.82	28.51	-35518	-56304	75012	56706	266.9
2600	36.91	28.60	-31831	-53448	78699	59561	268.3
2700	36.99	28.68	-28136	-50584	82394	62425	269.7
2800	37.07	28.75	-24433	-47712	86097	65297	271.0
2900	37.14	28.82	-20722	-44833	89808	68176	272.3
3000	37.20	28.89	-17005	-41948	93525	71061	273.6
3100	37.26	28.95	-13282	-39056	97248	73953	274.8
3200	37.32	29.01	-9553	-36158	100978	76851	276.0
3300	37.38	29.06	-5818	-33254	104713	79755	277.2
3400	37.43	29.11	-2077	-30345	108453	82664	278.3
3500	37.48	29.16	1668	-27431	112198	85578	279.4

Table 17.11: Thermodynamic properties for CO_2

T	c_p	c_v	h	u	$h_T - h_{298}$	$u_T - u_{298}$	s°
K	J/mol-K	J/mol-K	J/mol	J/mol	J/mol	J/mol	J/mol-K
298.15	37.13	28.82	-393492	-395971	0	0	213.8
300	37.22	28.90	-393424	-395918	69	53	214.0
400	41.32	33.01	-389489	-392815	4003	3157	225.3
500	44.62	36.31	-385186	-389343	8307	6628	234.9
600	47.32	39.01	-380584	-385573	12908	10398	243.3
700	49.56	41.25	-375737	-381557	17755	14414	250.7
800	51.43	43.12	-370685	-377336	22808	18635	257.5
900	53.00	44.68	-365461	-372944	28031	23027	263.6
1000	54.31	45.99	-360094	-368408	33398	27563	269.3
1100	55.42	47.10	-354606	-363752	38886	32220	274.5
1200	56.34	48.03	-349017	-358994	44476	36978	279.4
1300	57.13	48.81	-343342	-354150	50150	41821	283.9
1400	57.80	49.48	-337595	-349235	55897	46736	288.2
1500	58.37	50.06	-331786	-344257	61706	51714	292.2
1600	58.87	50.56	-325923	-339226	67569	56745	296.0
1700	59.31	50.99	-320014	-334148	73479	61823	299.6
1800	59.69	51.38	-314064	-329029	79429	66942	303.0
1900	60.03	51.72	-308077	-323874	85415	72097	306.2
2000	60.33	52.02	-302059	-318687	91434	77284	309.3
2100	60.60	52.29	-296012	-313472	97481	82500	312.2
2200	60.84	52.53	-289940	-308231	103553	87741	315.1
2300	61.06	52.75	-283844	-302967	109648	93005	317.8
2400	61.26	52.95	-277728	-297682	115765	98290	320.4
2500	61.44	53.13	-271593	-292378	121900	103593	322.9
2600	61.61	53.29	-265440	-287057	128052	108914	325.3
2700	61.76	53.44	-259272	-281720	134220	114251	327.6
2800	61.90	53.58	-253089	-276369	140403	119603	329.9
2900	62.03	53.72	-246893	-271004	146600	124968	332.0
3000	62.15	53.84	-240683	-265626	152809	130345	334.1
3100	62.27	53.96	-234462	-260236	159030	135735	336.2
3200	62.38	54.07	-228230	-254835	165263	141137	338.2
3300	62.49	54.17	-221986	-249423	171506	146549	340.1
3400	62.59	54.28	-215732	-244000	177761	151971	341.9
3500	62.69	54.38	-209467	-238567	184025	157404	343.8

Table 17.12: Thermodynamic properties for H_2

T	c_p	c_v	h	u	$h_T - h_{298}$	$u_T - u_{298}$	s°
K	J/mol-K	J/mol-K	J/mol	J/mol	J/mol	J/mol	J/mol-K
298	28.84	20.52	0	-2479	0	0	130.7
300	28.85	20.53	53	-2441	53	38	130.9
400	29.19	20.87	2960	-366	2960	2113	139.2
500	29.25	20.94	5882	1725	5882	4204	145.7
600	29.32	21.00	8810	3822	8810	6301	151.1
700	29.44	21.13	11748	5928	11748	8407	155.6
800	29.63	21.31	14701	8050	14701	10528	159.5
900	29.87	21.56	17675	10193	17675	12671	163.0
1000	30.20	21.89	20678	12364	20678	14843	166.2
1100	30.57	22.25	23716	14570	23716	17049	169.1
1200	30.98	22.67	26793	16816	26793	19295	171.8
1300	31.42	23.11	29913	19105	29913	21584	174.3
1400	31.86	23.55	33077	21437	33077	23916	176.6
1500	32.30	23.99	36286	23814	36286	26293	178.8
1600	32.73	24.42	39538	26235	39538	28714	180.9
1700	33.14	24.83	42831	28697	42831	31176	182.9
1800	33.54	25.22	46165	31200	46165	33679	184.8
1900	33.91	25.60	49538	33741	49538	36220	186.7
2000	34.27	25.96	52948	36320	52948	38798	188.4
2100	34.62	26.30	56392	38933	56392	41412	190.1
2200	34.94	26.63	59871	41579	59871	44058	191.7
2300	35.25	26.94	63380	44258	63380	46737	193.3
2400	35.55	27.23	66921	46967	66921	49446	194.8
2500	35.83	27.52	70490	49704	70490	52183	196.2
2600	36.10	27.79	74086	52470	74086	54948	197.6
2700	36.36	28.04	77709	55261	77709	57740	199.0
2800	36.61	28.29	81358	58078	81358	60557	200.3
2900	36.85	28.53	85031	60920	85031	63398	201.6
3000	37.08	28.76	88727	63784	88727	66263	202.9
3100	37.30	28.99	92446	66672	92446	69151	204.1
3200	37.52	29.20	96186	69581	96186	72060	205.3
3300	37.73	29.41	99949	72512	99949	74991	206.4
3400	37.93	29.62	103732	75464	103732	77942	207.6
3500	38.13	29.82	107535	78435	107535	80914	208.7

Table 17.13: Thermodynamic properties for H_2O

T	c_p	c_v	h	u	$h_T - h_{298}$	$u_T - u_{298}$	$s°$
K	J/mol-K	J/mol-K	J/mol	J/mol	J/mol	J/mol	J/mol-K
298	33.59	25.27	-241815	-244294	0	0	188.8
300	33.59	25.28	-241753	-244247	62	47	189.0
400	34.26	25.95	-238364	-241689	3452	2605	198.8
500	35.22	26.91	-234891	-239048	6924	5246	206.5
600	36.32	28.01	-231314	-236303	10501	7991	213.0
700	37.50	29.18	-227624	-233444	14191	10850	218.7
800	38.73	30.41	-223813	-230464	18002	13830	223.8
900	40.00	31.68	-219877	-227360	21938	16934	228.4
1000	41.29	32.98	-215813	-224127	26002	20167	232.7
1100	42.57	34.26	-211620	-220766	30195	23528	236.7
1200	43.84	35.53	-207299	-217276	34516	27018	240.5
1300	45.06	36.75	-202853	-213662	38962	30632	244.0
1400	46.22	37.91	-198289	-209928	43527	34366	247.4
1500	47.32	39.00	-193611	-206082	48204	38212	250.6
1600	48.34	40.02	-188828	-202130	52988	42164	253.7
1700	49.29	40.98	-183946	-198080	57870	46214	256.7
1800	50.17	41.86	-178972	-193937	62843	50357	259.5
1900	50.99	42.68	-173913	-189710	67902	54584	262.3
2000	51.75	43.44	-168775	-185403	73040	58891	264.9
2100	52.46	44.14	-163564	-181024	78251	63270	267.4
2200	53.11	44.79	-158286	-176577	83529	67717	269.9
2300	53.71	45.39	-152945	-172067	88870	72227	272.3
2400	54.26	45.95	-147546	-167500	94269	76794	274.6
2500	54.77	46.46	-142094	-162879	99721	81415	276.8
2600	55.25	46.93	-136592	-158209	105223	86085	279.0
2700	55.69	47.37	-131045	-153493	110770	90801	281.1
2800	56.09	47.78	-125456	-148736	116359	95559	283.1
2900	56.47	48.16	-119827	-143938	121988	100356	285.1
3000	56.82	48.51	-114163	-139105	127653	105189	287.0
3100	57.15	48.83	-108464	-134238	133351	110056	288.8
3200	57.45	49.14	-102734	-129339	139081	114955	290.7
3300	57.73	49.42	-96975	-124411	144841	119883	292.4
3400	58.00	49.69	-91188	-119456	150627	124838	294.2
3500	58.25	49.94	-85375	-114475	156440	129819	295.9

Table 17.14: Thermodynamic properties for N_2

T	c_p	c_v	h	u	$h_T - h_{298}$	$u_T - u_{298}$	s°
K	J/mol-K	J/mol-K	J/mol	J/mol	J/mol	J/mol	J/mol-K
298	29.12	20.81	0	-2479	0	0	191.6
300	29.12	20.81	54	-2440	54	38	191.8
400	29.25	20.93	2971	-355	2971	2124	200.2
500	29.58	21.27	5911	1753	5911	4232	206.7
600	30.11	21.79	8894	3905	8894	6384	212.2
700	30.75	22.44	11936	6116	11936	8595	216.9
800	31.43	23.12	15045	8394	15045	10873	221.0
900	32.09	23.77	18222	10739	18222	13218	224.7
1000	32.69	24.38	21461	13147	21461	15626	228.2
1100	33.24	24.93	24759	15613	24759	18092	231.3
1200	33.72	25.41	28107	18130	28107	20609	234.2
1300	34.15	25.83	31501	20693	31501	23172	236.9
1400	34.52	26.20	34935	23295	34935	25774	239.5
1500	34.84	26.53	38403	25932	38403	28411	241.9
1600	35.13	26.81	41901	28599	41901	31078	244.1
1700	35.38	27.06	45427	31293	45427	33772	246.3
1800	35.60	27.28	48976	34010	48976	36489	248.3
1900	35.79	27.48	52545	36748	52545	39227	250.2
2000	35.97	27.65	56134	39505	56134	41984	252.1
2100	36.13	27.81	59738	42279	59738	44758	253.8
2200	36.27	27.95	63358	45067	63358	47546	255.5
2300	36.39	28.08	66991	47869	66991	50347	257.1
2400	36.51	28.19	70636	50682	70636	53161	258.7
2500	36.61	28.30	74292	53507	74292	55986	260.2
2600	36.71	28.40	77959	56342	77959	58821	261.6
2700	36.80	28.48	81634	59186	81634	61665	263.0
2800	36.88	28.57	85318	62038	85318	64517	264.3
2900	36.96	28.64	89010	64899	89010	67378	265.6
3000	37.03	28.71	92709	67766	92709	70245	266.9
3100	37.09	28.78	96415	70641	96415	73120	268.1
3200	37.15	28.84	100127	73522	100127	76001	269.3
3300	37.21	28.90	103845	76409	103845	78887	270.4
3400	37.27	28.95	107569	79301	107569	81780	271.5
3500	37.32	29.00	111298	82199	111298	84678	272.6

Table 17.15: Thermodynamic properties for O_2

T	c_p	c_v	h	u	$h_T - h_{298}$	$u_T - u_{298}$	$s°$
K	J/mol-K	J/mol-K	J/mol	J/mol	J/mol	J/mol	J/mol-K
298	29.38	21.06	0	-2479	0	0	205.1
300	29.39	21.07	54	-2440	54	39	205.3
400	30.11	21.80	3026	-300	3026	2179	213.9
500	31.09	22.78	6085	1928	6085	4407	220.7
600	32.09	23.77	9245	4256	9245	6735	226.4
700	32.99	24.67	12500	6680	12500	9159	231.5
800	33.74	25.43	15837	9186	15837	11665	235.9
900	34.36	26.05	19244	11761	19244	14240	239.9
1000	34.88	26.57	22706	14392	22706	16871	243.6
1100	35.33	27.02	26218	17072	26218	19551	246.9
1200	35.69	27.38	29770	19793	29770	22272	250.0
1300	36.00	27.69	33355	22546	33355	25025	252.9
1400	36.29	27.97	36970	25330	36970	27809	255.6
1500	36.55	28.24	40611	28140	40611	30619	258.1
1600	36.81	28.49	44279	30977	44279	33456	260.4
1700	37.06	28.74	47973	33839	47973	36317	262.7
1800	37.30	28.99	51690	36725	51690	39204	264.8
1900	37.54	29.23	55433	39636	55433	42115	266.8
2000	37.78	29.47	59199	42571	59199	45049	268.8
2100	38.02	29.70	62989	45529	62989	48008	270.6
2200	38.25	29.94	66803	48511	66803	50990	272.4
2300	38.48	30.17	70639	51517	70639	53996	274.1
2400	38.71	30.39	74499	54545	74499	57024	275.7
2500	38.93	30.62	78381	57596	78381	60074	277.3
2600	39.15	30.84	82285	60668	82285	63147	278.8
2700	39.36	31.05	86211	63763	86211	66241	280.3
2800	39.57	31.26	90158	66878	90158	69357	281.8
2900	39.78	31.46	94125	70014	94125	72493	283.2
3000	39.98	31.66	98113	73171	98113	75650	284.5
3100	40.17	31.86	102121	76347	102121	78826	285.8
3200	40.36	32.05	106147	79542	106147	82021	287.1
3300	40.55	32.23	110193	82756	110193	85235	288.3
3400	40.73	32.41	114257	85989	114257	88468	289.6
3500	40.90	32.59	118338	89239	118338	91718	290.7

17.2 Appendix 2: Equilibrium Combustion Products for Octane and Air

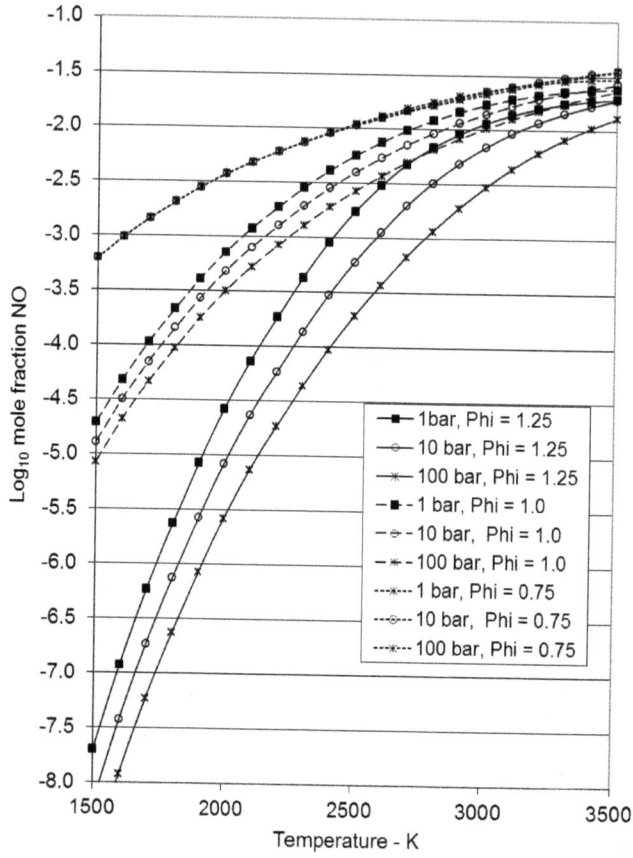

Figure 17.1: Equilibrium NO concentration in combustion products of octane and air.

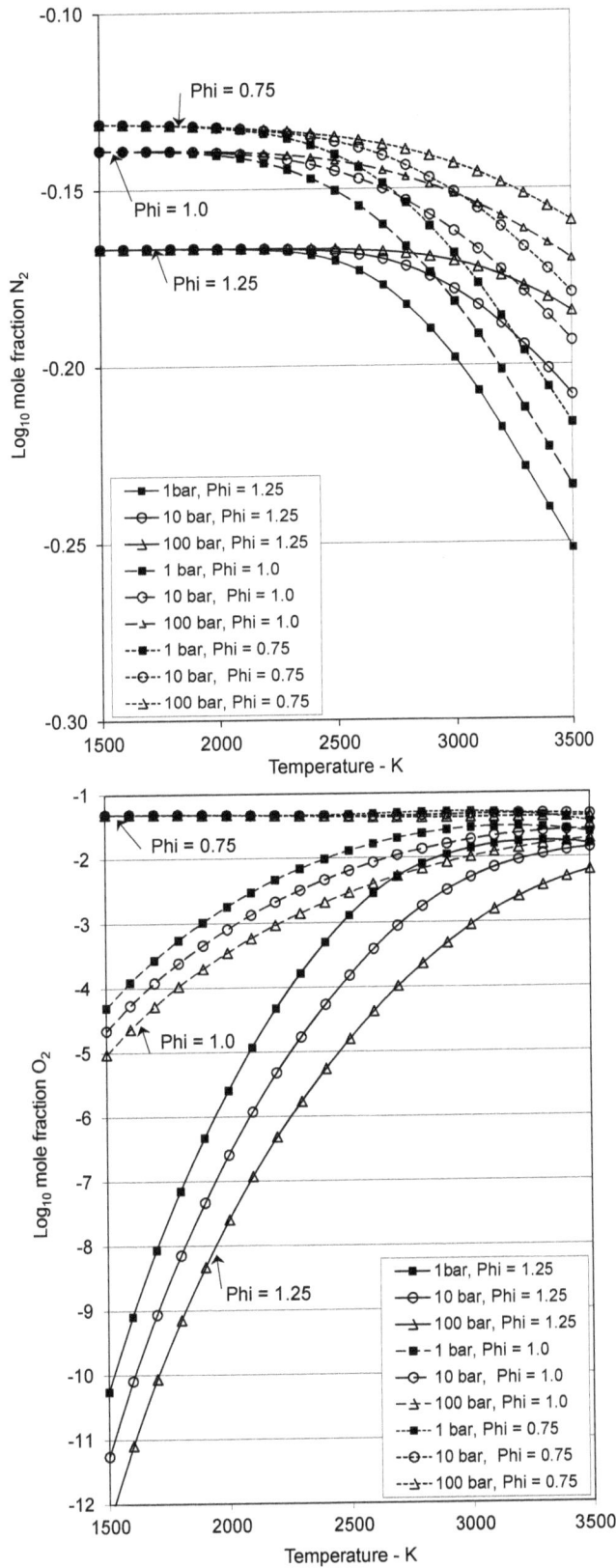

Figure 17.2: Equilibrium N_2 and O_2 concentrations in combustion products of octane and air.

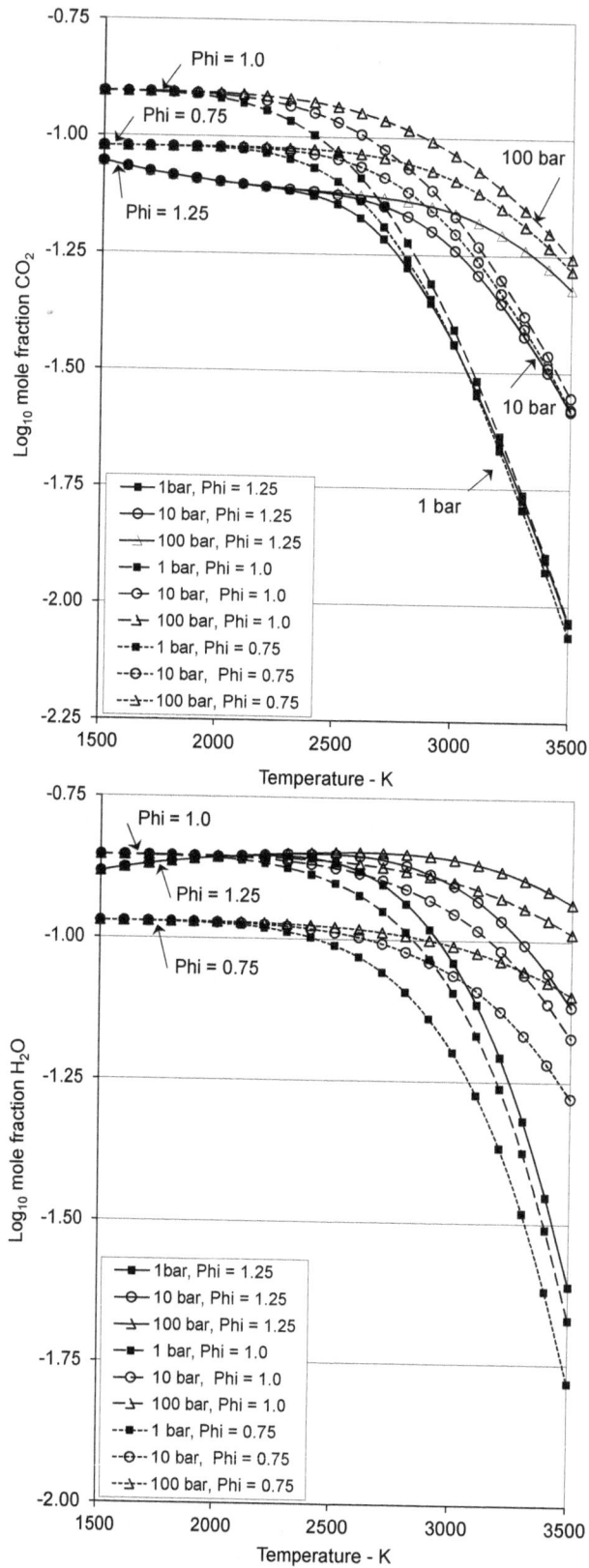

Figure 17.3: Equilibrium CO_2 and H_2O concentrations in combustion products of octane and air.

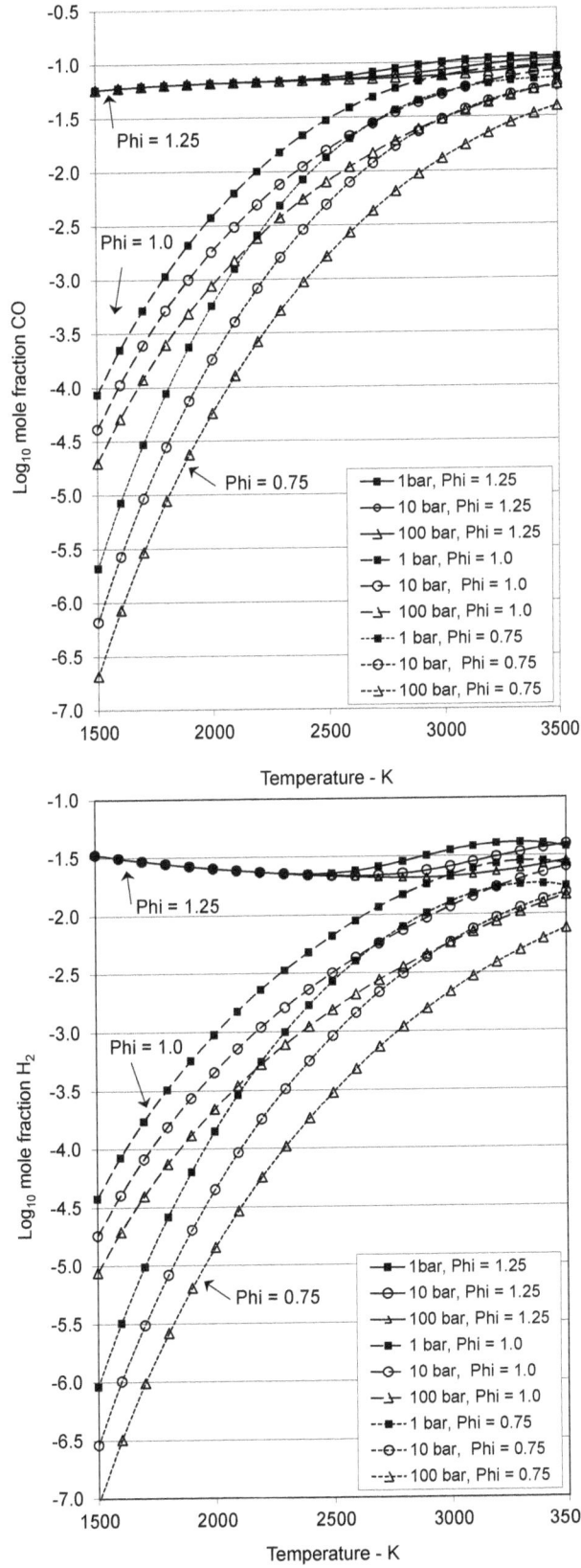

Figure 17.4: Equilibrium CO and H_2 concentrations in combustion products of octane and air.

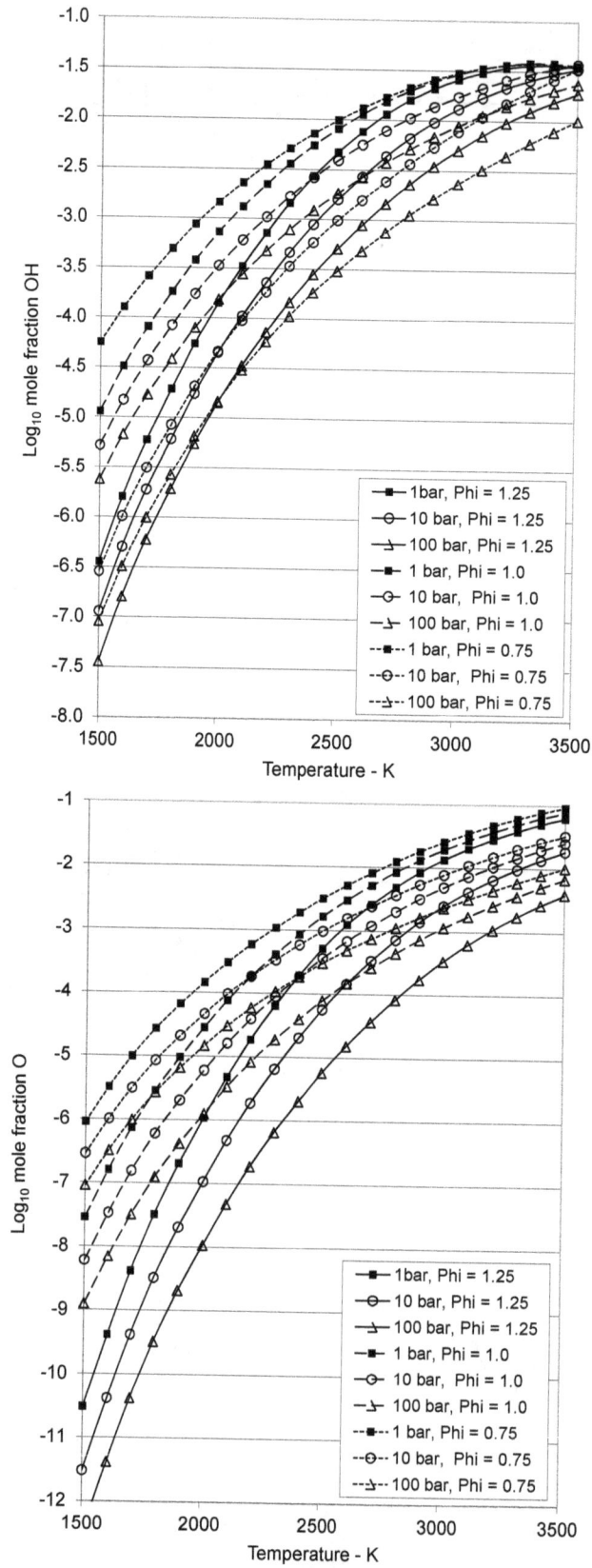

Figure 17.5: Equilibrium OH and O concentrations in combustion products of octane and air.

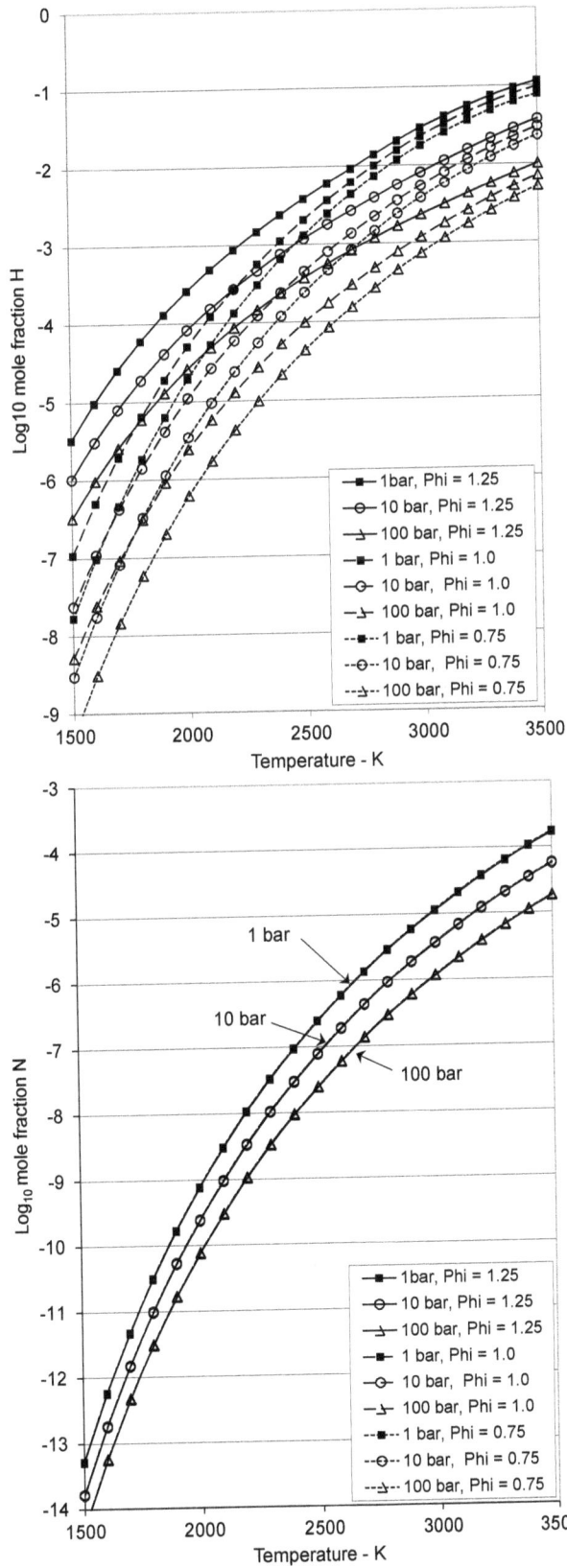

Figure 17.6: Equilibrium H and N concentrations in combustion products of octane and air.

Bibliography

[1] R. Klingmann, H. Brüggemann, A. Peters, and W. Pütz. Der Neue Vierzylinder-Dieselmotor OM 611 mit Common-Rail-Einspritzung Teil 1: Motorkonstruktion und mechanischer Aufbau. *MTZ*, 58 (12):pp 652–659, 1997.

[2] R. Klingmann, H. Brüggemann, A. Peters, and W. Pütz. Der Neue Vierzylinder-Dieselmotor OM 611 mit Common-Rail-Einspritzung Teil 2: Verbrennung und Motormanagement. *MTZ*, 58 - 12:pp 760–767, 1997.

[3] Pachernegg, S.J. A Closer Look at the Willans-Line. SAE paper 690182, 1969.

[4] E. D. Obert. *Internal Combustion Engines and Air Pollution*. International Textbook Co, 1972.

[5] C. R. Ferguson and A. T. Kirkpatrick. *Internal Combustion Engines: Applied Thermodynamics*. John Wiley and Sons, Inc., New York, NY, USA, 1986.

[6] C. F. Taylor and E. S. Taylor. *The Internal Combustion Engine*. International Textbook Co., Scranton Pa, 1961.

[7] J.B. Heywood. *Internal combustion engine fundamentals*. McGraw-Hill New York, 1988.

[8] S. R. Turns. *An introduction to combustion: Concepts and applications*. New York: McGraw-Hill, Inc, 1995.

[9] R. A. Roger A. Strehlow. *Combustion Fundamentals*. McGraw-Hill, New York, 1984.

[10] G.L. Borman and K.W. Ragland. *Combustion Engineering*. McGraw-Hill Boston, 1998.

[11] C. R. Ferguson. *Internal combustion engines. Applied thermosciences*. Wiley, 1986.

[12] Paul Blumberg and J. T. Kummer. Prediction of NO Formation in Spark-Ignited Engines - An Analysis of Methods of Control. *Combustion Science and Technology*, 4:73 – 95, 1971.

[13] D. R. Lancaster, R. B. Krieger, S. C. Sorenson, and W. L. Hull. Effects of Mixture Turbulence on Spark Ignition Engine Combustion. *SAE Transactions*, 85:937, 1976.

[14] G. P. Blair. *Design and Simulation of Two-Stroke Engines*. SAE International, 1996.

[15] G. P. Blair. *Design and Simulation of Four-Stroke Engines*. Society of Automotive Engineers, Inc, 1999.

[16] G. Eichelberg. Some New Investigations on Old Combustion Engine Problems - II. *Engineering*, 148:547–549, 1939.

[17] G. Woschni. A Universally Applicable Equation for the Instantaneous Heat Transfer Coeficient In the Internal Combustion Engine. *SAE Transactions*, 76:3065 – 3083, 1967.

[18] W. J. D. Annand. Heat Transfer in the Cylinders of Reciprocating Internal Combustion Engines. *Proc Instn Mech Engrs*, 177, No. 36:973 – 990, 1963.

[19] R. B. Krieger and G. L. Borman. The Computation of Apparent Heat Release for Internal Combustion Engines. ASME Paper 66-WA/DGP-4, 1966.

[20] T. A Huls, P. S. Myers, and O. A. Uyehara. Spark Ignition Engine Operation and Desgin for Minimum Exhaust Emission. *SAE Transactions*, 75, 1967.

[21] Ather A. Quader. Why intake charge dilution decreases nitric oxide emission from spark ignition engines. *SAE paper 710009*, 1971.

[22] David R. Lancaster. *Effects of Turbulence on Spark-Ignition Combustion*. PhD thesis, University of Illinois at Urbana-Champaign, 1975.

[23] B. G Groff and F. A Matekunas. The Nature of Turbulent Flame Propagation in a Homogeneous Spark-Ignited Engine. SAE paper 800133, 1980.

[24] James N. Mattavi, Edward G. Groff, John H. Lienesch, Frederic A Matekunas, and Robert N Noyes. Engine Improvements Through Combustion Modeling. *Combustion Modeling in Reciprocating Engines*, Plenum Press:537 – 579, 1980.

[25] S. D. Hires, R. J. Tabaczynski, and J. M. Novak. The Prediction of Ignition Delay and Combustion Intervals for a Homogeneous Charge Spark Ignition Engine. SAE paper 780232, 1978.

[26] Ather A. Quader. What limits lean operation in spark ignition engines - flame initiation or propagation. *SAE Paper 760760*, 1976.

[27] John E. Dec. Soot Distribution in a DI Diesel Engine Using 2D Imaging of Laser-Induced Incandescence Elastic Scattering and Flame Luminosity. *SAE Transactions*, Vol 101, No.4 (paper 922307):1642–1651, 1992.

[28] J. E. Dec and C. Espey. Ignition and early soot formation in a DI diesel engine using multiple 2D imaging diagnostics. *SAE Transactions*, 104 - 3:853–875, 1995.

[29] J. E. Dec. A Conceptual Model for DI Diesel Combustion Based on Laser-Sheet Imaging. *SAE Transactions*, 106, No. 3:1319–1348 Paper 970873, 1997.

[30] N. A. Chigier. *Energy, Combustion and Environment*. McGraw-Hill Boston, 1841.

[31] Yukio Matsui Yuzo Aoyagi, Takeyuki Kamimoto and Shin Matsuoka. A gas sampling study on the formation ,processes of soot and :no in a di diesel engine. SAE Paper 800254, 1980.

[32] A. Prothero M. P. Halstead, L. J. Kirsch and C. P. Quinn. A mathematical model for hydrocarbon autoignition at high pressures. *Proc. R. Soc. Lond. A.*, 346:515 – 538, 1975.

[33] V. Hamosfakidis and R. D. Reitz. Optimization of a hydrobarbon fuel ignition model for two single component surrogates of diesel fuel. *Combustion and Flame*, 132:433–450, 2003.

[34] L. J. Kirsch M. P. Halstead and C. P. Quinn. The autoignition of hydrocarbon fuels at high temperature and ppressure-fitting of a mathematical model. *Combustion and Flame*, 30:45 – 60, 1977.

[35] M. R. Heikal C Marooney S. S. Sazhin, E. M. Sazhine and S. V. Mikahalovsky. The shell autoignition model: A new mathematical formulation. *Combustion and Flame*, 117:529–540, 1999.

[36] William L. Brown. The Caterpillar imep Meter and Engine Friction. SAE Paper 730150, 1973.

[37] D. R. Lancaster, R. B. Krieger, and J. Lienesch. Measurements and Analysis of Engine Pressure Data. *SAE Transactions*, 86:paper 750026, 1984.

[38] M. Namazian, S. Hansen, E. Lyford-Pike, J. Sanchez-Barsse, J. B. Heywood, and J. Rife. Schileren Visulaization of the Flow and Density Fieds in the Cylinder of a Spark-Ignition Engine. SAE paper 800044, 1980.

[39] Michael B. Young and John H. Lienesch. An engine diagnostic package (edpac) - software for analysing pressure-time data. SAE Paper 780967, 1978.

[40] Bruce D. Peters and Gary L. Borman. Cyclic Variations and Average Burning Rates in an S. I. Engine. SAE paper 700064, 1970.

[41] J. A. Gatowksi, E. N. Balles, K. M. Chun, F. E. Nelson, J. A. Ekchian, and J. B. Heywood. Heat Release Analysis of Engine Pressure Data. *SAE Transactions*, 93:paper 841359, 1984.

[42] T. K. Hayes, L. D. Savage, and S. C. Sorenson. Cylinder Pressure Data Acquisition and Heat Release Analysis on a Personal Computer. SAE Paper 860029, 1986.

[43] Franz G. Chmela and Gerhard C. Orthaber. Rate of Heat Release Prediction for Direct Injection Diesel Engines Based on Purely Mixing Controlled Combustion. *SAE Special Publications*, Vol.1444:53–61, 1997 May 5-8 1999.

[44] G. Ferrari. *Motori A Combustione Interna*. Il Capitello,Turino, 1992.

[45] I. I. Vibe. *Brennverlauf und Kreisprozess von Verbrennungsmotoren*. Verlag Technik Berlin, 1970.

[46] Rolf Egnell. Comparison of Heat Release and NOx Formation in a DI Diesel Engine Running on DME and Diesel Fuel. SAE paper 2001-01-0651, 2001.

[47] W. Lovell. Knocking characteristics of hydrobarbons. *Ind. Eng. Chem.*, 40:2388–2438, 1948.

[48] F. W. Bowditch. Combustion Problems in Gasoline Engines. GMR Research Report GMR-260, 1960.

[49] K.Owen and T.Coley. *Automotive Fuels Reference Book - 2nd edition*. Society of Automotive Engineers, 1995.

[50] F. J. Ferfecki and S. C Sorenson. Performance of ethanol blends in gasoline engines. *Transactions of the ASAE*, 26(1):38, 1973.

[51] A. L. Humke and N. J. Barsic. Performance and Emission Characteristics of a Naturally Aspirated Diesel Engine with Vegetable Oil Fuels. SAE paper 810955, 1981.

[52] K. W. Scholl and S. C. Sorenson. Combustion of Soybean Oil Methyl Ester in a Direct Injection Diesel Engine. SAE paper 930934 in Alternatve Diesel Fuels, D. J. Holt Editer, SAE, pp 85-98., 2004.

[53] S. C. Sorenson and Svend-Erik Mikkelsen. Performance and Emissions of a 0.273 Liter Direct Injection Diesel Engine Fuelled with Neat Dimethyl Ether. *SAE Transactions*, 104:80 – 90, 1995.

[54] T. Fleisch, C. McCarthy, A. Basu, P. Charbonneau, Slodowske W., Svend-Erik Mikkelsen, and Jim McCandless. A new clean diesel technology: demonstration of ULEV emissions on a Navistar diesel engine fueled with dimethyl ether. *SAE Transactions*, 104:42 – 53, 1995.

[55] K. Fujimoto, editor. *DME Handbook*. Japan DME Forum, 2007.

[56] S. R. Thomas and S. C Sorenson. Characteristics of a four-cylinder hydrogen-fuelled internal combustion engine. SAE Paper, 760100, 1976.

[57] S Tullis and G Greeves. Optimising diesel combustion with an eui system for heavy-duty trucks. In *Diesel Fuel Injection Systems*, pages 119–129. Institution of Mechanical Engineers, 1992.

[58] Ray E. Bolz and George L. Tuve, editors. *Handbook of Tables for Applied Engineering Science, 2nd Ed., 1981,*. CRC Press, 1981.

[59] K. S. Varde. Bulk modulus of vegetable oil-diesel fuel blends. *Fuel*, 63:713–715, May 1984.

[60] N. A. Henein and D. J. Patterson. *Emissions From Combustion Engines and Their Control*. Ann Arbor Science Publishers inc, Ann Arbor, MI, USA, 1972.

[61] J. T. Wentworth and W. A. Daniel. Flame Photographs of LIght-Load Combustion Point the Way to Reduction of Hydrocarbons in Exhaust. In Vehicle Emissions, SAE Technical Progress Series, no 6, pp 121-136, 1955.

[62] E. Hendricks and S. C Sorenson. Mean Value Modelling of Spark Ignition Engines. SAE paper 90016, 1990.

[63] E. Hendricks, T. Vesterholm, and S. C Sorenson. Nonlinear, Closed Loop, SI Engine Control Observers. SAE Paper 920237, 1992.

[64] Spencer C Sorenson, E. Hendricks, S. Magnusson, and A. Bertelsen. Compact and Accurate Turbocharger Modelling for Engine Control. SAE paper 2005-01-1942, 2005.

[65] M. Müller, E. Hendricks, and S. C. Sorenson. Mean Value Modelling of Turbocharged SI Engines. SAE paper 980784, Also in SAE Special Publications SP-1330, Modeling of SI and Diesel Engines, 1998.

[66] E. Hendricks and S. C. Sorenson. SI Engine Controls and Mean Value Modelling. SAE Paper 910258, 1991.

[67] T. Vesterholm, E. Hendricks, and N. Houbak. Continuous SI Engine Observers. In *American Control Conference*, 1992.

[68] Alain Chevalier Michael Jensen Spencer C. Sorenson David K. Trumpy Hendricks, E. and Joseph R. Asik. Modelling of the manifold filling dy. *SAE Paper 960037*, 1996.

[69] D. L. Harrington and J. A. Bolt. Analysis and Digital Simulation of Carburetor Metering. SAE paper 700082, 1970.

[70] E. Hendricks, T. Vesterholm, and S. C. Sorenson. Nonlinear, Closed Loop, SI Engine Control Observers. SAE Paper 920237, 1992.

[71] P. S. Myers. Automotive Emissions - A Study in Environmental Benefits versus Technological Costs. SAE paper 700182, 1970.

[72] Jr H. W. Sigworth, P. S. Myers, and O. A. Uyehara. The disapprearance of ethylene, propylene, n-butane and 1- butane in spark-ignition engine ex. *SAE paper 700472*, 1970.

[73] M. C. Sellnau, G. S. G. S. Springer, and J. C. Keck. Measurements of Hydrocarbon Concentrations in the Exhaust Products From a Spherical Combustion Bomb. *SAE Paper 810148*, 1981.

[74] J. Schramm and S. C. Sorenson. A Model for HC Emissions from SI Engines. SAE Transaction, paper 902169, also in Automotive Emission Control, SAE SP-839, pg 163, 1990.

[75] M. C. G. S. Springer Sellnau and J. C. Keck. Measurements of hydrobarbon emissions in the exhaust products from a spherical combustion bomb. *SAE PSAE P810148*, 1981.

[76] H.K. Newhall and S. M. Shahed. Kinetics of Nitric Oxide Formation in High Pressure Flames. *Thirteenth Symposium (International) on Combustion Science and Technology*, pages 365–373, 1971.

[77] C. Olikara and G.L. Borman. A Computer Program for Calculating Properties of Equilibrium Combustion Products With Some Applications to IC Engines. SAE paper 750468, 1975.

[78] Robert H. Hammerle and C. H. Wu. Three-way catalyst performance characterization. *SAE Paper 810275*, 1981.

[79] C. A. Amann, D. L. Stivender, S. L. Plee, and J. S. MacDonald. Some Rudiments of Diesel Particulate Emissions. SAE Paper 800251, 1980.

[80] Tanvir Ahmad and Steven L. Plee. Application of flame temperature correlations to emissions from a direct-injection diesel engine. In *Combustion of Heterogeneous Mixtures, SP-557, papern 831734*. Society of Automotive Engineers, 1983.

[81] Denis Gill, Herwig Ofner, Daniel Schwarz, Eddie Sturman, and Marc Andrew Wolverton. The Performance of a Heavy Duty Diesel Engine with a Production Feasible DME Injection System. SAE paper 2001-01-3629, 2001.

[82] S. C. Sorenson, J. W. Hoej, and P.Stobbe. Flow Characteristics of SiC Diesel Particulate Filter Materials. *SAE Transactions*, 103:262, 1994.

[83] P. Stobbe, J. W. Hoej, H. Pedersen, and S. C. Sorenson. SiC as a Substrate for Diesel Particulate Filters. *SAE Transactions, Journal of Engines*, 102:2151, 1993.

[84] Hamilton L. Mabie and Fred V. Ocvirk. *Mechanisms and Dynamics of Machinery, 2nd ed.* John Wiley and Sons, Inc. New York, 1963.

[85] George Rudinger. *Nonsteady Duct Flow: Wave Diagram Analysis,*. Dover Publications, Inc, New York, 1969.

[86] S. N. Meisner and S. C. Sorenson. Computer Simulation of Intake and Exhuast Manifold Flow and Heat Transfer. SAE paper 860242, 1986.

[87] G. P. Blair. *The Basic Design of Two-Stroke Engines*. SAE International, 1990.

[88] Roy Douglas. Two-Stroke Engine Performance and Fuel Economy. Course notes, Spark Ignition Engine Emissions, Univeristy of Leeds, 1996.

[89] E. W. Huber. Measuring the Trapping Efficiency of Internal Combustion Engines Through Continuous Exhaust Gas Analysis. SAE paper 710144, 1971.

[90] Kim R. Hansen, Jakob D. Dolriis, Christopher Hansson, Claus S. Nielsen, Spencer C Sorenson, and Jesper Schramm. Optimizing the performance of a 50cc compression ignition two-stroke engine operating on dimethyl ether. *SAE paper 2011-01-0144*, 2011.

[91] I. M. Sivebaek, V. N. Samoilov, and B. N. J. Persson. Squeezing molecular thin alkane lubrication films between curved solid surfaces with long-range elasticity: Layering transitions and wear. *Journal of Chemical Physics*, 119, no 4:2314 – 2321, 2003.

[92] P. E. Vickery. Friction Losses in Automotive Plain Bearings - A Practical and Theoretical Study. SAE Paper 750052, 1975.

[93] Gary Borman and Kazuie Nishiwaki. Internal-Combustion Engine Heat Transfer. *Progress in Energy and Combustion Science*, 13:1–47, 1987.

[94] C. A. Finol and K. Robinson. Thermal modelling of modern engines: a review of empirical correlations to estimate the in-cylinder heat transfer coefficient. *Proc. IMechE, Part D: J. Automobile Engineering*, 22 issue 12:1765, 2006.

[95] Fredrik Ree Westlye. *Experimental study of liquid fuel spray combustion*. PhD thesis, Technical University of Denmark, 2017.

[96] Thomas LeFeuvre. *Instantaneous Metal Temperature and Heat Fluxes in a Diesel Eng*. PhD thesis, University of Wisconsin, 1968.

[97] T. LeFeuvre, P. S. Myers, and O.A. Uyehara. Experimental Instantaneous Heat Fluxes in a Diesel Engine and their Correlation. *SAE Transactions*, 78:Paper no. 690464, 1969.

[98] H. S. Carslaw and J. C. Jaeger. *Conduction of Heat in Solids, 2nd ed.* Oxford, 1959.

[99] Patrick F. Flynn. *An Experimental Determination of the Instantaneous Potential Radiant Heat Transfer Within an Operating Diesel Engine*. PhD thesis, University of Wisconsin, 1971.

[100] P. Flynn, Masatake Mizusawa, O. A. Uyehara, and P. S. Myers. An Experimental Determination of the Instantaneous Potential Radiant Heat Transfer Within an Operating Engine. *SAE Transactions*, 81(SAE paper 720022):95–126, 1972.

[101] J. A. Caton and J. B. Heywood. An Investigation and Analytical Study of Heat Transfer in an Engine Exhaust Port. *International Journal of Heat and Mass Transfer*, 24:581, 1981.

[102] M. Farrugia, A. C. Alkidas, and B. P. Sangeorzan. Cycle-Averaged Heat Flux Measurements in a Straight-Pipe Extension of the Exhaust Port of an SI Engine. SAE Paper 2006-01-1033, 2006.

[103] Gustav Winkler. Ein geschlossenes diagramm zur bestimmung der betriebspunkte von abgasturboladern an viertakmotoren. *MTZ*, 41:451 – 457, 1980.

[104] N. Watson and MS Janota. *Turbocharging the internal combustion engine*. Macmillan, London, 1982.

[105] Rowland S Benson and N.D Whitehouse. *Internal Combustion Engines*. Pergamon Press, 1979.

[106] K. Zinner. *Aufladung von Verbrennungsmotoren*. Springer-Verlag, 3 Aufl. Berlin, 1985.

[107] B. T. Pritchard E. R. R. Fuchs, K. G. Parker. Turbocharging the 3 litre v6 ford essex engine. *Proc Instn Mech Engrs*, 188(5/74):33 – 47, 1974.

[108] Horst Bergmann and Eberhard Mack. Die konstruktionsmerkmale des neuen nutzfahrzeug-dieselmotors om 904 la von mercedes benz. *Motortechnische Zeitschrift*, 57 no 2:74 – 80, 1996.

[109] M. Fortnagel, B. Hall, J. Giese, M.Mrwald, H. K. Weining, and P. Lückert. Technischder forschritt durch evolution - neue vierzylinder-ottomotoren von mercedes-benz af basis des erfolgreichen miii. *MTZ*, 16(9):582 – 588, 2000 2000.

[110] K.-H. Küttner. *Kolbenmaschinen*. B. G. Teubner, Stuttgart, 1974.

[111] Thomas Gillespie. *Fundamentals of Vehicle Dynamics*. Society of Automotive Engineers, 1992.

[112] Wolf-Heinrich Hucho, editor. *Aerodynamics of Road Vehicles, Fourth Edition*. Society of Automotive Engineers, 1998.

[113] N. S. Jackson, J. Stokes, P. A. Whitaker, and T. H. Lake. Stratified and homogeneous charge operation for the direct-injection gasoline engine-High power with low fuel consumption and emissions. SAE paper 970543, 1997.

Index

www.ingramcontent.com/pod-product-compliance
Lightning Source LLC
Chambersburg PA
CBHW082304210326
41598CB00028B/4434